Intermediate Algebra

Intermediate Algebra

Dennis Bila
Washtenaw Community College

Ralph Bottorff
Washtenaw Community College

Paul Merritt
Highland Park Community College

Donald Ross
Washtenaw Community College

WORTH PUBLISHERS, INC.

Intermediate Algebra

Printed in the United States of America

Library of Congress Catalog Card No. 74-84642

ISBN: 0-87901-038X

Second printing, August, 1978

WORTH PUBLISHERS, INC.

444 Park Avenue South

New York, New York 10016

To our students

Preface

For many years we have struggled with the problem of providing well-written programmed materials for those students who wish to learn mathematics. Many of these students have had unrewarding prior experiences with mathematics and are reluctant to take further courses. Our attempts to help them were hindered by the lack of adequate text material. As a result, we began writing semiprogrammed booklets, each covering a single concept, to supplement the textbook we were using. The positive reactions of students to these booklets were encouraging, and we decided to develop a series of semiprogrammed texts written specifically for students similar to those in our own classes. *Intermediate Algebra* is the third in a series which also includes *Core Mathematics* and *Introductory Algebra.* We sincerely hope that your experiences with these texts will be as rewarding as ours have been.

Intermediate Algebra is intended to be used in a one-semester course. Students who need to obtain a working knowledge of algebra for their later study of precalculus or other courses are provided here with a rigorously class-tested worktext which allows them to proceed at their own pace, and which reinforces them at each carefully graded step along the way.

The first two chapters offer students who have studied introductory algebra the opportunity to review and extend their knowledge. A major goal of Chapter 1 is to bring students to a thorough understanding of, and ability to deal with, the set of real numbers. We have found that a major shortcoming of many students in more advanced math courses is their inability to factor polynomials and perform operations with rational expressions; hence Chapter 2 covers these topics extensively. Naturally, students who have maintained their skills in these areas will be able to advance rapidly through these two chapters. Other students will have ample opportunity to master these topics without retaking a course in introductory algebra.

Chapters 3 through 7 develop topics that are frequently only introduced in introductory algebra. Chapter 3, for example, reviews and extends the student's knowledge of linear equations, inequalities, and absolute value. The chapter is written so that students who have not previously worked with inequalities or absolute value will not be at a disadvantage. Chapter 4 leads the student to a more complete understanding of the set of complex numbers, with special emphasis on operations with irrational numbers and expressions containing integer as well as rational exponents. Chapter 5 reviews solutions to quadratic equations with real roots, and extends this work to include equations with complex roots. Quadratic inequalities are solved by both geometric and analytic methods. Relations and functions are treated in some detail in Chapter 6 and no prior knowledge is assumed, since many students who are resuming their mathematical education do not have this background. Chapter 7 reviews the solution to systems of two linear equations in two variables by use of the addition and substitution methods. Cramer's rule is introduced and applied to systems of two equations in two variables, as well as systems of three equations in three variables.

Chapter 8 presents the important topics of exponential and logarithmic functions. The use of logarithms as an aid in computation has been somewhat de-emphasized because of the increasing availability of mini-calculators.

Sequences and series (Chapter 9) is a traditional topic in most intermediate algebra texts. Often it is not taught due to lack of time; however, many students find this material interesting and will want to study it independently.

Features of *Intermediate Algebra*

Format

Each unit of material (a single concept or a few closely related concepts) is presented in a section. Within a section, short, boxed, numbered frames contain all instructional material, including sample problems and sample solutions. Frames are followed by practice problems, with work space provided. Answers immediately follow, with numerous solutions and supplementary comments.

Exercises

Exercises at the end of each section are quite traditional in nature. However, they are generally shorter than those found in most texts because the student has already done numerous problems and answered many questions during the study of the section. Hence, the exercises serve as a review of the content of the section and as a guide that will help students to recognize whether they have mastered the material. Answers to the problems are provided immediately, permitting the student to advance through the text without having to turn to the back to check answers. Word problems are used throughout.

Sample Chapter Tests

At the end of each chapter we have provided a Sample Chapter Test. It is keyed to each section with the answers provided immediately for student convenience. It may also serve as a pretest for each chapter. However, if an instructor wishes to pretest without the availability of the answers (to the student), one form of the post-test for the chapter (provided in the *Test Manual*) could be used for this purpose.

Upon finishing the chapter, the student should complete the Chapter Sample Test, and the results should be shown to the instructor and discussed with the student. If the instructor and student are confident about mastery of the content of the chapter, a post-test can then be administered. Used in this manner, outstanding results can be anticipated on the first attempt for each post-test.

Objectives

The Sample Chapter Tests serve as objectives for both student and instructor. The instructor can readily ascertain the objectives of any chapter by examining the sample test at its end, and the student is in a good position to see what is expected of him by examining problems and questions that are going to be asked at the completion of the chapter. We believe that objectives stated verbally are of less benefit to the student than the statement of a problem that must be solved.

Glossary

The glossary at the back of the text provides the student with the pronunciation and the definition of all mathematical words and phrases used in the text. This is particularly important wherever the text is used in a laboratory situation where the student will not hear the words used in class discussions.

Acknowledgments

Reviews from the following teachers were particularly helpful in the writing of this text: Helen Hoke, Lansing Community College; Cecilia M. Cooper, Kendall College; and Richard Spangler, Tacoma Community College.

We would also like to express our appreciation to Janet Hastings at Washtenaw Community College for her assistance in developing the supplemental testing program, and to student reviewers John Dowding, Al Robbins, and Alan F. Turner. Our thanks also to our typists Phyllis Bostwich, Marilyn Myers, Kathy Stripp, and Carolyn Williams.

In addition, Bruce Bayha, Margaret Moy, Donna Reed, and Walter Steele provided valuable criticisms and suggestions during the class-testing.

Dennis Bila
Ralph Bottorff
Paul Merritt
Donald Ross

February, 1975

To the Student

You are about to brush up on some skills you may already have learned in a previous course; but they may be a bit rusty. You will learn some new techniques, too. This book is designed to help pinpoint those areas for which you need to sharpen your skills and to permit you to move quickly through those topics you know well.

If you follow the suggestions below, you will make best use of your time, proceeding through the course as quickly as possible, and you will have mastered all the material in this book.

Chapter Sample Tests

The Chapter Sample Test at the end of each chapter will help you to determine whether you can skip certain sections of the chapter. It is a cumulative review of the entire chapter, and questions are keyed to the individual sections of the chapter. To use this test to best advantage, we suggest this procedure:

1. Work the solutions to all problems neatly on your own paper to the best of your ability. You want to determine what you know about the material without help from someone else. If you receive help in completing this self-test, the result will show another person's knowledge rather than your own.
2. Once you have done as much as you can, check the test with the answers provided. On the basis of errors made, determine the most appropriate course of study. You may need to study all sections in the chapter or you may be able to skip one or two sections.
3. When you have completed the chapter, rework all problems originally missed on the Chapter Sample Test and review the entire test in preparation for the post-test that will cover the entire chapter.

Instructional Sections

To use each section most effectively:

1. Study the boxed frames carefully.
2. Following each frame are questions based on the information presented in the frame. When completing these questions, do the work in the spaces provided.
3. Use a blank piece of paper as a mask to cover the answers below the stop rule. For example:

Q1 How many eggs can 21 chickens lay in 1 hour, if 1 chicken can lay 7 eggs in $\frac{1}{2}$ hour?

(Work space)

STOP • STOP • STOP • STOP • STOP • STOP • STOP • STOP • STOP

Paper Mask

4. After you have written your calculations and response in the workspace, slide the paper mask down, uncovering the correct solution, and check your response with the correct answer given. In this way you can check your progress as you go without accidentally seeing the response before you have completed the necessary thinking or work.

Q1 How many eggs can 21 chickens lay in 1 hour, if 1 chicken can lay 7 eggs in $\frac{1}{2}$ hour?

$7 \times 2 = 14$

$$\begin{array}{r} 14 \\ \times 21 \\ \hline 14 \\ 28 \\ \hline 294 \end{array}$$

STOP • STOP • STOP • STOP • STOP • STOP • STOP • STOP • STOP

A1 294: because 1 chicken can lay 14 in 1 hour, 21 chickens can lay 21(14) in 1 hour.

Paper Mask

Notice that the answer is often followed by a colon (:). The information following the colon is one of the following:

a. The complete solution,
b. A partial solution, possibly a key step frequently missed by many students,
c. A remark to remind you of an important point, or
d. A comment about the solution.

If your answer is correct, advance immediately to the next step of instruction.

5. Space is provided for you to work in the text. However, you may prefer to use the paper mask to work out your solutions. If you do, be sure to show complete solutions to all problems, clearly numbered, on the paper mask. When both sides of the paper are full, file it in a notebook for future reference.
6. Make all necessary corrections before continuing to the next frame or problem.
7. If something is not clear, talk to your instructor immediately.
8. If difficulty arises when you are studying outside of class, be sure to note the difficulty so that you can ask about it at your earliest convenience.
9. When preparing for a chapter post-test, the frames of each section serve as an excellent review of the chapter.

Section Exercises

The exercises at the end of each section are provided for additional practice on the content of the section. For your convenience, answers are given immediately following the exercises. However, detailed solutions to these problems are not shown.

You should:

1. Work all problems neatly on your own paper.
2. Check your responses against the answers given.
3. Rework any problem with which you disagree. If you cannot verify the response given, discuss the result with your instructor.
4. In the process of completing the exercise, use the instructional material in the section for review when necessary.
5. Problems marked with an asterisk (*) are considered more difficult than examples in the instructional section. They are intended to challenge the interested student. Problems of this difficulty will not appear on the Chapter Sample Test or the post-test.

Contents

Intermediate Algebra

Chapter 1

The Real Number System

1.1 Set Ideas

Section 1.1 is designed to provide the student with the background necessary to understand the role of sets in the development of number ideas throughout the text.

1	A set is a well-defined collection of things. The things that make up a set are called *elements* of the set. The letters of the English alphabet form a set of 26 elements. The letter b is an element of this set. The months of the year form a set of 12 elements. July is an element of this set.

Q1 The days of the week form a ____ of seven elements.

STOP • STOP • STOP • STOP • STOP • STOP • STOP • STOP • STOP

A1 set

Q2 Name an element of the set in Q1. ________

STOP • STOP • STOP • STOP • STOP • STOP • STOP • STOP • STOP

A2 Any day of the week: for example, Wednesday.

2	A set may contain elements that have a common property, such as the Presidents of the United States or the residents of a particular city. A set may also consist of unrelated objects, such as a chair, a pencil, and a book. However, a set must be *well defined.* This means that membership in the set is clear. The set of men who have been President of the United States is a well-defined collection because it is clear exactly which men are included in this set. A collection of any three letters of the English alphabet is not a well-defined collection because many possibilities exist and you would not be sure which is intended.

Q3 Choose the sets from the following collections: ________

a. the collection of states in the United States bordered by the Great Lakes
b. the collection of three consecutive warm months
c. the collection of beautiful people
d. the collection of women who have been President of the United States

STOP • STOP • STOP • STOP • STOP • STOP • STOP • STOP • STOP

A3 a and d. b is not well defined because temperature is dependent upon location and would vary in different parts of the world. c is not well defined because there would not be agreement on who is beautiful.

3 The use of braces, { }, to indicate a set is called *set notation.* The elements of the set are enclosed in braces and are separated by commas. For example, the set of one-digit numbers* between 1 and 7 is denoted as {2, 3, 4, 5, 6}. The braces are read "the set containing," so {1, 2, 5} is read "the set containing 1, 2, and 5."

*The "digits" are 0, 1, 2, 3, 4, 5, 6, 7, 8, and 9.

Q4 Use set notation to indicate the set of one-digit numbers between 2 and 9. ________

STOP • STOP • STOP • STOP • STOP • STOP • STOP • STOP • STOP

A4 {3, 4, 5, 6, 7, 8}

4 It is not always convenient to use set notation. A description is used when it would be difficult or impractical to list the elements. Such an example would be the set of U.S. Congressmen.

Q5 Where convenient, use set notation to indicate the set:

a. the set of vowels in the English alphabet ________

b. the set of different kinds of flowers found in Ontario, Canada ________

c. the set of digits in the number 100,537,375 ________

STOP • STOP • STOP • STOP • STOP • STOP • STOP • STOP • STOP

A5 **a.** {a, e, i, o, u} (*Note:* It is common, and preferred, to arrange letter elements of a set alphabetically.)

b. Although this information could be listed, it would not be convenient to do so in set notation.

c. {0, 1, 3, 5, 7} (*Note:* The order in which the elements are arranged does not affect the set. That is, {0, 1, 3, 5, 7} = {1, 0, 5, 3, 7}. Also, duplicate elements need be listed only once.)

5 The set that contains no elements is called the *empty set.* It is written using braces with no elements, { }. The symbol ∅ also denotes the empty set. The set of one-digit numbers between 4 and 5 is an example of the empty set.

Q6 The set of living 200-year-old people represents an ________ set.

STOP • STOP • STOP • STOP • STOP • STOP • STOP • STOP • STOP

A6 empty

Q7 Use set notation to indicate the set of two-digit numbers greater than 99. ________

STOP • STOP • STOP • STOP • STOP • STOP • STOP • STOP • STOP

A7 { } or ∅

6 How many men had walked on the moon by 1960? The answer to this question is zero. That is, the set of men who had walked on the moon by 1960 is empty. You should observe that the number of elements in the empty set is zero; that is, there are no elements in the empty set.

Q8 a. How many elements are there in $\{0\}$? ________

b. How many elements are there in $\{\ \}$? ________

c. Is $\{0\} = \{\ \}$? ________

STOP • STOP • STOP • STOP • STOP • STOP • STOP • STOP • STOP

A8 a. one, the element 0 b. zero c. no

---

7 If each element of one set is also an element of a second set, the first set is a *subset* of the second set. The set of dogs is a subset of the set of animals.

Q9 The set of New England states is a ________ of the set of states in the United States.

STOP • STOP • STOP • STOP • STOP • STOP • STOP • STOP • STOP

A9 subset

Q10 Determine whether the first set is a subset of the second set:

a. {Jane, Mary} {June, Judy, Jane} ________

b. $\{1, 2, 5\}$ $\{0, 1, 2, 3, 4, 5\}$ ________

c. $\{2, 4, 6\}$ $\{1, 2, 3, 4, 5\}$ ________

d. $\{1, 2, 3\}$ $\{1, 2, 3\}$ ________

e. set of vowels set of all letters in English alphabet ________

STOP • STOP • STOP • STOP • STOP • STOP • STOP • STOP • STOP

A10 a. no b. yes c. no d. yes e. yes

---

8 The mathematical symbol $\subseteq$ means "is a subset of."

Examples:

"is a subset of"
↓
{Jane, Mary} $\subseteq$ {June, Judy, Jane, Mary}

$$\{1, 2, 5\} \subseteq \{0, 1, 2, 3, 4, 5, 6\}$$
$$\{1, 2, 3\} \subseteq \{1, 2, 3\}$$
$$\{\ \} \subseteq \{1, 2, 3\}$$

(*Note:* The third example illustrates that a set is a subset of itself. The fourth example indicates that the empty set is a subset of $\{1, 2, 3\}$.
$\not\subseteq$ means "is *not* a subset of.")

Examples:

$$\{a, c, d\} \not\subseteq \{a, e, i, o, u\}$$
$$\{0, 1, 2, 3\} \not\subseteq \{1, 2, 3\}$$
$$\left\{2\frac{1}{2}\right\} \not\subseteq \{0, 1, 2, 3, 4\}$$

Q11 Insert $\subseteq$ or $\not\subseteq$ to form a true statement:

a. $\{1\}$ ____ $\{1, 2\}$ b. $\{0\}$ ____ $\{1, 2, 3\}$

c. $\{1, 3, 5\}$ ____ $\{1, 3, 5, 7\}$ d. $\{0, 2\}$ ____ $\{0, 2\}$

STOP • STOP • STOP • STOP • STOP • STOP • STOP • STOP • STOP

A11 a. $\subseteq$ b. $\not\subseteq$ c. $\subseteq$ d. $\subseteq$

9

If a first set is a subset of a second set and *at least one* element appears in the second set that does not appear in the first set, the first set may be called a *proper* subset of the second. The symbol $\subset$ means "is a proper subset of." Formally, for sets A and B, if $A \subseteq B$, but $A \neq B$, then $A \subset B$.*

Examples:

"is a proper subset of"
↓
{Jane, Mary} $\subset$ {June, Jane, Judy, Mary}
$\{a, b, d\} \subset \{a, b, c, d, e\}$
$\{1\} \subset \{1, 2\}$
$\{1, 2, 3\} \not\subset \{1, 2, 3\}$

In the fourth example, $\{1, 2, 3\}$ is not a proper subset of $\{1, 2, 3\}$ because the sets are equal.

*It is common to let sets be denoted by capital letters.

Q12 Insert $\subset$ or $\not\subset$ to form a true statement:

a. $\{1\}$ ____ $\{1, 2, 3\}$
b. $\{2\}$ ____ $\{1, 2, 3\}$
c. $\{3\}$ ____ $\{1, 2, 3\}$
d. $\{1, 2\}$ ____ $\{1, 2, 3\}$
e. $\{1, 3\}$ ____ $\{1, 2, 3\}$
f. $\{2, 3\}$ ____ $\{1, 2, 3\}$
g. $\{1, 2, 3\}$ ____ $\{1, 2, 3\}$

STOP • STOP • STOP • STOP • STOP • STOP • STOP • STOP • STOP

A12 a. $\subset$ b. $\subset$ c. $\subset$ d. $\subset$ e. $\subset$ f. $\subset$ g. $\not\subset$

10

Let $A = \{1, 2, 3\}$ and $B = \{1, 2\}$. Is $A \subset B$? No, because there is an element, 3, in A that is not in B. Is $B \subset A$? Yes, because every element of B is an element of A and there is an element, 3, in A, that is not an element of B.

Let $C = \{\ \}$ and $B = \{1, 2\}$. Is $B \subset C$? No, because there are two elements, 1 and 2, in B that are not in C. Is $C \subset B$? Yes, because there are no elements in C and at least one element in B that is not in C.

The above argument illustrates that the empty set is a proper subset of all sets except itself.

Examples:

1. $\{\ \} \subset \{1\}$
2. $\{\ \} \subset \{a, c, d, f\}$
3. $\{\ \} \subset$ {June, Jane, Mary}
4. $\{\ \} \subset \{0, 1, 2, 3, \ldots\}$
5. $\{\ \} \not\subset \{\ \}$ or $\varnothing \not\subset \varnothing$
6. $\{\ \} \subseteq \{\ \}$ or $\varnothing \subseteq \varnothing$

The fifth example says that "the empty set is not a proper set of the empty set." The sixth example says that "the empty set is a subset of itself."

Q13 Insert $\subset$ or $\subseteq$ to form a true statement:

a. $\{1\}$ ____ $\{1\}$
b. $\{\ \}$ ____ $\{\ \}$

STOP • STOP • STOP • STOP • STOP • STOP • STOP • STOP • STOP

A13 a. $\subseteq$ b. $\subseteq$

Q14 True or false? (Use T or F.)

a. $\{1\} \subseteq \{1, 2, 3\}$ ______ b. $\{1\} \subset \{1, 2, 3\}$ ______
c. $\{2\} \subseteq \{1, 2, 3\}$ ______ d. $\{2\} \subset \{1, 2, 3\}$ ______
e. $\{3\} \subseteq \{1, 2, 3\}$ ______ f. $\{3\} \subset \{1, 2, 3\}$ ______
g. $\{1, 2\} \subseteq \{1, 2, 3\}$ ______ h. $\{1, 2\} \subset \{1, 2, 3\}$ ______
i. $\{1, 3\} \subseteq \{1, 2, 3\}$ ______ j. $\{1, 3\} \subset \{1, 2, 3\}$ ______
k. $\{2, 3\} \subseteq \{1, 2, 3\}$ ______ l. $\{2, 3\} \subset \{1, 2, 3\}$ ______
m. $\{1, 2, 3\} \subseteq \{1, 2, 3\}$ ______ n. $\{1, 2, 3\} \subset \{1, 2, 3\}$ ______
o. $\{\ \} \subseteq \{1, 2, 3\}$ ______ p. $\{\ \} \subset \{1, 2, 3\}$ ______

STOP • STOP • STOP • STOP • STOP • STOP • STOP • STOP • STOP

A14 a. T b. T c. T d. T e. T f. T g. T
h. T i. T j. T k. T l. T m. T n. F
o. T p. T

11 You should observe that the symbol $\subseteq$ permits the possibility that the two sets may be equal (although they don't have to be). The symbol $\subseteq$ says "is a subset of." The symbol $\subset$ says "is a proper subset of." The symbol $\subseteq$ also says "is a proper subset of or is equal to."

Q15 Insert $\subset$ whenever possible to form a true statement. Use $\subseteq$ otherwise.

a. $\{a\}$ ______ $\{a, b, c\}$ b. $\{b\}$ ______ $\{a, b, c\}$
c. $\{c\}$ ______ $\{a, b, c\}$ d. $\{a, b\}$ ______ $\{a, b, c\}$
e. $\{a, c\}$ ______ $\{a, b, c\}$ f. $\{b, c\}$ ______ $\{a, b, c\}$
g. $\{a, b, c\}$ ______ $\{a, b, c\}$ h. $\{\ \}$ ______ $\{a, b, c\}$

STOP • STOP • STOP • STOP • STOP • STOP • STOP • STOP • STOP

A15 a. $\subset$ b. $\subset$ c. $\subset$ d. $\subset$ e. $\subset$ f. $\subset$ g. $\subseteq$
h. $\subset$

12 All subsets of the set $\{x, y, z\}$ are formed by taking all subsets of one element, all subsets of two elements, and so on. For example, let $A = \{x, y, z\}$:

$\{x\} \subseteq A$ $\{x, y\} \subseteq A$ $\{x, y, z\} \subseteq A$
$\{y\} \subseteq A$ $\{x, z\} \subseteq A$ $\{\ \} \subseteq A$
$\{z\} \subseteq A$ $\{y, z\} \subseteq A$

Therefore, all subsets of $\{x, y, z\}$ are $\{x\}$, $\{y\}$, $\{z\}$, $\{x, y\}$, $\{x, z\}$, $\{y, z\}$, $\{x, y, z\}$, and $\{\ \}$.

Q16 Form all the subsets of $\{1, 2\}$. ______________________

STOP • STOP • STOP • STOP • STOP • STOP • STOP • STOP • STOP

A16 $\{1\}$, $\{2\}$, $\{1, 2\}$, $\{\ \}$ or $\varnothing$ (any order)

Q17 Indicate all the proper subsets of $\{1, 2\}$. ______________________

STOP • STOP • STOP • STOP • STOP • STOP • STOP • STOP • STOP

A17 {1}, {2}, { } or ∅ (any order)

Q18 Indicate all the proper subsets of {1, 2, 3}. ______

STOP • STOP • STOP • STOP • STOP • STOP • STOP • STOP • STOP

A18 {1}, {2}, {3}, {1, 2}, {1, 3}, {2, 3}, { } or ∅

Q19 **a.** How many proper subsets are there of {1, 2, 3}? ____

b. How many subsets are there of {1, 2, 3}? ____

STOP • STOP • STOP • STOP • STOP • STOP • STOP • STOP • STOP

A19 **a.** 7 **b.** 8

Q20 Form all subsets of {1, 2, 3, 4}.

______

______

STOP • STOP • STOP • STOP • STOP • STOP • STOP • STOP • STOP

A20 {1}, {2}, {3}, {4}, {1, 2}, {1, 3}, {1, 4}, {2, 3}, {2, 4}, {3, 4}, {1, 2, 3}, {1, 2, 4}, {1, 3, 4}, {2, 3, 4}, {1, 2, 3, 4}, { } or ∅

---

*Q21 **a.** A set with one element has ____ subsets, which is 2 to the ______ power.

b. A set with two elements has ____ subsets, which is 2 to the ______ power.

c. A set with three elements has ____ subsets, which is 2 to the ______ power.

d. A set with four elements has ____ subsets, which is 2 to the ______ power.

e. If this pattern continues, a set with n elements has ____ subsets.

f. A set with 5 elements has ____ subsets.

STOP • STOP • STOP • STOP • STOP • STOP • STOP • STOP • STOP

*A21 **a.** 2, first **b.** 4, second **c.** 8, third
d. 16, fourth **e.** 2^n **f.** 32

---

13 The symbol ∈ is used to abbreviate "is an element of." $x \in \{x, y, z\}$ is read "x is an element of the set containing x, y, and z." The slant bar, /, is used as a negation symbol. The symbol ∉ means "is *not* an element of." $1 \notin \{0, 2, 4, 6\}$ is read "1 is not an element of the set containing 0, 2, 4, and 6."

Q22 Insert the correct symbol, ∈ or ∉, in the blanks:

a. r ____ {a, e, i, o, u} **b.** 5 ____ {1, 3, 5, 7, 9}

STOP • STOP • STOP • STOP • STOP • STOP • STOP • STOP • STOP

A22 **a.** ∉ **b.** ∈

---

14 Since the letters in the set comprising the English alphabet are numerous, this set may be indicated as {a, b, c, . . . , z}. The three dots, read "and so on," indicate that all elements of the set have not been listed. Additional elements of the set can be found by following the pattern established in the elements preceding the dots.

Q23 Insert the correct symbol, $\in$ or $\notin$, in the blanks:

a. 17 ______ $\{1, 2, 3, \ldots, 20\}$ b. 12 ______ $\{1, 3, 5, \ldots, 19$

c. 10 ______ $\{2, 4, 6, \ldots, 20\}$

STOP • STOP • STOP • STOP • STOP • STOP • STOP • STOP • STOP

A23 a. $\in$ b. $\notin$ c. $\in$

15	If the elements of a set can be counted and the count has a last number, the set is a *finite* set. A set whose count is unending is an *infinite* set. For example, $\{a, b, c, d\}$ is a finite set, whereas $\{1, 2, 3, 4, \ldots\}$ is an infinite set.

Q24 Describe each of the following sets as finite or infinite:

a. the set of letters in the word "infinite" ______

b. $\{1, 2, 3, 4, \ldots, 15\}$ ______

c. $\{5, 10, 15, \ldots\}$ ______

d. the set of provincial parks in Canada ______

e. $\varnothing$ ______

STOP • STOP • STOP • STOP • STOP • STOP • STOP • STOP • STOP

A24 a. finite b. finite c. infinite
d. finite e. finite (the empty set has a count of zero)

16 The *intersection* of two sets is a third set that contains those elements, and only those elements, that belong to *both* (one and the other) of the original sets. Let

$A = \{a, e, i, o, u\}$

and

$B = \{a, r, i, s, t, o, c, e\}$

The intersection of sets A and B is $\{a, e, i, o\}$.

Q25 Let $C = \{1, 2, 3, 4\}$ and $D = \{2, 4, 6\}$. Determine the intersection of sets C and D.

______

STOP • STOP • STOP • STOP • STOP • STOP • STOP • STOP • STOP

A25 $\{2, 4\}$

17 The symbol $\cap$ is used to express the set operation of intersection. For example, if

$A = \{1, 2, 3, 5, 7\}$

and

$B = \{2, 3, 7, 8, 9\}$

the intersection of A and B may be shown as

$A \cap B = \{2, 3, 7\}$

Q26 Given $A = \{a, b, c\}$ and $B = \{b, c, d\}$, determine $A \cap B$. ________

STOP • STOP • STOP • STOP • STOP • STOP • STOP • STOP • STOP

A26 $\{b, c\}$

18 Two sets are *disjoint* sets when they have no common elements. The sets $\{1, 3, 5\}$ and $\{0, 2, 4\}$ are disjoint. These sets have no elements in common; hence the intersection of these sets has no elements and is, therefore, the empty set. If

$$C = \{1, 3, 5\}$$

and

$$D = \{0, 2, 4\}$$

then

$$C \cap D = \varnothing$$

If the intersection of two sets is empty, the sets are disjoint.

Q27 $A = \{2, 4, 6, 8, 10, 12\}$ $D = \{4, 8, 12\}$
$B = \{1, 3, 5, 7, 9, 11\}$ $E = \varnothing$
$C = \{3, 6, 9, 12\}$
Given the sets above, find:

a. $A \cap B =$ ________ **b.** $B \cap C =$ ________
c. $A \cap C =$ ________ **d.** $A \cap D =$ ________
e. $B \cap D =$ ________ **f.** $C \cap D =$ ________
g. $B \cap E =$ ________

STOP • STOP • STOP • STOP • STOP • STOP • STOP • STOP • STOP

A27 **a.** $\varnothing$ **b.** $\{3, 9\}$ **c.** $\{6, 12\}$ **d.** $\{4, 8, 12\}$
e. $\varnothing$ **f.** $\{12\}$ **g.** $\varnothing$

Q28 Consider sets A, B, C, D, and E in Q27, and indicate the pairs of sets that are disjoint. ________________

STOP • STOP • STOP • STOP • STOP • STOP • STOP • STOP • STOP

A28 A and B, A and E, B and D, B and E, C and E, D and E

19 The *union* of two sets is a third set that contains those elements that belong to *either* (one or the other) of the original sets. If

$$A = \{a, c, t\}$$

and

$$B = \{a, e, r, t\}$$

the union of sets A and B is $\{a, c, e, r, t\}$.

Q29 Let $C = \{1, 2, 3, 4\}$ and $D = \{2, 4, 6\}$. Determine the union of sets C and D.
____

STOP • STOP • STOP • STOP • STOP • STOP • STOP • STOP • STOP

A29 $\{1, 2, 3, 4, 6\}$

20 The symbol $\cup$ is used to express the set operation of union. For example, if

$A = \{1, 2, 3, 5, 7\}$

and

$B = \{2, 3, 7, 8, 9\}$

the union of A and B may be shown as

$A \cup B = \{1, 2, 3, 5, 7, 8, 9\}$

Q30 Given $A = \{a, b, c\}$ and $B = \{b, c, d\}$, determine $A \cup B$. ____________

STOP • STOP • STOP • STOP • STOP • STOP • STOP • STOP • STOP

A30 $\{a, b, c, d\}$

Q31 $A = \{2, 4, 6, 8, 10, 12\}$ $C = \{3, 6, 9, 12\}$ $E = \varnothing$
$B = \{1, 3, 5, 7, 9, 11\}$ $D = \{4, 8, 12\}$
Given the sets A–E, find:

a. $A \cup B =$ ____________ **b.** $B \cup C =$ ____________

c. $A \cup C =$ ____________ **d.** $A \cup D =$ ____________

e. $B \cup D =$ ____________ **f.** $B \cup E =$ ____________

STOP • STOP • STOP • STOP • STOP • STOP • STOP • STOP • STOP

A31 **a.** $\{1, 2, 3, \ldots, 12\}$ **b.** $\{1, 3, 5, 6, 7, 9, 11, 12\}$
c. $\{2, 3, 4, 6, 8, 9, 10, 12\}$ **d.** $\{2, 4, 6, 8, 10, 12\}$
e. $\{1, 3, 4, 5, 7, 8, 9, 11, 12\}$ **f.** $\{1, 3, 5, 7, 9, 11\}$

This completes the instruction for this section.

1.1 Exercises

1. Give the word or phrase that best describes each of the following:
   a. a well-defined collection
   b. the set that contains no elements
   c. a set in which the elements can be counted and the count has a last number
   d. a set whose count is unending
2. Specify the elements of each of the following sets:
   a. the set of states in the United States bordered by the Pacific Ocean
   b. the set of the last three letters of the English alphabet
   c. $\{1, 2, 3\}$
3. Which of the following are not sets:
   a. $\{1, 3, 5, 7, 9, 11, \ldots\}$
   b. the collection of ex-presidents of the United States
   c. the collection of consonants in the English alphabet
   d. the collection of large numbers
4. Where convenient, use set notation to indicate the following sets:
   a. the set of letters in the word "college"
   b. the set of digits in the number 10,274

c. the set of five-letter words beginning with C
d. the days of the week beginning with the letter R
e. the set of digits between 0 and 9

5. True or false? (Use T or F.)
   a. $\{a, c, d\}$ is a subset of $\{a, b, c, d\}$.
   b. The set of cities that are provincial capitals in Canada is a subset of the set of all cities in Canada.
   c. $\{0\}$ is a subset of $\{1, 2, 3, \ldots\}$.
   d. {Jane, Mary, Sue, Ann} is a subset of {Mary, Sue, Jane, Ann}.
6. a. Write the symbol for "is a subset of."
   b. Write the symbol for "is a proper subset of."
7. True or false? (Use T or F.)
   a. $\{5, 7\} \subset \{0, 1, 2, 3, \ldots\}$
   b. $\{1, 2\} \subset \{1, 2\}$
   c. $\varnothing \subseteq \varnothing$
   d. $\{4, 5, 6\} \subset \{1, 3, 5, 7\}$
   e. $\varnothing \subseteq \{0, 2, 4, 6, \ldots\}$
   f. $\{a, b, c, d\} \subseteq \{a, b, c, d\}$
   g. $\{5, 7\} \subseteq \{0, 1, 2, 3, \ldots\}$
   h. $\{1, 2, 3\} \subset \{1, 2, 3\}$
   i. $\{1, 2, 3\} \not\subset \{1, 2, 3\}$
   j. $\varnothing \subseteq X$, where X represents any set

*8. How many subsets are there of the set $\{0, 1, 2, 3, 4, 5, 6\}$?

9. Insert the correct symbol, $\in$ or $\notin$, in each of the following:
   a. h ____ {a, e, i, o, u}
   b. 0 ____ $\{1, 2, 3, \ldots, 12\}$
   c. 32 ____ $\{0, 2, 4, 6, \ldots\}$
   d. 2,000 ____ $\{0, 1, 2, 3, \ldots\}$
10. Describe each of the following sets as finite or infinite:
   a. the set of consonants in the English alphabet
   b. $\{0, 1, 2, 3, 4, \ldots\}$
   c. the set of digits between 5 and 6
   d. $\{0, 2, 4, 6, \ldots, 20\}$
11. Write the correct symbol for:
   a. set union
   b. set intersection
12. $A = \{1, 3, 5, 7, \ldots\}$ $D = \{2, 5, 7, 8\}$
   $B = \{0, 2, 4, 6, \ldots\}$ $E = \varnothing$
   $C = \{3, 7, 8, 9\}$
   Given sets A–E, find:
   a. $B \cap C$
   b. $C \cup D$
   c. $A \cup B$
   d. $A \cap C$
   e. $B \cap D$
   f. $C \cap D$
   g. $B \cup E$
   h. $A \cap E$
   i. $A \cap B$
   j. $(A \cup C) \cap D$
13. Which pairs of sets are disjoint?
   $A = \{1, 3, 5, 7, 9, \ldots\}$ $C = \{3, 7, 8, 12\}$
   $B = \{0, 2, 4, 6, \ldots\}$ $D = \varnothing$

1.1 Exercise Answers

1. a. set b. empty set c. finite set d. infinite set
2. a. Alaska, California, Hawaii, Oregon, and Washington
   b. x, y, and z c. 1, 2, and 3
3. d
4. a. {c, e, g, l, o} b. $\{0, 1, 2, 4, 7\}$ c. not convenient d. $\varnothing$
   e. $\{1, 2, 3, 4, 5, 6, 7, 8\}$ or $\{1, 2, 3, \ldots, 8\}$
5. a. T b. T c. F d. T
6. a. $\subseteq$ b. $\subset$

7. **a.** T **b.** F **c.** T **d.** F **e.** T **f.** T **g.** T
h. F **i.** T **j.** T

***8.** 128: 2^7

9. **a.** $\notin$ **b.** $\notin$ **c.** $\in$ **d.** $\in$

10. **a.** finite **b.** infinite **c.** finite **d.** finite

11. **a.** $\cup$ **b.** $\cap$

12. **a.** $\{8\}$ **b.** $\{2, 3, 5, 7, 8, 9\}$ **c.** $\{0, 1, 2, 3, \ldots\}$ **d.** $\{3, 7, 9\}$
e. $\{2, 8\}$ **f.** $\{7, 8\}$ **g.** $\{0, 2, 4, 6, \ldots\}$ **h.** $\varnothing$
i. $\varnothing$ **j.** $\{5, 7, 8\}$

13. A and B, A and D, B and D, C and D

1.2 The Set of Whole Numbers

1 A basic set of numbers with which you should be familiar is the set of counting numbers:

$\{1, 2, 3, 4, 5, \ldots\}$

Mathematicians usually refer to this set as the set of *natural numbers,* denoted by N. That is,

$N = \{1, 2, 3, 4, 5, \ldots\}$

Q1 Insert $\in$ or $\notin$ correctly:

a. 4 ____ N **b.** 9 ____ N **c.** 0 ____ N **d.** 12 ____ N

e. $5\frac{1}{2}$ ____ N **f.** 0.23 ____ N **g.** $\sqrt{2}$ ____ N **h.** $1{,}001$ ____ N

STOP • STOP • STOP • STOP • STOP • STOP • STOP • STOP • STOP

A1 **a.** $\in$ **b.** $\in$ **c.** $\notin$ **d.** $\in$
e. $\notin$ **f.** $\notin$ **g.** $\notin$ **h.** $\in$

2 The set of *whole numbers,* W, is the set of natural numbers together with the set containing zero.

$W = N \cup \{0\}$

That is,

$W = \{0, 1, 2, 3, 4, \ldots\}$

Q2 True or false? (Use T or F.)

a. $N \subset W$ ____

b. $0 \in W$ but $0 \notin N$ ____

c. $5 \in N$ and $5 \in W$ ____

d. The set of whole numbers is an infinite set. ____

e. $\{0, 1, 2, 3, 4, \ldots\}$ is a well-defined collection. ____

STOP • STOP • STOP • STOP • STOP • STOP • STOP • STOP • STOP

A2 **a.** T **b.** T **c.** T **d.** T **e.** T

3 The phrase $x + 2$ is called an *open expression* or *algebraic expression.* The algebraic expression $x + 2$ may be evaluated for various values of x. For this reason, x is called a *variable.* The value 2 will remain constantly 2. Such a value is called a *constant.* Some examples:

1. In $x - 3$, x is a variable and 3 is a constant.
2. In $8x + 7$, x is a variable and 8 and 7 are constants. The number factor 8 is called the *numerical coefficient* of $8x$; the letter factor (variable) is called the *literal coefficient* of $8x$.
3. In $xy + 5 - z$, x, y, and z are variables and 5 is a constant.

Q3 In each algebraic expression, identify the (1) variable(s), and (2) the constant(s):

a. $5(x + 2)$ (1) ________ (2) ________

b. $7xyz$ (1) ________ (2) ________

c. $3x$ (1) ________ (2) ________

d. $4m + 7n - 2p$ (1) ________ (2) ________

e. $15r - 9t - 4uv$ (1) ________ (2) ________

STOP • STOP • STOP • STOP • STOP • STOP • STOP • STOP • STOP

A3 **a.** (1) x (2) 5 and 2 **b.** (1) x, y, and z (2) 7
c. (1) x (2) 3 **d.** (1) m, n, and p (2) 4, 7, and 2
e. (1) r, t, u, and v (2) 15, 9, and 4

4 Any letter may be used as a variable in an algebraic expression. However, mathematicians usually use letters from the end of the alphabet, such as m, n, o, . . . , z. Once the variable is replaced by a numerical value, the resulting expression may be evaluated.* Evaluate for the given value of the following variables:

1. $x + 2$ when $x = 7$.

$$\begin{aligned} x + 2 &= 7 + 2 \\ &= 9 \end{aligned}$$

2. $x + 2$ when $x = 12$.

$$\begin{aligned} x + 2 &= 12 + 2 \\ &= 14 \end{aligned}$$

3. $3y - 4$ when $y = 8$.

$$\begin{aligned} 3y - 4 &= 3(8) - 4 \\ &= 24 - 4 \\ &= 20 \end{aligned}$$

(*Note:* $3y$ means $3 \cdot y$, where the "raised dot" indicates multiplication.) Other conventions for showing multiplication include $3(8) = (3)(8) = (3)8 = 3 \cdot 8$.

*Recall the *order of operations:*

Step 1: Perform all work within grouping symbols first.

Step 2: Next, perform multiplications and divisions from left to right.

Step 3: Finally, perform additions and subtractions from left to right.

4. $5(2y + 7)$ when $y = 3$.

$$\begin{aligned} 5(2y + 7) &= 5(2 \cdot 3 + 7) \\ &= 5(6 + 7) \\ &= 5(13) \\ &= 65 \end{aligned}$$

5. $(6w - 2)(10 - w)$ when $w = 7$.

$$\begin{aligned} (6w - 2)(10 - w) &= (7 \cdot 6 - 2)(10 - 7) \\ &= (42 - 2)3 \\ &= 40 \cdot 3 \\ &= 120 \end{aligned}$$

Q4 Evaluate:

a. $2r - 5$ when $r = 9$

b. $4(3x + 5)$ when $x = 5$

c. $3(3y - 3)$ when $y = 3$

d. $x(x + 7)$ when $x = 10$

e. $4 + 3(t - 2)$ when $t = 6$

f. $(m + 2)(m - 2)$ when $m = 7$

STOP • STOP • STOP • STOP • STOP • STOP • STOP • STOP • STOP

A4

a. 13: $\begin{aligned} 2r - 5 &= 2 \cdot 9 - 5 \\ &= 18 - 5 \end{aligned}$

b. 80: $\begin{aligned} 4(3x + 5) &= 4(3 \cdot 5 + 5) \\ &= 4(15 + 5) \\ &= 4 \cdot 20 \end{aligned}$

c. 18: $\begin{aligned} 3(3y - 3) &= 3(3 \cdot 3 - 3) \\ &= 3(9 - 3) \\ &= 3 \cdot 6 \end{aligned}$

d. 170: $\begin{aligned} x(x + 7) &= 10(10 + 7) \\ &= 10 \cdot 17 \end{aligned}$

e. 16: $\begin{aligned} 4 + 3(t - 2) &= 4 + 3(6 - 2) \\ &= 4 + 3 \cdot 4 \\ &= 4 + 12 \end{aligned}$

f. 45: $\begin{aligned} (m + 2)(m - 2) &= (7 + 2)(7 - 2) \\ &= 9 \cdot 5 \end{aligned}$

5

The algebraic expressions $x + 7$ and $7 + x$ are *equivalent* because they have the same evaluation for all replacements of the variable. Hence $x + 7 = 7 + x$ is true for all whole-number replacements of x. For example:

$x + 7 = 7 + x$
$0 + 7 = 7 + 0$
$1 + 7 = 7 + 1$
$2 + 7 = 7 + 2$ etc.

The algebraic expressions $x + y$ and $y + x$ are equivalent because they have the same evaluation for all replacements of the variable. Hence $x + y = y + x$ is true for all whole-number replacements of x and y.

Q5 Verify that $x + y = y + x$ is true when $x = 5$ and $y = 4$.

STOP • STOP • STOP • STOP • STOP • STOP • STOP • STOP • STOP

A5
$$x + y = y + x$$
$$5 + 4 = 4 + 5$$
$$9 = 9 \quad \text{(true statement)}$$

6 The assumption that $x + y = y + x$ is true for all whole-number replacements of x and y is called the *commutative property of addition*. It is concerned with the *order* in which two numbers are added. Addition is called a commutative operation because the order in which two numbers are added does not affect the sum. Examples of this property are:

$$0 + 5 = 5 + 0$$
$$2 + 7 = 7 + 2$$
$$52 + 1 = 1 + 52$$
$$547{,}245 + 249{,}346 = 249{,}346 + 547{,}245$$

Q6 $15 + 7 = 7 + 15$ is an example of the ____________ property of addition.

STOP • STOP • STOP • STOP • STOP • STOP • STOP • STOP • STOP

A6 commutative

Q7 Use the commutative property of addition to complete a true statement:

a. $17 + 8 =$ __________ **b.** $0 + 9 =$ __________

c. $1 + 12 =$ __________ **d.** $92 + 18 =$ __________

STOP • STOP • STOP • STOP • STOP • STOP • STOP • STOP • STOP

A7 **a.** $8 + 17$ **b.** $9 + 0$ **c.** $12 + 1$ **d.** $18 + 92$

Q8 The commutative property of addition indicates that the __________ of two terms in an addition problem may be reversed without affecting the sum.

STOP • STOP • STOP • STOP • STOP • STOP • STOP • STOP • STOP

A8 order

7 The commutative property of addition includes an infinite number of specific cases. The statement "$x + y = y + x$ is true for all possible whole-number replacements of x and y" illustrates the power of making one statement that covers an infinite number of cases. A similar property is true for multiplication. It is true that $3 \cdot 5 = 5 \cdot 3$. In general, the *order* of the factors in a multiplication does not affect the product. The *commutative property of multiplication* assumes that $xy = yx$ is true for all whole-number replacements of x and y. Other examples of this property are:

$$2 \cdot 17 = 17 \cdot 2$$
$$93 \cdot 6 = 6 \cdot 93$$
$$0 \cdot 15 = 15 \cdot 0$$

Q9 Use the commutative property of multiplication to complete a true statement:

a. $1 \cdot 15 =$ ______ b. $107 \cdot 49 =$ ______

c. $7 \cdot 12 =$ ______ d. $50 \cdot 75 =$ ______

STOP • STOP • STOP • STOP • STOP • STOP • STOP • STOP • STOP

A9 a. $15 \cdot 1$ b. $49 \cdot 107$ c. $12 \cdot 7$ d. $75 \cdot 50$

Q10 a. Use the commutative property of addition to change $2x + 5$ into an equivalent expression. ______

b. Use the commutative property of multiplication to change $3x \cdot 5$ into an equivalent expression. ______

STOP • STOP • STOP • STOP • STOP • STOP • STOP • STOP • STOP

A10 a. $5 + 2x$ b. $5 \cdot 3x$

Q11 Verify that $1 + (2 + 3) = (1 + 2) + 3$ is a true statement by showing that $1 + (2 + 3)$ and $(1 + 2) + 3$ are equivalent expressions.

STOP • STOP • STOP • STOP • STOP • STOP • STOP • STOP • STOP

A11 $1 + (2 + 3) = 1 + 5 = 6$
$(1 + 2) + 3 = 3 + 3 = 6$ or

$$1 + (2 + 3) = (1 + 2) + 3$$
$$1 + 5 = 3 + 3$$
$$6 = 6 \quad \text{(true)}$$

Q12 a. Are $2 + (m + 3)$ and $(2 + m) + 3$ equivalent for all whole-number replacements of m? ______

b. Is $2 + (t + 3) = (2 + t) + 3$ true for all whole-number replacements of t? ______

STOP • STOP • STOP • STOP • STOP • STOP • STOP • STOP • STOP

A12 a. yes (assumption) b. yes

Q13 a. Are $r + (s + t)$ and $(r + s) + t$ equivalent for all whole-number replacements of r, s, and t? ______

b. Is $p + (q + r) = (p + q) + r$ true for all whole-number replacements of p, q, and r? ______

STOP • STOP • STOP • STOP • STOP • STOP • STOP • STOP • STOP

A13 a. yes (assumption) b. yes

8 The assumption that $x + (y + z) = (x + y) + z$ is true for all whole-number replacements of x, y, and z is called the *associative property of addition*. It indicates that, in addition, the way three numbers are *grouped* does not affect the sum. Other examples of this property are:

$$(5 + 6) + 2 = 5 + (6 + 2)$$
$$12 + (5 + 0) = (12 + 5) + 0$$

Q14 The associative property of addition indicates that the ______________ of three numbers in an addition problem can be changed without affecting the sum.

STOP • STOP • STOP • STOP • STOP • STOP • STOP • STOP • STOP

A14 grouping

Q15 Use the associative property of addition to complete the following into true statements:

a. $7 + (103 + 28) =$ ______________

b. $(9 + 12) + 8 =$ ______________

c. $(8 + 4) + 6 =$ ______________

d. $9 + (31 + 12) =$ ______________

STOP • STOP • STOP • STOP • STOP • STOP • STOP • STOP • STOP

A15 a. $(7 + 103) + 28$ b. $9 + (12 + 8)$ c. $8 + (4 + 6)$ d. $(9 + 31) + 12$

---

9 A similar property is true for multiplication. In multiplication, the way three factors are grouped does not affect the product. This assumption is the *associative property of multiplication,* which states: $x(yz) = (xy)z$ is true for all whole-number replacements of x, y, and z. Examples of this property are:

$7(3 \cdot 2) = (7 \cdot 3)2$
$(3 \cdot 4)5 = 3(4 \cdot 5)$

Q16 The associative property of multiplication indicates that the grouping of three ______________ in a multiplication problem may be changed without affecting the product.

STOP • STOP • STOP • STOP • STOP • STOP • STOP • STOP • STOP

A16 factors

Q17 Use the associative property of multiplication to complete the following into true statements:

a. $(72 \cdot 5)2 =$ ______________ b. $4(25 \cdot 7) =$ ______________

c. $125(8 \cdot 9) =$ ______________ d. $(19 \cdot 8)5 =$ ______________

STOP • STOP • STOP • STOP • STOP • STOP • STOP • STOP • STOP

A17 a. $72(5 \cdot 2)$ b. $(4 \cdot 25)7$ c. $(125 \cdot 8)9$ d. $19(8 \cdot 5)$

Q18 a. Change $(x + 3) + 2$ into an equivalent expression using the associative property of addition. ______________

b. Change $2(3x)$ into an equivalent expression using the associative property of multiplication. ______________

STOP • STOP • STOP • STOP • STOP • STOP • STOP • STOP • STOP

A18 a. $x + (3 + 2)$ b. $(2 \cdot 3)x$

---

10 You should note that subtraction and division are not associative operations.

Q19 Verify the above fact by showing that each of the following are false statements:

a. $(15 - 7) - 3 = 15 - (7 - 3)$ **b.** $60 \div (15 \div 3) = (60 \div 15) \div 3$

STOP • STOP • STOP • STOP • STOP • STOP • STOP • STOP • STOP

A19 **a.** $8 - 3 = 15 - 4$ (false)
$8 - 3 \neq 15 - 4$ (true)

b. $60 \div 5 = 4 \div 3$ (false)
$60 \div 5 \neq 4 \div 3$ (true)

Q20 Give the name of the property used to change the expression on the left to the equivalent expression on the right:

a. $1x = x \cdot 1$ ______________________

b. $2 + (x + 5) = 2 + (5 + x)$ ______________________

c. $(x - 2)5 = 5(x - 2)$ ______________________

d. $(3x + 7) + 9 = 3x + (7 + 9)$ ______________________

e. $x + 0 = 0 + x$ ______________________

f. $2(5x) = (2 \cdot 5)x$ ______________________

STOP • STOP • STOP • STOP • STOP • STOP • STOP • STOP • STOP

A20 **a.** commutative property of multiplication
b. commutative property of addition
c. commutative property of multiplication
d. associative property of addition
e. commutative property of addition
f. associative property of multiplication

---

11 The following properties are much easier to recognize and remember. The *addition property of zero* (additive identity property) states that $x + 0 = 0 + x = x$ is true for all whole-number replacements of x. If zero is added to any whole number, the result is always the identical whole number.

Examples:

$5 + 0 = 0 + 5 = 5$
$17 + 0 = 0 + 17 = 17$

The *multiplication property of zero* states that $x \cdot 0 = 0x = 0$ is true for all whole-number replacements of x. If zero is multiplied by any whole number, the result is always zero.

Examples:

$5 \cdot 0 = 0 \cdot 5 = 0$
$32 \cdot 0 = 0 \cdot 32 = 0$

The *multiplication property of 1* (multiplicative identity property) states that $x \cdot 1 = 1x = x$ is true for all whole-number replacements of x. If 1 is multiplied by any whole number, the result is always the identical whole number.

Examples:

$5 \cdot 1 = 1 \cdot 5 = 5$
$108 \cdot 1 = 1 \cdot 108 = 108$

The *division property of 1* states that $\frac{x}{1} = x$ is true for all whole-number replacements of x.

If any whole number is divided by 1, the result is the same whole number.

Examples:

$\frac{5}{1} = 5$

$\frac{297}{1} = 297$

$\frac{0}{1} = 0$

Q21 Indicate the name of the property used to change the expression on the left to the equivalent expression on the right:

a. $\frac{5x}{1} = 5x$ __________

b. $(2x + 1)1 = 2x + 1$ __________

c. $(5x + 0) + 3 = 5x + 3$ __________

d. $\frac{2}{3}x \cdot 0 = 0$ __________

STOP • STOP • STOP • STOP • STOP • STOP • STOP • STOP • STOP

A21 **a.** division property of 1
b. multiplication property of 1 or multiplicative identity property
c. addition property of zero or additive identity property
d. multiplication property of zero

12 Consider the following evaluations*:

$2(3 + 4)$	and	$2(3 + 4)$
$2(7)$		$2 \cdot 3 + 2 \cdot 4$
14		$6 + 8$
		14

$2(3 + 4)$ and $2 \cdot 3 + 2 \cdot 4$ are equivalent expressions, because both expressions have the same evaluation. Therefore, $2(3 + 4) = 2 \cdot 3 + 2 \cdot 4$ is a true statement.

*When steps of an evaluation are written one under the other without equal signs, each line is assumed to be equal to the one above it.

Q22 Verify that $5 \cdot 6 + 5 \cdot 2 = 5(6 + 2)$ is a true statement by showing that $5 \cdot 6 + 5 \cdot 2$ and $5(6 + 2)$ are equivalent expressions.

STOP • STOP • STOP • STOP • STOP • STOP • STOP • STOP • STOP

A22 $5 \cdot 6 + 5 \cdot 2$ and $5(6 + 2)$
$30 + 10$ $5 \cdot 8$
40 40

Q23 Complete the following to form true statements:

a. $3(2 + 5)$
$3 \cdot 2 + 3 \cdot$ _____

b. $7(8 + 6)$
$7 \cdot$ _____ $+ 7 \cdot 6$

STOP • STOP • STOP • STOP • STOP • STOP • STOP • STOP • STOP

A23 **a.** $3 \cdot 2 + 3 \cdot \underline{5}$ **b.** $7 \cdot \underline{8} + 7 \cdot 6$

13 The true statement $9(10 + 3) = 9 \cdot 10 + 9 \cdot 3$ is an example of the *left distributive property of multiplication over addition.* You should observe that $9(10 + 3)$ is a *product* of 9 times 13. $9 \cdot 10 + 9 \cdot 3$ is a *sum* of 90 and 27. This property changes a product to a sum. It also changes a sum to a product.

$$\underbrace{2 \cdot 6 + 2 \cdot 4}_{\text{sum}} = \underbrace{2(6 + 4)}_{\text{product}}$$

Q24 Use the left distributive property of multiplication over addition to change each sum to a product or each product to a sum (parts a and b are given as examples):

a. $5 \cdot 3 + 5 \cdot 7 =$ $\underline{5(3 + 7)}$
b. $6(4 + 5) =$ $\underline{6 \cdot 4 + 6 \cdot 5}$
c. $9(2 + 1) =$ __________
d. $17(5 + 6) =$ __________
e. $10 \cdot 2 + 10 \cdot 3 =$ __________
f. $8 \cdot 12 + 8 \cdot 8 =$ __________

STOP • STOP • STOP • STOP • STOP • STOP • STOP • STOP • STOP

A24 **c.** $9 \cdot 2 + 9 \cdot 1$ **d.** $17 \cdot 5 + 17 \cdot 6$
e. $10(2 + 3)$ **f.** $8(12 + 8)$

14 In general, the *left distributive property of multiplication over addition* states that $x(y + z) = xy + xz$ is true for all whole-number replacements of x, y, and z.

Q25 Use the left distributive property of multiplication over addition to complete the evaluation of each numerical expression:

a. $4 \cdot 7 + 4 \cdot 3$
4(_____ + _____)
$4 \cdot$ _____
_____

b. $15(10 + 2)$
$15 \cdot$ _____ $+ 15 \cdot$ _____
_____ + _____
_____

STOP • STOP • STOP • STOP • STOP • STOP • STOP • STOP • STOP

A25 **a.** $4(\underline{7} + \underline{3})$
$4 \cdot \underline{10}$
$\underline{40}$

b. $15 \cdot \underline{10} + 15 \cdot \underline{2}$
$\underline{150} + \underline{30}$
$\underline{180}$

15 Consider these evaluations:

$(3 + 4)5$	and	$3 \cdot 5 + 4 \cdot 5$
$(7)5$		$15 + 20$
35		35

Hence $(3 + 4)5 = 3 \cdot 5 + 4 \cdot 5$. This is an example of the *right distributive property of multiplication over addition.* This property also changes products to sums or sums to products.

Q26 Use the right distributive property of multiplication over addition to change each sum to a product or each product to a sum:

a. $(4 + 2)7 =$ ____________

b. $2 \cdot 3 + 21 \cdot 3 =$ ____________

c. $18 \cdot 12 + 2 \cdot 12 =$ ____________

d. $(13 + 19)10 =$ ____________

STOP • STOP • STOP • STOP • STOP • STOP • STOP • STOP • STOP

A26 a. $4 \cdot 7 + 2 \cdot 7$ b. $(2 + 21)3$
c. $(18 + 2) \cdot 12$ d. $13 \cdot 10 + 19 \cdot 10$

16 In general, the *right distributive property of multiplication over addition* states that $(x + y)z = xz + yz$ is true for all whole-number replacements of x, y, and z.

Q27 Complete the following evaluations:

a. $14 \cdot 9 + 6 \cdot 9$

(____ + ____) $\cdot$ 9

____ $\cdot$ 9

____

b. $(17 + 8)12$

$17 \cdot$ ____ $+ 8 \cdot$ ____

____ + ____

____

STOP • STOP • STOP • STOP • STOP • STOP • STOP • STOP • STOP

A27 a. $(\underline{14} + \underline{6}) \cdot 9$

$\underline{20} \cdot 9$

$\underline{180}$

b. $17 \cdot \underline{12} + 8 \cdot \underline{12}$

$\underline{204} + \underline{96}$

$\underline{300}$

17 Notice that

$7(5 - 2)$	and	$7 \cdot 5 - 7 \cdot 2$
$7 \cdot 3$		$35 - 14$
21		21

Hence $7(5 - 2) = 7 \cdot 5 - 7 \cdot 2$. This is an example of the *left distributive property of multiplication over subtraction.* Observe that $7(5 - 2)$ is a product while $7 \cdot 5 - 7 \cdot 2$ is a difference. This property also changes a difference to a product. For example, $7 \cdot 5 - 7 \cdot 2 = 7(5 - 2)$.

Q28 Use the left distributive property of multiplication over subtraction to change each product to a difference or each difference to a product (parts a and b are given as examples):

a. $5 \cdot 7 - 5 \cdot 3 =$ $\underline{5(7 - 3)}$

b. $10(19 - 7) =$ $\underline{10 \cdot 19 - 10 \cdot 7}$

c. $12(13 - 6) =$ ______________

d. $8 \cdot 7 - 8 \cdot 3 =$ ______________

e. $24 \cdot 12 - 24 \cdot 2 =$ ______________

f. $3(36 - 27) =$ ______________

STOP • STOP • STOP • STOP • STOP • STOP • STOP • STOP • STOP

A28 **c.** $12 \cdot 13 - 12 \cdot 6$ **d.** $8(7 - 3)$
e. $24(12 - 2)$ **f.** $3 \cdot 36 - 3 \cdot 27$

18	In general, the *left distributive property of multiplication over subtraction* states that $x(y - z) = xy - xz$ is true for all whole-number replacements of x, y, and z.

Q29 Complete the following evaluations:

a. $24(12 - 2)$

$24 \cdot$ _____ $- 24 \cdot$ _____

_____ $-$ _____

_____

b. $7 \cdot 23 - 7 \cdot 3$

$7($_____ $-$ _____$)$

$7 \cdot$ _____

_____

STOP • STOP • STOP • STOP • STOP • STOP • STOP • STOP • STOP

A29 **a.** $24 \cdot \underline{12} - 24 \cdot \underline{2}$

$\underline{288} - \underline{48}$

$\underline{240}$

b. $7(\underline{23} - \underline{3})$

$7 \cdot \underline{20}$

$\underline{140}$

19	The true statement $(12 - 9)2 = 12 \cdot 2 - 9 \cdot 2$ is an example of the *right distributive property of multiplication over subtraction*. This property also changes products to differences or differences to products.

Q30 Use the right distributive property of multiplication over subtraction to change each product to a difference or each difference to a product:

a. $(12 - 2)6 =$ ______________

b. $7 \cdot 4 - 5 \cdot 4 =$ ______________

STOP • STOP • STOP • STOP • STOP • STOP • STOP • STOP • STOP

A30 **a.** $12 \cdot 6 - 2 \cdot 6$ **b.** $(7 - 5)4$

20	In general, the *right distributive property of multiplication over subtraction* states that $(x - y)z = xz - yz$ is true for all whole-number replacements of x, y, and z.

Q31 Complete the following evaluations:

a. $(45 - 7)100$

_____ $\cdot 100 -$ _____ $\cdot 100$

_______ $-$ _______

____

b. $35 \cdot 9 - 5 \cdot 9$

$($_____ $-$ _____$) \cdot 9$

_____ $\cdot 9$

_____

STOP • STOP • STOP • STOP • STOP • STOP • STOP • STOP • STOP

A31

a. $\underline{45} \cdot 100 - \underline{7} \cdot 100$

$\underline{4{,}500} - \underline{700}$

$\underline{3{,}800}$

b. $(\underline{35} - \underline{5}) \cdot 9$

$\underline{30} \cdot \underline{9}$

$\underline{270}$

21

Consider the following examples of the distributive properties:

$$5(7 + 9) = 5 \cdot 7 + 5 \cdot 9$$
$$5(x + 9) = 5x + 5 \cdot 9$$
$$(12 - 5)6 = 12 \cdot 6 - 5 \cdot 6$$
$$(y - 5)6 = y \cdot 6 - 5 \cdot 6$$
$$10(15 - 7) = 10 \cdot 15 - 10 \cdot 7$$
$$10(t - 7) = 10t - 10 \cdot 7$$
$$(32 + 11)8 = 32 \cdot 8 + 11 \cdot 8$$
$$(32 + m)8 = 32 \cdot 8 + m \cdot 8$$

In each example a product was changed to a sum or difference. Furthermore, it is said that the parentheses have been removed. When changing products to sums, the process of removing the parentheses requires the application of one of the distributive properties.

Q32

Remove the parentheses from each of the following expressions and complete the evaluation (part a is given as an example):

a. $2(7 - 3) =$ $\underline{2 \cdot 7 - 2 \cdot 3 = 14 - 6 = 8}$

b. $18(5 + 3) =$ ____________________

c. $(12 - 8)6 =$ ____________________

d. $20(9 + 7) =$ ____________________

e. $35(2 - 1) =$ ____________________

STOP • STOP • STOP • STOP • STOP • STOP • STOP • STOP • STOP

A32

b. $18 \cdot 5 + 18 \cdot 3 = 90 + 54 = 144$
c. $12 \cdot 6 - 8 \cdot 6 = 72 - 48 = 24$
d. $20 \cdot 9 + 20 \cdot 7 = 180 + 140 = 320$
e. $35 \cdot 2 - 35 \cdot 1 = 70 - 35 = 35$

22

When you perform an operation on *any two* numbers of a set and get another number in the set, the set is said to be *closed* under that operation. Consider $\{1, 2, 3\}$. $1 \in \{1, 2, 3\}$. $3 \in \{1, 2, 3\}$. Is $1 + 3 \in \{1, 2, 3\}$? Since $4 \notin \{1, 2, 3\}$, $\{1, 2, 3\}$ is *not* closed under addition.

Q33

Is the set $\{2, 4, 6\}$ closed under addition? ________

STOP • STOP • STOP • STOP • STOP • STOP • STOP • STOP • STOP

A33

no: $2 + 2$ and $2 + 4$ are the only sums that give results which are still in the set; $2 + 6$, $4 + 4$, $4 + 6$, and $6 + 6$ are not in the set

23

The set of natural numbers is closed under addition because the sum of any two natural numbers is another natural number. That is, given $N = \{1, 2, 3, \ldots\}$, if $x \in N$ and $y \in N$, then $x + y \in N$.

Q34

Is the set of whole numbers, $W = \{0, 1, 2, 3, \ldots\}$, closed under addition? ________

Why? ____________________

STOP • STOP • STOP • STOP • STOP • STOP • STOP • STOP • STOP

A34 yes, because the sum of any two whole numbers is another whole number

---

Q35 Is the set of whole numbers, W, closed under multiplication? ______

Why? ______

STOP • STOP • STOP • STOP • STOP • STOP • STOP • STOP • STOP

A35 yes, the product of any two whole numbers is another whole number

---

Q36 Complete the following: If $r \in W$ and $s \in W$, then $rs \in$ ______.

STOP • STOP • STOP • STOP • STOP • STOP • STOP • STOP • STOP

A36 W

24 The set of whole numbers, W, is not closed under subtraction. For example, $5 \in W$ and $6 \in W$, but $5 - 6$ or $-1 \notin W$.

Q37 Illustrate that the set of whole numbers is not closed under division.

STOP • STOP • STOP • STOP • STOP • STOP • STOP • STOP • STOP

A37 Many examples may be given. One such example is: $2 \in W$ and $4 \in W$, but $2 \div 4$ or $\frac{1}{2} \notin W$.

25 It is said that the set of whole numbers is closed under the operations of addition and multiplication. The *closure* properties for whole numbers state:

Closure Property of Addition of Whole Numbers

If $x \in W$ and $y \in W$, then $x + y \in W$.

Closure Property of Multiplication of Whole Numbers

If $x \in W$ and $y \in W$, then $xy \in W$.

The set of whole numbers is not closed under the operations of subtraction and division.

This completes the instruction for this section.

1.2 Exercises

1. One of the properties presented in this section was used to change the expression on the left to the equivalent expression on the right. Name the property used (assume whole-number replacements for all variables):
   a. $(xy)4 = 4(xy)$ b. $7z + 0 = 7z$
   c. $1(x + y) = x + y$ d. $(2m + n) + 3n = 2m + (n + 3n)$
   e. $(5 + t) + 3 = (t + 5) + 3$ f. $4(5x) = (4 \cdot 5)x$
   g. $5x \cdot 0 = 0$ h. $\dfrac{2x - 1}{1} = 2x - 1$
2. Indicate the number of the general statement that corresponds to its name. (Assume that general statements are truc for all whole-number replacements of x, y, and z.)

a. commutative property of addition
b. commutative property of multiplication
c. associative property of addition
d. associative property of multiplication
e. addition property of zero
f. multiplication property of zero
g. multiplication property of 1
h. division property of 1

1. $x \cdot 1 = 1x = x$
2. $x(yz) = (xy)z$
3. $x \cdot 0 = 0x = 0$
4. $x + y = y + x$
5. $\frac{x}{1} = x$
6. $(x + y) + z = x + (y + z)$
7. $xy = yx$
8. $x + 0 = 0 + x = x$

3. Use the property given to write a true statement:
   a. By the associative property of addition, $2 + (3 + 4) =$ ________.
   b. By the multiplication property of 1, $32 \cdot 1 =$ ________.
   c. By the commutative property of multiplication, $1 \cdot 15 =$ ________.
   d. By the addition property of zero, $(72 - 5) + 0 =$ ________.
   e. By the multiplication property of zero, $0(15 \cdot 3) =$ ________.
   f. By the associative property of multiplication, $1(4 \cdot 7) =$ ________.
   g. By the division property of 1, $\frac{42}{1} =$ ________.
   h. By the commutative property of addition, $2(3 + 4) =$ ________.
4. Change each product to a sum or each sum to a product:
   a. $9(6 + 7)$ b. $2 \cdot 4 + 2 \cdot 7$
   c. $(15 + 9)2$ d. $6 \cdot 8 + 10 \cdot 8$
5. Change each product to a difference or each difference to a product:
   a. $7 \cdot 9 - 7 \cdot 4$ b. $3(12 - 8)$
   c. $(17 - 2)5$ d. $27 \cdot 8 - 7 \cdot 8$
6. Use one of the distributive properties to evaluate each expression (remove the parentheses):
   a. $2(31 - 6)$ b. $(15 + 7)3$
   c. $(18 - 3)4$ d. $12(10 - 8)$
7. Name the distributive property that was used to change the expression on the left to the equivalent expression on the right (assume whole-number replacements for all variables):
   a. $3(x + 5) = 3x + 3 \cdot 5$ b. $4u - 1u = (4 - 1)u$
   c. $5y - 5 \cdot 3 = 5(y - 3)$ d. $7(p - 2) = 7p - 7 \cdot 2$
   e. $5x + 3x = (5 + 3)x$
8. Complete the following general statements assumed true for all whole-number replacements of x, y, and z:
   a. $x(y - z) =$ ____________ b. $(x + y)z =$ ____________
   c. $(x - y)z =$ ____________ d. $x(y + z) =$ ____________
9. State the closure properties for the set of whole numbers.
10. Is the set of natural numbers closed under subtraction? Why?
11. Give one example which illustrates that the set of natural numbers is not closed under division.

1.2 Exercise Answers

1. a. commutative property of multiplication
   b. addition property of zero (additive identity property)

- **c.** multiplication property of 1 (multiplicative identity property)
- **d.** associative property of addition
- **e.** commutative property of addition
- **f.** associative property of multiplication
- **g.** multiplication property of zero
- **h.** division property of 1

2. **a.** 4 **b.** 7 **c.** 6 **d.** 2 **e.** 8 **f.** 3 **g.** 1 **h.** 5

3. **a.** $(2 + 3) + 4$ **b.** 32 **c.** $15 \cdot 1$ **d.** $72 - 5$ **e.** 0 **f.** $(1 \cdot 4)7$ **g.** 42 **h.** $2(4 + 3)$

4. **a.** $9 \cdot 6 + 9 \cdot 7$ **b.** $2(4 + 7)$ **c.** $15 \cdot 2 + 9 \cdot 2$ **d.** $(6 + 10)8$

5. **a.** $7(9 - 4)$ **b.** $3 \cdot 12 - 3 \cdot 8$ **c.** $17 \cdot 5 - 2 \cdot 5$ **d.** $(27 - 7)8$

6. **a.** 50 **b.** 66 **c.** 60 **d.** 24

7. **a.** left distributive property of multiplication over addition
b. right distributive property of multiplication over subtraction
c. left distributive property of multiplication over subtraction
d. left distributive property of multiplication over subtraction
e. right distributive property of multiplication over addition

8. **a.** $xy - xz$ **b.** $xz + yz$ **c.** $xz - yz$ **d.** $xy + xz$

9. If $x \in W$ and $y \in W$, then $x + y \in W$. If $x \in W$ and $y \in W$, then $xy \in W$. (Variables other than x and y may be used.)

10. no, because the difference between any two natural numbers is not always another natural number (or any equivalent statement or example)

11. $2 \in N$ and $4 \in N$, but $2 \div 4 \notin N$. (This is one of many possible examples.)

1.3 The Set of Integers

1 In Section 1.2 the set of whole numbers was denoted:

$W = \{0, 1, 2, 3, 4, \ldots\}$

In this section the set of *integers* will be reviewed. In set notation, the set of integers is

$I = \{\ldots, {}^-3, {}^-2, {}^-1, 0, {}^+1, {}^+2, {}^+3, \ldots\}$

Each integer is considered to have two parts, *distance* and *direction*. Integers may be displayed on a *number line* as follows:

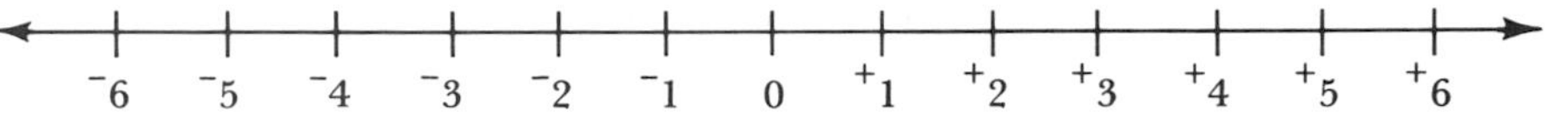

Zero indicates a reference point. $^+1$ (positive 1) represents a distance of *one* unit to the *right* of zero. $^-1$ (negative 1) represents a distance of *one* unit to the *left* of zero; and so on.

Q1 **a.** $^-5$ is read "__________ (positive/negative) five" and is located ____ units to the ________ (right/left) of zero.

b. $^{+}5$ is read "________ five" and is located ____ units to the ________ of zero.
(positive/negative) (right/left)

STOP • STOP • STOP • STOP • STOP • STOP • STOP • STOP • STOP

A1 **a.** negative, 5, left **b.** positive, 5, right

---

2 Zero is an integer but is considered to be neither positive nor negative. The set of integers can be thought of as being made up of three subsets:

$$I = \{\underbrace{\ldots, {}^{-}3, {}^{-}2, {}^{-}1}_{\text{negative integers}}, \overset{\text{zero}}{0}, \underbrace{{}^{+}1, {}^{+}2, {}^{+}3, \ldots}_{\text{positive integers}}\}$$

It is commonly agreed that positive integers may be written without the "+" sign. That is, $^{+}1$ may be written as 1, $^{+}2$ as 2, $^{+}3$ as 3, and so on. Even though 0 is written without a sign; it is not considered to be positive. In fact, zero is neither positive nor negative. (*Note:* The set of natural numbers, N, is a proper subset of I. That is, $N \subset I$.)

Q2 True or false? (Use T or F.)

a. $^{+}17$ may be written as 17. ____

b. $^{-}3$ may be written as 3. ____

c. $\{1, 2, 3, \ldots\}$ may be thought of as the set of positive integers. ____

d. $\{^{-}1, {}^{-}2, {}^{-}3, \ldots\}$ represents the set of negative integers. ____

e. 0 is neither positive nor negative. ____

f. $N \subset I$. ____

STOP • STOP • STOP • STOP • STOP • STOP • STOP • STOP • STOP

A2 **a.** T **b.** F **c.** T **d.** T **e.** T **f.** T

Q3 Insert $\in$ or $\notin$ to form a true statement:

a. 5 ____ N **b.** 5 ____ W **c.** 5 ____ I

d. 0 ____ N **e.** 0 ____ W **f.** 0 ____ I

g. $^{-}8$ ____ N **h.** $^{-}8$ ____ W **i.** $^{-}8$ ____ I

STOP • STOP • STOP • STOP • STOP • STOP • STOP • STOP • STOP

A3 **a.** $\in$ **b.** $\in$ **c.** $\in$
d. $\notin$ **e.** $\in$ **f.** $\in$
g. $\notin$ **h.** $\notin$ **i.** $\in$

---

3 Each integer may be paired with a second integer which is the same distance from zero. The paired integers are called *opposites.** That is, 7 is the opposite of $^{-}7$ and $^{-}7$ is the opposite of 7. Zero is its own opposite.

*Some mathematicians refer to opposites as "additive inverses."

Q4 Write the opposite of each integer:

a. 9 ____ **b.** $^{-}3$ ____ **c.** 0 ____

d. 4 _____ **e.** $^-12$ _____ **f.** $^-73$ _____

STOP • STOP • STOP • STOP • STOP • STOP • STOP • STOP • STOP

A4 **a.** $^-9$ **b.** 3 **c.** 0
d. $^-4$ **e.** 12 **f.** 73

4 The sum of integers may be pictured on the number line as follows (notice that each problem starts at zero):

Examples:

1. $4 + 5 = ?$

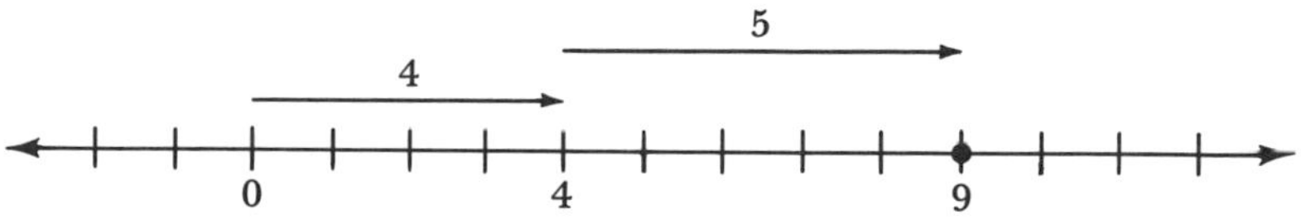

Therefore, $4 + 5 = 9$.

2. $^-4 + 5 = ?$

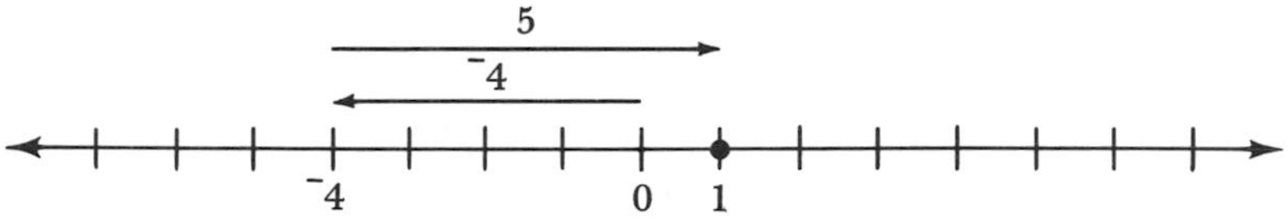

Therefore, $^-4 + 5 = 1$.

3. $^-7 + 3 = ?$

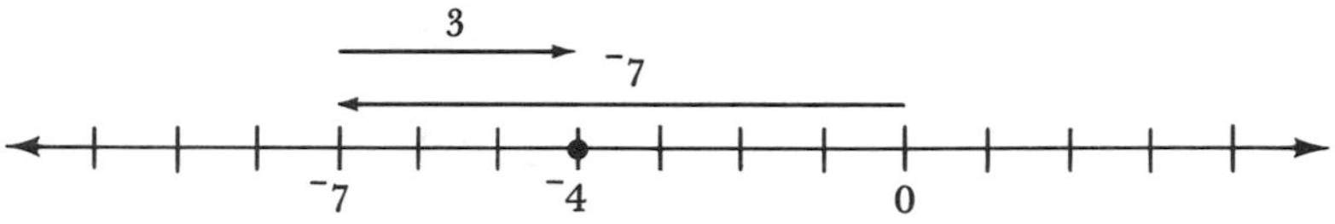

Therefore, $^-7 + 3 = {}^-4$.

4. $9 + {}^-2 = ?$

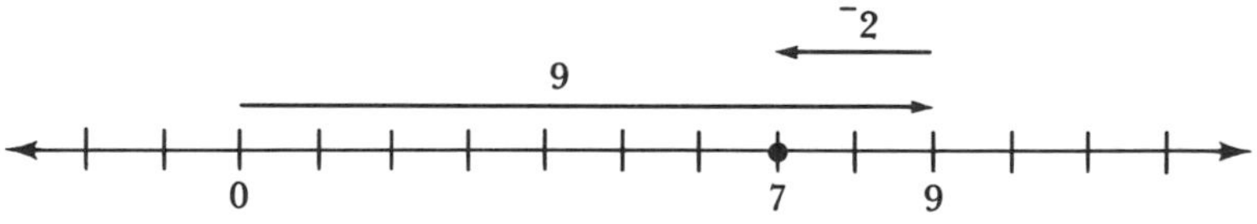

Therefore, $9 + {}^-2 = 7$.

5. $^-3 + {}^-4 = ?$

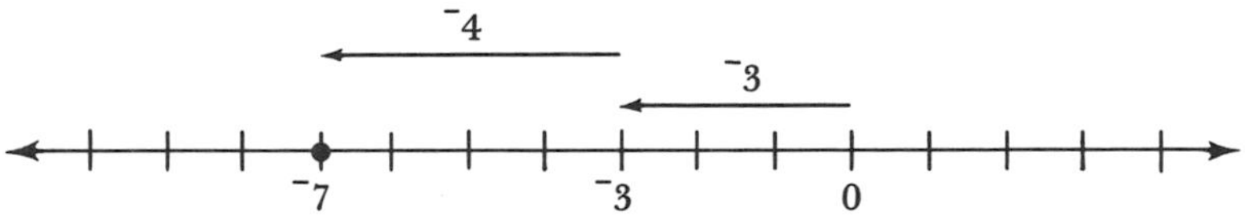

Therefore, $^-3 + {}^-4 = {}^-7$.

Q5 Use a number line (if necessary) to find:

a. $^-4 + {}^-5 =$ _____ **b.** $3 + 2 =$ _____

c. $^-6 + 4 =$ _____ **d.** $6 + {}^-6 =$ _____

e. $^-5 + 9 =$ _____ **f.** $^-2 + {}^-7 =$ _____

g. $2 + {}^-9 =$ _____ **h.** $^-5 + 5 =$ _____

i. $9 + {}^-12 =$ _____ **j.** $^-3 + 12 =$ _____

STOP • STOP • STOP • STOP • STOP • STOP • STOP • STOP • STOP

A5 a. $^-9$ b. 5 c. $^-2$ d. 0 e. 4 f. $^-9$ g. $^-7$
h. 0 i. $^-3$ j. 9

5 The properties developed for the set of whole numbers are also true for the set of integers. That is, the commutative properties of addition and multiplication, the associative properties of addition and multiplication, and the distributive properties of multiplication over addition and subtraction are all true for the set of integers. The set of integers is likewise closed under addition and multiplication.

In addition, you should observe that the sum of opposites is zero.

Examples:

$7 + {}^-7 = {}^-7 + 7 = 0$
${}^-8 + 8 = 8 + {}^-8 = 0$

The fact that the sum of any integer and its opposite is zero is generalized as:

$x + {}^-x = {}^-x + x = 0$ for any integer replacement of x

(Read "x plus the opposite of x equals the opposite of x plus x equals 0.") That is, ^-x says "the *opposite* of x."

If $x = 2$, ${}^-x = {}^-(2) = {}^-2$
If $x = {}^-2$, ${}^-x = {}^-({}^-2) = 2$

From this point on, the raised negative sign or the raised opposing sign will only be used to avoid confusion over the interpretation of a "$-$" sign or to avoid excessive use of grouping symbols. For example, $^-3$ will be written as -3 and ^-x as $-x$. Also, $3 - {}^-5$ will be written rather than $3 - ({}^-5)$ and $-2(3 + {}^-5)$ rather than $-2[3 + ({}^-5)]$.

Q6 Verify that $x + {}^-x = 0$ is true when:

a. $x = 3$ b. $x = -5$

STOP • STOP • STOP • STOP • STOP • STOP • STOP • STOP • STOP

A6 a. $x + {}^-x$
$3 + {}^-(3)$
$3 + {}^-3$
0

b. $x + {}^-x$
$-5 + {}^-(-5)$
$-5 + 5$
0

Q7 Find the replacement for x that converts each open sentence into a true statement:

a. $2 + {}^-2 = x$ _____ b. $x + 7 = 0$ _____

c. $-15 + x = 0$ _____ d. $-x = 5$ _____

STOP • STOP • STOP • STOP • STOP • STOP • STOP • STOP • STOP

A7 a. 0 b. -7 c. 15 d. -5

6 The fact that $x + {}^-x = -x + x = 0$ is true for all integer replacements of x is called the *addition property of opposites.*

Q8 According to the addition property of opposites, $x +$ ____ $= 0$.

STOP • STOP • STOP • STOP • STOP • STOP • STOP • STOP • STOP

A8 $-x$ (read "opposite of x")

Q9 If $x + y = 0$, then $x =$ ____ or $y =$ ____ for all integer replacements of x and y.

STOP • STOP • STOP • STOP • STOP • STOP • STOP • STOP • STOP

A9 $-y$, $-x$

7	"If $x + y = 0$, then $x = -y$ and $y = -x$ for all integer replacements of x and y" means that "If the sum of two integers is zero, one integer is the opposite of the other."

Q10 **a.** If $x + 3 = 0$, $x = -(\quad)$ or ____.

b. If $x + {}^{-}5 = 0$, $x = -(\quad)$ or ____.

STOP • STOP • STOP • STOP • STOP • STOP • STOP • STOP • STOP

A10 **a.** $-(3)$ or $\underline{-3}$ (read "the opposite of 3 or negative 3")

b. $-(-5)$ or $\underline{5}$ (read "the opposite of negative 5 or 5")

Q11 Verify that:

a. $x + y = y + x$ is true when $x = -5$ and $y = 8$.

b. $(x + y) + z = x + (y + z)$ is true when $x = -3$, $y = 4$, and $z = -5$.

c. $x + 0 = 0 + x = x$ is true when $x = -2$.

d. $x + {}^{-}x = -x + x = 0$ is true when $x = -12$.

e. $x + {}^{-}x = -x + x = 0$ is true when $x = 20$.

STOP • STOP • STOP • STOP • STOP • STOP • STOP • STOP • STOP

A11 **a.**
$$x + y = y + x$$
$$-5 + 8 = 8 + {}^{-}5$$
$$3 = 3$$

b.
$$(x + y) + z = x + (y + z)$$
$$(-3 + 4) + {}^{-}5 = -3 + (4 + {}^{-}5)$$
$$1 + {}^{-}5 = -3 + {}^{-}1$$
$$-4 = -4$$

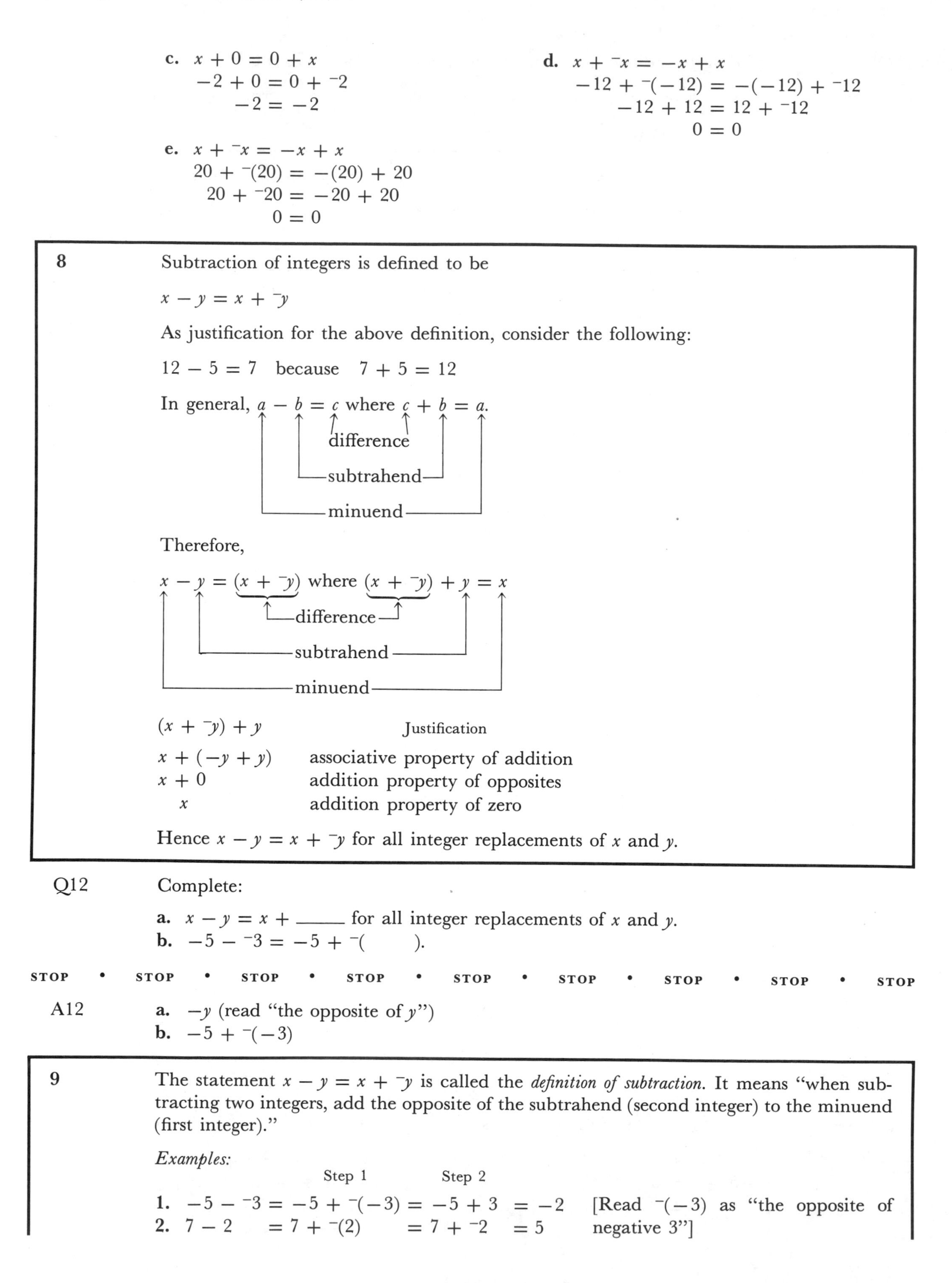

c. $x + 0 = 0 + x$
$-2 + 0 = 0 + {}^{-}2$
$-2 = -2$

d. $x + {}^{-}x = -x + x$
$-12 + {}^{-}(-12) = -(-12) + {}^{-}12$
$-12 + 12 = 12 + {}^{-}12$
$0 = 0$

e. $x + {}^{-}x = -x + x$
$20 + {}^{-}(20) = -(20) + 20$
$20 + {}^{-}20 = -20 + 20$
$0 = 0$

8 Subtraction of integers is defined to be

$x - y = x + {}^{-}y$

As justification for the above definition, consider the following:

$12 - 5 = 7$ because $7 + 5 = 12$

In general, $a - b = c$ where $c + b = a$.

Therefore,

$x - y = (x + {}^{-}y)$ where $(x + {}^{-}y) + y = x$

$(x + {}^{-}y) + y$	Justification
$x + (-y + y)$	associative property of addition
$x + 0$	addition property of opposites
x	addition property of zero

Hence $x - y = x + {}^{-}y$ for all integer replacements of x and y.

Q12 Complete:

a. $x - y = x +$ ____ for all integer replacements of x and y.
b. $-5 - {}^{-}3 = -5 + {}^{-}(\quad)$.

STOP • STOP • STOP • STOP • STOP • STOP • STOP • STOP • STOP

A12 a. $-y$ (read "the opposite of y")
b. $-5 + {}^{-}(-3)$

9 The statement $x - y = x + {}^{-}y$ is called the *definition of subtraction*. It means "when subtracting two integers, add the opposite of the subtrahend (second integer) to the minuend (first integer)."

Examples:

	Step 1	Step 2	
1. $-5 - {}^{-}3$	$= -5 + {}^{-}(-3)$	$= -5 + 3$	$= -2$
2. $7 - 2$	$= 7 + {}^{-}(2)$	$= 7 + {}^{-}2$	$= 5$

[Read ${}^{-}(-3)$ as "the opposite of negative 3"]

3. $9 - {}^{-}5 \quad = 9 + {}^{-}(-5) \quad = 9 + 5 \quad = 14$
4. $-2 - {}^{-}6 = -2 + {}^{-}(-6) = -2 + 6 = 4$
5. $-3 - 8 \quad = -3 + {}^{-}(8) \quad = -3 + {}^{-}8 = -11$

(*Note:* Step 1 is usually done mentally.)

6. $-2 - {}^{-}3 = -2 + 3 \quad = 1$
7. $5 - {}^{-}6 \quad = 5 + 6 \quad = 11$

Q13 Complete:

a. $-2 - {}^{-}5 = -2 + (\quad) =$ _____.

b. $7 - {}^{-}5 = 7 + (\quad) =$ _____.

c. $1 - 9 = 1 + (\quad) =$ _____.

d. $-12 - 3 = -12 + (\quad) =$ _____.

STOP • STOP • STOP • STOP • STOP • STOP • STOP • STOP • STOP

A13 a. 5, 3 b. 5, 12 c. $-9, -8$ d. $-3, -15$

10 When mixed operations of addition and subtraction are present, you should apply the definition of subtraction first. After all subtractions have been changed to equivalent additions, the associative property of addition can be used.

Examples:

1. $3 + {}^{-}7 - 5$

$3 + {}^{-}7 + {}^{-}5$	or	$3 + (-7 + {}^{-}5)$
$-4 + {}^{-}5$		$3 + {}^{-}12$
-9		-9

2. $-4 - {}^{-}2 + 3$

$-4 + 2 + 3$	or	$-4 + (2 + 3)$
$-2 + 3$		$-4 + 5$
1		1

3. $-3 - 4 - 6 + 7$

$-3 + {}^{-}4 + {}^{-}6 + 7$

$-7 + {}^{-}6 + 7$

$-13 + 7$

-6

Q14 Evaluate:

a. ${}^{-}3 + 5 - 7$

b. $1 - 5 - 6$

c. $4 + {}^{-}3 - 7 - {}^{-}6$

d. $7 - {}^{-}2 - 8 + {}^{-}3$

STOP • STOP • STOP • STOP • STOP • STOP • STOP • STOP • STOP

A14 a. -5: $-3 + 5 + {}^{-}7$ b. -10: $1 + {}^{-}5 + {}^{-}6$

c. 0: $4 + {}^{-}3 + {}^{-}7 + 6$ d. -2: $7 + 2 + {}^{-}8 + {}^{-}3$

(*Note:* Since every subtraction of two integers can be converted into an addition of two integers by the definition of subtraction, the set of integers is closed under the operation of subtraction.)

11

The rules for multiplication of integers can be justified by applying a fundamental idea from arithmetic. That is,

$5 \cdot 4 = 4 + 4 + 4 + 4 + 4$

$5 \cdot 4$ may be interpreted as five terms of 4. Therefore,

$$5 \cdot 4 = 4 + 4 + 4 + 4 + 4$$
$$= 20$$

Consider xy where both x and y are positive.

$$xy = \underbrace{y + y + y + \cdots + y}_{x \text{ terms of a positive integer } y}$$

Since y is positive, the sum will be positive. Hence a *positive* integer times a *positive* integer will always produce a *positive* result. Two examples are:

$5 \cdot 3 = 15$ and $5 \cdot 7 = 35$

Q15 Complete:

a. $2 \cdot 3 =$ _____ b. $7 \cdot 9 =$ _____

STOP • STOP • STOP • STOP • STOP • STOP • STOP • STOP • STOP

A15 a. 6 b. 63

12

Consider $4 \cdot {}^-7$. Use the same interpretation as in Frame 11:

$$4 \cdot {}^-7 = \underbrace{{}^-7 + {}^-7 + {}^-7 + {}^-7}_{4 \text{ terms of } -7} = -28$$

Consider xy where x is positive and y is negative.

$$xy = \underbrace{y + y + y + \cdots + y}_{x \text{ terms of a negative integer } y}$$

Since y is negative, the sum will be negative. Hence a *positive* integer times a *negative* integer will always produce a *negative* result. Two examples are:

$8 \cdot {}^-3 = -24$ and $7 \cdot {}^-4 = -28$

Q16 Complete:

a. $3 \cdot {}^-2 =$ _____ b. $4 \cdot {}^-9 =$ _____

c. $5 \cdot {}^-4 =$ _____ d. $6 \cdot {}^-8 =$ _____

e. $0 \cdot {}^-3 =$ _____ f. $8 \cdot {}^-5 =$ _____

STOP • STOP • STOP • STOP • STOP • STOP • STOP • STOP • STOP

A16 a. -6 b. -36 c. -20 d. -48 e. 0 f. -40

13

To find the product $-5 \cdot 4$ requires a different interpretation. Since multiplication is a commutative operation in the set of integers:

$$-5 \cdot 4 = 4 \cdot {}^-5$$
$$= -20$$

The product $-5 \cdot 4$ could be interpreted to mean subtraction of 4 five times. That is, $0 - 4 - 4 - 4 - 4 - 4$. This would produce

$$0 + {}^-4 + {}^-4 + {}^-4 + {}^-4 + {}^-4 = -20$$

This result agrees with the previous result and leads us to believe that a *negative* integer times a *positive* integer produces a *negative* result.

Q17 Complete:

a. $-2 \cdot 5 =$ ____ b. $-9 \cdot 4 =$ ____

c. $-4 \cdot 5 =$ ____ d. $-8 \cdot 6 =$ ____

e. $-3 \cdot 0 =$ ____ f. $-8 \cdot 7 =$ ____

STOP • STOP • STOP • STOP • STOP • STOP • STOP • STOP • STOP

A17 a. -10 b. -36 c. -20 d. -48 e. 0 f. -56

14 In Section 1.2 it was stated that $x \cdot 0 = 0x = 0$ is true for all whole-number replacements of x. This idea was called the multiplication property of zero. It is likewise true for all integer replacements of the variable. At that time it was also stated that $1x = x \cdot 1 = x$ for all whole-number replacements of x. This idea was called the multiplication property of 1. It is likewise true for integer replacements of the variable.

$$-5 \cdot 0 = 0 \cdot {}^-5 = 0$$
$$1 \cdot {}^-3 = -3 \cdot 1 = -3$$

Q18 Complete:

a. $1 \cdot {}^-5 =$ ____ b. $0 \cdot {}^-3 =$ ____

c. $-4 \cdot 5 =$ ____ d. $12 \cdot -5 =$ ____

e. $9 \cdot 0 =$ ____ f. $1 \cdot 4 =$ ____

g. $7 \cdot {}^-5 =$ ____ h. $-6 \cdot 9 =$ ____

STOP • STOP • STOP • STOP • STOP • STOP • STOP • STOP • STOP

A18 a. -5 b. 0 c. -20 d. -60 e. 0 f. 4 g. -35 h. -54

15 The product of two negative integers will now be developed.

$$\begin{aligned} 4 \cdot 5 &= 5 + 5 + 5 + 5 = 20 \\ 4 \cdot {}^-5 &= -5 + {}^-5 + {}^-5 + {}^-5 = -20 \\ -4 \cdot 5 &= 0 - 5 - 5 - 5 - 5 \\ &= 0 + {}^-5 + {}^-5 + {}^-5 + {}^-5 = -20 \end{aligned}$$

Using the same interpretation,

$$\begin{aligned} -4 \cdot {}^-5 &= 0 - {}^-5 - {}^-5 - {}^-5 - {}^-5 \\ &= 0 + 5 + 5 + 5 + 5 = 20 \end{aligned}$$

Hence it can be reasoned that a *negative* integer times a *negative* integer is a *positive* integer. This result may be produced in an algebraic manner. Since the left distributive property of multiplication over addition is valid for the set of integers,

$$-4(5 + {}^-5) = -4 \cdot 5 + \underbrace{{}^-4 \cdot {}^-5}$$
$$-4(0) = -20 + x$$
$$0 = -20 + x$$

The product $-4 \cdot {}^-5$ is a number, x, such that $0 = -20 + x$. Since $x = 20$, the product $-4 \cdot {}^-5 = 20$.

Q19 The product of a negative integer times a negative integer produces a ______ (positive/negative) result.

STOP • STOP • STOP • STOP • STOP • STOP • STOP • STOP • STOP

A19 positive

16 Summary: The product of

1. A positive integer and a positive integer is a positive integer.
2. A negative integer and a negative integer is a positive integer.
3. A positive integer and a negative integer is a negative integer.

Q20 Find the following products:

a. $-3 \cdot 4 =$ ______ b. $-4 \cdot 9 =$ ______

c. $-3 \cdot {}^-5 =$ ______ d. $9 \cdot {}^-8 =$ ______

e. $0 \cdot {}^-6 =$ ______ f. $-7 \cdot 1 =$ ______

g. $6 \cdot {}^-1 =$ ______ h. $-1 \cdot {}^-5 =$ ______

STOP • STOP • STOP • STOP • STOP • STOP • STOP • STOP • STOP

A20 a. -12 b. -36 c. 15 d. -72 e. 0 f. -7 g. -6 h. 5

17 Study the following examples.

$-1 \cdot 4 = -4$ (4 and -4: opposites) $-1 \cdot {}^-5 = 5$ ($^-5$ and 5: opposites)

The fact that -1 times an integer produces its opposite is called the *multiplication property of -1*. In general, $-1x = -x$ is true for all integer replacements of x.

Examples:

$-1x = -x$

$-1 \cdot 3 = -(3) = -3$ $-1 \cdot {}^-3 = -(-3) = 3$

$-1 \cdot 0 = -(0) = 0$ $-1 \cdot {}^-1 = -(-1) = 1$

Q21 Use the multiplication property of -1 to find:

a. $-1 \cdot {}^-7 = -(\quad) =$ ______ b. $-1 \cdot 9 = -(\quad) =$ ______

c. $-1 \cdot {}^-12 = -(\quad) =$ ______

STOP • STOP • STOP • STOP • STOP • STOP • STOP • STOP • STOP

A21 a. $-(-7) = 7$ b. $-(9) = -9$ c. $-(-12) = 12$

18 To find the product of more than two integers:

Step 1: Do all work within parentheses first.

Step 2: Find the product of two numbers at a time in any order desired. (Justified by commutative and associative properties of multiplication.)

Examples:

$-2 \cdot 4 \cdot {}^{-}5$	$3(-6 \cdot {}^{-}5) \cdot 9$
$-2 \cdot {}^{-}5 \cdot 4$	$3 \cdot 30 \cdot 9$
$10 \cdot 4$	$90 \cdot 9$
40	810

Q22 Determine the product:

a. $-2 \cdot {}^{-}3 \cdot 5$ **b.** $(-1 \cdot 3)(-4 \cdot {}^{-}7)$

STOP • STOP • STOP • STOP • STOP • STOP • STOP • STOP • STOP

A22 **a.** 30 **b.** -84

19 Division of integers follows from multiplication of integers and the idea that

If $\frac{x}{y} = z, y \neq 0,$ then $x = yz$

That is, the quotient, z, is the value which when multiplied by the divisor, y, equals the dividend, x. For example,

$\frac{10}{2} = 5$ because $10 = 2 \cdot 5$

The restriction, $y \neq 0$, will be discussed in Frame 21.

Q23 Complete:

$\frac{10}{2} = 5$ because $10 =$ _____ $\cdot$ _____.

STOP • STOP • STOP • STOP • STOP • STOP • STOP • STOP • STOP

A23 $2 \cdot 5$

20 The quotient of two positive integers is a positive result.

Examples:

$\frac{15}{3} = 5$ because $15 = 3 \cdot 5$

$\frac{49}{7} = 7$ because $49 = 7 \cdot 7$

Q24 Complete:

a. $\frac{72}{9} =$ _____ **b.** $\frac{102}{3} =$ _____

c. $\dfrac{90}{15} =$ ______ d. $\dfrac{0}{7} =$ ______

STOP • STOP • STOP • STOP • STOP • STOP • STOP • STOP • STOP

A24 **a.** 8 **b.** 34 **c.** 6 **d.** 0

21 Consider:

$\dfrac{7}{0} =$ __?__ because $7 = 0 \cdot$ __?__

Since there is *no number* times 0 that equals 7, the result of $\dfrac{7}{0}$ does not exist. Hence, *division by zero* is said to be *undefined.* Therefore, in order for $\dfrac{x}{y}$ to be a rational number, the restriction $y \neq 0$ is necessary.

Q25 Complete:

a. $\dfrac{100}{5} =$ ____________ b. $\dfrac{36}{12} =$ ____________

c. $\dfrac{0}{19} =$ ____________ d. $\dfrac{23}{0} =$ ____________

STOP • STOP • STOP • STOP • STOP • STOP • STOP • STOP • STOP

A25 **a.** 20 **b.** 3 **c.** 0 **d.** undefined

22 The quotient of two integers with different signs (neither zero) is illustrated in the same way as in Frame 20.

Examples:

$\dfrac{12}{-3} = -4$ because $12 = -3 \cdot {}^{-}4$

$\dfrac{-12}{3} = -4$ because $-12 = 3 \cdot {}^{-}4$

The above examples illustrate that the quotient of two integers with *different signs* (neither zero) is a *negative result.*

Q26 Complete:

a. $\dfrac{40}{-8} =$ ____________ b. $\dfrac{-92}{2} =$ ____________

c. $\dfrac{{}^{-}5}{0} =$ ____________ d. $\dfrac{0}{-5} =$ ____________

e. $\dfrac{12}{-6} =$ ____________ f. $\dfrac{-15}{5} =$ ____________

g. $\dfrac{18}{9} =$ ____________ h. $\dfrac{-81}{3} =$ ____________

STOP • STOP • STOP • STOP • STOP • STOP • STOP • STOP • STOP

A26 **a.** -5 **b.** -46 **c.** undefined **d.** 0
e. -2 **f.** -3 **g.** 2 **h.** -27

---

23 The quotient of two negative integers is a *positive result.*

Examples:

$\frac{-12}{-3} = 4$ because $-12 = -3 \cdot 4$

$\frac{-20}{-5} = 4$ because $-20 = -5 \cdot 4$

Q27 Complete:

a. $\frac{-16}{-4} =$ ____________ **b.** $\frac{-24}{-8} =$ ____________

c. $\frac{-32}{-16} =$ ____________ **d.** $\frac{-2}{0} =$ ____________

e. $\frac{0}{-3} =$ ____________ **f.** $\frac{64}{4} =$ ____________

g. $\frac{64}{-16} =$ ____________ **h.** $\frac{-64}{-8} -$ ____________

STOP • STOP • STOP • STOP • STOP • STOP • STOP • STOP • STOP

A27 **a.** 4 **b.** 3 **c.** 2 **d.** undefined
e. 0 **f.** 16 **g.** -4 **h.** 8

Q28 Complete:

a. The quotient of two positive integers is a ____________ result.
positive/negative

b. The quotient of two integers with different signs (neither zero) is a ____________ result.
positive/negative

c. The quotient of two negative integers is a ____________ result.
positive/negative

d. The quotient of a nonzero integer and zero is ____________.

e. The quotient of zero and a nonzero integer is ____________.

STOP • STOP • STOP • STOP • STOP • STOP • STOP • STOP • STOP

A28 **a.** positive **b.** negative **c.** positive **d.** undefined
e. zero

---

24 Summary: The quotient of:

1. Integers with the same sign is positive.
2. Integers with different signs is negative.
3. Zero by a nonzero integer is zero.
4. An integer by zero is undefined.

Q29 Complete:

a. $\frac{-4}{-1} =$ ______ b. $\frac{-54}{9} =$ ______

c. $\frac{56}{-7} =$ ______ d. $\frac{14}{7} =$ ______

e. $\frac{0}{-15} =$ ______ f. $\frac{0}{0} =$ ______

g. $\frac{-30}{-6} =$ ______ h. $\frac{48}{12} =$ ______

STOP • STOP • STOP • STOP • STOP • STOP • STOP • STOP • STOP

A29 a. 4 b. -6 c. -8 d. 2
e. 0 f. undefined g. 5 h. 4

25 You should observe that the division property of 1 is valid for the set of integers. That is,

$\frac{x}{1} = x$ for all integer replacements of x

Examples:

$\frac{5}{1} = 5$ $\frac{-7}{1} = -7$ $\frac{0}{1} = 0$ $\frac{12}{1} = 12$

Consider these examples:

$\frac{5}{-1} = -5$ $\frac{-7}{-1} = 7$ $\frac{0}{-1} = 0$ $\frac{12}{-1} = -12$

Note that dividing by -1 produces the opposite of the original dividend (numerator).
The *division property of* -1 states:

$\frac{x}{-1} = -x$ for all integer replacements of x

Both properties could be shortened to (I, the set of integers):

$\frac{x}{1} = x$ $x \in I$

$\frac{x}{-1} = -x$ $x \in I$

(*Recall:* $-x$ means the opposite of x.)

Q30 a. $-x$ is read "the ______ of x."
b. The multiplication property of -1 states that -1 times any integer is that integer's ______.
c. The division property of -1 states that any integer divided by -1 is that integer's ______.

STOP • STOP • STOP • STOP • STOP • STOP • STOP • STOP • STOP

A30 **a.** opposite **b.** opposite **c.** opposite

Q31 Complete:

a. $\frac{-12}{1} =$ ________ **b.** $\frac{-12}{-1} =$ ________

c. $\frac{-12}{0} =$ ________ **d.** $\frac{0}{-12} =$ ________

STOP • STOP • STOP • STOP • STOP • STOP • STOP • STOP • STOP

A31 **a.** -12 **b.** 12 **c.** undefined **d.** 0

This completes the instruction for this section.

1.3 Exercises

1. The set $\{\ldots, -2, -1, 0, 1, 2, 3, \ldots\}$ is called the set of ________.
2. Given that

   $N = \{1, 2, 3, \ldots\}$
   $W = \{0, 1, 2, 3, \ldots\}$
   $I = \{\ldots, -2, -1, 0, 1, 2, \ldots\}$

   insert $\in$ or $\notin$ to form a true statement:
   a. -17 ____ W **b.** -17 ____ N **c.** -17 ____ I
   d. 0 ____ N **e.** 0 ____ W **f.** 0 ____ I
   g. 3 ____ N **h.** 3 ____ I
3. Write the opposite of each integer:
   a. 0 **b.** -17 **c.** 5 **d.** 8
4. Find:
   a. $-3 + {}^{-}5$ **b.** $5 + {}^{-}9$ **c.** $7 + {}^{-}6$ **d.** $-5 + 5$
   e. $-7 + 0$ **f.** $-3 + {}^{-}7 + 3$ **g.** $9 + {}^{-}2 + {}^{-}5 + 6$
   h. $-8 + 5 + {}^{-}5 + 8$
5. Find:
   a. $-2 - 4$ **b.** $-5 - {}^{-}2$ **c.** $7 - {}^{-}3$ **d.** $0 - {}^{-}4$
   e. $3 - 9$ **f.** $-12 - {}^{-}12$ **g.** $0 - 5$ **h.** $-3 - {}^{-}9$
   i. $4 - 5$ **j.** $9 - {}^{-}3 + {}^{-}4 - {}^{-}3$
6. Find:
   a. $4 \cdot {}^{-}9$ **b.** $-3 \cdot {}^{-}4$ **c.** $-2 \cdot 0$ **d.** $-1 \cdot {}^{-}8$
   e. $-1 \cdot 9$ **f.** $-5 \cdot 7$ **g.** $12 \cdot 10$ **h.** $9 \cdot {}^{-}8$
   i. $11 \cdot {}^{-}11$ **j.** $0 \cdot 5$
7. Find:
   a. $72 \div -9$ **b.** $\frac{-72}{-9}$ **c.** $\frac{72}{9}$ **d.** $\frac{-72}{9}$
   e. $\frac{0}{9}$ **f.** $\frac{-9}{0}$ **g.** $\frac{-8}{-8}$ **h.** $12 \div 12$
   i. $\frac{52}{-1}$ **j.** $\frac{-17}{1}$

8. Find:

   a. $3 + {}^-2 - 5 - {}^-3$
   b. $3 \cdot {}^-5 + {}^-2$
   c. $(-3 - {}^-2) \cdot {}^-3$
   d. $-7 + (-3 + {}^-5)$
   e. $2(-5 + 9)$
   f. $(2 - 5)(5 - 2)$
   g. $\dfrac{-9 - {}^-3}{-2}$
   h. $\dfrac{-8}{-2 - {}^-2}$
   i. $-7 - {}^-3 \cdot {}^-5 - {}^-1$
   j. $5 \cdot {}^-2 - {}^-3 \cdot 4$

9. Insert $\in$ or $\notin$ correctly:
   a. If $a \in I$ and $b \in I$, then $a + b$ ____ I.
   b. If $a \in I$ and $b \in I$, then ab ____ I.
   c. If $a \in I$ and $b \in I$, then $a - b$ ____ I.

10. True or false? (Use T or F.)
    a. $-7 - 9 = 9 - {}^-7$.
    b. $xy = yx$ is true when $x \in I$ and when $y \in I$.
    c. $(x + y) + z = x + (y + z)$ is true for all $x \in I$, $y \in I$, and $z \in I$.
    d. According to the multiplication property of -1 and the division property of -1, any integer multiplied or divided by -1 will produce the opposite of the original integer.
    e. The fact that $x + {}^-x = -x + x = 0$ is true for all integer replacements of x is called the addition property of opposites.

1.3 Exercise Answers

1. integers
2. a. $\notin$ b. $\notin$ c. $\in$ d. $\notin$ e. $\in$ f. $\in$ g. $\in$ h. $\in$
3. a. 0 b. 17 c. -5 d. -8
4. a. -8 b. -4 c. 1 d. 0 e. -7 f. -7 g. 8 h. 0
5. a. -6 b. -3 c. 10 d. 4 e. -6 f. 0 g. -5 h. 6 i. -1 j. 11
6. a. -36 b. 12 c. 0 d. 8 e. -9 f. -35 g. 120 h. -72 i. -121 j. 0
7. a. -8 b. 8 c. 8 d. -8 e. 0 f. undefined g. 1 h. 1 i. -52 j. -17
8. a. -1 b. -17 c. 3 d. -15 e. 8 f. -9 g. 3 h. undefined i. -21 j. 2
9. a. $\in$ b. $\in$ c. $\in$
10. a. F b. T c. T d. T e. T

1.4 The Set of Rational Numbers

1 Because it is impossible to list the set of rational numbers, a verbal description is required.

A rational number is any number that can be written in the form $\frac{p}{q}$, *where* $p \in I$ *and* $q \in I$, $q \neq 0$.

The restriction that $q \neq 0$ is required because division by zero is undefined. The set of rational numbers is a collection of all rational numbers. It is common to denote the set of rational numbers by the letter Q. (R will be used later to represent the set of real numbers.)

Some examples of rational numbers are:

$\frac{2}{3}, \frac{-7}{2}, 5, -3, 0, 4\frac{2}{3}$

The numbers $\frac{2}{3}$ and $\frac{-7}{2}$ are obviously in the $\frac{p}{q}$ form, where $p \in I$ and $q \in I$, $q \neq 0$. The numbers 5, -3, and 0 can be put in the $\frac{p}{q}$ form by selecting q to be 1. That is, $5 = \frac{5}{1}$, $-3 = \frac{-3}{1}$, and $0 = \frac{0}{1}$. The mixed number $4\frac{2}{3}$ can be changed to the improper fraction $\frac{14}{3}$ and hence can be put in the $\frac{p}{q}$ form.

Q1 Put the following in the $\frac{p}{q}$ form:

a. $-3 =$ _____ b. $3\frac{1}{7} =$ _____

c. $0 =$ _____ d. $9 =$ _____

STOP • STOP • STOP • STOP • STOP • STOP • STOP • STOP • STOP

A1 a. $\frac{-3}{1}$ b. $\frac{22}{7}$ c. $\frac{0}{1}$: the denominator can be any nonzero integer

d. $\frac{9}{1}$

Q2 Why is $\frac{5}{0}$ not a rational number? _____

STOP • STOP • STOP • STOP • STOP • STOP • STOP • STOP • STOP

A2 Because the denominator is zero and division by zero is undefined.

Q3 a. Is every integer a rational number? _____

b. Is the set of integers a proper subset of the set of rational numbers? _____

c. Is it true that $I \subset Q$? _____

STOP • STOP • STOP • STOP • STOP • STOP • STOP • STOP • STOP

A3 a. yes b. yes c. yes

2 It is also possible to describe a rational number in terms of its decimal representation as either a *terminating* decimal or an *infinite-repeating* decimal.

Examples:

$\frac{1}{2} = 0.5$ (terminating)

$\frac{1}{3} = 0.333\ldots = 0.\overline{3}$ (infinite-repeating)

(*Note:* The bar above the 3 indicates that 3 repeats endlessly.)

$\frac{-5}{8} = -0.625$ (terminating)

$\frac{2}{7} = 0.285714285714\ldots = 0.\overline{285714}$ (infinite-repeating)

(*Note:* The bar above the 285714 indicates that this block of digits repeats endlessly.)

$-4 = -4.0$ (terminating)

In each case the decimal representation of a rational number may be found by dividing the numerator by the denominator.

Example: Determine the decimal representation of $\frac{5}{33}$.

Solution

```
     0.1515 . . .   (repeating 15s)
33)5.0000
   3 3
   1 70
   1 65
      50
      33
      170
      165
        5   etc.
```

Hence $\frac{5}{33} = 0.1515\ldots = 0.\overline{15}$. Therefore, the decimal representation of $\frac{5}{33}$ is an infinite-repeating decimal.

Q4 **a.** Find the decimal representation for the rational number $\frac{3}{8}$.

b. Is it terminating or infinite-repeating? __________

STOP • STOP • STOP • STOP • STOP • STOP • STOP • STOP • STOP

A4 **a.** 0.375 **b.** terminating

Q5 **a.** Find the decimal representation for the rational number $\frac{-5}{11}$.

b. Is it terminating or infinite-repeating? ______________

STOP • STOP • STOP • STOP • STOP • STOP • STOP • STOP • STOP

A5 **a.** $-0.454545\ldots$ or $-0.\overline{45}$ **b.** infinite-repeating

---

Q6 The decimal representation of a rational number is either

a. ______________, or

b. ______________.

STOP • STOP • STOP • STOP • STOP • STOP • STOP • STOP • STOP

A6 **a.** terminating **b.** infinite-repeating (either order)

3 It is possible to show that any infinite repeating decimal can be put in $\frac{p}{q}$ form.

Examples:

1. $0.\overline{3} = 0.333\ldots$

$$\begin{aligned} \text{Let } n &= 0.\overline{3} = 0.333\ldots \\ 10n &= 3.333\ldots && \text{(equation 1)} \\ n &= 0.333\ldots && \text{(equation 2)} \\ 9n &= 3 && \text{(subtracting equation 2 from equation 1)} \\ n &= \frac{1}{3} \end{aligned}$$

Therefore, $0.\overline{3} = \frac{1}{3}$.

2. $0.\overline{142857} = 0.142857142857\ldots$

$$\begin{aligned} \text{Let } n &= 0.\overline{142857} \\ 1{,}000{,}000n &= 142{,}857.\overline{142{,}857} \\ n &= 0.\overline{142857} \\ 999{,}999n &= 142{,}857 \\ n &= \frac{142{,}857}{999{,}999} = \frac{1}{7} \end{aligned}$$

$(142{,}857 \cdot 7 = 999{,}999)$

Therefore, $0.\overline{142857} = \frac{1}{7}$.

3. $5.24\overline{6} = 5.24666\ldots$

$$\begin{aligned} \text{Let } n &= 5.24\overline{6} \\ 1{,}000n &= 5{,}246.\overline{6} \\ 100n &= 524.\overline{6} \\ 900n &= 4{,}722 \\ n &= \frac{4{,}722}{900} = \frac{787}{150} \end{aligned}$$

Therefore, $5.24\overline{6} = \frac{787}{150}$.

Q7 Is $0.\overline{6}$ a rational number? ______

STOP • STOP • STOP • STOP • STOP • STOP • STOP • STOP • STOP

A7 yes: $0.\overline{6} = \frac{2}{3}$

4 The remaining portion of this section will review operations with rational numbers. The multiplication property of 1 is valid for the set of rational numbers. It states that

$1a = a \cdot 1 = a$ when $a \in Q$

Examples:

$\frac{2}{3} \cdot 1 = 1 \cdot \frac{2}{3} = \frac{2}{3}$ $\frac{-5}{7} \cdot 1 = 1 \cdot \frac{-5}{7} = \frac{-5}{7}$

$0.\overline{14} \cdot 1 = 1 \cdot 0.\overline{14} = 0.\overline{14}$ $17.25 \cdot 1 = 1 \cdot 17.25 = 17.25$

Q8 Complete:

a. $\frac{-9}{2} \cdot 1 =$ ______ **b.** $1 \cdot 0.\overline{142573} =$ ______

STOP • STOP • STOP • STOP • STOP • STOP • STOP • STOP • STOP

A8 **a.** $\frac{-9}{2}$ **b.** $0.\overline{142573}$

5 The quotient of a nonzero number divided by itself is 1. In general,

$\frac{x}{x} = 1 \quad x \in I, \quad x \neq 0$

(*Note:* This idea is true when $x \in Q$, $x \neq 0$.)

Examples:

$1 = \frac{1}{1} = \frac{2}{2} = \frac{3}{3} = \frac{4}{4}$ etc. $1 = \frac{-1}{-1} = \frac{-2}{-2} = \frac{-3}{-3} = \frac{-4}{-4}$ etc.

Q9 Complete:

a. $1 = \frac{(\quad)}{-7}$ **b.** $1 = \frac{(\quad)}{15}$ **c.** $1 = \frac{(\quad)}{0}$

STOP • STOP • STOP • STOP • STOP • STOP • STOP • STOP • STOP

A9 **a.** -7 **b.** 15
c. no number: because division by zero is undefined

6 The product of two rational numbers may be generalized as:

$\frac{r}{s} \cdot \frac{t}{u} = \frac{rt}{su}$

Examples:

$\frac{2}{3} \cdot \frac{4}{5} = \frac{2 \cdot 4}{3 \cdot 5} = \frac{8}{15}$

$$\frac{-3}{7} \cdot \frac{4}{11} = \frac{-3 \cdot 4}{7 \cdot 11} = \frac{-12}{77}$$

$$\frac{-2}{5} \cdot \frac{-8}{3} = \frac{-2 \cdot {}^{-}8}{5 \cdot 3} = \frac{16}{15}$$

In algebra, rational numbers are usually left in $\frac{p}{q}$ form.

Q10 Find the following products (leave answers in $\frac{p}{q}$ form):

a. $\frac{-7}{9} \cdot \frac{4}{5}$ b. $\frac{-2}{3} \cdot \frac{-5}{9}$

c. $\frac{6}{13} \cdot \frac{-2}{5}$ d. $\frac{7}{2} \cdot \frac{3}{4}$

STOP • STOP • STOP • STOP • STOP • STOP • STOP • STOP • STOP

A10 a. $\frac{-28}{45}$ b. $\frac{10}{27}$ c. $\frac{-12}{65}$ d. $\frac{21}{8}$

7 Rational numbers may be rewritten as equivalent rational numbers over a new denominator.

Examples:

$$\frac{2}{3} = \frac{2}{3} \cdot 1 = \frac{2}{3} \cdot \frac{2}{2} = \frac{2 \cdot 2}{3 \cdot 2} = \frac{4}{6}$$

$$\frac{-4}{5} = \frac{-4}{5} \cdot 1 = \frac{-4}{5} \cdot \frac{3}{3} = \frac{-4 \cdot 3}{5 \cdot 3} = \frac{-12}{15}$$

This procedure is often called "raising a rational number to higher terms." It is usually done mentally. That is,

$\frac{2}{3} = \frac{4}{6}$ (mentally multiplying the numerator and the denominator of $\frac{2}{3}$ by 2)

$\frac{-4}{5} = \frac{-12}{15}$ (mentally multiplying the numerator and the denominator of $\frac{-4}{5}$ by 3)

Q11 Perform the following mentally. Multiply the numerator and the denominator of the original rational number by the necessary value to produce an equivalent rational number:

a. $\frac{-5}{7} = \frac{(\quad)}{21}$ b. $\frac{-4}{5} = \frac{(\quad)}{20}$

c. $\frac{9}{4} = \frac{(\quad)}{16}$ d. $5 = \frac{5}{1} = \frac{(\quad)}{7}$

STOP • STOP • STOP • STOP • STOP • STOP • STOP • STOP • STOP

A11 a. -15 b. -16 c. 36 d. 35

8 *Reducing* a rational number may be accomplished in the following manner.

$$\frac{2}{4} = \frac{1 \cdot 2}{2 \cdot 2} = \frac{1}{2} \cdot \frac{2}{2} = \frac{1}{2} \cdot 1 = \frac{1}{2}$$

$$\frac{-6}{9} = \frac{-2 \cdot 3}{3 \cdot 3} = \frac{-2}{3} \cdot \frac{3}{3} = \frac{-2}{3} \cdot 1 = \frac{-2}{3}$$

However, the same result is obtained by simply dividing numerator and denominator by their *greatest common factor.* A rational number is said to be reduced to lowest terms when the numerator and denominator no longer contain a common factor (other than 1).

Examples:

$\frac{2}{4} = \frac{\overset{1}{\cancel{2}}}{\underset{2}{\cancel{4}}} = \frac{1}{2}$ (dividing numerator and denominator by 2, their greatest common factor)

$\frac{-6}{9} = \frac{\overset{-2}{\cancel{-6}}}{\underset{3}{\cancel{9}}} = \frac{-2}{3}$ (dividing numerator and denominator by 3, their greatest common factor)

Q12 Reduce the following rational numbers to lowest terms:

a. $\frac{-10}{15}$ b. $\frac{36}{72}$ c. $\frac{-12}{18}$ d. $\frac{26}{39}$

STOP • STOP • STOP • STOP • STOP • STOP • STOP • STOP • STOP

A12 a. $\frac{-2}{3}$: $\frac{-10}{15} = \frac{\overset{-2}{\cancel{-10}}}{\underset{3}{\cancel{15}}}$ (dividing numerator and denominator by 5)

b. $\frac{1}{2}$: $\frac{36}{72} = \frac{\overset{1}{\cancel{36}}}{\underset{2}{\cancel{72}}}$ (dividing numerator and denominator by 36)

(*Note:* Dividing numerator and denominator by the greatest common factor produces the rational number in lowest terms in one step. However, the numerator and denominator may be divided by any common factor repeatedly until the rational number is in lowest terms.)

c. $\frac{-2}{3}$: $\frac{-12}{18} = \frac{\overset{-2}{\cancel{-12}}}{\underset{3}{\cancel{18}}}$ (dividing numerator and denominator by 6)

d. $\frac{2}{3}$: $\frac{26}{39} = \frac{\overset{2}{\cancel{26}}}{\underset{3}{\cancel{39}}}$ (dividing numerator and denominator by 13)

(*Note:* Reducing $\frac{26}{39}$ could be aided by rewriting it as $\frac{2 \cdot 13}{3 \cdot 13}$. The common factor 13 would now be obvious.)

9 Consider the following example:

$$\frac{3}{5} \cdot \frac{10}{13} = \frac{3 \cdot 10}{5 \cdot 13} = \frac{30}{65} = \frac{6}{13}$$

A common procedure when multiplying rational numbers is to divide any numerator and denominator by a common factor first.

Examples:

$$\frac{3}{5}\cdot\frac{10}{13}=\frac{3}{\underset{1}{\cancel{5}}}\cdot\frac{\overset{2}{\cancel{10}}}{13}=\frac{6}{13}$$

$$\frac{-4}{7}\cdot\frac{-14}{20}=\frac{\overset{-1}{\cancel{-4}}}{\underset{1}{\cancel{7}}}\cdot\frac{\overset{-2}{\cancel{-14}}}{\underset{5}{\cancel{20}}}=\frac{2}{5}\qquad\text{(common factors of 4 and 7)}$$

$$\frac{-9}{12}\cdot\frac{5}{7}=\frac{\overset{-3}{\cancel{-9}}}{\underset{4}{\cancel{12}}}\cdot\frac{5}{7}=\frac{-15}{28}$$

Q13 Find the product by first dividing out common factors:

a. $\frac{-4}{8}\cdot\frac{3}{7}$ b. $\frac{-15}{39}\cdot\frac{-13}{10}$

STOP • STOP • STOP • STOP • STOP • STOP • STOP • STOP • STOP

A13 a. $\frac{-3}{14}$: $\frac{\overset{-1}{\cancel{-4}}}{\underset{2}{\cancel{8}}}\cdot\frac{3}{7}$ b. $\frac{1}{2}$: $\frac{\overset{-1}{\cancel{\overset{-3}{\cancel{-15}}}}}{\underset{1}{\cancel{\underset{3}{\cancel{39}}}}}\cdot\frac{\overset{-1}{\cancel{-13}}}{\underset{2}{\cancel{10}}}$

10 The numbers $\frac{2}{-3}$ and $\frac{-2}{3}$ are equivalent forms for the same rational number. Both represent the same negative rational number because they involve a quotient of integers with different signs. When writing a negative rational number, it is customary to place the negative sign on the number in the numerator. For example, $\frac{5}{-8}$ is written $\frac{-5}{8}$.

Q14 Find each product:

a. $\frac{-12}{15}\cdot\frac{3}{6}$ b. $\frac{-4}{6}\cdot\frac{-3}{18}$

c. $\frac{-5}{7}\cdot\frac{6}{-13}$ d. $\frac{7}{12}\cdot\frac{9}{-11}$

e. $\frac{-4}{7} \cdot \frac{7}{4}$

f. $\frac{5}{-28} \cdot \frac{-16}{17}$

STOP • STOP • STOP • STOP • STOP • STOP • STOP • STOP • STOP

A14 a. $\frac{-2}{5}$ b. $\frac{1}{9}$ c. $\frac{30}{91}$ d. $\frac{-21}{44}$ e. -1 f. $\frac{20}{119}$

11 Consider the following examples, which illustrate the sum of two rational numbers.

$$\frac{1}{2} + \frac{1}{2} = \frac{1+1}{2} = \frac{2}{2} = 1 \qquad \frac{3}{5} + \frac{1}{5} = \frac{3+1}{5} = \frac{4}{5}$$

$$\frac{-1}{7} + \frac{4}{7} = \frac{-1+4}{7} = \frac{3}{7} \qquad \frac{-2}{9} + \frac{5}{9} = \frac{-2+5}{9} = \frac{3}{9} = \frac{1}{3}$$

These examples demonstrate the following general statement:

$$\frac{r}{u} + \frac{s}{u} = \frac{r+s}{u} \qquad u \neq 0$$

Q15 Find the sum:

a. $\frac{9}{13} + \frac{4}{13}$

b. $\frac{-5}{9} + \frac{-2}{9}$

c. $\frac{-3}{5} + \frac{2}{5}$

d. $\frac{7}{12} + \frac{-4}{12}$

STOP • STOP • STOP • STOP • STOP • STOP • STOP • STOP • STOP

A15 a. 1 b. $\frac{-7}{9}$ c. $\frac{-1}{5}$ d. $\frac{1}{4}$

12 When finding the sum of two rational numbers with unlike denominators, the rational numbers must be rewritten over a common denominator. To simplify the calculation, the *least* (smallest) *common denominator* (LCD) is usually chosen.

Example: Find $\frac{1}{2} + \frac{1}{3}$.

Solution

The LCD is 6.

$$\frac{1}{2} + \frac{1}{3}$$

$$\left.\begin{array}{c} \dfrac{1}{2} \cdot 1 + \dfrac{1}{3} \cdot 1 \\ \dfrac{1}{2} \cdot \dfrac{3}{3} + \dfrac{1}{3} \cdot \dfrac{2}{2} \\ \dfrac{3}{6} + \dfrac{2}{6} \end{array}\right\} \text{raising the original rational numbers to higher terms}$$

$$\frac{5}{6}$$

The above solution is usually shortened to

$$\frac{1}{2} + \frac{1}{3} \qquad \text{or} \qquad \frac{1}{2} + \frac{1}{3} = \frac{3}{6} + \frac{2}{6}$$

$$\frac{3}{6} + \frac{2}{6} \qquad\qquad = \frac{5}{6}$$

$$\frac{5}{6}$$

This procedure is called *finding the sum of two rational numbers by the LCD method.*

Q16 Find the sum by the LCD method.

a. $\frac{2}{3} + \frac{4}{5}$ **b.** $\frac{-3}{7} + \frac{1}{2}$

STOP • STOP • STOP • STOP • STOP • STOP • STOP • STOP • STOP

A16 **a.** $\frac{22}{15}$: $\frac{2}{3} + \frac{4}{5}$

$$\frac{10}{15} + \frac{12}{15}$$

b. $\frac{1}{14}$: $\frac{-3}{7} + \frac{1}{2}$

$$\frac{-6}{14} + \frac{7}{14}$$

13 To avoid the necessity of using the LCD method, an alternative method will now be developed. Consider:

$$\frac{r}{s} + \frac{u}{v} \qquad s \neq 0 \text{ and } v \neq 0$$

$$\frac{r}{s} \cdot 1 + \frac{u}{v} \cdot 1$$

$$\frac{r}{s} \cdot \frac{v}{v} + \frac{u}{v} \cdot \frac{s}{s}$$

$$\frac{rv}{sv} + \frac{us}{sv}$$

$$\frac{rv + us}{sv}$$

Hence

$$\frac{r}{s} + \frac{u}{v} = \frac{rv + su}{sv} \qquad sv \neq 0$$

(*Note:* "$sv \neq 0$" and "$s \neq 0$ and $v \neq 0$" are equivalent statements.)

$$\frac{r}{s} + \frac{u}{v} = \frac{rv + su}{sv}$$

will be called the *cross-product rule for the addition of rational numbers.*

Examples:

$$\frac{1}{2} + \frac{2}{3} = \frac{1 \cdot 3 + 2 \cdot 2}{2 \cdot 3} = \frac{3 + 4}{6} = \frac{7}{6}$$

$$\frac{-1}{4} + \frac{5}{6} = \frac{-1 \cdot 6 + 4 \cdot 5}{4 \cdot 6} = \frac{-6 + 20}{24} = \frac{14}{24} = \frac{7}{12}$$

$$\frac{-1}{2} + \frac{-3}{5} = \frac{-1 \cdot 5 + 2 \cdot {}^-3}{10} = \frac{-5 + {}^-6}{10} = \frac{-11}{10}$$

Q17 Use the cross-product rule for addition of rational numbers to find:

a. $\frac{2}{3} + \frac{4}{5}$ **b.** $\frac{5}{8} + \frac{-1}{2}$

STOP • STOP • STOP • STOP • STOP • STOP • STOP • STOP • STOP

A17 **a.** $\frac{22}{15}$: $\frac{2 \cdot 5 + 3 \cdot 4}{3 \cdot 5}$

$\frac{10 + 12}{15}$

b. $\frac{1}{8}$: $\frac{5 \cdot 2 + 8 \cdot {}^-1}{8 \cdot 2}$

$\frac{10 + {}^-8}{16}$

$\frac{2}{16}$

14 To find the sum of mixed rational numbers, the following procedure may be used.

$$-3\frac{1}{2} + \frac{5}{7} = \frac{-7}{2} + \frac{5}{7} = \frac{-49 + 10}{14} = \frac{-39}{14}$$

$$\left[\textit{Note: } -3\frac{1}{2} = -\left(3\frac{1}{2}\right) = -\left(\frac{3 \cdot 2 + 1}{2}\right) = \frac{-7}{2}\right]$$

$$4\frac{1}{3} + {}^-2\frac{1}{2} = \frac{13}{3} + \frac{-5}{2} = \frac{26 + {}^-15}{6} = \frac{11}{6}$$

$$-1\frac{1}{5} + {}^-3\frac{1}{4} = \frac{-6}{5} + \frac{-13}{4} = \frac{-24 + {}^-65}{20} = \frac{-89}{20}$$

Q18 Find:

a. $-5\frac{1}{2} + \frac{2}{3}$

b. $\frac{-4}{5} + 2\frac{2}{3}$

STOP • STOP • STOP • STOP • STOP • STOP • STOP • STOP • STOP

A18 a. $\frac{-29}{6}$: $\frac{-11}{2} + \frac{2}{3}$

$\frac{-33 + 4}{6}$

b. $\frac{28}{15}$: $\frac{-4}{5} + \frac{8}{3}$

$\frac{-12 + 40}{15}$

Q19 Find:

a. $\frac{-1}{2} + {}^{-}1\frac{1}{4}$

b. $-2\frac{1}{8} + {}^{-}1\frac{3}{16}$

c. $\frac{5}{8} + {}^{-}1\frac{1}{2}$

d. $-4\frac{1}{4} + 3\frac{5}{6}$

STOP • STOP • STOP • STOP • STOP • STOP • STOP • STOP • STOP

A19 a. $\frac{-7}{4}$: $\frac{-1}{2} + \frac{-5}{4}$

$\frac{-4 + {}^{-}10}{8}$

$\frac{-14}{8}$

b. $\frac{-53}{16}$: $\frac{-17}{8} + \frac{-19}{16}$

$\frac{-272 + {}^{-}152}{128}$

$\frac{-424}{128}$

c. $\frac{-7}{8}$

d. $\frac{-5}{12}$

15 To find the product of any two rational numbers, change all mixed numbers to improper fractions and continue as before.

Examples:

$$-2\frac{1}{3} \cdot \frac{-5}{14} = \frac{\overset{-1}{\cancel{-7}}}{3} \cdot \frac{-5}{\underset{2}{\cancel{14}}} = \frac{5}{6}$$

$$5\frac{1}{2} \cdot \frac{-2}{3} = \frac{11}{\underset{1}{\cancel{2}}} \cdot \frac{\overset{-1}{\cancel{-2}}}{3} = \frac{-11}{3}$$

$$-4\frac{1}{3} \cdot 2\frac{5}{6} = \frac{-13}{3} \cdot \frac{17}{6} = \frac{-221}{18}$$

Q20 Find:

a. $5\frac{1}{8} \cdot \frac{-4}{9}$ b. $-1\frac{2}{3} \cdot {}^{-}4\frac{2}{7}$

STOP • STOP • STOP • STOP • STOP • STOP • STOP • STOP • STOP

A20 a. $\frac{-41}{18}$ b. $\frac{50}{7}$

16 When finding a product of a rational number and an integer, write the integer with a denominator of 1 before continuing as before. For example,

$$5\frac{2}{9} \cdot {}^{-}3 = \frac{47}{\cancel{9}_{3}} \cdot \frac{\cancel{-3}^{-1}}{1}$$

$$= \frac{-47}{3}$$

Q21 Find:

a. $2 \cdot {}^{-}1\frac{2}{5}$ b. $\frac{-1}{5} \cdot {}^{-}5$

STOP • STOP • STOP • STOP • STOP • STOP • STOP • STOP • STOP

A21 a. $\frac{-14}{5}$ b. 1

17 The multiplication property of zero is valid for the set of rational numbers. It states:

$x \cdot 0 = 0x = 0$ when $x \in Q$

This property can be stated:

$\frac{p}{q} \cdot 0 = 0 \cdot \frac{p}{q} = 0$ for all rational numbers $\frac{p}{q}$

Examples:

$\frac{-2}{5} \cdot 0 = 0$ $0 \cdot {}^{-}3\frac{1}{3} = 0$

Q22 Find the product:

a. $4\frac{1}{2} \cdot 0 =$ _____ b. $0 \cdot \frac{-5}{8} =$ _____

STOP • STOP • STOP • STOP • STOP • STOP • STOP • STOP • STOP

A22 **a.** 0 **b.** 0

18 When finding the product of more than two rational numbers, write all rational numbers in $\frac{p}{q}$ form, perform all possible reductions, and place the product of the numerators over the product of the denominators.

Example:

$$\frac{4}{7} \cdot \frac{-3}{8} \cdot {}^{-}2\frac{1}{3} = \frac{\overset{1}{\cancel{4}}}{\underset{1}{\cancel{7}}} \cdot \frac{\overset{-1}{\cancel{-3}}}{\underset{2}{\cancel{8}}} \cdot \frac{\overset{-1}{\cancel{-7}}}{\underset{1}{\cancel{3}}}$$

$$= \frac{1}{2}$$

Q23 Find:

a. $-3\frac{1}{3} \cdot \frac{7}{20} \cdot \frac{-9}{14}$ **b.** $2\frac{3}{4} \cdot 0 \cdot {}^{-}6$

STOP • STOP • STOP • STOP • STOP • STOP • STOP • STOP • STOP

A23 **a.** $\frac{3}{4}$ **b.** 0

19 Two rational numbers whose product is 1 are called *reciprocals* of each other.* Some examples are:

1. 5 is the reciprocal of $\frac{1}{5}$, because $5 \cdot \frac{1}{5} = 1$
2. $\frac{-1}{5}$ is the reciprocal of -5, because $\frac{-1}{5} \cdot {}^{-}5 = 1$
3. $\frac{-5}{8}$ and $\frac{-8}{5}$ are reciprocals of each other, because $\frac{-5}{8} \cdot \frac{-8}{5} = 1$

In general, x and $\frac{1}{x}$ are reciprocals, because $x \cdot \frac{1}{x} = 1$. The above statement is true only if $x \neq 0$. That is, 0 does not have a reciprocal.

*Some mathematicians refer to "reciprocals" as "multiplicative inverses."

Q24 Determine the reciprocal:

a. $\frac{2}{3}$ ____________ **b.** -9 ____________ **c.** 15 ____________

d. 0 ____________ **e.** $4\frac{1}{2}$ ____________ $\left(\textit{Hint: } 4\frac{1}{2} = \frac{9}{2}\right)$

f. $-2\frac{1}{3}$ ____________ $\left(\textit{Hint: } -2\frac{1}{3} = \frac{-7}{3}\right)$ **g.** $\frac{-1}{2}$ ____________

STOP • STOP • STOP • STOP • STOP • STOP • STOP • STOP • STOP

A24

a. $\frac{3}{2}$ b. $\frac{-1}{9}$ c. $\frac{1}{15}$ d. does not exist

e. $\frac{2}{9}$ f. $\frac{-3}{7}$ g. -2

20 Consider the quotient of two rational numbers $\frac{w}{x}$ and $\frac{y}{z}$, $\frac{y}{z} \neq 0$.

$$\frac{w}{x} \div \frac{y}{z} = \frac{\frac{w}{x}}{\frac{y}{z}} = \frac{\frac{w}{x}}{\frac{y}{z}} \cdot 1 = \frac{\frac{w}{x}}{\frac{y}{z}} \cdot \frac{\frac{z}{y}}{\frac{z}{y}} \qquad \left(\textit{Note: } 1 = \frac{\frac{z}{y}}{\frac{z}{y}}\right)$$

$$= \frac{\frac{w}{x} \cdot \frac{z}{y}}{\frac{y}{z} \cdot \frac{z}{y}}$$

$$= \frac{\frac{w}{x} \cdot \frac{z}{y}}{1}$$

$$= \frac{w}{x} \cdot \frac{z}{y}$$

Therefore, if $\frac{w}{x}$ and $\frac{y}{z}$ are rational numbers,

$\frac{y}{z} \neq 0$ $\quad \frac{w}{x} \div \frac{y}{z} = \frac{w}{x} \cdot \frac{z}{y}$

reciprocals

This general statement is called the *definition of division for rational numbers.* In words, the quotient of two rational numbers (with a nonzero divisor) is equal to the first (dividend) multiplied by the reciprocal of the second (divisor).

$$\frac{2}{3} \div \frac{4}{9} = \frac{\overset{1}{\cancel{2}}}{\underset{1}{\cancel{3}}} \cdot \frac{\overset{3}{\cancel{9}}}{\underset{2}{\cancel{4}}} = \frac{3}{2}$$

$$\frac{4}{5} \div \frac{-6}{7} = \frac{\overset{2}{\cancel{4}}}{5} \cdot \frac{-7}{\underset{3}{\cancel{6}}} = \frac{-14}{15}$$

$$-1\frac{1}{2} \div 4\frac{1}{3} = \frac{-3}{2} \div \frac{13}{3} = \frac{-3}{2} \cdot \frac{3}{13} = \frac{-9}{26}$$

Q25 Find:

a. $\frac{-2}{3} \div \frac{3}{4}$ b. $\frac{5}{14} \div -3\frac{2}{7}$

STOP • STOP • STOP • STOP • STOP • STOP • STOP • STOP • STOP

A25

a. $\frac{-8}{9}$: $\frac{-2}{3} \cdot \frac{4}{3}$

b. $\frac{-5}{46}$: $\frac{5}{14} \div \frac{-23}{7}$

$\frac{5}{\cancel{14}_{2}} \cdot \frac{\cancel{-7}^{-1}}{23}$

Q26 Find:

a. $\frac{-7}{18} \div \frac{1}{3}$

b. $0 \div \frac{-5}{9}$

c. $\frac{-5}{8} \div \frac{-5}{8}$

d. $8\frac{1}{2} \div {}^{-}1\frac{3}{4}$

e. $\frac{-1}{2} \div 0$

f. $\frac{-11}{15} \div {}^{-}4\frac{2}{5}$

STOP • STOP • STOP • STOP • STOP • STOP • STOP • STOP • STOP

A26

a. $\frac{-7}{6}$ b. 0 c. 1 d. $\frac{-34}{7}$ e. undefined

f. $\frac{1}{6}$

Q27 Find:

a. $\frac{-5}{8} \div 12$

b. $2\frac{2}{3} \div {}^{-}8$

STOP • STOP • STOP • STOP • STOP • STOP • STOP • STOP • STOP

A27

a. $\frac{-5}{96}$: $\frac{-5}{8} \cdot \frac{1}{12}$

b. $\frac{-1}{3}$: $\frac{\cancel{8}^{1}}{3} \cdot \frac{-1}{\cancel{8}_{1}}$

Q28 Find:

a. $-9 \div {}^{-}4\frac{2}{7}$

b. $-8 \div \frac{-1}{8}$

STOP • STOP • STOP • STOP • STOP • STOP • STOP • STOP • STOP

A28

a. $\frac{21}{10}$: $\frac{-9}{1} \div \frac{-30}{7}$

$\frac{\overset{-3}{\cancel{-9}}}{1} \cdot \frac{7}{\underset{-10}{\cancel{-30}}}$

$\frac{21}{10}$

b. 64: $\frac{-8}{1} \cdot \frac{-8}{1}$

21 The *opposite* of $\frac{2}{3}$ is $\frac{-2}{3}$. Recall that the sum of two integers which are opposites is zero. That is, $x + {}^{-}x = 0$, where $x \in I$. It is also true that $x + {}^{-}x = 0$, where $x \in Q$.

Examples:

$$\frac{2}{3} + \frac{-2}{3} = \frac{2 + {}^{-}2}{3} = \frac{0}{3} = 0$$

$$-2\frac{1}{2} + 2\frac{1}{2} = \frac{-5}{2} + \frac{5}{2} = \frac{-5 + 5}{2} = \frac{0}{2} = 0$$

$$0.\overline{14} + {}^{-}0.\overline{14} = 0$$

$$5.125 + {}^{-}5.125 = 0$$

Q29 Complete the following, to form true statements:

a. $\frac{-1}{2} + (\quad\quad) = 0$

b. $\frac{7}{5} + (\quad\quad) = 0$

c. $3\frac{1}{3} + (\quad\quad) = 0$

d. $-5 + (\quad\quad) = 0$

e. $(\quad\quad) + 5.1 = 0$

f. $(\quad\quad) + 2.\overline{3} = 0$

STOP • STOP • STOP • STOP • STOP • STOP • STOP • STOP • STOP

A29

a. $\frac{1}{2}$

b. $\frac{-7}{5}$

c. $-3\frac{1}{3}$ or $\frac{-10}{3}$

d. 5

e. -5.1 or $-5\frac{1}{10}$ or $\frac{-51}{10}$

f. $-2.\overline{3}$ or $-2\frac{1}{3}\left(0.\overline{3} = .333\ldots = \frac{1}{3}\right)$ or $\frac{-7}{3}$

22 When subtracting two integers, the definition of subtraction stated that

$x - y = x + {}^{-}y$ where $x \in I$ and $y \in I$

x: minuend

y: subtrahend

$-y$: opposite of y (the opposite of the subtrahend)

This idea is likewise valid for subtraction of rational numbers. That is,

$x - y = x + {}^{-}y \qquad \text{where } x \in Q, y \in Q$

Examples:

$$\frac{1}{2} - \frac{2}{3} = \frac{1}{2} + \frac{-2}{3}$$

(opposites)

$$\frac{-2}{3} - \frac{-3}{4} = \frac{-2}{3} + \frac{3}{4}$$

(opposites)

Q30 Use the definition of subtraction in Frame 22 to rewrite each subtraction problem as an addition problem (*do no further evaluating*):

a. $-4 - \frac{4}{5} =$ __________

b. $9\frac{1}{4} - {}^{-}2\frac{1}{2} =$ __________

c. $-1\frac{1}{3} - 4 =$ __________

d. $-1\frac{5}{6} - \frac{-4}{5} =$ __________

STOP • STOP • STOP • STOP • STOP • STOP • STOP • STOP • STOP

A30 **a.** $-4 + \frac{-4}{5}$

b. $9\frac{1}{4} + 2\frac{1}{2}$

c. $-1\frac{1}{3} + {}^{-}4$

d. $-1\frac{5}{6} + \frac{4}{5}$

23 Completing the subtraction of the problems in Q30 would require the use of either the LCD method of adding rational numbers or the cross-product rule for addition of rational numbers.

Examples:

Cross-product rule

$$-4 - \frac{4}{5} = \frac{-4}{1} + \frac{-4}{5} = \frac{-20 + {}^{-}4}{5} = \frac{-24}{5}$$

$$9\frac{1}{4} - {}^{-}2\frac{1}{2} = 9\frac{1}{4} + 2\frac{1}{2} = \frac{37}{4} + \frac{5}{2} = \frac{74 + 20}{8} = \frac{94}{8} = \frac{47}{4}$$

LCD method

$$-4 - \frac{4}{5} = \frac{-4}{1} + \frac{-4}{5} = \frac{-20}{5} + \frac{-4}{5} = \frac{-24}{5}$$

$$9\frac{1}{4} - {}^{-}2\frac{1}{2} = 9\frac{1}{4} + 2\frac{1}{2} = 9\frac{1}{4} + 2\frac{2}{4} = 11\frac{3}{4} = \frac{47}{4}$$

Q31 Complete the evaluation by (1) the cross-product rule, and (2) the LCD method:

a. $-1\frac{1}{3} - 4$

(1) (2)

b. $-1\frac{5}{6} - \frac{-4}{5}$

(1) (2)

STOP • STOP • STOP • STOP • STOP • STOP • STOP • STOP • STOP

A31 a. (1) $-1\frac{1}{3} - 4 = -1\frac{1}{3} + {}^{-}4$

$= \frac{-4}{3} + \frac{-4}{1}$

$= \frac{-4 + {}^{-}12}{3}$

$= \frac{-16}{3}$

(2) $-1\frac{1}{3} - 4 = -1\frac{1}{3} + {}^{-}4$

$= -5\frac{1}{3}$

$= \frac{-16}{3}$

b. (1) $-1\frac{5}{6} - \frac{-4}{5} = -1\frac{5}{6} + \frac{4}{5}$

$= \frac{-11}{6} + \frac{4}{5}$

$= \frac{-55 + 24}{30}$

$= \frac{-31}{30}$

(2) $-1\frac{5}{6} - \frac{-4}{5} = -1\frac{5}{6} + \frac{4}{5}$

$= \frac{-11}{6} + \frac{4}{5}$

$= \frac{-55}{30} + \frac{24}{30}$

$= \frac{-31}{30}$

Q32 Complete (use the method of your choice):

a. $2\frac{1}{4} - 3\frac{3}{4}$

b. $-4\frac{3}{7} - 6\frac{1}{7}$

c. $-2\frac{1}{9} - 3\frac{2}{5}$

d. $2\frac{1}{3} - \frac{9}{11}$

e. $\frac{-6}{13} - 4\frac{1}{2}$

f. $-15\frac{1}{6} - {}^{-}2\frac{7}{15}$

STOP • STOP • STOP • STOP • STOP • STOP • STOP • STOP • STOP

A32 a. $\frac{-3}{2}$ b. $\frac{-74}{7}$ c. $\frac{-248}{45}$ d. $\frac{50}{33}$ e. $\frac{-129}{26}$

f. $\frac{-127}{10}$: $-15\frac{1}{6} - {}^{-}2\frac{7}{15} = \frac{-91}{6} + \frac{37}{15}$

$$= \frac{-455}{30} + \frac{74}{30}$$

$$= \frac{-381}{30}$$

24 The following properties are valid for the set of rational numbers, that is, when w, x, y, $z \in Q$.

1. Commutative property of addition

   $x + y = y + x$

2. Associative property of addition

   $(x + y) + z = x + (y + z)$

3. Commutative property of multiplication

   $xy = yx$

4. Associative property of multiplication

   $(xy)z = x(yz)$

   (*Note:* The set of rational numbers is not commutative nor associative with respect to subtraction or division.)

5. Addition property of zero

   $x + 0 = 0 + x = x$

6. Multiplication property of zero

   $0x = x \cdot 0 = 0$

7. Multiplication property of 1

$1x = x \cdot 1 = x$

8. Multiplication property of $^{-}1$

$-1x = -x$

9. Division property of 1

$\frac{x}{1} = x$

10. Division property of -1

$\frac{x}{-1} = -x$

11. Addition property of opposites

$x + {}^{-}x = -x + x = 0$

12. Property of reciprocals

$x \cdot \frac{1}{x} = 1 \qquad x \neq 0$

13. Distributive properties:
(1) $x(y + z) = xy + xz$ (left over addition)
(2) $x(y - z) = xy - xz$ (left over subtraction)
(3) $(x + y)z = xz + yz$ (right over addition)
(4) $(x - y)z = xz - yz$ (right over subtraction)

14. Definition of subtraction

$x - y = x + {}^{-}y$

In the following properties $\frac{w}{x}$ and $\frac{y}{z} \in Q$.

15. Cross-product rule for addition (of rational numbers)

$\frac{w}{x} + \frac{y}{z} = \frac{wz + xy}{xz}$

16. Definition of multiplication of rational numbers

$\frac{w}{x} \cdot \frac{y}{z} = \frac{wy}{xz}$

17. Definition of division of rational numbers

$\frac{w}{x} \div \frac{y}{z} = \frac{w}{x} \cdot \frac{z}{y}$

18. Closure properties:
(1) If $x \in Q$ and $y \in Q$, then $x + y \in Q$.
(2) If $x \in Q$ and $y \in Q$, then $x - y \in Q$.
(3) If $x \in Q$ and $y \in Q$, then $xy \in Q$.
(4) If $x \in Q$ and $y \in Q$, $y \neq 0$, then $\frac{x}{y} \in Q$.

This completes the instruction for this section.

1.4 Exercises

1. Insert $\in$ or $\notin$ to form a true statement:

N = set of natural numbers
W = set of whole numbers
I = set of integers
Q = set of rational numbers

a. 0 ____ N **b.** 0 ____ W **c.** 0 ____ I
d. 0 ____ Q **e.** 15 ____ N **f.** 15 ____ W
g. 15 ____ I **h.** 15 ____ Q **i.** -15 ____ N
j. -15 ____ W **k.** -15 ____ I **l.** -15 ____ Q
m. $\frac{1}{3}$ ____ N **n.** $\frac{1}{3}$ ____ W **o.** $\frac{1}{3}$ ____ I
p. $\frac{1}{3}$ ____ Q **q.** $\frac{-1}{3}$ ____ I **r.** $\frac{-1}{3}$ ____ Q
s. $0.\overline{15}$ ____ I **t.** $0.\overline{15}$ ____ Q **u.** -3.75 ____ I
v. -3.75 ____ Q **w.** $\frac{-7}{5}$ ____ I **x.** $\frac{-7}{5}$ ____ Q
y. $-7.\overline{3}$ ____ I **z.** $-7.\overline{3}$ ____ Q

2. (1) Find the decimal representation for each rational number, and (2) state whether each is terminating or infinite repeating:

a. $\frac{-2}{3}$ **b.** $2\frac{1}{5}$ **c.** $4\frac{7}{9}$ **d.** $\frac{5}{8}$ **e.** $\frac{-7}{11}$

3. Find the sum:

a. $-3\frac{3}{4} + 1\frac{1}{4}$ **b.** $-4\frac{3}{7} + {}^-4\frac{5}{7}$ **c.** $1\frac{1}{3} + {}^-7\frac{2}{3}$

d. $15 + {}^-2\frac{4}{11}$ **e.** $-7 + \frac{4}{5}$ **f.** $\frac{-3}{5} + 0$

g. $\frac{3}{10} + \frac{-7}{6} + \frac{5}{12}$ **h.** $\frac{-5}{12} + \frac{7}{8} + \frac{-7}{12}$

4. Find the difference:

a. $\frac{-5}{8} - \frac{1}{8}$ **b.** $\frac{4}{5} - 1$ **c.** $\frac{-5}{6} - \frac{-10}{12}$

d. $-3 - 2\frac{3}{7}$ **e.** $2\frac{1}{3} - \frac{5}{6} + \frac{2}{9}$ **f.** $4\frac{1}{5} - 2\frac{2}{5} - 1\frac{4}{5}$

5. Find the product:

a. $-2\frac{1}{3} \cdot \frac{6}{7} \cdot \frac{-1}{2}$ **b.** $4\frac{3}{19} \cdot {}^-5 \cdot 0$ **c.** $4\frac{3}{8} \cdot {}^-2\frac{1}{5} \cdot 1\frac{1}{7}$

d. $-4\frac{2}{3} \cdot \frac{2}{5} \cdot \frac{-3}{14}$

6. Find the quotient:

a. $\frac{-4}{15} \div \frac{3}{2}$ **b.** $-2 \div \frac{-6}{11}$ **c.** $-2\frac{1}{2} \div 5$

d. $-3\frac{2}{3} \div -3\frac{1}{3}$ **e.** $0 \div \frac{-1}{2}$ **f.** $-5\frac{2}{3} \div 0$

7. True or false? (Use T or F.)
(*Recall:* $\subset$ says "is a proper subset of.")
a. $N \subset I$ **b.** $N \subset Q$ **c.** $W \subset Q$ **d.** $I \subset Q$

8. Show by an example that there is no:
a. commutative property of subtraction for the set of rational numbers.
b. associative property of subtraction for the set of rational numbers.
c. commutative property of division for the set of rational numbers.
d. associative property of division for the set of rational numbers.

9. a. Verify that $x(y - z) = xy - xz$ is true for $x = \frac{-1}{2}$, $y = \frac{3}{4}$, and $z = \frac{7}{8}$.
b. Verify that $(x + y)z = xz + yz$ is true for $x = \frac{1}{8}$, $y = \frac{-4}{5}$, and $z = 2\frac{1}{4}$.

***10.** Perform the indicated operations:
a. $\frac{-2}{3} + \frac{-3}{4} \cdot \frac{5}{7}$
b. $\frac{-1}{2} \cdot \frac{2}{3} - \frac{-4}{5} \cdot \frac{1}{3}$
c. $\frac{2}{3} \div \frac{-1}{2} \div \frac{4}{5}$
d. $\frac{-5}{6} - \frac{-2}{3} \cdot \frac{9}{10} + \frac{-1}{5}$
e. $\frac{4}{5} \div \frac{2}{5} \cdot \frac{1}{2}$
f. $\frac{1}{2}\left(\frac{-3}{4} + \frac{-7}{8}\right)$

1.4 Exercise Answers

1. **a.** $\notin$ **b.** $\in$ **c.** $\in$ **d.** $\in$ **e.** $\in$ **f.** $\in$ **g.** $\in$
h. $\in$ **i.** $\notin$ **j.** $\notin$ **k.** $\in$ **l.** $\in$ **m.** $\notin$ **n.** $\notin$
o. $\notin$ **p.** $\in$ **q.** $\notin$ **r.** $\in$ **s.** $\notin$ **t.** $\in$ **u.** $\notin$
v. $\in$ **w.** $\notin$ **x.** $\in$ **y.** $\notin$ **z.** $\in$

2. **a.** (1) $-0.\overline{6}$ (2) infinite-repeating
b. (1) 2.2 (2) terminating
c. (1) $4.\overline{7}$ (2) infinite-repeating
d. (1) 0.625 (2) terminating
e. (1) $-0.\overline{63}$ (2) infinite-repeating

3. **a.** $\frac{-5}{2}$ **b.** $\frac{-64}{7}$
c. $\frac{-19}{3}$ **d.** $\frac{139}{11}$
e. $\frac{-31}{5}$ **f.** $\frac{-3}{5}$
g. $\frac{-9}{20}$ **h.** $\frac{-1}{8}$

4. **a.** $\frac{-3}{4}$ **b.** $\frac{-1}{5}$
c. 0 **d.** $\frac{-38}{7}$
e. $\frac{31}{18}$ **f.** 0

5. **a.** 1 **b.** 0 **c.** -11 **d.** $\frac{2}{5}$

6. **a.** $\frac{-8}{45}$ **b.** $\frac{11}{3}$

c. $\frac{-1}{2}$ **d.** $\frac{11}{10}$

e. 0 **f.** undefined

7. **a.** T **b.** T **c.** T **d.** T

8. **a.** $1 - 5 \neq 5 - 1$ (one of many examples)
b. $(5 - 4) - 2 \neq 5 - (4 - 2)$ (one of many examples)
c. $1 \div 2 \neq 2 \div 1$ (one of many examples)
d. $(20 \div 10) \div 2 \neq 20 \div (10 \div 2)$ (one of many examples)

9. **a.**

$$\frac{-1}{2}\left(\frac{3}{4} - \frac{7}{8}\right) \stackrel{?}{=} \frac{-1}{2} \cdot \frac{3}{4} - \frac{-1}{2} \cdot \frac{7}{8}$$

$$\frac{-1}{2} \cdot \frac{-1}{8} \stackrel{?}{=} \frac{-3}{8} - \frac{-7}{16}$$

$$\frac{1}{16} \stackrel{?}{=} \frac{-3}{8} + \frac{7}{16}$$

$$\frac{1}{16} \stackrel{?}{=} \frac{-6}{16} + \frac{7}{16}$$

$$\frac{1}{16} = \frac{1}{16}$$

b.

$$\left(\frac{1}{8} + \frac{-4}{5}\right) \cdot 2\frac{1}{4} \stackrel{?}{=} \frac{1}{8} \cdot 2\frac{1}{4} + \frac{-4}{5} \cdot 2\frac{1}{4}$$

$$\frac{-27}{40} \cdot \frac{9}{4} \stackrel{?}{=} \frac{1}{8} \cdot \frac{9}{4} + \frac{\overset{-1}{\cancel{-4}}}{5} \cdot \frac{9}{\underset{1}{\cancel{4}}}$$

$$\frac{-243}{160} \stackrel{?}{=} \frac{9}{32} + \frac{-9}{5}$$

$$\frac{-243}{160} = \frac{-243}{160}$$

***10.** **a.** $\frac{-101}{84}$:

$$\frac{-2}{3} + \frac{-3}{4} \cdot \frac{5}{7} = \frac{-2}{3} + \frac{-15}{28}$$

$$= \frac{-56 + -45}{84}$$

$$= \frac{-101}{84}$$

b. $\frac{-1}{15}$:

$$\frac{-1}{2} \cdot \frac{2}{3} - \frac{-4}{5} \cdot \frac{1}{3} = \frac{-1}{3} - \frac{-4}{15}$$

$$= \frac{-1}{3} + \frac{4}{15}$$

$$= \frac{-5}{15} + \frac{4}{15}$$

c. $\frac{-5}{3}$:

$$\frac{2}{3} \div \frac{-1}{2} \div \frac{4}{5} = \frac{\overset{1}{\cancel{2}}}{3} \cdot \frac{\overset{-1}{\cancel{-2}}}{1} \cdot \frac{5}{\underset{\underset{1}{\cancel{2}}}{\cancel{4}}}$$

d. $\frac{-13}{30}$: $\frac{-5}{6} - \frac{\overset{-1}{\cancel{-2}}}{\underset{1}{\cancel{3}}} \cdot \frac{\overset{3}{\cancel{9}}}{\underset{5}{\cancel{10}}} + \frac{-1}{5} = \frac{-5}{6} - \frac{-3}{5} + \frac{-1}{5}$

$$= \frac{-5}{6} + \left(\frac{3}{5} + \frac{-1}{5}\right)$$

$$= \frac{-5}{6} + \frac{2}{5}$$

$$= \frac{-25 + 12}{30}$$

e. 1: $\frac{4}{5} \div \frac{2}{5} \cdot \frac{1}{2} = \frac{\overset{\overset{1}{\cancel{2}}}{\cancel{4}}}{\underset{1}{\cancel{5}}} \cdot \frac{\overset{1}{\cancel{5}}}{\underset{1}{\cancel{2}}} \cdot \frac{1}{\underset{1}{\cancel{2}}}$

f. $\frac{-13}{16}$: $\frac{1}{2}\left(\frac{-3}{4} + \frac{-7}{8}\right) = \frac{1}{2}\left(\frac{-6}{8} + \frac{-7}{8}\right)$

$$= \frac{1}{2} \cdot \frac{-13}{8}$$

1.5 The Set of Irrational Numbers and the Set of Real Numbers

1 In Section 1.4 the set of rational numbers, Q, was defined to be numbers of the form

$$\frac{p}{q} \quad \text{where } p \in I,\ q \in I,\ q \neq 0$$

Numbers of this form were (1) terminating decimals, or (2) infinite-repeating decimals.

Examples:

$\frac{-1}{2} = -0.5$ (terminating)

$\frac{1}{3} = 0.333\ldots = 0.\overline{3}$ (infinite-repeating)

$3\frac{1}{7} = \frac{22}{7} = 3.\overline{142857}$ (infinite-repeating)

$-5\frac{1}{8} = \frac{-41}{8} = -5.125$ (terminating)

The following decimal is neither terminating nor infinite-repeating and thus is *not* a rational number.

5.01001000100001 . . .

It is *infinite, nonrepeating*. A number whose decimal representation is infinite, nonrepeating is called an *irrational number*. The set of irrational numbers is denoted by L.

Q1 The decimal representation of every rational number is either:

a. ____________________, or **b.** ____________________

STOP • STOP • STOP • STOP • STOP • STOP • STOP • STOP • STOP

A1 **a.** terminating **b.** infinite-repeating (either order)

Q2 The decimal representation of every ________ number is infinite-nonrepeating.

STOP • STOP • STOP • STOP • STOP • STOP • STOP • STOP • STOP

A2 irrational

2 Irrational numbers may be generated in many ways. Such numbers as 1, 4, 9, 16, 25, 36, 49, 64, 81, 100, 121, 144, 169, 196, 225, . . . are referred to as perfect squares. *Square roots** of perfect squares are rational numbers.

Examples:

$\sqrt{1} = 1$ $\sqrt{4} = 2$ $\sqrt{9} = 3$

$\sqrt{25} = 5$ $\sqrt{36} = 6$ $\sqrt{49} = 7$ etc.

$-\sqrt{64} = -8$ $-\sqrt{81} = -9$ $-\sqrt{121} = -11$ etc.

Square roots of positive nonperfect squares are irrational numbers.

Examples:

$\sqrt{2}$, $\sqrt{3}$, $\sqrt{5}$, $\sqrt{6}$, $\sqrt{7}$, $\sqrt{8}$, $\sqrt{10}$, etc.

*The square root of a number n is a number x such that $x \cdot x = n$. The positive square root of n is written $\sqrt{n}$.

Q3 Answer "rational" or "irrational":

a. 196 is a perfect square; hence $\sqrt{196}$ is a(n) ________ number.

b. 50 is not a perfect square; hence $\sqrt{50}$ is a(n) ________ number.

STOP • STOP • STOP • STOP • STOP • STOP • STOP • STOP • STOP

A3 **a.** rational **b.** irrational

Q4 Indicate whether each of the following numbers represents a rational or an irrational number:

a. $\sqrt{32}$ ________ **b.** $\sqrt{400}$ ________

c. $-\sqrt{25}$ ________ **d.** $\sqrt{37}$ ________

e. $\sqrt{0}$ ________ **f.** $-\sqrt{120}$ ________

STOP • STOP • STOP • STOP • STOP • STOP • STOP • STOP • STOP

A4 **a.** irrational **b.** rational: (20)

c. rational: (-5) **d.** irrational

e. rational: (0) **f.** irrational

3 You should not get the idea that irrational numbers are generated only by taking square roots of positive nonperfect square *natural* numbers. 1.21 is a perfect square (rational number); hence $\sqrt{1.21}$ is a rational number. [*Note:* $1.21 = (1.1)(1.1)$ or $(1.1)^2$.] The number 12.1 is not a perfect square; hence $\sqrt{12.1}$ is an irrational number.

Q5 Answer "rational" or "irrational":

a. 0.04 is a perfect square; hence $\sqrt{0.04}$ is a(n) ________ number.

b. 0.1 is not a perfect square; hence $\sqrt{0.1}$ is a(n) ______________ number.

c. $\frac{1}{4}$ is a perfect square; hence $\sqrt{\frac{1}{4}}$ is a(n) ______________ number.

d. $\frac{9}{25}$ $\underset{\text{is/is not}}{\underline{\hspace{3em}}}$ a perfect square.

e. $\sqrt{\frac{9}{25}}$ is a(n) $\underset{\text{rational/irrational}}{\underline{\hspace{6em}}}$ number.

f. $\frac{2}{5}$ $\underset{\text{is/is not}}{\underline{\hspace{3em}}}$ a perfect square.

g. $\sqrt{\frac{2}{5}}$ is a(n) $\underset{\text{rational/irrational}}{\underline{\hspace{6em}}}$ number.

STOP • STOP • STOP • STOP • STOP • STOP • STOP • STOP • STOP

A5 a. rational b. irrational c. rational d. is
e. rational f. is not g. irrational

4 The set of rational numbers, Q, is an infinite set. The set of irrational numbers, L, is an infinite set. In fact, it can be shown that there are an infinite number of rational numbers between any two consecutive integers. There are also an infinite number of irrational numbers between any two consecutive integers.

Q6 Insert $\in$ or $\notin$ to form a true statement:

a. 1.1 ____ Q b. 1.1 ____ L
c. $\sqrt{1.1}$ ____ Q d. $\sqrt{1.1}$ ____ L
e. 1.44 ____ Q f. 1.44 ____ L
g. $\sqrt{1.44}$ ____ Q h. $\sqrt{1.44}$ ____ L
i. 1.69 ____ Q j. 1.69 ____ L
k. $\sqrt{1.69}$ ____ Q l. $\sqrt{1.69}$ ____ L
m. $\sqrt{1.71}$ ____ Q n. $\sqrt{1.71}$ ____ L
o. $\sqrt{3.9}$ ____ Q p. $\sqrt{3.9}$ ____ L
q. $-\sqrt{4}$ ____ Q r. $-\sqrt{4}$ ____ L
s. $5.\overline{236}$ ____ Q t. $5.\overline{236}$ ____ L
u. $\sqrt{\frac{2}{3}}$ ____ Q v. $\sqrt{\frac{2}{3}}$ ____ L
w. $\sqrt{105}$ ____ Q x. $\sqrt{105}$ ____ L
y. $\sqrt{225}$ ____ Q z. $\sqrt{225}$ ____ L

STOP • STOP • STOP • STOP • STOP • STOP • STOP • STOP • STOP

A6 a. $\in$ b. $\notin$ c. $\notin$ d. $\in$ e. $\in$ f. $\notin$ g. $\in$
h. $\notin$ i. $\in$ j. $\notin$ k. $\in$ l. $\notin$ m. $\notin$ n. $\in$
o. $\notin$ p. $\in$ q. $\in$ r. $\notin$ s. $\in$ t. $\notin$ u. $\notin$
v. $\in$ w. $\notin$ x. $\in$ y. $\in$ z. $\notin$

5 It should be clear that if a number is rational, it cannot be irrational also. Or, if a number is irrational, it cannot be rational. Symbolically, $Q \cap L = \varnothing$. That is, the set of rational

numbers and the set of irrational numbers are disjoint. *Recall:* If $Q \cap L = \varnothing$, Q and L are said to be disjoint sets. The union of the set of rational numbers, Q, and the set of irrational numbers, L, is the set of *real numbers*, R. That is,

$Q \cup L = R$

The real numbers may be shown in the following diagram, where

$N =$ set of natural numbers
$W =$ set of whole numbers
$I =$ set of integers
$Q =$ set of rational numbers
$L =$ set of irrational numbers
$R =$ set of real numbers

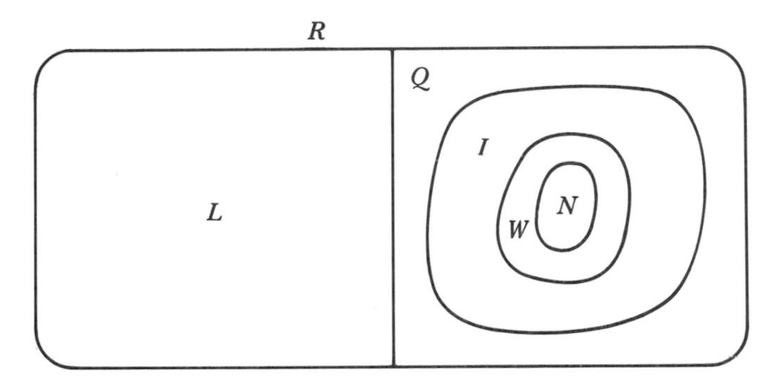

Each set is infinite.

Q7 True or false? (Use T or F.)

a. $N \subset W$ ____ b. $W \subset I$ ____
c. $I \subset Q$ ____ d. $I \subset L$ ____
e. $Q \subset R$ ____ f. $L \subset R$ ____
g. $Q \cap L = \varnothing$ ____ h. $Q \cup L = R$ ____
i. $W \subset L$ ____ j. $I \subset L$ ____
k. $W \subset R$ ____ l. $I \subset R$ ____
m. $N \subset I \subset Q \subset R$ ____

STOP • STOP • STOP • STOP • STOP • STOP • STOP • STOP • STOP

A7 a. T b. T c. T
d. F: If $x \in I$, then $x \in Q$. Since $Q \cap L = \varnothing$, $x \notin L$. Hence $I \not\subset L$.
e. T f. T g. T h. T
i. F: If $x \in W$, then $x \in Q$. Since $Q \cap L = \varnothing$, $x \notin L$. Hence $W \not\subset L$.
j. F: If $x \in I$, then $x \in Q$. Since $Q \cap L = \varnothing$, $x \notin L$. Hence $I \not\subset L$.
k. T l. T m. T

6 This section has presented only one way of generating irrational numbers. Chapter 4 will expand your knowledge of operations with irrational numbers. Many logarithms are also irrational numbers. You will encounter numerous other classes of irrational numbers as you advance in your mathematical education. The arithmetic of irrational numbers will be developed over many years. You should realize, however, that all the properties listed in Frame 24 of Section 1.4 are valid for the set of real numbers.

This completes the instruction for this section.

1.5 Exercises

1. a. Indicate whether the decimal representation of any rational number is terminating, infinite-repeating, and/or infinite-nonrepeating.
   b. Indicate whether the decimal representation of any irrational number is terminating, infinite-repeating, and/or infinite-nonrepeating.

2. Given N = set of natural numbers
   W = set of whole numbers
   I = set of integers
   Q = set of rational numbers
   L = set of irrational numbers
   R = set of real numbers

   True or false? (Use T or F.)

   a. $1 \in R$
   b. $W \subset N$
   c. $-0.\overline{1} \in Q$
   d. $3.14 \notin L$
   e. $Q \cap L \neq R$
   f. $L \cap Q = \varnothing$
   g. $L \cup Q = R$
   h. $L \not\subset R$
   i. $Q \subset R$
   j. $I \subset Q$
   k. $I \subset L$
   l. $Q \subset L$
   m. $L \subset R$
   n. $\{0\} \cup N = W$
   o. $-5 \in Q$
   p. $-5 \in R$
   q. $\frac{-1}{7} \in L$
   r. $5.\overline{636} \in Q$
   s. $\pi \in L$*
   t. $3\frac{5}{9} \in R$
   u. $\frac{-11}{2} \notin Q$
   v. $N \cup W = W$
   w. $N \cap W = N$
   x. $-0.15625 \in Q$
   y. $7.\overline{6} \in L$
   z. $7.\overline{6} \in R$

3. Insert $\in$ or $\notin$ to form a true statement:

   a. $\sqrt{5}$ ____ Q
   b. $\sqrt{144}$ ____ Q
   c. $\sqrt{0.05}$ ____ L
   d. $-\sqrt{100}$ ____ L
   e. $\sqrt{\frac{1}{16}}$ ____ Q
   f. $10\sqrt{36}$ ____ Q
   *g. $\frac{-\sqrt{4}}{9}$ ____ L
   *h. $\frac{\sqrt{3}}{3}$ ____ L

*4. For each given number, indicate all the sets to which it belongs, using N, W, I, Q, L, and R (part a is given as an example).

   a. 9: N, W, I, Q, R
   b. $\sqrt{2}$
   c. $-\sqrt{4}$
   d. $\sqrt{100}$
   e. $\frac{-2}{7}$
   f. $1\frac{1}{3}$
   g. 5.3125
   h. $-12.\overline{7}$
   i. $\sqrt{122}$
   j. -101
   k. 0
   l. π

5. Why can't a number be both a rational and an irrational number?

*π represents an infinite-nonrepeating decimal (3.14159. . .).

1.5 Exercise Answers

1. a. terminating or infinite-repeating (either order)
   b. infinite-nonrepeating
2. a. T b. F c. T d. T e. T f. T g. T
   h. F i. T j. T k. F l. F m. T n. T
   o. T p. T q. F r. T s. T t. T u. F
   v. T w. T x. T y. F z. T
3. a. $\notin$ b. $\in$ c. $\in$ d. $\notin$ e. $\in$ f. $\in$
   *g. $\notin$: $\frac{-\sqrt{4}}{9} = \frac{-2}{9}$ *h. $\in$

*4. b. L, R c. I, Q, R d. N, W, I, Q, R e. Q, R
   f. Q, R g. Q, R h. Q, R i. L, R
   j. I, Q, R k. W, I, Q, R l. L, R
5. Possible responses:
   1. $Q \cap L = \varnothing$
   2. The intersection between the two sets is empty.
   3. The sets are disjoint.

Chapter 1 Sample Test

At the completion of Chapter 1 it is expected that you will be able to work the following problems.

1.1 Set Ideas

1. Define:
   a. set b. empty set
   c. finite set d. infinite set
   e. subset f. proper subset
   g. intersection of two sets h. union of two sets
2. Indicate the mathematical symbol used for each of the following:
   a. empty set b. "is a subset of"
   c. "is a proper subset of" d. set intersection
   e. set union f. "is an element of"
3. True or false? (Use T or F.)
   a. $\{1, 2, 3\} \subset \{1, 2, 3\}$ b. $\varnothing \subseteq \{1, 2, 3, \ldots\}$
   c. $5 \in \{1, 2, 3, \ldots\}$ d. $\{2, 4, 6\} \subset \{0, 1, 2, \ldots\}$
   e. $\{2, 3, 5\} \subset \{1, 3, 5, 7, \ldots\}$ f. $17 \notin \{1, 2, 3, \ldots\}$
4. Determine the intersection or union of the following sets:
   $A = \{1, 2, 4, 7, 9\}$, $B = \{2, 3, 4, 8, 9\}$, $C = \{3, 5, 6, 8, 10\}$, $D = \varnothing$
   a. $A \cup B$ b. $B \cap C$
   c. $A \cap C$ d. $A \cup D$
   e. $B \cap D$ f. $A \cap B$
5. Of the four sets in problem 4, which pairs are disjoint?

1.2 The Set of Whole Numbers

Assume whole-number replacements for all variables.

6. Use set notation to denote the set of:
   a. natural numbers
   b. whole numbers
7. Name the property used to change the expression on the left to the equivalent expression on the right:
   a. $7(3x) = (7 \cdot 3)x$
   b. $(t + 6) + 5 = t + (6 + 5)$
   c. $1(2x - 1) = 2x - 1$
   d. $5x \cdot 0 = 0$
   e. $7y + 2 = 2 + 7y$
   f. $\frac{19c}{1} = 19c$
   g. $x \cdot 9 = 9x$
   h. $0 + 3m = 3m$
   i. $3(y + 2) = 3y + 3 \cdot 2$
   j. $9c - 7c = (9 - 7)c$
8. Use the property given to write a true statement: By the
   a. commutative property of addition, $17 + 3y =$ ______________.
   b. associative property of addition, $5 + (3 + 2x) =$ ______________.
   c. left distributive property of multiplication over subtraction, $5(x - 2) =$ ______________.
   d. right distributive property of multiplication over addition, $18y + 5y =$ ______________.
   e. division property of 1, $\frac{99z}{1} =$ ______________.
9. True or false? (Use T or F.)
   The set of whole numbers is closed under:
   a. addition
   b. subtraction
   c. multiplication
   d. division

1.3 The Set of Integers

10. Write the set of integers in set notation.
11. Given that

$$N = \{1, 2, 3, \ldots\}$$
$$W = \{0, 1, 2, \ldots\}$$
$$I = \{\ldots, -2, -1, 0, 1, 2, \ldots\}$$

insert $\in$ or $\notin$ to form a true statement:

   a. -101 ______ N
   b. -101 ______ W
   c. -101 ______ I
   d. 0 ______ N
   e. 0 ______ W
   f. 0 ______ I
   g. 72 ______ N
   h. 72 ______ W
   i. 72 ______ I
12. Find:
   a. $-12 + 17$
   b. $-12 + {}^{-}13$
   c. $-18 + 18$
   d. $-9 + 12 + {}^{-}3$
   e. $-5 - {}^{-}8$
   f. $-10 - 2$
   g. $13 - 15$
   h. $9 - {}^{-}8$
   i. $-5 - {}^{-}5$
   j. $-7 \cdot {}^{-}9$
   k. $14 \cdot {}^{-}5$
   l. $-1 \cdot {}^{-}19$
   m. $0 \cdot {}^{-}5$
   n. $\frac{-42}{-3}$
   o. $\frac{28}{-4}$
   p. $\frac{0}{-12}$
   q. $\frac{-84}{7}$
   r. $\frac{17}{0}$
   s. $\frac{-32}{-1}$
   t. $\frac{-19}{-19}$

13. Find:

a. $-3 + {}^{-}2 - {}^{-}3 - 5$ b. $-2 - {}^{-}3 \cdot 5$ c. $-2 \cdot {}^{-}8 - 5 \cdot {}^{-}6$

d. $-4(-7 + 9)$ e. $(4 - 7)(7 - 4)$ f. $\dfrac{-12}{7 - 7}$

g. $5 - (-4 + {}^{-}3)$ h. $(9 - {}^{-}2) \div {}^{-}11$

14. True or false? (Use T or F.)

a. If $a \in I$ and $b \in I$, then $a + b \in I$.
b. If $a \in I$ and $b \in I$, then $a - b \in I$.
c. If $a \in I$ and $b \in I$, then $ab \in I$.
d. If $a \in I$ and $b \in I$, then $\dfrac{a}{b} \in I$.
e. The addition property of opposites states that $x + {}^{-}x = 0$ is true if $x \in I$.
f. Multiplying or dividing an integer by -1 results in the opposite of the original integer.
g. $x - y = y - x$ is true for all integer replacements of x and y.
h. The commutative and associative properties of addition are valid for the set of integers.
i. The commutative property of multiplication is valid for the set of integers but the associative property of multiplication is not.
j. $-2(-3 - {}^{-}4) = -2 \cdot {}^{-}3 - {}^{-}2 \cdot {}^{-}4$.

1.4 The Set of Rational Numbers

15. True or false? (Use T or F.)
Q: The set of rational numbers

a. $0 \in Q$ b. $32 \in Q$ c. $98 \in Q$

d. $\dfrac{1}{3} \in Q$ e. $\dfrac{-2}{5} \in Q$ f. $-0.8 \in Q$

g. $5.2\overline{3} \in Q$ h. $5\dfrac{1}{2} \in Q$ i. $\dfrac{-19}{5} \in Q$

16. (1) Find the decimal representation, and (2) state whether it is terminating or infinite-repeating:

a. $\dfrac{-4}{5}$ b. $2\dfrac{5}{6}$ c. $\dfrac{5}{12}$ d. $\dfrac{-11}{40}$

17. Find $\left(\text{leave answer in } \dfrac{p}{q} \text{ form}\right)$:

a. $-2\dfrac{1}{2} + {}^{-}1\dfrac{3}{4}$ b. $-12 + \dfrac{15}{16}$ c. $0 + \dfrac{-2}{3}$

d. $\dfrac{-7}{8} - \dfrac{2}{3}$ e. $-4\dfrac{2}{3} \cdot {}^{-}5\dfrac{1}{8} \cdot 0$ f. $9 \cdot \dfrac{-2}{3} \cdot \dfrac{5}{18}$

g. $\dfrac{-5}{8} \cdot {}^{-}1\dfrac{1}{2} \cdot {}^{-}4\dfrac{2}{3}$ h. $\dfrac{5}{9} \div \dfrac{-2}{3}$ i. $0 \div 9\dfrac{5}{8}$

j. $\dfrac{7}{9} \div 0$

1.5 The Set of Irrational Numbers and the Set of Real Numbers

18. Use "terminating," "infinite-repeating," and "infinite-nonrepeating" to complete the following statement correctly; The decimal representation of any rational number is either a(n) ______(a)______ decimal, or a(n) ______(b)______ decimal; an irrational number is a(n) ______(c)______ decimal.

19. Insert $\in$, $\notin$, $\cup$, $\cap$, $\subset$, or $\not\subset$ to form a true statement:

N = set of natural numbers
W = set of whole numbers
I = set of integers
Q = set of rational numbers
L = set of irrational numbers
R = set of real numbers

a. -15 ____ I
b. $\varnothing$ ____ R
c. N ____ W
d. W ____ I
e. I ____ Q
f. Q ____ R
g. L ____ R
h. Q ____ $L = R$
i. Q ____ $L = \varnothing$
j. Q ____ L
k. -7 ____ W
l. $-0.\bar{5}$ ____ Q
m. π ____ L
n. $\sqrt{15}$ ____ Q
o. $-\sqrt{\frac{1}{25}}$ ____ Q
p. $\frac{-7}{2}$ ____ R
q. $0.101001000\ldots$ ____ L
r. $\frac{-7}{2}$ ____ L
s. 72 ____ I
t. 72 ____ Q
u. 72 ____ L
v. 72 ____ R
w. $\sqrt{72}$ ____ L
x. $\sqrt{72}$ ____ Q
y. $\sqrt{72}$ ____ R

Chapter 1 Sample Test Answers

1. (*Note:* Statements to answers 1a–1h do not have to be word for word.)
   a. a well-defined collection
   b. a set with no elements
   c. a set whose count has a last number
   d. a set whose count is unending
   e. a set whose elements are also in another set
   f. a set whose elements are also in another set and at least one additional element is in the other set
   g. a third set that contains elements, and only those elements, that belong to both original sets
   h. a third set that contains elements that belong to either of the original sets
2. **a.** { } or $\varnothing$ **b.** $\subseteq$ **c.** $\subset$ **d.** $\cap$ **e.** $\cup$ **f.** $\in$
3. **a.** F **b.** T **c.** T **d.** T **e.** F **f.** F
4. **a.** $\{1, 2, 3, 4, 7, 8, 9\}$ **b.** $\{3, 8\}$ **c.** { } or $\varnothing$ **d.** A **e.** D or { } or $\varnothing$ **f.** $\{2, 4, 9\}$
5. A and D, B and D, C and D, A and C
6. **a.** $\{1, 2, 3, \ldots\}$ **b.** $\{0, 1, 2, 3, \ldots\}$
7. **a.** associative property of multiplication
   b. associative property of addition
   c. multiplication property of 1 or multiplicative identity property
   d. multiplication property of zero
   e. commutative property of addition
   f. division property of 1
   g. commutative property of multiplication

h. addition property of zero or additive identity property
i. left distributive property of multiplication over addition
j. right distributive property of multiplication over subtraction

8. **a.** $3y + 17$ **b.** $(5 + 3) + 2x$ or $8 + 2x$
c. $5x - 5 \cdot 2$ or $5x - 10$ **d.** $(18 + 5)y$ or $23y$
e. $99z$

9. **a.** T **b.** F **c.** T **d.** F

10. $\{\ldots, -2, -1, 0, 1, 2, \ldots\}$

11. **a.** $\notin$ **b.** $\notin$ **c.** $\in$ **d.** $\notin$ **e.** $\in$ **f.** $\in$ **g.** $\in$
h. $\in$ **i.** $\in$

12. **a.** 5 **b.** -25 **c.** 0 **d.** 0 **e.** 3 **f.** -12 **g.** -2
h. 17 **i.** 0 **j.** 63 **k.** -70 **l.** 19 **m.** 0 **n.** 14
o. -7 **p.** 0 **q.** -12 **r.** undefined **s.** 32 **t.** 1

13. **a.** -7 **b.** 13 **c.** 46 **d.** -8 **e.** -9 **f.** undefined
g. 12 **h.** -1

14. **a.** T **b.** T **c.** T **d.** F **e.** T **f.** T **g.** F
h. T **i.** F **j.** T

15. **a.** T **b.** T **c.** T **d.** T **e.** T **f.** T **g.** T
h. T **i.** T

16. **a.** (1) -0.8 (2) terminating
b. (1) $2.8\overline{3}$ (2) infinite-repeating
c. (1) $0.41\overline{6}$ (2) infinite-repeating
d. (1) -0.275 (2) terminating

17. **a.** $\frac{-17}{4}$ **b.** $\frac{-177}{16}$ **c.** $\frac{-2}{3}$
d. $\frac{-37}{24}$ **e.** 0 **f.** $\frac{-5}{3}$
g. $\frac{-35}{8}$ **h.** $\frac{-5}{6}$ **i.** 0
j. undefined

18. **a.** terminating
b. infinite-repeating (answers to parts a and b may be reversed)
c. infinite-nonrepeating

19. **a.** $\in$ **b.** $\subset$ **c.** $\subset$ **d.** $\subset$ **e.** $\subset$ **f.** $\subset$ **g.** $\subset$
h. $\cup$ **i.** $\cap$ **j.** $\not\subset$ **k.** $\notin$ **l.** $\in$ **m.** $\in$ **n.** $\notin$
o. $\in$ **p.** $\in$ **q.** $\in$ **r.** $\notin$ **s.** $\in$ **t.** $\in$ **u.** $\notin$
v. $\in$ **w.** $\in$ **x.** $\notin$ **y.** $\in$

Chapter 2

Polynomials and Rational Expressions

This chapter will review and extend algebraic manipulation of polynomials and rational algebraic expressions.

2.1 Introduction

1 In the expression x^n, n is called an *exponent* and x is called the *base*. Symbols such as 2^3 and x^5 are called *powers*. The exponent of a variable such as x is understood to be 1. That is, $x = x^1$ and is read "x" or "the first power of x." The symbol x^2 is read "x squared" or "the second power of x." The symbol x^3 is read "x cubed" or "the third power of x." Other powers are read according to the number used as an exponent. That is, x^4 is read "the fourth power of x."

Q1 Read (write) each of the following:

a. x^5 ______________________

b. y^3 ______________________

c. $-z$ ______________________

d. x^2y ______________________

STOP • STOP • STOP • STOP • STOP • STOP • STOP • STOP • STOP

A1 **a.** the fifth power of x*
b. y cubed or the third power of y
c. the opposite of z or the opposite of the first power of z
d. x squared y or the second power of x times y

*Other acceptable ways of saying x^5 are: (1) x raised to the fifth power, (2) x to the fifth power, or simply (3) x to the fifth.

2 Positive integer exponents indicate the number of times that the base appears as a factor. Some examples are:

1. x means x^1 or one factor of x.
2. x^2 means $x \cdot x$ or two factors of x.
3. x^3 means $x \cdot x \cdot x$ or three factors of x.

In general,

$$x^m = \underbrace{x \cdot x \cdot x \cdot \ldots \cdot x}_{m \text{ factors of } x} \qquad \text{where } x \in N$$

4. x^5y^2 means $x \cdot x \cdot x \cdot x \cdot x \cdot y \cdot y$ or five factors of x and two factors of y.

Q2 Complete to form a true statement:

a. y^3 means ________ or ________ factors of y.

b. z^2 means ________ or two ________ of z.

c. m^1 means ________ or ________ factor of m.

d. x^3y^2 means ________.

STOP • STOP • STOP • STOP • STOP • STOP • STOP • STOP • STOP

A2 **a.** $y \cdot y \cdot y$, three **b.** $z \cdot z$, factors **c.** m, one
d. $x \cdot x \cdot x \cdot y \cdot y$

3 A *polynomial* is defined as an algebraic expression made up of sums, differences, and products of variables and numbers. Examples of polynomials are

$$2x - 3,\ 5y^2 - 3y + 7,\ -5,\ x^2 - xy - y^2,\ \frac{-2}{3}x^2y$$

where the variables represent real numbers.* You should observe that all constants, such as -5, $\frac{1}{2}$, and $\frac{-2}{3}$, are polynomials.

Expressions that contain quotients where the variable is in the denominator are not polynomials. Examples of expressions that are not polynomials are

$$1 + \frac{3}{x} \qquad \frac{5}{3x - 7} \qquad 4y^2 + 3y + \frac{1}{y}$$

Expressions of this type are called *rational algebraic expressions.* Operations with rational algebraic expressions will be discussed later in the chapter.

*For any real-number replacement of the variables, a polynomial represents a real number. Hence all properties that are valid for the set of real numbers can be used when performing operations with polynomials. Unless stated otherwise, the replacements for all variables will come from the set of real numbers.

Q3 Write "yes" if the algebraic expression is a polynomial. Otherwise, write "no":

a. 5 ______ **b.** $5 - \frac{3}{x}$ ______

c. $5x$ ______ **d.** $5y - 5$ ______

e. $\frac{-4}{7}x^2 - y$ ______ **f.** $\frac{5}{x} + \frac{5}{y}$ ______

g. $2x^2 + 3x - 5$ ______ **h.** $\frac{7x}{9 - 2x}$ ______

i. 0 ______ **j.** $4x^5 + 6x^2 - 2x + 7$ ______

STOP • STOP • STOP • STOP • STOP • STOP • STOP • STOP • STOP

A3 **a.** yes **b.** no **c.** yes **d.** yes **e.** yes **f.** no **g.** yes
h. no **i.** yes **j.** yes

4 When a polynomial shows only additions, the *terms* are the parts separated by plus signs. Subtraction signs that appear in a polynomial can be converted to addition signs by use of the definition of subtraction.

	Terms
$5y - 5 = 5y + {}^{-}5$	$5y$ and -5
$5x^2 - 3x + 7 = 5x^2 + {}^{-}3x + 7$	$5x^2$, $-3x$, and 7
$x^3 - y^3 = x^3 + {}^{-}y^3$	x^3 and $-y^3$
$x^2 - xy + y^2 = x^2 + {}^{-}xy + y^2$	x^2, $-xy$, and y^2
$x^4 - x^2y^2 - 6y^4 = x^4 + {}^{-}x^2y^2 + {}^{-}6y^4$	x^4, $-x^2y^2$, and $-6y^4$

Q4 Identify the terms of each polynomial:

a. $-7 + x$ ________________

b. $5x^3 - 3$ ________________

c. $7x^5 - 3x^2 + x - 2$ ________________

d. $8x^3 - y^3$ ________________

STOP • STOP • STOP • STOP • STOP • STOP • STOP • STOP • STOP

A4 **a.** -7 and x **b.** $5x^3$ and -3 **c.** $7x^5$, $-3x^2$, x, and -2 **d.** $8x^3$ and $-y^3$

5 Polynomials that contain only *one* term are called *monomials*. Polynomials that have exactly *two* terms are called *binomials*. Polynomials that have exactly *three* terms are called *trinomials*. No special names are associated with polynomials with more than three terms. You should observe that a monomial, a binomial, and a trinomial are *all* polynomials.

Q5 Identify each polynomial as a monomial, binomial, or trinomial:

a. $5x^2 - 2xy - 3y^2$ __________

b. $x^2 - y^2$ __________

c. $x^3 + y^3$ __________

d. -12 __________

e. $5x^2y^2$ __________

STOP • STOP • STOP • STOP • STOP • STOP • STOP • STOP • STOP

A5 **a.** trinomial **b.** binomial **c.** binomial **d.** monomial **e.** monomial

6 The *degree* of a term containing variables is the sum of the exponents present on the variables in the term. The degree of all real numbers is zero, except for the real number zero (0), which has no degree.

	Degree
y	1, because $y = y^1$
5	0
xy	2, because $xy = x^1y^1$ and $1 + 1 = 2$
$-3x^2$	2
$5x^2y$	3, because $x^2y = x^2y^1$ and $2 + 1 = 3$
0	no degree

Q6 Indicate the degree of each term:

a. $-3t$ ________ b. $9xyz$ ________

c. -8 ________ d. $12xy^2z^3$ ________

e. 0 ________ f. $-5m^2$ ________

STOP • STOP • STOP • STOP • STOP • STOP • STOP • STOP • STOP

A6 **a.** 1 **b.** 3 **c.** 0 **d.** 6 **e.** none **f.** 2

7 The degree of a polynomial is the greatest degree of any of its terms.

	Degree of the polynomial
$x^2 - xy - y^2$	2
$x^3 + y^3$	3
$7x - 2$	1
$x^2yz^3 - 3xy^2$	6, because $x^2yz^3 = x^2y^1z^3$ and $2 + 1 + 3 = 6$

Q7 Indicate the degree of each polynomial:

a. $-3x^2 + x - 2$ _____ b. $x^2 - y^2$ _____

c. $y^2 + 2yz + z^2$ _____ d. $125a^3 - 27b^3$ _____

e. $-x + 9$ _____ f. 5 _____

STOP • STOP • STOP • STOP • STOP • STOP • STOP • STOP • STOP

A7 **a.** 2 **b.** 2 **c.** 2 **d.** 3 **e.** 1 **f.** 0

8 The *numerical coefficient* (number factor) of $-3x^2$ is -3. The *literal coefficient* (letter factor) of $-3x^2$ is x^2. (*Note:* In this text, the use of "coefficient" alone will mean "numerical coefficient.") A term with no variable is called a *constant term.* That is, in $-3x^2 + 5$, 5 is called a constant term and is also referred to as the coefficient of that term.

Q8 True or false? (Use T or F.)

a. -2 is the coefficient of $-2x^2y$. _____

b. x^2 is a literal coefficient of $-2x^2y$. _____

c. x^2 and y are literal coefficients of $-2x^2y$. _____

d. The coefficient of $-x$ is -1. _____

STOP • STOP • STOP • STOP • STOP • STOP • STOP • STOP • STOP

A8 **a.** T **b.** T **c.** T

d. T: by the multiplication property of -1, $-x = -1x$

Q9 Indicate (in order from left to right) the coefficient of each term:

a. $x^3 - 3x^2 + x - 1$ ____________

b. $2x^3 - x^2y + 5xy^2 - 8y^3$ ____________

STOP • STOP • STOP • STOP • STOP • STOP • STOP • STOP • STOP

A9

a. 1, −3, 1, −1

b. 2, −1, 5, −8:
$2x^3 - x^2y + 5xy^2 - 8y^3$
$2x^3 - 1x^2y + 5xy^2 - 8y^3$
$2x^3 + {}^-1x^2y + 5xy^2 + {}^-8y^3$

9

It is common (and sometimes necessary) to rearrange the order of the terms of a polynomial. The commutative and associative properties of addition, together with the definition of subtraction, permit the following rule: *The terms of a polynomial can be rearranged in any order as long as the original sign of each term is left unchanged.* Some examples are:

$-2 + x = x - 2$

$-3 - 5x = -5x - 3$

$5x^2 - 3 + 2x = 5x^2 + 2x - 3$

$9x^3 - y^3 + x^2y - xy^2 = 9x^3 + x^2y - xy^2 - y^3$

$-x + 4 = 4 - x$

Q10

Complete:

a. $y - x =$ ______ $+ y$

b. $-3 + 5x = 5x$ ______

c. $2x^2 - 7 + 9x = 2x^2$ ______ $- 7$

STOP • STOP • STOP • STOP • STOP • STOP • STOP • STOP • STOP

A10

a. $-x$

b. -3

c. $+9x$

10

Polynomials in one variable are usually arranged so that the exponents decrease. When this is done, the polynomial is said to be in *descending order.*

Descending order

$$2x^2 - 7 + 9x = 2x^2 + 9x - 7$$
$$5 - 2x = -2x + 5$$
$$-2 + 5y - y^2 = -y^2 + 5y - 2$$

Polynomials in one variable can be arranged so that the exponents increase. If so, the polynomial is said to be in *ascending order.*

Q11

Arrange in descending order:

a. $3x^4 - x + 7x^3 + 2x^2 - 3 =$ ______

b. $9y^2 + 2 - 6y^3 + 5y =$ ______

STOP • STOP • STOP • STOP • STOP • STOP • STOP • STOP • STOP

A11

a. $3x^4 + 7x^3 + 2x^2 - x - 3$

b. $-6y^3 + 9y^2 + 5y + 2$

11

Polynomials in several variables can be arranged with respect to any variable present in either ascending or descending order.

Example 1: Arrange $4xy - x^2 + y^2$ in descending order with respect to x.

Solution

$4xy - x^2 + y^2 = -x^2 + 4xy + y^2$

(*Note:* The result is in ascending order with respect to y.)

Example 2: Arrange $4x^2y + 5xy^2 - 3x^3y^3$ in descending order with respect to y.

Solution

$4x^2y + 5xy^2 - 3x^3y^3 = -3x^3y^3 + 5xy^2 + 4x^2y$

Q12 Arrange $5xyz - 4x^2y^3z^2 + y^2z^3$ in:

a. ascending order with respect to x. ____________

b. descending order with respect to z. ____________

c. descending order with respect to y. ____________

STOP • STOP • STOP • STOP • STOP • STOP • STOP • STOP • STOP

A12 a. $y^2z^3 + 5xyz - 4x^2y^3z^2$
b. $y^2z^3 - 4x^2y^3z^2 + 5xyz$
c. $-4x^2y^3z^2 + y^2z^3 + 5xyz$

This completes the instruction for this section.

2.1 Exercises

1. Read (write) each of the following:
   a. z^3 b. xy^2z c. y^7 d. $-t$
2. a. What does x^2y^3 mean?
   b. How many factors of s are present in $-2r^3st^2$?
3. Indicate the terms of each polynomial:
   a. $h - 3$ b. $y^3 - 4y^2 + y - 8$
   c. $m^2 + 7mn + 6n^2$ d. $z^3 - 64$
   e. $-12 - t$
4. Identify each expression in problem 3 as a monomial, binomial, or trinomial. Otherwise, write "polynomial."
5. What is the degree of each term?
   a. 0 b. 5 c. $-x$ d. x^2y^3 e. t^7 f. $-3xy^2z$
6. Indicate the degree of each polynomial:
   a. $3x^2 + xy - 4y^2$ b. $x^3 - x^2 + x - 1$
   c. $9 - x$ d. $9 - x^2$
   e. $x - x^2$ f. $1 - x - x^2$
   g. $x^2y^2 - 16$ h. $x^6 - y^6$
7. Indicate:
   a. the coefficient of xy.
   b. the literal coefficients of $2xy$.
   c. all coefficients in $2x^2 - xy + 3y^2$.
8. Arrange in descending order:
   a. $12x - 5x^2 + 4$ b. $2 + x$
   c. $-4x + 5 + 9x^2$ d. $6 - y - y^2$
9. Arrange $12y^4 - 4x^2y^2 + xy^3 - 2x^3y + 3x^4$ in:
   a. descending order with respect to y.
   b. descending order with respect to x.

2.1 Exercise Answers

1. a. z cubed
   b. x y-squared z
   c. y to the seventh (power) or the seventh power of y
   d. opposite of t
2. a. $x \cdot x \cdot y \cdot y \cdot y$ b. 1
3. a. h and -3 b. y^3, $-4y^2$, y, and -8
   c. m^2, $7mn$, and $6n^2$ d. z^3 and -64 e. -12 and $-t$
4. a. binomial b. polynomial c. trinomial
   d. binomial e. binomial
5. a. none b. 0 c. 1 d. 5 e. 7 f. 4
6. a. 2 b. 3 c. 1 d. 2 e. 2 f. 2 g. 4
   h. 6
7. a. 1 b. x and y c. 2, -1, and 3
8. a. $-5x^2 + 12x + 4$ b. $x + 2$
   c. $9x^2 - 4x + 5$ d. $-y^2 - y + 6$
9. a. $12y^4 + xy^3 - 4x^2y^2 - 2x^3y + 3x^4$
   b. $3x^4 - 2x^3y - 4x^2y^2 + xy^3 + 12y^4$

2.2 Addition and Subtraction of Polynomials

1

Terms are called *like* if they have the same literal coefficients (variables) as well as the same exponents on each variable. $2x$ and $-3x$ are like terms. The expressions $2x$ and $-3x^2$ are not like terms. Constant terms such as 5 and -3 are considered to be like terms. The sum or difference of like terms can be combined or simplified. In the examples below, each line is understood to be equivalent to the line above it.* In each, simplify (combine like terms):

	Justification
1. $5x + 2x$	
$(5 + 2)x$	right distributive property of multiplication over addition
$7x$	
2. $3y - 4y$	
$(3 - 4)y$	right distributive property of multiplication over subtraction
$-1y$	
$-y$	multiplication property of -1
3. $5x^3 + 4x^2 + 2x^3 - 7x^2$	
$5x^3 + 2x^3 + 4x^2 - 7x^2$	
$(5 + 2)x^3 + (4 - 7)x^2$	
$7x^3 + {}^{-}3x^2$	
$7x^3 - 3x^2$	definition of subtraction

*Two algebraic expressions are equivalent if they have the same evaluation for all replacements of the variable.

Q1 Simplify:

a. $4x - 7 + 2x + 3$ b. $5x^2y + 7xy - 8x^2y$

c. $-3st + 9uv - 3uv + 10$

STOP • STOP • STOP • STOP • STOP • STOP • STOP • STOP • STOP

A1 **a.** $6x - 4$ **b.** $-3x^2y + 7xy$ or $7xy - 3x^2y$
c. $-3st + 6uv + 10$ (any order)

---

2 The sum of two or more polynomials is found as follows:

$(x^2 + x - 7) + (2x^2 - 3x + 2)$
$x^2 + x - 7 + 2x^2 - 3x + 2$
$x^2 + 2x^2 + x - 3x - 7 + 2$ (*Note:* $x^2 = 1x^2$
$3x^2 - 2x - 5$ $x = 1x$)

The same calculations could be performed by arranging like terms in the same column.

Example: Find $(x^2 + x - 7) + (2x^2 - 3x + 2)$.

Solution

$$\begin{array}{r} x^2 + x - 7 \\ 2x^2 - 3x + 2 \\ \hline 3x^2 - 2x - 5 \end{array}$$

Do as much work mentally as possible.

Q2 Simplify (find the sum):
a. $(2x^3 - 5x^2 + 3x - 2) + (4x^3 - 4x + 5)$

b. $(9r^2t^2 - 2rt + 8) + (-2rt - 9)$

STOP • STOP • STOP • STOP • STOP • STOP • STOP • STOP • STOP

A2 **a.** $6x^3 - 5x^2 - x + 3$ **b.** $9r^2t^2 - 4rt - 1$

---

3 The addition property of opposites states that $x + {}^-x = 0$ for all real-number replacements of x. It is also true that if $x + y = 0$, then $x = -y$ and $y = -x$. That is, if the sum of two values is zero, one value is the opposite of the other. For example, since $(2x - 5) + (-2x + 5) = 0$, $2x - 5 = -(-2x + 5)$ and $-2x + 5 = -(2x - 5)$. In general, $(a + b) + (-a - b) = 0$. Hence $-(a + b) = -a - b$. Therefore, to form the opposite of a polynomial it is possible to simply change the sign of each term of the polynomial.

Q3 Find:
a. $-(5x^2 - 2x + 4)$ **b.** $-(x^2y^2 - 16)$ **c.** $-(2 + 3r - 2r^2)$

STOP • STOP • STOP • STOP • STOP • STOP • STOP • STOP • STOP

A3 a. $-5x^2 + 2x - 4$ b. $-x^2y^2 + 16$ c. $-2 - 3r + 2r^2$

4 The definition of subtraction states that

$x - y = x + {}^-y$ where $x \in R, y \in R$

That is, to subtract two values, add the opposite of the subtrahend to the minuend. The difference between two polynomials can be found by applying this definition. For example,

$(2x + 5) - (3x - 7)$
$2x + 5 + {}^-(3x - 7)$ definition of subtraction
$2x + 5 + {}^-3x + 7$
$-x + 12$ or $12 - x$

Q4 Find the difference (simplify):
a. $(7x^2 - 4) - (6x^2 + 2)$ b. $(12x^2y^2 + 7xy - 3) - (-8x^2y^2 + 4)$

c. $-15q^3 + 11q^2 - 7q - 1 - (4q^2 + 8q - 5)$

STOP • STOP • STOP • STOP • STOP • STOP • STOP • STOP • STOP

A4 a. $x^2 - 6$ b. $20x^2y^2 + 7xy - 7$
c. $-15q^3 + 7q^2 - 15q + 4$

5 Subtraction of polynomials can also be done using columns.

Example: Find $5r^2 - 3r + 6 - (-3r^2 - 3r + 4)$.

Solution

$$\begin{array}{l} 5r^2 - 3r + 6 \\ \underline{3r^2 + 3r - 4} \quad \text{(change signs of each term of the subtrahend)} \\ 8r^2 \qquad\; + 2 \end{array}$$

Hence the difference is $8r^2 + 2$.

Q5 Simplify:
a. $32y^3 - 19y^2x + 12yx^2 - 9x^3 - (15y^2x + 7x^3)$

b. $13x^2 + 7x - (5x + 3)$

STOP • STOP • STOP • STOP • STOP • STOP • STOP • STOP • STOP

A5 a. $32y^3 - 34y^2x + 12yx^2 - 16x^3$ b. $13x^2 + 2x - 3$

This completes the instruction for this section.

2.2 Exercises

1. Simplify (combine like terms):
   a. $12x + 3 + 9x - 5$
   b. $8xy - 4 - 10xy$
   c. $17mn + 8mn - 3$
   d. $12x^3 + 9x^2 - 5x^3 - 4x^2$
2. Simplify (find the sum):
   a. $(-3x^4 - 7x^3 + 8x^2 + 7) + (-5x^3 - 11)$
   b. $(-2y^2 - 3y + 5) + (3y^2 - 7)$
3. Find:
   a. $-(2x - 7)$
   b. $-(-3x^2 + 12x - 6)$
4. Simplify (find the difference):
   a. $(2x^3 + 4x - 5) - (x - 4)$
   b. $23xy^2 + 10x^2y - (15x^2y + 24xy^2)$

2.2 Exercise Answers

1. **a.** $21x - 2$
   b. $-2xy - 4$
   c. $25mn - 3$
   d. $7x^3 + 5x^2$
2. **a.** $-3x^4 - 12x^3 + 8x^2 - 4$
   b. $y^2 - 3y - 2$
3. **a.** $-2x + 7$ or $7 - 2x$
   b. $3x^2 - 12x + 6$
4. **a.** $2x^3 + 3x - 1$
   b. $-xy^2 - 5x^2y$ or $-5x^2y - xy^2$

2.3 Multiplication of Polynomials

1

Multiplication of monomial factors is the simplest product involving polynomials. Three examples are:

$5 \cdot x = 5x \qquad x \cdot y = xy \qquad (-2x)(5y) = (-2)(5) \cdot x \cdot y = -10xy$

When exponents are present, certain properties of exponents can be applied. One such property is

$x^m \cdot x^n = x^{m+n}$ where $x \in R, m,n \in N$

Two examples are:

$x^2 \cdot x^3 \qquad (-3x^2yz^3)(-4xy^3z^2)$
$x^{2+3} \qquad (-3)(-4)x^2 \cdot x \cdot y \cdot y^3 \cdot z^3 \cdot z^2$
$x^5 \qquad 12x^3y^4z^5$

[*Note:* $x^2 \cdot x^3 = (x \cdot x)(x \cdot x \cdot x) = x^5$.]

Q1 Find the product:

a. $(3x^3)(-4x^5) =$ ________
b. $(-7x^4y^5)(-2xy^3) =$ ________
c. $(-5m^2n)(-8mn^3) =$ ________
d. $(-xy^2)(6xy^7) =$ ________

STOP • STOP • STOP • STOP • STOP • STOP • STOP • STOP • STOP

A1

a. $-12x^8$
b. $14x^5y^8$
c. $40m^3n^4$
d. $-6x^2y^9$: $(-xy^2)(6xy^7) = (-1xy^2)(6xy^7)$

Q2 When multiplying monomials, the exponents of like bases can be ________.

STOP • STOP • STOP • STOP • STOP • STOP • STOP • STOP • STOP

A2 added

---

2 When raising a monomial to a power, two additional properties can be used. One is:

$(xy)^m = x^m y^m$ where $x \in R$, $m \in N$

Examples:

$(xy)^2$	$(5xy)^2$	$(-2x)^3$
x^2y^2	$5^2x^2y^2$	$(-2)^3x^3$
	$25x^2y^2$	$-8x^3$

[*Note:* $(xy)^2 = (xy)(xy) = x^2y^2$.]

The other is:

$(x^m)^n = x^{mn}$ where $x \in R$, $m,n \in N$

Examples:

$(x^2)^3$	$(-5x^2y^3z)^3$
$x^{2\cdot 3}$	$(-5)^3(x^2)^3(y^3)^3z^3$
x^6	$-125x^6y^9z^3$

[*Note:* $(x^2)^3$ means $x^2 \cdot x^2 \cdot x^2 = x^6$.]

Q3 Simplify:

a. $(2x)^3 =$ ________ **b.** $(-3xyz)^2 =$ ________
c. $(mn)^4 =$ ________ **d.** $(x^3y^2)^3 =$ ________
e. $(x^2)^7 =$ ________ **f.** $(-4p^4)^3 =$ ________
g. $(-x^5)^2 =$ ________ **h.** $(-t)^5 =$ ________

STOP • STOP • STOP • STOP • STOP • STOP • STOP • STOP • STOP

A3 **a.** $8x^3$ **b.** $9x^2y^2z^2$ **c.** m^4n^4
d. x^9y^6 **e.** x^{14} **f.** $-64p^{12}$
g. x^{10}: $(-x^5)^2 = (-1x^5)^2$ **h.** $-t^5$

Q4 **a.** When raising a monomial to a power, each ________ of the monomial is raised to that power.
b. When raising a monomial to a power, the exponent on each variable can be ________ by the power.

STOP • STOP • STOP • STOP • STOP • STOP • STOP • STOP • STOP

A4 **a.** factor **b.** multiplied

3

The simplification of a monomial times a polynomial requires the application of the distributive properties. Following are five examples:

1. $2(x + 5)$
$2 \cdot x + 2 \cdot 5$
$2x + 10$
left distributive property of multiplication over addition

2. $(x - 3)5x$
$x \cdot 5x - 3 \cdot 5x$
$5x^2 - 15x$
right distributive property of multiplication over subtraction

3. $-x(2x + 5)$
$-1x(2x + 5)$ — multiplication property of -1
$-1x \cdot 2x + {}^-1x \cdot 5$
$-2x^2 + {}^-5x$
$-2x^2 - 5x$ — definition of subtraction

4. $xy(x^2 + 2x - 3)$
$xy \cdot x^2 + xy \cdot 2x - xy \cdot 3$*
$x^3y + 2x^2y - 3xy$

5. $-3x^2(x^2 - y^2)$
$-3x^2 \cdot x^2 - {}^-3x^2 \cdot y^2$
$-3x^4 - {}^-3x^2y^2$
$-3x^4 + 3x^2y^2$

*An extended distributive property can be proved. A simple example would be:

$m(x + y - z)$
$m[(x + y) - z]$ — order of operations
$m(x + y) - mz$ — left distributive property of multiplication over subtraction
$mx + my - mz$ — left distributive property of multiplication over addition

This idea can be applied to any product involving two polynomials.

Q5 Simplify (do as much mentally as possible):

a. $-3(x^2 - 7x + 6)$ **b.** $(x + 4)7x$ **c.** $-xy(x^2 + x - 5)$

d. $-2y^2(3y^2 + 4y - 1)$

STOP • STOP • STOP • STOP • STOP • STOP • STOP • STOP • STOP

A5 **a.** $-3x^2 + 21x - 18$ **b.** $7x^2 + 28x$ **c.** $-x^3y - x^2y + 5xy$
d. $-6y^4 - 8y^3 + 2y^2$

4

The product of two binomials may be completed as follows:

$(2x + 3)(x - 4)$
$2x(x - 4) + 3(x - 4)$ — right distributive property of multiplication over addition

$2x^2 - 8x + 3x - 12$ — left distributive property of multiplication over subtraction
$2x^2 - 5x - 12$

Another example is:

$(3x - y)(2x - 3y)$
$3x(2x - 3y) - y(2x - 3y)$ — right distributive property of multiplication over subtraction

$6x^2 - 9xy - 2xy + 3y^2$ — left distributive property of multiplication over subtraction
$6x^2 - 11xy + 3y^2$

Q6 Complete the following simplification:

$(2x + 5)(x - 6)$

a. $2x(\quad\quad) + 5(\quad\quad)$ — right distributive property of multiplication over addition

b. ______________________ — left distributive property of multiplication over subtraction

c. ______________________ — final simplification

STOP • STOP • STOP • STOP • STOP • STOP • STOP • STOP • STOP

A6 **a.** $x - 6$ (both blanks) **b.** $2x^2 - 12x + 5x - 30$
c. $2x^2 - 7x - 30$

Q7 Complete the following simplification: $(3x - y)(2x + 5y)$

a. $3x(\quad\quad) - y(\quad\quad)$ — right distributive property of multiplication over subtraction

b. ______________________ — left distributive property of multiplication over addition

c. ______________________ — final simplification

STOP • STOP • STOP • STOP • STOP • STOP • STOP • STOP • STOP

A7 **a.** $2x + 5y$ (both blanks) **b.** $6x^2 + 15xy - 2xy - 5y^2$
c. $6x^2 + 13xy - 5y^2$

Q8 Simplify by use of the above procedure:

a. $(x + 5)(x - 2)$ **b.** $(x - 3y)(x + 7y)$ **c.** $(x - 2y)(x - 3y)$

d. $(2x + 1)(x + 4)$ **e.** $(x - 5y)(2x + 3y)$ **f.** $(4x + 3)(x - 5)$

g. $(3x - 2)(2x - 5)$ **h.** $(5x + y)(5x - y)$

STOP • STOP • STOP • STOP • STOP • STOP • STOP • STOP • STOP

A8 **a.** $x^2 + 3x - 10$ **b.** $x^2 + 4xy - 21y^2$ **c.** $x^2 - 5xy + 6y^2$
d. $2x^2 + 9x + 4$ **e.** $2x^2 - 7xy - 15y^2$ **f.** $4x^2 - 17x - 15$
g. $6x^2 - 19x + 10$ **h.** $25x^2 - y^2$

5 Consider the following product between any two binomials:

$(a + b)(c + d)$
$a(c + d) + b(c + d)$
$ac + ad + bc + bd$

This simplification illustrates a method (called the *foil method*) by which you can multiply two binomials mentally:

1. ac is the product of the *f*irst terms of each binomial.
2. ad is the product of the *o*uter terms of each binomial.
3. bc is the product of the *i*nner terms of each binomial.
4. bd is the product of the *l*ast terms of each binomial.
5. The final result is a simplification of the *sum* of these individual products.

Q9 Given $(x + 2)(x - 7)$, indicate the product of:

a. the first terms __________

b. the outer terms __________

c. the inner terms __________

d. the last terms __________

e. the simplification of the sum of the four terms in a–d __________

STOP • STOP • STOP • STOP • STOP • STOP • STOP • STOP • STOP

A9 **a.** x^2 **b.** $-7x$ **c.** $2x$ **d.** -14
e. $x^2 - 5x - 14$

6 The work is usually displayed:

$(x + 2)(x - 7)$
$x^2 - 7x + 2x - 14$
$x^2 - 5x - 14$

Often the product of the outer terms and the product of the inner terms are like terms. With practice their sum can be obtained mentally. Example:

$$(x + 3)(x + 7) = \underset{(1)}{x^2} + \underset{(2)}{10x} + \underset{(3)}{21}$$

(1) Product of first terms
(2) Sum of the products of the outer and inner terms $(7x + 3x)$
(3) Product of the last terms

Q10 Use the technique of Frame 6 to simplify the following products:

a. $(x + 4)(x + 5)$ **b.** $(x - 2)(x - 3)$ **c.** $(x + 7)(x - 4)$

d. $(x - 3)(x + 5)$ **e.** $(2x + 5)(x - 2)$ **f.** $(3x - 2)(2x - 3)$

g. $(x + 4)(x - 4)$

h. $(2x + 3)^2$
$(2x + 3)(2x + 3)$

STOP • STOP • STOP • STOP • STOP • STOP • STOP • STOP • STOP

A10

a. $x^2 + 9x + 20$
b. $x^2 - 5x + 6$
c. $x^2 + 3x - 28$
d. $x^2 + 2x - 15$
e. $2x^2 + x - 10$
f. $6x^2 - 13x + 6$
g. $x^2 - 16$
h. $4x^2 + 12x + 9$

7 The foil method (Frames 5 and 6) can be used when finding the product of binomials containing more than one variable.

Examples:

1. $(x - y)(2x + 3y)$
   $2x^2 + 3xy - 2xy - 3y^2$
   $2x^2 + xy - 3y^2$
2. $(3x + 5y)(3x - 5y)$
   $9x^2 - 15xy + 15xy - 25y^2$
   $9x^2 - 25y^2$

Do as much mentally as possible.

Q11 Simplify:

a. $(x - 2y)(2x - y)$
b. $(3x + 2y)(5x + 3y)$
c. $(x - 5y)(x + 3y)$
d. $(2x + y)(2x - y)$
e. $(x + y)^2$
f. $(2x - y)^2$

STOP • STOP • STOP • STOP • STOP • STOP • STOP • STOP • STOP

A11

a. $2x^2 - 5xy + 2y^2$
b. $15x^2 + 19xy + 6y^2$
c. $x^2 - 2xy - 15y^2$
d. $4x^2 - y^2$
e. $x^2 + 2xy + y^2$
f. $4x^2 - 4xy + y^2$

8 The product of a binomial and a trinomial or any two polynomials can be obtained by use of the distributive properties.

Examples:

1. $(2x - 3)(x^2 + 2x - 4)$
   $2x(x^2 + 2x - 4) - 3(x^2 + 2x - 4)$ — right distributive property of multiplication over subtraction
   $2x^3 + 4x^2 - 8x - 3x^2 - 6x + 12$ — left distributive properties
   $2x^3 + x^2 - 14x + 12$

2. $(x^2 - 2x + 3)(2x^2 + 3x - 5)$

$x^2(2x^2 + 3x - 5) - 2x(2x^2 + 3x - 5) + 3(2x^2 + 3x - 5)$
$2x^4 + 3x^3 - 5x^2 - 4x^3 - 6x^2 + 10x + 6x^2 + 9x - 15$
$2x^4 + 3x^3 - 4x^3 - 5x^2 - 6x^2 + 6x^2 + 10x + 9x - 15$
$2x^4 - x^3 - 5x^2 + 19x - 15$

Q12 Find the product (simplify):

a. $(x - 3)(x^2 - 2x + 3)$

b. $(2x^2 - x - 3)(x^2 + 2x - 4)$

STOP • STOP • STOP • STOP • STOP • STOP • STOP • STOP • STOP

A12 **a.** $x^3 - 5x^2 + 9x - 9$: $(x - 3)(x^2 - 2x + 3)$
$x(x^2 - 2x + 3) - 3(x^2 - 2x + 3)$
$x^3 - 2x^2 + 3x - 3x^2 + 6x - 9$
$x^3 - 5x^2 + 9x - 9$

b. $2x^4 + 3x^3 - 13x^2 - 2x + 12$:
$(2x^2 - x - 3)(x^2 + 2x - 4)$
$2x^2(x^2 + 2x - 4) - x(x^2 + 2x - 4) - 3(x^2 + 2x - 4)$
$2x^4 + 4x^3 - 8x^2 - x^3 - 2x^2 + 4x - 3x^2 - 6x + 12$
$2x^4 + 4x^3 - x^3 - 8x^2 - 2x^2 - 3x^2 + 4x - 6x + 12$

9 The product of two polynomials containing more than one variable deserves special care, owing to the numerous possibilities for error.

Example: Simplify $(x + y)(x^2 - xy + y^2)$.

Solution

$(x + y)(x^2 - xy + y^2)$
$x(x^2 - xy + y^2) + y(x^2 - xy + y^2)$
$x^3 - x^2y + xy^2 + x^2y - xy^2 + y^3$
$x^3 - x^2y + x^2y + xy^2 - xy^2 + y^3$
$x^3 + y^3$

Q13 Simplify $(3x + y)(9x^2 - 3xy + y^2)$.

STOP • STOP • STOP • STOP • STOP • STOP • STOP • STOP • STOP

A13 $27x^3 + y^3$: $(3x + y)(9x^2 - 3xy + y^2)$
$3x(9x^2 - 3xy + y^2) + y(9x^2 - 3xy + y^2)$
$27x^3 - 9x^2y + 3xy^2 + 9x^2y - 3xy^2 + y^3$

Q14 Simplify $(x - y)(x^2 + xy + y^2)$.

STOP • STOP • STOP • STOP • STOP • STOP • STOP • STOP • STOP

A14 $x^3 - y^3$: $(x - y)(x^2 + xy + y^2)$
$x(x^2 + xy + y^2) - y(x^2 + xy + y^2)$
$x^3 + x^2y + xy^2 - x^2y - xy^2 - y^3$

Q15 Simplify $(2x - 3y)(4x^2 + 6xy + 9y^2)$.

STOP • STOP • STOP • STOP • STOP • STOP • STOP • STOP • STOP

A15 $8x^3 - 27y^3$: $(2x - 3y)(4x^2 + 6xy + 9y^2)$
$2x(4x^2 + 6xy + 9y^2) - 3y(4x^2 + 6xy + 9y^2)$
$8x^3 + 12x^2y + 18xy^2 - 12x^2y - 18xy^2 - 27y^3$

10 When multiplying a monomial times two or more binomials, multiply two factors and then multiply this result times the third factor.

Example 1: Simplify $3(x + 1)(x - 5)$.

Solution

$3(x + 1)(x - 5)$
$3(x^2 - 4x - 5)$
$3x^2 - 12x - 15$

Example 2: Simplify $2x(x + y)(2x - 3y)$.

Solution

$2x(x + y)(2x - 3y)$
$2x(2x^2 - xy - 3y^2)$
$4x^3 - 2x^2y - 6xy^2$

In both examples the foil method was used first on the product of the two binomials.

Q16 Simplify $-3x(x - y)(x - 2y)$.

STOP • STOP • STOP • STOP • STOP • STOP • STOP • STOP • STOP

A16 $-3x^3 + 9x^2y - 6xy^2$: $-3x(x - y)(x - 2y)$
$-3x(x^2 - 3xy + 2y^2)$

11 Study the following example carefully.

Example: Simplify $5x^2(x - y)(x^2 + xy + y^2)$.

Solution

$5x^2(x - y)(x^2 + xy + y^2)$
$5x^2[x(x^2 + xy + y^2) - y(x^2 + xy + y^2)]$
$5x^2(x^3 + x^2y + xy^2 - x^2y - xy^2 - y^3)$
$5x^2(x^3 - y^3)$
$5x^5 - 5x^2y^3$

Q17 Simplify $-3x(x + 2y)(x^2 - 2xy + 4y^2)$.

STOP • STOP • STOP • STOP • STOP • STOP • STOP • STOP • STOP

A17 $-3x^4 - 24xy^3$: $-3x(x + 2y)(x^2 - 2xy + 4y^2)$
$-3x[x(x^2 - 2xy + 4y^2) + 2y(x^2 - 2xy + 4y^2)]$
$-3x(x^3 - 2x^2y + 4xy^2 + 2x^2y - 4xy^2 + 8y^3)$
$-3x(x^3 + 8y^3)$

This completes the instruction for this section.

2.3 Exercises

1. Find the product:
   a. $(5x^2)(-3x^4)$
   b. $(-4m^2n)(-8mn^2)$
   c. $9xy^2z \cdot 7x^2yz^3$
   d. $(-r^3s^2t)(11rst^3)$
   e. $(-2x^2y^3)^4$
   f. $(-3mn^5)^2$
   g. $\left(\frac{2}{3}p^2q^3\right)^2$
   h. $(-0.1r^2st^3)^3$

   In parts i–l, $x,y,z \in R$ and $a,b,c \in N$.
   ***i.** $(y^a \cdot y^b)^c$
   ***j.** $(az^b)^c$
   ***k.** $x \cdot x^a$
   ***l.** $(bx^{a+1}y^b)^c$
2. Simplify:
   a. $-4(x^2 - 5x + 2)$
   b. $5x(x - 3)$
   c. $-mn(m^2 + m + 1)$
   d. $-3y(x^2 - xy + y^2)$
   e. $6x^2(2x^2 + 5x - 2)$
   f. $-4xy(x^3 - y^3)$

3. Simplify:
   a. $(2x - 1)(3x - 2)$
   b. $(5x + 3y)(2x - 7y)$
   c. $(m + 3)(4m + 5)$
   d. $(2x + 3y)(2x - 3y)$
   e. $(x + 3y)^2$
   f. $(2x - 3y)^2$
4. Simplify:
   a. $(x^2 - 2x + 5)(2x - 1)$
   b. $(y^2 + 2y - 1)(y^2 - y + 2)$
5. Simplify:
   a. $(2m + 3n)(m^2 - 3mn - n^2)$
   *b. $(m^2 + 2mn - n^2)(m^2 - mn - 3n^2)$
   c. $(x + 3y)(x^2 - 3xy + 9y^2)$
   d. $(2y - 1)(4y^2 + 2y + 1)$
6. Simplify:
   a. $2(x + 2)(x - 4)$
   b. $-3(y - 2)(2y + 1)$
   c. $x(2x + y)(x - y)$
   d. $-2y(x + 2y)^2$
   e. $-xy(x - 3y)(x + 3y)$
   f. $5x^2(x + 1)(x^2 - 3x + 2)$
   g. $-(x - 1)(x^2 + x + 1)$
   *h. $-3(x^2 + xy - y^2)(x^2 - xy + y^2)$

2.3 Exercise Answers

1. a. $-15x^6$ b. $32m^3n^3$ c. $63x^3y^3z^4$ d. $-11r^4s^3t^4$
   e. $16x^8y^{12}$ f. $9m^2n^{10}$ g. $\frac{4}{9}p^4q^6$ h. $-0.001r^6s^3t^9$
   *i. y^{ac+bc} *j. a^cz^{bc} *k. x^{a+1} *l. $b^cx^{ac+c}y^{bc}$
2. a. $-4x^2 + 20x - 8$
   b. $5x^2 - 15x$
   c. $-m^3n - m^2n - mn$
   d. $-3x^2y + 3xy^2 - 3y^3$
   e. $12x^4 + 30x^3 - 12x^2$
   f. $-4x^4y + 4xy^4$
3. a. $6x^2 - 7x + 2$
   b. $10x^2 - 29xy - 21y^2$
   c. $4m^2 + 17m + 15$
   d. $4x^2 - 9y^2$
   e. $x^2 + 6xy + 9y^2$
   f. $4x^2 - 12xy + 9y^2$
4. a. $2x^3 - 5x^2 + 12x - 5$
   b. $y^4 + y^3 - y^2 + 5y - 2$
5. a. $2m^3 - 3m^2n - 11mn^2 - 3n^3$
   *b. $m^4 + m^3n - 6m^2n^2 - 5mn^3 + 3n^4$
   c. $x^3 + 27y^3$
   d. $8y^3 - 1$
6. a. $2x^2 - 4x - 16$
   b. $-6y^2 + 9y + 6$
   c. $2x^3 - x^2y - xy^2$
   d. $-2x^2y - 8xy^2 - 8y^3$
   e. $-x^3y + 9xy^3$
   f. $5x^5 - 10x^4 - 5x^3 + 10x^2$
   g. $-x^3 + 1$ or $1 - x^3$
   *h. $-3x^4 + 3x^2y^2 - 6xy^3 + 3y^4$

2.4 Special Products

1

The foil method may always be used when finding the product of two binomials. However, in special cases, other shortcuts will help. Consider the *square of a binomial*.

$$\begin{aligned}(a + b)^2 &= (a + b)(a + b)\\ &= a^2 + ab + ab + b^2\\ &= a^2 + 2ab + b^2\end{aligned}$$

Hence $(a + b)^2 = a^2 + 2ab + b^2$, where a is the *first* term and b is the *second* term.

The square of any binomial is the square of the first term plus twice the product of the two terms plus the square of the second term.

Examples:

$$(2x + 3y)^2 = (2x)^2 + 2(2x \cdot 3y) + (3y)^2$$
$$= 4x^2 + 12xy + 9y^2$$

$$(y - 2z)^2 = (y + {}^-2z)^2$$
$$= (y)^2 + 2(y \cdot {}^-2z) + (-2z)^2$$
$$= y^2 - 4yz + 4z^2$$

Q1 Complete the following:

$(a - b)^2 = (a + {}^-b)^2$

a. $= (\quad)^2 + 2(\quad) + (\quad)^2$

b. $=$ ______________

STOP • STOP • STOP • STOP • STOP • STOP • STOP • STOP • STOP

A1 **a.** $(a)^2 + 2(a \cdot {}^-b) + ({}^-b)^2$
b. $a^2 - 2ab + b^2$

Q2 Use the technique of Frame 1 to simplify each product (write only the result if possible):

a. $(2x + 5)^2$

b. $(x - 7)^2$

c. $(2x - 5y)^2$

d. $(x + 7y)^2$

STOP • STOP • STOP • STOP • STOP • STOP • STOP • STOP • STOP

A2 **a.** $4x^2 + 20x + 25$: $(2x + 5)^2 = (2x)^2 + 2(2x)(5) + (5)^2 = 4x^2 + 20x + 25$

b. $x^2 - 14x + 49$

c. $4x^2 - 20xy + 25y^2$: $(2x - 5y)^2 = (2x)^2 + 2(2x)(-5y) + (-5y)^2 = 4x^2 - 20xy + 25y^2$

d. $x^2 + 14xy + 49y^2$

Q3 Practice writing only the answer:

a. $(x - 2)^2 =$ ______________

b. $(y + 5)^2 =$ ______________

c. $(3m - 2)^2 =$ ______________

d. $(x - 3y)^2 =$ ______________

e. $(5r + 2t)^2 =$ ______________

STOP • STOP • STOP • STOP • STOP • STOP • STOP • STOP • STOP

A3 **a.** $x^2 - 4x + 4$
b. $y^2 + 10y + 25$
c. $9m^2 - 12m + 4$
d. $x^2 - 6xy + 9y^2$
e. $25r^2 + 20rt + 4t^2$

---

2 Powers of binomials greater than 2 can be found by applying the technique of Frame 1. For example,

$$(a + b)^3 = (a + b)^2(a + b)$$
$$= (a^2 + 2ab + b^2)(a + b)$$
$$= a^2(a + b) + 2ab(a + b) + b^2(a + b)$$

$$= a^3 + a^2b + 2a^2b + 2ab^2 + ab^2 + b^3$$
$$= a^3 + 3a^2b + 3ab^2 + b^3$$

Hence $(a + b)^3 = a^3 + 3a^2b + 3ab^2 + b^3$.

Q4 Find:

a. $(2x - 1)^3$ **b.** $(3x + 2y)^3$

STOP • STOP • STOP • STOP • STOP • STOP • STOP • STOP • STOP

A4 **a.** $8x^3 - 12x^2 + 6x - 1$: $(2x - 1)^2(2x - 1)$
$(4x^2 - 4x + 1)(2x - 1)$

b. $27x^3 + 54x^2y + 36xy^2 + 8y^3$: $(3x + 2y)^2(3x + 2y)$
$(9x^2 + 12xy + 4y^2)(3x + 2y)$

3 The equation $(a + b)^3 = a^3 + 3a^2b + 3ab^2 + b^3$ can be used to expand the cube of a binomial where a is the first term and b is the second term.

Example 1: Expand $(2x - 1)^3$.

Solution

$$(2x - 1)^3 = (2x + {}^-1)^3$$
$$(a + b)^3 = a^3 + 3a^2b + 3ab^2 + b^3$$
$$(2x + {}^-1)^3 = (2x)^3 + 3(2x)^2(-1) + 3(2x)(-1)^2 + (-1)^3$$
$$= 8x^3 - 12x^2 + 6x - 1$$

Hence $(2x - 1)^3 = 8x^3 - 12x^2 + 6x - 1$.

Example 2: Expand $(3x + 2y)^3$.

Solution

$$(a + b)^3 = a^3 + 3a^2b + 3ab^2 + b^3$$
$$(3x + 2y)^3 = (3x)^3 + 3(3x)^2(2y) + 3(3x)(2y)^2 + (2y)^3$$
$$= 27x^3 + 54x^2y + 36xy^2 + 8y^3$$

Hence $(3x + 2y)^3 = 27x^3 + 54x^2y + 36xy^2 + 8y^3$.

Q5 Use the equation $(a + b)^3 = a^3 + 3a^2b + 3ab^2 + b^3$ to expand:

a. $(x + 5)^3$ **b.** $(2x - 3)^3$

STOP • STOP • STOP • STOP • STOP • STOP • STOP • STOP • STOP

A5

a. $x^3 + 15x^2 + 75x + 125$:
$(x + 5)^3$
$(x)^3 + 3(x)^2(5) + 3(x)(5)^2 + (5)^3$

b. $8x^3 - 36x^2 + 54x - 27$:
$(2x - 3)^3$
$(2x)^3 + 3(2x)^2(-3) + 3(2x)(-3)^2 + (-3)^3$

4

The product $(a + b)(a - b)$ is the *product of the sum and the difference of the same two values.**

$$(a + b)(a - b) = a^2 - ab + ab - b^2$$
$$= a^2 - b^2$$

The result, $a^2 - b^2$, is the *difference of two squares*. You should observe that in this case the sum of the products of the inner and outer terms will always be zero. The product of the sum and the difference of the same two values is the square of the first term minus the square of the second term.

Examples:

$$(x + 7)(x - 7) = (x)^2 - (7)^2$$
$$= x^2 - 49$$

$$(x - 2y)(x + 2y) = (x)^2 - (2y)^2$$
$$= x^2 - 4y^2$$

Such products can easily be done mentally.

*The expressions $a + b$ and $a - b$ are called *conjugates* of each other. Conjugates are two binomials that differ only in the sign of the second term.

Q6

Simplify the following products mentally:

a. $(2x - 1)(2x + 1) =$ ____________

b. $(3y + 2)(3y - 2) =$ ____________

c. $(5m + 7)(5m - 7) =$ ____________

d. $(7x - 3y)(7x + 3y) =$ ____________

STOP • STOP • STOP • STOP • STOP • STOP • STOP • STOP • STOP

A6

a. $4x^2 - 1$ b. $9y^2 - 4$ c. $25m^2 - 49$ d. $49x^2 - 9y^2$

Q7

Use the special techniques presented or the foil method where necessary:

a. $(x + y)(x - y) =$ ____________

b. $(x + y)^2 =$ ____________

c. $(x - y)^2 =$ ____________

d. $(w + x)(y - z) =$ ____________

STOP • STOP • STOP • STOP • STOP • STOP • STOP • STOP • STOP

A7

a. $x^2 - y^2$ b. $x^2 + 2xy + y^2$ c. $x^2 - 2xy + y^2$
d. $wy - wz + xy - xz$

5 The following special product will be useful in Section 2.5. However, great care must be observed in its use.

same signs

opposite signs

$$(a + b)(a^2 - ab + b^2) = a^3 + b^3$$

1 2 3 4 5 6 7

1: first term
2: second term
3: square of first term
4: product of first and second terms
5: square of second term
6: cube of first term
7: cube of second term

The result of such a product is the *sum of two cubes.*

Examples:

$(2x + 1)(4x^2 - 2x + 1) = 8x^3 + 1$
$(x + 3y)(x^2 - 3xy + 9y^2) = x^3 + 27y^3$

Q8 Simplify:

a. $(2x + 3y)(4x^2 - 6xy + 9y^2) =$ ________

b. $(m + n)(m^2 - mn + n^2) =$ ________

STOP • STOP • STOP • STOP • STOP • STOP • STOP • STOP • STOP

A8 **a.** $8x^3 + 27y^3$ **b.** $m^3 + n^3$

6 The pattern of the product in Frame 5 is difficult to recognize. However, when you do, the simplification becomes easy. A similar special product is

same signs

opposite signs

$$(a - b)(a^2 + ab + b^2) = a^3 - b^3$$

You should observe the same pattern as in Frame 5 except for the arrangement of the signs. The product in this case is the *difference of two cubes.*

Examples:

$(x - 2)(x^2 + 2x + 4) = x^3 - 8$
$(2x - y)(4x^2 + 2xy + y^2) = 8x^3 - y^3$

Q9 Simplify:

a. $(x - 2y)(x^2 + 2xy + 4y^2) =$ ________

b. $(m - n)(m^2 + mn + n^2) =$ ________

STOP • STOP • STOP • STOP • STOP • STOP • STOP • STOP • STOP

A9 **a.** $x^3 - 8y^3$ **b.** $m^3 - n^3$

Q10 Complete:

a. $(x - y)(\qquad\qquad) = x^3 - y^3$

b. $(x + y)(\qquad\qquad) = x^3 + y^3$

STOP • STOP • STOP • STOP • STOP • STOP • STOP • STOP • STOP

A10 a. $x^2 + xy + y^2$ b. $x^2 - xy + y^2$

7	Although special products allow a savings of time (if you recognize them), care must always be observed to apply them correctly.

This completes the instruction for this section.

2.4 Exercises

Some "nonspecial" products are included for practice as well as to keep you alert.

1. Match the column on the right with the column on the left:

a. $(a + b)(a - b)$	1. difference of two cubes
b. $a^3 - b^3$	2. product of any two binomials
c. $(a + b)^2$	3. cube of a binomial
d. $a^3 + b^3$	4. product of the sum and the difference of the same two values
e. $(a + b)(c + d)$	5. square of a binomial
f. $(a + b)^3$	6. sum of two cubes

2. Indicate the product of each of the following:
   a. $(a + b)(a - b)$ b. $(a - b)(a^2 + ab + b^2)$
   c. $(a + b)^2$ d. $(a + b)(c + d)$
   e. $(a + b)(a^2 - ab + b^2)$ f. $(a - b)^2$
   g. $(a + b)^3$
3. Simplify:
   a. $(3x - 2)^2$ b. $(7x - 4y)(7x + 4y)$
   c. $(5x + 2)(25x^2 - 10x + 4)$ d. $(3x + 1)(x - 4)$
   e. $-3(x + 2)^2$ f. $(4x - 3y)^2$
   g. $(2x - 3y)(4x^2 + 6xy + 9y^2)$ h. $x(2x - 1)$
   i. $(x - 1)(x^2 - 2x + 1)$ j. $-2(x + y)(x - y)$
   k. $(x^2 - 2y^2)(x^2 + 2y^2)$ l. $(x - 2y)(3x - y)$
   m. $(3 - 2x)^3$ n. $(5x + 2y)^3$

*4. Simplify:
   a. $(x + y)(x^2 - xy + y^2)(x^2 + xy + y^2)(x - y)$
   b. $(x - 1)(x^2 - 1)(x^2 + 1)$
   c. $(x + 5)(x - 5)^2$
   d. $(x + 2y)^2(x - 2y)^2$
   e. $2x(x + y)(x - 2y)^2$
   f. $(x - 5y)(x - 5y)^2$

2.4 Exercise Answers

1. a. 4 b. 1 c. 5 d. 6 e. 2 f. 3
2. a. $a^2 - b^2$ b. $a^3 - b^3$ c. $a^2 + 2ab + b^2$

d. $ac + ad + bc + bd$ (any order) e. $a^3 + b^3$
f. $a^2 - 2ab + b^2$ g. $a^3 + 3a^2b + 3ab^2 + b^3$

3. a. $9x^2 - 12x + 4$ b. $49x^2 - 16y^2$
c. $125x^3 + 8$ d. $3x^2 - 11x - 4$
e. $-3x^2 - 12x - 12$ f. $16x^2 - 24xy + 9y^2$
g. $8x^3 - 27y^3$ h. $2x^2 - x$
i. $x^3 - 3x^2 + 3x - 1$ [This was not a special product. $(x - 1)(x^2 + x + 1)$ would have been $x^3 - 1$.]
j. $-2x^2 + 2y^2$ or $2y^2 - 2x^2$ k. $x^4 - 4y^4$ l. $3x^2 - 7xy + 2y^2$
m. $27 - 54x + 36x^2 - 8x^3$ n. $125x^3 + 150x^2y + 60xy^2 + 8y^3$

*4. a. $x^6 - y^6$ b. $x^5 - x^4 - x + 1$
c. $x^3 - 5x^2 - 25x + 125$ d. $x^4 - 8x^2y^2 + 16y^4$
e. $2x^4 - 6x^3y + 8xy^3$ f. $x^3 - 15x^2y + 75xy^2 - 125y^3$

2.5 Factoring Polynomials

1

When we use one of the distributive properties to simplify a product, it is often called "expanding" or "removing the parentheses."

Examples:

1. $2(x - 3) = 2x - 6$
2. $5x(x + y) = 5x^2 + 5xy$
3. $(2x - 1)x = 2x^2 - x$
4. $(x + 3)(x - 2) = x(x - 2) + 3(x - 2)$
$= x^2 - 2x + 3x - 6$
$= x^2 + x - 6$

In each example the expression on the left of the equal sign is a *product*. The result on the right of the equal sign is a *sum*. Recall that the distributive properties change products to sums.

Q1

a. $-5(x - 7)$ is a ________ (product/sum); $-5x + 35$ is a ________ (product/sum).

b. Expand $2x(x - y)$. ________

c. Remove the parentheses from $(x + 7)(x - 3)$.

d. Simplify $x(x - 2) - 3(x - 2)$.

STOP • STOP • STOP • STOP • STOP • STOP • STOP • STOP • STOP

A1

a. product, sum b. $2x^2 - 2xy$
c. $x^2 + 4x - 21$: $(x + 7)(x - 3) = x(x - 3) + 7(x - 3)$
$= x^2 - 3x + 7x - 21$
$= x^2 + 4x - 21$
d. $x^2 - 5x + 6$: $x(x - 2) - 3(x - 2)$
$x^2 - 2x - 3x + 6$

2 The distributive properties also change *sums* to *products*.

Examples:

1. $5x - 10 = 5(x - 2)$

The expression $5x - 10$ is a sum because, by the definition of subtraction, $5x - 10 = 5x + {}^{-}10$.

2. $x^2 + x = x^2 + 1x$
$= x(x + 1)$
3. $x^2y + x^3y^2 = x^2y(1 + xy)$

When the distributive properties are used to change sums to products, the process is called *factoring*. If each term of a polynomial contains the same monomial factor, the polynomial can be written as a product of the common factor times another polynomial. The common factor is said to be "factored from" or "removed from" the polynomial. For example,

$$3x + 12 = 3x + 3 \cdot 4$$
$$= 3(x + 4)$$

In this example the common factor 3 has been factored (removed) from $3x + 12$. It is wise to check the factoring by expanding. That is, $3(x + 4) = 3x + 12$.

Q2 a. In $2x + 2 \cdot 5$, what is the common factor? __________

b. Factor the common factor from $2x + 2 \cdot 5$. __________

STOP • STOP • STOP • STOP • STOP • STOP • STOP • STOP • STOP

A2 a. 2 b. $2(x + 5)$

3 When factoring a common monomial from a polynomial, always consider the *greatest common factor*. The expression $4x^2 + 8x$ has common factors of 2, $2x$, 4, and $4x$. $4x$ would be the greatest common factor.* Hence

$4x^2 + 8x$
$4x \cdot x + 4x \cdot 2$
$4x(x + 2)$

The middle step can be done mentally. The above result "checks," because $4x(x + 2) = 4x^2 + 8x$.

*Any rational number is a common factor of any polynomial. That is, $4x - 5 = \frac{1}{2}(8x - 10) = \frac{-2}{3}(-6x + 15)$, and so on. However, when the phrase "greatest common factor" is used, only positive integer values and variables are considered.

Q3 Complete the following factorizations:
a. $6y^2 - 3y = 3y(\qquad)$
b. $12x^2y + 18xy^2 = 6xy(\qquad)$
c. $20mn^2t^3 - 15m^3nt^2 + 25m^2n^2t = 5mnt(\qquad)$

STOP • STOP • STOP • STOP • STOP • STOP • STOP • STOP • STOP

A3 a. $2y - 1$ b. $2x + 3y$ c. $4nt^2 - 3m^2t + 5mn$

Q4 Remove the greatest common factor (check your results):

a. $8x + 12x^2 =$ ____________

b. $18xy^2 - 24x^2y^3 =$ ____________

c. $12ab^2c^3 + 36a^3b^2c - 30a^2b^2c^2 =$ ____________

STOP • STOP • STOP • STOP • STOP • STOP • STOP • STOP • STOP

A4 a. $4x(2 + 3x)$ b. $6xy^2(3 - 4xy)$ c. $6ab^2c(2c^2 + 6a^2 - 5ac)$

4 In many cases it is necessary to factor monomials from a polynomial in which the first term of the polynomial is negative.

Example 1: Factor $-15x^2 + 10x$.

Solution

Choose the common factor to be $-5x$. Hence

$-15x^2 + 10x = -5x(3x - 2)$

This choice was made so that the first term of the polynomial within the parentheses would be positive.

Example 2: Factor $-5 - 15x$.

Solution

$-5 - 15x = -5(1 + 3x)$

The expression $-4x + 3$ has a common factor of -1. Therefore, it is possible to factor -1 from both terms. That is,

$-4x + 3 = -1(4x - 3)$

By the multiplication property of -1, $-1(4x - 3) = -(4x - 3)$. You should observe that -1 is a factor of any polynomial. However, -1 is factored from a polynomial only when it is necessary to make the first term of the polynomial (within the parentheses) positive.

Q5 Factor (make the first term of the polynomial within the parentheses positive where necessary):

a. $-3x + 6 =$ ____________ b. $9x^2 - 3x =$ ____________

c. $-10x - 15x^2 =$ ____________ d. $-xy + x =$ ____________

e. $-2x - 7 =$ ____________ f. $-12x^2 + 5x - 3 =$ ____________

STOP • STOP • STOP • STOP • STOP • STOP • STOP • STOP • STOP

A5 a. $-3(x - 2)$ b. $3x(3x - 1)$
c. $-5x(2 + 3x)$ d. $-x(y - 1)$
e. $-(2x + 7)$ f. $-(12x^2 - 5x + 3)$

5 When removing a common factor, the quantity within the parentheses can be said to be the "sum of the remaining factors."

Common factor — Sum of the remaining factors

$2x + 10 = 2x + 2 \cdot 5 \quad = 2\overbrace{(x + 5)}$

$xy^2 - 4xy = xy \cdot y - xy \cdot 4 = xy(y - 4)$

In the following expression, $(x + 3)$ is a common factor:

$(x + 3)x + (x + 3)5$

By the left distributive property of multiplication over addition,

$(x + 3)x + (x + 3)5$ can be factored $(x + 3)(x + 5)$

where $(x + 5)$ is the sum of the remaining factors.

Q6 a. What is the common factor in $(x - 2)x + (x - 2)4$? ________

b. What is the sum of the remaining factors? ________

STOP • STOP • STOP • STOP • STOP • STOP • STOP • STOP • STOP

A6 a. $(x - 2)$ b. $(x + 4)$

Q7 Complete the factorization: $(x - 2)x + (x - 2)4$
$(x - 2)(\qquad)$

STOP • STOP • STOP • STOP • STOP • STOP • STOP • STOP • STOP

A7 $(x - 2)(x + 4)$

6 In $(x + 3)x - (x + 3)5$, the common factor $x + 3$ appears on the left. Hence, by the left distributive property of multiplication over subtraction,

$(x + 3)x - (x + 3)5 = (x + 3)(x - 5)$

where the common factor, $x + 3$, has been placed on the left and the sum of the remaining factors, $x - 5$, has been placed on the right. You should observe that it makes no difference whether the common factor appears on the left or the right. For example, by the right distributive property of multiplication over subtraction, $x(x + 3) - 5(x + 3) = (x - 5)(x + 3)$.

Q8 $3x(x + 6) - 5(x + 6) = (\qquad)(\qquad)$

STOP • STOP • STOP • STOP • STOP • STOP • STOP • STOP • STOP

A8 $(3x - 5)(x + 6)$

Q9 Factor:

a. $x(2x - 1) - 3(2x - 1) =$ ____________

b. $2x(x + 1) - 3(x + 1) =$ ____________

c. $x(2x - 3) + 1(2x - 3) =$ ____________

STOP • STOP • STOP • STOP • STOP • STOP • STOP • STOP • STOP

A9 a. $(x - 3)(2x - 1)$ b. $(2x - 3)(x + 1)$ c. $(x + 1)(2x - 3)$

7 The factoring of second-degree polynomials of the form $ax^2 + bx + c$, where $a = 1$ and $b,c \in I$, will now be considered. Recall that:

$(x + 2)(x + 3)$
$x(x + 3) + 2(x + 3)$
$x^2 + 3x + 2x + 6$
$x^2 + 5x + 6$

The process of factoring $x^2 + 5x + 6$ will reverse the above process to obtain the two binomials whose product is $x^2 + 5x + 6$. To make the process easier, the terms of $x^2 + 5x + 6$ will be examined to see how they were obtained. Note that 6 (the constant term) is the product of 2 and 3, the last terms of the original binomials. Also note that the first-degree term, $5x$, is the sum of $2x$ and $3x$.

Q10

$(x - 2)(x - 3)$
$x(x - 3) - 2(x - 3)$
$x^2 - 3x - 2x + 6$
$x^2 - 5x + 6$

a. The constant term, 6, is the ________ (product/sum) of -2 and -3, the last terms of the original binomials.

b. The first-degree term, $-5x$, is the ________ (product/sum) of $-3x$ and $-2x$.

STOP • STOP • STOP • STOP • STOP • STOP • STOP • STOP • STOP

A10 **a.** product **b.** sum

Q11

$(x + 2)(x - 3)$
$x(x - 3) + 2(x - 3)$
$x^2 - 3x + 2x - 6$
$x^2 - x - 6$

a. -6 is the ________ (product/sum) of 2 and -3.

b. $-x$ is the ________ (product/sum) of $-3x$ and $2x$.

STOP • STOP • STOP • STOP • STOP • STOP • STOP • STOP • STOP

A11 **a.** product **b.** sum

8

To factor $x^2 + 5x + 6$ you must first find a pair of integers whose product is 6 and whose sum is 5, the coefficient of $5x$. The possible integers whose product is 6 are

$(1)(6)$, $(-1)(-6)$, $(2)(3)$, $(-2)(-3)$

The combination that gives the proper sum is $2 \cdot 3$, because $2 + 3 = 5$. These integers are used as coefficients of x terms whose sum is $5x$. That is, $2x + 3x = 5x$. The factoring now proceeds as follows:

$x^2 + 5x + 6$
$x^2 + 2x + 3x + 6$ ($5x$ is replaced by $2x + 3x$)
$(x^2 + 2x) + (3x + 6)$
$x(x + 2) + 3(x + 2)$
$(x + 3)(x + 2)$

$(x + 3)$ ↑ sum of remaining factors; $(x + 2)$ ↑ common factor

The order in which $2x$ and $3x$ are placed in the second step does not affect the factorization (except in the order of the final factors). For example,

$x^2 + 5x + 6$
$x^2 + 3x + 2x + 6$
$(x^2 + 3x) + (2x + 6)$
$x(x + 3) + 2(x + 3)$
$(x + 2)(x + 3)$

Q12 To factor $x^2 - 5x + 6$, first find two integers whose product is 6 and whose sum is ____.

STOP • STOP • STOP • STOP • STOP • STOP • STOP • STOP • STOP

A12 -5

Q13 The possible pairs of integers whose product is 6 are 1( ), -1( ), 2( ), and -2( ).

STOP • STOP • STOP • STOP • STOP • STOP • STOP • STOP • STOP

A13 $6, -6, 3, -3$

Q14 The two integers whose product is 6 and whose sum is -5 are ____ and ____.

STOP • STOP • STOP • STOP • STOP • STOP • STOP • STOP • STOP

A14 $-2, -3$ (either order): because $-2 + {}^{-}3 = -5$.

9 To factor $x^2 - 5x + 6$:

Step 1: Find two integers whose product is 6 and whose sum is -5. They are -2 and -3.

Step 2: Use these integers as coefficients of x terms to form a sum equivalent to $-5x$: $-2x - 3x$ or $-3x - 2x$.

Step 3: Then proceed as follows:

	or	
$x^2 - 5x + 6$		$x^2 - 5x + 6$
$x^2 - 2x - 3x + 6$		$x^2 - 3x - 2x + 6$
$x(x - 2) - 3(x - 2)$		$x(x - 3) - 2(x - 3)$
$(x - 3)(x - 2)$		$(x - 2)(x - 3)$

Q15 Factor $x^2 + 12x + 27$ by completing the following:

a. The integers whose product is 27 and whose sum is 12 are ____ and ____. (*Note:* The combinations that involve negative integers would be ignored, because the sum must be positive.)

$x^2 + 12x + 27$

b. $x^2 + 3x +$ ____ $+ 27$

c. $x(x + 3) +$ ________

d. ( )( )

STOP • STOP • STOP • STOP • STOP • STOP • STOP • STOP • STOP

A15 **a.** 3, 9 (either order) **b.** $9x$ **c.** $9(x + 3)$
d. $(x + 9)(x + 3)$

Q16 Complete the following factorization:

$x^2 + 12x + 27$
$x^2 + 9x + 3x + 27$
$x(x + 9)$ ____ ( )
 b **a**

c. ( )( )

STOP • STOP • STOP • STOP • STOP • STOP • STOP • STOP • STOP

A16 **a.** $x + 9$ **b.** $+3$ **c.** $(x + 3)(x + 9)$

Q17 Factor $x^2 - 8x + 15$.

STOP • STOP • STOP • STOP • STOP • STOP • STOP • STOP • STOP

A17 $(x - 3)(x - 5)$ or $(x - 5)(x - 3)$:

$$\begin{array}{lcl} x^2 - 8x + 15 & & \\ x^2 - 3x - 5x + 15 & \text{or} & x^2 - 5x - 3x + 15 \\ x(x - 3) - 5(x - 3) & & x(x - 5) - 3(x - 5) \\ (x - 5)(x - 3) & & (x - 3)(x - 5) \end{array}$$

Q18 Given $x^2 - x - 6$, find the pair of integers whose product is -6 and whose sum is -1.

____

STOP • STOP • STOP • STOP • STOP • STOP • STOP • STOP • STOP

A18 -3 and 2 (either order)

10 To factor $x^2 - x - 6$, first find a pair of integers whose product is -6 and whose sum is -1. Since the product is negative, the signs of the integers must be different, one positive and one negative. The possible choices that would produce a negative product and a negative sum are:

$(1)(-6) = -6$ and $1 + {}^{-}6 = -5$
$(2)(-3) = -6$ and $2 + {}^{-}3 = -1$

Hence the proper combination is 2 and -3. The factoring would be completed as follows:

$$\begin{array}{lcl} x^2 - x - 6 & \text{or} & x^2 - x - 6 \\ x^2 + 2x - 3x - 6 & & x^2 - 3x + 2x - 6 \\ x(x + 2) - 3(x + 2) & & x(x - 3) + 2(x - 3) \\ (x - 3)(x + 2) & & (x + 2)(x - 3) \end{array}$$

Q19 Factor $x^2 + 4x - 32$.

STOP • STOP • STOP • STOP • STOP • STOP • STOP • STOP • STOP

A19 $(x + 8)(x - 4)$ or $(x - 4)(x + 8)$:

$$\begin{array}{l} x^2 + 4x - 32 \\ x^2 + 8x - 4x - 32 \\ x(x + 8) - 4(x + 8) \\ (x - 4)(x + 8) \end{array}$$

(other possibility not shown)

Q20 Factor $x^2 - 9x - 22$.

STOP • STOP • STOP • STOP • STOP • STOP • STOP • STOP • STOP

A20 $(x - 11)(x + 2)$ or $(x + 2)(x - 11)$

11 Many polynomials of several variables can also be factored.

Examples:

1. $x^2 - xy - 6y^2$ $\quad -6 = (-3)(2)$ and $-3 + 2 = -1$
   $x^2 - 3xy + 2xy - 6y^2$
   $x(x - 3y) + 2y(x - 3y)$
   $(x + 2y)(x - 3y)$

2. $x^2 - 11xy + 24y^2$ $\quad 24 = (-3)(-8)$ and $-3 + {}^-8 = -11$
   $x^2 - 3xy - 8xy + 24y^2$
   $x(x - 3y) - 8y(x - 3y)$
   $(x - 8y)(x - 3y)$

Q21 Factor:

a. $x^2 + 11xy + 30y^2$

b. $x^2 - 14xy + 45y^2$

c. $x^2 - 3xy - 28y^2$

d. $x^2 + 6xy - 16y^2$

STOP • STOP • STOP • STOP • STOP • STOP • STOP • STOP • STOP

A21

a. $(x + 5y)(x + 6y)$ or $(x + 6y)(x + 5y)$:
   $x^2 + 5xy + 6xy + 30y^2$
   $x(x + 5y) + 6y(x + 5y)$

b. $(x - 5y)(x - 9y)$ or $(x - 9y)(x - 5y)$:
   $x^2 - 9xy - 5xy + 45y^2$
   $x(x - 9y) - 5y(x - 9y)$

c. $(x - 7y)(x + 4y)$ or $(x + 4y)(x - 7y)$:
   $x^2 - 7xy + 4xy - 28y^2$
   $x(x - 7y) + 4y(x - 7y)$

d. $(x + 8y)(x - 2y)$ or $(x - 2y)(x + 8y)$:
   $x^2 + 8xy - 2xy - 16y^2$
   $x(x + 8y) - 2y(x + 8y)$

12 When factoring the previous polynomials, you may have noticed that the result could be written immediately once you found the correct factors which give the correct sum. Some examples are:

$x^2 + 5x + 6 = (x + 2)(x + 3)$
$x^2 - 5x + 6 = (x - 2)(x - 3)$
$x^2 - x - 6 = (x - 3)(x + 2)$
$x^2 + 4x - 32 = (x + 8)(x - 4)$
$x^2 - 9x - 22 = (x - 11)(x + 2)$
$x^2 + 11xy + 30y^2 = (x + 5y)(x + 6y)$
$x^2 - 14xy + 45y^2 = (x - 5y)(x - 9y)$
$x^2 - 3xy - 28y^2 = (x - 7y)(x + 4y)$
$x^2 + 6xy - 16y^2 = (x + 8y)(x - 2y)$

Q22 Write the result of the factoring immediately:

a. $x^2 - 9x + 8 =$ ______

b. $x^2 - 7xy - 18y^2 =$ ______

c. $x^2 + 2x - 24 =$ ______

d. $x^2 + 5xy - 36y^2 =$ ______

e. $x^2 + 20x + 36 =$ ______

f. $x^2 + 2x + 1 =$ ______

g. $x^2 - xy - 6y^2 =$ ______

h. $x^2 - 2xy + y^2 =$ ______

i. $x^2 + 6x + 9 =$ ______

j. $x^2 + 11xy + 10y^2 =$ ______

STOP • STOP • STOP • STOP • STOP • STOP • STOP • STOP • STOP

A22 Factors may be commuted.

a. $(x - 8)(x - 1)$
b. $(x - 9y)(x + 2y)$
c. $(x + 6)(x - 4)$
d. $(x + 9y)(x - 4y)$
e. $(x + 18)(x + 2)$
f. $(x + 1)(x + 1)$ or $(x + 1)^2$
g. $(x - 3y)(x + 2y)$
h. $(x - y)(x - y)$ or $(x - y)^2$
i. $(x + 3)(x + 3)$ or $(x + 3)^2$
j. $(x + 10y)(x + y)$

13 Although many trinomials may be factored mentally, an adaptation of the technique presented earlier is important if we are to factor more difficult second-degree polynomials of the form $ax^2 + bx + c$, where $a \neq 1$, $a,b,c \in I$. For example, when factoring $6x^2 + 7x - 3$, you cannot immediately write the result. To factor $6x^2 + 7x - 3$:

Step 1: Multiply 6 (the coefficient of the second-degree term) by -3 (the constant term), which is -18.

Step 2: Find two integers whose product is -18 and whose sum is 7 (the coefficient of the first-degree term). They are 9 and -2, because $(9)(-2) = -18$ and $9 + {}^{-}2 = 7$.

Step 3: Use these integers as coefficients of x terms whose sum is equivalent to $7x$: $9x - 2x$ or $-2x + 9x$.

Step 4: Proceed as follows:

$6x^2 + 7x - 3$	or	$6x^2 + 7x - 3$
$6x^2 + 9x - 2x - 3$		$6x^2 - 2x + 9x - 3$
$3x(2x + 3) - 1(2x + 3)$		$2x(3x - 1) + 3(3x - 1)$
$(3x - 1)(2x + 3)$		$(2x + 3)(3x - 1)$

Q23 To factor $2x^2 + 7x + 6$, multiply 2 by 6, which gives 12.

a. Now find the pair of integers whose product is 12 and whose sum is ______.

b. They are ______.

STOP • STOP • STOP • STOP • STOP • STOP • STOP • STOP • STOP

A23 a. 7 b. 4 and 3 (either order)

Q24 Complete the factorization: $2x^2 + 7x + 6$
$2x^2 + 4x + 3x + 6$

a. ____________ + ____________

b. (   )(   )

STOP • STOP • STOP • STOP • STOP • STOP • STOP • STOP • STOP

A24 **a.** $2x(x + 2) + 3(x + 2)$ **b.** $(2x + 3)(x + 2)$

14 The justification of the procedure of Frame 13 is given below. Consider the product of the two binomials $ax + b$ and $cx + d$, where $a,b,c,d \in I$.

$(ax + b)(cx + d)$
$ax(cx + d) + b(cx + d)$
$acx^2 + adx + bcx + bd$
$acx^2 + (adx + bcx) + bd$
$acx^2 + (ad + bc)x + bd$

Notice that the *sum*, $ad + bc$ (the coefficient of the first-degree term), is a combination of the factors $acbd$ (the *product* of the coefficient of the second-degree term and the constant term).

In general, the second-degree polynomial $ax^2 + bx + c$, $a,b,c \in I$, $a \neq 0$, $c \neq 0$, is factorable into the product of two binomials with integral coefficients if two integers exist whose product is ac and whose sum is b.

To factor $5x^2 - 12x + 4$:

Step 1: Multiply 5 by 4, which is 20.

Step 2: Find the pair of integers whose product is 20 and whose sum is -12. They are -2 and -10.

Hence $5x^2 - 12x + 4$ is factorable into the product of two binomials with integral coefficients.

Q25 Complete the factorization: $5x^2 - 12x + 4$
$5x^2 - 10x - 2x + 4$

STOP • STOP • STOP • STOP • STOP • STOP • STOP • STOP • STOP

A25 $(5x - 2)(x - 2)$: $5x(x - 2) - 2(x - 2)$
$(5x - 2)(x - 2)$

15 **Example:** Factor $4x^2 + 12x - 7$.

Solution

Step 1: 4 times -7 is -28.

Step 2: $(-2)(14) = -28$ and $-2 + 14 = 12$.

Now proceed as follows:

$4x^2 + 12x - 7$	or	$4x^2 + 12x - 7$
$4x^2 - 2x + 14x - 7$		$4x^2 + 14x - 2x - 7$
$2x(2x - 1) + 7(2x - 1)$		$2x(2x + 7) - 1(2x + 7)$
$(2x + 7)(2x - 1)$		$(2x - 1)(2x + 7)$

Q26 Factor $3x^2 - 8x - 3$.

STOP • STOP • STOP • STOP • STOP • STOP • STOP • STOP • STOP

A26

$(3x + 1)(x - 3)$:	$3x^2 - 8x - 3$
$(3)(-3) = -9$	$3x^2 - 9x + 1x - 3$
$(-9)(1) = -9$	$3x(x - 3) + 1(x - 3)$
and $-9 + 1 = -8$	$(3x + 1)(x - 3)$
	(other possibility not shown)

Q27 Use the technique developed above to factor the following trinomials:

a. $6x^2 + 17x + 5$

b. $3x^2 - 14x + 8$

c. $6x^2 + 5x - 4$

d. $4x^2 - 9$
(*Hint:* $4x^2 - 9 = 4x^2 + 0x - 9$.)

e. $4x^2 - 20x + 25$

f. $8x^2 + 14x - 15$

STOP • STOP • STOP • STOP • STOP • STOP • STOP • STOP • STOP

A27

a. $(2x + 5)(3x + 1)$

b. $(x - 4)(3x - 2)$

c. $(3x + 4)(2x - 1)$

d. $(2x + 3)(2x - 3)$

e. $(2x - 5)(2x - 5)$ or $(2x - 5)^2$

f. $(4x - 3)(2x + 5)$

16 Not all polynomials of the form $ax^2 + bx + c$, $a,b,c \in I$, $a \neq 0$, $c \neq 0$, are factorable into the product of two binomials. Examine $3x^2 - 20x - 4$. The possible pairs of integers whose product is -12 which produce a negative sum are 1 and -12, 2 and -6, and 3 and -4. In no case is the sum of these integers -20. Hence $3x^2 - 20x - 4$ is not factorable over the set of integers into the product of two binomials. Also, since there is no integer factor (other than 1 or -1) or variable common to each term, $3x^2 - 20x - 4$ is said to be *prime* over the set of integers.

In general, any polynomial is prime over the set of integers if the polynomial does not contain an integer factor (other than 1 or -1) common to each term, and the polynomial cannot be factored into two other polynomials with integer coefficients. In this text, when a polynomial is said to be prime, it will be understood to mean over the set of integers.

Assuming that there is no common factor, the second-degree polynomial $ax^2 + bx + c$, $a,b,c \in I$, $a \neq 0$, $c \neq 0$, is prime if there are no integers whose product is ac and whose sum is b.

Q28 Is $3x^2 - 2x + 7$ prime? ______

STOP • STOP • STOP • STOP • STOP • STOP • STOP • STOP • STOP

A28 yes: there is no common monomial factor and there are no integers whose product is 21 and whose sum is -2.

Q29 Is $2x^2 + 7x + 5$ factorable? ______

STOP • STOP • STOP • STOP • STOP • STOP • STOP • STOP • STOP

A29 yes: $(2)(5) = 10$ and $2 + 5 = 7$

Q30 Is $3x^2 + 17x - 9$ factorable? ______

STOP • STOP • STOP • STOP • STOP • STOP • STOP • STOP • STOP

A30 no: there is no common monomial factor and there are no integers whose product is -27 and whose sum is 17.

17 A trinomial such as $x^2 + 2x + 3$ is of the form $ax^2 + bx + c$, where $a = 1$. Therefore, the test to determine whether a trinomial of this form is factorable into the product of two binomials applies. Since $(1)(3) = 3$ and since there are no integers whose product is 3 and whose sum is 2, $x^2 + 2x + 3$ is not factorable into the product of two binomials. Also, there is no common monomial factor. Hence $x^2 + 2x + 3$ is prime.

Q31
a. Is $x^2 - 3x - 5$ factorable? ______
b. Is $x^2 - 3x - 5$ prime? ______

STOP • STOP • STOP • STOP • STOP • STOP • STOP • STOP • STOP

A31 **a.** no **b.** yes

Q32
a. Is $3x^2 + 4x + 1$ prime? ______
b. Is $2x^2 - 3x - 5$ prime? ______

STOP • STOP • STOP • STOP • STOP • STOP • STOP • STOP • STOP

A32 **a.** no **b.** no

Q33 Factor where possible. Otherwise, write "prime."
a. $3x^2 + x - 2$
b. $4x^2 - 12x + 9$
c. $x^2 + 9x - 20$

d. $4x^2 - 25$ (*Hint:* $4x^2 - 25 = 4x^2 + 0x - 25$.)

STOP • STOP • STOP • STOP • STOP • STOP • STOP • STOP • STOP

A33 **a.** $(3x - 2)(x + 1)$ **b.** $(2x - 3)(2x - 3)$ or $(2x - 3)^2$
c. prime **d.** $(2x + 5)(2x - 5)$

18 The procedures discussed previously also apply to polynomials that contain more than one variable.

Example: Factor $6x^2 - xy - 12y^2$.

Solution

$6(-12) = {}^-72$

$(8)(-9) = {}^-72$ and $8 + {}^-9 = -1$

	or	
$6x^2 - xy - 12y^2$		$6x^2 - xy - 12y^2$
$6x^2 + 8xy - 9xy - 12y^2$		$6x^2 - 9xy + 8xy - 12y^2$
$2x(3x + 4y) - 3y(3x + 4y)$		$3x(2x - 3y) + 4y(2x - 3y)$
$(2x - 3y)(3x + 4y)$		$(3x + 4y)(2x - 3y)$

Q34 Factor $6x^2 + 7xy - 3y^2$.

STOP • STOP • STOP • STOP • STOP • STOP • STOP • STOP • STOP

A34 $(3x - y)(2x + 3y)$: $(6)(-3) = -18$

$(9)(-2) = -18$ and $9 + {}^-2 = 7$

$6x^2 + 9xy - 2xy - 3y^2$

$3x(2x + 3y) - y(2x + 3y)$

Q35 Factor $4x^2 - 16xy + 7y^2$.

STOP • STOP • STOP • STOP • STOP • STOP • STOP • STOP • STOP

A35 $(2x - y)(2x - 7y)$

Q36 Factor $2x^2 + 5xy - y^2$.

STOP • STOP • STOP • STOP • STOP • STOP • STOP • STOP • STOP

A36 prime: because $(2)(-1) = -2$ and there are no integers whose product is -2 and whose sum is 5; also, there is no common monomial factor.

19 Many polynomials possess a common factor. If so, it should be factored from the polynomial before attempting any other type of factoring.

Examples:

1. $2x^2 + x = x(2x + 1)$

2. $3x^2 - 6x - 9 = 3(x^2 - 2x - 3)$
$= 3(x - 3)(x + 1)$

3. $2x^3 - 7x^2 + 5x = x(2x^2 - 7x + 5)$
$= x(2x^2 - 2x - 5x + 5)$
$= x[2x(x - 1) - 5(x - 1)]$
$= x(2x - 5)(x - 1)$

4. $-9x^3 - 3x^2 + 12x = -3x(3x^2 + x - 4)$
$= -3x(3x^2 + 4x - 3x - 4)$
$= -3x[x(3x + 4) - 1(3x + 4)]$
$= -3x(x - 1)(3x + 4)$

5. $-x^2 + 10xy - 25y^2 = -(x^2 - 10xy + 25y^2)$
$= -(x - 5y)(x - 5y)$ or $-(x - 5y)^2$

6. $2x^3y - 6x^2y + 10xy = 2xy(x^2 - 3x + 5)$

(*Note:* $x^2 - 3x + 5$ is prime.)

7. $5x^2 - 3xy - 7y^2$ is prime

(*Note:* There is no common factor and there are no integers whose product is -35 and whose sum is -3.)

Remember: Always remove a common factor *first* (if it exists) before attempting any other type of factoring.

Q37 Factor:

a. $-5x^2 - 15x + 50$ **b.** $6x^2 - 3x - 45$ **c.** $2x^2 - 26xy + 80y^2$

d. $8x^3 - 17x^2 + 2x$ **e.** $2x^3y + 5x^2y - 2xy$ **f.** $-2x^2 - xy + 15y^2$

STOP • STOP • STOP • STOP • STOP • STOP • STOP • STOP • STOP

A37 **a.** $-5(x + 5)(x - 2)$ **b.** $3(2x + 5)(x - 3)$ **c.** $2(x - 5y)(x - 8y)$
d. $x(8x - 1)(x - 2)$ **e.** $xy(2x^2 + 5x - 2)$ **f.** $-(2x - 5y)(x + 3y)$

20 This frame will illustrate a method of simplifying that only works occasionally. Nevertheless, it is a very valuable technique.

Example:

$(x - 3)(x - 5) - (x - 3)(2x + 5)$
$(x - 3)[(x - 5) - (2x + 5)]$
$(x - 3)(-x - 10)$
$-x^2 - 7x + 30$

You should note that $(x - 3)$ is a common factor of the original expression. By removing it (factoring it from the polynomial), the resulting work was simplified considerably.

Q38 Simplify (use the technique of Frame 20):

a. $(x + 2)(3x - 1) + (x + 2)(x + 3)$

b. $(3x + 2y)(x - y) + (x - 7y)(x - y)$

STOP • STOP • STOP • STOP • STOP • STOP • STOP • STOP • STOP

A38

a. $4x^2 + 10x + 4$: $(x + 2)[(3x - 1) + (x + 3)]$
$(x + 2)(4x + 2)$

b. $4x^2 - 9xy + 5y^2$: $[(3x + 2y) + (x - 7y)](x - y)$
$(4x - 5y)(x - y)$

Q39 Simplify:

a. $(x + 2)(x - 3) - (x - 2)(x + 3)$

b. $(x + 5)^2 + (x + 5)(x - 5)$

STOP • STOP • STOP • STOP • STOP • STOP • STOP • STOP • STOP

A39

a. $-2x$: $(x^2 - x - 6) - (x^2 + x - 6)$
$x^2 - x - 6 - x^2 - x + 6$

b. $2x^2 + 10x$: $(x + 5)(x + 5) + (x + 5)(x - 5)$
$(x + 5)[(x + 5) + (x - 5)]$
$(x + 5)2x$

21 Many polynomials of higher degree than 2 are factorable applying the technique of *factoring by parts.*

Examples:

1. $3m^3 + 9m^2 + 8m + 24$
$3m^2(m + 3) + 8(m + 3)$
$(3m^2 + 8)(m + 3)$

2. $y^3 + 5y^2 - 2y - 10$
$y^2(y + 5) - 2(y + 5)$
$(y^2 - 2)(y + 5)$

3. $x^5 - x^4 + 3x^3 - 3x^2 - x + 1$
$x^4(x - 1) + 3x^2(x - 1) - 1(x - 1)$
$(x^4 + 3x^2 - 1)(x - 1)$

This method is not a general method for factoring polynomials as it only works in special cases.

[*Note:* An intermediate step in the earlier procedure for factoring trinomials is actually factoring by parts. For example,

$2x^2 - 7x + 3$
$2x^2 - x - 6x + 3$
$x(2x - 1) - 3(2x - 1)$ (factoring by parts)
$(x - 3)(2x - 1)$]

Q40 Factor by parts:
a. $10x^2 - 15x + 4x - 6$ **b.** $y^3 - 2y^2 + 8y - 16$

STOP • STOP • STOP • STOP • STOP • STOP • STOP • STOP • STOP

A40 **a.** $(5x + 2)(2x - 3)$: $5x(2x - 3) + 2(2x - 3)$
b. $(y^2 + 8)(y - 2)$: $y^2(y - 2) + 8(y - 2)$

Q41 Factor by parts:
a. $x^4 + x^3 - 2x^2 - 2x + 3x + 3$

b. $2x^5 + 4x^4 - x^3 - 2x^2 - 3x - 6$

STOP • STOP • STOP • STOP • STOP • STOP • STOP • STOP • STOP

A41 **a.** $(x^3 - 2x + 3)(x + 1)$: $x^3(x + 1) - 2x(x + 1) + 3(x + 1)$
b. $(2x^2 - 3)(x^2 + 1)(x + 2)$: $2x^4(x + 2) - x^2(x + 2) - 3(x + 2)$

This completes the instruction for this section.

2.5 Exercises

1. Factor:
   a. $5x - 10$ **b.** $3x^2 - 6x$
   c. $9x^2 - 3x + 15$ **d.** $x^3 + 4x^2 - 3x$
   e. $x(x - 2) + 2(x - 2)$ **f.** $x(2x + 3) - 2(2x + 3)$
   g. $3x(3x - 1) - 1(3x - 1)$ **h.** $-x - 1$
   i. $-12x^2y^2 + 18x^2y^3$ **j.** $-x^2y + y^2$
2. Factor:
   a. $x^2 + 10x + 21$ **b.** $x^2 - 12x + 35$
   c. $x^2 - 18x - 40$ **d.** $x^2 + 5x - 36$
   e. $x^2 - 100$ **f.** $x^2 - 5x$

g. $x^2 - 14x + 24$
h. $x^2 + 9x + 12$
i. $x^2 + 4x + 4$
j. $x^2 + x - 42$

3. Factor:
   a. $3x^2 - 11x + 6$
   b. $y^2 + 19y + 84$
   c. $x^3 + 5x^2 + 6x$
   d. $4x^2 - 3x + 9$
   e. $5x^2 - 180$
   f. $3x^2 + 6x - 24$
   g. $2x^2 - 5x - 12$
   h. $-2x^2 + 3x + 20$
   i. $3x^3 + 5x^2 - 2x$
   j. $5x^3 - 20x^2 + 20x$
4. Factor:
   a. $4x^2 + 4xy + y^2$
   b. $x^2 - 2xy - 8y^2$
   c. $m^2 + 13mn + 22n^2$
   d. $-x^2 - 5xy + 3y^2$
   e. $2x^2 + 2xy - 4y^2$
   f. $x^2 + 11xy - 26y^2$
   g. $2x^3 - 4x^2y + 2xy^2$
   h. $8x^2 + 2xy - 15y^2$
   i. $5x^2 - xy + 3y^2$
   j. $3x^2 + 30xy + 75y^2$
5. Factor (if possible) and simplify:
   a. $(2x - 1)(3x + 2) + (2x - 1)(x - 3)$
   b. $(x + 5)(x - 4) - (2x + 3)(x - 1)$
   c. $(x + 1)^3 - (x + 1)^2$
   d. $(x + 5)(x + 1) - (x + 3)(x + 1)$
   e. $(x - 5)(x - 7) + 2(x - 5)(2x - 3)$
6. Factor by parts:
   a. $2x^2 - 3x + 10x - 15$
   b. $2x^5 - 2x^4 + 5x - 5$
   c. $2x^5 - 6x^4 - 5x^3 + 15x^2 + 4x - 12$
   d. $5 - 5y + y^2 - y^3$

2.5 Exercise Answers

(Binomial factors can be commuted.)

1. a. $5(x - 2)$
   b. $3x(x - 2)$
   c. $3(3x^2 - x + 5)$
   d. $x(x^2 + 4x - 3)$
   e. $(x + 2)(x - 2)$
   f. $(x - 2)(2x + 3)$
   g. $(3x - 1)(3x - 1)$ or $(3x - 1)^2$
   h. $-(x + 1)$
   i. $-6x^2y^2(2 - 3y)$
   j. $-y(x^2 - y)$
2. a. $(x + 3)(x + 7)$
   b. $(x - 5)(x - 7)$
   c. $(x - 20)(x + 2)$
   d. $(x + 9)(x - 4)$
   e. $(x + 10)(x - 10)$
   f. $x(x - 5)$
   g. $(x - 12)(x - 2)$
   h. prime
   i. $(x + 2)(x + 2)$ or $(x + 2)^2$
   j. $(x + 7)(x - 6)$
3. a. $(3x - 2)(x - 3)$
   b. $(y + 7)(y + 12)$
   c. $x(x + 2)(x + 3)$
   d. prime
   e. $5(x + 6)(x - 6)$
   f. $3(x + 4)(x - 2)$
   g. $(2x + 3)(x - 4)$
   h. $-(2x + 5)(x - 4)$
   i. $x(x + 2)(3x - 1)$
   j. $5x(x - 2)(x - 2)$ or $5x(x - 2)^2$
4. a. $(2x + y)(2x + y)$ or $(2x + y)^2$
   b. $(x + 2y)(x - 4y)$
   c. $(m + 2n)(m + 11n)$
   d. $-(x^2 + 5xy - 3y^2)$
   e. $2(x - y)(x + 2y)$
   f. $(x + 13y)(x - 2y)$
   g. $2x(x - y)^2$
   h. $(2x + 3y)(4x - 5y)$
   i. prime
   j. $3(x + 5y)^2$
5. a. $8x^2 - 6x + 1$
   b. $-x^2 - 17$
   c. $x^3 + 2x^2 + x$
   d. $2x + 2$
   e. $5x^2 - 38x + 65$

6. **a.** $(x + 5)(2x - 3)$ **b.** $(2x^4 + 5)(x - 1)$
c. $(2x^4 - 5x^2 + 4)(x - 3)$ **d.** $(1 - y)(5 + y^2)$

2.6 Perfect Squares and Completing the Square

1

Perfect square integers are numbers obtained by squaring an integer. *Perfect square rational numbers* are numbers obtained by squaring a rational number.

Perfect square integers	Perfect square rational numbers
$(1)^2 = 1$	$\left(\frac{2}{3}\right)^2 = \frac{4}{9}$
$(-2)^2 = 4$	$(0.1)^2 = 0.01$
$7^2 = 49$	$\left(\frac{-4}{7}\right)^2 = \frac{16}{49}$
$(-10)^2 = 100$	$(2.3)^2 = 5.29$
etc.	etc.

Q1 List the perfect square integers less than or equal to 400. ____________________

____

STOP • STOP • STOP • STOP • STOP • STOP • STOP • STOP • STOP

A1 1, 4, 9, 16, 25, 36, 49, 64, 81, 100, 121, 144, 169, 196, 225, 256, 289, 324, 361, 400

2

Examples of perfect square monomials are:

$(a)^2 = a^2$ $(xy)^2 = x^2y^2$

$(-2x)^2 = 4x^2$ $(x^2)^2 = x^4$

$\left(\frac{2}{3}y^2\right)^2 = \frac{4}{9}y^4$ $(-0.5m^3)^2 = 0.25m^6$

For a monomial to be a perfect square, the coefficient must be a perfect square integer or perfect square rational number and all variables must be raised to an even power.

Q2 Write "yes" if the monomial is a perfect square; otherwise, write "no."

a. x^2y ______ **b.** x^2y^4 ______ **c.** $-x^2$ ______

d. $1.21m^6$ ______ **e.** $16x^3$ ______ **f.** $\frac{25}{64}y^2$ ______

g. $225x^2y^2$ ______ **h.** $0.1y^4$ ______

STOP • STOP • STOP • STOP • STOP • STOP • STOP • STOP • STOP

A2 **a.** no **b.** yes: $(xy^2)^2$ **c.** no

d. yes: $(1.1m^3)^2$ **e.** no **f.** yes: $\left(\frac{5}{8}y\right)^2$

g. yes: $(15xy)^2$ **h.** no

Q3 Complete each expression to form a true statement:

a. $(\qquad)^2 = 144x^2$ b. $(\qquad)^2 = \frac{1}{9}x^4$

c. $(\qquad)^2 = 0.81x^2y^2$ d. $(\qquad)^2 = \frac{49}{100}x^2y^4$

e. $(\qquad)^2 = 196y^4$ f. $(\qquad)^2 = 1.69x^6$

STOP • STOP • STOP • STOP • STOP • STOP • STOP • STOP • STOP

A3 a. $12x$ or $-12x$ b. $\frac{1}{3}x^2$ or $\frac{-1}{3}x^2$ c. $0.9xy$ or $-0.9xy$

d. $\frac{7}{10}xy^2$ or $\frac{-7}{10}xy^2$ e. $14y^2$ or $-14y^2$ f. $1.3x^3$ or $-1.3x^3$

3 The square of a binomial is always a trinomial. From Section 2.4:

$(a + b)^2 = a^2 + 2ab + b^2$
$(a - b)^2 = a^2 - 2ab + b^2$

Trinomials obtained from the square of a binomial are called *perfect square trinomials*. To determine whether a trinomial is a perfect square, perform the following test:

1. First and third terms must be perfect square monomials.
2. If so, express the first and third terms as squares of a monomial.
3. If twice the product of the two monomials thus formed equals the middle term (disregarding its sign), the original trinomial is a perfect square.

Examples:

1. $x^2 + 14x + 49 = (x)^2 + 14x + (7)^2$ *Test:* $2(x)(7) = 14x$

Hence $x^2 + 14x + 49$ is a perfect square trinomial.

2. $x^2 - 12x + 36 = (x)^2 - 12x + (6)^2$ *Test:* $2(x)(6) = 12x$

Hence $x^2 - 12x + 36$ is a perfect square trinomial.

3. $x^2 + 10x - 25$

Not a perfect square trinomial because the third term, -25, is not a perfect square monomial.

4. $4x^2 - 20xy + 25y^2 = (2x)^2 - 20xy + (5y)^2$ *Test:* $2(2x)(5y) = 20xy$

Hence $4x^2 - 20xy + 25y^2$ is a perfect square trinomial.

5. $x^2 + 12x + 25 = (x)^2 + 12x + (5)^2$ *Test:* $2(x)(5) \neq 12x$

Hence $x^2 + 12x + 25$ is not a perfect square trinomial.

Q4 Determine whether each trinomial is a perfect square. If it is, write "yes"; otherwise, write "no."

a. $y^2 + 20y - 100$ ______ b. $x^2 - 20x + 100$ ______

c. $x^2 - 22x + 121$ ______ d. $9x^2 - 40x + 49$ ______

e. $9m^2 - 30mn + 25n^2$ ______ f. $16y^2 + 72xy + 81x^2$ ______

g. $196x^2 - 28xy + y^2$ ______ h. $50x^2 + 14x + 1$ ______

STOP • STOP • STOP • STOP • STOP • STOP • STOP • STOP • STOP

A4

a. no: -100 is not a perfect square monomial
b. yes: $(x)^2 - 20x + (10)^2$ and $2(x)(10) = 20x$
c. yes: $(x)^2 - 22x + (11)^2$ and $2(x)(11) = 22x$
d. no: $(3x)^2 + 40x + (7)^2$ and $2(3x)(7) \neq 40x$
e. yes: $(3m)^2 - 30mn + (5n)^2$ and $2(3m)(5n) = 30mn$
f. yes: $(4y)^2 + 72xy + (9x)^2$ and $2(4y)(9x) = 72xy$
g. yes: $(14x)^2 - 28xy + (y)^2$ and $2(14x)(y) = 28xy$
h. no: $50x^2$ is not a perfect square monomial

4

Recall that x^4 is a perfect square monomial since $(x^2)^2 = x^4$. Trinomials involving greater powers than 2 can be perfect square trinomials.

Examples:

1. $x^4 + 2x^2y^2 + y^4 = (x^2)^2 + 2x^2y^2 + (y^2)^2$ *Test:* $2(x^2)(y^2) = 2x^2y^2$

Hence $x^4 + 2x^2y^2 + y^4$ is a perfect square trinomial.

2. $36y^6 - 12y^3 + 1 = (6y^3)^2 - 12y^3 + (1)^2$ *Test:* $2(6y^3)(1) = 12y^3$

Hence $36y^6 - 12y^3 + 1$ is a perfect square trinomial.

Q5

Write "yes" if the trinomial is a perfect square; otherwise, write "no."

a. $4x^4 - 5x^2y^2 + y^4$ ______ **b.** $x^6 + 2x^3y^2 + y^4$ ______

STOP • STOP • STOP • STOP • STOP • STOP • STOP • STOP • STOP

A5

a. no: $2(2x^2)(y^2) \neq 5x^2y^2$ **b.** yes

5

If a trinomial is a perfect square, it can be factored rapidly.

$$a^2 + 2ab + b^2 \qquad a^2 - 2ab + b^2$$
$$(a)^2 + 2ab + (b)^2 \qquad (a)^2 - 2ab + (b)^2$$
$$(a + b)^2 \qquad (a - b)^2$$

Consider the factoring of the following trinomials.

Examples:

1. $x^2 + 14x + 49$
$(x)^2 + 14x + (7)^2$ *Test:* $2(x)(7) = 14x$
$(x + 7)^2$

2. $x^2 - 12x + 36$
$(x)^2 - 12x + (6)^2$ *Test:* $2(x)(6) = 12x$
$(x - 6)^2$

3. $x^2 + 10x - 25$
prime

4. $4x^2 - 20xy + 25y^2$
$(2x)^2 - 20xy + (5y)^2$ *Test:* $2(2x)(5y) = 20xy$
$(2x - 5y)^2$

5. $x^4 + 2x^2y^2 + y^4$
$(x^2)^2 + 2x^2y^2 + (y^2)^2$ *Test:* $2(x^2)(y^2) = 2x^2y^2$
$(x^2 + y^2)^2$

Q6 Factor:

a. $x^2 - 20x + 100$ b. $x^2 - 22x + 121$ c. $9x^2 + 42x + 49$

d. $9m^2 - 30mn + 25n^2$ e. $16y^2 + 72xy + 81x^2$ f. $196x^2 - 28xy + y^2$

g. $36y^6 - 12y^3 + 1$ h. $x^6 + 2x^3y^2 + y^4$

STOP • STOP • STOP • STOP • STOP • STOP • STOP • STOP • STOP

A6 a. $(x - 10)^2$ b. $(x - 11)^2$ c. $(3x + 7)^2$
d. $(3m - 5n)^2$ e. $(4y + 9x)^2$ f. $(14x - y)^2$
g. $(6y^3 - 1)^2$ h. $(x^3 + y^2)^2$

6 Polynomials can be arranged in either descending or ascending order of the powers of the variables. For example,

$36x^2 - 12x + 1 = 1 - 12x + 36x^2$
$= (1)^2 - 12x + (6x)^2$ *Test:* $2(1)(6x) = 12x$
$= (1 - 6x)^2$

The result can always be checked by expanding.

Q7 Factor:

a. $1 + 12x^3 + 36x^6$ b. $49 - 84y^2 + 36y^4$

STOP • STOP • STOP • STOP • STOP • STOP • STOP • STOP • STOP

A7 a. $(1 + 6x^3)^2$ b. $(7 - 6y^2)^2$

7 Naturally all trinomials are not perfect squares. However, it is wise to test each trinomial where two of the three terms are perfect square monomials. If the trinomial is a perfect square, the factoring is then easy. If it is not, other means of factoring must be tried. Regardless, remove common factors first if they are present. It is also possible that the trinomial might be prime. Study the following examples.

1. $-8x^2 - 24xy - 18y^2$
$-2(4x^2 + 12xy + 9y^2)$ *Test:* $2(2x)(3y) = 12xy$
$-2(2x + 3y)^2$

2. $-y^2 + 29y - 100$
$-(y^2 - 29y + 100)$ *Test:* $2(y)(10) \neq 29y$

Hence $y^2 - 29y + 100$ is not a perfect square trinomial. Howcver, it is factorable into the product of two binomials.
$-(y - 25)(y - 4)$

3. $4x^2 - 5x + 1$ *Test:* $2(2x)(1) \neq 5x$

Hence $4x^2 - 5x + 1$ is not a perfect square trinomial. Therefore, try the factoring process discussed in Section 2.5. $4(1) = 4$, $(-1)(-4) = 4$, and $-1 + {}^-4 = -5$.

$4x^2 - 5x + 1$
$4x^2 - 1x - 4x + 1$
$x(4x - 1) - 1(4x - 1)$
$(x - 1)(4x - 1)$

4. $xy^4 + 6x^3y^2 - 16x^5$
$x(y^4 + 6x^2y^2 - 16x^4)$

$-16x^4$ is not a perfect square monomial. However, $y^4 + 6x^2y^2 - 16x^4$ is factorable into the product of two binomials.

$x(y^4 + 8x^2y^2 - 2x^2y^2 - 16x^4)$
$x[y^2(y^2 + 8x^2) - 2x^2(y^2 + 8x^2)]$
$x(y^2 - 2x^2)(y^2 + 8x^2)$

5. $8x^3 - 28x^2 + 18x$
$2x(4x^2 - 14x + 9)$ *Test:* $2(2x)(3) \neq 14x$

Hence $4x^2 - 14x + 9$ is not a perfect square trinomial. Also,

$4(9) = 36$ and
$(-1)(-36) = 36$
$(-2)(-18) = 36$
$(-3)(-12) = 36$
$(-4)(-9) = 36$
$(-6)(-6) = 36$

Since no combination of the above factors sum to -14, $4x^2 - 14x + 9$ is prime and the factorization is complete.

Q8 Factor:

a. $-x^2 - 2xy - y^2$

b. $2y^5 - 12x^2y^3 + 18x^4y$

c. $m^4 - 24m^2 - 52$

d. $-6 - 12y + 3y^2$

e. $6x^2 + xy - 15y^2$ f. $m^2 - 7mn + 9n^2$

STOP • STOP • STOP • STOP • STOP • STOP • STOP • STOP • STOP

A8 a. $-(x + y)^2$ b. $2y(y^2 - 3x^2)^2$ c. $(m^2 + 2)(m^2 - 26)$
d. $-3(2 + 4y - y^2)$ (*Note:* $2 + 4y - y^2$ is prime because there are no integers whose product is -2 and whose sum is 4.)
e. $(2x - 3y)(3x + 5y)$ f. prime

8 As you have seen, the square of a binomial is a perfect square trinomial. An important relationship exists between the terms of a perfect square trinomial. For example,

$(x + 5)^2 = x^2 + 10x + 25$

Notice that $\left[\frac{1}{2}(10)\right]^2 = (5)^2 = 25$. This indicates that the square of one-half of 10 (the coefficient of the first-degree term) equals 25 (the constant term). In general, $(x + b)^2 = x^2 + 2bx + b^2$. Observe that

$\left[\frac{1}{2}(2b)\right]^2 = (b)^2 = b^2$ regardless of the value of b

Q9 a. $x^2 + 18x + 81$ is a perfect square trinomial. Is $\left[\frac{1}{2}(18)\right]^2 = 81$? ______

b. $x^2 - 14x + 49$ is a perfect square trinomial. Is $\left[\frac{1}{2}(-14)\right]^2 = 49$? ______

STOP • STOP • STOP • STOP • STOP • STOP • STOP • STOP • STOP

A9 a. yes b. yes

9 Consider the polynomial $x^2 - 6x$. By adding 9, $\left[\frac{1}{2}(-6)\right]^2$, a perfect square trinomial, $x^2 - 6x + 9$, is formed. Now, $x^2 - 6x + 9 = (x - 3)^2$. This process is called *completing the square.**

*In Chapter 5 the process of "completing the square" will be applied to the solution of quadratic (second-degree) equations.

Q10 Complete the square of $x^2 - 16x$ by performing the following steps:

a. $\left[\frac{1}{2}(-16)\right]^2 =$ ______

b. $x^2 - 16x +$ ______ $= (\qquad)^2$

STOP • STOP • STOP • STOP • STOP • STOP • STOP • STOP • STOP

A10 a. 64: $(-8)^2$ b. $x^2 - 16x + 64 = (x - 8)^2$

10 Study the following examples of completing the square and the factoring of the resulting perfect square trinomial:

$x^2 - 5x + \underline{\ ?\ }$

$\left[\frac{1}{2}(-5)\right]^2 = \left(\frac{-5}{2}\right)^2 = \frac{25}{4}$

$x^2 - 5x + \frac{25}{4}$

$(x)^2 - 5x + \left(\frac{5}{2}\right)^2$

$\left(x - \frac{5}{2}\right)^2$

$x^2 + \frac{1}{2}x + \underline{\ ?\ }$

$\left[\frac{1}{2}\left(\frac{1}{2}\right)\right]^2 = \left(\frac{1}{4}\right)^2 = \frac{1}{16}$

$x^2 + \frac{1}{2}x + \frac{1}{16}$

$(x)^2 + \frac{1}{2}x + \left(\frac{1}{4}\right)^2$

$\left(x + \frac{1}{4}\right)^2$

These polynomials involve rational-number coefficients. The factoring has been completed over the set of rational numbers.

Q11 (1) Complete the square and (2) factor:

a. (1) $x^2 - 4x + \_\_\_\_$ (2) $\_\_\_\_\_\_$

b. (1) $x^2 - \frac{2}{5}x + \_\_\_\_$ (2) $\_\_\_\_\_\_$

c. (1) $x^2 + \frac{1}{3}x + \_\_\_\_$ (2) $\_\_\_\_\_\_$

d. (1) $m^2n^2 + 7mn + \_\_\_\_$ (2) $\_\_\_\_\_\_$

STOP • STOP • STOP • STOP • STOP • STOP • STOP • STOP • STOP

A11 **a.** (1) 4 (2) $(x - 2)^2$

b. (1) $\frac{1}{25}$ (2) $\left(x - \frac{1}{5}\right)^2$

c. (1) $\frac{1}{36}$ (2) $\left(x + \frac{1}{6}\right)^2$

d. (1) $\frac{49}{4}$ (2) $\left(mn + \frac{7}{2}\right)^2$

Q12 **a.** $\left[\frac{1}{2}\left(\frac{b}{a}\right)\right]^2 = \left(\frac{b}{2a}\right)^2 = \_\_\_\_$

b. (1) Complete the square and (2) factor.

(1) $x^2 + \frac{b}{a}x + \_\_\_\_$ (2) $\_\_\_\_\_\_$

STOP • STOP • STOP • STOP • STOP • STOP • STOP • STOP • STOP

A12 **a.** $\frac{b^2}{4a^2}$ **b.** (1) $x^2 + \frac{b}{a}x + \frac{b^2}{4a^2}$ (2) $\left(x + \frac{b}{2a}\right)^2$

11 You should realize that if a term is added to a polynomial, the value of that polynomial has been changed. To compensate, the same value added to the polynomial must be subtracted in order to maintain the original value.

Examples:

$x^2 + 10x$
$x^2 + 10x + \underline{25} - \underline{25}$
$(x + 5)^2 - 25$

All expressions above are equivalent. However, $x^2 + 10x \neq x^2 + 10x + 25$.

$x^4 + 64$
$x^4 + \underline{16x^2} + 64 - \underline{16x^2}$
$(x^2 + 8)^2 - 16x^2$

All expressions above are equivalent. However, $x^4 + 64 \neq x^4 + 16x^2 + 64$. This technique will be used in Section 2.7.

This completes the instruction for this section.

2.6 Exercises

1. Write "yes" if the trinomial is a perfect square; otherwise, write "no."
   a. $x^2 + 2x + 1$ **b.** $y^2 - 4xy + 4x^2$
   c. $x^2 - 26x + 25$ **d.** $25m^2 + 20mn + 4n^2$
   e. $y^2 + 9y - 36$ **f.** $16x^2 - 24x + 9$
   g. $n^2 - 11n + 36$ **h.** $4y^2 + 28yz + 49z^2$
   i. $100r^4 + 20r^2s^2 + s^4$ **j.** $1 - 18y^3 + 81y^6$
2. Factor each trinomial in problem 1. Write "prime" where appropriate.
3. Factor:
   a. $-4x^2 - 4x + 24$ **b.** $4x^3 - 4x^2y + xy^2$
   c. $-25x^2 - 10x - 1$ **d.** $60x^3 + 25x^2 - 10x$
   e. $18x^2 - 40x + 8$ **f.** $9x^2 + 15xy - 4y^2$
4. (1) Replace k by the value that will make the trinomial a perfect square and (2) factor the resulting trinomial over the set of rational numbers:
   a. $y^2 - 18y + k$ **b.** $m^2 + 4m + k$ **c.** $x^2 + 9x + k$
   d. $x^2 - \frac{4}{5}x + k$ **e.** $x^2 - \frac{1}{5}x + k$ **f.** $x^2 - \frac{b}{a}x + k$

2.6 Exercise Answers

1. **a.** yes **b.** yes **c.** no
   d. yes **e.** no **f.** yes
   g. no **h.** yes **i.** yes
   j. yes
2. **a.** $(x + 1)^2$ **b.** $(y - 2x)^2$ **c.** $(x - 25)(x - 1)$
   d. $(5m + 2n)^2$ **e.** $(y + 12)(y - 3)$ **f.** $(4x - 3)^2$
   g. prime **h.** $(2y + 7z)^2$ **i.** $(10r^2 + s^2)^2$
   j. $(1 - 9y^3)^2$
3. **a.** $-4(x + 3)(x - 2)$ **b.** $x(2x - y)^2$ **c.** $-(5x + 1)^2$
   d. $5x(4x - 1)(3x + 2)$ **e.** $2(9x - 2)(x - 2)$ **f.** prime

4. **a.** (1) $y^2 - 18y + 81$
(2) $(y - 9)^2$

b. (1) $m^2 + 4m + 4$
(2) $(m + 2)^2$

c. (1) $x^2 + 9x + \frac{81}{4}$
(2) $\left(x + \frac{9}{2}\right)^2$

d. (1) $x^2 - \frac{4}{5}x + \frac{4}{25}$
(2) $\left(x - \frac{2}{5}\right)^2$

e. (1) $x^2 - \frac{1}{5}x + \frac{1}{100}$
(2) $\left(x - \frac{1}{10}\right)^2$

f. (1) $x^2 - \frac{b}{a}x + \frac{b^2}{4a^2}$
(2) $\left(x - \frac{b}{2a}\right)^2$

2.7 Sums or Differences of Two Squares and Two Cubes

1

A special product from Section 2.4 was

$(a + b)(a - b) = a^2 - b^2$

The expression $a^2 - b^2$ is called the *difference of two squares.*

Examples:

$(3x + 4)(3x - 4) = 9x^2 - 16$

$(x + 2y)(x - 2y) = x^2 - 4y^2$

The equation $a^2 - b^2 = (a + b)(a - b)$ can be used to factor the difference of two squares.

Examples:

$x^2 - 9$
$(x)^2 - (3)^2$
$(x + 3)(x - 3)$

$25y^2 - 16$
$(5y)^2 - (4)^2$
$(5y + 4)(5y - 4)$

$4x^2 - 49y^2$
$(2x)^2 - (7y)^2$
$(2x + 7y)(2x - 7y)$

$1 - 36m^2$
$(1)^2 - (6m)^2$
$(1 + 6m)(1 - 6m)$

Do as much mentally as possible.

Q1 Factor:

a. $4 - y^2$ **b.** $64x^2 - 121$ **c.** $9x^2 - 100$

d. $x^4 - 16y^2$

STOP • STOP • STOP • STOP • STOP • STOP • STOP • STOP • STOP

A1 **a.** $(2 + y)(2 - y)$ **b.** $(8x + 11)(8x - 11)$ **c.** $(3x + 10)(3x - 10)$
d. $(x^2 + 4y)(x^2 - 4y)$

2 When factoring polynomials, always remove a common factor first (if it exists) before attempting any other type of factoring.

Examples:

$-18 + 2x^2$	$x^3y^2 - 81x$
$-2(9 - x^2)$	$x(x^2y^2 - 81)$
$-2(3 + x)(3 - x)$	$x(xy + 9)(xy - 9)$

Q2 Factor:

a. $y^3 - 25x^2y$ b. $9x^2 - 144$

c. $-4x^3y^2 + 100xz^2$ d. $x^2 - x$

STOP • STOP • STOP • STOP • STOP • STOP • STOP • STOP • STOP

A2 a. $y(y + 5x)(y - 5x)$ b. $9(x + 4)(x - 4)$
c. $-4x(xy + 5z)(xy - 5z)$ d. $x(x - 1)$

3 Occasionally it will be possible to factor again. For example,

$x^4 - y^4$
$(x^2 + y^2)(x^2 - y^2)$
$(x^2 + y^2)(x + y)(x - y)$

$x^2 + y^2$ is prime. The only two possibilities, $(x + y)^2$ and $(x - y)^2$, are not factors. *Recall:*

$(x + y)^2 = x^2 + 2xy + y^2$
$(x - y)^2 = x^2 - 2xy + y^2$

Q3 Factor:

a. $16x^4 - 81y^4$ b. $x^8 - 1 = (x^4)^2 - 1$

STOP • STOP • STOP • STOP • STOP • STOP • STOP • STOP • STOP

A3 a. $(4x^2 + 9y^2)(2x + 3y)(2x - 3y)$ b. $(x^4 + 1)(x^2 + 1)(x + 1)(x - 1)$

4 The expression $x^2 + y^2$ (the sum of two squares) is prime. However, in special cases, the sum of two squares is factorable. $4x^4 + 1$ is an example of the sum of two squares that can be factored. Factoring such expressions requires a special technique.

Example: Factor $4x^4 + 1$.

Solution

Step 1: Determine the value of h that would make $4x^4 + h + 1$ a perfect square trinomial.

$4x^4 + h + 1$
$(2x^2)^2 + h + (1)^2$
$h = 2(2x^2)(1) = 4x^2$

Step 2: Add and subtract $4x^2$ from $4x^4 + 1$.

$4x^4 + 4x^2 + 1 - 4x^2$

(*Note:* Since $4x^2 - 4x^2 = 0$, the value of $4x^4 + 1$ is unchanged.)

Step 3: Factor the first three terms.

$(2x^2 + 1)^2 - 4x^2$

Step 4: Factor $(2x^2 + 1)^2 - 4x^2$ as the difference of two squares.

$(2x^2 + 1)^2 - (2x)^2$
$[(2x^2 + 1) + 2x][(2x^2 + 1) - 2x]$

Step 5: Rearrange terms of both factors in descending order and simplify.

$(2x^2 + 2x + 1)(2x^2 - 2x + 1)$

The above result can be verified by expanding. *This factorization is presented not for mastery but to dispel any notion you might have that all sums of two squares are prime.*

Q4 Factor $16x^8 - 1$. [*Hint:* Factor $16x^8 - 1$ as the difference of two squares, $(4x^4)^2 - (1)^2$, and use the result of Frame 4.]

STOP • STOP • STOP • STOP • STOP • STOP • STOP • STOP • STOP

A4 $(2x^2 + 2x + 1)(2x^2 - 2x + 1)(2x^2 + 1)(2x^2 - 1)$: $(4x^4 + 1)(4x^4 - 1)$

Q5 Factor:

a. $x^{16} - y^{64}$

b. $x^4 - 2x^2y^2 + y^4$

STOP • STOP • STOP • STOP • STOP • STOP • STOP • STOP • STOP

A5 **a.** $(x^8 + y^{32})(x^4 + y^{16})(x^2 + y^8)(x + y^4)(x - y^4)$

b. $(x + y)^2(x - y)^2$: $x^4 - 2x^2y^2 + y^4$
$(x^2 - y^2)^2$
$(x^2 - y^2)(x^2 - y^2)$
$(x + y)(x - y)(x + y)(x - y)$

5 Two other special products were presented in Section 2.4. They were:

$(a + b)(a^2 - ab + b^2) = a^3 + b^3$
$(a - b)(a^2 + ab + b^2) = a^3 - b^3$

$a^3 + b^3$ is called the *sum of two cubes.* $a^3 - b^3$ is called the *difference of two cubes.* When factoring sums or differences of two cubes, you must recognize *perfect cube monomials.* You should first be familiar with *perfect cube integers.* $(1)^3 = 1$, $(2)^3 = 8$, $(3)^3 = 27$, $(4)^3 = 64$, $(5)^3 = 125$, $(6)^3 = 216$, $(7)^3 = 343$, $(8)^3 = 512$, $(9)^3 = 729$, $(10)^3 = 1{,}000$, and so on. [*Note:* Although $(-1)^3 = -1$, $(-2)^3 = -8, \ldots$, the negative values are not important to this discussion.]

A perfect cube monomial will have a perfect cube integer coefficient and the variables will be to a power that is a multiple of 3 (3, 6, 9, 12, . . .).

Q6 Complete the following:

a. $8 = (\quad)^3$ b. $27x^3 = (\quad)^3$ c. $64x^6 = (\quad)^3$

STOP • STOP • STOP • STOP • STOP • STOP • STOP • STOP • STOP

A6 a. $(2)^3$ b. $(3x)^3$ c. $(4x^2)^3$

---

6 The following equation is helpful when factoring the sum of two cubes.

$$a^3 + b^3 = (a)^3 + (b)^3$$
$$= (a + b)(a^2 - ab + b^2)$$

(1) (2)(3) (4) (5)

opposite sign

1: first term
2: second term
3: square of the first term
4: product of the first term and the second term
5: square of the second term

Example 1: Factor $8x^3 + 1$.

Solution

$8x^3 + 1$
$(2x)^3 + (1)^3$
$(2x + 1)[(2x)^2 - (2x)(1) + (1)^2]$
$(2x + 1)(4x^2 - 2x + 1)$

Example 2: Factor $x^6 + y^6$.

Solution

$x^6 + y^6$
$(x^2)^3 + (y^2)^3$
$(x^2 + y^2)[(x^2)^2 - (x^2)(y^2) + (y^2)^2]$
$(x^2 + y^2)(x^4 - x^2y^2 + y^4)$

Example 3: Factor $x^3 + 27y^9$.

Solution

$x^3 + 27y^9$
$(x)^3 + (3y^3)^3$
$(x + 3y^3)[(x)^2 - (x)(3y^3) + (3y^3)^2]$
$(x + 3y^3)(x^2 - 3xy^3 + 9y^6)$

With practice many of the above steps can be done mentally. In each example, the trinomial factor is prime.

Q7 Complete the factorization:

$8 + 27x^3$

a. $(\quad)^3 + (\quad)^3$
b. $(\_\_\_ + \_\_\_)[(\quad)^2 - (\quad)(\quad) + (\quad)^2]$
c. $(\quad)(\quad)$

STOP • STOP • STOP • STOP • STOP • STOP • STOP • STOP • STOP

A7

a. $(2)^3 + (3x)^3$
b. $(2 + 3x)[(2)^2 - (2)(3x) + (3x)^2]$
c. $(2 + 3x)(4 - 6x + 9x^2)$

Q8

Factor:

a. $27 + y^3$

b. $8x^3 + y^3$

c. $z^3 + \frac{1}{27}$ $\left[\textit{Note: } \left(\frac{1}{3}\right)^3 = \frac{1}{27}.\right]$

d. $1.331 + m^3$ [*Hint:* $(1.1)^3 = 1.331$.]

STOP • STOP • STOP • STOP • STOP • STOP • STOP • STOP • STOP

A8

a. $(3 + y)(9 - 3y + y^2)$

b. $(2x + y)(4x^2 - 2xy + y^2)$

c. $\left(z + \frac{1}{3}\right)\left(z^2 - \frac{1}{3}z + \frac{1}{9}\right)$

d. $(1.1 + m)(1.21 - 1.1m + m^2)$

7

The equation $a^3 - b^3 = (a - b)(a^2 + ab + b^2)$ can be used as a pattern for factoring the difference of two cubes.

$$a^3 - b^3 = (a)^3 - (b)^3$$
$$= (a - b)(a^2 + ab + b^2)$$

opposite sign

Example 1: Factor $y^3 - 64$.

Solution

$y^3 - 64$
$(y)^3 - (4)^3$
$(y - 4)[(y)^2 + (y)(4) + (4)^2]$
$(y - 4)(y^2 + 4y + 16)$

Example 2: Factor $8m^6 - 27n^9$.

Solution

$8m^6 - 27n^9$
$(2m^2)^3 - (3n^3)^3$
$(2m^2 - 3n^3)[(2m^2)^2 + (2m^2)(3n^3) + (3n^3)^2]$
$(2m^2 - 3n^3)(4m^4 + 6m^2n^3 + 9n^6)$

Q9

Factor:

a. $x^3 - 1$

*b. $1 - y^6$

c. $x^3 - 8y^3$

d. $125m^6 - p^3q^3$

STOP • STOP • STOP • STOP • STOP • STOP • STOP • STOP • STOP

A9

a. $(x - 1)(x^2 + x + 1)$

*b. $(1 + y)(1 - y)(1 - y + y^2)(1 + y + y^2)$: $(1 - y^2)(1 + y^2 + y^4)$

c. $(x - 2y)(x^2 + 2xy + 4y^2)$

d. $(5m^2 - pq)(25m^4 + 5m^2pq + p^2q^2)$

8

The equations for factoring the sum or difference of two squares and two cubes can be summarized:

$a^2 + b^2$: Only factorable in special cases.
$a^2 - b^2 = (a + b)(a - b)$
$a^3 + b^3 = (a + b)(a^2 - ab + b^2)$
$a^3 - b^3 = (a - b)(a^2 + ab + b^2)$

You should observe that $x^6 + 1$ is the sum of two squares, $(x^3)^2 + (1)^2$. This is another special case where the sum of two squares is factorable. However, it is easier to factor $x^6 + 1$ as the sum of two cubes, $(x^2)^3 + (1)^3$. That is, $x^6 + 1 = (x^2 + 1)(x^4 - x^2 + 1)$. To factor $x^6 + 1$ as the sum of two squares requires the technique of Frame 4.

Q10

a. Is $x^6 + y^6$ the sum of two squares? ________

b. Is $x^6 + y^6$ factorable as the sum of two cubes? ________

c. Factor: $x^6 + y^6 =$ ________________________________

STOP • STOP • STOP • STOP • STOP • STOP • STOP • STOP • STOP

A10

a. yes b. yes c. $(x^2 + y^2)(x^4 - x^2y^2 + y^4)$

9

Q9b asked you to factor $1 - y^6$. You might have noticed that it could have been factored as the difference of two squares.

$$\begin{aligned} 1 - y^6 &= (1)^2 - (y^3)^2 \\ &= (1 + y^3)(1 - y^3) \end{aligned}$$

Now, the two factors are the sum and difference of two cubes:

$= (1 + y)(1 - y + y^2)(1 - y)(1 + y + y^2)$

Often, it is possible to factor expressions as either the difference of two squares or the difference of two cubes. You should choose the method you prefer.

Q11

a. Is $x^6 - y^6$ the difference between two squares? ________

b. Is $x^6 - y^6$ the difference between two cubes? ________

c. Is $x^8 - y^8$ the difference between two squares? ________

d. Is $x^8 - y^8$ the difference between two cubes? ________

e. Is $x^9 - y^9$ the difference between two squares? ________

f. Is $x^9 - y^9$ the difference between two cubes? ________

STOP • STOP • STOP • STOP • STOP • STOP • STOP • STOP • STOP

A11

a. yes b. yes c. yes d. no e. no f. yes

10

To be both the difference of two squares and the difference of two cubes, all variables must be to a power that is a multiple of 6. Furthermore, all coefficients must be both squares and cubes. $(1)^2 = 1$ and $(1)^3 = 1$, $(8)^2 = 64$ and $(4)^3 = 64$, and so on.

Q12 **a.** Factor $x^6 - y^6$ as the difference of two squares.

***b.** Factor $x^6 - y^6$ as the difference of two cubes. (Look at solution below if you have difficulty.)

STOP • STOP • STOP • STOP • STOP • STOP • STOP • STOP • STOP

A12 **a.** $(x + y)(x - y)(x^2 - xy + y^2)(x^2 + xy + y^2)$:
$x^6 - y^6$
$(x^3)^2 - (y^3)^2$
$(x^3 + y^3)(x^3 - y^3)$
$(x + y)(x^2 - xy + y^2)(x - y)(x^2 + xy + y^2)$

***b.** Same as part a: $x^6 - y^6$
$(x^2)^3 - (y^2)^3$
$(x^2 - y^2)(x^4 + x^2y^2 + y^4)$
$(x + y)(x - y)(x^4 + x^2y^2 + y^4)$

You may have stopped here not knowing what to do with $x^4 + x^2y^2 + y^4$. Recall the procedure of Frame 4.

$x^4 + x^2y^2 + \underline{x^2y^2} + y^4 - \underline{x^2y^2}$
$(x^4 + 2x^2y^2 + y^4) - x^2y^2$
$(x^2 + y^2)^2 - x^2y^2$
$[(x^2 + y^2) + xy][(x^2 + y^2) - xy]$
$(x^2 + xy + y^2)(x^2 - xy + y^2)$

11	Generally, to avoid the situation that arose in Q12b when you have a choice, factor such an expression as $x^6 - y^6$ as the difference of two squares.

Q13 Factor (as the difference of two squares) $x^6 - 64y^6$.

STOP • STOP • STOP • STOP • STOP • STOP • STOP • STOP • STOP

A13 $(x + 2y)(x - 2y)(x^2 - 2xy + 4y^2)(x^2 + 2xy + 4y^2)$: $x^6 - 64y^6$
$(x^3)^2 - (8y^3)^2$
$(x^3 + 8y^3)(x^3 - 8y^3)$

This completes the instruction for this section.

2.7 Exercises

Factor (remove common factors first if they exist):

1. **a.** $9x^2 - 25$ **b.** $6 - 6x^2$ **c.** $121x^2 - 169y^2$ **d.** $m^2n^2 - 100$
   e. $98x^2 - 2y^2z^2$ **f.** $4ax^2 - 36ay^2$ **g.** $64x^4 - 81y^2$ **h.** $4x^6 - 225y^6$

i. $16x^4 - 1$ j. $x^4 - 625y^4$ $[(25)^2 = 625]$

2. a. $x^3 + y^3$ b. $x^3 - y^3$ c. $x^3 + 8$ d. $8x^3 + 27y^3$
e. $8x^3 - 27y^3$ f. $x^3 - y^6$ g. $27x^6 + y^6$ h. $64x^3 - 125$
i. $8 - y^9$ j. $1 + 8x^3$

3. a. $x^6 + y^6$ (*Hint:* Factor as the sum of two cubes.)
b. $64x^6 - y^6$ (*Hint:* Factor as the difference of two squares.)
*c. $x^9 + 1$ *d. $x^9 - 1$
*e. $x^{12} + 1$ *f. $x^{12} - 1$

The following problems will involve factoring techniques included in the entire chapter. Watch for polynomials that are prime.

4. a. $2x^2 - 11x + 12$ b. $144 - y^2$ c. $z^3 + 1$
d. $m^2 + 12m + 36$ e. $x^2 - 2x - 24$ f. $x^2 - xy$
g. $4x^2 - 12xy + 9y^2$ h. $8 - m^6$ i. $9x^2 + 15x + 4$
j. $x - 2y$ k. $x^2 + 6xy - 27y^2$ l. $x^3y - 6x^2y - 16xy$
m. $3x^8 - 3y^8$ n. $20x - 5xy^2$ o. $y^2 - 3y + 4$
p. $-x^2 - 4x - 4$ q. $-x^2 + 1$ r. $40 - 18y + 2y^2$
s. $x^6 + 6x^3 - 16$ t. $x^5 - 16xy^4$
u. $x^5 + 6x^4 - 2x^3 - 12x^2 - 7x - 42$
v. $4y^3 - 7y^2 + 16y - 28$

2.7 Exercise Answers

1. a. $(3x + 5)(3x - 5)$ b. $6(1 + x)(1 - x)$
c. $(11x + 13y)(11x - 13y)$ d. $(mn + 10)(mn - 10)$
e. $2(7x + yz)(7x - yz)$ f. $4a(x + 3y)(x - 3y)$
g. $(8x^2 + 9y)(8x^2 - 9y)$ h. $(2x^3 + 15y^3)(2x^3 - 15y^3)$
i. $(4x^2 + 1)(2x + 1)(2x - 1)$ j. $(x^2 + 25y^2)(x + 5y)(x - 5y)$

2. a. $(x + y)(x^2 - xy + y^2)$ b. $(x - y)(x^2 + xy + y^2)$
c. $(x + 2)(x^2 - 2x + 4)$ d. $(2x + 3y)(4x^2 - 6xy + 9y^2)$
e. $(2x - 3y)(4x^2 + 6xy + 9y^2)$ f. $(x - y^2)(x^2 + xy^2 + y^4)$
g. $(3x^2 + y^2)(9x^4 - 3x^2y^2 + y^4)$ h. $(4x - 5)(16x^2 + 20x + 25)$
i. $(2 - y^3)(4 + 2y^3 + y^6)$ j. $(1 + 2x)(1 - 2x + 4x^2)$

3. a. $(x^2 + y^2)(x^4 - x^2y^2 + y^4)$
b. $(2x + y)(2x - y)(4x^2 - 2xy + y^2)(4x^2 + 2xy + y^2)$
*c. $(x + 1)(x^2 - x + 1)(x^6 - x^3 + 1)$
*d. $(x - 1)(x^2 + x + 1)(x^6 + x^3 + 1)$
*e. $(x^4 + 1)(x^8 - x^4 + 1)$
*f. $(x^2 + 1)(x^4 - x^2 + 1)(x + 1)(x^2 - x + 1)(x - 1)(x^2 + x + 1)$

4. a. $(2x - 3)(x - 4)$ b. $(12 + y)(12 - y)$
c. $(z + 1)(z^2 - z + 1)$ d. $(m + 6)^2$
e. $(x - 6)(x + 4)$ f. $x(x - y)$
g. $(2x - 3y)^2$ h. $(2 - m^2)(4 + 2m^2 + m^4)$
i. $(3x + 1)(3x + 4)$ j. prime
k. $(x + 9y)(x - 3y)$ l. $xy(x - 8)(x + 2)$
m. $3(x^4 + y^4)(x^2 + y^2)(x + y)(x - y)$ n. $5x(2 + y)(2 - y)$
o. prime p. $-(x + 2)^2$
q. $-(x + 1)(x - 1)$ or $(1 + x)(1 - x)$ r. $2(5 - y)(4 - y)$
s. $(x^3 - 2)(x + 2)(x^2 - 2x + 4)$ t. $x(x^2 + 4y^2)(x + 2y)(x - 2y)$
u. $(x^4 - 2x^2 - 7)(x + 6)$ v. $(y^2 + 4)(4y - 7)$

2.8 Reducing Rational Expressions

1

A rational number is a number that can be expressed in the form $\frac{p}{q}$, where $p \in I$ and $q \in I$, $q \neq 0$.* A *rational expression* is an expression that can be written as a fraction in which both numerator and denominator are polynomials. All possible values of the variables that would produce a zero denominator are excluded as possible replacements for the variable. Examples of rational expressions are:

$-3 \qquad 0$

$\frac{2}{3} \qquad 5x^3$

$\frac{1}{y^2},\ y \neq 0 \qquad \frac{1}{x-2},\ x \neq 2$

$\frac{x}{x-y},\ x \neq y \qquad \frac{5x}{(x+5)(2x-1)},\ x \neq -5 \text{ or } x \neq \frac{1}{2}$

$5x^2 - 3x + 2 \qquad \frac{-4}{5}x$

A rational expression is said to be *undefined* for any value that makes the denominator zero.

*Division by zero is impossible.

Q1

a. The value of $\frac{1}{x+2}$ is __________ for $x = -2$.

b. $\frac{x+3}{x(x-5)}$ is undefined for $x = 0$ or $x =$ _____.

STOP • STOP • STOP • STOP • STOP • STOP • STOP • STOP • STOP

A1

a. undefined **b.** 5

2

Reducing rational numbers requires division of both the numerator and denominator by a common factor. Simplifying rational expressions by reducing also requires division by a common factor.

Examples:

$$\frac{4}{10} = \frac{\overset{2}{\cancel{4}}}{\underset{5}{\cancel{10}}} = \frac{2}{5} \quad \left(\frac{4}{10} = \frac{\overset{1}{\cancel{2}} \cdot 2}{\underset{1}{\cancel{2}} \cdot 5} = \frac{2}{5}\right)$$

$$\frac{2x}{10} = \frac{\overset{1}{\cancel{2}}x}{\underset{5}{\cancel{10}}} = \frac{x}{5} \quad \left(\frac{2x}{10} = \frac{\overset{1}{\cancel{2}} \cdot x}{\underset{1}{\cancel{2}} \cdot 5} = \frac{x}{5}\right)$$

$$\frac{32x^7}{20x^5} = \frac{\overset{8}{\cancel{32}}\overset{x^2}{\cancel{x^7}}}{\underset{5}{\cancel{20}}\underset{1}{\cancel{x^5}}} = \frac{8x^2}{5} \quad \left(\frac{32x^7}{20x^5} = \frac{\overset{1}{\cancel{4x^5}} \cdot 8x^2}{\underset{7}{\cancel{4x^5}} \cdot 5} = \frac{8x^2}{5}\right)^*$$

$$\frac{15x^2}{24x^3} = \frac{\overset{5}{\cancel{15}}\overset{1}{\cancel{x^2}}}{\underset{8}{\cancel{24}}\underset{x}{\cancel{x^3}}} = \frac{5}{8x} \quad \left(\frac{15x^2}{24x^3} = \frac{\overset{1}{\cancel{3x^2}} \cdot 5}{\underset{1}{\cancel{3x^2}} \cdot 8x} = \frac{5}{8x}\right)$$

*The restriction $x \neq 0$ has been omitted. Henceforth it will be assumed in all examples that the expression is defined for all permissible replacements of the variable.

A rational expression is said to be in *lowest terms* when both the numerator and denominator do not contain a common factor.

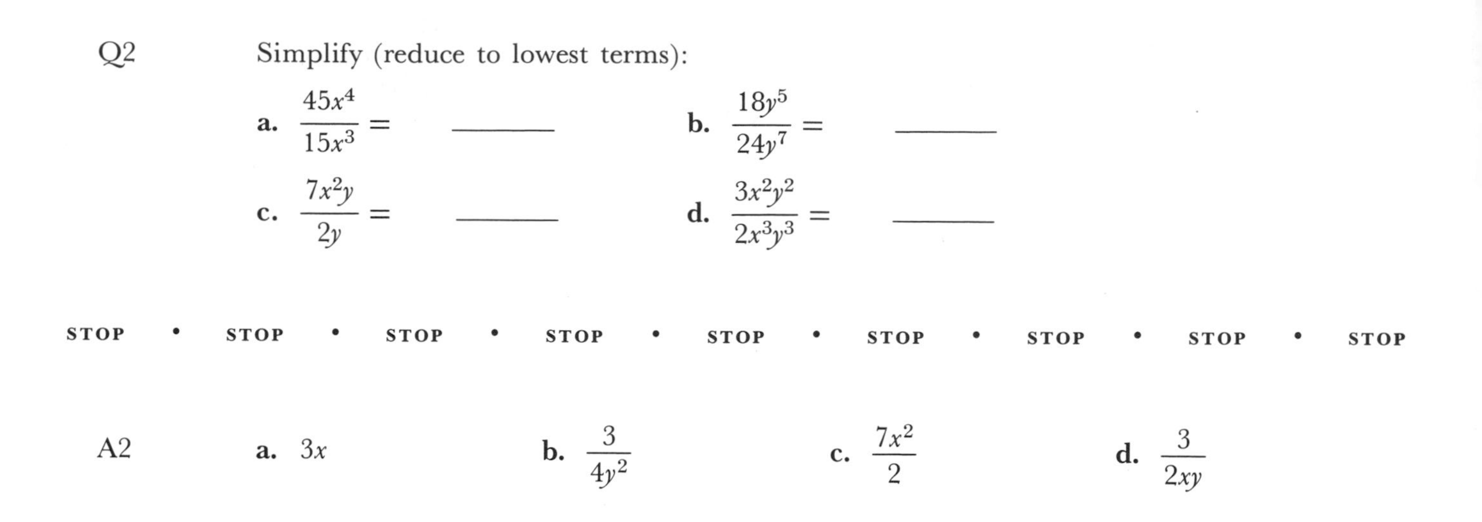

Q2 Simplify (reduce to lowest terms):

a. $\dfrac{45x^4}{15x^3} =$ ______ b. $\dfrac{18y^5}{24y^7} =$ ______

c. $\dfrac{7x^2y}{2y} =$ ______ d. $\dfrac{3x^2y^2}{2x^3y^3} =$ ______

STOP • STOP • STOP • STOP • STOP • STOP • STOP • STOP • STOP

A2 a. $3x$ b. $\dfrac{3}{4y^2}$ c. $\dfrac{7x^2}{2}$ d. $\dfrac{3}{2xy}$

3 The following properties of exponents can be used when simplifying rational expressions where $x,y \in R$, $m,n \in N$, $x \neq 0$, $y \neq 0$:

1. $\dfrac{x^m}{x^n} = x^{m-n}$ when m is greater than n

2. $\dfrac{x^m}{x^n} = \dfrac{1}{x^{n-m}}$ when n is greater than m

3. $\dfrac{x^m}{x^m} = 1$

4. $\left(\dfrac{x}{y}\right)^m = \dfrac{x^m}{y^m}$

Some examples are:

1. $\dfrac{x^5}{x^2} = x^{5-2} = x^3$

$\left(\textit{Note: } \dfrac{x^5}{x^2} = \dfrac{x \cdot x \cdot x \cdot x \cdot x}{x \cdot x} = x^3\right)$

2. $\dfrac{x^3}{x^7} = \dfrac{1}{x^{7-3}} = \dfrac{1}{x^4}$

$\left(\textit{Note: } \dfrac{x^3}{x^7} = \dfrac{x \cdot x \cdot x}{x \cdot x \cdot x \cdot x \cdot x \cdot x \cdot x} = \dfrac{1}{x^4}\right)$

3. $\dfrac{x^3}{x^3} = 1$

4. $\dfrac{(x+7)^3}{(x+7)^2} = (x+7)^{3-2}$
$= (x+7)^1 = x + 7$

5. $\dfrac{2x-1}{(2x-1)^2} = \dfrac{(2x-1)^1}{(2x-1)^2} = \dfrac{1}{(2x-1)^{2-1}} = \dfrac{1}{2x-1}$

6. $\dfrac{x-3}{x-3} = 1$

7. $\left(\dfrac{2x^3}{3y}\right)^2 = \dfrac{(2x^3)^2}{(3y)^2} = \dfrac{4x^6}{9y^2}$

Do as much mentally as possible.

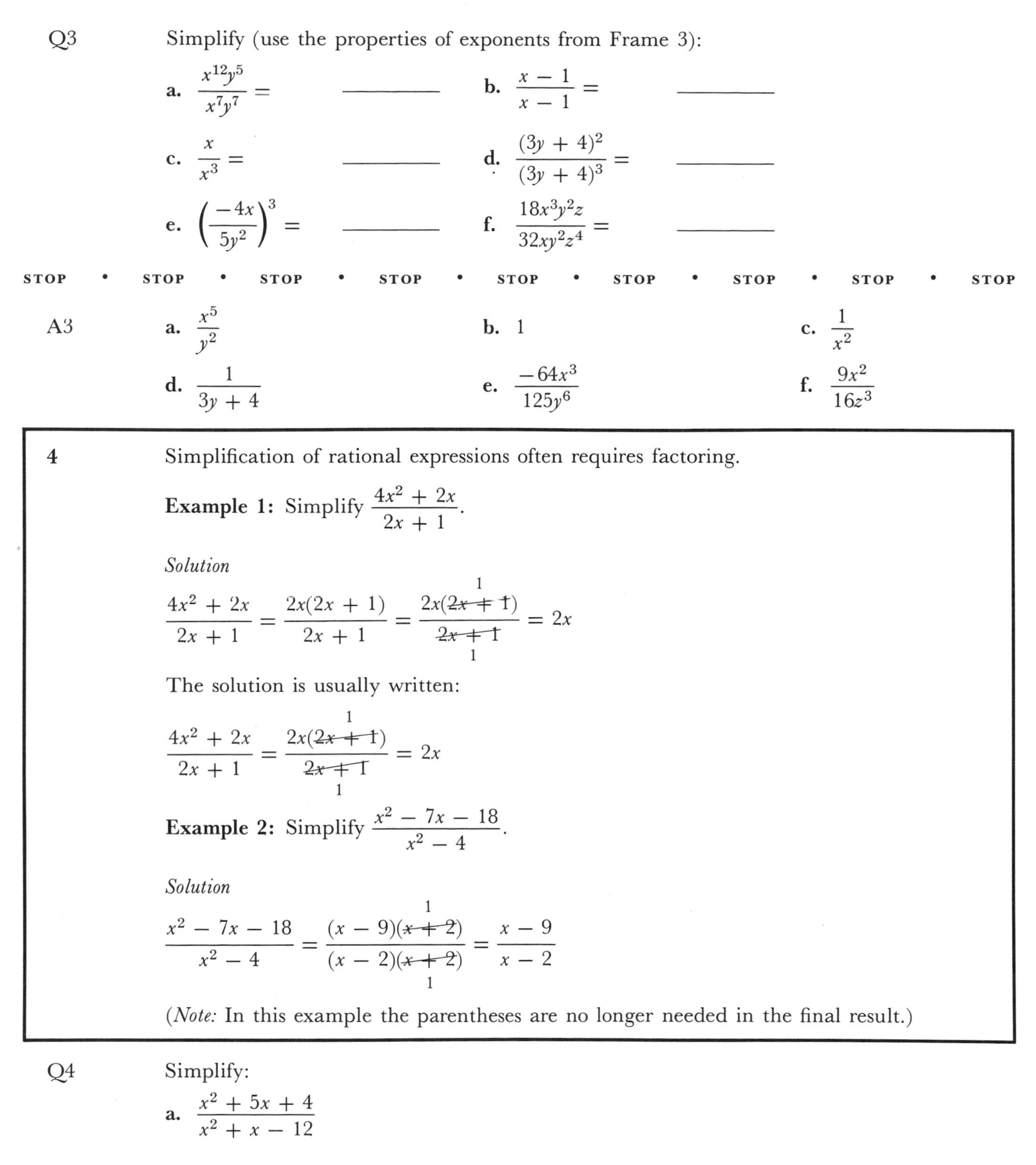

Q3 Simplify (use the properties of exponents from Frame 3):

a. $\frac{x^{12}y^5}{x^7y^7} =$ ________ b. $\frac{x-1}{x-1} =$ ________

c. $\frac{x}{x^3} =$ ________ d. $\frac{(3y+4)^2}{(3y+4)^3} =$ ________

e. $\left(\frac{-4x}{5y^2}\right)^3 =$ ________ f. $\frac{18x^3y^2z}{32xy^2z^4} =$ ________

STOP • STOP • STOP • STOP • STOP • STOP • STOP • STOP • STOP

A3 a. $\frac{x^5}{y^2}$ b. 1 c. $\frac{1}{x^2}$

d. $\frac{1}{3y+4}$ e. $\frac{-64x^3}{125y^6}$ f. $\frac{9x^2}{16z^3}$

4 Simplification of rational expressions often requires factoring.

Example 1: Simplify $\frac{4x^2+2x}{2x+1}$.

Solution

$$\frac{4x^2+2x}{2x+1} = \frac{2x(2x+1)}{2x+1} = \frac{2x\overset{1}{\cancel{(2x+1)}}}{\underset{1}{\cancel{2x+1}}} = 2x$$

The solution is usually written:

$$\frac{4x^2+2x}{2x+1} = \frac{2x\overset{1}{\cancel{(2x+1)}}}{\underset{1}{\cancel{2x+1}}} = 2x$$

Example 2: Simplify $\frac{x^2-7x-18}{x^2-4}$.

Solution

$$\frac{x^2-7x-18}{x^2-4} = \frac{(x-9)\overset{1}{\cancel{(x+2)}}}{(x-2)\underset{1}{\cancel{(x+2)}}} = \frac{x-9}{x-2}$$

(*Note:* In this example the parentheses are no longer needed in the final result.)

Q4 Simplify:

a. $\frac{x^2+5x+4}{x^2+x-12}$

b. $\frac{5x-15}{x^2-9x+18}$

c. $\frac{3x^2-3y^2}{x^3+y^3}$

d. $\dfrac{-6x - 15}{4x^2 + 20x + 25}$

e. $\dfrac{-x - 3}{x + 3}$

STOP • STOP • STOP • STOP • STOP • STOP • STOP • STOP • STOP

A4 a. $\dfrac{x+1}{x-3}$: $\dfrac{(x+1)(x+4)}{(x-3)(x+4)}$ b. $\dfrac{5}{x-6}$: $\dfrac{5(x-3)}{(x-6)(x-3)}$

c. $\dfrac{3(x-y)}{x^2 - xy + y^2}$: $\dfrac{3(x+y)(x-y)}{(x+y)(x^2 - xy + y^2)}$ (*Note:* It is common to leave answers in factored form. However, $\dfrac{3x - 3y}{x^2 - xy + y^2}$ is in simplest form.)

d. $\dfrac{-3}{2x+5}$: $\dfrac{-3(2x+5)}{(2x+5)^2}$ e. -1: $\dfrac{-1(x+3)}{x+3}$

5 The multiplication property of -1 states:

$-1x = -x \quad$ for all $x \in R$

You should observe its application to rational expressions.

$-1(a - b) = -(a - b) = -a + b = b + {}^{-}a = b - a$

Therefore, $a - b$ and $b - a$ are opposites of each other. Since the quotient of opposites is -1,

$$\frac{a-b}{b-a} = \frac{b-a}{a-b} = -1, \qquad a \neq b$$

The above fact can be shown algebraically.

$$\frac{b-a}{a-b} = \frac{-a+b}{a-b} = \frac{-(a-b)}{a-b} = \frac{-1(a-b)}{a-b} = -1$$

Examples:

$\dfrac{x-2}{2-x} = -1$ $\qquad$ $\dfrac{m-n}{n-m} = -1$

$\dfrac{2x-y}{y-2x} = -1$ $\qquad$ $\dfrac{-(x-2)}{x-2} = -1$

$\dfrac{-x-2}{x+2} = \dfrac{-(x+2)}{x+2} = -1$ $\qquad$ $\dfrac{-(5+x)}{x+5} = -1$

Q5 Simplify:

a. $\dfrac{5-2x}{2x-5} =$ ______ b. $\dfrac{x+2}{2+x} =$ ______

c. $\dfrac{-(x-y)}{x-y} =$ ______ d. $\dfrac{-9-2y}{9+2y} =$ ______

e. $\dfrac{n-m}{m-n} =$ ______ f. $\dfrac{-m-n}{m+n} =$ ______

STOP • STOP • STOP • STOP • STOP • STOP • STOP • STOP • STOP

A5

a. -1 b. 1: $x + 2 = 2 + x$ c. -1

d. -1: $-9 - 2y = -(9 + 2y)$ e. -1 f. -1

6

Confusion often arises concerning operations with expressions like $a - b$, $a + b$, and $b - a$. The expressions $a - b$ and $a + b$ are *conjugates* of each other. The product of conjugates is the difference of two squares. That is,

$$(a + b)(a - b) = a^2 - b^2$$

The quotient of conjugates, $\frac{a - b}{a + b}$ or $\frac{a + b}{a - b}$, cannot be simplified. The expressions $a - b$ and $b - a$ are *opposites* of each other. The sum of opposites is zero. That is,

$$(a - b) + (b - a) = 0$$

The quotient of opposites is -1. That is,

$$\frac{a - b}{b - a} = \frac{b - a}{a - b} = -1$$

Other operations that involve conjugates or opposites must be calculated in the same manner as with any other polynomials.

Q6

Simplify, if possible; otherwise, write "cannot be simplified."

a. $\frac{x^2 - 3x - 10}{-x^2 + 3x + 10} =$ ______________

b. $\frac{x - 2y}{2y - x} =$ ______________

c. $\frac{x + 2y}{x - 2y} =$ ______________

d. $\frac{x + 2y}{2y + x} =$ ______________

STOP • STOP • STOP • STOP • STOP • STOP • STOP • STOP • STOP • STOP

A6

a. -1 b. -1

c. cannot be simplified d. 1

Q7

Simplify:

a. $\frac{2x^2 - x - 1}{x^3 - 1}$

b. $\frac{x^2 + x}{x}$

c. $\frac{25x^2 - 49y^2}{25x^2 - 70xy + 49y^2}$

d. $\frac{m^3 + n^3}{m^2 - 2mn - 3n^2}$

e. $\dfrac{2y^2 - 7y + 6}{3y - 6}$

f. $\dfrac{x^4 + 4x^2y^2 - 21y^4}{x^4 - 9y^4}$

STOP • STOP • STOP • STOP • STOP • STOP • STOP • STOP • STOP

A7 a. $\dfrac{2x + 1}{x^2 + x + 1}$ b. $x + 1$ c. $\dfrac{5x + 7y}{5x - 7y}$ d. $\dfrac{m^2 - mn + n^2}{m - 3n}$

e. $\dfrac{2y - 3}{3}$ f. $\dfrac{x^2 + 7y^2}{x^2 + 3y^2}$

7 Each answer above is reduced to *lowest terms;* that is, *no common factor* remains in both the numerator and denominator. Each answer is also in simplest form; that is, no parentheses are present. In this text, answers involving rational expressions may be left with parentheses as long as the result is in lowest terms.

This completes the instruction for this section.

2.8 Exercises

1. The process of reducing means to ____________ both the numerator and the denominator by a common ____________.
2. If neither the numerator nor the denominator contains the same common factor, a rational expression is in ____________. (two words)
3. A rational expression is undefined for any value of a variable that produces a denominator of ____________, because division by ____________ is impossible.

(*Note:* In this exercise, all rational expressions are understood to be defined for permissible replacements of the variables.)

4. Simplify:

a. $\dfrac{54x^2}{10x^5}$ b. $\dfrac{32x^2y^3}{128x^3y^4}$ c. $\dfrac{24x^2yz^3}{36x^2y^2z}$

d. $\dfrac{15mn}{4m}$

In the following problems $a,b,c \in R$ and $x,y,z \in N$ ($>$: is greater than):

e. $\dfrac{a^x}{a^y}$, $x > y$ f. $\dfrac{a^x}{a^y}$, $y > x$ g. $\dfrac{b^x}{b^y}$, $x = y$

*h. $\dfrac{a^x a^y}{a^z}$, $x > y > z$ *i. $\dfrac{(c^x)^y}{c^z}$, $x > y > z$

*j. $\dfrac{a^x b^y c}{ab^z c^y}$, $x > 1, z > y, y > 1$

5. Simplify:

a. $\dfrac{(5x - 2)^2}{(5x - 2)^3}$ b. $\dfrac{(x + y)^3}{(x + y)(x - y)}$

6. Simplify:

a. $\dfrac{x^2 - 11x + 30}{x^2 - 2x - 15}$ b. $\dfrac{5x + 15}{x^2 - 9}$ c. $\dfrac{-2x^2 - 18x - 28}{8 + x^3}$

d. $\dfrac{y^3 - 27}{y - 3}$ e. $\dfrac{9m^2 + 12mn + 4n^2}{9m^2 - 4n^2}$ f. $\dfrac{z - 1}{1 - z}$

g. $\dfrac{-t - 1}{t + 1}$ h. $\dfrac{x^2 - y^2}{2y - 2x}$ i. $\dfrac{2x^2 - 5xy - 12y^2}{4x^2 + 12xy + 9y^2}$

j. $\dfrac{x^4 - x^2y^2 - 2y^4}{x^4 - y^4}$

2.8 Exercise Answers

1. divide, factor
2. lowest terms
3. zero, zero

4. a. $\dfrac{27}{5x^3}$ b. $\dfrac{1}{4xy}$ c. $\dfrac{2z^2}{3y}$

d. $\dfrac{15n}{4}$ e. a^{x-y} f. $\dfrac{1}{a^{y-x}}$

g. 1 *h. a^{x+y-z} *i. c^{xy-z}

*j. $\dfrac{a^{x-1}}{b^{z-y}c^{y-1}}$

5. a. $\dfrac{1}{5x - 2}$ b. $\dfrac{(x + y)^2}{x - y}$

6. a. $\dfrac{x - 6}{x + 3}$ b. $\dfrac{5}{x - 3}$

c. $\dfrac{-2x - 14}{4 - 2x + x^2}$ or $\dfrac{-2(x + 7)}{x^2 - 2x + 4}$ d. $y^2 + 3y + 9$

e. $\dfrac{3m + 2n}{3m - 2n}$ f. -1

g. -1 h. $\dfrac{-x - y}{2}$ or $\dfrac{-(x + y)}{2}$ or $\dfrac{-1}{2}(x + y)$

i. $\dfrac{x - 4y}{2x + 3y}$ j. $\dfrac{x^2 - 2y^2}{x^2 - y^2}$ or $\dfrac{x^2 - 2y^2}{(x + y)(x - y)}$

2.9 Multiplication and Division of Rational Expressions

1

Multiplication of the rational numbers $\frac{r}{s}$ and $\frac{u}{v}$ is defined to be

$$\frac{r}{s} \cdot \frac{u}{v} = \frac{ru}{sv}$$

Common factors to both numerator and denominator may be divided out first.

For example,

$$\frac{4}{15}\cdot\frac{25}{32} = \frac{\overset{1}{\cancel{4}}}{\underset{3}{\cancel{15}}}\cdot\frac{\overset{5}{\cancel{25}}}{\underset{8}{\cancel{32}}} = \underbrace{\frac{1\cdot 5}{3\cdot 8}} = \frac{5}{24}$$

This step is usually done mentally

The final result should be in lowest terms. The same principles apply when multiplying rational expressions.

Example: Simplify $\frac{2x}{5}\cdot\frac{x+7}{4}$.

Solution

$$\frac{2x}{5}\cdot\frac{x+7}{4} = \frac{\overset{1}{\cancel{2}x}}{5}\cdot\frac{x+7}{\underset{2}{\cancel{4}}} = \frac{x(x+7)}{10} = \frac{x^2+7x}{10}$$

$\frac{x(x+7)}{10}$ is in lowest terms; however, it is said to be in factored form

$\frac{x^2+7x}{10}$ is also in lowest terms, but the parentheses have been removed

A rational expression is said to be in simplest form when it is in lowest terms (reduced) and the parentheses have been removed. However, it is common practice to leave rational expressions in factored form. This procedure will be followed in this text.

Q1 Simplify (leave in factored form):

a. $\frac{5x^2}{8}\cdot\frac{x-2}{10}$

b. $\frac{3x+5}{24}\cdot\frac{8x}{15}$

STOP • STOP • STOP • STOP • STOP • STOP • STOP • STOP • STOP

A1 a. $\frac{x^2(x-2)}{16}$: $\frac{\overset{1}{\cancel{5}x^2}}{8}\cdot\frac{x-2}{\underset{2}{\cancel{10}}}$ b. $\frac{x(3x+5)}{45}$: $\frac{3x+5}{\underset{3}{\cancel{24}}}\cdot\frac{\overset{1}{\cancel{8}x}}{15}$

2 Rational expressions involving quantities that are common factors can also be reduced.

Example 1: Simplify $\frac{x-2}{x+2}\cdot\frac{x+2}{(x+2)(x-2)}$.

Solution

$$\frac{\overset{1}{\cancel{x-2}}}{x+2}\cdot\frac{\overset{1}{\cancel{x+2}}}{\underset{1}{\cancel{(x+2)}}\underset{1}{\cancel{(x-2)}}} = \frac{1}{x+2}$$

Example 2: Simplify $\frac{2y(1-y)}{(y-1)(y+1)}\cdot\frac{(y+1)^2}{4y^2}$.

Solution

$$\frac{\overset{1}{\cancel{2y}}\overset{-1}{\cancel{(1-y)}}}{\underset{1}{\cancel{(y-1)}}\underset{1}{\cancel{(y+1)}}} \cdot \frac{\overset{y+1}{\cancel{(y+1)^2}}}{\underset{2y}{\cancel{4y^2}}} = \frac{-(y+1)}{2y} \quad \text{or} \quad \frac{-y-1}{2y}$$

$\left[\textit{Note: } \frac{1-y}{y-1} = -1 \text{ (quotient of opposites).}\right]$

Q2 Simplify:

a. $\frac{x+y}{y-x} \cdot \frac{x-y}{(x+y)^2}$

b. $\frac{(m+2n)(m-2n)}{16m^2n^2} \cdot \frac{m^3n(m^2-2mn+4n^2)}{(m+2n)(m^2-2mn+4n^2)}$

STOP • STOP • STOP • STOP • STOP • STOP • STOP • STOP • STOP

A2 a. $\frac{-1}{x+y}$: $\frac{\overset{1}{\cancel{x+y}}}{\underset{1}{\cancel{y-x}}} \cdot \frac{\overset{-1}{\cancel{x-y}}}{\underset{x+y}{\cancel{(x+y)^2}}}$ b. $\frac{m(m-2n)}{16n}$

3 Factoring is often required before the simplification of the product of two rational expressions can be completed.

Example 1: Simplify $\frac{x^2-y^2}{x^3+y^3} \cdot \frac{2x^2-2xy+2y^2}{2x^2-xy-y^2}$.

Solution

$$\frac{x^2-y^2}{x^3+y^3} \cdot \frac{2x^2-2xy+2y^2}{2x^2-xy-y^2} = \frac{\overset{1}{\cancel{(x+y)}}\overset{1}{\cancel{(x-y)}}}{\underset{1}{\cancel{(x+y)}}\underset{1}{\cancel{(x^2-xy+y^2)}}} \cdot \frac{2\overset{1}{\cancel{(x^2-xy+y^2)}}}{\underset{1}{\cancel{(x-y)}}(2x+y)}$$

$$= \frac{2}{2x+y}$$

Example 2: Simplify $\frac{y^2-14y+49}{7xy-xy^2} \cdot \frac{5y-35}{y^2-10y+21}$.

Solution

$$\frac{\overset{\overset{-1}{\cancel{(y-7)}}(y-7)}{\cancel{(y-7)^2}}}{xy\underset{1}{\cancel{(7-y)}}} \cdot \frac{5\overset{1}{\cancel{(y-7)}}}{\underset{1}{\cancel{(y-7)}}(y-3)} = \frac{-5(y-7)}{xy(y-3)}$$

Other acceptable forms of the above answer are $\frac{5(7-y)}{xy(y-3)}$ and $\frac{5(y-7)}{xy(3-y)}$.

Q3 Simplify:
(*Suggestion:* Do factoring work on separate paper and transfer results to this work space.)

a. $\dfrac{x^2 - 2xy + y^2}{x^2 - y^2} \cdot \dfrac{x^3 + y^3}{x^2 - 3xy + 2y^2}$

b. $\dfrac{y^2 - y - 12}{y^3 - 2y^2 + 5y - 10} \cdot \dfrac{10 - 5y}{y^2 - 9}$

c. $\dfrac{x^4 - y^4}{x^3 - y^3} \cdot \dfrac{x^3 + x^2y + xy^2}{x^2 + xy}$

d. $\dfrac{4 - x^2}{15 - 8x + x^2} \cdot \dfrac{25 - 10x + x^2}{x^2 - 4}$

STOP • STOP • STOP • STOP • STOP • STOP • STOP • STOP • STOP

A3

a. $\dfrac{x^2 - xy + y^2}{x - 2y}$: $\dfrac{\overset{\overset{1}{\cancel{x-y}}}{\cancel{(x-y)^2}}}{\underset{1}{\cancel{(x+y)}}\underset{1}{\cancel{(x-y)}}} \cdot \dfrac{\overset{1}{\cancel{(x+y)}}(x^2 - xy + y^2)}{\underset{1}{\cancel{(x-y)}}(x - 2y)}$

b. $\dfrac{-5(y-4)}{(y^2+5)(y-3)}$ or $\dfrac{5(4-y)}{(y^2+5)(y-3)}$ or $\dfrac{5(y-4)}{(y^2+5)(3-y)}$:

$\dfrac{\overset{1}{\cancel{(y+3)}}(y-4)}{(y^2+5)\underset{1}{\cancel{(y-2)}}} \cdot \dfrac{5\overset{-1}{\cancel{(2-y)}}}{\underset{1}{\cancel{(y+3)}}(y-3)}$

c. $x^2 + y^2$: $\dfrac{(x^2+y^2)\overset{1}{\cancel{(x+y)}}\overset{1}{\cancel{(x-y)}}}{\underset{1}{\cancel{(x-y)}}\underset{1}{\cancel{(x^2+xy+y^2)}}} \cdot \dfrac{\overset{1}{\cancel{x}}\overset{1}{\cancel{(x^2+xy+y^2)}}}{\underset{1}{\cancel{x}}\underset{1}{\cancel{(x+y)}}}$

d. $\dfrac{x-5}{3-x}$: $\dfrac{\overset{1}{\cancel{(2+x)}}\overset{-1}{\cancel{(2-x)}}}{\underset{1}{\cancel{(5-x)}}(3-x)} \cdot \dfrac{\overset{5-x}{\cancel{(5-x)^2}}}{\underset{1}{\cancel{(x+2)}}\underset{1}{\cancel{(x-2)}}}$

---

4 The definition of division of real numbers states: If $r,s,u,v \in R$:

$$\frac{r}{s} \div \frac{u}{v} = \frac{r}{s} \cdot \frac{v}{u} \qquad suv \neq 0$$

(*Note:* "$suv \neq 0$" is the same as "$s \neq 0$ and $u \neq 0$ and $v \neq 0$." $\frac{v}{u}$ is the reciprocal of $\frac{u}{v}$, $\frac{u}{v} \neq 0$.)

The division of rational expressions also involves the definition of division of real numbers. That is, to divide two rational expressions, multiply the dividend (first expression) by the reciprocal of the divisor (second expression).

Example 1: Simplify $\dfrac{x-2}{3-x} \div \dfrac{4-x^2}{x^2-9}$.

Solution

$$\frac{x-2}{3-x}\cdot\frac{x^2-9}{4-x^2} = \frac{\overset{-1}{\cancel{x-2}}}{\underset{1}{\cancel{3-x}}}\cdot\frac{(x+3)\overset{-1}{\cancel{(x-3)}}}{(2+x)\underset{1}{\cancel{(2-x)}}} = \frac{x+3}{x+2}$$

Example 2: Simplify $\dfrac{15x^2y^3}{x^3+8y^3} \div \dfrac{60xy^5}{x^2+4xy+4y^2}$.

Solution

$$\frac{\overset{1\ x\ 1}{\cancel{15x^2y^3}}}{\underset{1}{\cancel{(x+2y)}}(x^2-2xy+4y^2)}\cdot\frac{\overset{x+2y}{\cancel{(x+2y)^2}}}{\underset{4\ 1\ y^2}{\cancel{60xy^5}}} = \frac{x(x+2y)}{4y^2(x^2-2xy+4y^2)}$$

(*Note:* It is common practice to form the reciprocal and factor in the same step.)

Q4 Simplify:

a. $\dfrac{y^3-1}{y-1} \div \dfrac{y^2+y+1}{y^2}$

b. $\dfrac{m^2-n^2}{m^2+2mn+n^2} \div \dfrac{m-n}{m+n}$

c. $\dfrac{6x^2+7x-20}{3x^2-10x+8} \div \dfrac{2x+5}{2-x}$

d. $\dfrac{3x-3y}{x^2y^2} \div \dfrac{x^2-xy}{xy}$

STOP • STOP • STOP • STOP • STOP • STOP • STOP • STOP • STOP

A4 a. y^2: $\dfrac{\overset{1}{\cancel{(y-1)}}\overset{1}{\cancel{(y^2+y+1)}}}{\underset{1}{\cancel{y-1}}}\cdot\dfrac{y^2}{\underset{1}{\cancel{y^2+y+1}}}$

b. 1: $\dfrac{\overset{1}{\cancel{(m+n)}}\overset{1}{\cancel{(m-n)}}}{\underset{\underset{1}{m+n}}{\cancel{(m+n)^2}}} \cdot \dfrac{\overset{1}{\cancel{m+n}}}{\underset{1}{\cancel{m-n}}}$

c. -1: $\dfrac{\overset{1}{\cancel{(2x+5)}}\overset{1}{\cancel{(3x-4)}}}{\underset{1}{\cancel{(x-2)}}\underset{1}{\cancel{(3x-4)}}} \cdot \dfrac{\overset{-1}{\cancel{2-x}}}{\underset{1}{\cancel{2x+5}}}$

d. $\dfrac{3}{x^2y}$: $\dfrac{3\overset{1}{\cancel{(x-y)}}}{\underset{xy}{\cancel{x^2y^2}}} \cdot \dfrac{\overset{1}{\cancel{xy}}}{x\underset{1}{\cancel{(x-y)}}}$

5 Many rational expressions cannot be simplified. For example, $\dfrac{x+1}{x}$ is already in simplest form. Occasionally, in more advanced mathematics, it is appropriate (and necessary) to write expressions such as $\dfrac{x+1}{x}$ as the *sum of its individual terms.*

Examples:

1. $\dfrac{x+1}{x} = \underbrace{(x+1) \div x}_{\text{this step is usually done mentally}} = (x+1)\cdot\dfrac{1}{x} = x\cdot\dfrac{1}{x} + 1\cdot\dfrac{1}{x} = 1 + \dfrac{1}{x}$

2. $$\begin{aligned}\frac{x^3 - 10x^2 + 6x - 1}{2x} &= (x^3 - 10x^2 + 6x - 1)\cdot\frac{1}{2x} \\ &= x^3\cdot\frac{1}{2x} - 10x^2\cdot\frac{1}{2x} + 6x\cdot\frac{1}{2x} - 1\cdot\frac{1}{2x} \\ &= \frac{x^2}{2} - 5x + 3 - \frac{1}{2x} \quad \text{or} \quad \frac{1}{2}x^2 - 5x + 3 - \frac{1}{2x}\end{aligned}$$

The work in these examples may be shortened to:

1. $\dfrac{x+1}{x} = \dfrac{x}{x} + \dfrac{1}{x} = 1 + \dfrac{1}{x}$

2. $$\begin{aligned}\frac{x^3 - 10x^2 + 6x - 1}{2x} &= \frac{x^3}{2x} - \frac{10x^2}{2x} + \frac{6x}{2x} - \frac{1}{2x} \\ &= \frac{x^2}{2} - 5x + 3 - \frac{1}{2x}\end{aligned}$$

Q5 Write each rational expression as the sum of its individual terms.

a. $\dfrac{2x+5}{2x}$

b. $\dfrac{x^2 + 7x - 9}{x}$

c. $\dfrac{15y^3 - 6y^2 + 18y + 3}{3y^2}$

STOP • STOP • STOP • STOP • STOP • STOP • STOP • STOP • STOP

A5 **a.** $1 + \dfrac{5}{2x}$ **b.** $x + 7 - \dfrac{9}{x}$ **c.** $5y - 2 + \dfrac{6}{y} + \dfrac{1}{y^2}$

6

The rational expression $\dfrac{x^2 + 7x - 8}{x + 8}$ can be simplified by factoring and reducing.

$$\frac{x^2 + 7x - 8}{x + 8} = \frac{(x + 8)(x - 1)}{x + 8} = x - 1$$

However, $\dfrac{x^2 + 7x - 9}{x + 8}$ is in simplest form. To write $\dfrac{x^2 + 7x - 9}{x + 8}$ as the sum of its individual terms requires a process similar to long division in arithmetic.

Step 1: $x \leftarrow \dfrac{x^2 \text{ (first term of } x^2 + 7x - 9)}{x \text{ (first term of } x + 8)}$

$$x + 8\overline{)x^2 + 7x - 9}$$

Step 2: $x^2 + 8x \leftarrow x(x + 8)$

Step 3: $-x \leftarrow (x^2 + 7x) - (x^2 + 8x)$

$$\begin{array}{r} x \phantom{{}+7x-9} \\ x + 8\overline{)x^2 + 7x - 9} \\ \underline{x^2 + 8x} \phantom{{}-9} \\ -x - 9 \end{array}$$

Step 4: $-x - 9 \leftarrow$ bring down next term of dividend

Step 5: $x - 1 \leftarrow -1 = \dfrac{-x \text{ (first term of } -x - 9)}{x \text{ (first term of } x + 8)}$

$$\begin{array}{r} x - 1 \phantom{{}-9} \\ x + 8\overline{)x^2 + 7x - 9} \\ \underline{x^2 + 8x} \phantom{{}-9} \\ -x - 9 \end{array}$$

Step 6: $-x - 8 \leftarrow -1(x + 8)$

Step 7: $-1 \leftarrow (-x - 9) - (-x - 8)$

Step 8:

$$\begin{array}{r} x - 1 + \dfrac{-1}{x + 8} \\ x + 8\overline{)x^2 + 7x - 9} \\ \underline{x^2 + 8x} \phantom{{}-9} \\ -x - 9 \\ \underline{-x - 8} \\ -1 \end{array}$$

$x - 1$: quotient; -1: remainder; $x + 8$: divisor; $x^2 + 7x - 9$: dividend; divisor $\rightarrow x + 8$

Note that each partial quotient is formed by dividing by the *first term* of the divisor. Applying the definition of subtraction,

$$x - 1 + \frac{-1}{x + 8} = x - 1 - \frac{1}{x + 8}$$

Hence

$$\frac{x^2 + 7x - 9}{x + 8} = x - 1 - \frac{1}{x + 8}$$

This result can be checked by multiplying the quotient times the divisor and adding the remainder (if any) to obtain the dividend.

$$(x + 8)(x - 1) + {}^{-}1 = (x^2 + 7x - 8) + {}^{-}1$$
$$= x^2 + 7x - 9$$

Q6 **a.** Use the long-division process of Frame 6 to determine the individual terms of $\frac{x^2 - 3x + 5}{x - 2}$.

b. Check the result from Q6a.

STOP • STOP • STOP • STOP • STOP • STOP • STOP • STOP • STOP

A6 **a.** $x - 1 + \frac{3}{x - 2}$:

$$\begin{array}{r} x - 1 \\ x - 2\overline{)x^2 - 3x + 5} \\ \underline{x^2 - 2x} \\ -x + 5 \\ \underline{-x + 2} \\ 3 \end{array}$$

b. $(x - 2)(x - 1) + 3$
$(x^2 - 3x + 2) + 3$
$x^2 - 3x + 5$

Q7 Determine the individual terms of:

a. $\frac{x^2 + 9x - 3}{x + 5}$ **b.** $\frac{4x^2 - 3x - 10}{4x + 5}$

STOP • STOP • STOP • STOP • STOP • STOP • STOP • STOP • STOP

A7

a. $x + 4 - \dfrac{23}{x + 5}$:

$$\begin{array}{r} x + 4 \\ x + 5\overline{)x^2 + 9x - 3} \\ \underline{x^2 + 5x} \\ 4x - 3 \\ \underline{4x + 20} \\ -23 \end{array}$$

b. $x - 2$:

$$\begin{array}{r} x - 2 \\ 4x + 5\overline{)4x^2 - 3x - 10} \\ \underline{4x^2 + 5x} \\ -8x - 10 \\ \underline{-8x - 10} \\ 0 \end{array}$$

(*Note:* The fact that there is a zero remainder indicates that the divisor is a factor of the dividend.)

7

To write the individual terms of $\dfrac{x^3 + 2x^2 - 5}{x + 3}$ requires that space for missing terms be provided for during the division process.

$$\begin{array}{r} x^2 - x + 3 \\ x + 3\overline{)x^3 + 2x^2 + 0x - 5} \\ \underline{x^3 + 3x^2} \\ -x^2 + 0x \\ \underline{-x^2 - 3x} \\ 3x - 5 \\ \underline{3x + 9} \\ -14 \end{array}$$

(*Note:* Adding $0x$ does not change the value of the dividend.)

The division continues until all terms of the dividend have been used. Hence,

$$\frac{x^3 + 2x^2 - 5}{x + 3} = x^2 - x + 3 - \frac{14}{x + 3}$$

Q8

Determine the individual terms of:

a. $\dfrac{y^3 - 2y^2 + 5y - 10}{y^2 + 5}$

b. $\dfrac{8y^3 - 27}{2y - 3}$

STOP • STOP • STOP • STOP • STOP • STOP • STOP • STOP • STOP

A8

a. $y - 2$:

$$\begin{array}{r} y - 2 \\ y^2 + 5\overline{)y^3 - 2y^2 + 5y - 10} \\ \underline{y^3 \qquad\quad + 5y \quad\;\;} \\ -2y^2 \qquad\; - 10 \\ \underline{-2y^2 \qquad\; - 10} \\ 0 \end{array}$$

b. $4y^2 + 6y + 9$

This completes the instruction for this section.

2.9 Exercises

1. Simplify:

a. $\frac{32x^2}{44} \cdot \frac{10x - 4}{24x}$

b. $\frac{x^2 - 3x + 2}{x^2 - 1} \cdot \frac{x^2 + 4x + 3}{4x - 8}$

c. $\frac{4x^2 + 5}{x^2 - 12x + 20} \cdot \frac{x^2 - 10x}{4x^2 + 5}$

d. $\frac{x - 5}{5 - x} \cdot \frac{x + 2}{2 + x}$

e. $\frac{2x}{x + 4} \div \frac{8x^2}{x^2 - 16}$

f. $\frac{x + 2y}{x - 3y} \cdot \frac{x^2 + 2xy - 15y^2}{x^2 + 6xy + 8y^2}$

g. $\frac{m^2 - 9n^2}{m + 4n} \cdot \frac{m - 4n}{m^2 - 6mn + 9n^2}$

h. $\frac{1 - x^3}{x - 1} \div \frac{1 + x^3}{x + 1}$

i. $\frac{x^4 - 16}{x^2 - 2x} \cdot \frac{8x^3}{x^2 + 4}$

j. $\frac{x^2 + 2x + 1}{x^3 + x^2 - 2x - 2} \div \frac{x^2 + 2}{x^4 - 4}$

k. $\frac{6y^2 + 7y - 3}{1 - 3y} \div \frac{9y^2 - 6y + 1}{9y^2 - 1}$

l. $\frac{x + 1}{x} \cdot \frac{y}{y - 1}$

2. Determine the individual terms of each rational expression:

a. $\frac{x^2 - x + 1}{x}$

b. $\frac{3x + 7}{x}$

c. $\frac{16y^3 + 8y^2 - 24y + 9}{4y^2}$

d. $\frac{5r^2 - 4r - 3}{2r^2}$

e. $\frac{2x^2 + 5}{x + 2}$

f. $\frac{3x^3 - 4x^2 - 5x + 6}{3x - 1}$

g. $\frac{x^2 - 17x + 49}{x - 7}$

h. $\frac{27y^3 - 125}{3y - 5}$

i. $\frac{25m^2 + 80m + 64}{5m + 8}$

*j. $\frac{121 - 100x^2}{11 + 8x}$

k. $\frac{x^3 + 2xy^2 + 3y^3}{x + y}$

*l. $\frac{x^4 + 4y^4}{x^2 + 2xy + 2y^2}$

2.9 Exercise Answers

1. a. $\frac{2x(5x - 2)}{33}$ or $\frac{10x^2 - 4x}{33}$

b. $\frac{x + 3}{4}$

c. $\frac{x}{x - 2}$

d. -1

e. $\frac{x - 4}{4x}$

f. $\frac{x + 5y}{x + 4y}$

g. $\frac{m^2 - mn - 12n^2}{m^2 + mn - 12n^2}$

h. $\frac{-(1 + x + x^2)}{1 - x + x^2}$ or $\frac{-1 - x - x^2}{1 - x + x^2}$

i. $8x^2(x + 2)$ or $8x^3 + 16x^2$

j. $x + 1$

k. $\frac{(2y + 3)(3y + 1)}{1 - 3y}$ or $\frac{6y^2 + 11y + 3}{1 - 3y}$

l. $\frac{y(x + 1)}{x(y - 1)}$ or $\frac{xy + y}{xy - x}$

2. a. $x - 1 + \frac{1}{x}$

b. $3 + \frac{7}{x}$

c. $4y + 2 - \frac{6}{y} + \frac{9}{4y^2}$

d. $\frac{5}{2} - \frac{2}{r} - \frac{3}{2r^2}$

e. $2x - 4 + \frac{13}{x + 2}$

f. $x^2 - x - 2 + \frac{4}{3x - 1}$

g. $x - 10 - \frac{21}{x - 7}$

h. $9y^2 + 15y + 25$

i. $5m + 8$

*j. $11 - 8x - \frac{36x^2}{11 + 8x}$

k. $x^2 - xy + 3y^2$

*l. $x^2 - 2xy + 2y^2$

2.10 Addition and Subtraction of Rational Expressions

1

To find the sum or difference of two rational numbers with the same denominator you can use either of the following generalizations where $x, y, z \in I$, $z \neq 0$:

1. $\frac{x}{z} + \frac{y}{z} = \frac{x + y}{z}$

2. $\frac{x}{z} - \frac{y}{z} = \frac{x - y}{z}$

Examples:

$$\frac{1}{7} + \frac{4}{7} = \frac{1 + 4}{7} = \frac{5}{7} \qquad \frac{-2}{5} - \frac{3}{5} = \frac{-2 - 3}{5} = \frac{-5}{5} = -1$$

The same principles apply to the sum or difference of rational expressions with like denominators.

Examples:

$$\frac{x}{9} + \frac{2x}{9} = \frac{x + 2x}{9} = \frac{3x}{9} = \frac{x}{3}$$

$$\frac{5}{x} - \frac{3}{x} = \frac{5 - 3}{x} = \frac{2}{x}$$

$$\frac{2x + 1}{y} + \frac{x - 3}{y} = \frac{2x + 1 + x - 3}{y} = \frac{3x - 2}{y}$$

$$\frac{4x + 7}{x + 2} - \frac{2x + 3}{x + 2} = \frac{4x + 7 - (2x + 3)}{x + 2} = \frac{4x + 7 - 2x - 3}{x + 2}$$

$$= \frac{2x + 4}{x + 2} = \frac{2\overset{1}{\cancel{(x + 2)}}}{\underset{1}{\cancel{x + 2}}} = 2$$

Q1

Find:

a. $\frac{5x}{12} + \frac{7}{12}$

b. $\frac{11}{y} - \frac{6}{y}$

c. $\dfrac{3m-2}{n} - \dfrac{m+3}{n}$

d. $\dfrac{x^2+y^2}{x+y} + \dfrac{2xy}{x+y}$

e. $\dfrac{x^2}{x(x-2)} - \dfrac{x}{x(x-2)}$

STOP • STOP • STOP • STOP • STOP • STOP • STOP • STOP • STOP

A1 a. $\dfrac{5x+7}{12}$ b. $\dfrac{5}{y}$

c. $\dfrac{2m-5}{n}$: $\dfrac{3m-2-(m+3)}{n} = \dfrac{3m-2-m-3}{n}$

d. $x+y$: $\dfrac{x^2+y^2+2xy}{x+y} = \dfrac{x^2+2xy+y^2}{x+y} = \dfrac{(x+y)^2}{x+y}$

e. $\dfrac{x-1}{x-2}$: $\dfrac{x^2-x}{x(x-2)} = \dfrac{x(x-1)}{x(x-2)}$

2 To find the sum or difference of two rational numbers with unlike denominators you could use either:

1. the least common denominator (LCD) method, or
2. the cross-product rule for addition or subtraction of rational numbers.

That is, where $x,y,u,v \in I$, $uv \neq 0$.

1. $\dfrac{x}{u} + \dfrac{y}{v} = \dfrac{xv+uy}{uv}$

2. $\dfrac{x}{u} - \dfrac{y}{v} = \dfrac{xv-uy}{uv}$

If necessary, see Section 1.5 for a more complete review.

Example: Find $\dfrac{2}{3} + \dfrac{-4}{9}$.

Solution

LCD method (the LCD is 9)	Cross-product rule
$\dfrac{2}{3} + \dfrac{-4}{9}$	$\dfrac{2}{3} + \dfrac{-4}{9}$
$\dfrac{2}{3}\cdot\dfrac{3}{3} + \dfrac{-4}{9}$	$\dfrac{2\cdot 9 + 3\cdot {}^-4}{3\cdot 9}$
$\dfrac{6}{9} + \dfrac{^-4}{9}$	$\dfrac{18 + {}^-12}{27}$
$\dfrac{6 + {}^-4}{9}$	$\dfrac{6}{27}$
$\dfrac{2}{9}$	$\dfrac{2}{9}$

The cross-product rule is not always the most convenient method. It is most convenient when the product of the denominators is the least common denominator.

Q2 Use the method of your choice to find:

a. $\frac{-5}{6} + \frac{-3}{10}$ b. $\frac{7}{9} + \frac{-1}{2}$

STOP • STOP • STOP • STOP • STOP • STOP • STOP • STOP • STOP

A2 a. $\frac{-17}{15}$: $\frac{-25}{30} + \frac{-9}{30}$ b. $\frac{5}{18}$: $\frac{7 \cdot 2 + 9 \cdot {}^{-}1}{9 \cdot 2}$

(*Note:* The LCD method would be used, because $6 \cdot 10$ or 60 is not the least common denominator.)

3 Finding the sum or difference of two rational expressions requires the same arithmetic procedures as the sum or difference of two rational numbers.

Example 1: Find $\frac{2x}{3} + \frac{4x}{5}$.

Solution

LCD method (the LCD is 15)	Cross-product rule
$\frac{2x}{3} + \frac{4x}{5}$	$\frac{2x}{3} + \frac{4x}{5}$
$\frac{2x}{3} \cdot \frac{5}{5} + \frac{4x}{5} \cdot \frac{3}{3}$	$\frac{10x + 12x}{15}$
$\frac{10x}{15} + \frac{12x}{15}$	$\frac{22x}{15}$
$\frac{22x}{15}$	

The advantage of the cross-product rule in this example is evident if the second step is done mentally. That is, $\frac{2x}{3} + \frac{4x}{5} = \frac{22x}{15}$. Think: $10x + 12x = 22x$.

Example 2: Find $\frac{5x}{7} - \frac{3x}{4}$.

Solution

LCD method (the LCD is 28)

$$\frac{5x}{7} - \frac{3x}{4}$$

$$\frac{5x}{7} \cdot \frac{4}{4} - \frac{3x}{4} \cdot \frac{7}{7}$$

$$\frac{20x}{28} - \frac{21x}{28}$$

$$\frac{-x}{28}$$

Cross-product rule

$$\frac{5x}{7} - \frac{3x}{4}$$

$$\frac{-x}{28}$$

Think: $\frac{20x - 21x}{28}$

Again, the cross-product rule would permit the result to be obtained mentally.

Q3 Use the cross-product rule to find:

a. $\frac{4y}{7} + \frac{2y}{5} =$ ______________

b. $\frac{7x}{8} - \frac{4x}{5} =$ ______________

STOP • STOP • STOP • STOP • STOP • STOP • STOP • STOP • STOP

A3 a. $\frac{34y}{35}$: think: $\frac{20y + 14y}{35}$ b. $\frac{3x}{40}$: think: $\frac{35x - 32x}{40}$

4 The cross-product rule for adding or subtracting rational numbers is particularly useful in problems such as those that follow.

Example 1: Find $\frac{x - 2}{4} + \frac{x + 2}{5}$.

Solution

$$\frac{x - 2}{4} + \frac{x + 2}{5}$$

$$\frac{5(x - 2) + 4(x + 2)}{4 \cdot 5}$$

$$\frac{5x - 10 + 4x + 8}{20}$$

$$\frac{9x - 2}{20}$$

Example 2: Find $\frac{2x - y}{2} - \frac{3x + 2y}{3}$.

Solution

$$\frac{3(2x - y) - 2(3x + 2y)}{2 \cdot 3}$$

$$\frac{6x - 3y - 6x - 4y}{6}$$

$$\frac{-7y}{6}$$

As much as possible may be done mentally. However, be careful to avoid errors with signs.

Q4 Find:

a. $\frac{x+3y}{7}+\frac{x-5y}{8}$ b. $\frac{x-7}{3}-\frac{2x+5}{4}$

STOP • STOP • STOP • STOP • STOP • STOP • STOP • STOP • STOP

A4 a. $\frac{15x-11y}{56}$: $\frac{8x+24y+7x-35y}{56}$ b. $\frac{-2x-43}{12}$: $\frac{4x-28-6x-15}{12}$

5 As mentioned, if the product of the denominators is the LCD, the cross-product rule is usually the most convenient method.

Example: Find $\frac{x+3}{12}+\frac{x-2}{18}$.

Solution

LCD method (the LCD is 36)

$$\frac{x+3}{12}+\frac{x-2}{18}$$

$$\frac{x+3}{12}\cdot\frac{3}{3}+\frac{x-2}{18}\cdot\frac{2}{2}$$

$$\frac{3x+9}{36}+\frac{2x-4}{36}$$

$$\frac{5x+5}{36}$$

Cross-product rule

$$\frac{x+3}{12}+\frac{x-2}{18}$$

$$\frac{18(x+3)+12(x-2)}{12\cdot 18}$$

$$\frac{18x+54+12x-24}{216}$$

$$\frac{30x+30}{216}$$

$$\frac{\overset{5}{\cancel{30}}(x+1)}{\underset{36}{\cancel{216}}}=\frac{5(x+1)}{36}$$

$\left[\textit{Note: } \frac{5x+5}{36}=\frac{5(x+1)}{36}\text{, which indicates that } \frac{5x+5}{36} \text{ is in lowest terms.}\right]$ In the event that the product of the denominators is not the LCD, the cross-product rule method requires an additional reducing step.

Q5 Use the LCD method to find:

a. $\dfrac{y-4}{10}+\dfrac{y+6}{15}$ b. $\dfrac{2m+1}{6}-\dfrac{m-5}{4}$

STOP • STOP • STOP • STOP • STOP • STOP • STOP • STOP • STOP

A5 a. $\dfrac{y}{6}$: $\dfrac{y-4}{10}\cdot\dfrac{3}{3}+\dfrac{y+6}{15}\cdot\dfrac{2}{2}$

$\dfrac{3y-12}{30}+\dfrac{2y+12}{30}$

b. $\dfrac{m+17}{12}$: $\dfrac{2m+1}{6}\cdot\dfrac{2}{2}-\dfrac{m-5}{4}\cdot\dfrac{3}{3}$

$\dfrac{4m+2}{12}-\dfrac{3m-15}{12}$

$\dfrac{4m+2}{12}+{}^-\left(\dfrac{3m-15}{12}\right)$

$\dfrac{4m+2}{12}+\dfrac{-3m+15}{12}$

6 Procedures discussed thus far are adequate if the LCD is obvious. However, for more difficult examples, another method of obtaining the LCD is helpful. The LCD of two or more rational expressions is the *least common multiple* (LCM) of the denominators.

Example 1: Find the LCD for $\frac{1}{2}+\frac{1}{3}+\frac{1}{5}$.

Solution

The LCM of 2, 3, and 5 is $2\cdot3\cdot5$ or 30. Therefore, the LCD for $\frac{1}{2}+\frac{1}{3}+\frac{1}{5}$ is 30.

Example 2: Find the LCD for $\frac{1}{12}+\frac{1}{15}$.

Solution

Step 1: Determine the prime factorization of both 12 and 15.

$12 = 2\cdot2\cdot3$
$15 = 3\cdot5$

Step 2: The LCM is the *product* of *different prime factors* that appear the *greatest* number of times in any *one* factorization.

$12 = (2\cdot2)\cdot(3)$ ← 2 appears *twice* as a factor of 12; 3 appears *once* as a factor in both 12 and 15 (circle only one factor of 3)

$15 = 3\cdot(5)$ ← 5 appears *once* as a factor of 15

The LCM of 12 and 15 must contain *two* factors of 2, *one* factor of 3, and *one* factor of 5. Therefore, the LCM of 12 and 15 is $2 \cdot 2 \cdot 3 \cdot 5$ or 60. Hence the LCD for $\frac{1}{12} + \frac{1}{15}$ is 60.

Q6 Complete:

a. The LCD of two or more rational expressions is the ________ of the denominators.

b. The LCM of two or more numbers is the ________ of the different ________ factors that appear the ________ number of times in any ________ factorization.

STOP • STOP • STOP • STOP • STOP • STOP • STOP • STOP • STOP

A6 **a.** LCM **b.** product, prime, greatest, one

Q7 Use the method of Frame 6 to find the LCD for $\frac{x+5}{20} + \frac{x-2}{35}$. Determine the prime factorization:

a. 20 ________

b. 35 ________

c. The LCM of 20 and 35 is the product of ________ (word) factors of 2, ________ (word) factor of 5, and ________ (word) factor of 7.

d. The LCM of 20 and 35 is ________.

e. The LCD for $\frac{x+5}{20} + \frac{x-2}{35}$ is ________.

STOP • STOP • STOP • STOP • STOP • STOP • STOP • STOP • STOP

A7 **a.** $2 \cdot 2 \cdot 5$ **b.** $5 \cdot 7$ **c.** two, one, one
d. 140: $2 \cdot 2 \cdot 5 \cdot 7$ **e.** 140

Q8 Find the LCD for:

a. $\frac{5x}{24} + \frac{7x}{36}$

b. $\frac{x-6}{18} - \frac{x+7}{40}$

STOP • STOP • STOP • STOP • STOP • STOP • STOP • STOP • STOP

A8 **a.** 72: $24 = 2 \cdot 2 \cdot 2 \cdot 3$
$36 = 2 \cdot 2 \cdot 3 \cdot 3$
$\text{LCM} = 2 \cdot 2 \cdot 2 \cdot 3 \cdot 3$

b. 360: $18 = 2 \cdot 3 \cdot 3$
$40 = 2 \cdot 2 \cdot 2 \cdot 5$
$\text{LCM} = 2 \cdot 2 \cdot 2 \cdot 3 \cdot 3 \cdot 5$

7

When variables appear in the denominator, the same method can be used to find the LCD.

Example 1: Find the LCD for $\frac{7}{10x} - \frac{5}{6x}$.

Solution

Find the LCM of $10x$ and $6x$.

Step 1: Determine the prime factorization of both terms.

$$10x = 2 \cdot 5 \cdot x$$
$$6x = 2 \cdot 3 \cdot x$$

Step 2: The LCM is the product of the different prime factors that appear the greatest number of times in any one factorization.

Therefore, the LCM of $10x$ and $6x$ is $2 \cdot 3 \cdot 5 \cdot x$ or $30x$. Hence the LCD for $\frac{7}{10x} - \frac{5}{6x}$ is $30x$.

Example 2: Find the LCD for $\frac{5}{2x^2} + \frac{3}{4x}$.

Solution

$$2x^2 = 2 \cdot x \cdot x$$
$$4x = 2 \cdot 2 \cdot x$$

The LCM of $2x^2$ and $4x$ is $2 \cdot 2 \cdot x \cdot x$ or $4x^2$. Hence the LCD for $\frac{5}{2x^2} + \frac{3}{4x}$ is $4x^2$.

Q9

Find the LCD for:

a. $\frac{7}{15x} + \frac{23}{21x}$

b. $\frac{3y + 1}{3y} - \frac{y + 2}{9y^2}$

c. $\frac{1}{6x} + \frac{x - 2}{10x^2}$

d. $\frac{1}{4x} + \frac{1}{3y}$

e. $\frac{2}{x^2y} - \frac{3}{xy^3}$

f. $\frac{m + 9}{12m^2} - \frac{2m + 7}{14m^3}$

STOP • STOP • STOP • STOP • STOP • STOP • STOP • STOP • STOP

A9

a. $105x$: $15x = 3 \cdot 5 \cdot x$
$21x = 3 \cdot 7 \cdot x$
$\text{LCM} = 3 \cdot 5 \cdot 7 \cdot x$

b. $9y^2$: $3y = 3 \cdot y$
$9y^2 = 3 \cdot 3 \cdot y \cdot y$
$\text{LCM} = 3 \cdot 3 \cdot y \cdot y$
(*Note:* $9y^2$ is a multiple of $3y$.)

c. $30x^2$: $6x = 2 \cdot 3 \cdot x$
$10x^2 = 2 \cdot 5 \cdot x \cdot x$
$\text{LCM} = 2 \cdot 3 \cdot 5 \cdot x \cdot x$

d. $12xy$: $4x = 2 \cdot 2 \cdot x$
$3y = 3 \cdot y$
$\text{LCM} = 2 \cdot 2 \cdot 3 \cdot x \cdot y$
(*Note:* $12xy$ is the product of the original denominators.)

e. x^2y^3: $x^2y = x \cdot x \cdot y$
$xy^3 = x \cdot y \cdot y \cdot y$
$\text{LCM} = x \cdot x \cdot y \cdot y \cdot y$
(*Note:* The largest exponent on each variable determines the number of factors needed.)

f. $84m^3$: $12m^2 = 2 \cdot 2 \cdot 3m^2$
$14m^3 = 2 \cdot 7m^3$
$\text{LCM} = 2 \cdot 2 \cdot 3 \cdot 7 \cdot m^3$

---

8 **Example:** Find the LCD for $\frac{x+2}{3x-6} + \frac{x+3}{2x-4}$.

Solution

Step 1: Factor both denominators:

$3x - 6 = 3(x - 2)$
$2x - 4 = 2(x - 2)$

Step 2: Form the LCM by indicating the product of the different prime factors that appear the greatest number of times in any one factorization.

$\text{LCM} = 2 \cdot 3(x - 2)$

Hence the LCD for $\frac{x+2}{3x-6} + \frac{x+3}{2x-4}$ is $6(x - 2)$.

Q10 Find the LCD for $\frac{5}{2x+2} - \frac{7}{3x+3}$.

STOP • STOP • STOP • STOP • STOP • STOP • STOP • STOP • STOP

A10 $6(x + 1)$: $2x + 2 = 2(x + 1)$
$3x + 3 = 3(x + 1)$

---

9 **Example:** Find the LCD for $\frac{5x}{x^2-16} + \frac{7x}{x^2+4x}$.

Solution

Step 1: Factor both denominators:

$x^2 - 16 = (x + 4)(x - 4)$

$x^2 + 4x = x(x + 4)$

Step 2: Form the LCM by indicating the product of the different prime factors that appear the greatest number of times in any one factorization.

LCM $= x(x + 4)(x - 4)$

Hence the LCD for $\dfrac{5x}{x^2 - 16} + \dfrac{7x}{x^2 + 4x}$ is $x(x + 4)(x - 4)$. The binomial factors may be commuted.

Q11 Find the LCD for $\dfrac{1}{x^2 - 25} - \dfrac{3x}{5x - 20}$.

STOP • STOP • STOP • STOP • STOP • STOP • STOP • STOP • STOP

A11 $5(x - 4)(x + 5)(x - 5)$: $x^2 - 25 = (x + 5)(x - 5)$
$5x - 20 = 5(x - 4)$

10 **Example:** Find the LCD for $\dfrac{x + 5}{x^2 + 5x + 6} + \dfrac{x + 3}{x^2 - x - 6}$.

Solution

Step 1: Factor both denominators:

$x^2 + 5x + 6 = (x + 2)(x + 3)$
$x^2 - x - 6 = (x - 3)(x + 2)$

Step 2: Form the LCM of the denominators.

LCM $= (x + 2)(x + 3)(x - 3)$

Hence the LCD for $\dfrac{x + 5}{x^2 + 5x + 6} + \dfrac{x + 3}{x^2 - x - 6}$ is $(x + 2)(x + 3)(x - 3)$. The binomial factors may be rearranged in any order.

Q12 Find the LCD for:

a. $\dfrac{x + 1}{x - 1} + \dfrac{x - 1}{x + 1}$

b. $\dfrac{x + 5}{x^2 + 4x + 4} - \dfrac{x - 2}{2x^2 + 5x + 2}$

c. $\dfrac{5}{3x} - \dfrac{2}{5y}$

d. $\dfrac{x + y}{x^2 - 8xy + 15y^2} + \dfrac{x - y}{x^2 - 25y^2}$

STOP • STOP • STOP • STOP • STOP • STOP • STOP • STOP • STOP

A12 a. $(x - 1)(x + 1)$ b. $(x + 2)^2(2x + 1)$
c. $15xy$ d. $(x - 3y)(x + 5y)(x - 5y)$

11 The LCD for $\frac{x+5}{20} + \frac{x-2}{35}$ is 140 (see Q7). To find the sum, proceed as follows:

Step 1: Rewrite each rational expression over the LCD, 140.

$$\frac{x+5}{20} + \frac{x-2}{35}$$

$$\frac{x+5}{20} \cdot \frac{7}{7} + \frac{x-2}{35} \cdot \frac{4}{4}$$

$$\frac{7(x+5)}{140} + \frac{4(x-2)}{140}$$

Step 2: Simplify the numerators of each term.

$$\frac{7x+35}{140} + \frac{4x-8}{140}$$

Step 3: Write the sum (or difference) of the numerators over the LCD.

$$\frac{7x+35+4x-8}{140}$$

Step 4: Simplify the numerator over the LCD.

$$\frac{11x+27}{140}$$

Step 5: Determine whether the result is in simplest form which requires factoring (if possible) and reducing.

$\frac{11x+27}{140}$ is in simplest form

Hence $\frac{x+5}{20} + \frac{x-2}{35} = \frac{11x+27}{140}$.

Q13 Find:

a. $\frac{5x}{24} + \frac{7x}{36}$ (see Q8a)

b. $\frac{7}{15x} + \frac{23}{21x}$ (see Q9a)

STOP • **STOP** • **STOP** • **STOP** • **STOP** • **STOP** • **STOP** • **STOP** • **STOP**

A13 **a.** $\frac{29x}{72}$: $\frac{5x}{24} \cdot \frac{3}{3} + \frac{7x}{36} \cdot \frac{2}{2}$

$$\frac{15x}{72} + \frac{14x}{72}$$

b. $\frac{164}{105x}$: $\frac{7}{15x} \cdot \frac{7}{7} + \frac{23}{21x} \cdot \frac{5}{5}$

$$\frac{49}{105x} + \frac{115}{105x}$$

$$\frac{164}{105x}$$

(*Note:* $164 = 2 \cdot 2 \cdot 41$
$105 = 3 \cdot 5 \cdot 7$
Hence 164 and 105 have no common factors.)

12 When subtracting, special care must be taken to avoid errors with the signs.

Example: Find $\frac{x-6}{18} - \frac{x+7}{40}$.

Solution

The LCD is 360 (see Q8b).

$$\frac{x-6}{18} - \frac{x+7}{40}$$

$$\frac{x-6}{18} \cdot \frac{20}{20} - \frac{x+7}{40} \cdot \frac{9}{9}$$

It is suggested at this point that the expression be rewritten.

$$\frac{20(x-6)}{360} - \frac{9(x+7)}{360}$$

Now, complete the simplification.

$$\frac{20(x-6) - 9(x+7)}{360}$$

$$\frac{20x - 120 - 9x - 63}{360}$$

$$\frac{11x - 183}{360}$$

Since the numerator and denominator have no common factors, the expression is in simplest form. Hence,

$$\frac{x-6}{18} - \frac{x+7}{40} = \frac{11x-183}{360}$$

Q14 Find $\frac{3y+1}{3y} - \frac{y+2}{9y^2}$ (see Q9b).

STOP • STOP • STOP • STOP • STOP • STOP • STOP • STOP • STOP

A14 $\frac{9y^2 + 2y - 2}{9y^2}$: $\frac{3y+1}{3y} \cdot \frac{3y}{3y} - \frac{y+2}{9y^2}$

$$\frac{3y(3y+1)}{9y^2} - \frac{(y+2)}{9y^2}$$

$$\frac{3y(3y+1) - (y+2)}{9y^2}$$

Q15 Find:

a. $\frac{1}{6x} + \frac{x+2}{10x^2}$

b. $\frac{1}{4x} + \frac{1}{3y}$

c. $\frac{2}{x^2y} - \frac{3}{xy^3}$

d. $\frac{m+9}{12m^2} - \frac{2m+7}{14m^3}$

STOP • STOP • STOP • STOP • STOP • STOP • STOP • STOP • STOP

A15 a. $\frac{4x+3}{15x^2}$: $\frac{5x}{30x^2} + \frac{3(x+2)}{30x^2}$

$\frac{5x+3x+6}{30x^2}$

$\frac{8x+6}{30x^2}$

$\frac{2(4x+3)}{30x^2}$

b. $\frac{4x+3y}{12xy}$ (*Note:* The cross-product rule method works well in this example, because the LCD is $12xy$.)

c. $\frac{2y^2-3x}{x^2y^3}$: $\frac{2y^2}{x^2y^3} - \frac{3x}{x^2y^3}$

d. $\frac{7m^2+51m-42}{84m^3}$: $\frac{7m(m+9)}{84m^3} - \frac{6(2m+7)}{84m^3}$

$\frac{7m(m+9)-6(2m+7)}{84m^3}$

$\frac{7m^2+63m-12m-42}{84m^3}$

13

Example: Find $\dfrac{1}{3x+6} - \dfrac{5}{3x-6}$.

Solution

Step 1: Determine the LCD. Recall that the LCD is the LCM of the denominators.

$$3x + 6 = 3(x + 2)$$
$$3x - 6 = 3(x - 2)$$
$$\text{LCD} = 3(x + 2)(x - 2)$$

Step 2: Rewrite original rational expressions over the LCD.

$$\frac{1}{3(x+2)} \cdot \frac{x-2}{x-2} - \frac{5}{3(x-2)} \cdot \frac{x+2}{x+2}$$

Step 3: Rewrite as follows:

$$\frac{x-2}{3(x+2)(x-2)} - \frac{5(x+2)}{3(x+2)(x-2)}$$

Step 4: Continue simplification:

$$\frac{x-2-5(x+2)}{3(x+2)(x-2)}$$

$$\frac{x-2-5x-10}{3(x+2)(x-2)}$$

$$\frac{-4x-12}{3(x+2)(x-2)}$$

Step 5: Check to see whether the expression is in simplest form (factor).

$$\frac{-4(x+3)}{3(x+2)(x-2)}$$

No common factor is present in both numerator and denominator; hence the expression is in lowest terms. The result may be left in factored form.

Q16 Find:

a. $\dfrac{5}{2x+2} - \dfrac{7}{3x+3}$ **b.** $\dfrac{x+2}{3x-6} + \dfrac{x+3}{2x-4}$

STOP • STOP • STOP • STOP • STOP • STOP • STOP • STOP • STOP

A16

a. $\dfrac{1}{6(x+1)}$: $2x + 2 = 2(x+1)$
$3x + 3 = 3(x+1)$
$\text{LCD} = 6(x+1)$

$$\frac{5}{2(x+1)} - \frac{7}{3(x+1)}$$

$$\frac{5}{2(x+1)} \cdot \frac{3}{3} - \frac{7}{3(x+1)} \cdot \frac{2}{2}$$

$$\frac{15}{6(x+1)} - \frac{14}{6(x+1)}$$

b. $\dfrac{5x+13}{6(x-2)}$: $3x - 6 = 3(x-2)$
$2x - 4 = 2(x-2)$
$\text{LCD} = 6(x-2)$

$$\frac{x+2}{3(x-2)} + \frac{x+3}{2(x-2)}$$

$$\frac{x+2}{3(x-2)} \cdot \frac{2}{2} + \frac{x+3}{2(x-2)} \cdot \frac{3}{3}$$

$$\frac{2(x+2)}{6(x-2)} + \frac{3(x+3)}{6(x-2)}$$

14

Example: Find $\dfrac{5x}{x^2 - 16} + \dfrac{7x}{x^2 + 4x}$.

Solution

$x^2 - 16 = (x+4)(x-4)$

$x^2 + 4x = x(x+4)$

$\text{LCD} = x(x+4)(x-4)$

$$\frac{5x}{(x+4)(x-4)} + \frac{7x}{x(x+4)}$$

$$\frac{5x}{(x+4)(x-4)} \cdot \frac{x}{x} + \frac{7x}{x(x+4)} \cdot \frac{x-4}{x-4}$$

$$\frac{5x^2}{x(x+4)(x-4)} + \frac{7x(x-4)}{x(x+4)(x-4)}$$

$$\frac{5x^2 + 7x^2 - 28x}{x(x+4)(x-4)}$$

$$\frac{12x^2 - 28x}{x(x+4)(x-4)}$$

$$\frac{4x(3x-7)}{x(x+4)(x-4)}$$

$$\frac{4(3x-7)}{(x+4)(x-4)}$$

The final result is left in factored form to emphasize the fact that it is in lowest terms.

Q17

Find $\dfrac{1}{x^2 - 25} - \dfrac{3x}{4x - 20}$.

STOP • STOP • STOP • STOP • STOP • STOP • STOP • STOP • STOP

A17 $\frac{-(3x^2 + 15x - 4)}{4(x + 5)(x - 5)}$: $\frac{1}{x^2 - 25} - \frac{3x}{4x - 20}$

$$\frac{1}{(x + 5)(x - 5)} - \frac{3x}{4(x - 5)}$$

$$\text{LCD} = 4(x + 5)(x - 5)$$

$$\frac{1}{(x + 5)(x - 5)} \cdot \frac{4}{4} - \frac{3x}{4(x - 5)} \cdot \frac{(x + 5)}{(x + 5)}$$

$$\frac{4 - 3x(x + 5)}{4(x + 5)(x - 5)}$$

$$\frac{4 - 3x^2 - 15x}{4(x + 5)(x - 5)}$$

$$\frac{-3x^2 - 15x + 4}{4(x + 5)(x - 5)}$$

15 **Example:** Find $\frac{x + 5}{x^2 + 5x + 6} - \frac{x + 3}{x^2 - x - 6}$.

Solution

$$\frac{x + 5}{(x + 2)(x + 3)} - \frac{x + 3}{(x - 3)(x + 2)}$$

$$\text{LCD} = (x + 2)(x + 3)(x - 3)$$

$$\frac{x + 5}{(x + 2)(x + 3)} \cdot \frac{x - 3}{x - 3} - \frac{x + 3}{(x - 3)(x + 2)} \cdot \frac{x + 3}{x + 3}$$

$$\frac{(x + 5)(x - 3) - (x + 3)^2}{(x + 2)(x + 3)(x - 3)}$$

At this point, care must be taken to obtain the proper signs.

$$\frac{x^2 + 2x - 15 - (x^2 + 6x + 9)}{(x + 2)(x + 3)(x - 3)}$$

$$\frac{x^2 + 2x - 15 - x^2 - 6x - 9}{(x + 2)(x + 3)(x - 3)}$$

$$\frac{-4x - 24}{(x + 2)(x + 3)(x - 3)}$$

$$\frac{-4(x + 6)}{(x + 2)(x + 3)(x - 3)}$$

Q18 Find:

a. $\frac{x + 1}{x - 1} + \frac{x - 1}{x + 1}$

b. $\frac{x + 1}{x - 1} - \frac{x - 1}{x + 1}$

c. $\dfrac{x+5}{x^2+4x+4} - \dfrac{x-2}{2x^2+5x+2}$

d. $\dfrac{5}{3x} + \dfrac{2}{5y}$

*e. $\dfrac{x+y}{x^2+2xy-15y^2} + \dfrac{x-y}{x^2-25y^2}$

f. $\dfrac{5}{x^2-6x+8} - \dfrac{2}{3x-12}$

STOP • STOP • STOP • STOP • STOP • STOP • STOP • STOP • STOP

A18 a. $\dfrac{2(x^2+1)}{(x-1)(x+1)}$: $\dfrac{(x+1)^2+(x-1)^2}{(x-1)(x+1)}$

$$\frac{x^2+2x+1+x^2-2x+1}{(x-1)(x+1)}$$

$$\frac{2x^2+2}{(x-1)(x+1)}$$

(*Note:* The cross-product rule was used since the product of $x - 1$ and $x + 1$ is the LCD.)

b. $\dfrac{4x}{(x+1)(x-1)}$: $\dfrac{(x+1)^2-(x-1)^2}{(x-1)(x+1)}$

$$\frac{x^2+2x+1-x^2+2x-1}{(x-1)(x+1)}$$

c. $\dfrac{x^2 + 11x + 9}{(x + 2)^2(2x + 1)}$:

$$\frac{x + 5}{(x + 2)^2} - \frac{x - 2}{(2x + 1)(x + 2)}.$$

$$\frac{x + 5}{(x + 2)^2} \cdot \frac{2x + 1}{2x + 1} - \frac{x - 2}{(2x + 1)(x + 2)} \cdot \frac{x + 2}{x + 2}$$

$$\frac{(x + 5)(2x + 1) - (x - 2)(x + 2)}{(x + 2)^2(2x + 1)}$$

$$\frac{2x^2 + 11x + 5 - (x^2 - 4)}{(x + 2)^2(2x + 1)}$$

$$\frac{2x^2 + 11x + 5 - x^2 + 4}{(x + 2)^2(2x + 1)}$$

d. $\dfrac{25y + 6x}{15xy}$

*e. $\dfrac{2(x^2 - 4xy - y^2)}{(x - 3y)(x + 5y)(x - 5y)}$:

$$\frac{x + y}{(x - 3y)(x + 5y)} + \frac{x - y}{(x + 5y)(x - 5y)}$$

$$\frac{x + y}{(x - 3y)(x + 5y)} \cdot \frac{x - 5y}{x - 5y} + \frac{x - y}{(x + 5y)(x - 5y)} \cdot \frac{x - 3y}{x - 3y}$$

$$\frac{(x + y)(x - 5y) + (x - y)(x - 3y)}{(x - 3y)(x + 5y)(x - 5y)}$$

$$\frac{x^2 - 4xy - 5y^2 + x^2 - 4xy + 3y^2}{(x - 3y)(x + 5y)(x - 5y)}$$

$$\frac{2x^2 - 8xy - 2y^2}{(x - 3y)(x + 5y)(x - 5y)}$$

f. $\dfrac{19 - 2x}{3(x - 4)(x - 2)}$:

$$\frac{5}{(x - 4)(x - 2)} - \frac{2}{3(x - 4)}$$

$$\frac{5}{(x - 4)(x - 2)} \cdot \frac{3}{3} - \frac{2}{3(x - 4)} \cdot \frac{x - 2}{x - 2}$$

$$\frac{15}{3(x - 4)(x - 2)} - \frac{2(x - 2)}{3(x - 4)(x - 2)}$$

$$\frac{15 - 2(x - 2)}{3(x - 4)(x - 2)}$$

16

If opposites are present in the denominators, a simple change can be made to simplify the work.

Example 1: Find $\dfrac{1}{x - 1} + \dfrac{3}{1 - x}$.

Solution

$$\frac{1}{x - 1} + \frac{3}{1 - x} \cdot \frac{-1}{-1}$$

$$\frac{1}{x - 1} + \frac{-3}{x - 1}$$

$$\frac{-2}{x - 1}$$

$\left(\textit{Note: } \dfrac{-2}{x - 1} = \dfrac{-2}{x - 1} \cdot \dfrac{-1}{-1} = \dfrac{2}{1 - x}\right)$

Example 2: Find $\frac{x-3}{5-x} - \frac{x+7}{x-5}$.

Solution

$$\frac{x-3}{5-x} \cdot \frac{-1}{-1} - \frac{x+7}{x-5}$$

$$\frac{3-x}{x-5} - \frac{x+7}{x-5}$$

$$\frac{3-x-(x+7)}{x-5}$$

$$\frac{3-x-x-7}{x-5}$$

$$\frac{-2x-4}{x-5}$$

$$\frac{-2(x+2)}{x-5} \text{ or } \frac{2(x+2)}{5-x}$$

Example 3: Find $\frac{x+2}{x-2} - \frac{x-3}{2-x}$.

Solution

$$\frac{x+2}{x-2} - \frac{x-3}{2-x} \cdot \frac{-1}{-1}$$

$$\frac{x+2}{x-2} - \frac{3-x}{x-2}$$

$$\frac{x+2-(3-x)}{x-2}$$

$$\frac{x+2-3+x}{x-2}$$

$$\frac{2x-1}{x-2} \text{ or } \frac{1-2x}{2-x}$$

Q19 Find:

a. $\frac{x^2+4}{x-7} - \frac{x^2-5}{7-x}$ **b.** $\frac{4}{2-x} + \frac{5}{x-2}$

STOP • STOP • STOP • STOP • STOP • STOP • STOP • STOP • STOP

A19 **a.** $\dfrac{2x^2 - 1}{x - 7}$ or $\dfrac{1 - 2x^2}{7 - x}$ **b.** $\dfrac{1}{x - 2}$

17 More complicated problems can be done using the techniques presented in this section.

Example: Find $\dfrac{1}{x} + \dfrac{x + 2}{x^2 - 1} - \dfrac{5}{x + 1}$.

Solution

$$\left(\frac{1}{x} + \frac{x + 2}{(x + 1)(x - 1)}\right) - \frac{5}{x + 1}$$

Apply the cross-product rule to the first two expressions:

$$\frac{(x + 1)(x - 1) + x(x + 2)}{x(x + 1)(x - 1)} - \frac{5}{x + 1}$$

Rewrite the second expression over the LCD, $x(x + 1)(x - 1)$:

$$\frac{(x + 1)(x - 1) + x(x + 2)}{x(x + 1)(x - 1)} - \frac{5}{x + 1} \cdot \frac{x(x - 1)}{x(x - 1)}$$

Simplify:

$$\frac{(x + 1)(x - 1) + x(x + 2) - 5x(x - 1)}{x(x + 1)(x - 1)}$$

$$\frac{x^2 - 1 + x^2 + 2x - 5x^2 + 5x}{x(x + 1)(x - 1)}$$

$$\frac{-3x^2 + 7x - 1}{x(x + 1)(x - 1)}$$

$$\frac{-(3x^2 - 7x + 1)}{x(x + 1)(x - 1)}$$

$3x^2 - 7x + 1$ is prime; hence the simplification is complete.

A few examples of this degree of difficulty will be starred problems in the exercises.

This completes the instruction for this section.

2.10 Exercises

(All rational expressions are assumed to be defined for permissible replacements of the variables.)

1. Find:

a. $\dfrac{4}{x} + \dfrac{3}{x}$ **b.** $\dfrac{7}{5y} - \dfrac{2}{5y}$

c. $\dfrac{x - 2}{x - 3} - \dfrac{x - 1}{x - 3}$ **d.** $\dfrac{3x + 4}{2x} + \dfrac{x - 3}{2x}$

e. $\dfrac{5x}{x^2 + 2x + 3} + \dfrac{x - 2}{x^2 + 2x + 3}$ **f.** $\dfrac{x^2 + 3x + 1}{x - 3} - \dfrac{x^2 + 2x + 4}{x - 3}$

g. $\dfrac{2x}{x^2 + 2x + 1} - \dfrac{x - 1}{x^2 + 2x + 1}$ **h.** $\dfrac{x}{x^2 - y^2} + \dfrac{y}{x^2 - y^2}$

2. The least common denominator (LCD) is the ______________ (LCM) of the denominators.
3. Find the LCM for:
   a. 12 and 18
   b. 15 and 50
   c. $3x$ and $7y$
   d. $4x$ and $5x$
   e. $x^2 - 4$ and $3x + 6$
   f. $5x^2$ and $15x$
   g. $x^2 - 3x$ and $x^2 - 6x + 9$
   h. $x^2 - 25$ and $x^2 + x - 6$
   i. $x^2 + 7x + 12$ and $x^2 + x - 6$
   j. $3x^2 - 5x - 12$ and $9x^2 - 16$
   *k. $x^3 + y^3$ and $x^2 - y^2$
   *l. $x + 1$, $x^2 - 1$, and $x^2 - 2x + 1$
4. Find:
   a. $\frac{x+5}{5} + \frac{2x-7}{6}$
   b. $\frac{x+4}{3} - \frac{x-3}{4}$
   c. $\frac{1}{5x^2} + \frac{7}{15x}$
   d. $\frac{x-1}{4x} - \frac{x+6}{5x}$
   e. $\frac{5}{3x} + \frac{4}{7y}$
   f. $\frac{x}{y} - \frac{a}{b}$
   g. $\frac{2x}{y} + 5$
   h. $9 - \frac{4y}{5}$
   i. $\frac{x+3}{x^2-4} - \frac{5}{3x+6}$
   j. $\frac{5}{x^2-16} + \frac{3}{x^2+4x}$
   k. $\frac{x-1}{x^2+7x+12} - \frac{x+2}{x^2+x-6}$
   l. $\frac{x+3}{x^2+7x+10} + \frac{x-4}{x^2-x-6}$
   m. $\frac{3x}{3x^2-5x-12} - \frac{5x}{9x^2-16}$
   n. $\frac{x+y}{x^2-y^2} + \frac{x-y}{x^2+2xy+y^2}$
   o. $\frac{x-2}{x+3} - \frac{x+3}{x-2}$
   p. $\frac{x-y}{x+y} + \frac{x+y}{x-y}$
   q. $\frac{x+2}{x-3} + \frac{5}{3-x}$
   r. $\frac{x}{x-1} - \frac{1-3x}{1-x^2}$
   *s. $\frac{x-y}{x^3+y^3} + \frac{5}{x^2-y^2}$
   *t. $\frac{7x}{4x^2-12xy+9y^2} - \frac{3x}{4x^2-9y^2}$
*5. Find:
   a. $\frac{1}{x} + \frac{1}{xy} - \frac{1}{x+y}$
   b. $\frac{5}{x^2y} - \frac{7}{x^3y^2} + \frac{3}{xy^3}$
   c. $\frac{x}{x+1} - \frac{3}{x^2-1} - \frac{x-2}{x^2-2x+1}$
   d. $\frac{-2}{1-x^3} + \frac{x}{x^3+x^2+x} - \frac{1}{x-1}$

2.10 Exercise Answers

1. a. $\frac{7}{x}$
   b. $\frac{1}{y}$
   c. $\frac{-1}{x-3}$ or $\frac{1}{3-x}$
   d. $\frac{4x+1}{2x}$
   e. $\frac{2(3x-1)}{x^2+2x+3}$
   f. 1
   g. $\frac{1}{x+1}$
   h. $\frac{1}{x-y}$
2. least common multiple

3. a. 36 b. 150 c. $21xy$ d. $20x$ e. $3(x+2)(x-2)$
f. $15x^2$ g. $x(x-3)^2$ h. $(x+5)(x-5)(x+3)(x-2)$
i. $(x+4)(x+3)(x-2)$ j. $(3x+4)(3x-4)(x-3)$
*k. $(x+y)(x^2-xy+y^2)(x-y)$
*l. $(x+1)(x-1)^2$ or $(x+1)(x-1)(x-1)$

4. a. $\dfrac{16x-5}{30}$ b. $\dfrac{x+25}{12}$
c. $\dfrac{7x+3}{15x^2}$ d. $\dfrac{x-29}{20x}$
e. $\dfrac{12x+35y}{21xy}$ f. $\dfrac{bx-ay}{by}$
g. $\dfrac{2x+5y}{y}$ h. $\dfrac{45-4y}{5}$
i. $\dfrac{19-2x}{3(x+2)(x-2)}$ j. $\dfrac{4(2x-3)}{x(x+4)(x-4)}$
k. $\dfrac{-3(3x+2)}{(x+4)(x+3)(x-2)}$ or $\dfrac{3(3x+2)}{(x+4)(x+3)(2-x)}$
l. $\dfrac{2x^2+x-29}{(x+2)(x+5)(x-3)}$ m. $\dfrac{x(4x+3)}{(3x+4)(3x-4)(x-3)}$
n. $\dfrac{2(x^2+y^2)}{(x+y)^2(x-y)}$
o. $\dfrac{-5(2x+1)}{(x+3)(x-2)}$ or $\dfrac{5(2x+1)}{(x+3)(2-x)}$: $\dfrac{(x-2)^2-(x+3)^2}{(x+3)(x-2)}$
p. $\dfrac{2(x^2+y^2)}{(x+y)(x-y)}$ q. 1
r. $\dfrac{x-1}{x+1}$ *s. $\dfrac{6x^2-7xy+6y^2}{(x+y)(x^2-xy+y^2)(x-y)}$
*t. $\dfrac{2x(4x+15y)}{(2x-3y)^2(2x+3y)}$

*5. a. $\dfrac{y^2+x+y}{xy(x+y)}$ b. $\dfrac{5xy^2-7y+3x^2}{x^3y^3}$
c. $\dfrac{x^3-3x^2-x+5}{(x+1)(x-1)^2}$: $\dfrac{x(x-1)^2-3(x-1)-(x-2)(x+1)}{(x+1)(x-1)^2}$
d. $\dfrac{-x^2}{(x-1)(x^2+x+1)}$ or $\dfrac{x^2}{(1-x)(1+x+x^2)}$
$\left(\textit{Note: } \dfrac{x}{x^3+x^2+x} = \dfrac{1}{x^2+x+1}\right)$

2.11 Complex Fractions

1 A *complex fraction* is a rational expression that has a fraction in its numerator or its denominator or both. Examples of complex fractions are:

$$\frac{\frac{1}{2}}{5}, \quad \frac{\frac{2x}{3}}{\frac{4}{5}}, \quad \frac{x+\frac{1}{3}}{2}, \quad \frac{x}{\frac{2}{3}-x}, \quad \frac{5-\frac{x}{y}}{5+\frac{x}{y}}, \quad \frac{\frac{1}{x}-\frac{1}{y}}{\frac{1}{x}+\frac{1}{y}}$$

Q1 Write "yes" if the expression is a complex fraction; otherwise, write "no":

a. $x + 2$ ______ **b.** $\dfrac{x+\frac{1}{2}}{5}$ ______

c. $\dfrac{2}{3}$ ______ **d.** $\dfrac{\frac{2}{3}}{6}$ ______

e. $x^2 + x - 6$ ______ **f.** $\dfrac{x^2-\frac{1}{4}}{\frac{2}{3}}$ ______

g. $\dfrac{x+y}{x-y}$ ______ **h.** $\dfrac{x+\frac{1}{y}}{x-\frac{1}{y}}$ ______

i. $\dfrac{2-\frac{4}{5}}{\frac{7}{8}}$ ______ **j.** $\dfrac{\frac{4x}{9}-3}{3-\frac{4x}{9}}$ ______

STOP • STOP • STOP • STOP • STOP • STOP • STOP • STOP • STOP

A1 **a.** no **b.** yes **c.** no **d.** yes **e.** no **f.** yes **g.** no
h. yes **i.** yes **j.** yes

2 Any complex fraction is not in simplest form. There are two methods for simplifying a complex fraction.

Method 1: Simplify both the numerator and the denominator (if necessary) and use the definition of division.

Method 2: Multiply both the numerator and denominator of the complex fraction by the LCM of all denominators of individual terms of the complex fraction.

Example 1: Simplify $\dfrac{\frac{1}{2}}{5}$.

Solution

Method 1

$$\frac{\frac{1}{2}}{5} = \frac{1}{2} \div \frac{5}{1}$$
$$= \frac{1}{2} \cdot \frac{1}{5}$$
$$= \frac{1}{10}$$

Method 2

$$\frac{\frac{1}{2}}{5} = \frac{\frac{1}{2}\cdot 2}{\frac{5}{1}\cdot 2}$$
$$= \frac{1}{10}$$

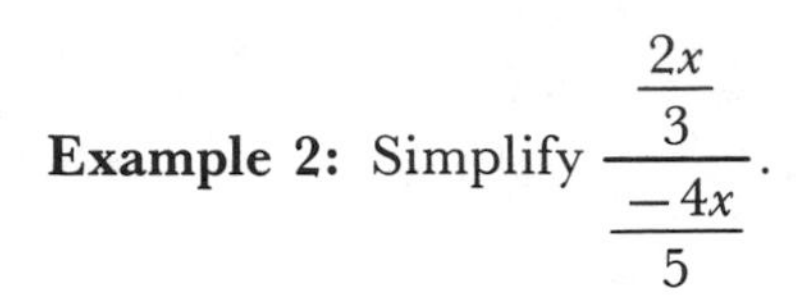

Example 2: Simplify $\dfrac{\frac{2x}{3}}{\frac{-4x}{5}}$.

Solution

Method 1

$$\frac{\frac{2x}{3}}{\frac{-4x}{5}} = \frac{2x}{3} \div \frac{-4x}{5}$$

$$= \frac{\overset{1}{\cancel{2x}}}{3} \cdot \frac{-5}{\underset{2}{\cancel{4x}}}$$

$$= \frac{-5}{6}$$

Method 2

$$\frac{\frac{2x}{3}}{\frac{-4x}{5}} = \frac{\frac{2x}{\underset{1}{\cancel{3}}} \cdot \overset{5}{\cancel{15}}}{\frac{-4x}{\underset{1}{\cancel{5}}} \cdot \overset{3}{\cancel{15}}}$$

$$= \frac{\overset{5\ 1}{\cancel{10x}}}{\underset{-6\ 1}{\cancel{-12x}}}$$

$$= \frac{-5}{6}$$

Q2 Simplify (1) by Method 1 and (2) by Method 2:

a. $\dfrac{\frac{2}{3}}{6}$ (1) (2)

b. $\dfrac{\frac{-5x^2}{6}}{\frac{2x}{3}}$ (1) (2)

STOP • STOP • STOP • STOP • STOP • STOP • STOP • STOP • STOP

A2 **a.** (1) $\frac{1}{9}$: $\dfrac{\frac{2}{3}}{6} = \frac{2}{3} \div \frac{6}{1}$

$$= \frac{\overset{1}{\cancel{2}}}{3} \cdot \frac{1}{\underset{3}{\cancel{6}}}$$

(2) $\frac{1}{9}$: $\dfrac{\frac{2}{3} \cdot 3}{\frac{6}{1} \cdot 3} = \dfrac{\overset{1}{\cancel{2}}}{\underset{9}{\cancel{18}}}$

b. (1) $\frac{-5x}{4}$: $\dfrac{\frac{-5x^2}{6}}{\frac{2x}{3}} = \frac{-5x^2}{6} \div \frac{2x}{3}$

$= \frac{-5\cancel{x^2}^{\,x}}{\cancel{6}_{2}} \cdot \frac{\cancel{3}^{\,1}}{2\cancel{x}_{1}}$

(2) $\frac{-5x}{4}$: $\dfrac{\frac{-5x^2}{\cancel{6}_{1}} \cdot \cancel{6}^{\,1}}{\frac{2x}{\cancel{3}_{1}} \cdot \cancel{6}^{\,2}} = \frac{-5\cancel{x^2}^{\,x}}{4\cancel{x}_{1}}$

3 From experience you will find that one method or the other works best in certain situations.

Example 1: Simplify $x + \frac{1}{3}$.

Solution

Method 1

$$x + \frac{1}{3} = \frac{x}{1} + \frac{1}{3} \quad {}^{*}$$
$$= \frac{3x + 1}{3}$$

Method 2

$$x + \frac{1}{3} = \frac{\left(x + \frac{1}{3}\right)3}{1 \cdot 3}$$
$$= \frac{3x + 1}{3}$$

*Cross-product rule.

(*Note:* In this example either method is equally difficult. However, Method 1 is usually preferred.)

Example 2: Simplify $\dfrac{x}{\frac{2}{3} - x}$.

Solution

Method 1

$$\frac{x}{\frac{2}{3} - x} = \frac{x}{\frac{2}{3} - \frac{x}{1}} \quad {}^{*}$$
$$= \frac{x}{\frac{2 - 3x}{3}}$$
$$= x \div \frac{2 - 3x}{3}$$
$$= x \cdot \frac{3}{2 - 3x}$$
$$= \frac{3x}{2 - 3x}$$

Method 2

$$\frac{x}{\frac{2}{3} - x} = \frac{\frac{x}{1} \cdot 3}{\left(\frac{2}{3} - x\right)3}$$
$$= \frac{3x}{2 - 3x}$$

*Cross-product rule.

(*Note:* In this example Method 2 is considerably easier.)

Example 3: Simplify $\dfrac{5-\frac{x}{y}}{5+\frac{y}{x}}$.

Solution

Method 1

$$\frac{5-\frac{x}{y}}{5+\frac{y}{x}} = \frac{\frac{5y-x}{y}}{\frac{5x+y}{x}} = \frac{5y-x}{y}\cdot\frac{x}{5x+y} = \frac{x(5y-x)}{y(5x+y)}$$

Method 2

$$\frac{5-\frac{x}{y}}{5+\frac{y}{x}} = \frac{\left(5-\frac{x}{y}\right)xy}{\left(5+\frac{y}{x}\right)xy} = \frac{5xy-x^2}{5xy+y^2} = \frac{x(5y-x)}{y(5x+y)}$$

(*Note:* In this example Method 1 appears to be slightly less confusing, for two reasons:

1. Reducing during the distributive step of Method 2 is not required.
2. The result is seen to be in lowest terms immediately without the necessity of factoring.)

You should practice both methods until it is possible to anticipate the most efficient method with respect to a given problem.

Q3 Simplify (use both methods):

a. $\dfrac{\frac{1}{x}-\frac{1}{y}}{\frac{1}{x}+\frac{1}{y}}$

b. $\dfrac{1-\frac{m}{n}}{\frac{m^2}{n^2}}$

STOP • STOP • STOP • STOP • STOP • STOP • STOP • STOP • STOP

A3 a. $\dfrac{y-x}{x+y}$: Method 1: $\dfrac{\frac{y-x}{xy}}{\frac{y+x}{xy}} = \dfrac{y-x}{\cancel{xy}_1}\cdot\dfrac{\overset{1}{\cancel{xy}}}{y+x} = \dfrac{y-x}{y+x}$

Method 2: $\dfrac{\left(\frac{1}{x}-\frac{1}{y}\right)xy}{\left(\frac{1}{x}+\frac{1}{y}\right)xy} = \dfrac{y-x}{y+x}$

b. $\frac{n(n-m)}{m^2}$: Method 1: $\frac{\frac{n-m}{n}}{\frac{m^2}{n^2}} = \frac{n-m}{\cancel{n}_1} \cdot \frac{\cancel{n^2}^{\,n}}{m^2} = \frac{n(n-m)}{m^2}$

Method 2: $\frac{\left(1-\frac{m}{n}\right)n^2}{\frac{m^2}{n^2} \cdot n^2} = \frac{n^2 - mn}{m^2} = \frac{n(n-m)}{m^2}$

Q4 Simplify (use the method of your choice):

a. $x + \frac{1}{2}$ **b.** $\frac{\frac{2x}{3}}{6}$ **c.** $\frac{1}{4} - x^2$ **d.** $\frac{x + \frac{1}{y}}{x - \frac{1}{y}}$

e. $\frac{2 - \frac{4x}{5}}{\frac{7}{8}}$ **f.** $\frac{\frac{4x}{9} - 3}{3 - \frac{4x}{9}}$

STOP • STOP • STOP • STOP • STOP • STOP • STOP • STOP • STOP

A4 **a.** $\frac{2x+1}{2}$ **b.** $\frac{x}{9}$ **c.** $\frac{1-4x^2}{4}$ or $\frac{(1+2x)(1-2x)}{4}$

d. $\frac{xy+1}{xy-1}$ **e.** $\frac{80-32x}{35}$ or $\frac{16(5-2x)}{35}$

f. -1 (quotient of opposites)

Q5 Simplify:

a. $\frac{\frac{x^2-y^2}{8}}{\frac{x+y}{4}}$ **b.** $\frac{\frac{1}{x}+\frac{1}{y}}{\frac{x^2-y^2}{xy}}$ **c.** $\frac{\frac{27}{y^2}+y}{1+\frac{3}{y}}$ **d.** $\frac{\frac{1}{m^3}-m}{\frac{1}{m^2}-1}$

STOP • STOP • STOP • STOP • STOP • STOP • STOP • STOP • STOP

A5

a. $\dfrac{x-y}{2}$: $\dfrac{\dfrac{x^2-y^2}{8}\cdot 8}{\dfrac{x+y}{4}\cdot 8} = \dfrac{\overset{1}{\cancel{(x+y)}}(x-y)}{2\underset{1}{\cancel{(x+y)}}}$

b. $\dfrac{1}{x-y}$

c. $\dfrac{9-3y+y^2}{y}$: $\dfrac{\left(\dfrac{27}{y^2}+y\right)y^2}{\left(1+\dfrac{3}{y}\right)y^2} = \dfrac{27+y^3}{y^2+3y} = \dfrac{\overset{1}{\cancel{(3+y)}}(9-3y+y^2)}{y\underset{1}{\cancel{(y+3)}}}$

d. $\dfrac{1+m^2}{m}$: $\dfrac{\dfrac{1-m^4}{m^3}}{\dfrac{1-m^2}{m^2}} = \dfrac{(1+m^2)\overset{1}{\cancel{(1-m^2)}}}{\underset{m}{\cancel{m^3}}}\cdot\dfrac{\overset{1}{\cancel{m^2}}}{\underset{1}{\cancel{1-m^2}}}$

4	In the process of simplifying complex fractions, remember to factor both the numerator and denominator to determine whether the final expression is in lowest terms.

This completes the instruction for this section.

2.11 Exercises

Simplify:

1. $\dfrac{\dfrac{7x}{12}}{\dfrac{5x}{18}}$

2. $y - \dfrac{2}{3}$

3. $x^2 - \dfrac{1}{16}$

4. $\dfrac{1-\dfrac{1}{x}}{1+\dfrac{1}{x}}$

5. $\dfrac{\dfrac{2x}{3}+5}{\dfrac{4x}{9}-5}$

6. $\dfrac{\dfrac{1}{2}-\dfrac{2}{3}}{\dfrac{2}{3}-\dfrac{1}{2}}$

7. $\dfrac{\dfrac{2}{m}+\dfrac{3}{n}}{\dfrac{4}{m}-\dfrac{6}{n}}$

8. $\dfrac{\dfrac{x-y}{5}}{\dfrac{x^2-y^2}{10}}$

9. $\dfrac{\dfrac{1}{x^2}-\dfrac{1}{y^2}}{\dfrac{1}{x}-\dfrac{1}{y}}$

10. $\dfrac{7+\dfrac{x}{y}}{\dfrac{x}{y}-7}$

11. $\dfrac{x^3-8}{1-\dfrac{x}{2}}$

12. $\dfrac{\dfrac{x+y}{x}}{\dfrac{x+y}{y}}$

13. $\dfrac{\dfrac{x^2+14x+49}{x+2}}{\dfrac{x^2+9x+14}{x^2-4}}$

14. $\dfrac{\dfrac{x^2-25y^2}{10x^2}}{\dfrac{x^2+10xy+25y^2}{25x}}$

2.11 Exercise Answers

1. $\frac{21}{10}$
2. $\frac{3y-2}{3}$
3. $\frac{(4x+1)(4x-1)}{16}$
4. $\frac{x-1}{x+1}$
5. $\frac{3(2x+15)}{4x-45}$
6. -1
7. $\frac{2n+3m}{2(2n-3m)}$
8. $\frac{2}{x+y}$
9. $\frac{x+y}{xy}$
10. $\frac{x+7y}{x-7y}$
11. $-2x^2-4x-8$
12. $\frac{y}{x}$
13. $\frac{(x+7)(x-2)}{x+2}$
14. $\frac{5(x-5y)}{2x(x+5y)}$

[*Note:* Answers, except 11, were left in factored form. You may expand each result to indicate the expression in simplest form. Expressions (like 11) that are polynomials are usually expanded because it is obvious that the expression is in lowest terms.]

Chapter 2 Sample Test

At the completion of Chapter 2 it is expected that you will be able to work the following problems.

2.1 Introduction

1. Identify each expression as a monomial, binomial, or trinomial. Otherwise, write "polynomial."
   a. x^3-3x^2+5x+9 **b.** $x^3y-x^2y^2+y^4$ **c.** x^2yz^3
   d. x^6+y^6 **e.** 4
2. What is the degree of each term?
   a. -7 **b.** xy **c.** 0
   d. x^3 **e.** $-xy^2z^2$
3. What is the degree of each polynomial?
   a. $5+x-2x^2+x^3$ **b.** $x-1$ **c.** $-3x^2+9x$
   d. $x^3y-x^2y^2+xy^3$ **e.** x^2-xy+y^2
4. Arrange:
   a. $-3+4x^2-7x$ in descending order
   b. y^2-4y+3 in ascending order
   c. $x^3-x^2y+5xy^2-2y^3$ in ascending order with respect to x

2.2 Addition and Subtraction of Polynomials

5. Simplify:
   a. $7x-4-9x+8$ **b.** $-(2x^2-3x+7)$
   c. $3x^2y+5xy^2-4y^3-(-2x^2y+5y^3-3xy^2)$
   d. $3y+12+(y^2-4y-9)$
6. Subtract:
   a. $\begin{array}{r} -7x-3 \\ \underline{-7x+4} \end{array}$ **b.** $\begin{array}{r} 5x^2y^2-5xy^3 \\ \underline{5x^2y^2-7xy^3} \end{array}$

2.3 Multiplication of Polynomials

7. Simplify:
 a. $(-5x^2)^3$ b. $(-8x^2y^2)(7xy^3)$ c. $(-mn)^4$
 d. $(4xy^2)^3(-3x^2y)^2$ e. $-x(x^2 + 3x - 2)$
 f. $7y^2(3y^2 + xy - 2x^2)$ g. $(3x - 5)(2x - 3)$ h. $(y + 5)(3y - 4)$
 i. $(5m + 2n)(5m - 2n)$ j. $(x^2 + 3x - 4)(x - 3)$
 k. $(x - 2)(x^2 + 3x + 4)$ l. $(x^2 - x + 2)(x^2 + 3x - 1)$
 m. $-5(x - 2y)(x + 3y)$ n. $3x(x - 1)(x + 6)$

2.4 Special Products

8. Simplify:
 a. $(x + y)(x^2 - xy + y^2)$ b. $(m + n)(p + q)$ c. $(x - y)^2$
 d. $(x + y)^2$ e. $(x + y)(x - y)$
 f. $(x - y)(x^2 + xy + y^2)$ g. $(m + 2n)(m - 2n)$ h. $(5p + 2q)^2$
 i. $(3x - 7y)^2$ j. $(y - 3)(y^2 + 3y + 9)$
 k. $(p + 2q)(p^2 - 2pq + 4q^2)$ l. $(1 - 2x)^3$

2.5 Factoring Polynomials

9. Factor:
 a. $-3x + 9$ b. $y(y + 2) - 5(y + 2)$ c. $4x^3 - 2x^2 + 6x$
 d. $-x^3y^2 - x^2y$ e. $y^2 + 4y - 12$ f. $2x^2 + 7x - 4$
 g. $x^2 + 2xy - 3y^2$ h. $x^2 + 5x + 12$
 i. $4x^2 - 11xy + 6y^2$ j. $x^3 + 4x^2y + 4xy^2$
10. Factor (if possible) and simplify:
 a. $(x + 2)(x - 1) - (x + 2)(x + 3)$
 b. $(x - 1)^2 + (x - 1)^3$
11. Factor by parts:
 a. $x^3 + x^2 - x - 1$ b. $4 - 2y - 2y^2 + y^3$

2.6 Perfect Squares and Completing the Square

12. Factor:
 a. $x^2 - 14x + 49$ b. $25y^2 + 40y + 16$ c. $x^2 + 20x + 64$
 d. $m^2 - 6m - 9$ e. $72 - 24y + 2y^2$ f. $x^2 + 6x - 16$
13. (1) Replace k by the value that will make the trinomial a perfect square and (2) factor:
 a. $x^2 + 8x + k$ b. $y^2 - 3y + k$ c. $m^2 - \frac{2}{3}m + k$
 d. $x^2 + \frac{b}{a}x + k$

2.7 Sums or Differences of Two Squares and Two Cubes

14. Factor:
 a. $64 - x^2$ b. $3x^2 - 27y^2$ c. $x^4 - 16$
 d. $m^3 + n^3$ e. $m^3 - n^3$ f. $y^3 - 64$

2.8 Reducing Rational Expressions

15. Answer the following:
    a. Explain what is meant by "reducing a rational expression."
    b. Under what conditions is a rational expression undefined?

In the remaining problems all rational expressions are understood to be defined for permissible replacements of the variables.

16. Simplify:

a. $\dfrac{15x^2yz}{20xy^3z}$ b. $\left(\dfrac{-6x^3}{20x}\right)^2$ c. $\dfrac{x+y}{x^2+2xy+y^2}$

d. $\dfrac{3x-9}{x^2-9}$ e. $\dfrac{x^2-2x-8}{x^2-7x+12}$ f. $\dfrac{-t+1}{t-1}$

g. $\dfrac{x^3+1}{2x^2+5x+3}$ h. $\dfrac{xy-2x}{8-y^3}$

2.9 Multiplication and Division of Rational Expressions

17. Simplify:

a. $\dfrac{45x^3}{16}\cdot\dfrac{4x-8}{15x^4}$ b. $\dfrac{x^2+7x+10}{x^2+10x+25}\cdot\dfrac{x^2+3x-10}{x^2-4}$

c. $\dfrac{x^2-4x+4}{x^2+x-6}\div\dfrac{16-x^4}{x^3+2x^2+4x+8}$

d. $\dfrac{x^2-2xy-15y^2}{x^2-6xy+8y^2}\div\dfrac{x^2+3xy}{xy-2y^2}$

18. Determine the individual terms of each rational expression:

a. $\dfrac{x^3-2x^2+4x-6}{x^2}$ b. $\dfrac{2x-9}{x}$ c. $\dfrac{4x^2-3x+6}{x-3}$

d. $\dfrac{15x^3+4x^2-x+2}{3x+2}$

2.10 Addition and Subtraction of Rational Expressions

19. Find:

a. $\dfrac{x-2}{12}-\dfrac{x+4}{18}$ b. $\dfrac{5}{7x^2}+\dfrac{3}{14x}$ c. $\dfrac{4}{3x}+\dfrac{6}{5y}$

d. $7-\dfrac{3y}{5}$ e. $\dfrac{x+3}{x^2-1}+\dfrac{9}{2x-2}$

f. $\dfrac{x-1}{x^2+3x-10}-\dfrac{x+2}{x^2-5x+6}$

g. $\dfrac{7}{2-x}-\dfrac{-3}{x-2}$ h. $\dfrac{2x}{3-x}+\dfrac{x+5}{x^2-9}$

2.11 Complex Fractions

20. Simplify:

a. $\dfrac{\dfrac{5x}{9}}{\dfrac{10x}{27}}$ b. $m - \dfrac{4}{5}$ c. $\dfrac{\dfrac{1}{y} + 2}{\dfrac{1}{y} - 2}$

d. $\dfrac{\dfrac{5}{x} + \dfrac{2}{y}}{\dfrac{10}{x} + \dfrac{4}{y}}$ e. $\dfrac{1 - \dfrac{5}{p}}{\dfrac{5}{p} - 1}$

f. $\dfrac{\dfrac{x^2 - 16xy + 64y^2}{30x^2y^2}}{\dfrac{x^2 - 10xy + 16y^2}{75xy}}$

Chapter 2 Sample Test Answers

1. **a.** polynomial **b.** trinomial **c.** monomial
d. binomial **e.** monomial

2. **a.** 0 **b.** 2 **c.** none **d.** 3 **e.** 5

3. **a.** 3 **b.** 1 **c.** 2 **d.** 4 **e.** 2

4. **a.** $4x^2 - 7x - 3$ **b.** $3 - 4y + y^2$
c. $-2y^3 + 5xy^2 - x^2y + x^3$

5. **a.** $-2x + 4$ or $4 - 2x$ **b.** $-2x^2 + 3x - 7$
c. $5x^2y + 8xy^2 - 9y^3$ **d.** $y^2 - y + 3$

6. $a - 7$ **b.** $2xy^3$

7. **a.** $-125x^6$ **b.** $-56x^3y^5$ **c.** m^4n^4
d. $576x^7y^8$ **e.** $-x^3 - 3x^2 + 2x$ or $2x - 3x^2 - x^3$
f. $21y^4 + 7xy^3 - 14x^2y^2$ **g.** $6x^2 - 19x + 15$ **h.** $3y^2 + 11y - 20$
i. $25m^2 - 4n^2$ **j.** $x^3 - 13x + 12$ **k.** $x^3 + x^2 - 2x - 8$
l. $x^4 + 2x^3 - 2x^2 + 7x - 2$ **m.** $-5x^2 - 5xy + 30y^2$ **n.** $3x^3 + 15x^2 - 18x$

8. **a.** $x^3 + y^3$ **b.** $mp + mq + np + nq$ (any order)
c. $x^2 - 2xy + y^2$ **d.** $x^2 + 2xy + y^2$ **e.** $x^2 - y^2$
f. $x^3 - y^3$ **g.** $m^2 - 4n^2$
h. $25p^2 + 20pq + 4q^2$ **i.** $9x^2 - 42xy + 49y^2$ **j.** $y^3 - 27$
k. $p^3 + 8q^3$ **l.** $1 - 6x + 12x^2 - 8x^3$

9. **a.** $-3(x - 3)$ **b.** $(y - 5)(y + 2)$ **c.** $2x(2x^2 - x + 3)$
d. $-x^2y(xy + 1)$ **e.** $(y + 6)(y - 2)$ **f.** $(2x - 1)(x + 4)$
g. $(x + 3y)(x - y)$ **h.** prime **i.** $(4x - 3y)(x - 2y)$
j. $x(x + 2y)^2$ or $x(x + 2y)(x + 2y)$

10. **a.** $-4x - 8$ **b.** $x^3 - 2x^2 + x$

11. **a.** $(x - 1)(x + 1)^2$ **b.** $(2 - y^2)(2 - y)$

12. **a.** $(x - 7)^2$ **b.** $(5y + 4)^2$ **c.** $(x + 16)(x + 4)$
d. prime **e.** $2(6 - y)^2$ **f.** $(x + 8)(x - 2)$

13. **a.** (1) $x^2 + 8x + 16$ **b.** (1) $y^2 - 3y + \frac{9}{4}$

(2) $(x + 4)^2$ (2) $\left(y - \frac{3}{2}\right)^2$

c. (1) $m^2 - \frac{2}{3}m + \frac{1}{9}$ **d.** (1) $x^2 + \frac{b}{a}x + \frac{b^2}{4a^2}$

(2) $\left(m - \frac{1}{3}\right)^2$ (2) $\left(x + \frac{b}{2a}\right)^2$

14. **a.** $(8 + x)(8 - x)$ **b.** $3(x + 3y)(x - 3y)$

c. $(x^2 + 4)(x + 2)(x - 2)$ **d.** $(m + n)(m^2 - mn + n^2)$

e. $(m - n)(m^2 + mn + n^2)$ **f.** $(y - 4)(y^2 + 4y + 16)$

15. **a.** dividing both numerator and denominator by some common factor

b. whenever the value of any variable produces a zero denominator (since division by zero is impossible)

16. **a.** $\frac{3x}{4y^2}$ **b.** $\frac{9x^4}{100}$ **c.** $\frac{1}{x + y}$

d. $\frac{3}{x + 3}$ **e.** $\frac{x + 2}{x - 3}$ **f.** -1

g. $\frac{x^2 - x + 1}{2x + 3}$ **h.** $\frac{-x}{4 + 2y + y^2}$ or $\frac{-x}{y^2 + 2y + 4}$

17. **a.** $\frac{3(x - 2)}{4x}$ **b.** 1 **c.** $\frac{-1}{x + 3}$

d. $\frac{y(x - 5y)}{x(x - 4y)}$

18. **a.** $x - 2 + \frac{4}{x} - \frac{6}{x^2}$ **b.** $2 - \frac{9}{x}$ **c.** $4x + 9 + \frac{33}{x - 3}$

d. $5x^2 - 2x + 1$

19. **a.** $\frac{x - 14}{36}$ **b.** $\frac{3x + 10}{14x^2}$ **c.** $\frac{2(9x + 10y)}{15xy}$

d. $\frac{35 - 3y}{5}$ **e.** $\frac{11x + 15}{2(x + 1)(x - 1)}$

f. $\frac{-(11x + 7)}{(x + 5)(x - 2)(x - 3)}$ **g.** $\frac{-4}{x - 2}$ or $\frac{4}{2 - x}$

h. $\frac{-(2x^2 + 5x - 5)}{(x + 3)(x - 3)}$

20. **a.** $\frac{3}{2}$ **b.** $\frac{5m - 4}{5}$ **c.** $\frac{1 + 2y}{1 - 2y}$

d. $\frac{1}{2}$ **e.** -1 **f.** $\frac{5(x - 8y)}{2xy(x - 2y)}$

Chapter 3

Solution of Linear Equations, Inequalities, and Absolute Value

Much of algebra centers around finding the solution set for various types of equations and inequalities. The procedures for solving equations and inequalities will be similar to those developed in beginning algebra; however, they will be somewhat condensed.

The expression "linear equation" will be defined and explained in Chapter 6. A lack of understanding of the meaning of "linear" should not hinder the solution of equations in this chapter.

3.1 Solution of Linear Equations

1

An *equation* is a mathematical statement of equality. The statement

$10 + 13 = 23$

is a true statement of equality, whereas

$7 - 3 = 3 - 7$

is a false statement of equality.

The open sentence

$x + 7 = 13$

is neither true nor false. The sentence will be true if $x = 6$ and false if x equals any other real number. Finding the replacement for the variable that makes the statement true is called *solving the equation.* The set of replacements that makes the equation true is called the *solution set* or the *truth set*. The solution set for $x + 7 = 13$ is $\{6\}$. The solution set can sometimes be found by observation.

Example 1: Solve $x - 8 = -2$.

Solution

Think: "some number minus 8 equals negative 2." Hence $x = 6$ because $6 - 8 = -2$. Therefore, the solution set is $\{6\}$.

Example 2: Solve $\frac{1}{2}x = 15$.

Solution

Think: "$\frac{1}{2}$ of some number equals 15." Hence $x = 30$ because $\frac{1}{2}(30) = 15$. Solution set $= \{30\}$.

Q1 Find the solution set by observation:

a. $x - 15 = 3$ ______

b. $12 + x = 0$ ______

c. $2.4x = 0$ ______

d. $\frac{1}{3}x = 6$ ______

STOP • STOP • STOP • STOP • STOP • STOP • STOP • STOP • STOP

A1 **a.** $\{18\}$ **b.** $\{-12\}$

c. $\{0\}$ **d.** $\{18\}$

2 Solving equations by observation is often difficult; however, equations of the form

$ax + b = 0, \quad a \neq 0, \quad a, b \in R$

can be solved using certain properties of equality.

Property 1: Addition property of equality

$a = b$

is equivalent to

$a + c = b + c \qquad a, b, c \subset R$

That is, an equivalent equation is formed by adding a real number to both sides of a given equation. *Equivalent equations* have the same solution set.

Example 1: Find the solution set for $x - 18 = 32$ by use of the addition property of equality.

Solution

$$\begin{aligned} x - 18 &= 32 \\ x - 18 + 18 &= 32 + 18 \quad \text{(adding 18 to both sides)} \\ x + 0 &= 50 \\ x &= 50 \qquad \text{Check: } 50 - 18 = 32 \\ \text{solution set} &= \{50\} \end{aligned}$$

Example 2: Find the solution set for $x + 8 = -5$ by use of the addition property of equality.

Solution

$$\begin{aligned} x + 8 &= -5 \\ x + 8 + {}^{-}8 &= -5 + {}^{-}8 \quad \text{(adding } -8 \text{ to both sides)} \\ x &= -13 \qquad \text{Check: } -13 + 8 = -5 \\ \text{solution set} &= \{-13\} \end{aligned}$$

Q2 Find the solution set for the following equations by use of the addition property of equality and check the solution:

a. $x + 13 = 6$

b. $y - \frac{1}{3} = \frac{1}{6}$

c. $12 = -20 + x$

d. $z + 7.2 = 9.8$

STOP • STOP • STOP • STOP • STOP • STOP • STOP • STOP • STOP

A2 **a.** $\{-7\}$: $x + 13 = 6$

$$x + 13 + (-13) = 6 + (-13)$$
$$x = -7$$

Check: $-7 + 13 = 6$

b. $\left\{\frac{1}{2}\right\}$: $y - \frac{1}{3} = \frac{1}{6}$

$$y - \frac{1}{3} + \frac{1}{3} = \frac{1}{6} + \frac{1}{3}$$
$$y = \frac{1}{2}$$

Check: $\frac{1}{2} - \frac{1}{3} = \frac{1}{6}$

c. $\{32\}$: $12 = -20 + x$

$$12 + 20 = -20 + x + 20$$
$$32 = x$$

Check: $12 = -20 + 32$

d. $\{2.6\}$: $z + 7.2 = 9.8$

$$z + 7.2 + (-7.2) = 9.8 + (-7.2)$$
$$z = 2.6$$

Check: $2.6 + 7.2 = 9.8$

Q3 Solve* the following equations by use of the addition property of equality:

a. $x + 6 = 15$ **b.** $x + 9 = -11$ **c.** $x - 4 = -9$ **d.** $x - 8 = 12$

STOP • STOP • STOP • STOP • STOP • STOP • STOP • STOP • STOP

A3 **a.** $x = 9$ **b.** $x = -20$ **c.** $x = -5$ **d.** $x = 20$

*Note that the solution is presented without braces, { }. This is done for convenience; however, whenever the solution set or truth set is asked for, place the answer in braces. When asked to solve an equation in x, find an equivalent equation in the form $x = a$ or $a = x$. Also, even though it is important to check the solution to all problems, the check will usually be omitted in the answer.

3

Another important property for solving equations follows.

Property 2: Multiplication property of equality

$a = b$

is equivalent to

$ac = bc \qquad a, b, c \in R, c \neq 0$

That is, an equivalent equation is formed by multiplying both sides of an equation by a nonzero real number.

Example 1: Solve $3x = 18$ by use of the multiplication property of equality.

Solution

When the equation is solved, the coefficient of x will equal 1. Hence both sides of the equation should be multiplied by the *reciprocal** of 3, which is $\frac{1}{3}$.

$$\begin{aligned} 3x &= 18 \\ \frac{1}{3}(3x) &= \frac{1}{3}(18) \\ 1x &= 6 \\ x &= 6 \end{aligned}$$

Example 2: Solve $\frac{2}{3}x = 10$ by use of the multiplication property of equality.

Solution

$$\begin{aligned} \frac{2}{3}x &= 10 \\ \frac{3}{2}\left(\frac{2}{3}x\right) &= \frac{3}{2}(10) \\ 1x &= 15 \\ x &= 15 \end{aligned}$$

*The reciprocal of a nonzero real number x is $\frac{1}{x}$.

Q4 Determine the reciprocals of the following numbers:

a. $\frac{3}{4}$ **b.** -5 **c.** 0 **d.** 1

e. -1 **f.** $2\frac{1}{2}$

STOP • STOP • STOP • STOP • STOP • STOP • STOP • STOP • STOP

A4 **a.** $\frac{4}{3}$: $\frac{1}{\frac{3}{4}} = \frac{4}{3}$ **b.** $\frac{1}{-5}$ or $\frac{-1}{5}$ **c.** does not exist: $\frac{1}{0}$ is undefined

d. 1: $\frac{1}{1} = 1$ **e.** -1 **f.** $\frac{2}{5}$

Q5 Solve the following equations by use of the multiplication property of equality:

a. $6x = 30$

b. $\frac{1}{3}x = 5$

c. $-4y = \frac{1}{8}$

d. $\frac{8}{9}x = -16$

STOP • STOP • STOP • STOP • STOP • STOP • STOP • STOP • STOP

A5 a. $x = 5$: $\frac{1}{6}(6x) = \frac{1}{6}(30)$

b. $x = 15$: $\frac{3}{1}\left(\frac{1}{3}x\right) = \frac{3}{1}(15)$

c. $y = \frac{-1}{32}$: $\frac{-1}{4}(-4y) = \frac{-1}{4}\left(\frac{1}{8}\right)$

d. $x = -18$: $\frac{9}{8}\left(\frac{8}{9}x\right) = \frac{9}{8}(-16)$

Q6 Solve the following equations:

a. $\frac{2x}{3} = 22$

b. $\frac{x}{0.7} = 1.2$

STOP • STOP • STOP • STOP • STOP • STOP • STOP • STOP • STOP

A6 a. $x = 33$: $\frac{3}{2}\left(\frac{2x}{3}\right) = \frac{3}{2}(22)$

b. $x = 0.84$: $\frac{0.7}{1}\left(\frac{x}{0.7}\right) = \frac{0.7}{1}(1.2)$

Q7 Solve the following equations:

a. $\frac{4x}{7} = -12$

b. $\frac{2}{3}x = 0$

c. $-x = 17$

d. $18y = 6$

STOP • STOP • STOP • STOP • STOP • STOP • STOP • STOP • STOP

A7 a. $x = -21$

b. $x = 0$

c. $x = -17$:
$$-x = 17$$
$$-1(-x) = -1(17)$$
$$x = -17$$

d. $y = \frac{1}{3}$

4 It will often be necessary to use both the addition and multiplication properties of equality to solve equations. Consider the equation

$3x - 6 = 30$

The equation is solved when it is simplified to the form

$x = a \qquad a \in R$

The equation could be solved by first adding 6 to both sides of the equation. That is,

$$\begin{aligned} 3x - 6 &= 30 \\ 3x - 6 + 6 &= 30 + 6 \\ 3x &= 36 \\ \frac{1}{3}(3x) &= \frac{1}{3}(36) \\ x &= 12 \end{aligned}$$

The equation could also be solved by first multiplying both sides of the equation by the reciprocal of the coefficient of x. That is,

$$\begin{aligned} 3x - 6 &= 30 \\ \frac{1}{3}(3x - 6) &= \frac{1}{3}(30) \\ \frac{1}{3}(3x) - \frac{1}{3}(6) &= \frac{1}{3}(30) \\ x - 2 &= 10 \\ x - 2 + 2 &= 10 + 2 \\ x &= 12 \end{aligned}$$

The procedure followed in solving equations is often a matter of choice. However, adding first usually provides a simpler procedure.

Q8 Solve $3x - 2 = 19$ by first adding 2 to both sides of the equation.

STOP • STOP • STOP • STOP • STOP • STOP • STOP • STOP • STOP

A8 $x = 7$:

$$\begin{aligned} 3x - 2 &= 19 \\ 3x - 2 + 2 &= 19 + 2 \\ 3x &= 21 \\ \frac{1}{3}(3x) &= \frac{1}{3}(21) \\ x &= 7 \end{aligned}$$

Q9 Solve $3x - 2 = 19$ by first multiplying both sides by $\frac{1}{3}$, the reciprocal of 3.

STOP • STOP • STOP • STOP • STOP • STOP • STOP • STOP • STOP

A9 $x = 7$:

$$3x - 2 = 19$$
$$\frac{1}{3}(3x - 2) = \frac{1}{3}(19)$$
$$\frac{1}{3}(3x) - \frac{1}{3}(2) = \frac{19}{3}$$
$$x - \frac{2}{3} = \frac{19}{3}$$
$$x - \frac{2}{3} + \frac{2}{3} = \frac{19}{3} + \frac{2}{3}$$
$$x = \frac{21}{3}$$
$$x = 7$$

Q10 **a.** Solve $2x + 5 = 17$. **b.** Check the solution.

STOP • STOP • STOP • STOP • STOP • STOP • STOP • STOP • STOP

A10 **a.** $x = 6$:

$$2x + 5 = 17$$
$$2x + 5 + {}^{-}5 = 17 + {}^{-}5$$
$$2x = 12$$
$$\frac{1}{2}(2x) = \frac{1}{2}(12)$$
$$x = 6$$

b.

$$2x + 5 = 17$$
$$2(6) + 5 \stackrel{?}{=} 17$$
$$12 + 5 \stackrel{?}{=} 17$$
$$17 = 17$$

Q11 Solve the following equations:

a. $-2x + 6 = 12$

b. $5n - 4 = 16$

STOP • STOP • STOP • STOP • STOP • STOP • STOP • STOP • STOP

A11

a. $x = -3$:

$$\begin{aligned} -2x + 6 &= 12 \\ -2x + 6 + {}^{-}6 &= 12 + {}^{-}6 \\ -2x &= 6 \\ \frac{-1}{2}(-2x) &= \frac{-1}{2}(6) \\ x &= -3 \end{aligned}$$

b. $n = 4$:

$$\begin{aligned} 5n - 4 &= 16 \\ 5n - 4 + 4 &= 16 + 4 \\ 5n &= 20 \\ \frac{1}{5}(5n) &= \frac{1}{5}(20) \\ n &= 4 \end{aligned}$$

Q12 Solve the following equations:

a. $-8m - 3 = 21$

b. $2x + 7 = 12$

STOP • STOP • STOP • STOP • STOP • STOP • STOP • STOP • STOP

A12 a. $m = -3$ b. $x = \frac{5}{2}$

---

5 Before applying the addition and multiplication properties of equality, it may be necessary to simplify one or both sides of the equation.

Example 1: Solve $5x - 7 - 2x = 13$.

Solution

First simplify the left side.

$$\begin{aligned} 5x - 7 - 2x &= 13 \\ 3x - 7 &= 13 \\ 3x &= 20 && \text{(adding 7 to both sides)} \\ x &= \frac{20}{3} && \left(\text{multiplying both sides by } \frac{1}{3}\right) \end{aligned}$$

Example 2: Solve $2x + 7 - x = 4x - 2$.

Solution

$$\begin{aligned} x + 7 &= 4x - 2 && (\text{adding } 2x + {}^{-}x) \\ -4x + x + 7 &= -4x + 4x - 2 && (\text{adding } -4x \text{ to both sides}) \\ -3x + 7 &= -2 \\ -3x &= -9 \\ x &= 3 \end{aligned}$$

In the second example it was necessary to get the x terms on one side of the equation; hence $-4x$ was added to both sides to place the x terms on the left side of the equation. The x terms could have been placed on the right side of the equation by adding x to both sides.

Q13 Solve $2x + 7 - x = 4x - 2$ by placing the x terms on the right side of the equation.

STOP • STOP • STOP • STOP • STOP • STOP • STOP • STOP • STOP

A13 $x = 3$:

$$\begin{aligned} 2x + 7 - x &= 4x - 2 \\ x + 7 &= 4x - 2 \\ -x + x + 7 &= -x + 4x - 2 \quad \text{(adding } -x \text{ to both sides)} \\ 7 &= 3x - 2 \\ 9 &= 3x \\ 3 &= x \end{aligned}$$

Q14 Solve the following equations:

a. $14x - 9x + 7 = -23$ **b.** $5x + 3 = 3x - 1$

STOP • STOP • STOP • STOP • STOP • STOP • STOP • STOP • STOP

A14 **a.** $x = -6$:

$$\begin{aligned} 14x - 9x + 7 &= -23 \\ 5x + 7 &= -23 \\ 5x &= -30 \\ x &= -6 \end{aligned}$$

b. $x = -2$:

$$\begin{aligned} 5x + 3 &= 3x - 1 \\ 2x + 3 &= -1 \\ 2x &= -4 \\ x &= -2 \end{aligned}$$

Q15 Solve the following equations:

a. $8y - 4 = 3y - 1$ **b.** $a - 9 + 4a = 6$

STOP • STOP • STOP • STOP • STOP • STOP • STOP • STOP • STOP

A15 **a.** $y = \frac{3}{5}$ **b.** $a = 3$

---

6 Equations involving parentheses can be solved by first applying a distributive property and then proceeding as before.

Example: Solve $4(x - 1) + 2(x + 3) = 6$.

Solution

$$\begin{aligned} 4(x-1)+2(x+3) &= 6 \\ 4x-4+2x+6 &= 6 \\ 6x+2 &= 6 \\ 6x &= 4 \\ x &= \frac{2}{3} \end{aligned}$$

Q16 Solve the following equations:

a. $6(2x-5)=18$

b. $5(2n-4)=-2(2n-2)$

STOP • STOP • STOP • STOP • STOP • STOP • STOP • STOP • STOP

A16 **a.** $x=4$: $\begin{aligned} 6(2x-5) &= 18 \\ 12x-30 &= 18 \\ 12x &= 48 \\ x &= 4 \end{aligned}$

b. $n=\frac{12}{7}$: $\begin{aligned} 5(2n-4) &= -2(2n-2) \\ 10n-20 &= -4n+4 \\ 14n-20 &= 4 \\ 14n &= 24 \\ n &= \frac{24}{14} \end{aligned}$

Q17 Solve the following equations:

a. $3(2x-7)=11$

b. $x+3(x-1)=7x+1$

STOP • STOP • STOP • STOP • STOP • STOP • STOP • STOP • STOP

A17 **a.** $x=\frac{16}{3}$ **b.** $x=\frac{-4}{3}$

7 When solving an equation, one must be aware of the numbers under consideration. That is, what numbers will be considered as possible values for the solution set. This set is often called the *universal* or *replacement set*. If both sides of an equation simplify to identical expressions, the solution set is all values from the universal set for which the equation is defined.

Example: Given the universal set to be R, solve $2(x-3)=x+x-6$.

Solution

$$2(x - 3) = x + x - 6$$
$$\underbrace{2x - 6}_{} = \underbrace{2x - 6}_{}$$

identical expressions

The solution set is R, because any real number substituted for x will result in a true statement.

If an equation is simplified to a point where a false statement occurs, then the solution set is empty.

Example: Solve $3(x - 2) + x = 4x - 1$.

Solution

$$\begin{aligned} 3x - 6 + x &= 4x - 1 \\ 4x - 6 &= 4x - 1 \\ -6 &= -1 \end{aligned}$$

The false statement $-6 = -1$ indicates that there is no solution. The solution set $= \{\ \}$ or $\varnothing$.

Q18 Solve the following equations:

a. $7(x - 1) - x = 6x - 7$

b. $-8(x + 3) = 4(-2x + 1)$

STOP • STOP • STOP • STOP • STOP • STOP • STOP • STOP • STOP

A18 a. R:
$$\begin{aligned} 7(x - 1) - x &= 6x - 7 \\ 7x - 7 - x &= 6x - 7 \\ \underbrace{6x - 7}_{} &= \underbrace{6x - 7}_{} \end{aligned}$$

identical expressions

b. $\varnothing$:
$$\begin{aligned} -8(x + 3) &= 4(-2x + 1) \\ -8x - 24 &= -8x + 4 \\ -24 &= 4 \end{aligned}$$

(a false statement)

8 Equations that involve fractions can be solved by first multiplying both sides of the equation by the least common denominator (LCD) of the fractions.

Example 1: Solve $\frac{x}{2} - \frac{1}{4} = \frac{1}{3}$.

Solution

The LCD for $\frac{x}{2}$, $\frac{1}{3}$, and $\frac{1}{4}$ is 12; hence multiply both sides of the equation by 12.

$$\begin{aligned} 12\left(\frac{x}{2} - \frac{1}{4}\right) &= 12\left(\frac{1}{3}\right) \\ 12\left(\frac{x}{2}\right) - 12\left(\frac{1}{4}\right) &= 4 \\ 6x - 3 &= 4 \\ 6x &= 7 \\ x &= \frac{7}{6} \end{aligned}$$

Example 2: Solve $\frac{x+1}{x} = \frac{x-1}{2x}$, $x \neq 0$.

Solution

The LCD for $\frac{x+1}{x}$ and $\frac{x-1}{2x}$ is $2x$; hence multiply both sides of the equation by $2x$.

$$2x\left(\frac{x+1}{x}\right) = 2x\left(\frac{x-1}{2x}\right)$$
$$2(x+1) = x - 1$$
$$2x + 2 = x - 1$$
$$x + 2 = -1$$
$$x = -3$$

Q19 Solve the following equations:

a. $\frac{2x}{5} + \frac{1}{2} = \frac{9}{10}$

b. $\frac{2x+1}{4} = \frac{x-1}{8}$

STOP • STOP • STOP • STOP • STOP • STOP • STOP • STOP • STOP

A19 a. $x = 1$: $10\left(\frac{2x}{5} + \frac{1}{2}\right) = 10\left(\frac{9}{10}\right)$, LCD = 10

$$10\left(\frac{2x}{5}\right) + 10\left(\frac{1}{2}\right) = 9$$
$$4x + 5 = 9$$
$$4x = 4$$
$$x = 1$$

b. $x = -1$: $8\left(\frac{2x+1}{4}\right) = 8\left(\frac{x-1}{8}\right)$

$$2(2x+1) = x - 1$$
$$4x + 2 = x - 1$$
$$3x = -3$$
$$x = -1$$

Q20 Solve the following equations:

a. $\frac{2m - 1}{3m} = \frac{m + 1}{2m}, \ m \neq 0$

b. $\frac{1}{x + 1} = \frac{x}{2(x + 1)}, \ x \neq -1$

STOP • STOP • STOP • STOP • STOP • STOP • STOP • STOP • STOP

A20 a. $m = 5$: $6m\left(\frac{2m - 1}{3m}\right) = 6m\left(\frac{m + 1}{2m}\right)$

$$2(2m - 1) = 3(m + 1)$$

b. $x = 2$: $2(x + 1)\left(\frac{1}{x + 1}\right) = 2(x + 1)\left(\frac{x}{2(x + 1)}\right)$

$$2 = x$$

9 Formulas can be solved for certain variables by use of the addition and multiplication properties of equality.

Example 1: Solve $P = QR - S$ for R.

Solution

The solution will be of the form $R = ?$; hence solve for R by treating all other variables as constants.

$$P = QR - S$$

$$P + S = QR - S + S \quad \text{(adding } S \text{ to both sides)}$$

$$P + S = QR$$

$$\frac{1}{Q}(P + S) = \frac{1}{Q}(QR) \quad \left(\text{multiplying both sides by } \frac{1}{Q}\right)$$

$$\frac{P + S}{Q} = R$$

Example 2: Solve $A = \frac{1}{2}bh$ for h.

Solution

$$2A = 2\left(\frac{1}{2}bh\right)$$

$$2A = bh$$

$$\frac{1}{b}(2A) = \frac{1}{b}(bh)$$

$$\frac{2A}{b} = h$$

Q21 Solve $i = prt$ for r.

STOP • STOP • STOP • STOP • STOP • STOP • STOP • STOP • STOP

A21 $r = \frac{i}{pt}$:

$$i = prt$$

$$\left(\frac{1}{pt}\right)i = \left(\frac{1}{pt}\right)ptr$$

$$\frac{i}{pt} = r$$

Q22 Solve $P = 2w + 2l$ for l.

STOP • STOP • STOP • STOP • STOP • STOP • STOP • STOP • STOP

A22 $l = \frac{P - 2w}{2}$:

$$P = 2w + 2l$$

$$P - 2w = 2l$$

$$\frac{P - 2w}{2} = l$$

Q23 Solve for the variable indicated:

a. $P = a + b + c$ (for c) **b.** $d = rt$ (for r)

STOP • STOP • STOP • STOP • STOP • STOP • STOP • STOP • STOP

A23 **a.** $c = P - a - b$ **b.** $r = \frac{d}{t}$

---

10 Linear equations can be used to solve verbal problems by translating English sentences into mathematical sentences. Recognizing certain key words as mathematical operations will aid in this task. In the following examples the key word is underlined in the English phrase. The corresponding operation is represented in the mathematical phrase.

English phrase	Mathematical phrase
a number increased by two	$n + 2$
twelve more than a number	$n + 12$
seven less than a number	$n - 7$
a number decreased by seven	$n - 7$
seven minus a number	$7 - n$
the difference between seven and a number	$7 - n$
the product of three and a number	$3n$

eight times a number	$8n$
fifteen percent of a number	$0.15n$
the quotient of twelve and a number	$\frac{12}{n}$
a number divided by three	$\frac{n}{3}$
twice a number decreased by fifteen	$2n - 15$

Q24 Translate each English phrase into a mathematical phrase (use n to represent the number):

a. a number added to seven ________

b. three increased by a number ________

c. three times a number, increased by twelve ________

d. a number less five ________

e. the difference between twice a number and nine ________

f. the quotient of a number and seven ________

STOP • STOP • STOP • STOP • STOP • STOP • STOP • STOP • STOP

A24 **a.** $7 + n$ **b.** $3 + n$ **c.** $3n + 12$ **d.** $n - 5$

e. $2n - 9$ **f.** $\frac{n}{7}$

11 English sentences can now be translated into mathematical sentences. For example,

1. Thirty-five is seven less than a number.

$$35 = n - 7$$

2. Three times a number plus two more than the number is 18.

$$3n + n + 2 = 18$$

Q25 Express each English sentence as a mathematical sentence; use x as the variable:

a. The sum of a number and eight is four. ________

b. If a number is decreased by ten, twice the difference is thirty-two. ________

c. Three more than twice a certain number is seven less than that number. ________

STOP • STOP • STOP • STOP • STOP • STOP • STOP • STOP • STOP

A25 **a.** $x + 8 = 4$ **b.** $2(x - 10) = 32$ **c.** $2x + 3 = x - 7$

12 A verbal problem can now be solved with the aid of the following steps.

Step 1: List the parts of the problem mathematically; use a variable for the unknown quantity. Give a pictorial representation if possible.

Step 2: Write a mathematical sentence (equation or inequality).

Step 3: Solve the resulting equation or inequality.

Step 4: List and check the solutions.

Example: Five plus twice a number is seven times the number. Find the number.

Solution

Step 1: $5 + 2n$ (five plus twice a number)
$7n$ (seven times the number)

Step 2: $5 + 2n = 7n$

Step 3: $5 = 5n$
$1 = n$

Step 4: Check: five plus twice one is seven times one.

Q26 The sum of three consecutive integers is 36. Find the integers.

Step 1: x (first integer)
_______ (second integer)
_______ (third integer)

Step 2:

Step 3:

Step 4: (check):

STOP • STOP • STOP • STOP • STOP • STOP • STOP • STOP • STOP

A26 11, 12, 13:

Step 1: x
$x + 1$
$x + 2$

Step 2: $x + x + 1 + x + 2 = 36$

Step 3: $3x + 3 = 36$
$3x = 33$
$x = 11$

Step 4 (check): $11 + 12 + 13 = 36$

Q27 The perimeter of a rectangle is 26 feet. If the width is 2 feet less than half the length, find the dimensions (draw a picture).

STOP • STOP • STOP • STOP • STOP • STOP • STOP • STOP • STOP

A27 $l = 10$ feet and $w = 3$ feet:

Step 1: $w = \frac{1}{2}l - 2$ $\left(2 \text{ feet less than } \frac{1}{2} \text{ the length}\right)$

l = length

Step 2: $P = 2l + 2w$

$26 = 2l + 2\left(\frac{1}{2}l - 2\right)$

Step 3: $26 = 2l + l - 4$

$26 = 3l - 4$

$30 = 3l$

$10 = l$

Step 4: (check): $l = 10$ feet

$w = \frac{1}{2}(10) - 2 = 3$ feet

$P = 2(10) + 2(3)$

$P = 26$ feet

Q28 Bill and Glee have just made a canoe trip that lasted 5 hours. They took turns at paddling and kept a record of their time. Glee paddled only $\frac{2}{3}$ as long as Bill. How long did each paddle?

STOP • STOP • STOP • STOP • STOP • STOP • STOP • STOP • STOP

A28 Glee 2 hours, Bill 3 hours:

Step 1: t (time Bill paddled)

$\frac{2}{3}t$ (time Glee paddled)

Step 2: $t + \frac{2}{3}t = 5$

Step 3: $3t + 2t = 15$

$5t = 15$

$t = 3$

Step 4 (check): Bill = 3 hours

Glee $= \frac{2}{3}(3) = 2$ hours

3 hours + 2 hours = 5 hours

Q29 A plane leaves an airport going 150 mph. Two hours later a second plane starts after the first one and goes 250 mph. How many hours will it take the second plane to catch the first one? (*Hint:* $d = rt$; let t = time of the first plane.)

STOP • STOP • STOP • STOP • STOP • STOP • STOP • STOP • STOP

A29 3 hours:

Step 1: t (time first plane); $t - 2$ (time second plane)
150 (rate first plane); 250 (rate second plane)

Step 2: Since they will have traveled the same distance and $d = rt$, $150t = 250(t - 2)$

Step 3:
$$150t = 250t - 500$$
$$-100t = -500$$
$$t = 5$$

Step 4: $t - 2 = 3$ hours (time second plane traveled)

(the check has been omitted)

(*Note:* The problem could have been solved by letting t = time of the second plane and $t + 2$ = the time of the first plane.)

Q30 A house and lot together cost $26,000. If the house costs $2,000 more than twice the lot, how much did each cost?

STOP • STOP • STOP • STOP • STOP • STOP • STOP • STOP • STOP

A30 lot \$8,000; house \$18,000:

Step 1: x (cost of lot)
$2x + 2{,}000$ (cost of house)

Step 2: $x + 2x + 2{,}000 = 26{,}000$

Step 3:
$$3x + 2{,}000 = 26{,}000$$
$$3x = 24{,}000$$
$$x = 8{,}000$$

Step 4 (check): lot = \$8,000
house = \$18,000
\$8,000 + \$18,000 = \$26,000

This completes the instruction for this section.

3.1 Exercises

Solve the following:

1. $10x - 8 = 7x + 4$
2. $7x - 3 = 2x + 27$
3. $y + 3(y - 3) = y - 3$
4. $5(2y - 9) - 3y = 4y$
5. $3n - 5 = 2(4 - 5n) + n - 1$
6. $2(5x - 7) = 4(5 - 3x) + 32$
7. $x + x + 2(2x - 2) = 6 - 9x$
8. $5(3x - 2) = 6(2 - 3x)$
9. $6(3x + 7) + 4(6 - 3x) = 3(x + 30)$
10. $4t - 2(3t + 5) + 3(t - 6) = 3(4 - 2t) + 9$
11. $4(3n - 2) + 6(n + 6) = 5(2n - 7) - 1$
12. $12 + 18m + 42m + 42 = 16m + 35m - 9$
13. $2(3k - 15) - k - 50 = 10$
14. $27(2x + 3) - 20(x + 4) = 6(7 + 3x) + 7$
15. $\dfrac{2x}{3} = \dfrac{4(6 - x)}{9} + \dfrac{5x + 9}{6}$
16. $\dfrac{2(x - 3)}{3} + \dfrac{x + 2}{2} = \dfrac{3x + 4}{6} + \dfrac{1}{3}$
17. Solve $A = p + prt$ for t.
18. Solve $p = 2a + 3c$ for c.
19. The perimeter of a rectangle is 68 feet. If the width is 2 feet less than $\frac{1}{3}$ of the length, what is the length?
20. If a number is doubled and 17 is added, the result is 51. What is the number?
21. The sum of three consecutive even integers is 276. Find the integers.
22. The length of a rectangle is 12 meters less than five times its width. The perimeter of the rectangle is 60 meters. Find the width.

3.1 Exercise Answers

1. $x = 4$
2. $x = 6$
3. $y = 2$
4. $y = 15$
5. $n = 1$
6. $x = 3$
7. $x = \dfrac{2}{3}$
8. $x = \dfrac{2}{3}$
9. $x = 8$
10. $t = 7$
11. $n = -8$
12. $m = -7$

13. $k = 18$ **14.** $x = 3$ **15.** $x = 15$ **16.** $x = 3$

17. $t = \dfrac{A - p}{pr}$ **18.** $c = \dfrac{p - 2a}{3}$ **19.** 27 feet **20.** 17

21. 90, 92, 94 **22.** 7 meters

3.2 Linear Inequalities

1

Before inequalities can be solved, an understanding of the order of the real numbers is necessary. The order of the real numbers can be illustrated by use of the number line.

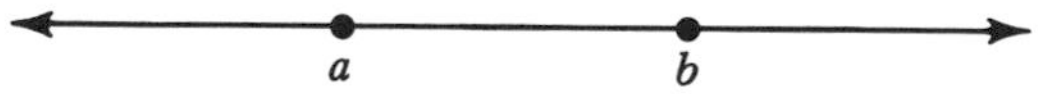

Given two real numbers a and b, a is less than b if a is to the left of b on the number line. For example, -2 is less than 3 because -2 is to the left of 3 on the number line.

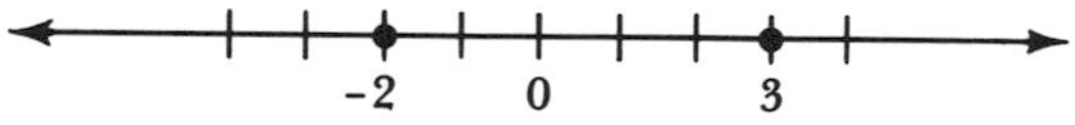

The mathematical symbol for "is less than" is $<$. The symbol for "is greater than" is $>$.

$3 < 4.2$ is read "3 is less than 4.2"
$0 > -12$ is read "0 is greater than -12"

Q1 Write the following pairs of numbers in the proper order:

a. $-2, 6$ _____ $<$ _____ b. $-3, -8$ _____ $>$ _____

c. $0, -3.2$ _____ $<$ _____ d. $-1.81, -1.811$ _____ $>$ _____

STOP • STOP • STOP • STOP • STOP • STOP • STOP • STOP • STOP

A1 a. $-2 < 6$ b. $-3 > -8$ c. $-3.2 < 0$ d. $-1.81 > -1.811$
(*Explanation:* The number to the left on the number line is of lesser value.)

Q2 Insert $<$ or $>$ between the following pairs of numbers to form a true statement:

a. -5 _____ $\dfrac{5}{2}$ b. -3 _____ -10 c. 0.001 _____ 0.01

d. $\dfrac{2}{3}$ _____ $\dfrac{3}{2}$ e. $\dfrac{-3}{2}$ _____ $\dfrac{-2}{3}$ f. $\dfrac{10}{11}$ _____ $\dfrac{8}{9}$

STOP • STOP • STOP • STOP • STOP • STOP • STOP • STOP • STOP

A2 a. $<$ b. $>$ c. $<$ d. $<$ e. $<$ f. $>$

2

The statement $a < b$ could also be interpreted as $b > a$. For example, if $0 < 5$, then $5 > 0$. The symbol $a \leqslant b$ means $a < b$ *or* $a = b$ and is read "a is less than or equal to b." The symbol $a \geqslant b$ means $a > b$ *or* $a = b$.

Examples:

$-2 \geqslant -10$ means $-2 > -10$ or $-2 = -10$
$x \leqslant 3$ means $x < 3$ or $x = 3$

Q3 Rewrite the statement by use of the word "or":

a. $x \geqslant -2$ ______ b. $x - 1 \leqslant 5$ ______

STOP • STOP • STOP • STOP • STOP • STOP • STOP • STOP • STOP

A3 a. $x > -2$ or $x = -2$ b. $x - 1 < 5$ or $x - 1 = 5$ (either order)

Q4 Rewrite the statement without the word "or":

a. $x = 5$ or $x < 5$ ______ b. $-5 < x$ or $x = -5$ ______

STOP • STOP • STOP • STOP • STOP • STOP • STOP • STOP • STOP

A4 a. $x \leqslant 5$
b. $x \geqslant -5$ ($-5 \leqslant x$ is an equivalent statement.)

Q5 Give the meaning of the following symbols:

a. $<$ ______ b. $\geqslant$ ______

STOP • STOP • STOP • STOP • STOP • STOP • STOP • STOP • STOP

A5 a. is less than b. is greater than or equal to

---

3 Positive numbers are greater than zero and can be represented as $\{x|x > 0\}$.* Negative numbers are represented as $\{x|x < 0\}$.

positive numbers $x > 0$

negative numbers $x < 0$

0

The number zero is neither positive nor negative.

Example: Under what conditions is $-x$ positive?

Solution

If $x < 0$, then $-x > 0$. This statement is read "if x is negative, then the opposite of x is positive."

*The notation $\{x|x > 0\}$ is called set builder notation and is read "the set of all x such that x is greater than 0." Note that the vertical bar is read "such that."

Q6 Insert $<$ or $>$ between the following pairs of numbers:

a. -8 ____ 0 b. $-(-8)$ ____ 0

c. if $x = -2$, then $-x$ ____ 0 d. if $x < 0$, then $-x$ ____ 0

e. if $x > 0$, then $-x$ ____ 0 f. if $-x < 0$, then x ____ 0.

STOP • STOP • STOP • STOP • STOP • STOP • STOP • STOP • STOP

A6 a. $<$ b. $>$ c. $>$ d. $>$ e. $<$ f. $>$

Q7 a. Choose a number for x such that $-x > 0$. ____

b. Choose a number x such that $-x < 0$. ____

STOP • STOP • STOP • STOP • STOP • STOP • STOP • STOP • STOP

A7 a. -3: $-(-3) = 3$, any negative real number would make it true
b. 2: any positive real number would make it true

4 If the numbers a, b, and c are arranged as follows on the number line,

a b c

b is said to be between a and c. Hence

$b > a$ *and* $b < c$ (1)

If $b > a$, it follows that $a < b$ and statement (1) could be written

$a < b$ *and* $b < c$ (2)

A shortened method for writing statement (2) is

$a < b < c$ (3)

Examples:

The statement $-1 < 0 < 1$ means that $0 > -1$ and $0 < 1$.
The statement $-5 < x < 3$ means that $x > -5$ and $x < 3$.

Statements such as $-5 < x < 3$ are called *combined inequalities* and are usually read from the middle part. For example, the statement $0 < x < 2$ should be read "x is greater than 0 and less than 2."

Q8 Rewrite the combined inequalities as two inequalities reading from the middle part:

a. $-3 < 1 < 5$ ________ **b.** $-6 < x < 6$ ________

STOP • STOP • STOP • STOP • STOP • STOP • STOP • STOP • STOP

A8 **a.** $1 > -3$ and $1 < 5$ **b.** $x > -6$ and $x < 6$

Q9 Rewrite the two inequalities as a combined inequality:

a. $1 < 3$ and $3 < 5$ ________ **b.** $0 > -1$ and $0 < 4$ ________

c. $-5 < x$ and $x < 5$ ________ **d.** $x < \frac{1}{2}$ and $x > \frac{-1}{2}$ ________

STOP • STOP • STOP • STOP • STOP • STOP • STOP • STOP • STOP

A9 **a.** $1 < 3 < 5$ **b.** $-1 < 0 < 4$

c. $-5 < x < 5$ **d.** $\frac{-1}{2} < x < \frac{1}{2}$

Q10 Is the statement $5 > x > -5$ equivalent to $-5 < x < 5$? ________

STOP • STOP • STOP • STOP • STOP • STOP • STOP • STOP • STOP

A10 yes: $5 > x > -5$ means $x < 5$ and $x > -5$ or $-5 < x$ and $x < 5$ or $-5 < x < 5$

Q11 Rewrite the combined inequalities as two inequalities, reading from the middle:

a. $-6 < x + 2 < 6$ ________

b. $0 < 2m - 3 < 8$ ________

STOP • STOP • STOP • STOP • STOP • STOP • STOP • STOP • STOP

A11 **a.** $x + 2 > -6$ and $x + 2 < 6$
b. $2m - 3 > 0$ and $2m - 3 < 8$

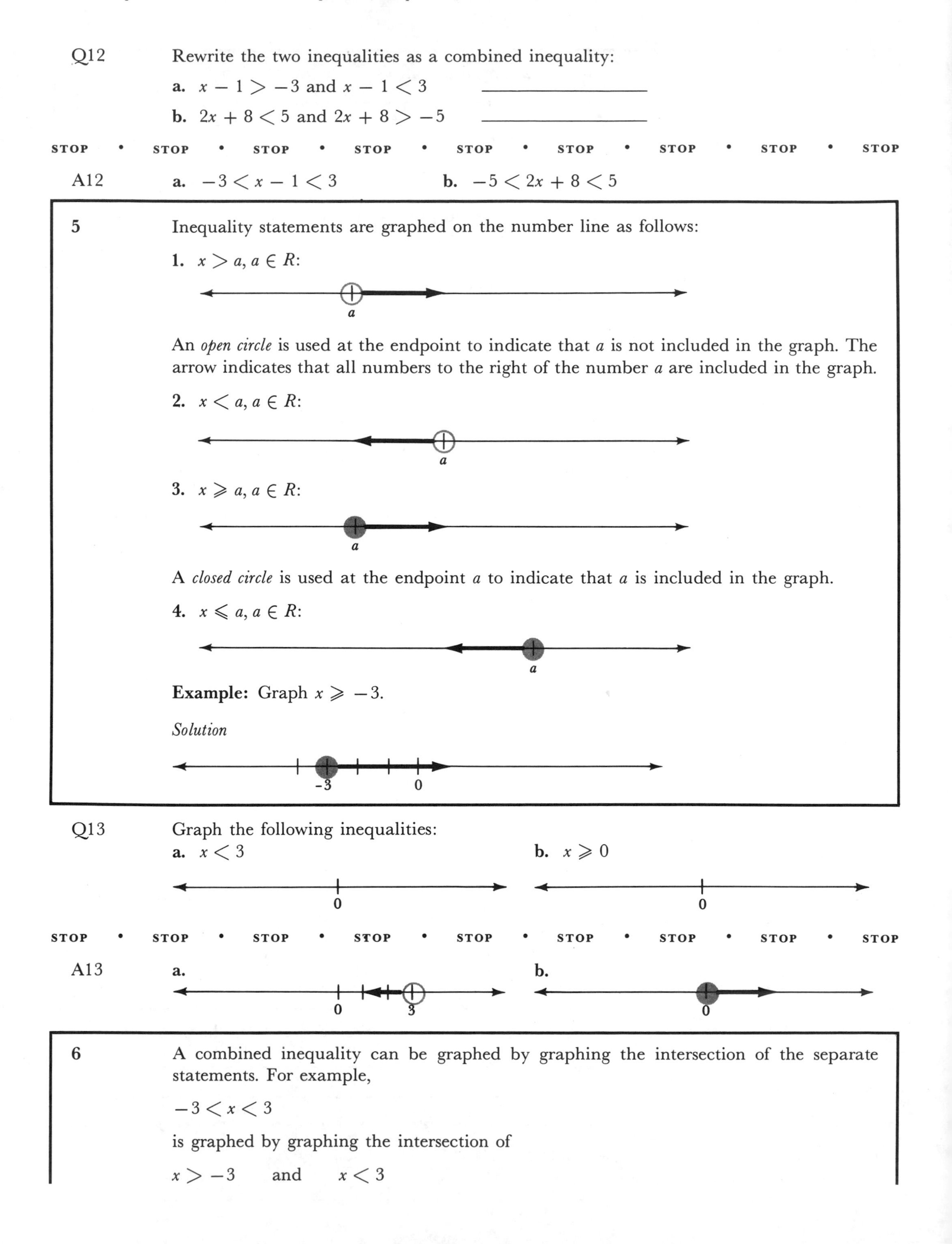

Q12 Rewrite the two inequalities as a combined inequality:

a. $x - 1 > -3$ and $x - 1 < 3$ ______________

b. $2x + 8 < 5$ and $2x + 8 > -5$ ______________

STOP • STOP • STOP • STOP • STOP • STOP • STOP • STOP • STOP

A12 a. $-3 < x - 1 < 3$ b. $-5 < 2x + 8 < 5$

5 Inequality statements are graphed on the number line as follows:

1. $x > a, a \in R$:

An *open circle* is used at the endpoint to indicate that a is not included in the graph. The arrow indicates that all numbers to the right of the number a are included in the graph.

2. $x < a, a \in R$:

3. $x \geqslant a, a \in R$:

A *closed circle* is used at the endpoint a to indicate that a is included in the graph.

4. $x \leqslant a, a \in R$:

Example: Graph $x \geqslant -3$.

Solution

Q13 Graph the following inequalities:

a. $x < 3$

b. $x \geqslant 0$

STOP • STOP • STOP • STOP • STOP • STOP • STOP • STOP • STOP

A13 a.

b.

6 A combined inequality can be graphed by graphing the intersection of the separate statements. For example,

$-3 < x < 3$

is graphed by graphing the intersection of

$x > -3$ and $x < 3$

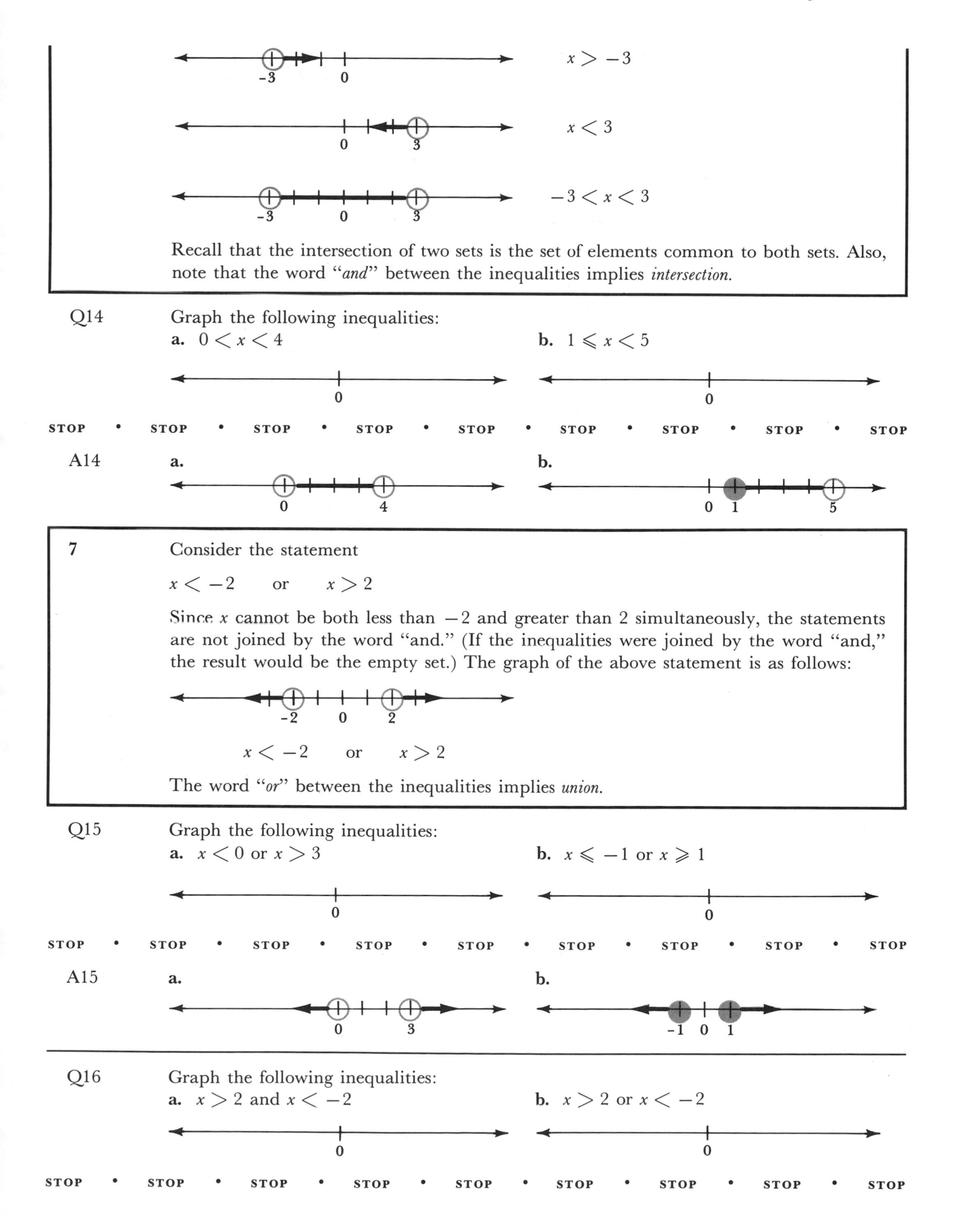

Recall that the intersection of two sets is the set of elements common to both sets. Also, note that the word "*and*" between the inequalities implies *intersection.*

Q14 Graph the following inequalities:

a. $0 < x < 4$

b. $1 \leqslant x < 5$

STOP • STOP • STOP • STOP • STOP • STOP • STOP • STOP • STOP

A14 a. b.

7 Consider the statement

$x < -2$ or $x > 2$

Since x cannot be both less than -2 and greater than 2 simultaneously, the statements are not joined by the word "and." (If the inequalities were joined by the word "and," the result would be the empty set.) The graph of the above statement is as follows:

$x < -2$ or $x > 2$

The word "*or*" between the inequalities implies *union.*

Q15 Graph the following inequalities:

a. $x < 0$ or $x > 3$

b. $x \leqslant -1$ or $x \geqslant 1$

STOP • STOP • STOP • STOP • STOP • STOP • STOP • STOP • STOP

A15 a. b.

Q16 Graph the following inequalities:

a. $x > 2$ and $x < -2$

b. $x > 2$ or $x < -2$

STOP • STOP • STOP • STOP • STOP • STOP • STOP • STOP • STOP

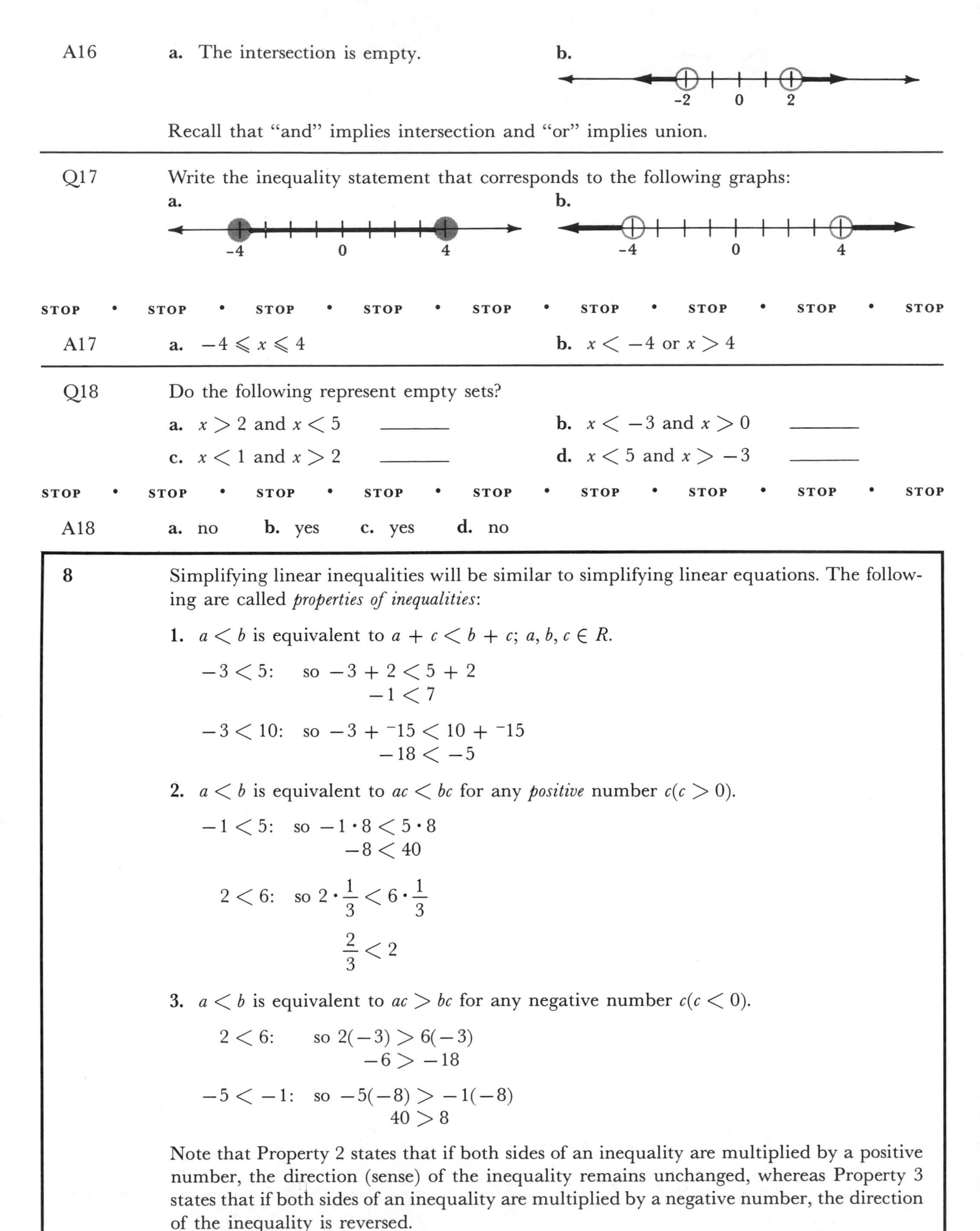

A16 **a.** The intersection is empty. **b.**

Recall that "and" implies intersection and "or" implies union.

Q17 Write the inequality statement that corresponds to the following graphs:

a. **b.**

STOP • STOP • STOP • STOP • STOP • STOP • STOP • STOP • STOP

A17 **a.** $-4 \leqslant x \leqslant 4$ **b.** $x < -4$ or $x > 4$

Q18 Do the following represent empty sets?

a. $x > 2$ and $x < 5$ ______ **b.** $x < -3$ and $x > 0$ ______

c. $x < 1$ and $x > 2$ ______ **d.** $x < 5$ and $x > -3$ ______

STOP • STOP • STOP • STOP • STOP • STOP • STOP • STOP • STOP

A18 **a.** no **b.** yes **c.** yes **d.** no

---

8 Simplifying linear inequalities will be similar to simplifying linear equations. The following are called *properties of inequalities*:

1. $a < b$ is equivalent to $a + c < b + c$; $a, b, c \in R$.

$-3 < 5$: so $-3 + 2 < 5 + 2$
$-1 < 7$

$-3 < 10$: so $-3 + {}^{-}15 < 10 + {}^{-}15$
$-18 < -5$

2. $a < b$ is equivalent to $ac < bc$ for any *positive* number $c(c > 0)$.

$-1 < 5$: so $-1 \cdot 8 < 5 \cdot 8$
$-8 < 40$

$2 < 6$: so $2 \cdot \frac{1}{3} < 6 \cdot \frac{1}{3}$
$\frac{2}{3} < 2$

3. $a < b$ is equivalent to $ac > bc$ for any negative number $c(c < 0)$.

$2 < 6$: so $2(-3) > 6(-3)$
$-6 > -18$

$-5 < -1$: so $-5(-8) > -1(-8)$
$40 > 8$

Note that Property 2 states that if both sides of an inequality are multiplied by a positive number, the direction (sense) of the inequality remains unchanged, whereas Property 3 states that if both sides of an inequality are multiplied by a negative number, the direction of the inequality is reversed.

These three properties would also hold if the symbol $<$ were replaced by the symbol $>$, $\leqslant$, or $\geqslant$.

Q19 Insert $<$ or $>$ in the blank to form a true statement:

a. $3 < 8$ is equivalent to $3 + {}^{-}7$ _____ $8 + {}^{-}7$

b. $-3 < 2$ is equivalent to $-3\left(\frac{1}{2}\right)$ _____ $2\left(\frac{1}{2}\right)$

c. $-3 < 2$ is equivalent to $-3(-6)$ _____ $2(-6)$

d. $-10 < -2$ is equivalent to $-10(5)$ _____ $-2(5)$

STOP • STOP • STOP • STOP • STOP • STOP • STOP • STOP • STOP

A19 **a.** $<$ **b.** $<$ **c.** $>$ **d.** $<$

Q20 Insert the proper symbol in the blank to form a true statement:

a. $5 > -1$ is equivalent to $5 + 3$ _____ $-1 + 3$

b. $5 > -1$ is equivalent to $5(-3)$ _____ $-1(-3)$

c. $\frac{1}{2} \leqslant \frac{2}{3}$ is equivalent to $\frac{1}{2} + \frac{-3}{5}$ _____ $\frac{2}{3} + \frac{-3}{5}$

d. $\frac{1}{2} \leqslant \frac{2}{3}$ is equivalent to $\frac{1}{2}\left(\frac{-1}{5}\right)$ _____ $\frac{2}{3}\left(\frac{-1}{5}\right)$

STOP • STOP • STOP • STOP • STOP • STOP • STOP • STOP • STOP

A20 **a.** $>$ **b.** $<$ **c.** $\leqslant$ **d.** $\geqslant$

Q21 Insert the proper symbol in the blank to form a true statement:

a. $a > b$ is equivalent to ac _____ bc, $c > 0$

b. $a \geqslant b$ is equivalent to ac _____ bc, $c < 0$

c. $x < y$ is equivalent to $x + c$ _____ $y + c$, $c < 0$

STOP • STOP • STOP • STOP • STOP • STOP • STOP • STOP • STOP

A21 **a.** $>$ **b.** $\leqslant$ **c.** $<$

9 Linear inequalities can now be simplified by use of one or more of the properties of inequalities.

Example 1: Solve $x + 2 < -5$.

Solution

$x + 2 < -5$

Add -2 to both sides (Property 1):

$x + 2 + {}^{-}2 < -5 + {}^{-}2$

$x < -7$

Example 2: Solve $3x > 12$.

Solution

Multiply both sides by $\frac{1}{3}$ (Property 2):

$$\frac{1}{3}(3x) > \frac{1}{3}(12)$$

$$x > 4$$

Example 3: Solve and graph the solution to $2 - 3x \leqslant -5$.

Solution

$2 - 3x \leqslant -5$

Add -2 to both sides (Property 1):

$-3x \leqslant -7$

Multiply both sides by $\frac{-1}{3}$ (Property 3):

$$x \geqslant \frac{7}{3}$$

0 $\frac{7}{3}$

Q22 Give the property that justifies the solution to the following inequality: $-7x + 3 < 17$

$-7x < 14$ (Property ____)

$x > -2$ (Property ____)

STOP • STOP • STOP • STOP • STOP • STOP • STOP • STOP • STOP

A22 1, 3

Q23 Solve the following inequalities:

a. $x - 7 > 4$ **b.** $x + 3 < -2$

STOP • STOP • STOP • STOP • STOP • STOP • STOP • STOP • STOP

A23 **a.** $x > 11$: add 7 to both sides

b. $x < -5$: add -3 to both sides

Q24 Solve the following inequalities:

a. $3x < 45$ **b.** $-7x \geqslant 42$

STOP • STOP • STOP • STOP • STOP • STOP • STOP • STOP • STOP

A24 **a.** $x < 15$: multiply both sides by $\frac{1}{3}$

b. $x \leqslant -6$: multiply both sides by $\frac{-1}{7}$

Q25 Solve the following inequalities:

a. $-2x + 3 \leqslant 13$ **b.** $x + 2 < 7x - 1$

STOP • **STOP** • **STOP** • **STOP** • **STOP** • **STOP** • **STOP** • **STOP** • **STOP**

A25 **a.** $x \geqslant -5$: $-2x + 3 \leqslant 13$

$$-2x \leqslant 10$$
$$x \geqslant -5$$

b. $x > \frac{1}{2}$: $x + 2 < 7x - 1$

$$x < 7x - 3$$
$$-6x < -3$$
$$x > \frac{1}{2}$$

Q26 Solve and graph $3x + 2 < -7$.

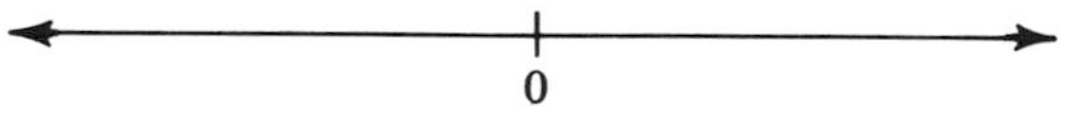

0

STOP • **STOP** • **STOP** • **STOP** • **STOP** • **STOP** • **STOP** • **STOP** • **STOP**

A26 $x < -3$

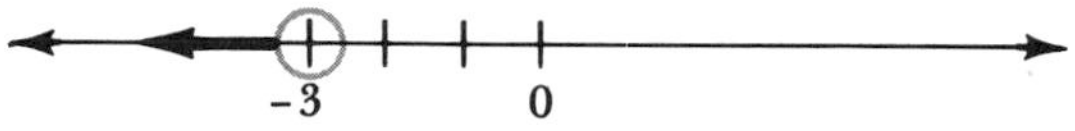

-3 0

Q27 Solve the following inequalities:

a. $\frac{2x - 3}{2} \leqslant 6$ **b.** $\frac{x - 6x}{2} < -20$

STOP • **STOP** • **STOP** • **STOP** • **STOP** • **STOP** • **STOP** • **STOP** • **STOP**

A27 **a.** $x \leqslant \frac{15}{2}$: $\frac{2x - 3}{2} \leqslant 6$

$$2x - 3 \leqslant 12$$
$$2x \leqslant 15$$
$$x \leqslant \frac{15}{2}$$

b. $x > 8$: $\frac{x - 6x}{2} < -20$

$$x - 6x < -40$$
$$-5x < -40$$
$$x > 8$$

Q28 Solve $\frac{x}{2} + 1 < \frac{x}{3} - x$.

STOP • **STOP** • **STOP** • **STOP** • **STOP** • **STOP** • **STOP** • **STOP** • **STOP**

A28 $x < \frac{-6}{7}$: $\frac{x}{2} + 1 < \frac{x}{3} - x$

$$3x + 6 < 2x - 6x$$
$$3x + 6 < -4x$$
$$7x < -6$$
$$x < \frac{-6}{7}$$

10 Since combined inequalities are actually two inequalities written as one statement, they can be solved using the properties presented in Frame 8.

Example 1: Solve $-3 < x - 5 < 3$.

Solution

Add 5 to each member of the inequality:

$$-3 + 5 < x - 5 + 5 < 3 + 5$$
$$2 < x < 8$$

Example 2:

$$-7 \leqslant 2x + 5 \leqslant 7$$
$$-7 + {}^{-}5 \leqslant 2x + 5 + {}^{-}5 \leqslant 7 + {}^{-}5$$
$$-12 \leqslant 2x \leqslant 2$$
$$\frac{1}{2}(-12) \leqslant \frac{1}{2}(2x) \leqslant \frac{1}{2}(2)$$
$$-6 \leqslant x \leqslant 1$$

Q29 Solve the following inequalities:

a. $-8 < x + 4 \leqslant 10$ **b.** $4 \leqslant 3x - 2 \leqslant 10$

STOP • STOP • STOP • STOP • STOP • STOP • STOP • STOP • STOP

A29 **a.** $-12 < x \leqslant 6$: add -4 to all members

b. $2 \leqslant x \leqslant 4$: $4 \leqslant 3x - 2 \leqslant 10$

$$6 \leqslant 3x \leqslant 12$$
$$2 \leqslant x \leqslant 4$$

Q30 Solve and graph the following inequalities:

a. $3 \geqslant 3 - 4x > -19$ **b.** $-12 \leqslant 2 - 5x \leqslant 7$

STOP • STOP • STOP • STOP • STOP • STOP • STOP • STOP • STOP

A30 **a.** $0 \leqslant x < \frac{11}{2}$ **b.** $\frac{14}{5} \geqslant x \geqslant -1$ (graphs not provided)

Q31 Rewrite the answer to Q30b so that it reads $\leqslant$. ____________

STOP • STOP • STOP • STOP • STOP • STOP • STOP • STOP • STOP

A31 $-1 \leqslant x \leqslant \frac{14}{5}$

11 The solution set for inequalities could be the set of real numbers or the empty set. Consider the following examples:

Example 1: Solve $3 + x > -1 + x$.

Solution

Add 1 to both sides:

$4 + x > x$

Add $-x$ to both sides:

$4 > 0$

The true statement $4 > 0$ indicates that the solution is all real numbers. The left side will be 4 greater than the right side for any real number x; hence the solution set is R.

Example 2: Solve $-2x + 3 < x - 3x - 1$.

Solution

Add 1 to both sides and simplify:

$-2x + 4 < -2x$

Add $2x$ to both sides:

$4 < 0$

The false statement $4 < 0$ indicates that there is no solution. The left side will be 4 greater than the right side; however, the inequality says otherwise. Hence the solution set is $\varnothing$.

Q32 Solve the following inequalities:

a. $4 - 3x < 2x - 5x + 13$ **b.** $4x > x + 3x + 7$

STOP • STOP • STOP • STOP • STOP • STOP • STOP • STOP • STOP

A32* **a.** R:
$$4 - 3x < 2x - 5x + 13$$
$$4 - 3x < -3x + 13$$
$$-3x < -3x + 9$$
$$0 < 9 \text{ (true)}$$

b. $\varnothing$:
$$4x > x + 3x + 7$$
$$4x > 4x + 7$$
$$0 > 7 \text{ (false)}$$

*Note that Q32 said "solve." According to the agreement in Section 3.1, the answer would not be written as a set. However, for convenience, whenever the solution to an equation is all real numbers, the answer will be given as R, the set of real numbers. Similarly, for convenience, if there is no solution to an equation, the answer will be given as $\varnothing$, the empty set. This practice will be followed throughout the rest of the text.

Q33 John and his helper, Kevin, work together on a job. John earns twice as much as Kevin. If they receive at least $840 for doing the job, find the least amount that each would receive.

STOP • STOP • STOP • STOP • STOP • STOP • STOP • STOP • STOP

A33 John earns at least $560 and Kevin earns at least $280:
Let x = Kevin's earnings and $2x$ = John's

$$x + 2x \geqslant 840$$
$$3x \geqslant 840$$
$$x \geqslant 280$$

Q34 Mr. Cole can drive from his home on a trip at the rate of 30 mph and return home at the rate of 40 mph. What is the greatest total distance that he can drive out and return home if he has at most 7 hours to spend on the trip?

STOP • STOP • STOP • STOP • STOP • STOP • STOP • STOP • STOP

A34 240 miles: $d = rt$

$30t$ = distance out, $40(7 - t)$ = distance back

$$d = 30t + 40(7 - t)$$

Since the trip out equals the trip back:

$$30t = 40(7 - t)$$
$$30t = 280 - 40t$$
$$70t = 280$$
$$t = 4 \text{ hours}$$
$$d = 30(4) + 40(3)$$

This completes the instruction for this section.

3.2 Exercises

1. Give the meaning of the following symbols:
   a. $\leqslant$ **b.** $>$ **c.** $<$ **d.** $\geqslant$

2. Given $n \in R$, when is:
   a. $-n > 0$? b. $-n < 0$? c. $-n = 0$?
3. Insert the correct inequality symbol:
   a. $x < y$ is equivalent to $x \cdot c$ ____ $y \cdot c$, where $c > 0$.
   b. $x > y$ is equivalent to $x \cdot c$ ____ $y \cdot c$, where $c < 0$.
   c. $x \leqslant y$ is equivalent to $x + c$ ____ $y + c$, where $c < 0$.
4. Express the following statements using inequality symbols when $x \in R$:
   a. positive real numbers b. negative real numbers
   c. nonnegative real numbers
5. Solve the following inequalities:
   a. $7x < 7$
   b. $-x > 12$
   c. $5x + 6 < 13$
   d. $\frac{x}{3} + 2 < \frac{x}{4} - 2x$
   e. $-3x + 1 > 0$
   f. $x + 6 \geqslant 4 - 3x$
   g. $4 + x > x - 1$
   h. $3x - 4 > 7x + 5$
   i. $12 - 3x < 1 + 2x + 8$
   j. $7x + 1 + 3x \geqslant 2x - 15$
   k. $5x - 1 \leqslant 6$
   l. $-4x + 8 \geqslant -12$
   m. $6x - 9 > -9$
   n. $3(x + 2) - 5x < x$
   o. $\frac{1}{2}(x - 1) + \frac{3}{4}(x + 1) < 0$
   p. $5 \leqslant 2x - 3 \leqslant 13$
   q. $8 < 5x + 4 \leqslant 10$
   r. $-7 \leqslant 2 - 3x \leqslant 5$
   s. $5 - x < -x + 3$
   t. $2 > -3 - 3x \geqslant -7$
6. Graph the solution to the inequalities for Q5m, p, and r.
7. A chemist has 60 quarts of an acid solution that is 20 percent pure acid. At least, how many quarts of pure acid must be added to this solution to obtain a solution that is at least 50 percent pure acid? (*Hint:* Let x = amount of acid added and write an equation in which both members represent the amount of pure acid.)

3.2 Exercise Answers

1. a. is less than or equal to b. is greater than c. is less than
   d. is greater than or equal to
2. a. $n < 0$ b. $n > 0$ c. $n = 0$
3. a. $<$ b. $<$ c. $\leqslant$
4. a. $x > 0$ b. $x < 0$ c. $x \geqslant 0$
5. a. $x < 1$ b. $x < -12$ c. $x < \frac{7}{5}$ d. $x < \frac{-24}{25}$
   e. $x < \frac{1}{3}$ f. $x \geqslant \frac{-1}{2}$ g. R h. $x < \frac{-9}{4}$
   i. $x > \frac{3}{5}$ j. $x \geqslant -2$ k. $x \leqslant \frac{7}{5}$ l. $x \leqslant 5$
   m. $x > 0$ n. $x > 2$ o. $x < \frac{-1}{5}$ p. $4 \leqslant x \leqslant 8$
   q. $\frac{4}{5} < x \leqslant \frac{6}{5}$ r. $-1 \leqslant x \leqslant 3$ or $3 \geqslant x \geqslant -1$
   s. $\varnothing$ t. $\frac{-5}{3} < x \leqslant \frac{4}{3}$

6.

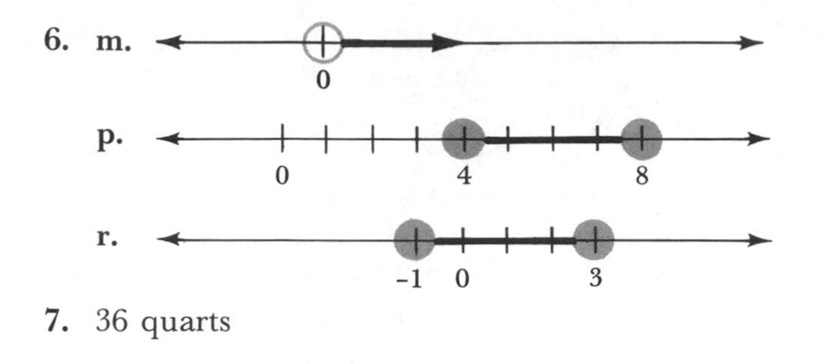

7. 36 quarts

3.3 Absolute Value

---

1 A number expressing both directional and quantitative values is called a *signed number*. On the number line signed numbers indicate *direction* and *distance* from zero. For example, $+2$ indicates the number 2 units to the right of zero; -3 is a number 3 units to the left of zero.

The distance between a number and zero will be defined as the *absolute value* of the number. For example, the absolute value of $+8$ is 8, because $+8$ is a distance of 8 units from zero. The absolute value of -8 is 8, because -8 is a distance of 8 units from zero.

-8 ← 8 units → 0 ← 8 units → +8

It is important to note that distance is a nonnegative quantity.

Q1 What is the absolute value (distance from zero) of the following signed numbers?

a. $+10$ ____ **b.** -5 ____ **c.** $-\sqrt{3}$ ____ **d.** 0 ____

STOP • STOP • STOP • STOP • STOP • STOP • STOP • STOP • STOP

A1 **a.** 10 **b.** 5 **c.** $\sqrt{3}$
d. 0: since 0 is zero units from zero, the absolute value of 0 is 0

---

2 The absolute value of a real number x is the distance between x and 0. The absolute value of x is abbreviated $|x|$.

Examples:

$|+100| = 100$ $|-10| = 10$ $|0| = 0$

Q2 $|-3|$ is read "______________________________."

STOP • STOP • STOP • STOP • STOP • STOP • STOP • STOP • STOP

A2 the absolute value of negative three

Q3 Evaluate the following expressions:

a. $|-13| =$ ____ **b.** $\left|\frac{-3}{2}\right| =$ ____ **c.** $|9-13| =$ ____

d. $|-7-(-7)| =$ ____ **e.** $-|-12+2| =$ ____

f. $|-2| + |-8| =$ ____

STOP • STOP • STOP • STOP • STOP • STOP • STOP • STOP • STOP

A3 a. 13 b. $\frac{3}{2}$ c. 4 d. 0 e. -10 f. 10

Q4 If $x = -3$ and $y = 2$, evaluate the following expressions:

a. $|x| + |y|$ b. $|x + y|$ c. $|x - y|$
d. $|x| - |y|$ e. $|xy|$ f. $2|x| - 4|y|$

STOP • STOP • STOP • STOP • STOP • STOP • STOP • STOP • STOP

A4 a. 5 b. 1 c. 5 d. 1 e. 6 f. -2

---

3 The absolute value of a number is defined algebraically as follows:

$|x| = x$ if $x \geqslant 0$ (read "the absolute value of a number is the number if the number is positive or zero")

$|x| = -x$ if $x < 0$ (read "the absolute value of a number is the *opposite* of the number if the number is negative")

Examples:

$|-3| = -(-3) = 3$

$|n| = -n$ if $n < 0$

Q5 Find:

a. $|x| =$ _____ if $x > 0$ b. $|a| =$ _____ if $a < 0$

c. $|x| =$ _____ if $x = 0$ d. $|-m| =$ _____ if $-m > 0$

STOP • STOP • STOP • STOP • STOP • STOP • STOP • STOP • STOP

A5 a. x b. $-a$ c. x d. $-m$

Q6 When is $|n| = -n$? (write in words) ____________________

STOP • STOP • STOP • STOP • STOP • STOP • STOP • STOP • STOP

A6 When n is negative.

---

4 One method of solving absolute-value equations is to recall that absolute value means distance from zero. Solving the equation $|x| = c$, $c > 0$, means finding the values of x that are c units from zero.

Example: Solve $|x| = 4$.

Solution

x = −4   0   x = 4
←4 units→←4 units→

$x = -4$ or $x = 4$

The solution set is $\{-4, 4\}$; that is, two values satisfy the equation $|x| = 4$.

Check: Let $x = -4$: $|-4| \stackrel{?}{=} 4$, $4 = 4$

Let $x = 4$: $|4| \stackrel{?}{=} 4$, $4 = 4$

Q7 **a.** Solve the absolute-value equation $|x| = 6$.

b. Check the solution.

STOP • STOP • STOP • STOP • STOP • STOP • STOP • STOP • STOP

A7 **a.** $x = -6$ or $x = 6$: -6 and 6 are 6 units from zero

b. Check: Let $x = -6$: $|-6| \stackrel{?}{=} 6$, $6 = 6$ Let $x = 6$: $|6| \stackrel{?}{=} 6$, $6 = 6$

Q8 Solve the following equations:

a. $|x| = 17$ **b.** $|x| = -5$

STOP • STOP • STOP • STOP • STOP • STOP • STOP • STOP • STOP

A8 **a.** $x = -17$ or $x = 17$

b. $\varnothing$: the absolute value of a number is always nonnegative

Q9 Solve the following equations:

a. $|x| = \sqrt{3}$ **b.** $|x| = 0$

STOP • STOP • STOP • STOP • STOP • STOP • STOP • STOP • STOP

A9 **a.** $x = -\sqrt{3}$ or $x = \sqrt{3}$ **b.** $x = 0$

5 More complicated absolute-value equations can be solved by also expressing the quantity between the absolute-value symbols as a distance from zero. Solving the equation $|x + a| = c$, $c > 0$, means that the expression $x + a$ is c units from zero.

Example: Solve $|x - 3| = 5$.

Solution

-5 0 5

← x - 3 → ← x - 3 →

The expression $x - 3$ is 5 units from zero; hence

$x - 3 = -5$ or $x - 3 = 5$

Each equation can now be solved for x,

$x = -2$ or $x = 8$

Check: Let $x = -2$: $|-2 - 3| \stackrel{?}{=} 5$ Let $x = 8$: $|8 - 3| \stackrel{?}{=} 5$

$|-5| \stackrel{?}{=} 5$ $|5| \stackrel{?}{=} 5$

$5 = 5$ $5 = 5$

Q10 Solve and check the equation $|x + 3| = 7$.

STOP • STOP • STOP • STOP • STOP • STOP • STOP • STOP • STOP

A10 $x = -10$ or $x = 4$: the expression $x + 3$ is 7 units from zero; hence $x + 3 = -7$ or $x + 3 = 7$.

Check: Let $x = -10$: $|-10 + 3| \stackrel{?}{=} 7$ Let $x = 4$: $|4 + 3| \stackrel{?}{=} 7$

$|-7| \stackrel{?}{=} 7$ $|7| \stackrel{?}{=} 7$

$7 = 7$ $7 = 7$

Q11 Solve the following equations:

a. $|x + 2| = 10$ **b.** $|x - 7| = 1$

STOP • STOP • STOP • STOP • STOP • STOP • STOP • STOP • STOP

A11 **a.** $x = -12$ or $x = 8$:

$x + 2 = -10$ or $x + 2 = 10$

b. $x = 6$ or $x = 8$:

$x - 7 = -1$ or $x - 7 = 1$

Q12 Solve the following equations:

a. $|3x + 2| = 5$ **b.** $|2x + 5| = 7$

STOP • STOP • STOP • STOP • STOP • STOP • STOP • STOP • STOP

A12 **a.** $x = 1$ or $x = \frac{-7}{3}$:

$3x + 2 = -5$ or $3x + 2 = 5$

$3x = -7$ $3x = 3$

$x = \frac{-7}{3}$ $x = 1$

b. $x = -6$ or $x = 1$:

$2x + 5 = -7$ or $2x + 5 = 7$

Q13 Solve:

a. $|3x - 1| = 5$ **b.** $\left|\frac{2x}{3} - 1\right| = 2$

STOP • STOP • STOP • STOP • STOP • STOP • STOP • STOP • STOP

A13 **a.** $x = \frac{-4}{3}$ or $x = 2$ **b.** $x = \frac{-3}{2}$ or $x = \frac{9}{2}$

Q14 Solve:

a. $|2 - 4x| = 6$ **b.** $|7 - \frac{3}{2}x| = -12$

STOP • STOP • STOP • STOP • STOP • STOP • STOP • STOP • STOP

A14 **a.** $x = -1$ or $x = 2$

$2 - 4x = -6$ or $2 - 4x = 6$

$-4x = -8$ $\quad -4x = 4$

$x = 2$ $\quad x = -1$

b. $\varnothing$: absolute values must be nonnegative

6 Absolute-value inequalities can be solved utilizing the techniques developed for solving absolute-value equations. Solving the inequality $|x| < c$, $c > 0$, means finding the values of x that are less than c units from zero.

Example: Solve $|x| < 3$.

Solution

x must be less than 3 units from zero; hence $x > -3$ and $x < 3$. These inequalities can be combined as

$-3 < x < 3$

Geometrically, the solution is

x → ← x

-3 0 3

$x > -3$ and $x < 3$

Q15 Solve and graph the inequality $|x| < 5$.

0

STOP • STOP • STOP • STOP • STOP • STOP • STOP • STOP • STOP

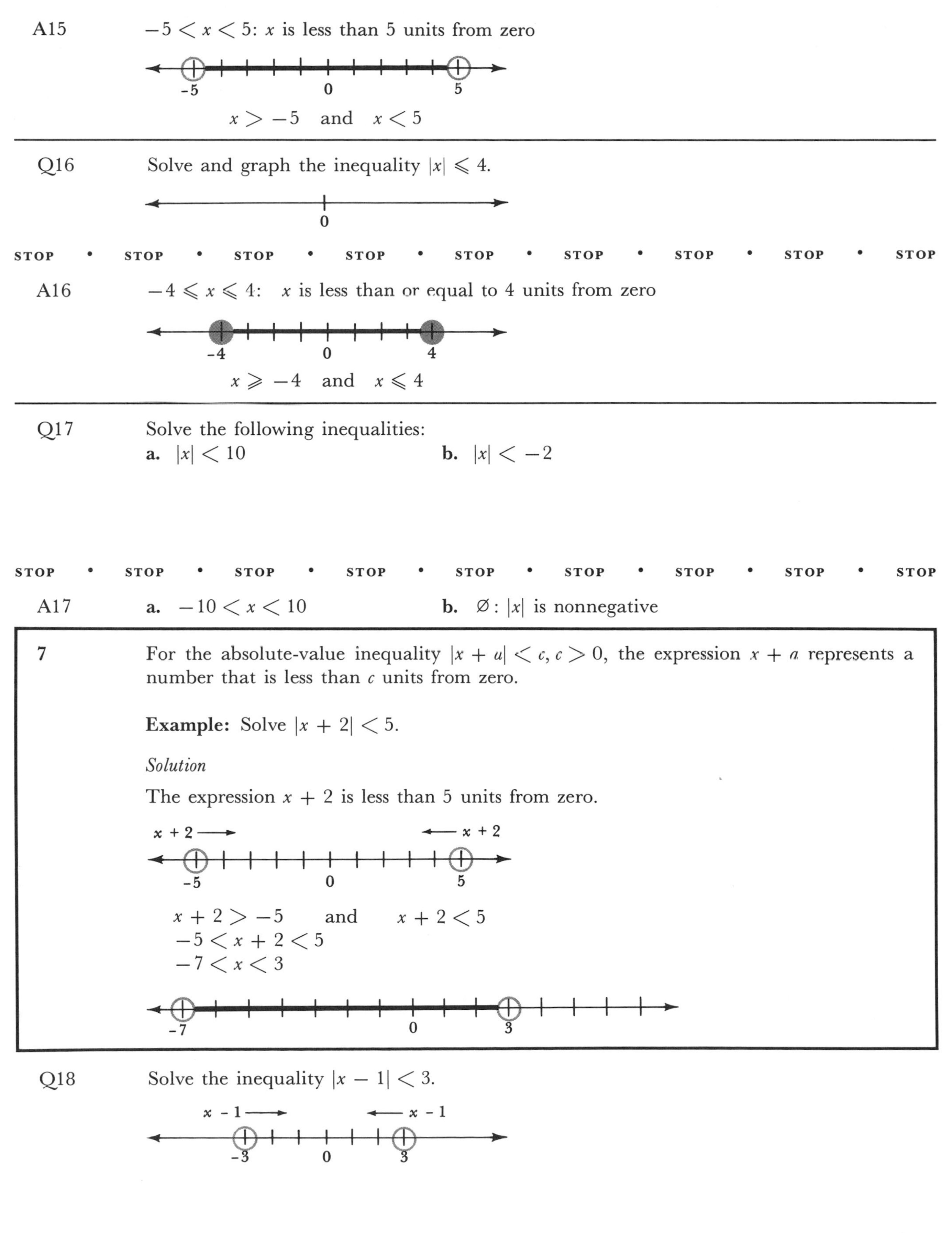

A15 $-5 < x < 5$: x is less than 5 units from zero

$x > -5$ and $x < 5$

Q16 Solve and graph the inequality $|x| \leqslant 4$.

STOP • STOP • STOP • STOP • STOP • STOP • STOP • STOP • STOP

A16 $-4 \leqslant x \leqslant 4$: x is less than or equal to 4 units from zero

$x \geqslant -4$ and $x \leqslant 4$

Q17 Solve the following inequalities:

a. $|x| < 10$ **b.** $|x| < -2$

STOP • STOP • STOP • STOP • STOP • STOP • STOP • STOP • STOP

A17 **a.** $-10 < x < 10$ **b.** $\varnothing$: $|x|$ is nonnegative

7 For the absolute-value inequality $|x + a| < c$, $c > 0$, the expression $x + a$ represents a number that is less than c units from zero.

Example: Solve $|x + 2| < 5$.

Solution

The expression $x + 2$ is less than 5 units from zero.

$x + 2 > -5$ and $x + 2 < 5$

$-5 < x + 2 < 5$

$-7 < x < 3$

Q18 Solve the inequality $|x - 1| < 3$.

STOP • STOP • STOP • STOP • STOP • STOP • STOP • STOP • STOP

A18 $-2 < x < 4$: $-3 < x - 1 < 3$
$-2 < x < 4$

Q19 Solve the following inequalities:
a. $|x + 2| < 7$ **b.** $|3 - x| < 1$

STOP • STOP • STOP • STOP • STOP • STOP • STOP • STOP • STOP

A19 **a.** $-9 < x < 5$: $-7 < x + 2 < 7$
$-9 < x < 5$

b. $2 < x < 4$: $-1 < 3 - x < 1$
$-4 < -x < -2$
$4 > x > 2$
or $2 < x < 4$

Q20 Solve:
a. $|x - 1| \leqslant 3$ **b.** $|3 - 2x| < 5$

STOP • STOP • STOP • STOP • STOP • STOP • STOP • STOP • STOP

A20 **a.** $-2 \leqslant x \leqslant 4$: $-3 \leqslant x - 1 \leqslant 3$
$-2 \leqslant x \leqslant 4$

b. $-1 < x < 4$: $-5 < 3 - 2x < 5$
$-8 < -2x < 2$
$4 > x > -1$
or $-1 < x < 4$

---

8 Solving the absolute-value inequality $|x| > c$, $c > 0$, means finding the values of x that are more than c units from zero.

Example: Solve $|x| > 3$.

Solution

x must be more than 3 units from zero; hence $x < -3$ or $x > 3$. Geometrically, the solution is

← x x →
-3 0 3

$x < -3$ or $x > 3$

Note that the solution must be joined by the word "or." Combining the inequalities with "and" would yield the empty set.

Q21 Solve and graph the inequality $|x| > 5$.

0

STOP • STOP • STOP • STOP • STOP • STOP • STOP • STOP • STOP

A21 $x < -5$ or $x > 5$: x is more than 5 units from zero

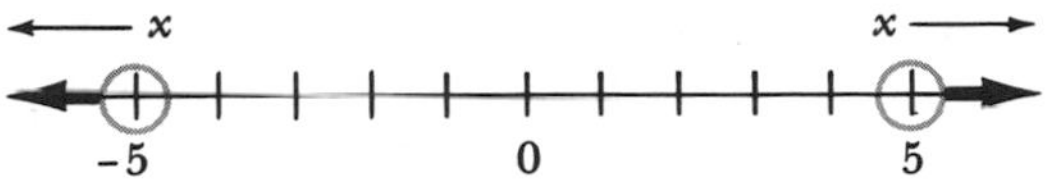

$x < -5$ or $x > 5$

Q22 Solve the following inequalities:

a. $|x| > 12$ b. $|x| > -3$

STOP • STOP • STOP • STOP • STOP • STOP • STOP • STOP • STOP

A22 a. $x < -12$ or $x > 12$: x is more than 12 units from zero

b. R: $|x|$ is nonnegative and always greater than -3

Q23 Solve the following inequalities:

a. $|x| \geqslant 1$ b. $|-x| > \sqrt{2}$

STOP • STOP • STOP • STOP • STOP • STOP • STOP • STOP • STOP

A23 a. $x \leqslant -1$ or $x \geqslant 1$

b. $x < -\sqrt{2}$ or $x > \sqrt{2}$:
$-x < -\sqrt{2}$ or $-x > \sqrt{2}$
$x > \sqrt{2}$ or $x < -\sqrt{2}$

9 For the absolute-value inequality $|x + a| > c, c > 0$, the expression $x + a$ represents a number that is more than c units from zero.

Example: Solve $|x - 1| > 4$.

Solution

The expression $x - 1$ is more than 4 units from zero.

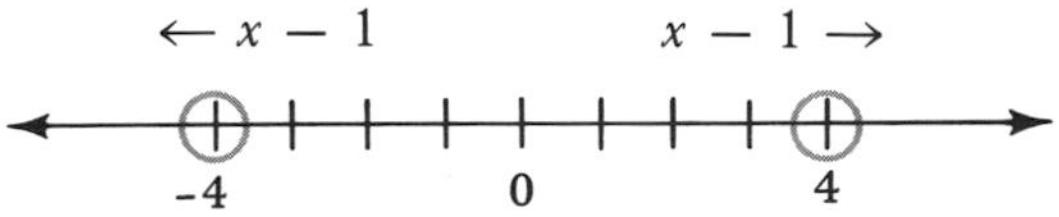

$x - 1 < -4$ or $x - 1 > 4$
$x < -3$ or $x > 5$

Geometrically the solution is as shown.

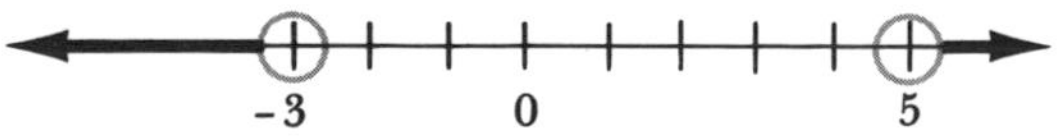

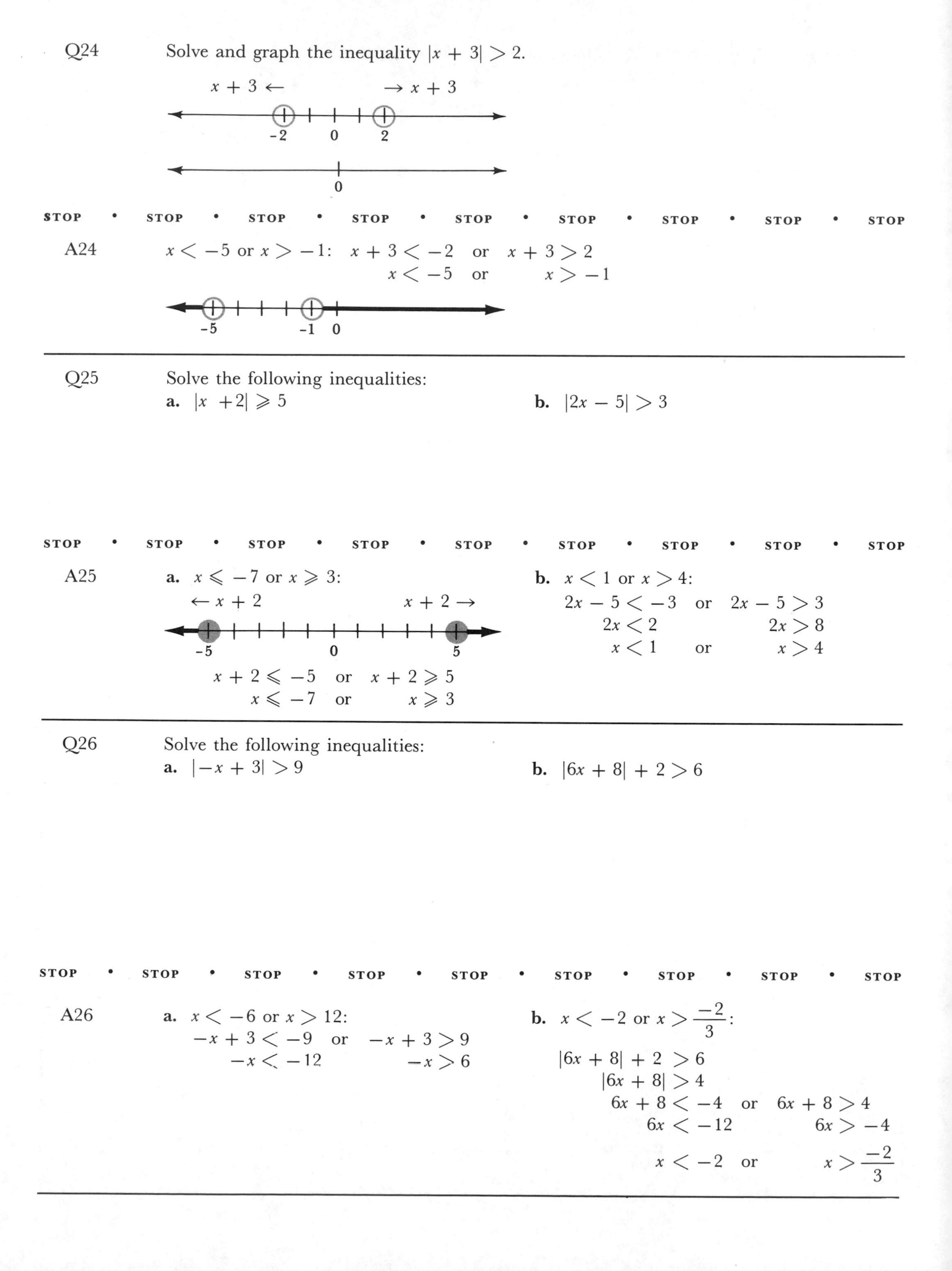

Q24 Solve and graph the inequality $|x + 3| > 2$.

$x + 3 \leftarrow$ $\rightarrow x + 3$

-2 0 2

0

STOP • STOP • STOP • STOP • STOP • STOP • STOP • STOP • STOP

A24 $x < -5$ or $x > -1$: $x + 3 < -2$ or $x + 3 > 2$

$x < -5$ or $x > -1$

-5 -1 0

Q25 Solve the following inequalities:

a. $|x + 2| \geqslant 5$

b. $|2x - 5| > 3$

STOP • STOP • STOP • STOP • STOP • STOP • STOP • STOP • STOP

A25 a. $x \leqslant -7$ or $x \geqslant 3$:

$\leftarrow x + 2$ $x + 2 \rightarrow$

-5 0 5

$x + 2 \leqslant -5$ or $x + 2 \geqslant 5$

$x \leqslant -7$ or $x \geqslant 3$

b. $x < 1$ or $x > 4$:

$2x - 5 < -3$ or $2x - 5 > 3$

$2x < 2$ $2x > 8$

$x < 1$ or $x > 4$

Q26 Solve the following inequalities:

a. $|-x + 3| > 9$

b. $|6x + 8| + 2 > 6$

STOP • STOP • STOP • STOP • STOP • STOP • STOP • STOP • STOP

A26 a. $x < -6$ or $x > 12$:

$-x + 3 < -9$ or $-x + 3 > 9$

$-x < -12$ $-x > 6$

b. $x < -2$ or $x > \frac{-2}{3}$:

$|6x + 8| + 2 > 6$

$|6x + 8| > 4$

$6x + 8 < -4$ or $6x + 8 > 4$

$6x < -12$ $6x > -4$

$x < -2$ or $x > \frac{-2}{3}$

Q27 Solve the following inequalities:

a. $|x| < k, k > 0$ b. $|x| > k, k > 0$

STOP • STOP • STOP • STOP • STOP • STOP • STOP • STOP • STOP

A27 a. $-k < x < k$ b. $x < -k$ or $x > k$

This completes the instruction for this section.

3.3 Exercises

1. a. When is the absolute value of a number k equal to $-k$?
   b. Read (write in words) $|x| = -x$ if $x < 0$.
2. Find:
   a. $|m| =$ ______ if $m > 0$ b. $|-k| =$ ______ if $k < 0$
   c. $|k| = -k$ if ______ d. $|y| = 0$ if ______
3. If $x = -6$ and $y = -1$, find the following:
   a. $|x| + |y|$ b. $|x - y|$ c. $\dfrac{|x|}{|y|}$ d. $|y| - |x|$
4. Solve the following equations:
   a. $|x| = 15$ b. $|-2x| = 7$ c. $|2x - 6| = 0$
   d. $|x - 7| = 2$ e. $|2x - 5| = 9$ f. $|2 - 3x| = 6$
   g. $|5x - 5| = 10$ h. $|4x + 6| = -8$ i. $\left|x - \dfrac{1}{2}\right| = 2$
5. Solve the following inequalities:
   a. $|x| < 23$ b. $|x| > 5$ c. $|x + 3| < 6$
   d. $|x - 1| \leqslant 7$ e. $|x + 4| > 1$ f. $|2x - 1| > -6$
   g. $|2x + 1| > 3$ h. $|-5x + 2| > 8$ i. $|-5x + 2| \leqslant 8$
   j. $|4 - x| \geqslant 10$ k. $|3 - x| \leqslant 0$ l. $\left|\dfrac{1}{2}x - 5\right| \leqslant 13$
6. Graph the solution to problems **5g** and **5i.**

3.3 Exercise Answers

1. a. When k is negative.
   b. The absolute value of x is equal to the opposite of x if x is negative.
2. a. m b. $-k$ c. $k < 0$ d. $y = 0$
3. a. 7 b. 5 c. 6 d. -5
4. a. $x = -15$ or $x = 15$ b. $x = \dfrac{-7}{2}$ or $x = \dfrac{7}{2}$ c. $x = 3$
   d. $x = 5$ or $x = 9$ e. $x = -2$ or $x = 7$
   f. $x = \dfrac{-4}{3}$ or $x = \dfrac{8}{3}$ g. $x = -1$ or $x = 3$ h. $\varnothing$
   i. $x = \dfrac{-3}{2}$ or $x = \dfrac{5}{2}$
5. a. $-23 < x < 23$ b. $x < -5$ or $x > 5$ c. $-9 < x < 3$

d. $-6 \leqslant x \leqslant 8$ **e.** $x < -5$ or $x > -3$ **f.** R

g. $x < -2$ or $x > 1$ **h.** $x < \frac{-6}{5}$ or $x > 2$ **i.** $\frac{-6}{5} \leqslant x \leqslant 2$

j. $x \leqslant -6$ or $x \geqslant 14$ **k.** $x = 3$ **l.** $-16 \leqslant x \leqslant 36$

6. **5g.**

5i.

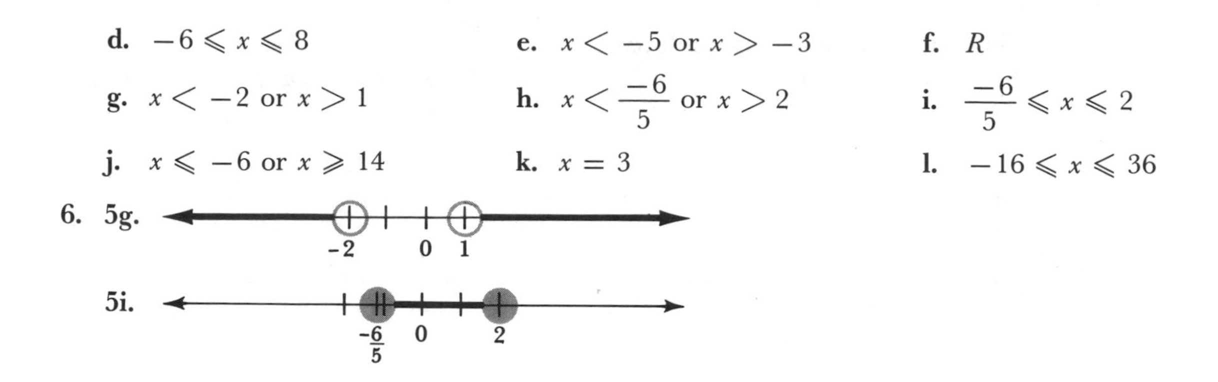

Chapter 3 Sample Test

At the completion of Chapter 3 it is expected that you will be able to work the following problems.

3.1 Solution of Linear Equations

1. Solve the following equations:

a. $12x = -48$ **b.** $-3 = x - 7$

c. $11x + 12 = 39$ **d.** $4n + 3n - 7 = 30$

e. $\frac{k}{3} - \frac{k}{4} = \frac{2}{3}$ **f.** $10 - y = 3 - 6(y - 2)$

2. **a.** Find three consecutive integers whose sum is 99.

b. Helen is 3 times as old as Sally. Ten years from now, Helen will be twice as old as Sally. How old is each now?

3.2 Linear Inequalities

3. Solve the following inequalities:

a. $x - 3 > 4$ **b.** $-3x \leqslant -12$

c. $4x - 3 < 17$ **d.** $3(2x - 1) < 21$

e. $-8 < 2x - 1 \leqslant 5$ **f.** $0 < \frac{3}{2}x + 5 < 15$

4. Graph the solution to problems **3b** and **3e.**

5. **a.** $x < y$ is equivalent to xc _____ yc, where $c < 0$.

b. $x > y$ is equivalent to $x + c$ _____ $y + c$, where $c < 0$.

3.3 Absolute Value

6. Find:

a. $|x| =$ _____ if $x < 0$ **b.** $|k| = k$ if _____

7. Evaluate the following expression if $x = -5$ and $y = 2$:

a. $|x + y|$ **b.** $|y| - |x|$

8. Solve the following equations:

a. $|x - 2| = 9$ **b.** $|4 - 3x| = 1$

9. Solve the following inequalities:
   a. $|x - 3| \leqslant 5$ b. $|-2x + 3| > 2$
10. Graph the solution to the inequalities of problem 9.

Chapter 3 Sample Test Answers

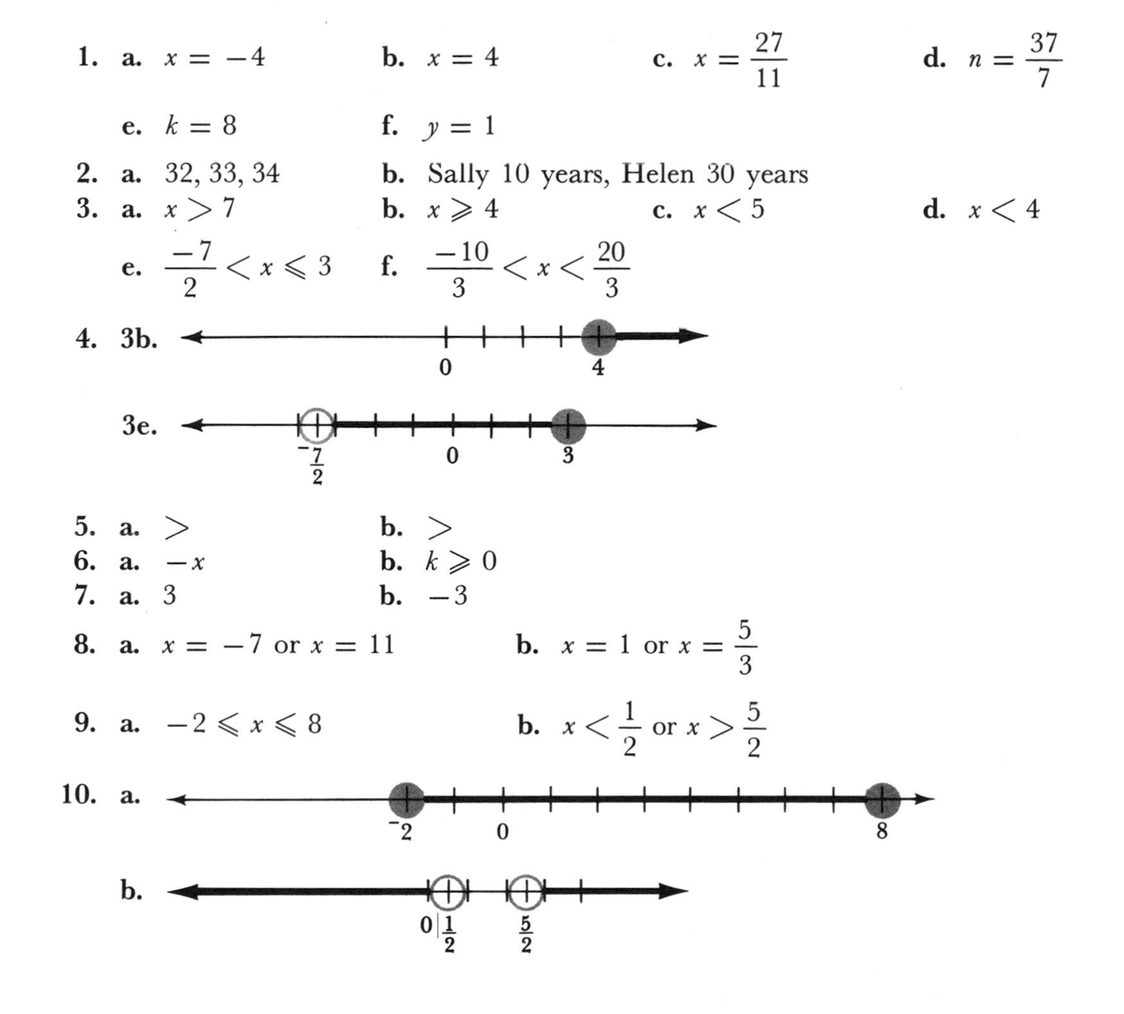

1. a. $x = -4$ b. $x = 4$ c. $x = \frac{27}{11}$ d. $n = \frac{37}{7}$
   e. $k = 8$ f. $y = 1$
2. a. 32, 33, 34 b. Sally 10 years, Helen 30 years
3. a. $x > 7$ b. $x \geqslant 4$ c. $x < 5$ d. $x < 4$
   e. $\frac{-7}{2} < x \leqslant 3$ f. $\frac{-10}{3} < x < \frac{20}{3}$
4. 3b.

   3e.

5. a. $>$ b. $>$
6. a. $-x$ b. $k \geqslant 0$
7. a. 3 b. -3
8. a. $x = -7$ or $x = 11$ b. $x = 1$ or $x = \frac{5}{3}$
9. a. $-2 \leqslant x \leqslant 8$ b. $x < \frac{1}{2}$ or $x > \frac{5}{2}$
10. a.

    b.

Chapter 4

Irrational and Complex Numbers

The set of rational numbers is large enough to provide solutions to most of the problems of arithmetic. However, there is no solution in the rational numbers to equations such as $x^2 = 2$. This gives a reason to create a new set of numbers that will provide solutions to equations which formerly had no solution. The new set of numbers is called the irrational numbers. Operations on irrational numbers will be discussed, and the chapter will conclude with a section on a still larger set of numbers, the complex numbers.

4.1 Introduction to Radicals and Simplest Radical Form

1 Rational numbers can be written as the quotient of two integers. Each rational number can be located on a number line. However, when each rational number is associated with a point on the number line, there are many points with no number associated with them. Numbers associated with the unused points on the number line are called *irrational numbers*. An irrational number cannot be written as the quotient of two integers. The number π, used in the formula for the area of a circle, is an example of an irrational number. Also, many roots of integers, such as $\sqrt{2}$, $\sqrt{10}$, and $\sqrt[3]{-3}$, are irrational numbers. The process of simplifying roots will be developed in this section. An irrational number written as a decimal goes on forever without repeating any block of digits.

There is no number that is both rational and irrational. The sets are said to be *disjoint* because their intersection is empty. The union of the rational and irrational numbers forms the set of real numbers. The real numbers can be put in a one-to-one correspondence with the points on a number line. This means that every point is associated with exactly one number and every number is associated with exactly one point. The real numbers will be represented by R, the irrational numbers by L, and the rational numbers by Q.

Q1 The real numbers can be separated into two sets, the __________ numbers and the __________ numbers.

STOP • STOP • STOP • STOP • STOP • STOP • STOP • STOP • STOP

A1 rational, irrational (any order)

Q2 The set of numbers that is in one-to-one correspondence with the number line is called the set of __________.

STOP • STOP • STOP • STOP • STOP • STOP • STOP • STOP • STOP

A2 real numbers

Q3 **a.** $Q \cap L =$ ______ **b.** $Q \cup L =$ ______

STOP • STOP • STOP • STOP • STOP • STOP • STOP • STOP • STOP

A3 **a.** { } or ∅ **b.** R

2

If n represents a positive real number, the *positive square root of* n, $\sqrt{n}$, is the positive number that when squared equals n.

Examples:

$\sqrt{16} = 4$ because $4^2 = 16$

$\sqrt{0.04} = 0.2$ because $(0.2)^2 = 0.04$

$\sqrt{\dfrac{9}{25}} = \dfrac{3}{5}$ because $\left(\dfrac{3}{5}\right)^2 = \dfrac{9}{25}$

The $\sqrt{\ }$ symbol is called the *radical sign,* and the expression under the radical sign is called the *radicand.*

Q4 Complete the equations to make true statements:

a. $\sqrt{49} =$ ______ **b.** $\sqrt{81} =$ ______

c. $\sqrt{\dfrac{4}{9}} =$ ______ **d.** $\sqrt{0.01} =$ ______

STOP • STOP • STOP • STOP • STOP • STOP • STOP • STOP • STOP

A4 **a.** 7 **b.** 9 **c.** $\dfrac{2}{3}$ **d.** 0.1

3

Some radical expressions cannot be simplified. For example, $\sqrt{7}$ cannot be written as a whole number. Decimals can be written that are close to $\sqrt{7}$, but none is exactly equal to it. If $\sqrt{7}$ is squared, the result is 7, because it represents the number that when squared equals 7.

Example 1: Simplify $(\sqrt{7})^2$.

Solution

$(\sqrt{7})^2 = 7$

Example 2: Simplify $(\sqrt{1{,}317})^2$.

Solution

$(\sqrt{1{,}317})^2 = 1{,}317$

Q5 Complete the equations to make true statements:

a. $(\sqrt{9})^2 =$ ______ **b.** $(\sqrt{11})^2 =$ ______

c. $(\sqrt{2{,}469})^2 =$ ______ **d.** $(\sqrt{0.216})^2 =$ ______

STOP • STOP • STOP • STOP • STOP • STOP • STOP • STOP • STOP

A5 **a.** 9 **b.** 11 **c.** 2,469 **d.** 0.216

Q6 What is the radicand in the expression $8x + \sqrt{21xy}$? ______

STOP • STOP • STOP • STOP • STOP • STOP • STOP • STOP • STOP

A6 $21xy$

4 Even though $\sqrt{9} = 3$, there is *another* square root of 9, -3, which when squared equals 9. When referring to the negative square root of a positive number n, use the notation $-\sqrt{n}$.

Example 1: Write the two square roots of 16.

Solution

$\sqrt{16} = 4$ and $-\sqrt{16} = -4$

Example 2: Write the two square roots of 5.

Solution

$\sqrt{5}$ and $-\sqrt{5}$

Q7 Complete the statements:

a. $\sqrt{64} =$ ____ **b.** $-\sqrt{64} =$ ____

c. The square roots of 49 are ________

d. The square roots of 23 are ________

STOP • STOP • STOP • STOP • STOP • STOP • STOP • STOP • STOP

A7 **a.** 8 **b.** -8 **c.** 7 and -7 **d.** $\sqrt{23}$ and $-\sqrt{23}$

5 When finding roots other than square roots it is helpful to know the first few powers of the numbers 2, 3, and 5.

$2^2 = 4$	$2^7 = 128$	$3^2 = 9$	$5^2 = 25$
$2^3 = 8$	$2^8 = 256$	$3^3 = 27$	$5^3 = 125$
$2^4 = 16$	$2^9 = 512$	$3^4 = 81$	$5^4 = 625$
$2^5 = 32$	$2^{10} = 1{,}024$	$3^5 = 243$	
$2^6 = 64$			

These facts should be memorized.

Q8 Fill in the following without referring to Frame 5.

a. $2^3 =$ ____ **b.** $3^2 =$ ____ **c.** $5^3 =$ ____ **d.** $2^5 =$ ____

e. $2^{10} =$ ____ **f.** $3^3 =$ ____ **g.** $3^4 =$ ____ **h.** $2^6 =$ ____

i. $5^2 =$ ____ **j.** $2^4 =$ ____ **k.** $2^9 =$ ____ **l.** $5^4 =$ ____

STOP • STOP • STOP • STOP • STOP • STOP • STOP • STOP • STOP

A8 **a.** 8 **b.** 9 **c.** 125 **d.** 32
e. 1,024 **f.** 27 **g.** 81 **h.** 64
i. 25 **j.** 16 **k.** 512 **l.** 625

6 A more general definition of roots, which also includes square roots, follows. The principal *n*th *root of a positive number x*, written $\sqrt[n]{x}$, is the positive number that when raised to the nth power is x. Symbolically it could be written $\sqrt[n]{x} = y$ if $y^n = x$. The third root of a number is commonly called the cube root. The number n is called the *index* of the radical. In summary:

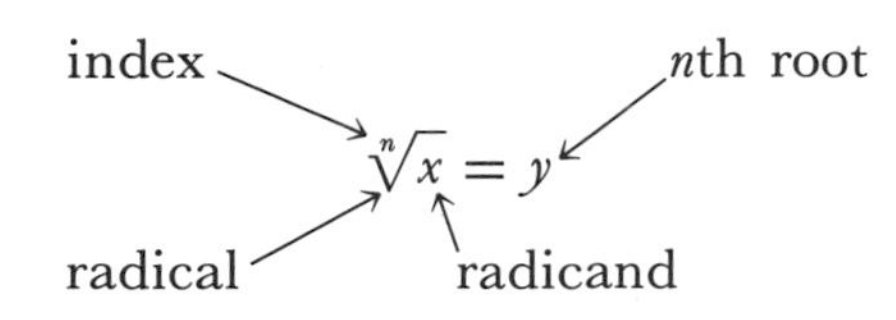

Examples:

$\sqrt[3]{8} = 2$ because $2^3 = 8$

$\sqrt[5]{243} = 3$ because $3^5 = 243$

$\sqrt[4]{16x^4y^8z^{12}} = 2xy^2z^3$ where $x > 0$, $y > 0$, $z > 0$

Q9 Write the roots below (assume all variables to be positive):

a. $\sqrt[3]{27} =$ ______ b. $\sqrt[5]{32a^5b^{15}} =$ ______

c. $\sqrt[8]{256} =$ ______ d. $\sqrt[3]{8x^3y^6} =$ ______

e. $\sqrt[10]{1{,}024} =$ ______ f. $\sqrt[5]{243x^{10}} =$ ______

STOP • STOP • STOP • STOP • STOP • STOP • STOP • STOP • STOP

A9 a. 3 b. $2ab^3$ c. 2 d. $2xy^2$
e. 2 f. $3x^2$

7 It is possible to find cube roots of negative numbers. For example, $\sqrt[3]{-27} = -3$, because $(-3)^3 = -27$. It is also possible to find fifth roots, seventh roots, or any other odd-number root of a negative number. This follows from the fact that an odd power of a negative number is negative. For example, $(-2)^7 = -128$; therefore, $\sqrt[7]{-128} = -2$. Generalizing, it is said that if n is odd, the principal nth root of a negative number x, $\sqrt[n]{x}$, is the negative number that when raised to the nth power equals x. Symbolically, if $x < 0$ and n odd, $\sqrt[n]{x} = y$ if $y^n = x$. Note that for odd roots the definition of $\sqrt[n]{x}$ is the same for both positive and negative real numbers in the radicand. For example, $\sqrt[9]{-512x^9y^{18}} = -2xy^2$ for all real numbers x and y.

Q10 Find the roots:

a. $\sqrt[3]{-8x^{12}y^3} =$ ______ b. $\sqrt[5]{-243z^5} =$ ______

c. $\sqrt[3]{-125} =$ ______ d. $\sqrt[5]{-32x^{20}} =$ ______

STOP • STOP • STOP • STOP • STOP • STOP • STOP • STOP • STOP

A10 a. $-2x^4y$ b. $-3z$ c. -5 d. $-2x^4$

8 Even roots of negative numbers present a different problem. There is no real number that when squared equals -4; therefore, $\sqrt{-4}$ is not a real number. In fact, if $a > 0$ and n is even, $\sqrt[n]{-a}$ is not a real number. This follows from the fact that even powers of both negative and positive numbers are always positive. Therefore, it is impossible to raise a real number to an even power and get a negative result.

Although $\sqrt{-4}$ is not a real number, do not say that $\sqrt{-4}$ is not a number. The number sets will be expanded in Section 4.5 to include numbers of this type. All expressions will be assumed to represent real numbers until the larger set is developed. Therefore, in this chapter all variables raised to an odd power (including 1) in a radicand are assumed to be positive when the index is even.

Q11 Simplify the radicals if possible. If a number is not a real number, write "not real."

a. $\sqrt[4]{-16} =$ ______ **b.** $\sqrt[5]{-243} =$ ______

c. $\sqrt[6]{-1} =$ ______ **d.** $\sqrt[3]{-1} =$ ______

e. $\sqrt[3]{-8} =$ ______ **f.** $\sqrt{-25} =$ ______

STOP • STOP • STOP • STOP • STOP • STOP • STOP • STOP • STOP

A11 **a.** not real: because the index is even and the radicand negative
b. -3 **c.** not real **d.** -1 **e.** -2
f. not real

9 In the simplification of expressions involving radicals, the radicands will be assumed to be positive. In the case of $\sqrt{x^2}$, x^2 is always positive, even when x is negative. For example,

$\sqrt{(5)^2} = \sqrt{25} = 5$ $\sqrt{(-3)^2} = \sqrt{9} = 3$

If x is positive, $\sqrt{x^2} = x$. If x is negative, $\sqrt{x^2} \neq x$, because $\sqrt{x^2}$ must be positive.

Q12 Evaluate the radicals:

a. $\sqrt{3^2}$ ______ **b.** $\sqrt{(-4)^2}$ ______

c. $\sqrt{(5)^2}$ ______ **d.** $\sqrt{(-8)^2}$ ______

STOP • STOP • STOP • STOP • STOP • STOP • STOP • STOP • STOP

A12 **a.** 3 **b.** 4 **c.** 5 **d.** 8

10 If x represents a negative number, $-x$ (the opposite of x) would be positive. Note this carefully: $-x$ is positive if you know that x represents a negative number. If x is negative, $\sqrt{x^2} \neq x$, because the principal square root must be positive. However, if x is negative, $\sqrt{x^2} = -x$, where $-x$ (the opposite of x) represents a positive number.

Example 1: Show that $\sqrt{x^2} = x$ is false when $x = -3$.

Solution

Substituting -3 for x in $\sqrt{x^2} = x$ gives

$\sqrt{(-3)^2} = -3$

which is false, because $\sqrt{(-3)^2} = 3$.

Example 2: Show that $\sqrt{x^2} = -x$ is true when $x = -5$.

Solution

Substituting -5 for x in $\sqrt{x^2} = -x$ gives

$$\begin{aligned}\sqrt{(-5)^2} &= -(-5)\\ &= 5\end{aligned}$$

which is true.

Q13 True or false? (Use T or F.)

a. When $x = 3$, $\sqrt{x^2} = x$. _____

b. When $x = -12$, $\sqrt{x^2} = x$. _____

c. When $x = -6$, $\sqrt{x^2} = -x$. _____

d. When $x = 7$, $\sqrt{x^2} = -x$. _____

e. When $x = -6$, $\sqrt{x^2} = |x|$. _____

f. When $x = 7$, $\sqrt{x^2} = |x|$. _____

STOP • STOP • STOP • STOP • STOP • STOP • STOP • STOP • STOP

A13 **a.** T **b.** F **c.** T **d.** F **e.** T **f.** T

11 From the results of Q13, it can be seen that both $\sqrt{x^2} = x$ and $\sqrt{x^2} = -x$ are sometimes false. To get out of the dilemma, the absolute value of x is used. Recall that the absolute value of a number is the distance the number is from zero and is always nonnegative. Since $\sqrt{x^2}$ is also always nonncgative, a statement that is always true for all real numbers is $\sqrt{x^2} = |x|$. The same problem exists for all roots with even index. Therefore, $\sqrt[4]{x^4} = |x|$, $\sqrt[6]{x^6} = |x|$, and so on.

Example 1: $\sqrt{(x + 3)^2} = ?$

Solution

$$\sqrt{(x + 3)^2} = |x + 3|$$

Example 2: $\sqrt[4]{(x - 2)^8} = ?$

Solution

$$\sqrt[4]{(x - 2)^8} = |(x - 2)^2|$$
$$= (x - 2)^2 \quad \text{because } (x - 2)^2 \text{ has to be positive}$$

Q14 Answer "yes" or "no" to the following statements:

a. Is $\sqrt{x^2} = x$ a true statement for all real-number replacements for x? _______

b. Is $\sqrt{x^2} = -x$ a true statement for all real-number replacements for x? _______

c. Is $\sqrt{x^2} = |x|$ a true statement for all real-number replacements for x? _______

STOP • STOP • STOP • STOP • STOP • STOP • STOP • STOP • STOP

A14 **a.** no **b.** no **c.** yes

Q15 Simplify the roots (you must guarantee that your answer is positive):

a. $\sqrt{y^2} =$ _________ b. $\sqrt{y^4} =$ _________

c. $\sqrt{(x - 3)^2} =$ _________ d. $\sqrt[4]{(x + 3)^4} =$ _________

e. $\sqrt{(x - 5)^4} =$ _________ f. $\sqrt[6]{x^6 y^{12}} =$ _________

STOP • STOP • STOP • STOP • STOP • STOP • STOP • STOP • STOP

A15 **a.** $|y|$ **b.** y^2

c. $|x - 3|$ **d.** $|x + 3|$

e. $(x - 5)^2$ **f.** $|x| y^2$

12 If a radical expression represents an irrational number, it can sometimes be simplified so that there is a smaller radicand. The property that allows this is:

$\sqrt[n]{ab} = \sqrt[n]{a} \cdot \sqrt[n]{b}$ where a and b are not negative if n is even

For example,

$\sqrt{4 \cdot 9} = \sqrt{4} \cdot \sqrt{9} = 2 \cdot 3 = 6$ and $\sqrt{4 \cdot 9} = \sqrt{36} = 6$

To simplify a radical expression, factor the radicand into a factor to the nth power times another factor, separate the factors in two radicals, and find the nth root of the expression that is an nth power.

Example 1: Simplify $\sqrt{45}$.

Solution

$\sqrt{45} = \sqrt{9 \cdot 5} = \sqrt{9} \cdot \sqrt{5} = 3\sqrt{5}$

Example 2: Simplify $\sqrt[3]{32}$.

Solution

$\sqrt[3]{32} = \sqrt[3]{8 \cdot 4} = \sqrt[3]{8} \cdot \sqrt[3]{4} = 2\sqrt[3]{4}$

Q16 Simplify the irrational numbers:

a. $\sqrt{50}$

b. $\sqrt{54}$

c. $\sqrt[3]{54}$

d. $\sqrt[4]{32}$

STOP • STOP • STOP • STOP • STOP • STOP • STOP • STOP • STOP

A16

a. $5\sqrt{2}$: $\sqrt{50} = \sqrt{25 \cdot 2} = \sqrt{25} \cdot \sqrt{2} = 5\sqrt{2}$

b. $3\sqrt{6}$: $\sqrt{54} = \sqrt{9 \cdot 6} = \sqrt{9} \cdot \sqrt{6} = 3\sqrt{6}$

c. $3\sqrt[3]{2}$: $\sqrt[3]{54} = \sqrt[3]{27 \cdot 2} = \sqrt[3]{27} \cdot \sqrt[3]{2} = 3\sqrt[3]{2}$

d. $2\sqrt[4]{2}$: $\sqrt[4]{32} = \sqrt[4]{16 \cdot 2} = \sqrt[4]{16} \cdot \sqrt[4]{2} = 2\sqrt[4]{2}$

13 If variables are in the radicand, they should also be included with the perfect nth power and simplified. Care must be taken with even roots of even powers where the variable might not be positive. You must guarantee that the final result is positive.

Example 1: Simplify $\sqrt{18a^3b}$. (Since a and b are to odd powers, assume that $a > 0$, $b > 0$.)

Solution

$\sqrt{18a^3b} = \sqrt{9a^2 \cdot 2ab} = \sqrt{9a^2} \cdot \sqrt{2ab} = 3a\sqrt{2ab}$

Example 2: Simplify $\sqrt[3]{-81a^8b^4}$.

Solution

$$\sqrt[3]{-81a^8b^4} = \sqrt[3]{-27a^6b^3 \cdot 3a^2b} = \sqrt[3]{-27a^6b^3} \cdot \sqrt[3]{3a^2b} = -3a^2b\sqrt[3]{3a^2b}$$

The negative sign is included with the perfect cube term.

Another example is

$$\sqrt{4x^6y} = \sqrt{4x^6} \cdot \sqrt{y} = 2|x^3|\sqrt{y}$$

Q17 Simplify the radical expressions:

a. $\sqrt{96x^2}$

b. $\sqrt{x^2y^3z^5}$

c. $\sqrt[3]{-8ab^4}$

d. $\sqrt[4]{64a^6b^4}$

STOP • STOP • STOP • STOP • STOP • STOP • STOP • STOP • STOP

A17 a. $4|x|\sqrt{6}$: $\sqrt{96x^2} = \sqrt{16x^2 \cdot 6} = \sqrt{16x^2} \cdot \sqrt{6} = 4|x|\sqrt{6}$

b. $|x|yz^2\sqrt{yz}$: $\sqrt{x^2y^3z^5} = \sqrt{x^2y^2z^4 \cdot yz} = \sqrt{x^2y^2z^4} \cdot \sqrt{yz} = |x|yz^2\sqrt{yz}$
Because y appeared originally to an odd power, it is assumed to be positive.

c. $-2b\sqrt[3]{ab}$: $\sqrt[3]{-8ab^4} = \sqrt[3]{-8b^3 \cdot ab} = \sqrt[3]{-8b^3} \cdot \sqrt[3]{ab} = -2b\sqrt[3]{ab}$

d. $2|ab|\sqrt[4]{4a^2}$: $\sqrt[4]{64a^6b^4} = \sqrt[4]{16a^4b^4 \cdot 4a^2} = \sqrt[4]{16a^4b^4} \cdot \sqrt[4]{4a^2} = 2|ab|\sqrt[4]{4a^2}$

14 In simplest radical form there should be no fractions in the radicand. To simplify such numbers use the property

$$\sqrt[n]{\frac{a}{b}} = \frac{\sqrt[n]{a}}{\sqrt[n]{b}}$$ where a and b are not negative if n is even

For example,

$$\sqrt{\frac{4}{9}} = \frac{\sqrt{4}}{\sqrt{9}} = \frac{2}{3} \quad \text{and} \quad \sqrt{\frac{4}{9}} = \sqrt{\left(\frac{2}{3}\right)^2} = \frac{2}{3}$$

Example 1: Simplify $\sqrt{\frac{8}{9}}$.

Solution

$$\sqrt{\frac{8}{9}} = \frac{\sqrt{8}}{\sqrt{9}} = \frac{\sqrt{4 \cdot 2}}{3} = \frac{2\sqrt{2}}{3}$$

Example 2: Simplify $\sqrt{\frac{7}{8}}$.

Solution

$$\sqrt{\frac{7}{8}} = \frac{\sqrt{7}}{\sqrt{8}} = \frac{\sqrt{7}}{\sqrt{8}} \cdot \frac{\sqrt{2}}{\sqrt{2}} = \frac{\sqrt{14}}{\sqrt{16}} = \frac{\sqrt{14}}{4}$$

Notice that the 8 which was the radicand in the denominator was not a perfect square. To remove the radical in the denominator, the fraction was multiplied by 1 in the form of $\frac{\sqrt{2}}{\sqrt{2}}$. The square root of 2 was chosen because the product of 8 and 2 is the perfect square 16. Another solution to the second example is

$$\sqrt{\frac{7}{8}} = \sqrt{\frac{7}{8} \cdot \frac{2}{2}} = \sqrt{\frac{14}{16}} = \frac{\sqrt{14}}{\sqrt{16}} = \frac{\sqrt{14}}{4}$$

Example 3: Simplify $\sqrt[3]{\frac{1}{3}}$.

Solution

$$\sqrt[3]{\frac{1}{3}} = \frac{\sqrt[3]{1}}{\sqrt[3]{3}} = \frac{1}{\sqrt[3]{3}} \cdot \frac{\sqrt[3]{9}}{\sqrt[3]{9}} = \frac{\sqrt[3]{9}}{\sqrt[3]{27}} = \frac{\sqrt[3]{9}}{3}$$

Notice that $\sqrt[3]{9}$ was chosen so that the new denominator would be a perfect cube. The process carried out in this frame is called *rationalizing the denominator.*

Q18 Simplify the irrational numbers (rationalize the denominator):

a. $\sqrt{\frac{3}{4}}$

b. $\sqrt{\frac{5}{6}}$

c. $\sqrt[3]{\frac{1}{2}}$

d. $\sqrt[3]{\frac{3}{4}}$

STOP • STOP • STOP • STOP • STOP • STOP • STOP • STOP • STOP

A18 **a.** $\frac{\sqrt{3}}{2}$: $\sqrt{\frac{3}{4}} = \frac{\sqrt{3}}{\sqrt{4}} = \frac{\sqrt{3}}{2}$

b. $\frac{\sqrt{30}}{6}$: $\sqrt{\frac{5}{6}} = \frac{\sqrt{5}}{\sqrt{6}} = \frac{\sqrt{5}}{\sqrt{6}} \cdot \frac{\sqrt{6}}{\sqrt{6}} = \frac{\sqrt{30}}{6}$

c. $\frac{\sqrt[3]{4}}{2}$: $\sqrt[3]{\frac{1}{2}} = \frac{\sqrt[3]{1}}{\sqrt[3]{2}} = \frac{1}{\sqrt[3]{2}} \cdot \frac{\sqrt[3]{4}}{\sqrt[3]{4}} = \frac{\sqrt[3]{4}}{\sqrt[3]{8}} = \frac{\sqrt[3]{4}}{2}$

d. $\frac{\sqrt[3]{6}}{2}$: $\sqrt[3]{\frac{3}{4}} = \frac{\sqrt[3]{3}}{\sqrt[3]{4}} \cdot \frac{\sqrt[3]{2}}{\sqrt[3]{2}} = \frac{\sqrt[3]{6}}{\sqrt[3]{8}} = \frac{\sqrt[3]{6}}{2}$

15 If radicands have fractions involving variables they may be simplified in a similar manner.

Example 1: Simplify $\sqrt{\dfrac{a}{3b}}$.

Solution

$$\sqrt{\frac{a}{3b}} = \frac{\sqrt{a}}{\sqrt{3b}} = \frac{\sqrt{a}}{\sqrt{3b}} \cdot \frac{\sqrt{3b}}{\sqrt{3b}} = \frac{\sqrt{3ab}}{3b}$$

Note that $\sqrt{3b} \cdot \sqrt{3b}$ is $3b$. It is not necessary to write $\sqrt{3b} \cdot \sqrt{3b} = \sqrt{9b^2} = 3b$. For example, since $\sqrt{15} \cdot \sqrt{15} = 15$ you can save yourself some time by not multiplying the radicands.

Example 2: Simplify $\sqrt{\dfrac{2x}{3y^2z}}$.

Solution

$$\sqrt{\frac{2x}{3y^2z}} = \frac{\sqrt{2x}}{\sqrt{3y^2z}} \cdot \frac{\sqrt{3z}}{\sqrt{3z}} = \frac{\sqrt{6xz}}{\sqrt{9y^2z^2}} = \frac{\sqrt{6xz}}{3|y|z}$$

(z is assumed to be positive since it appeared to an odd power)

Q19 Simplify the radical expressions:

a. $\sqrt{\dfrac{2a}{b}}$

b. $\sqrt{\dfrac{9a}{b^3}}$

c. $\sqrt{\dfrac{x}{yz^3}}$

d. $\sqrt[3]{\dfrac{2}{3x}}$

STOP • STOP • STOP • STOP • STOP • STOP • STOP • STOP • STOP

A19

a. $\dfrac{\sqrt{2ab}}{b}$: $\sqrt{\dfrac{2a}{b}} = \dfrac{\sqrt{2a}}{\sqrt{b}} \cdot \dfrac{\sqrt{b}}{\sqrt{b}} = \dfrac{\sqrt{2ab}}{b}$ (b is assumed to be positive since it originally appeared to an odd power)

b. $\dfrac{3\sqrt{ab}}{b^2}$: $\sqrt{\dfrac{9a}{b^3}} = \dfrac{3\sqrt{a}}{\sqrt{b^3}} = \dfrac{3\sqrt{a}}{\sqrt{b^3}} \cdot \dfrac{\sqrt{b}}{\sqrt{b}} = \dfrac{3\sqrt{ab}}{\sqrt{b^4}} = \dfrac{3\sqrt{ab}}{b^2}$

c. $\dfrac{\sqrt{xyz}}{yz^2}$: $\sqrt{\dfrac{x}{yz^3}} = \dfrac{\sqrt{x}}{\sqrt{yz^3}} = \dfrac{\sqrt{x}}{\sqrt{yz^3}} \cdot \dfrac{\sqrt{yz}}{\sqrt{yz}} = \dfrac{\sqrt{xyz}}{\sqrt{y^2z^4}} = \dfrac{\sqrt{xyz}}{yz^2}$

d. $\dfrac{\sqrt[3]{18x^2}}{3x}$: $\sqrt[3]{\dfrac{2}{3x}} = \dfrac{\sqrt[3]{2}}{\sqrt[3]{3x}} \cdot \dfrac{\sqrt[3]{9x^2}}{\sqrt[3]{9x^2}} = \dfrac{\sqrt[3]{18x^2}}{\sqrt[3]{27x^3}} = \dfrac{\sqrt[3]{18x^2}}{3x}$

16 In simplest radical form there should be no radicals in the denominator. To remove a radical of index n, multiply by a number that makes the radicand a perfect nth power.

The numerator and denominator must both be multiplied by the same number, because multiplying by 1 leaves the value of the fraction the same.

Example 1: Simplify $\frac{4x}{\sqrt{3x}}$.

Solution

$$\frac{4x}{\sqrt{3x}} = \frac{4x}{\sqrt{3x}} \cdot \frac{\sqrt{3x}}{\sqrt{3x}} = \frac{4x\sqrt{3x}}{3x} = \frac{4\sqrt{3x}}{3}$$

Example 2: Simplify $\frac{\sqrt{4y}}{\sqrt{8}}$.

Solution

$$\frac{\sqrt{4y}}{\sqrt{8}} = \frac{2\sqrt{y}}{\sqrt{8}} \cdot \frac{\sqrt{2}}{\sqrt{2}} = \frac{2\sqrt{2y}}{\sqrt{16}} = \frac{2\sqrt{2y}}{4} = \frac{\sqrt{2y}}{2}$$

Example 3: Simplify $\frac{\sqrt[3]{8y}}{\sqrt[3]{9}}$.

Solution

$$\frac{\sqrt[3]{8y}}{\sqrt[3]{9}} = \frac{\sqrt[3]{8y}}{\sqrt[3]{9}} \cdot \frac{\sqrt[3]{3}}{\sqrt[3]{3}} = \frac{2\sqrt[3]{3y}}{\sqrt[3]{27}} = \frac{2\sqrt[3]{3y}}{3}$$

Q20 Write the following in simplest radical form:

a. $\frac{\sqrt{27x^5y^6}}{\sqrt{32z^7}}$

b. $\frac{2\sqrt{xy}}{\sqrt{x^3y}}$

c. $\frac{\sqrt[4]{7a^3}}{\sqrt[4]{8b^6}}$

d. $\frac{\sqrt[3]{xy^2}}{\sqrt[3]{x^3y}}$

STOP • STOP • STOP • STOP • STOP • STOP • STOP • STOP • STOP

A20

a. $\frac{3x^2|y^3|\sqrt{6xz}}{8z^4}$: $\frac{\sqrt{27x^5y^6}}{\sqrt{32z^7}} \cdot \frac{\sqrt{2z}}{\sqrt{2z}} = \frac{\sqrt{9x^4y^6 \cdot 6xz}}{\sqrt{64z^8}} = \frac{3x^2|y^3|\sqrt{6xz}}{8z^4}$

b. $\frac{2}{x}$: $\frac{2\sqrt{xy}}{\sqrt{x^3y}} \cdot \frac{\sqrt{xy}}{\sqrt{xy}} = \frac{2xy}{\sqrt{x^4y^2}} = \frac{2xy}{x^2y} = \frac{2}{x}$

c. $\frac{\sqrt[4]{14a^3b^2}}{2b^2}$: $\frac{\sqrt[4]{7a^3}}{\sqrt[4]{8b^6}} \cdot \frac{\sqrt[4]{2b^2}}{\sqrt[4]{2b^2}} = \frac{\sqrt[4]{14a^3b^2}}{\sqrt[4]{16b^8}} = \frac{\sqrt[4]{14a^3b^2}}{2b^2}$

d. $\frac{\sqrt[3]{xy}}{x}$: $\frac{\sqrt[3]{xy^2}}{\sqrt[3]{x^3y}} \cdot \frac{\sqrt[3]{y^2}}{\sqrt[3]{y^2}} = \frac{\sqrt[3]{xy^4}}{\sqrt[3]{x^3y^3}} = \frac{y\sqrt[3]{xy}}{xy} = \frac{\sqrt[3]{xy}}{x}$

17 The properties of an expression that is in simplest radical form can be summarized as follows: An expression is in simplest radical form if the following are true:

1. The radicand contains no factor whose exponent is larger than the index.
2. There are no fractions in the radicand.
3. There is no radical in the denominator.

Q21 Write in simplest radical form (assume that variables are positive if they appear to an odd power under the radical of even index):

a. $\sqrt{75}$

b. $\sqrt[3]{-54}$

c. $\sqrt{8x^3y^4z^2}$

d. $\sqrt[3]{24x^4y^6}$

e. $\sqrt{\dfrac{24}{81}}$

f. $\sqrt{\dfrac{9}{12}}$

g. $\sqrt{\dfrac{3a}{9b^3}}$

h. $\sqrt[3]{\dfrac{8a^2}{3b^2}}$

i. $\sqrt{\dfrac{2a^2}{6a^3}}$

j. $\dfrac{\sqrt[3]{2a^2}}{\sqrt[3]{6a^3}}$

STOP • STOP • STOP • STOP • STOP • STOP • STOP • STOP • STOP

A21 **a.** $5\sqrt{3}$: $\sqrt{75} = \sqrt{25 \cdot 3} = 5\sqrt{3}$

b. $-3\sqrt[3]{2}$: $\sqrt[3]{-54} = \sqrt[3]{-27 \cdot 2} = -3\sqrt[3]{2}$

c. $2xy^2|z|\sqrt{2x}$: $\sqrt{8x^3y^4z^2} = \sqrt{4x^2y^4z^2 \cdot 2x} = 2xy^2|z|\sqrt{2x}$

d. $2xy^2\sqrt[3]{3x}$: $\sqrt[3]{24x^4y^6} = \sqrt[3]{8x^3y^6 \cdot 3x} = 2xy^2\sqrt[3]{3x}$

e. $\dfrac{2\sqrt{6}}{9}$: $\sqrt{\dfrac{24}{81}} = \dfrac{\sqrt{24}}{\sqrt{81}} = \dfrac{\sqrt{4 \cdot 6}}{9} = \dfrac{2\sqrt{6}}{9}$

f. $\frac{\sqrt{3}}{2}$: $\sqrt{\frac{9}{12}} = \frac{\sqrt{9}}{\sqrt{12}} = \frac{\sqrt{9}}{\sqrt{12}} \cdot \frac{\sqrt{3}}{\sqrt{3}} = \frac{3\sqrt{3}}{\sqrt{36}} = \frac{3\sqrt{3}}{6} = \frac{\sqrt{3}}{2}$

g. $\frac{\sqrt{3ab}}{3b^2}$: $\sqrt{\frac{3a}{9b^3}} = \frac{\sqrt{3a}}{\sqrt{9b^3}} \cdot \frac{\sqrt{b}}{\sqrt{b}} = \frac{\sqrt{3ab}}{\sqrt{9b^4}} = \frac{\sqrt{3ab}}{3b^2}$

h. $\frac{2\sqrt[3]{9a^2b}}{3b}$: $\sqrt[3]{\frac{8a^2}{3b^2}} = \frac{\sqrt[3]{8a^2}}{\sqrt[3]{3b^2}} \cdot \frac{\sqrt[3]{9b}}{\sqrt[3]{9b}} = \frac{\sqrt[3]{8}\sqrt[3]{9a^2b}}{\sqrt[3]{27b^3}} = \frac{2\sqrt[3]{9a^2b}}{3b}$

i. $\frac{\sqrt{3a}}{3a}$: $\frac{\sqrt{2a^2}}{\sqrt{6a^3}} = \frac{\sqrt{2a^2}}{\sqrt{6a^3}} \cdot \frac{\sqrt{6a}}{\sqrt{6a}} = \frac{\sqrt{12a^3}}{\sqrt{36a^4}} = \frac{\sqrt{4a^2 \cdot 3a}}{6a^2} = \frac{2a\sqrt{3a}}{6a^2} = \frac{\sqrt{3a}}{3a}$

j. $\frac{\sqrt[3]{9a^2}}{3a}$: $\frac{\sqrt[3]{2a^2}}{\sqrt[3]{6a^3}} = \frac{\sqrt[3]{2a^2}}{\sqrt[3]{6a^3}} \cdot \frac{\sqrt[3]{36}}{\sqrt[3]{36}} = \frac{\sqrt[3]{72a^2}}{\sqrt[3]{216a^3}} = \frac{\sqrt[3]{8 \cdot 9a^2}}{6a} = \frac{2\sqrt[3]{9a^2}}{6a} = \frac{\sqrt[3]{9a^2}}{3a}$

These answers do not all represent the shortest method. Try to find shorter solutions for parts f, i, and j.

This completes the instruction for this section.

4.1 Exercises

Write in simplest radical form. If a number is not a real number, write "not real." Assume that variables in the radicand are positive if they appear to an odd power under an even index.

1. $\sqrt{121}$
2. $\sqrt{-144}$
3. $\sqrt[3]{125}$
4. $\sqrt[7]{-128}$
5. $\sqrt[5]{-243}$
6. $\sqrt[8]{-256}$
7. $\sqrt{x^2}$
8. $\sqrt[3]{-1}$
9. $\sqrt[4]{x^4}$
10. if $y < 0$, $\sqrt{y^2}$
11. $\sqrt{(x-3)^2}$
12. if $z > 0$, $\sqrt{z^2}$
13. $\sqrt{125}$
14. $\sqrt{45}$
15. $\sqrt{80}$
16. $\sqrt[3]{48}$
17. $\sqrt[4]{48}$
18. $\sqrt[3]{-40x^3y^6}$
19. $\sqrt{128x^3y^2}$
20. $\sqrt[3]{250xy^5}$
21. $\sqrt{45x^4y^6}$
22. $\sqrt{\frac{12}{25}}$
23. $\sqrt[3]{\frac{-16}{27}}$
24. $\sqrt{\frac{14}{4xy}}$
25. $\sqrt{\frac{2a}{5b}}$
26. $\sqrt[3]{\frac{4x^3}{9y}}$
27. $\sqrt[3]{\frac{8x^4}{-25x}}$
28. $\sqrt{\frac{2a}{3a}}$
29. $\frac{\sqrt[3]{2x}}{\sqrt[3]{9x^4}}$
30. $\frac{1}{\sqrt{6y}}$

4.1 Exercise Answers

1. 11
2. not real
3. 5
4. -2
5. -3
6. not real
7. $|x|$
8. -1
9. $|x|$
10. $-y$
11. $|x-3|$
12. z
13. $5\sqrt{5}$
14. $3\sqrt{5}$
15. $4\sqrt{5}$
16. $2\sqrt[3]{6}$
17. $2\sqrt[4]{3}$
18. $-2xy^2\sqrt[3]{5}$
19. $8x|y|\sqrt{2x}$
20. $5y\sqrt[3]{2xy^2}$
21. $3x^2|y^3|\sqrt{5}$

22. $\dfrac{2\sqrt{3}}{5}$ 23. $\dfrac{-2\sqrt[3]{2}}{3}$ 24. $\dfrac{\sqrt{14xy}}{2xy}$

25. $\dfrac{\sqrt{10ab}}{5b}$ 26. $\dfrac{x\sqrt[3]{12y^2}}{3y}$ 27. $\dfrac{-2x\sqrt[3]{5}}{5}$

28. $\dfrac{\sqrt{6}}{3}$ 29. $\dfrac{\sqrt[3]{6}}{3x}$ 30. $\dfrac{\sqrt{6y}}{6y}$

4.2 Operations on Irrational Numbers

1 Many irrational numbers are represented by radical expressions. To add, subtract, multiply, and divide these numbers, techniques for manipulating radical expressions must be learned. These techniques will be the topic of this section.

An expression such as $3\sqrt{5} + 7\sqrt{5}$ can be simplified by the distributive property.

$$\begin{aligned} 3\sqrt{5} + 7\sqrt{5} &= (3 + 7)\sqrt{5} \\ &= 10\sqrt{5} \end{aligned}$$

Radical expressions with the same radicand and index are said to be *like* radicals. The sum or difference of like radicals may be simplified by using the distributive law. Radical expressions that are not like radicals cannot be simplified unless they can be changed into equivalent like radicals.

Example 1: Simplify $3\sqrt{6} + 2\sqrt{6}$.

Solution

$$3\sqrt{6} + 2\sqrt{6} = (3 + 2)\sqrt{6} = 5\sqrt{6}$$

Example 2: Simplify $3\sqrt{3} + 4\sqrt{2}$.

Solution

Actually, this sum cannot be simplified, because each radical is in simplest radical form and the terms are not like radicals.

Example 3: Simplify $\sqrt{12} + \sqrt{75}$.

Solution

$$\begin{aligned} \sqrt{12} + \sqrt{75} &= \sqrt{4 \cdot 3} + \sqrt{25 \cdot 3} \\ &= 2\sqrt{3} + 5\sqrt{3} \\ &= 7\sqrt{3} \end{aligned}$$

Q1 Simplify:

a. $3\sqrt{2} - \sqrt{2} + 5\sqrt{2}$

b. $-4\sqrt{2} + \sqrt{18}$

c. $\sqrt{50} + 2\sqrt{18} - \sqrt{3}$

d. $\sqrt{8} + 2\sqrt{12} - 4\sqrt{50} + \sqrt{3}$

STOP • STOP • STOP • STOP • STOP • STOP • STOP • STOP • STOP

A1

a. $7\sqrt{2}$: $3\sqrt{2} - \sqrt{2} + 5\sqrt{2} = (3 - 1 + 5)\sqrt{2} = 7\sqrt{2}$

b. $-\sqrt{2}$: $-4\sqrt{2} + \sqrt{18} = -4\sqrt{2} + \sqrt{9 \cdot 2} = -4\sqrt{2} + 3\sqrt{2} = -\sqrt{2}$

c. $11\sqrt{2} - \sqrt{3}$: $\sqrt{50} + 2\sqrt{18} - \sqrt{3} = \sqrt{25 \cdot 2} + 2\sqrt{9 \cdot 2} - \sqrt{3}$
$= 5\sqrt{2} + 6\sqrt{2} - \sqrt{3}$

d. $5\sqrt{3} - 18\sqrt{2}$: $\sqrt{8} + 2\sqrt{12} - 4\sqrt{50} + \sqrt{3} = 2\sqrt{2} + 4\sqrt{3} - 20\sqrt{2} + \sqrt{3}$
$= 5\sqrt{3} - 18\sqrt{2}$

2

Before simplifying an expression that contains radicals, write each radical in simplest radical form.

Example: Simplify $\sqrt{\frac{1}{2}} + \sqrt{98}$.

Solution

$$\sqrt{\frac{1}{2}} + \sqrt{98} = \frac{\sqrt{1}}{\sqrt{2}} \cdot \frac{\sqrt{2}}{\sqrt{2}} + \sqrt{49 \cdot 2}$$
$$= \frac{\sqrt{2}}{2} + 7\sqrt{2}$$
$$= \frac{\sqrt{2}}{2} + \frac{14\sqrt{2}}{2}$$
$$= \frac{15\sqrt{2}}{2}$$

Q2

Simplify:

a. $\sqrt{\frac{1}{2}} + \frac{1}{2}\sqrt{18}$

b. $\sqrt{8x^3} + 2x\sqrt{\frac{x}{2}}$

c. $y\sqrt{28x^3y} + x\sqrt{63xy^3} + \sqrt{175x^3y^3}$

d. $\sqrt[3]{2x^4y^2} + x\sqrt[3]{16xy^2}$

STOP • STOP • STOP • STOP • STOP • STOP • STOP • STOP • STOP

A2

a. $2\sqrt{2}$: $\sqrt{\frac{1}{2}} + \frac{1}{2}\sqrt{18} = \frac{\sqrt{1}}{\sqrt{2}} \cdot \frac{\sqrt{2}}{\sqrt{2}} + \frac{\sqrt{9 \cdot 2}}{2} = \frac{\sqrt{2}}{2} + \frac{3\sqrt{2}}{2} = \frac{4\sqrt{2}}{2} = 2\sqrt{2}$

b. $3x\sqrt{2x}$: $\sqrt{8x^3} + 2x\sqrt{\frac{x}{2}} = \sqrt{4x^2 2x} + 2x\frac{\sqrt{x}}{\sqrt{2}} \cdot \frac{\sqrt{2}}{\sqrt{2}} = 2x\sqrt{2x} + \frac{2x\sqrt{2x}}{2}$
$= 2x\sqrt{2x} + x\sqrt{2x} = 3x\sqrt{2x}$

c. $10xy\sqrt{7xy}$: $y\sqrt{4x^2 \cdot 7xy} + x\sqrt{9y^2 \cdot 7xy} + \sqrt{25x^2y^2 \cdot 7xy}$
$= 2xy\sqrt{7xy} + 3xy\sqrt{7xy} + 5xy\sqrt{7xy} = 10xy\sqrt{7xy}$

d. $3x\sqrt[3]{2xy^2}$: $\sqrt[3]{2x^4y^2} + x\sqrt[3]{16xy^2} = \sqrt[3]{x^3 \cdot 2xy^2} + x\sqrt[3]{8 \cdot 2xy^2}$
$= x\sqrt[3]{2xy^2} + 2x\sqrt[3]{2xy^2}$
$= 3x\sqrt[3]{2xy^2}$

3

Irrational numbers written with radicals can be multiplied by using the same property that was used to simplify radical expressions.

$\sqrt[n]{a} \cdot \sqrt[n]{b} = \sqrt[n]{ab}$ where a and b are not negative when n is even

Example 1: Multiply $\sqrt{3} \cdot \sqrt{18}$.

Solution

$\sqrt{3} \cdot \sqrt{18} = \sqrt{3 \cdot 18} = \sqrt{3 \cdot 9 \cdot 2} = \sqrt{9} \cdot \sqrt{6} = 3\sqrt{6}$

Example 2: Multiply $(2\sqrt{5})(3\sqrt{2})$.

Solution

$(2\sqrt{5})(3\sqrt{2}) = 2 \cdot 3\sqrt{5} \cdot \sqrt{2} = 6\sqrt{10}$

Example 3: Multiply $(5)(8\sqrt{7})$.

Solution

$(5)(8\sqrt{7}) = 5 \cdot 8\sqrt{7} = 40\sqrt{7}$

Notice that a number outside the radical *cannot* be multiplied times a number inside the radical. For example, $2\sqrt{3} \neq \sqrt{6}$.

Q3

Multiply the numbers:

a. $(4)(3\sqrt{6})$

b. $\sqrt{3}(4\sqrt{12})$

c. $\sqrt{18x} \cdot \sqrt{8x}$

d. $\sqrt{3xy^3} \cdot \sqrt{48xy}$

STOP • STOP • STOP • STOP • STOP • STOP • STOP • STOP • STOP

A3

a. $12\sqrt{6}$: $4(3\sqrt{6}) = 4 \cdot 3\sqrt{6} = 12\sqrt{6}$

b. 24: $\sqrt{3}(4\sqrt{12}) = 4\sqrt{3} \cdot \sqrt{12} = 4\sqrt{36} = 4 \cdot 6 = 24$

c. $12x$: $\sqrt{18x} \cdot \sqrt{8x} = \sqrt{18 \cdot 8 \cdot x^2} = \sqrt{9 \cdot 2 \cdot 2 \cdot 4x^2} = 3 \cdot 2 \cdot 2x = 12x$

d. $12xy^2$: $\sqrt{3xy^3} \cdot \sqrt{48xy} = \sqrt{3 \cdot 3 \cdot 16x^2y^4} = 3 \cdot 4xy^2 = 12xy^2$

Q4 Multiply and simplify:

a. $\sqrt[3]{2x^2} \cdot \sqrt[3]{4x^2}$

b. $\sqrt[4]{9x^3} \cdot \sqrt[4]{9x}$

c. $(y\sqrt{y^2})(y^2\sqrt{y})(y\sqrt{y^3})$

d. $2\sqrt[3]{-9} \cdot \sqrt[3]{2x} \cdot \sqrt[3]{-3x^4}$

STOP • STOP • STOP • STOP • STOP • STOP • STOP • STOP • STOP

A4 a. $2x\sqrt[3]{x}$: $\sqrt[3]{2x^2} \cdot \sqrt[3]{4x^2} = \sqrt[3]{8x^4} = \sqrt[3]{8x^3 \cdot x} = 2x\sqrt[3]{x}$

b. $3x$: $\sqrt[4]{9x^3} \cdot \sqrt[4]{9x} = \sqrt[4]{81x^4} = 3x$

c. y^7: $(y\sqrt{y^2})(y^2\sqrt{y})(y\sqrt{y^3}) = y^4\sqrt{y^6} = y^4 \cdot y^3 = y^7$

d. $6x\sqrt[3]{2x^2}$: $2\sqrt[3]{-9} \cdot \sqrt[3]{2x} \cdot \sqrt[3]{-3x^4} = 2\sqrt[3]{27 \cdot 2 \cdot x^5} = 6x\sqrt[3]{2x^2}$

4 To multiply a monomial times a binomial, use the distributive property.

$$\begin{aligned}\sqrt{3}(2+\sqrt{15}) &= \sqrt{3}\cdot 2 + \sqrt{3}\cdot\sqrt{15}\\ &= 2\sqrt{3} + \sqrt{3}\cdot\sqrt{3}\cdot\sqrt{5}\\ &= 2\sqrt{3} + 3\sqrt{5}\end{aligned}$$

Although complete solutions are shown in the frames and answers in this section, try to shorten the steps by doing as much mentally as possible.

Q5 Multiply the numbers:

a. $\sqrt{2}(\sqrt{3} + \sqrt{8})$

b. $(4 + \sqrt{5})\sqrt{3}$

c. $\sqrt{5}(\sqrt{15} + \sqrt{10})$

d. $\sqrt[3]{3}(\sqrt[3]{9} + \sqrt[3]{3})$

STOP • STOP • STOP • STOP • STOP • STOP • STOP • STOP • STOP

A5 a. $4 + \sqrt{6}$: $\sqrt{2}(\sqrt{3} + \sqrt{8}) = \sqrt{2}\cdot\sqrt{3} + \sqrt{2}\cdot\sqrt{8} = \sqrt{6} + \sqrt{16} = 4 + \sqrt{6}$

b. $4\sqrt{3} + \sqrt{15}$: $(4 + \sqrt{5})\sqrt{3} = 4\sqrt{3} + \sqrt{5}\cdot\sqrt{3} = 4\sqrt{3} + \sqrt{15}$

c. $5\sqrt{3} + 5\sqrt{2}$: $\sqrt{5}(\sqrt{15} + \sqrt{10}) = \sqrt{75} + \sqrt{50} = 5\sqrt{3} + 5\sqrt{2}$

d. $3 + \sqrt[3]{9}$: $\sqrt[3]{3}(\sqrt[3]{9} + \sqrt[3]{3}) = \sqrt[3]{27} + \sqrt[3]{9} = 3 + \sqrt[3]{9}$

5 The product of two binomial expressions containing radicals is found also by using the distributive properties.

Example 1: Multiply $(2 + \sqrt{3})(3 + \sqrt{5})$.

Solution

$$\begin{aligned}(2+\sqrt{3})(3+\sqrt{5}) &= 2(3+\sqrt{5}) + \sqrt{3}(3+\sqrt{5})\\ &= 6 + 2\sqrt{5} + 3\sqrt{3} + \sqrt{15}\end{aligned}$$

In this instance the final expression cannot be simplified because no two of the terms are like radicals.

Example 2: Multiply $(3 + \sqrt{2})(5 + \sqrt{2})$.

Solution

$$\begin{aligned}(3 + \sqrt{2})(5 + \sqrt{2}) &= 3(5 + \sqrt{2}) + \sqrt{2}(5 + \sqrt{2}) \\ &= 15 + 3\sqrt{2} + 5\sqrt{2} + \sqrt{4} \\ &= 15 + 8\sqrt{2} + 2 \\ &= 17 + 8\sqrt{2}\end{aligned}$$

Q6 Multiply:

a. $(2 + \sqrt{3})(3 - \sqrt{2})$ **b.** $(2 - \sqrt{2})(3 + \sqrt{5})$

STOP • STOP • STOP • STOP • STOP • STOP • STOP • STOP • STOP

A6 **a.** $6 - 2\sqrt{2} + 3\sqrt{3} - \sqrt{6}$: $(2 + \sqrt{3})(3 - \sqrt{2}) = 2(3 - \sqrt{2}) + \sqrt{3}(3 - \sqrt{2})$
$= 6 - 2\sqrt{2} + 3\sqrt{3} - \sqrt{6}$

b. $6 + 2\sqrt{5} - 3\sqrt{2} - \sqrt{10}$: $(2 - \sqrt{2})(3 + \sqrt{5}) = 2(3 + \sqrt{5}) - \sqrt{2}(3 + \sqrt{5})$
$= 6 + 2\sqrt{5} - 3\sqrt{2} - \sqrt{10}$

Q7 Find the products:

a. $(\sqrt{7} + 3)(\sqrt{7} - 2)$ **b.** $(2 + \sqrt{6})(3 - \sqrt{6})$

STOP • STOP • STOP • STOP • STOP • STOP • STOP • STOP • STOP

A7 **a.** $1 + \sqrt{7}$: $(\sqrt{7} + 3)(\sqrt{7} - 2)$
$\sqrt{7}(\sqrt{7} - 2) + 3(\sqrt{7} - 2)$
$7 - 2\sqrt{7} + 3\sqrt{7} - 6$
$1 + \sqrt{7}$

b. $\sqrt{6}$: $(2 + \sqrt{6})(3 - \sqrt{6})$
$2(3 - \sqrt{6}) + \sqrt{6}(3 - \sqrt{6})$
$6 - 2\sqrt{6} + 3\sqrt{6} - 6$
$\sqrt{6}$

Q8 Multiply using the methods of squaring a binomial:

a. $(2 + \sqrt{3})^2$ **b.** $(4 - \sqrt{5})^2$

STOP • STOP • STOP • STOP • STOP • STOP • STOP • STOP • STOP

A8 **a.** $7 + 4\sqrt{3}$: $(2 + \sqrt{3})^2$
$2^2 + 2 \cdot 2\sqrt{3} + (\sqrt{3})^2$
$4 + 4\sqrt{3} + 3$
$7 + 4\sqrt{3}$

b. $21 - 8\sqrt{5}$: $(4 - \sqrt{5})^2$
$4^2 - 2 \cdot 4\sqrt{5} + (\sqrt{5})^2$
$16 - 8\sqrt{5} + 5$
$21 - 8\sqrt{5}$

Q9 Find the products:

a. $(2 - \sqrt{5})(2 + \sqrt{5})$

b. $(3 + \sqrt{2})(3 - \sqrt{2})$

STOP • STOP • STOP • STOP • STOP • STOP • STOP • STOP • STOP

A9 a. -1: $(2 - \sqrt{5})(2 + \sqrt{5})$
$2(2 + \sqrt{5}) - \sqrt{5}(2 + \sqrt{5})$
$4 + 2\sqrt{5} - 2\sqrt{5} - 5$
-1

b. 7: $(3 + \sqrt{2})(3 - \sqrt{2})$
$3^2 - (\sqrt{2})^2$ (using the techniques for special products)
$9 - 2$
7

6 The binomial that differs only in the sign of the second term from a given binomial is called its *conjugate*. The conjugate of $2 + \sqrt{5}$ is $2 - \sqrt{5}$, and the conjugate of $2 - \sqrt{5}$ is $2 + \sqrt{5}$.

The multiplication of conjugates that contain square roots will result in rational products. This result will be used in the solution of division problems.

Example: Multiply $2 + \sqrt{6}$ times its conjugate.

Solution

$$\begin{aligned}(2 + \sqrt{6})(2 - \sqrt{6}) &= 2^2 - (\sqrt{6})^2 \\ &= 4 - 6 \\ &= -2\end{aligned}$$

Q10 Give the conjugate of each binomial:

a. $2 + \sqrt{10}$ _______

b. $\sqrt{3} - \sqrt{2}$ _______

c. $\sqrt{7} + 3$ _______

d. $a + b$ _______

STOP • STOP • STOP • STOP • STOP • STOP • STOP • STOP • STOP

A10 a. $2 - \sqrt{10}$ b. $\sqrt{3} + \sqrt{2}$ c. $\sqrt{7} - 3$ d. $a - b$

Q11 Find the product of each binomial below times its conjugate:

a. $\sqrt{2} + \sqrt{10}$

b. $\sqrt{3} - \sqrt{2}$

c. $\sqrt{7} + 3$

d. $a + b$

STOP • STOP • STOP • STOP • STOP • STOP • STOP • STOP • STOP

A11 a. -8 b. 1 c. -2 d. $a^2 - b^2$

7 To divide radical expressions the following property can sometimes be used.

$$\frac{\sqrt[n]{a}}{\sqrt[n]{b}} = \sqrt[n]{\frac{a}{b}} \quad \text{where } a \text{ and } b \text{ are not negative if } n \text{ is even}$$

Two examples are:

$$\frac{\sqrt{12}}{\sqrt{3}} = \sqrt{\frac{12}{3}} = \sqrt{4} = 2$$

$$\frac{\sqrt{7abc}}{\sqrt{bc}} = \sqrt{\frac{7abc}{bc}} = \sqrt{7a}$$

Answers should be written in simplest radical form. This will sometimes require that the numerator and denominator be multiplied by an expression to make the denominator rational.

Example: Write $\frac{\sqrt{20}}{\sqrt{3}}$ in simplest radical form.

Solution

$$\frac{\sqrt{20}}{\sqrt{3}} = \frac{\sqrt{20}}{\sqrt{3}} \cdot \frac{\sqrt{3}}{\sqrt{3}} = \frac{\sqrt{4} \cdot \sqrt{5} \cdot \sqrt{3}}{3} = \frac{2\sqrt{15}}{3}$$

Q12 Simplify:

a. $\frac{\sqrt{48}}{\sqrt{3}}$

b. $\frac{\sqrt{42xy}}{\sqrt{6x}}$

c. $\frac{\sqrt[3]{54}}{\sqrt[3]{2}}$

d. $\frac{4\sqrt{12}}{\sqrt{3}}$

STOP • STOP • STOP • STOP • STOP • STOP • STOP • STOP • STOP

A12 **a.** 4: $\frac{\sqrt{48}}{\sqrt{3}} = \sqrt{\frac{48}{3}} = \sqrt{16} = 4$

b. $\sqrt{7y}$: $\frac{\sqrt{42xy}}{\sqrt{6x}} = \sqrt{\frac{42xy}{6x}} = \sqrt{7y}$

c. 3: $\frac{\sqrt[3]{54}}{\sqrt[3]{2}} = \sqrt[3]{\frac{54}{2}} = \sqrt[3]{27} = 3$

d. 8: $\frac{4\sqrt{12}}{\sqrt{3}} = 4\sqrt{\frac{12}{3}} = 4\sqrt{4} = 4 \cdot 2 = 8$

Q13 Write in simplest radical form:

a. $\dfrac{7}{\sqrt{3}}$

b. $\dfrac{2\sqrt{6}}{\sqrt{2}}$

c. $\dfrac{\sqrt{2}}{\sqrt{6}}$

d. $\dfrac{\sqrt[3]{12}}{\sqrt[3]{4}}$

STOP • STOP • STOP • STOP • STOP • STOP • STOP • STOP • STOP

A13

a. $\dfrac{7\sqrt{3}}{3}$: $\dfrac{7}{\sqrt{3}} = \dfrac{7}{\sqrt{3}} \cdot \dfrac{\sqrt{3}}{\sqrt{3}} = \dfrac{7\sqrt{3}}{3}$

b. $2\sqrt{3}$: $\dfrac{2\sqrt{6}}{\sqrt{2}} = \dfrac{2\sqrt{6}}{\sqrt{2}} \cdot \dfrac{\sqrt{2}}{\sqrt{2}} = \dfrac{2\sqrt{12}}{2} = \sqrt{4 \cdot 3} = 2\sqrt{3}$;

alternatively, $\dfrac{2\sqrt{6}}{\sqrt{2}} = \dfrac{2\sqrt{3} \cdot \overset{1}{\cancel{\sqrt{2}}}}{\underset{1}{\cancel{\sqrt{2}}}} = 2\sqrt{3}$

c. $\dfrac{\sqrt{3}}{3}$: $\dfrac{\sqrt{2}}{\sqrt{6}} = \dfrac{\sqrt{2}}{\sqrt{6}} \cdot \dfrac{\sqrt{6}}{\sqrt{6}} = \dfrac{\sqrt{12}}{6} = \dfrac{\sqrt{4 \cdot 3}}{6} = \dfrac{2\sqrt{3}}{6} = \dfrac{\sqrt{3}}{3}$;

alternatively, $\dfrac{\sqrt{2}}{\sqrt{6}} = \sqrt{\dfrac{2}{6}} = \sqrt{\dfrac{1}{3}} = \dfrac{\sqrt{1}}{\sqrt{3}} \cdot \dfrac{\sqrt{3}}{\sqrt{3}} = \dfrac{\sqrt{3}}{3}$

d. $\sqrt[3]{3}$: $\dfrac{\sqrt[3]{12}}{\sqrt[3]{4}} = \sqrt[3]{\dfrac{12}{4}} = \sqrt[3]{3}$

---

8 A binomial divided by a monomial can sometimes be reduced after the binomial is factored. For example,

1. $\dfrac{6 + \sqrt{72}}{6} = \dfrac{6 + \sqrt{36 \cdot 2}}{6} = \dfrac{6 + 6\sqrt{2}}{6} = \dfrac{6(1 + \sqrt{2})}{6} = 1 + \sqrt{2}$

2. $\dfrac{\sqrt{3} + \sqrt{15}}{\sqrt{3}} = \dfrac{\sqrt{3} + \sqrt{3} \cdot \sqrt{5}}{\sqrt{3}} = \dfrac{\sqrt{3}(1 + \sqrt{5})}{\sqrt{3}} = 1 + \sqrt{5}$

Q14 Reduce the fractions:

a. $\dfrac{2 + \sqrt{32}}{2}$

b. $\dfrac{6 + \sqrt{12}}{2}$

c. $\dfrac{14 - \sqrt{98}}{7}$

d. $\dfrac{\sqrt{3} + \sqrt{18}}{\sqrt{3}}$

STOP • STOP • STOP • STOP • STOP • STOP • STOP • STOP • STOP

A14

a. $1 + 2\sqrt{2}$: $\dfrac{2 + \sqrt{32}}{2} = \dfrac{2 + 4\sqrt{2}}{2} = \dfrac{2(1 + 2\sqrt{2})}{2} = 1 + 2\sqrt{2}$

b. $3 + \sqrt{3}$: $\dfrac{6 + \sqrt{12}}{2} = \dfrac{6 + 2\sqrt{3}}{2} = \dfrac{2(3 + \sqrt{3})}{2} = 3 + \sqrt{3}$

c. $2 - \sqrt{2}$: $\dfrac{14 - \sqrt{98}}{7} = \dfrac{14 - 7\sqrt{2}}{7} = \dfrac{7(2 - \sqrt{2})}{7} = 2 - \sqrt{2}$

d. $1 + \sqrt{6}$: $\dfrac{\sqrt{3} + \sqrt{18}}{\sqrt{3}} = \dfrac{\sqrt{3} + \sqrt{3} \cdot \sqrt{6}}{\sqrt{3}} = \dfrac{\sqrt{3}(1 + \sqrt{6})}{\sqrt{3}} = 1 + \sqrt{6}$

9

To rationalize the denominator of a fraction with a binomial denominator that contains square roots, multiply both numerator and denominator times the conjugate of the denominator.

Example 1: Rationalize the denominator of $\dfrac{2 + \sqrt{3}}{1 - \sqrt{5}}$.

Solution

$$\begin{aligned}
\frac{2 + \sqrt{3}}{1 - \sqrt{5}} &= \frac{2 + \sqrt{3}}{1 - \sqrt{5}} \cdot \frac{1 + \sqrt{5}}{1 + \sqrt{5}} \\
&= \frac{(2 + \sqrt{3})(1 + \sqrt{5})}{(1 - \sqrt{5})(1 + \sqrt{5})} \\
&= \frac{2 + 2\sqrt{5} + \sqrt{3} + \sqrt{15}}{1 + \sqrt{5} - \sqrt{5} - 5} \\
&= \frac{2 + 2\sqrt{5} + \sqrt{3} + \sqrt{15}}{-4}
\end{aligned}$$

The final fraction obtained is considered to be in simplest radical form.

Example 2: Divide $4 + \sqrt{3}$ by $1 - \sqrt{3}$.

Solution

$$\begin{aligned}
\frac{4 + \sqrt{3}}{1 - \sqrt{3}} &= \frac{4 + \sqrt{3}}{1 - \sqrt{3}} \cdot \frac{1 + \sqrt{3}}{1 + \sqrt{3}} \\
&= \frac{(4 + \sqrt{3})(1 + \sqrt{3})}{(1 - \sqrt{3})(1 + \sqrt{3})} \\
&= \frac{4 + 4\sqrt{3} + \sqrt{3} + 3}{1 - 3} \\
&= \frac{7 + 5\sqrt{3}}{-2}
\end{aligned}$$

Q15 Rationalize the denominators in these fractions:

a. $\dfrac{2+\sqrt{5}}{7-\sqrt{5}}$ **b.** $\dfrac{3-\sqrt{2}}{4+\sqrt{3}}$

STOP • STOP • STOP • STOP • STOP • STOP • STOP • STOP • STOP

A15 **a.** $\dfrac{19+9\sqrt{5}}{44}$: $\dfrac{2+\sqrt{5}}{7-\sqrt{5}}\cdot\dfrac{7+\sqrt{5}}{7+\sqrt{5}}$

$$\frac{14+2\sqrt{5}+7\sqrt{5}+5}{49-5}$$

$$\frac{19+9\sqrt{5}}{44}$$

b. $\dfrac{12-3\sqrt{3}-4\sqrt{2}+\sqrt{6}}{13}$: $\dfrac{3-\sqrt{2}}{4+\sqrt{3}}\cdot\dfrac{4-\sqrt{3}}{4-\sqrt{3}}$

$$\frac{12-3\sqrt{3}-4\sqrt{2}+\sqrt{6}}{16-3}$$

$$\frac{12-3\sqrt{3}-4\sqrt{2}+\sqrt{6}}{13}$$

Q16 Divide the real numbers:

a. 6 by $2+\sqrt{7}$ **b.** $2+\sqrt{3}$ by $\sqrt{3}+\sqrt{5}$

STOP • STOP • STOP • STOP • STOP • STOP • STOP • STOP • STOP

A16 **a.** $-4+2\sqrt{7}$: $\dfrac{6}{2+\sqrt{7}}=\dfrac{6}{2+\sqrt{7}}\cdot\dfrac{2-\sqrt{7}}{2-\sqrt{7}}=\dfrac{12-6\sqrt{7}}{4-7}$

$$=\frac{-3(-4+2\sqrt{7})}{-3}=-4+2\sqrt{7}$$

b. $\dfrac{3+2\sqrt{3}-2\sqrt{5}-\sqrt{15}}{-2}$: $\dfrac{2+\sqrt{3}}{\sqrt{3}+\sqrt{5}}=\dfrac{2+\sqrt{3}}{\sqrt{3}+\sqrt{5}}\cdot\dfrac{\sqrt{3}-\sqrt{5}}{\sqrt{3}-\sqrt{5}}$

$$=\frac{2\sqrt{3}-2\sqrt{5}+3-\sqrt{15}}{3-5}$$

$$=\frac{3+2\sqrt{3}-2\sqrt{5}-\sqrt{15}}{-2}$$

---

10 Solutions to equations are sometimes found to be irrational numbers. The methods for solving these equations will not be discussed in this section, but possible solutions can be checked with the techniques just discussed.

Example: Is $5\sqrt{2}$ a solution to the following equation?

$x^2 - 50 = 0$

Solution

Substituting $5\sqrt{2}$ for x,

$$\begin{aligned} x^2 - 50 &= (5\sqrt{2})^2 - 50 \\ &= 25 \cdot 2 - 50 \\ &= 50 - 50 \\ &= 0 \end{aligned}$$

Therefore, $5\sqrt{2}$ is a solution to $x^2 - 50 = 0$.

Q17 Show that $-7\sqrt{2}$ is a solution to $x^2 - 98 = 0$.

STOP • STOP • STOP • STOP • STOP • STOP • STOP • STOP • STOP

A17 $x^2 - 98 = (-7\sqrt{2})^2 - 98 = 98 - 98 = 0$

Q18 Show that $1 - \sqrt{3}$ is a solution to $x^2 - 2x - 2 = 0$.

STOP • STOP • STOP • STOP • STOP • STOP • STOP • STOP • STOP

A18
$$\begin{aligned} x^2 - 2x - 2 &= (1 - \sqrt{3})^2 - 2(1 - \sqrt{3}) - 2 \\ &= 1 - 2\sqrt{3} + 3 - 2 + 2\sqrt{3} - 2 \\ &= 0 \end{aligned}$$

Q19 By substituting and evaluating the result, find the *two* solutions to the equation $x^2 - 4x - 7 = 0$ from the replacement set $\{2 + \sqrt{11},\ 1 - 3\sqrt{2},\ 2 - \sqrt{11},\ 3 - \sqrt{2}\}$.

STOP • STOP • STOP • STOP • STOP • STOP • STOP • STOP • STOP

A19 $2 + \sqrt{11}$ and $2 - \sqrt{11}$

11 The last topic considered in this section will be the solution of equations containing radicals. It is helpful to use the following property: If both the right and left sides of an equation are squared, the solution set of the original equation is a subset of the solution set of the resulting equation.

$$\sqrt{2x+3} = 4$$
$$(\sqrt{2x+3})^2 = (4)^2$$
$$2x + 3 = 16$$
$$2x = 13$$
$$x = \frac{13}{2}$$

The solution set of $\sqrt{2x+3} = 4$ is a subset of $\left\{\frac{13}{2}\right\}$. To find if $\frac{13}{2}$ is a solution, substitute in the original equation.

$$\sqrt{2x+3} = 4$$
$$\sqrt{2\left(\frac{13}{2}\right) + 3} = 4$$
$$\sqrt{13+3} = 4$$
$$\sqrt{16} = 4 \quad \text{(true)}$$

Therefore, the solution set is $\left\{\frac{13}{2}\right\}$.

Q20 **a.** Solve the equation $3 = \sqrt{x+2}$.

b. Check your solutions in the original equation.

STOP • STOP • STOP • STOP • STOP • STOP • STOP • STOP • STOP

A20 **a.** $x = 7$:
$$3 = \sqrt{x+2}$$
$$9 = (\sqrt{x+2})^2$$
$$9 = x + 2$$
$$7 = x$$

The solution set is $\{7\}$.

b.
$$3 = \sqrt{x+2}$$
$$3 = \sqrt{7+2}$$
$$3 = \sqrt{9} \text{ (true)}$$

Q21 Solve the equation $\sqrt{-7x-12} = -4$ and check the solution.

STOP • STOP • STOP • STOP • STOP • STOP • STOP • STOP • STOP

A21 $\varnothing$:
$$(\sqrt{-7x-12})^2 = (-4)^2$$
$$-7x - 12 = 16$$
$$-7x = 28$$
$$x = -4$$

Check:
$$\sqrt{-7(-4) - 12} = -4$$
$$\sqrt{+28 - 12} = -4$$
$$\sqrt{16} = -4 \quad \text{(false)}$$

The solution set is $\varnothing$. (See footnote, Section 3.2, A32.) A careful look at the original problem indicates that the solution set must be empty because no principal square root results in a negative number.

12 It is good policy to check all solutions to equations to discover errors in calculations. Equations that involve radicals need to be checked for other reasons. There may be no errors in computations, and yet some of the possible solutions will not check. The reason for this is that the process of squaring both sides of an equation does not always result in an equivalent equation. Fortunately, extra solutions are obtained—the process never loses solutions.

To illustrate this, consider the equation

$$x = 9$$

By inspection, the only number that will satisfy this equation is 9. However, when both sides are squared, there are two solutions to the new equation $x^2 = 81$, 9 and -9. Therefore, always check solutions in the *original* equation after having squared both sides of an equation.

Q22 Solve $\sqrt{2x + 3} = 3$.

STOP • STOP • STOP • STOP • STOP • STOP • STOP • STOP • STOP

A22 $x = 3$:

$$\sqrt{2x+3} = 3 \qquad \text{Check: } \sqrt{2(3)+3} = 3$$
$$2x + 3 = 9 \qquad \sqrt{6+3} = 3$$
$$2x = 6 \qquad \sqrt{9} = 3 \text{ (true)}$$
$$x = 3$$

This completes the instruction for this section.

4.2 Exercises

1. Perform any indicated operations and write the answer in simplest radical form:
   a. $2\sqrt{3} + 5\sqrt{3} - \sqrt{3}$ b. $5\sqrt{2} - \sqrt{2} - 6\sqrt{2}$ c. $2\sqrt{3} + 2\sqrt{12}$
   d. $\sqrt[3]{16} - 3\sqrt[3]{2}$ e. $\sqrt{12} + \sqrt{75}$ f. $-4\sqrt{3} - \sqrt{48}$
   g. $\sqrt{4x^3} - \sqrt{25x^3}$ h. $\sqrt{32} + 4\sqrt{\frac{1}{2}}$ i. $\sqrt{\frac{5}{4}} + \sqrt{45}$
   j. $\sqrt{8xy} + 2\sqrt{18xy}$ k. $\sqrt[3]{-81} + 2\sqrt[3]{3}$ l. $\sqrt[5]{64} - 3\sqrt[5]{2}$
2. Simplify:
   a. $\sqrt{2} \cdot \sqrt{8}$ b. $\sqrt{12} \cdot 3\sqrt{8}$ c. $4\sqrt[3]{24} \cdot \sqrt[3]{9}$
   d. $(3x\sqrt{x^2})(x\sqrt{x})(\sqrt{9x^2})$ e. $\sqrt{5}(2 + 2\sqrt{5})$ f. $3\sqrt{3}(\sqrt{3} + \sqrt{6})$
   g. $\sqrt{2}(\sqrt{3} + \sqrt{5})$ h. $(2 + \sqrt{3})(-1 + \sqrt{3})$ i. $(\sqrt{2} - 4)(\sqrt{2} - 1)$

j. $(3 + 2\sqrt{3})(3 - 2\sqrt{3})$ k. $(\sqrt{5} - 2)(\sqrt{5} + 2)$ l. $(3 + \sqrt{6})^2$

m. $\dfrac{\sqrt{12}}{\sqrt{3}}$ n. $\dfrac{\sqrt{15}}{\sqrt{5}}$ o. $\dfrac{\sqrt{3xy}}{\sqrt{2x}}$

p. $\dfrac{\sqrt[3]{4}}{\sqrt[3]{18}}$ q. $\dfrac{2 + 4\sqrt{5}}{2}$ r. $\dfrac{5 + \sqrt{50}}{5}$

s. $\dfrac{6 - \sqrt{27}}{9}$ t. $\dfrac{\sqrt{10} + \sqrt{20}}{\sqrt{2}}$ u. $\dfrac{2 + \sqrt{3}}{1 - \sqrt{3}}$

v. $\dfrac{5 + \sqrt{5}}{2 - \sqrt{5}}$ w. $\dfrac{6 + \sqrt{10}}{1 - 2\sqrt{10}}$ x. $\dfrac{\sqrt{3} + \sqrt{7}}{2\sqrt{3} + \sqrt{7}}$

3. Determine if the number given for the variable is a solution to the equation by substitution:
   a. $x = 2 + \sqrt{2}$ in the equation $x^2 - 4x + 2 = 0$
   b. $x = 2 - \sqrt{6}$ in the equation $x^2 - 4x - 2 = 0$
4. Find the solution sets:
   a. $\sqrt{x + 1} = 2$ b. $-3 = \sqrt{2x + 1}$

4.2 Exercise Answers

1. a. $6\sqrt{3}$ b. $-2\sqrt{2}$ c. $6\sqrt{3}$
   d. $-\sqrt[3]{2}$ e. $7\sqrt{3}$ f. $-8\sqrt{3}$
   g. $-3x\sqrt{x}$ h. $6\sqrt{2}$ i. $\dfrac{7\sqrt{5}}{2}$
   j. $8\sqrt{2xy}$ k. $-\sqrt[3]{3}$ l. $-\sqrt[5]{2}$
2. a. 4 b. $12\sqrt{6}$ c. 24
   d. $9x^4\sqrt{x}$ e. $2\sqrt{5} + 10$ f. $9 + 9\sqrt{2}$
   g. $\sqrt{6} + \sqrt{10}$ h. $1 + \sqrt{3}$ i. $6 - 5\sqrt{2}$
   j. -3 k. 1 l. $15 + 6\sqrt{6}$
   m. 2 n. $\sqrt{3}$ o. $\dfrac{\sqrt{6y}}{2}$
   p. $\dfrac{\sqrt[3]{6}}{3}$ q. $1 + 2\sqrt{5}$ r. $1 + \sqrt{2}$
   s. $\dfrac{2 - \sqrt{3}}{3}$ t. $\sqrt{5} + \sqrt{10}$ u. $\dfrac{5 + 3\sqrt{3}}{-2}$
   v. $-15 - 7\sqrt{5}$ w. $\dfrac{2 + \sqrt{10}}{-3}$ or $\dfrac{-2 - \sqrt{10}}{3}$ x. $\dfrac{-1 + \sqrt{21}}{5}$
3. a. yes b. yes
4. a. $x = 3$ b. $\varnothing$

4.3 Integer Exponents

1 The properties of exponents discussed in Chapter 2 will be extended to integer values of the exponents in this section. The properties can be summarized as:

1. $x^m \cdot x^n = x^{m+n}$ $\quad m \in N, n \in N, x \in R$
2. $(x^m)^n = x^{mn}$ $\quad m \in N, n \in N, x \in R$
3. $(xy)^m = x^m y^m$ $\quad m \in N, x \in R, y \in R$
4. $\left(\frac{x}{y}\right)^m = \frac{x^m}{y^m}$ $\quad m \in N, x \in R, y \in R, y \neq 0$
5. $\frac{x^m}{x^n} = x^{m-n}$ $\quad m \in N, n \in N, m > n, x \neq 0$

Q1 Review the properties of exponents by simplifying the expressions below:

a. $2^3 \cdot 2^7 =$ ________ **b.** $(5^2)^3 =$ ________

c. $(2xy^2)^3 =$ ________ **d.** $\left(\frac{4}{3x^2}\right)^3 =$ ________

e. $\frac{8x^5}{2x^3} =$ ________ **f.** $\left(\frac{2x^2y^3}{y}\right)^5 =$ ________

STOP • STOP • STOP • STOP • STOP • STOP • STOP • STOP • STOP

A1 **a.** 2^{10} **b.** 5^6

c. $2^3x^3y^6$ **d.** $\frac{4^3}{3^3x^6}$

e. $4x^2$ **f.** $2^5x^{10}y^{10}$

2 One requirement that will be imposed as the exponents are allowed to assume values from sets other than the natural numbers is that the five properties of Frame 1 will continue to hold. The definitions must then be made in a way consistent with these properties.

If Property 4 still holds when $m = n$, then $\frac{x^m}{x^m} = x^{m-m} = x^0$. However, a nonzero number divided by itself is 1, so $\frac{x^m}{x^m} = 1$. Therefore, x^0 is defined to be 1 for all values of x except zero. 0^0 is not defined.

Example 1: Simplify 5^0.

Solution

$5^0 = 1$

Example 2: Simplify $(2x^2y^3)^0$, $x \neq 0, y \neq 0$.

Solution

$(2x^2y^3)^0 = 1$

Q2 Simplify:

a. $22^0 =$ ________ **b.** $2x^0y^2 =$ ________

c. $(4xy^2)(2x)^0 =$ ________ **d.** $0^0 =$ ________

STOP • STOP • STOP • STOP • STOP • STOP • STOP • STOP • STOP

A2 **a.** 1 **b.** $2y^2$

c. $4xy^2$ **d.** undefined

3

All five properties of Frame 1 may be restated for exponents from the whole numbers. Study these examples.

Example 1: Simplify $3^5 \cdot 3^0 \cdot 3^4$.

Solution

$3^5 \cdot 3^0 \cdot 3^4 = 3^{5+0+4} = 3^9$

Example 2: Simplify $\dfrac{6x^6y^8}{2x^0y^5}$.

Solution

$\dfrac{6x^6y^8}{2x^0y^5} = 3x^{6-0}y^{8-5} = 3x^6y^3$

Q3 Simplify by applying the properties of exponents:

a. $5^2 \cdot 5^5 \cdot 5^0$

b. $(2x^0y^3z)^3$

c. $(2x^2y)(3x^0y^4)$

d. $\left(\dfrac{x^3}{y^0}\right)^4$

STOP • STOP • STOP • STOP • STOP • STOP • STOP • STOP • STOP

A3

a. 5^7: $5^2 \cdot 5^5 \cdot 5^0 = 5^{2+5+0} = 5^7$
b. $2^3y^9z^3$: $(2x^0y^3z)^3 = 2^3x^{0\cdot 3}y^{3\cdot 3}z^{1\cdot 3} = 2^3x^0y^9z^3 = 2^3y^9z^3$
c. $6x^2y^5$: $(2x^2y)(3x^0y^4) = 6x^{2+0}y^{1+4} = 6x^2y^5$
d. x^{12}: $\left(\dfrac{x^3}{y^0}\right)^4 = \dfrac{x^{3\cdot 4}}{y^{0\cdot 4}} = \dfrac{x^{12}}{y^0} = \dfrac{x^{12}}{1} = x^{12}$

4

Before writing the definition of what 4^{-2} will mean, the properties will be examined to see what logical meaning would be required in order for the properties to remain applicable.

Consider $4^{-2} \cdot 4^2$. Using Property 1, in which $x^m \cdot x^n = x^{m+n}$, the product would be

$4^{-2} \cdot 4^2 = 4^{-2+2} = 4^0 = 1$

In other words, 4^{-2} is a number that when multiplied times 4^2 results in 1. Previously this number has been called the reciprocal of 4^2.

$4^2 \cdot \dfrac{1}{4^2} = 1$

To be consistent with the above statement, x^{-m} is defined to be the reciprocal of x^m; that is, $x^{-m} = \dfrac{1}{x^m}$. Note that

$x^m \cdot x^{-m} = x^{m+^-m} = x^0 = 1$

Example 1: Simplify 9^{-2}.

Solution

$$9^{-2} = \frac{1}{9^2} = \frac{1}{81}$$

Example 2: Simplify $2^{-1} \cdot 3^{-2}$.

Solution

$$2^{-1} \cdot 3^{-2} = \frac{1}{2^1} \cdot \frac{1}{3^2} = \frac{1}{2} \cdot \frac{1}{9} = \frac{1}{18}$$

Example 3: Write $2x^{-2}yz^{-3}$ with positive exponents.

Solution

$$2x^{-2}yz^{-3} = 2 \cdot \frac{1}{x^2} \cdot y \cdot \frac{1}{z^3} = \frac{2y}{x^2z^3}$$

Q4 Write each of the following as a fraction:

a. $2^{-2} =$ ______ b. $10^{-3} =$ ______

c. $5^{-1} =$ ______ d. $4^{-3} =$ ______

STOP • STOP • STOP • STOP • STOP • STOP • STOP • STOP • STOP

A4 a. $\frac{1}{4}$ b. $\frac{1}{1{,}000}$ c. $\frac{1}{5}$ d. $\frac{1}{64}$

Q5 Write each expression by using positive exponents:

a. $x^{-2} =$ ______ b. $2x^2y^{-3} =$ ______

c. $(2x^3)^{-1} =$ ______ d. $2x^{-1}y^3z^{-2} =$ ______

STOP • STOP • STOP • STOP • STOP • STOP • STOP • STOP • STOP

A5 a. $\frac{1}{x^2}$ b. $\frac{2x^2}{y^3}$ c. $\frac{1}{2x^3}$ d. $\frac{2y^3}{xz^2}$

5 The fact that $x^{-2} = \frac{1}{x^2}$ implies that when a factor with a negative exponent is found in the numerator, the factor is moved to the denominator and the sign of the exponent is changed.

Consider the effect of a negative exponent in the denominator of a fraction.

$$\frac{1}{x^{-3}} = \frac{1}{\frac{1}{x^3}} = 1 \cdot \frac{x^3}{1} = x^3$$

The example implies that a factor in the denominator with a negative exponent may be moved to the numerator and the sign of the exponent changed.

Example 1: Simplify $\frac{2x}{y^{-2}}$.

Solution

$$\frac{2x}{y^{-2}} = 2xy^2$$

Example 2: Simplify $\frac{2x^{-2}z}{y^{-3}}$.

Solution

$$\frac{2x^{-2}z}{y^{-3}} = \frac{2y^3z}{x^2}$$

Q6 Write the following fractions with positive exponents:

a. $\frac{5}{x^{-3}} =$ ______ b. $\frac{2x^{-2}}{y^{-3}} =$ ______

STOP • STOP • STOP • STOP • STOP • STOP • STOP • STOP • STOP

A6 a. $5x^3$ b. $\frac{2y^3}{x^2}$

Q7 Simplify and write with positive exponents:

a. $\frac{2x^2y}{4x^{-2}y^2} =$ ______ b. $\frac{24x^{-2}y}{2x^{-3}y} =$ ______

STOP • STOP • STOP • STOP • STOP • STOP • STOP • STOP • STOP

A7 a. $\frac{x^4}{2y}$ b. $12x$

6 Now that negative exponents have been defined, the properties of exponents will be restated for integral exponents.

1. $x^m \cdot x^n = x^{m+n}$ $m \in I, n \in I, x \in R$
2. $(x^m)^n = x^{mn}$ $m \in I, n \in I, x \in R$
3. $(xy)^m = x^m y^m$ $m \in I, x \in R, y \in R$
4. $\left(\frac{x}{y}\right)^m = \frac{x^m}{y^m}$ $m \in I, x \in R, y \in R, y \neq 0$
5. $\frac{x^m}{x^n} = x^{m-n}$ $m \in I, n \in I, x \in R, x \neq 0$

Note that there is no longer a restriction on Property 5 that m must be greater than n. Hence n may be greater, less, or equal to m.

When operations are to be performed on expressions involving negative exponents, you have a choice of methods. The exponents can be changed to positive exponents before or after the operations are performed.

Example: Simplify the expression $(2x^{-2})^2(4x^{-3})$ in two ways.

Solution 1

$$
\begin{aligned}
(2x^{-2})^2(4x^{-3}) &= \left(\frac{2}{x^2}\right)^2\left(\frac{4}{x^3}\right) \\
&= \frac{4}{x^4} \cdot \frac{4}{x^3} \\
&= \frac{16}{x^7}
\end{aligned}
$$

Solution 2

$$
\begin{aligned}
(2x^{-2})^2(4x^{-3}) &= (4x^{-4})(4x^{-3}) \\
&= 16x^{-7} \\
&= \frac{16}{x^7}
\end{aligned}
$$

Q8 Perform the indicated operations and write with positive exponents:

a. $(2x^{-2})(3x^4)(2x^{-1})$

b. $(2x^3)(4x^{-4})$

c. $(x^{-2})^3$

d. $(2y^{-3})^{-2}$

STOP • STOP • STOP • STOP • STOP • STOP • STOP • STOP • STOP

A8 **a.** $12x$: $(2x^{\ 2})(3x^4)(2x^{-1}) = 2 \cdot 3 \cdot 2x^{-2+4-1} = 12x$

b. $\frac{8}{x}$: $(2x^3)(4x^{-4}) = 8x^{3-4} = 8x^{-1} = \frac{8}{x}$

c. $\frac{1}{x^6}$: $(x^{-2})^3 = x^{-6} = \frac{1}{x^6}$

d. $\frac{y^6}{4}$: $(2y^{-3})^{-2} = 2^{-2}y^{(-3)(-2)} = 2^{-2}y^6 = \frac{y^6}{4}$

Q9 Perform the indicated operations and write with positive exponents:

a. $(2x^{-2})(x^3)^{-1}(x^{-4})^{-1}$

b. $(x^{-2}y^3)^{-2}$

c. $(3x^0y^4)^{-2}$

d. $(x^{-2})^3(x^4)^{-1}$

STOP • STOP • STOP • STOP • STOP • STOP • STOP • STOP • STOP

A9 **a.** $\frac{2}{x}$: $(2x^{-2})(x^3)^{-1}(x^{-4})^{-1} = (2x^{-2})(x^{-3})(x^4) = 2x^{-2-3+4} = 2x^{-1} = \frac{2}{x}$

b. $\frac{x^4}{y^6}$: $(x^{-2}y^3)^{-2} = x^4y^{-6} = \frac{x^4}{y^6}$

c. $\frac{1}{9y^8}$: $(3x^0y^4)^{-2} = 3^{-2}x^0y^{-8} = \frac{1}{9y^8}$

d. $\frac{1}{x^{10}}$: $(x^{-2})^3(x^4)^{-1} = (x^{-6})(x^{-4}) = x^{-10} = \frac{1}{x^{10}}$

7

When simplifying fractions, the properties may be applied first or exponents may be made positive first.

Example: Simplify $\left(\frac{x^{-2}}{x^3}\right)^{-2}$.

Solution 1

$$\left(\frac{x^{-2}}{x^3}\right)^{-2} = \left(\frac{1}{x^2x^3}\right)^{-2} = \left(\frac{1}{x^5}\right)^{-2} = \frac{1}{\left(\frac{1}{x^5}\right)^2} = \frac{1}{\frac{1}{x^{10}}} = x^{10}$$

Solution 2

$$\left(\frac{x^{-2}}{x^3}\right)^{-2} = \frac{x^4}{x^{-6}} = x^4 \cdot x^6 = x^{10}$$

In comparing the solutions it is more convenient to apply the properties first.

Q10 Simplify the expressions:

a. $\left(\frac{x^{-2}}{y}\right)^{-3}$

b. $\left(\frac{2x}{4x^{-2}}\right)^{-1}$

c. $\frac{2x^{-3}}{x^{-4}}$

STOP • STOP • STOP • STOP • STOP • STOP • STOP • STOP • STOP

A10 a. x^6y^3: $\left(\frac{x^{-2}}{y}\right)^{-3} = \frac{x^6}{y^{-3}} = x^6y^3$

b. $\frac{2}{x^3}$: $\left(\frac{2x}{4x^{-2}}\right)^{-1} = \frac{2^{-1}x^{-1}}{4^{-1}x^2} = \frac{4}{2x^2x} = \frac{2}{x^3}$

c. $2x$: $\frac{2x^{-3}}{x^{-4}} = 2x^{-3-(-4)} = 2x^{-3+4} = 2x$

8

Be very careful when a negative exponent is used in an expression with sums or differences. The factor with a negative exponent is placed in the denominator of its own term, then the terms are added. For example,

$$\begin{aligned} 2^{-1} + 3^{-1} &= \frac{1}{2} + \frac{1}{3} \\ &= \frac{3}{6} + \frac{2}{6} \\ &= \frac{5}{6} \end{aligned}$$

(*Note:* $2^{-1} + 3^{-1} \neq \frac{1}{2+3}$.)

Example: Simplify $x^{-1} + 2x^{-2}$.

Solution

$$x^{-1} + 2x^{-2} = \frac{1}{x} + \frac{2}{x^2}$$
$$= \frac{x}{x^2} + \frac{2}{x^2}$$
$$= \frac{x + 2}{x^2}$$

Q11 Add the following expressions:

a. $3x^2 + 2x^{-1}$ **b.** $5 + x^{-2}$

STOP • STOP • STOP • STOP • STOP • STOP • STOP • STOP • STOP

A11 **a.** $\frac{3x^3 + 2}{x}$: $3x^2 + 2x^{-1}$

$3x^2 + \frac{2}{x}$

$\frac{3x^3}{x} + \frac{2}{x}$

b. $\frac{5x^2 + 1}{x^2}$: $5 + x^{-2}$

$5 + \frac{1}{x^2}$

$\frac{5x^2}{x^2} + \frac{1}{x^2}$

9 Very large and very small numbers are used in the fields of chemistry and physics. To more easily calculate with these numbers a notation has been developed called *scientific notation.* Scientific notation will be useful in this text when logarithms are studied.

A number is written in scientific notation if it is written as a number greater than or equal to 1 and less than 10 multiplied by 10 raised to a power. The following numbers are written in scientific notation:

2.36×10^2 1.72×10^8
7.8×10^{-3} 4.7×10^0

The following numbers are not written in scientific notation:

0.78×10^2 $(0.78 < 1)$ 56.8×10^6 $(56.8 > 10)$

Q12 Write "yes" if the number is written in scientific notation; otherwise, indicate "no."

a. 48.7×10^2 ______ **b.** 3.72×10^6 ______

c. 4.2×10^{-2} ______ **d.** 0.04×10^2 ______

e. 0.002×10^{-4} ______ **f.** 5×10 ______

STOP • STOP • STOP • STOP • STOP • STOP • STOP • STOP • STOP

A12 **a.** no **b.** yes **c.** yes **d.** no **e.** no **f.** yes

10

Recall that multiplying a number times a multiple of 10 moves the decimal point to the right the same number of places as there are zeros in the multiple of 10.

Example 1: Write 2.62×10^3 without the use of scientific notation.

Solution

$$2.62 \times 10^3 = 2.62 \times 1{,}000$$
$$= 2{,}620$$

Example 2: Write 4.6×10^6 in decimal notation.

Solution

$$4.6 \times 10^6 = 4.6 \times 1{,}000{,}000$$
$$= 4{,}600{,}000$$

Note that the decimal is moved the same number of positions that appears as the exponent of 10. With this in mind the middle step need not be included.

Example 3: Write 4.7×10^3 in decimal notation.

Solution

$$4.7 \times 10^3 = 4{,}700$$

Q13 Write in decimal notation:

a. $5.72 \times 10^4 =$ ____________

b. $6.24 \times 10 =$ ____________

c. $7 \times 10^7 =$ ____________

STOP • STOP • STOP • STOP • STOP • STOP • STOP • STOP • STOP

A13 **a.** 57,200 **b.** 62.4 **c.** 70,000,000

11

Numbers that include negative exponents can be converted from scientific notation to decimal notation by means of division by a power of 10. This moves the decimal to the left.

Example 1: Write 8.2×10^{-2} in decimal notation.

Solution

$$8.2 \times 10^{-2} = 8.2 \times \frac{1}{10^2}$$
$$= \frac{8.2}{100}$$
$$= 0.082$$

The intermediate steps may be omitted and the decimal point moved the appropriate number of positions to the left.

Example 2: Convert 5.34×10^{-4} to decimal notation.

Solution

$$5.34 \times 10^{-4} = 0.000534$$

Q14 Convert to decimal notation:

a. $4.6 \times 10^{-2} =$ ____________

b. $9.804 \times 10^{-4} =$ ____________

c. $4.7 \times 10^{0} =$ ____________

STOP • STOP • STOP • STOP • STOP • STOP • STOP • STOP • STOP

A14 **a.** 0.046 **b.** 0.0009804
c. 4.7 (recall that $10^0 = 1$)

---

12 A number, N, is written in scientific notation by writing N as the product of a decimal, d ($1 \leqslant d < 10$), and a power of 10, 10^j, where $j \in I$. That is,

$$N = d \times 10^j$$

Example 1: Write 2,420 in scientific notation.

Solution

$2{,}420 = 2.420 \times 10^3$

Example 2: Write 0.00402 in scientific notation.

Solution

$0.00402 = 4.02 \times 10^{-3}$

Q15 Write in scientific notation:

a. $42{,}640 =$ ____________

b. $8{,}000{,}000 =$ ____________

c. $53 =$ ____________

d. $0.046 =$ ____________

e. $0.0009 =$ ____________

STOP • STOP • STOP • STOP • STOP • STOP • STOP • STOP • STOP

A15 **a.** 4.264×10^4 **b.** 8.0×10^6 **c.** 5.3×10 **d.** 4.6×10^{-2}
e. $9. \times 10^{-4}$

---

13 Sometimes a number is written in a form that is similar to scientific notation, but the first number is not between 1 and 10. Such a number can be written in scientific notation.

Example: Write $28. \times 10^4$ in scientific notation.

Solution

$$\begin{aligned} 28. \times 10^4 &= 2.8 \times 10^1 \times 10^4 \\ &= 2.8 \times 10^5 \end{aligned}$$

Notice that 28. is first converted to scientific notation and then multiplied by 10^4.

Q16 Convert to scientific notation:

a. $46. \times 10^3 =$ ____________

b. $522. \times 10^2 =$ ____________

c. $0.420 \times 10^{-5} =$ ____________

STOP • STOP • STOP • STOP • STOP • STOP • STOP • STOP • STOP

A16 a. 4.6×10^4 b. 5.22×10^4 c. 4.20×10^{-6}

---

14 When multiplying numbers that are written in scientific notation, separate the powers of 10 from the other factors.

Example 1: Multiply $(4.0 \times 10^4)(2.0 \times 10^2)$.

Solution

$(4.0 \times 10^4)(2.0 \times 10^2)$
$4.0 \times 2.0 \times 10^4 \times 10^2$
8.0×10^6

Example 2: Multiply $(6.0 \times 10^{-3})(5.0 \times 10^5)$.

Solution

$(6.0 \times 10^{-3})(5.0 \times 10^5)$
$6.0 \times 5.0 \times 10^{-3} \times 10^5$
30.0×10^2
3.0×10^3

Q17 Write the product in scientific notation:
a. $(3.0 \times 10^3)(1.5 \times 10^2)$ b. $(5.0 \times 10^2)(7.0 \times 10^4)$

STOP • STOP • STOP • STOP • STOP • STOP • STOP • STOP • STOP

A17 a. 4.5×10^5 b. 3.5×10^7

Q18 Write the products in scientific notation:
a. $(5.0 \times 10^{-2})(1.6 \times 10^{-4})$ b. $(3.0 \times 10^{-5})(7.0 \times 10^2)$

STOP • STOP • STOP • STOP • STOP • STOP • STOP • STOP • STOP

A18 a. 8.0×10^{-6} b. 2.1×10^{-2}

---

15 When dividing numbers written in scientific notation, separate the powers of 10 from the other numbers.

Example 1: Divide $\dfrac{9.0 \times 10^5}{1.5 \times 10^2}$.

Solution

$$\frac{9.0 \times 10^5}{1.5 \times 10^2} = \frac{9.0}{1.5} \times \frac{10^5}{10^2}$$
$$= 6.0 \times 10^{5-2}$$
$$= 6.0 \times 10^3$$

Example 2: Divide $\frac{1.2 \times 10^{-2}}{4.0 \times 10^{-5}}$.

Solution

$$\frac{1.2 \times 10^{-2}}{4.0 \times 10^{-5}} = \frac{1.2}{4.0} \times \frac{10^{-2}}{10^{-5}}$$
$$= 0.3 \times 10^{-2-(-5)}$$
$$= 0.3 \times 10^3$$
$$= 3.0 \times 10^2$$

Q19 Write the quotients in scientific notation:

a. $\frac{4.8 \times 10^6}{1.2 \times 10}$

b. $\frac{3.00 \times 10^4}{2.50 \times 10^3}$

c. $\frac{2.7 \times 10^2}{9.0 \times 10^6}$

d. $\frac{6.25 \times 10^4}{2.5 \times 10^6}$

STOP • STOP • STOP • STOP • STOP • STOP • STOP • STOP • STOP

A19 a. 4.0×10^5 b. 1.2×10 c. 3.0×10^{-5} d. 2.5×10^{-2}

This completes the instruction for this section.

4.3 Exercises

1. Write each expression with positive exponents:
   a. 2^{-3} b. $4x^{-2}y^3$ c. x^{-1} d. $\frac{1}{x^{-3}}$
   e. $\frac{x^{-5}}{y^{-3}}$ f. $4x^{-2}y^{-3}z$
2. Simplify each of the following and write the result with positive exponents (assume that variables represent nonzero real numbers):
   a. $(4x^{-2})^2$ b. $(2x^{-3}y^2)^{-1}$ c. $2x^0 \cdot x \cdot x^2$ d. $(-3x^{-2})(-x^2)$
   e. $(2x^{-3} \cdot y^2)^0$ f. $4x^{-1} \cdot x^0 \cdot x^{-2}$

3. Write each fraction by use of positive exponents only:

a. $\frac{2x^2}{x^{-3}}$ b. $\frac{4x^{-2}}{(2x^{-2})^2}$ c. $(x^{-1})^{-2}$ d. $\frac{x^{-2}y}{x^2y^{-1}}$

4. Perform the indicated operations and write the result with positive exponents:

a. $(x^{-2})^3$ b. $(2x^3)^{-1}$ c. $\frac{3x^{-2}}{x^{-2} \cdot x^4}$ d. $\frac{4x^3}{(2x^{-1})^3}$

5. Simplify each fraction and write with positive exponents:

a. $\left(\frac{x^{-2}}{y^2}\right)^{-1}$ b. $\left(\frac{x^{-2}}{x^3}\right)^2$ c. $\left(\frac{2x^2}{x^{-1}}\right)^3$ d. $\frac{2x^{-2}}{4x^{-1}y^{-1}}$

6. Add the following expressions:

a. $3^{-2} + 6^{-1}$ b. $2x^{-1} + x^3$

7. Indicate whether the number is written in scientific notation by a yes or no:

a. 2.5×10^{-3} b. 0.46×10^{-2} c. $452. \times 10^4$ d. 5.8×10^6

8. Write each number in decimal notation:

a. 4.6×10^4 b. 3.2×10 c. 5.72×10^{-3} d. 4.3×10^0

9. Write each number in scientific notation:

a. 462,000 b. 27. c. 0.0462 d. 0.000073

10. Write each of the following in scientific notation:

a. 43.6×10^5 b. 0.012×10^5 c. 43.0×10^{-1} d. 0.74×10^{-6}

11. Multiply and write the product in scientific notation:

a. $(2.4 \times 10^3)(2.0 \times 10^5)$ b. $(4.2 \times 10^{-2})(1.2 \times 10^3)$

c. $(3.6 \times 10^3)(4.0 \times 10^2)$ d. $(8.0 \times 10^{-2})(5.1 \times 10^{-3})$

12. Divide and write the quotients in scientific notation:

a. $\frac{4.8 \times 10^3}{3.0 \times 10^2}$ b. $\frac{6.0 \times 10^4}{1.2 \times 10^{-2}}$ c. $\frac{1.5 \times 10^{-2}}{5.0 \times 10^{-2}}$ d. $\frac{1.6 \times 10^{-2}}{8.0 \times 10^2}$

13. Simplify and write in scientific notation:

$$\frac{(1.4 \times 10^3)(6.0 \times 10^{-2})}{(1.2 \times 10^5)(2.8 \times 10^{-1})}$$

4.3 Exercise Answers

1. a. $\frac{1}{8}$ b. $\frac{4y^3}{x^2}$ c. $\frac{1}{x}$ d. x^3

e. $\frac{y^3}{x^5}$ f. $\frac{4z}{x^2y^3}$

2. a. $\frac{16}{x^4}$ b. $\frac{x^3}{2y^2}$ c. $2x^3$ d. 3

e. 1 f. $\frac{4}{x^3}$

3. a. $2x^5$ b. x^2 c. x^2 d. $\frac{y^2}{x^4}$

4. a. $\frac{1}{x^6}$ b. $\frac{1}{2x^3}$ c. $\frac{3}{x^4}$ d. $\frac{x^6}{2}$

5. a. x^2y^2 b. $\frac{1}{x^{10}}$ c. $8x^9$ d. $\frac{y}{2x}$

6. a. $\frac{5}{18}$ b. $\frac{2 + x^4}{x}$

7. a. yes b. no c. no d. yes
8. a. 46,000 b. 32 c. 0.00572 d. 4.3
9. a. 4.62×10^5 b. 2.7×10 c. 4.62×10^{-2} d. 7.3×10^{-5}
10. a. 4.36×10^6 b. 1.2×10^3 c. 4.3×10^0 d. 7.4×10^{-7}
11. a. 4.8×10^8 b. 5.04×10 c. 1.44×10^6 d. 4.08×10^{-4}
12. a. 1.6×10 b. 5.0×10^6 c. 3.0×10^{-1} d. 2.0×10^{-5}
13. 2.5×10^{-3}

4.4 Rational Exponents

1 In Section 4.3 a definition of negative exponents was made in such a way that five properties which held for natural-number exponents were also true for integral exponents. In this section a meaning will be given to rational exponents. Again, the motivation for the definition will be the desire to keep all former properties of exponents applicable to the expressions with fractional exponents.

Q1 Review the properties of exponents by completing the following: The statements below are true for all integer values of m and n.

a. $x^m \cdot x^n =$ ______ $x \in R, m \in I, n \in I$

b. $(x^m)^n =$ ______ $x \in R, m \in I, n \in I$

c. $(xy)^m =$ ______ $x \in R, y \in R, m \in I$

d. $\left(\frac{x}{y}\right)^m =$ ______ $x \in R, y \in R, m \in I, y \neq 0$

e. $\frac{x^m}{x^n} =$ ______ $x \in R, m \in I, n \in I, x \neq 0$

STOP • STOP • STOP • STOP • STOP • STOP • STOP • STOP • STOP

A1 a. x^{m+n} b. x^{mn} c. $x^m y^m$ d. $\frac{x^m}{y^m}$ e. x^{m-n}

2 Before writing the definition of what $4^{1/2}$ will mean, the properties will be examined to see what logical meaning would be required in order for the properties to remain applicable.

Consider $4^{1/2} \cdot 4^{1/2}$. Using property $x^m \cdot x^n = x^{m+n}$, the product would be $4^{1/2} \cdot 4^{1/2} = 4^{(1/2)+(1/2)} = 4^1 = 4$. In other words, $4^{1/2}$ is a number that when multiplied times itself results in 4. Such a number is the square root of 4. Therefore, $4^{1/2}$ is defined to be the positive square root of 4; that is, $4^{1/2} = \sqrt{4}$. In general, for the positive integer n, $x^{1/n}$ is the principal nth root of x.

$x^{1/n} = \sqrt[n]{x} \qquad n \in I, n > 0, x \in R$

Example 1: Write $9^{1/2}$ in radical form and simplify.

Solution

$9^{1/2} = \sqrt{9} = 3$

Example 2: Write $24^{1/3}$ in radical form and simplify.

Solution

$24^{1/3} = \sqrt[3]{24} = \sqrt[3]{8 \cdot 3} = 2\sqrt[3]{3}$

Q2 Convert the following to radical form and simplify:

a. $7^{1/2}$ ______ b. $25^{1/2}$ ______

c. $a^{1/3}$ ______ d. $(a^3)^{1/3}$ ______

e. $(6x^4)^{1/2}$ ______ f. $8^{1/2}$ ______

STOP • STOP • STOP • STOP • STOP • STOP • STOP • STOP • STOP

A2 a. $\sqrt{7}$ b. 5

c. $\sqrt[3]{a}$ d. a

e. $x^2\sqrt{6}$ f. $2\sqrt{2}$

Q3 Write each radical expression by using fractional exponents:

a. $\sqrt{5}$ ______ b. $\sqrt{3b}$ ______

c. $\sqrt{x^2}$ ______ d. $\sqrt{48}$ ______

STOP • STOP • STOP • STOP • STOP • STOP • STOP • STOP • STOP

A3 a. $5^{1/2}$ b. $(3b)^{1/2}$

c. $(x^2)^{1/2}$ d. $(48)^{1/2}$

3 The fractional exponent may be combined with a negative exponent.

Example: Write $4^{-1/2}$ in radical form and simplify.

Solution

$$4^{-1/2} = \frac{1}{4^{1/2}} \quad \text{because} \quad x^{-n} = \frac{1}{x^n}$$

$$= \frac{1}{\sqrt{4}}$$

$$= \frac{1}{2}$$

Q4 Write each expression in radical form and simplify:

a. $9^{-1/2}$

b. $27^{-1/3}$

c. $49^{1/2}$

d. $(4)^{-3}$

STOP • STOP • STOP • STOP • STOP • STOP • STOP • STOP • STOP

A4 a. $\frac{1}{3}$: $9^{-1/2} = \frac{1}{9^{1/2}} = \frac{1}{\sqrt{9}} = \frac{1}{3}$

b. $\frac{1}{3}$: $27^{-1/3} = \frac{1}{27^{1/3}} = \frac{1}{\sqrt[3]{27}} = \frac{1}{3}$

c. 7: $49^{1/2} = \sqrt{49} = 7$

d. $\frac{1}{64}$: $4^{-3} = \frac{1}{4^3} = \frac{1}{64}$

Q5 Write each expression by use of a negative exponent:

a. $\frac{1}{3^2} =$ ____ b. $\frac{1}{\sqrt{5}} =$ ____ c. $\frac{1}{a^3} =$ ____

d. $\frac{1}{\sqrt[n]{a}} =$ ____

STOP • STOP • STOP • STOP • STOP • STOP • STOP • STOP • STOP

A5 a. 3^{-2} b. $5^{-1/2}$ c. a^{-3} d. $a^{-1/n}$

4 Notice that $x^{1/n}$ is not a real number when x is negative and n is even. For example, $(-4)^{1/2}$ is not a real number. Variables will be assumed to be positive if they appear as radicands of radicals with even index.

So far, only fractional exponents with numerators of 1 have been defined. To stay consistent with $x^{mn} = (x^m)^n$, the following chain of equations would have to be true.

$$4^{3/2} = 4^{1/2 \cdot 3} = (4^{1/2})^3$$

In general, $x^{m/n}$ is defined to be $(x^{1/n})^m$. That is,

$$x^{m/n} = (x^{1/n})^m \qquad m \in I, n \in I, x \in R, \text{ and } x \text{ is positive if } n \text{ is even}$$

Example 1: Evaluate $9^{3/2}$.

Solution

$$9^{3/2} = (9^{1/2})^3 = (\sqrt{9})^3 = (3)^3 = 27$$

Example 2: Evaluate $8^{2/3}$.

Solution

$$8^{2/3} = (8^{1/3})^2 = (\sqrt[3]{8})^2 = (2)^2 = 4$$

Q6 Evaluate:

a. $4^{3/2} = ($____$)^3 =$ ____ b. $25^{3/2} = ($____$)^3 =$ ____

c. $16^{3/4} = ($____$)^3 =$ ____ d. $8^{4/3} = ($____$)^4 =$ ____

STOP • STOP • STOP • STOP • STOP • STOP • STOP • STOP • STOP

A6 a. $\sqrt{4}$, 8 b. $\sqrt{25}$, 125 c. $\sqrt[4]{16}$, 8 d. $\sqrt[3]{8}$, 16

Q7 Evaluate:

a. $32^{3/5} =$ ________ b. $4^{5/2} =$ ________

c. $9^{3/2} =$ ________ d. $(-27)^{2/3} =$ ________

STOP • STOP • STOP • STOP • STOP • STOP • STOP • STOP • STOP

A7 **a.** 8 **b.** 32 **c.** 27 **d.** 9: $(\sqrt[3]{-27})^2$

Q8 Write with fractional exponents:

a. $(\sqrt{9})^3 =$ ____ **b.** $(\sqrt[5]{20})^3 =$ ____

c. $(\sqrt[3]{15})^4 =$ ____ **d.** $(\sqrt{17})^4 =$ ____

STOP • STOP • STOP • STOP • STOP • STOP • STOP • STOP • STOP

A8 **a.** $9^{3/2}$ **b.** $20^{3/5}$ **c.** $15^{4/3}$ **d.** $17^{4/2}$

5 The properties of exponents will be restated as they apply to rational exponents. Also included for summary purposes are three definitions for fractional exponents.

Definition: $x^{1/n} = \sqrt[n]{x}$ $n \in N$, $x \in R$, and $x > 0$ if n is even

Definition: $x^{m/n} = (x^{1/n})^m$ $m \in N$, $n \in N$, $x \in R$ and $x > 0$ if n is even

Definition: $x^{-m/n} = \dfrac{1}{x^{m/n}}$ $m \in N$, $n \in N$, $x \in R$, and $x > 0$ if n is even

For the following properties a, b, c, and d are integers, x and y are real numbers, and x and y are positive if there is an even root, $b \neq 0$, $d \neq 0$.

1. $x^{a/b} \cdot x^{c/d} = x^{(a/b)+(c/d)}$
2. $(x^{a/b})^{c/d} = x^{ac/bd}$
3. $(xy)^{a/b} = x^{a/b} \cdot y^{a/b}$
4. $\left(\dfrac{x}{y}\right)^{a/b} = \dfrac{x^{a/b}}{y^{a/b}}$
5. $\dfrac{x^{a/b}}{x^{c/d}} = x^{(a/b)-(c/d)}$

Q9 Use the definitions or properties of rational exponents to complete the following equations (complete only the first step of the simplification):

a. $5^{2/3} \cdot 5^{4/3} =$ ____ **b.** $(3^{1/5})^{5/4} =$ ____

c. $(3 \cdot 5)^{2/3} =$ ____ • ____ **d.** $\left(\dfrac{9}{25}\right)^{1/2} =$ ____

e. $\dfrac{5^{2/3}}{5^{1/3}} =$ ____ **f.** $9^{-3/2} =$ ____

STOP • STOP • STOP • STOP • STOP • STOP • STOP • STOP • STOP

A9 **a.** 5^2 **b.** $3^{1/4}$

c. $3^{2/3} \cdot 5^{2/3}$ **d.** $\dfrac{9^{1/2}}{25^{1/2}}$

e. $5^{1/3}$ **f.** $\dfrac{1}{9^{3/2}}$

6 According to the definition of fractional exponents, $8^{2/3}$ may be evaluated two ways:

$8^{2/3} = 8^{1/3 \cdot 2} = (8^{1/3})^2 = (\sqrt[3]{8})^2 = 2^2 = 4$

$8^{2/3} = 8^{2 \cdot 1/3} = (8^2)^{1/3} = 64^{1/3} = \sqrt[3]{64} = 4$

Notice that taking the root first is easier. As long as x is positive when n is even, the following is true:

$$x^{m/n} = (x^{1/n})^m = (x^m)^{1/n}$$

For the rest of this section all variables used in radicands or with a fractional exponent will be assumed to represent positive numbers.

Q10

a. Evaluate $4^{3/2}$ by the first method used in Frame 6.

b. Evaluate $4^{3/2}$ by the second method used in Frame 6.

STOP • STOP • STOP • STOP • STOP • STOP • STOP • STOP • STOP

A10

a. $4^{3/2} = (4^{1/2})^3 = (\sqrt{4})^3 = 2^3 = 8$

b. $4^{3/2} = (4^3)^{1/2} = (64)^{1/2} = 8$

7

The laws of exponents can be used to simplify expressions with fractional exponents.

Example 1: Simplify $(w^{1/3})^6$.

Solution

$$(w^{1/3})^6 = w^{6/3} = w^2$$

Example 2: Simplify $\left(\frac{x^{2/3}}{y^{1/3}}\right)^6$.

Solution

$$\left(\frac{x^{2/3}}{y^{1/3}}\right)^6 = \frac{x^{12/3}}{y^{6/3}} = \frac{x^4}{y^2}$$

Q11

Simplify the following by use of the laws of exponents:

a. $(9a^2b^4)^{1/2} =$ ______

b. $(a^{2/3}b^{1/3})^3 =$ ______

c. $\left(\frac{x^6}{x^3}\right)^{1/2} =$ ______

d. $(x^0y^6)^{1/3} =$ ______

STOP • STOP • STOP • STOP • STOP • STOP • STOP • STOP • STOP

A11

a. $3ab^2$

b. a^2b

c. $x^{3/2}$

d. y^2

Q12

Simplify:

a. $\left(\frac{x^4y^2}{x^6}\right)^{1/2} =$ ______

b. $a^{1/2} \cdot a^{1/2} =$ ______

STOP • STOP • STOP • STOP • STOP • STOP • STOP • STOP • STOP

A12

a. $\frac{y}{x}$

b. a: $a^{1/2} \cdot a^{1/2} = a^{1/2+1/2} = a^1 = a$

Q13

Simplify:

a. $x^{1/2} \cdot x^{1/3} =$ ______

b. $x^{1/2}(x^{1/2} + x^{1/4}) =$ ______

STOP • STOP • STOP • STOP • STOP • STOP • STOP • STOP • STOP

A13

a. $x^{5/6}$: $x^{1/2} \cdot x^{1/3} = x^{(1/2+1/3)} = x^{5/6}$

b. $x + x^{3/4}$

Q14 Multiply by use of the properties of exponents:

a. $m^{1/2}(m^2 + m + m^{1/2})$

b. $(a^{1/2} + b^{1/2})(a^{1/2} - b^{1/2})$

STOP • STOP • STOP • STOP • STOP • STOP • STOP • STOP • STOP

A14 a. $m^{5/2} + m^{3/2} + m$

b. $a - b$: $(a^{1/2} + b^{1/2})(a^{1/2} - b^{1/2}) = a^{(1/2)+(1/2)} - b^{(1/2)+(1/2)} = a - b$

---

8 The laws of radicals can be translated into notation using fractional exponents. For example,

$\sqrt[n]{ab} = \sqrt[n]{a} \cdot \sqrt[n]{b}$ $a \in R$, $b \in R$, $n \in N$, with a and b not both negative

can be written $(ab)^{1/n} = a^{1/n}b^{1/n}$, which is a restatement of a law of exponents.

Q15 Translate the following law for radicals to exponential form:

$$\sqrt[n]{\frac{x}{y}} = \frac{\sqrt[n]{x}}{\sqrt[n]{y}}$$

STOP • STOP • STOP • STOP • STOP • STOP • STOP • STOP • STOP

A15 $$\left(\frac{x}{y}\right)^{1/n} = \frac{x^{1/n}}{y^{1/n}}$$

---

9 Fractional exponents may be used to simplify radical expressions by the use of laws stated in Frame 5.

Example 1: Simplify $\sqrt[4]{x^4y^8}$.

Solution

$\sqrt[4]{x^4y^8} = (x^4y^8)^{1/4} = x^{4/4}y^{8/4} = xy^2$

Example 2: Simplify $\sqrt[3]{x^4y^6}$.

Solution

$\sqrt[3]{x^4y^6} = (x^3y^6 \cdot x)^{1/3} = x^{3/3}y^{6/3} \cdot x^{1/3} = xy^2\sqrt[3]{x}$

Q16 Convert to fractional exponents and simplify:

a. $\sqrt{x^2y^4}$

b. $\sqrt[4]{x^8y^{12}}$

c. $\sqrt[3]{m^2n^3}$

d. $\sqrt[3]{m^5n^9}$

STOP • STOP • STOP • STOP • STOP • STOP • STOP • STOP • STOP

A16

a. xy^2: $\sqrt{x^2y^4} = (x^2y^4)^{1/2} = x^{2/2}y^{4/2} = xy^2$
b. x^2y^3: $\sqrt[4]{x^8y^{12}} = (x^8y^{12})^{1/4} = x^{8/4}y^{12/4} = x^2y^3$
c. $n\sqrt[3]{m^2}$: $\sqrt[3]{m^2n^3} = (m^2n^3)^{1/3} = m^{2/3}n^{3/3} = n\sqrt[3]{m^2}$
d. $mn^3\sqrt[3]{m^2}$: $\sqrt[3]{m^5n^9} = (m^3n^9 \cdot m^2)^{1/3} = m^{3/3}n^{9/3} \cdot m^{2/3} = mn^3\sqrt[3]{m^2}$

10

A radical expression can sometimes be simplified to another radical with smaller index by use of fractional exponents.

Example: Simplify $\sqrt[8]{y^6}$.

Solution

$$\begin{aligned}\sqrt[8]{y^6} &= (y^6)^{1/8}\\ &= y^{6/8}\\ &= y^{3/4}\\ &= \sqrt[4]{y^3}\end{aligned}$$

Q17

Simplify by use of exponents (write the results in radical form):

a. $\sqrt[6]{a^4b^2}$ **b.** $\sqrt[6]{64x^6y^3}$

STOP • STOP • STOP • STOP • STOP • STOP • STOP • STOP • STOP

A17

a. $\sqrt[3]{a^2b}$: $\sqrt[6]{a^4b^2} = (a^4b^2)^{1/6} = a^{4/6}b^{2/6} = a^{2/3}b^{1/3} = (a^2b)^{1/3} = \sqrt[3]{a^2b}$
b. $2x\sqrt{y}$: $\sqrt[6]{64x^6y^3} = (64x^6y^3)^{1/6} = 64^{1/6}x^{6/6}y^{3/6} = 2xy^{1/2} = 2x\sqrt{y}$

11

In Section 4.2 it was seen that radical expressions can be multiplied when they have the same index. For example,

$$\sqrt{2x} \cdot \sqrt{2x^3} = \sqrt{4x^4} = 2x^2$$

Up to this time radicals with different index could not be multiplied. However, fractional exponents may now be used to find the product. Each factor must be converted to the same index.

Example: Multiply $\sqrt{3} \cdot \sqrt[3]{5}$.

Solution

$$\begin{aligned}\sqrt{3} \cdot \sqrt[3]{5} &= 3^{1/2} \cdot 5^{1/3}\\ &= 3^{3/6} \cdot 5^{2/6} \quad \text{6 is the least common denominator}\\ &= \sqrt[6]{3^3} \cdot \sqrt[6]{5^2}\\ &= \sqrt[6]{3^3 5^2}\\ &= \sqrt[6]{27 \cdot 25}\\ &= \sqrt[6]{675}\end{aligned}$$

Q18 Multiply (write the products with one radical):

a. $\sqrt[3]{5} \cdot \sqrt[6]{2}$ b. $\sqrt{x} \cdot \sqrt[3]{x}$ c. $\sqrt{2} \cdot \sqrt[3]{3}$ d. $\sqrt[3]{x} \cdot \sqrt[6]{2x^2}$

STOP • STOP • STOP • STOP • STOP • STOP • STOP • STOP • STOP

A18 a. $\sqrt[6]{50}$: $\sqrt[3]{5} \cdot \sqrt[6]{2} = 5^{1/3} \cdot 2^{1/6} = 5^{2/6} \cdot 2^{1/6} = (5^2 \cdot 2)^{1/6} = \sqrt[6]{50}$

b. $\sqrt[6]{x^5}$: $\sqrt{x} \cdot \sqrt[3]{x} = x^{1/2} \cdot x^{1/3} = x^{3/6} \cdot x^{2/6} = x^{5/6} = \sqrt[6]{x^5}$

c. $\sqrt[6]{72}$: $\sqrt{2} \cdot \sqrt[3]{3} = 2^{1/2} \cdot 3^{1/3} = 2^{3/6} \cdot 3^{2/6} = \sqrt[6]{2^3} \cdot \sqrt[6]{3^2} = \sqrt[6]{8 \cdot 9} = \sqrt[6]{72}$

d. $\sqrt[6]{2x^4}$: $\sqrt[3]{x} \cdot \sqrt[6]{2x^2} = x^{1/3} \cdot \sqrt[6]{2x^2} = x^{2/6} \cdot \sqrt[6]{2x^2} = \sqrt[6]{x^2} \cdot \sqrt[6]{2x^2} = \sqrt[6]{2x^4}$

12 A negative fraction may be used as an exponent. The negative value has the same meaning as before. The base is placed in the denominator of the fraction and the exponent changes its sign.

Example 1: Simplify $4^{-5/2}$.

Solution

$$4^{-5/2} = \frac{1}{4^{5/2}} = \frac{1}{(\sqrt{4})^5} = \frac{1}{2^5} = \frac{1}{32}$$

Example 2: Simplify $\left(\frac{4}{9}\right)^{-3/2}$.

Solution

$$\left(\frac{4}{9}\right)^{-3/2} = \frac{1}{\left(\frac{4}{9}\right)^{3/2}} = \frac{1}{\left(\sqrt{\frac{4}{9}}\right)^3} = \frac{1}{\left(\frac{2}{3}\right)^3} = \frac{1}{\frac{8}{27}} = \frac{27}{8}$$

Q19 Simplify:

a. $25^{-1/2} =$ ______ b. $27^{-1/3} =$ ______

c. $8^{-2/3} =$ ______ d. $16^{-3/4} =$ ______

STOP • STOP • STOP • STOP • STOP • STOP • STOP • STOP • STOP

A19 a. $\frac{1}{5}$ b. $\frac{1}{3}$ c. $\frac{1}{4}$

d. $\frac{1}{8}$

13 The laws of exponents may be applied to negative fractional exponents. In most cases it is best *not* to convert all exponents to positive values before the laws are applied.

Example 1: Simplify $(3x^{-3/2})^{-2}$.

Solution

$$(3x^{-3/2})^{-2} = 3^{-2}x^{-3/2 \cdot -2}$$

$$= \frac{x^3}{9}$$

Example 2: Simplify $y^{-2/3} \cdot y^{-4/3}$.

Solution

$$\begin{aligned} y^{-2/3} \cdot y^{-4/3} &= y^{(-2/3)+(-4/3)} \\ &= y^{-6/3} \\ &= y^{-2} \\ &= \frac{1}{y^2} \end{aligned}$$

Q20 Simplify:

a. $x^{-1/2} \cdot x^{3/2} =$ ______

b. $(2x^{-1/2})^4 =$ ______

c. $\left(\dfrac{2^{1/3}x^{1/2}y^{-1/3}}{x^2y^{2/3}}\right)^6 =$ ______

STOP • STOP • STOP • STOP • STOP • STOP • STOP • STOP • STOP

A20 a. x: $x^{-1/2} \cdot x^{3/2} = x^{(-1/2)+(3/2)} = x^{2/2} = x$

b. $\dfrac{16}{x^2}$: $(2x^{-1/2})^4 = 2^4x^{-4/2} = 16x^{-2} = \dfrac{16}{x^2}$

c. $\dfrac{4}{x^9y^6}$: $\left(\dfrac{2^{1/3}x^{1/2}y^{-1/3}}{x^2y^{2/3}}\right)^6 = \dfrac{2^{6/3}x^{6/2}y^{-6/3}}{x^{12}y^{12/3}} = \dfrac{2^2x^3y^{-2}}{x^{12}y^4} = \dfrac{4}{x^9y^6}$

This completes the instruction for this section.

4.4 Exercises

(Assume that all variables represent positive numbers.)

1. Write in simplest radical form:

a. $49^{1/2}$ b. $81^{1/4}$ c. $(-27)^{1/3}$ d. $a^{1/5}$

e. $(4a^4)^{1/2}$ f. $(32)^{1/2}$ g. $25^{-1/2}$ h. $4^{-1/2}$

i. $(-27)^{-1/3}$ j. $\left(\dfrac{4}{25}\right)^{1/2}$ k. $\left(\dfrac{27}{8}\right)^{-1/3}$ l. $\left(\dfrac{4}{100}\right)^{-1/2}$

m. $4^{3/2}$ n. $(-27)^{5/3}$ o. $(-8)^{2/3}$ p. $(10{,}000)^{3/4}$

q. $(0.01)^{5/2}$ r. $(100)^{6/2}$

2. Write in simplest radical form:

a. $(x^{1/3}y^{2/3})^3$ b. $(a^{1/2}b^{1/4})^4$ c. $(m^{-1/2}n^{1/4})^2$

d. $\dfrac{x^{1/3}}{x^{-2/3}}$ e. $\dfrac{a^{-1/2}b^2}{a^{1/2}b}$ f. $\left(\dfrac{a^{-2/3}b^{1/3}}{a^{1/3}}\right)^6$

g. $(48b^3)^{1/2}$ h. $(80x^5)^{1/3}$ i. $\left(\dfrac{1}{2}\right)^{1/2}$

j. $a^{1/2} \cdot a^{3/4} \cdot a^{-1/4}$ k. $x^{1/3} \cdot x^{1/6}$ l. $x^{1/2}(2x^{1/2} + x^{1/4})$

m. $(2 + 3^{1/2})(2 + 3^{1/2})$ n. $(a^{1/2} - 1)(a^{1/2} + 1)$ o. $a^{1/4}(a^{-1/4} + a^{3/4})$

p. $\left(\dfrac{x^{-1/2}y}{x^{1/4}y^{-1}}\right)^4$ q. $\left(\dfrac{x^3}{x^{1/2}}\right)^6$ r. $\left(\dfrac{2a^{-1}b^{1/2}}{a^0b^{-1/2}}\right)^3$

3. Write each number or expression in simplest form using fractional exponents:

a. $\sqrt[3]{27}$ b. $\sqrt[3]{x}$ c. $\sqrt{2x^3}$ d. $\sqrt[3]{x^2}$
e. $\sqrt[5]{x^3}$ f. $\sqrt[4]{2x^3}$

4. Compute with fractional exponents and write in simplest radical form:

a. $(\sqrt{x})^3$ b. $(\sqrt{27})^3$ c. $(\sqrt[4]{x^2})^4$ d. $\sqrt{x^3y^6}$
e. $\sqrt[3]{x^6y^7}$ f. $\sqrt[5]{32x^4y^{10}}$ g. $\sqrt[4]{x^2}$ h. $\sqrt[6]{z^3}$
i. $\sqrt[4]{16x^6}$ j. $\sqrt{2}\cdot\sqrt[4]{2}$ k. $\sqrt[8]{x^6}\cdot\sqrt[4]{x^3}$ l. $\sqrt[6]{x^4}\cdot\sqrt[3]{x}$

4.4 Exercise Answers

1. a. 7 b. 3 c. -3 d. $\sqrt[5]{a}$
e. $2a^2$ f. $4\sqrt{2}$ g. $\frac{1}{5}$ h. $\frac{1}{2}$
i. $\frac{-1}{3}$ j. $\frac{2}{5}$ k. $\frac{2}{3}$ l. 5
m. 8 n. -243 o. 4 p. 1,000
q. 0.00001 r. 1,000,000

2. a. xy^2 b. a^2b c. $\frac{\sqrt{n}}{m}$ d. x
e. $\frac{b}{a}$ f. $\frac{b^2}{a^6}$ g. $4b\sqrt{3b}$ h. $2x\sqrt[3]{10x^2}$
i. $\frac{\sqrt{2}}{2}$ j. a k. $\sqrt{x}$ l. $2x+\sqrt[4]{x^3}$
m. $7+4\sqrt{3}$ n. $a-1$ o. $1+a$ p. $\frac{y^8}{x^3}$
q. x^{15} r. $\frac{8b^3}{a^3}$

3. a. 3 b. $x^{1/3}$ c. $(2x^3)^{1/2}$ d. $x^{2/3}$
e. $x^{3/5}$ f. $(2x^3)^{1/4}$

4. a. $x\sqrt{x}$ b. $81\sqrt{3}$ c. x^2 d. $xy^3\sqrt{x}$
e. $x^2y^2\sqrt[3]{y}$ f. $2y^2\sqrt[5]{x^4}$ g. $\sqrt{x}$ h. $\sqrt{z}$
i. $2x\sqrt{x}$ j. $\sqrt[4]{2^3}$ k. $x\sqrt{x}$ l. x

4.5 Complex Numbers

1

The set of real numbers does not include square roots of negative numbers such as $\sqrt{-4}$. However, a set of numbers can be developed that gives such expressions meaning.

To create the *complex numbers* a number, i, is introduced whose square is -1; that is, $i^2 = -1$. Thus

$i=\sqrt{-1}$

Recall that $\sqrt{ab}=\sqrt{a}\cdot\sqrt{b}$ if a and b are not negative. This property can now be extended for a and b both not negative. For example,

$$\begin{aligned}\sqrt{-4} &= \sqrt{4(-1)} \\ &= \sqrt{4} \cdot \sqrt{-1} \\ &= 2\sqrt{-1} \\ &= 2i\end{aligned}$$

Using the same property, all square roots of negative numbers can be written as a real number times the number i.

Example 1: Simplify $\sqrt{-25}$.

Solution

$$\begin{aligned}\sqrt{-25} &= \sqrt{25(-1)} \\ &= \sqrt{25} \cdot \sqrt{-1} \\ &= 5i\end{aligned}$$

Example 2: Simplify $\sqrt{-27}$.

Solution

$$\begin{aligned}\sqrt{-27} &= \sqrt{27(-1)} \\ &= \sqrt{27} \cdot \sqrt{-1} \\ &= 3\sqrt{3}i \text{ or } 3i\sqrt{3}\end{aligned}$$

Q1 Write each number as a real number in simplest form times the number i:

a. $\sqrt{-49}$ **b.** $\sqrt{-81}$ **c.** $\sqrt{-5}$

d. $\sqrt{-7}$ **e.** $\sqrt{-75}$ **f.** $\sqrt{-8}$

STOP • STOP • STOP • STOP • STOP • STOP • STOP • STOP • STOP

A1 **a.** $7i$ **b.** $9i$ **c.** $\sqrt{5}i$ or $i\sqrt{5}$

d. $\sqrt{7}i$ or $i\sqrt{7}$ **e.** $5\sqrt{3}i$ or $5i\sqrt{3}$ **f.** $2\sqrt{2}i$ or $2i\sqrt{2}$

2 All expressions of the form $a + bi$, where a and b are real numbers, will be called *complex numbers*. The real number a is called the *real part*. The real number b is called the *imaginary part* because it is multiplied times the imaginary number i. The following tabulation shows the real and imaginary parts of several complex numbers.

Complex number	Real part	Imaginary part
$8 + 2i$	8	2
$-3 + \sqrt{5}i$	-3	$\sqrt{5}$
$\pi - 2i$	π	-2
$-3i$	0	-3
8	8	0
0	0	0
$2 + (3 - \sqrt{5})i$	2	$3 - \sqrt{5}$

Q2 Write the real part and the imaginary part of each complex number:

	Real part a	Imaginary part b
a. $2 + \sqrt{6}i$	________	________
b. $-\sqrt{6}i$	________	________
c. $2 + \sqrt{3}$	________	________
d. $(2 + \sqrt{3})i$	________	________
e. $\sqrt{7} + (2 + \sqrt{2})i$	________	________
f. $5 + \pi$	________	________

STOP • STOP • STOP • STOP • STOP • STOP • STOP • STOP • STOP

A2 **a.** $2, \sqrt{6}$ **b.** $0, -\sqrt{6}$ **c.** $2 + \sqrt{3}, 0$ **d.** $0, 2 + \sqrt{3}$
e. $\sqrt{7}, (2 + \sqrt{2})$ **f.** $5 + \pi, 0$

3 Notice that when $b = 0$, the complex number $a + bi$ is also the real number a. All real numbers can be written in the form of a complex number by letting $b = 0$.

$5 = 5 + 0i$

$0 = 0 + 0i$

$\frac{2}{3} = \frac{2}{3} + 0i$, etc.

If b is not zero in $a + bi$, the complex number is called an *imaginary number*. If $a = 0$ and $b \neq 0$, the number is said to be a *pure imaginary number*.

The following numbers are all complex numbers:

Real numbers	Imaginary numbers	Pure imaginary numbers
$\sqrt{7}$	$2 + 0.5i$	$5i$
0	$(2 + \sqrt{3})i$	$0.73i$
$4 + \sqrt{6}$	$\sqrt{2} - i$	$\sqrt{2}i$
$3 - \sqrt{4}$	$4i$	$(2 + \sqrt{3})i$

The relationships between these sets can be described in a diagram such as:

Complex numbers

Real numbers	Imaginary numbers (Pure imaginary numbers)

Q3 Let C = complex numbers R = real numbers
J = imaginary numbers P = pure imaginary numbers
List the sets to which each number belongs:

a. $\sqrt{6}$ ________ **b.** $4i$ ________

c. $\frac{2}{3} + 3i$ ________ **d.** 0 ________

e. $2 + \sqrt{7}$ ________ **f.** $(3 - \sqrt{7})i$ ________

STOP • STOP • STOP • STOP • STOP • STOP • STOP • STOP • STOP

A3

a. C, R **b.** C, J, P **c.** C, J

d. C, R **e.** C, R **f.** C, J, P

4

Numbers that contain an indicated square root of a negative number may be changed to $a + bi$ form.

Example 1: Write $2\sqrt{-25}$ in $a + bi$ form.

Solution

$$\begin{aligned} 2\sqrt{-25} &= 2\sqrt{25(-1)} \\ &= 2\sqrt{25} \cdot \sqrt{-1} \\ &= 2 \cdot 5 \cdot i \\ &= 10i \end{aligned}$$

Example 2: Write $\dfrac{-3 + \sqrt{-18}}{3}$ in $a + bi$ form.

Solution

$$\begin{aligned} \frac{-3 + \sqrt{-18}}{3} &= \frac{-3 + \sqrt{18} \cdot \sqrt{-1}}{3} \\ &= \frac{-3 + 3\sqrt{2}i}{3} \\ &= -1 + \sqrt{2}i \text{ or } -1 + i\sqrt{2} \end{aligned}$$

Q4

Write the following in $a + bi$ form:

a. $\sqrt{-100} =$ __________ **b.** $3\sqrt{-49} =$ __________

c. $\sqrt{25} + \sqrt{-25} =$ __________ **d.** $\sqrt{-16} + \sqrt{48} =$ __________

STOP • STOP • STOP • STOP • STOP • STOP • STOP • STOP • STOP

A4

a. $10i$ **b.** $21i$

c. $5 + 5i$ **d.** $4\sqrt{3} + 4i$

Q5

Write the following in $a + bi$ form:

a. $\dfrac{\sqrt{36} + \sqrt{-9}}{3}$ **b.** $\sqrt{98} - \sqrt{-1}$

c. $\dfrac{\sqrt{-8} + \sqrt{16}}{2}$ **d.** $\dfrac{\sqrt{27} - \sqrt{-12}}{6}$

STOP • STOP • STOP • STOP • STOP • STOP • STOP • STOP • STOP

A5

a. $2 + i$ **b.** $7\sqrt{2} - i$ **c.** $2 + \sqrt{2}i$

d. $\dfrac{\sqrt{3}}{2} - \dfrac{\sqrt{3}}{3}i$

5 The sum of two complex numbers may be found by adding their real parts and their imaginary parts. The difference is found in a similar way. The definitions are as follows:

$(a + bi) + (c + di) = (a + c) + (b + d)i$
$(a + bi) - (c + di) = (a - c) + (b - d)i$

Example 1: Add $(2 + 3i) + (-5 + i)$.

Solution

$$(2 + 3i) + (-5 + i) = (2 + {}^{-}5) + (3 + 1)i$$
$$= -3 + 4i$$

Example 2: Add $(-2 + \sqrt{-4}) + (3 + \sqrt{-9})$.

Solution

$$(-2 + \sqrt{-4}) + (3 + \sqrt{-9}) = (-2 + 2\sqrt{-1}) + (3 + 3\sqrt{-1})$$
$$= (-2 + 2i) + (3 + 3i)$$
$$= 1 + 5i$$

Q6 Add or subtract the complex numbers:

a. $(4 + 6i) + (-3 - 8i)$

b. $3 + (-2 + i) + (-3 - i)$

c. $(4 + \sqrt{3}i) - (-7 + 2\sqrt{3}i)$

d. $(2 + \sqrt{2}i) + (\sqrt{5} + 3\sqrt{2}i)$

STOP • STOP • STOP • STOP • STOP • STOP • STOP • STOP • STOP

A6 **a.** $1 - 2i$ **b.** -2 **c.** $11 - \sqrt{3}i$
d. $(2 + \sqrt{5}) + 4\sqrt{2}i$

Q7 Add or subtract the complex numbers:

a. $(-3 - \sqrt{-3}) + (6 - \sqrt{-48}) - (5 + \sqrt{-27})$

b. $\dfrac{12 + 2\sqrt{-48}}{4} + \dfrac{12 + 3\sqrt{-108}}{3}$

STOP • STOP • STOP • STOP • STOP • STOP • STOP • STOP • STOP

A7 a. $-2 - 8\sqrt{3}i$ b. $7 + 8\sqrt{3}i$

6 Recall that i is the square root of -1. Therefore, $i^2 = -1$. Another way of writing this is

$$\sqrt{-1} \cdot \sqrt{-1} = -1$$

It can now be seen why the restriction of "a and b not both negative" must be included with the property $\sqrt{ab} = \sqrt{a} \cdot \sqrt{b}$. Notice what happens if this restriction is ignored.

$$-2 = \sqrt{-2} \cdot \sqrt{-2} \stackrel{?}{=} \sqrt{(-2)(-2)} = \sqrt{4} = 2$$

If the equal sign with a question mark over it were truly equal, the above statement shows that $-2 = 2$.

To avoid this problem, change to the $a + bi$ form before multiplying.

Example: Multiply $\sqrt{-16} \cdot \sqrt{-9}$.

Solution

$$\begin{aligned}\sqrt{-16} \cdot \sqrt{-9} &= 4i \cdot 3i \\ &= 12i^2 \\ &= 12(-1) \\ &= -12\end{aligned}$$

Q8 Multiply:

a. $(3i)(5i) =$ ______ b. $(-2i)(10i) =$ ______

c. $(-3i)(2i)(i) =$ ______ d. $(10i)^2 =$ ______

STOP • STOP • STOP • STOP • STOP • STOP • STOP • STOP • STOP

A8 a. -15 b. 20

c. $6i$ d. -100

Q9 Multiply:

a. $\sqrt{-4} \cdot \sqrt{-1}$ b. $\sqrt{-36} \cdot \sqrt{-4}$

c. $(\sqrt{-7})^2$ d. $\sqrt{-4} \cdot \sqrt{-9} \cdot \sqrt{-49}$

STOP • STOP • STOP • STOP • STOP • STOP • STOP • STOP • STOP

A9

a. -2: $\sqrt{-4} \cdot \sqrt{-1} = 2i \cdot i = 2i^2 = 2(-1) = -2$

b. -12: $\sqrt{-36} \cdot \sqrt{-4} = 6i \cdot 2i = 12i^2 = -12$

c. -7: $(\sqrt{-7})^2 = \sqrt{-7} \cdot \sqrt{-7} = \sqrt{7}i \cdot \sqrt{7}i = 7i^2 = -7$

d. $-42i$: $\sqrt{-4} \cdot \sqrt{-9} \cdot \sqrt{-49} = 2i \cdot 3i \cdot 7i = 42i^2 \cdot i = -42i$

7

The distributive law may be used to multiply two complex numbers.

$$\begin{aligned}(a + bi)(c + di) &= a(c + di) + bi(c + di)\\ &= ac + adi + bci + bdi^2\\ &= ac + (ad + bc)i + bd(-1)\\ &= (ac - bd) + (ad + bc)i\end{aligned}$$

The same pattern is used each time a product is found. For example,

$$\begin{aligned}(2 + i)(5 - 2i) &= 2(5 - 2i) + i(5 - 2i)\\ &= 10 - 4i + 5i - 2i^2\\ &= 10 - 4i + 5i - 2(-1)\\ &= 10 + 2 + (-4 + 5)i\\ &= 12 + i\end{aligned}$$

Q10 Multiply $(3 + 2i)(4 + 3i)$.

STOP • STOP • STOP • STOP • STOP • STOP • STOP • STOP • STOP

A10 $6 + 17i$:
$$\begin{aligned}(3 + 2i)(4 + 3i) &= 3(4 + 3i) + 2i(4 + 3i)\\ &= 12 + 9i + 8i + 6i^2\\ &= 12 - 6 + (9 + 8)i\\ &= 6 + 17i\end{aligned}$$

Q11 Multiply:

a. $(1 + i)(4 - i)$ **b.** $(3 - 2i)(5 + 2i)$

STOP • STOP • STOP • STOP • STOP • STOP • STOP • STOP • STOP

A11 **a.** $5 + 3i$ **b.** $19 - 4i$

Q12 Multiply:

a. $(4 + 2i)^2$ **b.** $6i(5 - 3i)$

c. $(5 + i)(5 - i)$

d. $(3 - 4i)(3 + 4i)$

STOP • STOP • STOP • STOP • STOP • STOP • STOP • STOP • STOP

A12 a. $12 + 16i$ b. $18 + 30i$ c. 26 d. 25

8 Notice that in Q12, *c* and *d*, the result of multiplying two imaginary numbers, formed a real number. Note that $5 + i$ was multiplied by $5 - i$. These numbers are called *complex conjugates* of each other. Complex conjugates are two complex numbers that differ only in the sign of thcir imaginary parts. The product of complex conjugates is always a real number.

Q13 Write the complex conjugate of each number:

a. $2 + 5i$ __________

b. $2 - \sqrt{5}i$ __________

c. $2i$ __________

d. $3 + (3 + \sqrt{2})i$ __________

STOP • STOP • STOP • STOP • STOP • STOP • STOP • STOP • STOP

A13 a. $2 - 5i$

b. $2 + \sqrt{5}i$

c. $-2i$

d. $3 - (3 + \sqrt{2})i$

Q14 Find the product of the given number and its complex conjugate:

a. $2 - 7i$

b. $3 + \sqrt{2}i$

STOP • STOP • STOP • STOP • STOP • STOP • STOP • STOP • STOP

A14 a. 53: $(2 - 7i)(2 + 7i) = 4 - 14i + 14i - 49i^2 = 53$

b. 11: $(3 + \sqrt{2}i)(3 - \sqrt{2}i) = 9 - 3\sqrt{2}i + 3\sqrt{2}i - 2i^2 = 11$

9 The complex conjugate of a number is used to divide complex numbers. The quotient of $2 + 3i$ and $3 - i$ is a complex number written in $a + bi$ form. To find this number, multiply numerator and denominator by the complex conjugate of the denominator. The new number equals the original since you are multiplying by 1.

Example 1: Simplify $\frac{2 + 3i}{3 - i}$.

Solution

$$\frac{2+3i}{3-i} = \frac{2+3i}{3-i} \cdot \frac{3+i}{3+i}$$
$$= \frac{6-3+(2+9)i}{9-i^2}$$
$$= \frac{3+11i}{10}$$
$$= \frac{3}{10} + \frac{11}{10}i$$

Example 2: Simplify $\frac{2+3i}{2i}$.

Solution

$$\frac{2+3i}{2i} = \frac{2+3i}{2i} \cdot \frac{-2i}{-2i}$$
$$= \frac{-4i-6i^2}{-4i^2}$$
$$= \frac{6-4i}{4}$$
$$= \frac{3}{2} - i$$

Q15 Divide (indicate the result in $a + bi$ form):

a. $\frac{3-2i}{5+i}$

b. $\frac{6+i}{3-5i}$

STOP • STOP • STOP • STOP • STOP • STOP • STOP • STOP • STOP

A15 **a.** $\frac{1}{2} - \frac{1}{2}i$: $\frac{3-2i}{5+i} \cdot \frac{5-i}{5-i} = \frac{(15+2i^2)+(-10-3)i}{25+1}$

$$= \frac{13}{26} + \frac{-13}{26}i = \frac{1}{2} - \frac{1}{2}i$$

b. $\frac{13}{34} + \frac{33}{34}i$: $\frac{6+i}{3-5i} \cdot \frac{3+5i}{3+5i} = \frac{18-5+(3+30)i}{9+25} = \frac{13}{34} + \frac{33}{34}i$

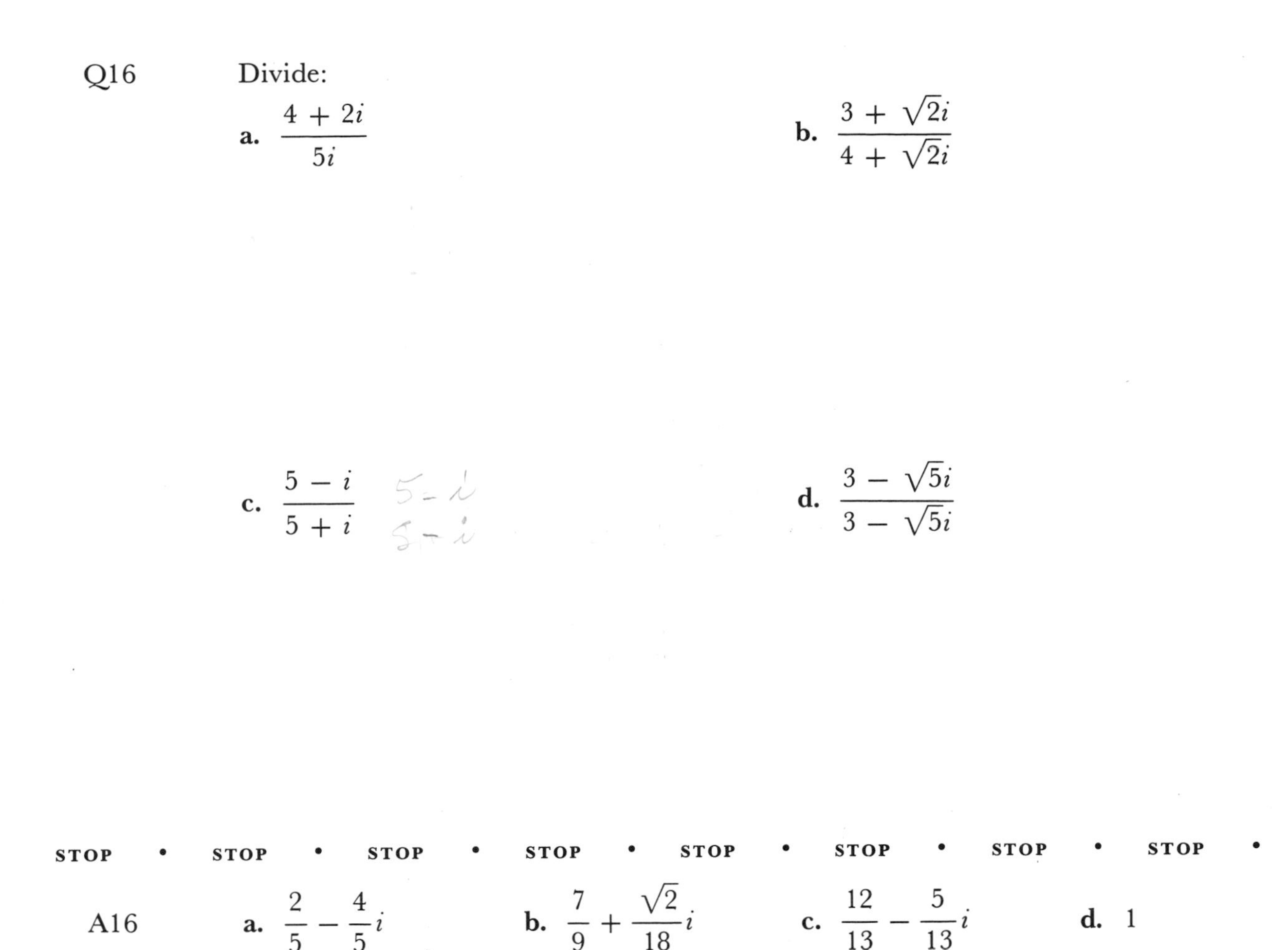

Q16 Divide:

a. $\dfrac{4 + 2i}{5i}$

b. $\dfrac{3 + \sqrt{2}i}{4 + \sqrt{2}i}$

c. $\dfrac{5 - i}{5 + i}$

d. $\dfrac{3 - \sqrt{5}i}{3 - \sqrt{5}i}$

STOP • STOP • STOP • STOP • STOP • STOP • STOP • STOP • STOP

A16 a. $\dfrac{2}{5} - \dfrac{4}{5}i$ b. $\dfrac{7}{9} + \dfrac{\sqrt{2}}{18}i$ c. $\dfrac{12}{13} - \dfrac{5}{13}i$ d. 1

This completes the instruction for this section.

4.5 Exercises

1. Consider the following complex numbers: $2 + i$, $\sqrt{-36}$, $2 - \sqrt{3}$, i^2, $-4i$, $\sqrt{2} + \sqrt{5}$, 0.
   - **a.** List the real numbers.
   - **b.** List the imaginary numbers.
   - **c.** List the pure imaginary numbers.
2. List the real part of each complex number:
   - **a.** $2 + 3i$ **b.** 0 **c.** $-\sqrt{3}i$ **d.** $2 + (3 + \sqrt{2})i$
   - **e.** $\sqrt{5} + \sqrt{3}i$
3. List the imaginary part of each number in problem 2.
4. Perform the indicated operations and write in $a + bi$ form:
   - **a.** $\sqrt{-48}$ **b.** $5\sqrt{-16}$
   - **c.** $\sqrt{-72} + \sqrt{9}$ **d.** $\dfrac{\sqrt{20} + \sqrt{-12}}{2}$
   - **e.** $\dfrac{\sqrt{-36} - \sqrt{12}}{4}$ **f.** $(2 + i) + (-3 - 2i)$
   - **g.** $(5 - i) - (-4 - 6i)$ **h.** $(2 + \sqrt{-4}) + (6 + \sqrt{-9})$

i. $(3 + \sqrt{2}) + (\sqrt{2} + \sqrt{3}i)$
j. $\sqrt{-48} - \sqrt{-75} + \sqrt{25}$
k. $(4i)(10i)$
l. $(3i)(-i)$
m. $\sqrt{-2} \cdot \sqrt{-6}$
n. $\sqrt{-4} \cdot \sqrt{-9}$
o. $\sqrt{-3}\,(\sqrt{-3} + \sqrt{-9})$
p. $\sqrt{-6}\,(6 + \sqrt{-12})$
q. $(2 - 4i)(5 + i)$
r. $(6 - 5i)(6 + 5i)$
s. $(2 - \sqrt{-6})(4 + \sqrt{-6})$
t. $(5 + 2i)^2$
u. $\dfrac{3 + i}{4 + i}$
v. $\dfrac{5 - 2i}{3 - i}$
w. $\dfrac{2 + \sqrt{-2}}{3 + \sqrt{-2}}$
x. $\dfrac{\sqrt{2} - \sqrt{-1}}{\sqrt{2} + \sqrt{-1}}$
y. $\dfrac{2i}{3i + 1}$
z. $\dfrac{1}{i}$

4.5 Exercise Answers

1. a. $2 - \sqrt{3}, i^2, \sqrt{2} + \sqrt{5}, 0$ b. $2 + i, \sqrt{-36}, -4i$ c. $\sqrt{-36}, -4i$
2. a. 2 b. 0 c. 0 d. 2 e. $\sqrt{5}$
3. a. 3 b. 0 c. $-\sqrt{3}$ d. $3 + \sqrt{2}$
   e. $\sqrt{3}$
4. a. $4\sqrt{3}i$ or $4i\sqrt{3}$ b. $20i$ c. $3 + 6\sqrt{2}i$
   d. $\sqrt{5} + \sqrt{3}i$ or $\sqrt{5} + i\sqrt{3}$ e. $\dfrac{-\sqrt{3}}{2} + \dfrac{3}{2}i$ f. $-1 - i$
   g. $9 + 5i$ h. $8 + 5i$ i. $(3 + 2\sqrt{2}) + \sqrt{3}i$
   j. $5 - \sqrt{3}i$ or $5 - i\sqrt{3}$ k. -40 l. 3
   m. $-2\sqrt{3}$ n. -6 o. $-3 - 3\sqrt{3}$
   p. $-6\sqrt{2} + 6\sqrt{6}i$ q. $14 - 18i$ r. 61
   s. $14 - 2\sqrt{6}i$ t. $21 + 20i$ u. $\dfrac{13}{17} + \dfrac{1}{17}i$
   v. $\dfrac{17}{10} - \dfrac{1}{10}i$ w. $\dfrac{8}{11} + \dfrac{\sqrt{2}}{11}i$ x. $\dfrac{1}{3} - \dfrac{2\sqrt{2}}{3}i$
   y. $\dfrac{3}{5} + \dfrac{1}{5}i$ z. $-i$

Chapter 4 Sample Test

At the completion of Chapter 4 it is expected that you will be able to work the following problems.

4.1 Introduction to Radicals and Simplest Radical Form

Simplify all real numbers. If a number is not a real number, write "not real." Assume that all variables represent positive numbers.

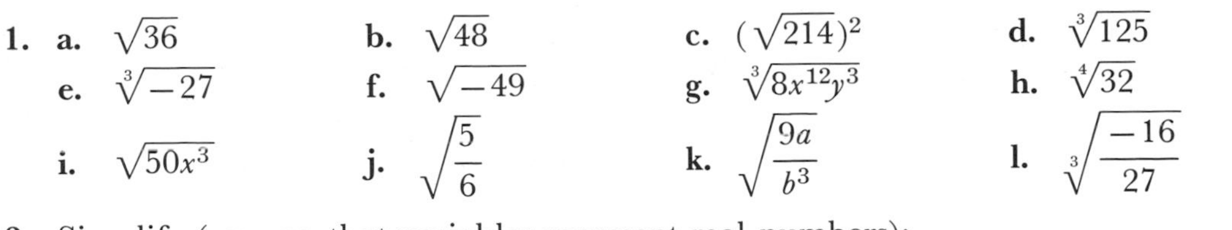

1. a. $\sqrt{36}$ b. $\sqrt{48}$ c. $(\sqrt{214})^2$ d. $\sqrt[3]{125}$
   e. $\sqrt[3]{-27}$ f. $\sqrt{-49}$ g. $\sqrt[3]{8x^{12}y^3}$ h. $\sqrt[4]{32}$
   i. $\sqrt{50x^3}$ j. $\sqrt{\dfrac{5}{6}}$ k. $\sqrt{\dfrac{9a}{b^3}}$ l. $\sqrt[3]{\dfrac{-16}{27}}$

2. Simplify (assume that variables represent real numbers):

   a. $\sqrt{4x^2y}$ b. $\sqrt[3]{-27x^3y^6}$ c. $\sqrt{27x^4y^6}$ d. $\sqrt{\dfrac{8x^2}{y^4}}$

   e. The index of $\sqrt[5]{3x^2}$ is ____. f. The radicand of $2x\sqrt{4y}$ is ____.

4.2 Operations on Irrational Numbers

3. Perform the indicated operation and write in simplest radical form:

   a. $2\sqrt{3} + 3\sqrt{9} + \sqrt{3}$ b. $-5\sqrt{3} - \sqrt{48}$ c. $\sqrt{27} + \sqrt{\dfrac{1}{3}}$

   d. $\sqrt[3]{-16} + 8\sqrt[3]{2}$ e. $\sqrt{3} \cdot \sqrt{12}$ f. $\sqrt[3]{-25} \cdot \sqrt[3]{-10}$

   g. $(2 + \sqrt{5})(2 - 3\sqrt{5})$ h. $(4 - \sqrt{3})(4 + \sqrt{3})$ i. $\dfrac{6 - \sqrt{27}}{9}$

   j. $\dfrac{4 + \sqrt{5}}{2 - \sqrt{5}}$ k. Solve $\sqrt{x - 1} = 5$.

4.3 Integer Exponents

4. Simplify, if possible, and write with positive exponents:

   a. 3^{-3} b. $\dfrac{x^{-5}}{x^{-3}}$ c. $4x^{-2} \cdot x^0 \cdot x$ d. $(2x^3)^{-1}(3x^2)$

   e. $3^{-1} + 3^{-2}$ f. $\left(\dfrac{2x^2}{x^{-1}}\right)^{-3}$

5. Write the following in scientific notation:

   a. 5,320,000 b. 0.0064 c. $426. \times 10^3$ d. 0.06×10^3

6. Perform the indicated operation and write the result in scientific notation:

   a. $(4.0 \times 10^3)(6 \times 10^2)$ b. $\dfrac{1.2 \times 10^3}{6.0 \times 10^5}$

4.4 Rational Exponents

7. Simplify and write answers in simplest radical form (assume that all variables represent positive real numbers):

   a. $81^{1/2}$ b. $(8a^6)^{1/3}$ c. $(25)^{3/2}$
   d. $(-27)^{2/3}$ e. $8^{-1/3}$ f. $100^{-3/2}$
   g. $(x^{-1/2}y^{1/4})^2$ h. $x^{1/2}(x^{1/2} + x^{3/2})$ i. $\left(\dfrac{x^{-1/2}y}{x^{1/2}}\right)^3$
   j. $x^{1/2}(x^{3/4})^2$

8. Compute with fractional exponents and write in simplest radical form (assume that all variables are positive):

 a. $\sqrt[4]{x^2}$ b. $\sqrt[4]{81x^6}$ c. $\sqrt{3}\cdot\sqrt[3]{2}$ d. $\sqrt[3]{x^2}\cdot\sqrt{x}$

4.5 Complex Numbers

9. Consider the set $\{0, 2 + 3i, \sqrt{2}, \sqrt{5}i, (2 + \sqrt{3})i, 5 + \sqrt{2}\}$.
 a. List the complex numbers.
 b. List the real numbers.
 c. List the imaginary numbers.
 d. List the pure imaginary numbers.

10. For each complex number, write (1) the real part and (2) the imaginary part:

 a. $2 + 7i$ b. $2 + \sqrt{5}$ c. $\sqrt{5}i$ d. 5

11. Perform the indicated operations and write in $a + bi$ form:

 a. $\sqrt{-36} + \sqrt{25}$ b. $(2 + 5i) + (-3 + {}^{-}7i)$ c. $(4i)(8i)$

 d. $2i(4 + 6i)$ e. $(2 + 3i)(4 - i)$ f. $(3 + 4i)(3 - 4i)$

 g. $\dfrac{4 + 6i}{3i}$ h. $\dfrac{6 + 2i}{1 - i}$

Chapter 4 Sample Test Answers

1. a. 6 b. $4\sqrt{3}$ c. 214 d. 5

 e. -3 f. not real g. $2x^4y$ h. $2\sqrt[4]{2}$

 i. $5x\sqrt{2x}$ j. $\dfrac{\sqrt{30}}{6}$ k. $\dfrac{3\sqrt{ab}}{b^2}$ l. $\dfrac{-2}{3}\sqrt[3]{2}$ or $\dfrac{-2\sqrt[3]{2}}{3}$

2. a. $2|x|\sqrt{y}$ b. $-3xy^2$ c. $3x^2|y^3|\sqrt{3}$ d. $\dfrac{2|x|\sqrt{2}}{y^2}$

 e. 5 f. $4y$

3. a. $9 + 3\sqrt{3}$ b. $-9\sqrt{3}$ c. $\dfrac{10\sqrt{3}}{3}$ d. $6\sqrt[3]{2}$

 e. 6 f. $5\sqrt[3]{2}$ g. $-11 - 4\sqrt{5}$ h. 13

 i. $\dfrac{2 - \sqrt{3}}{3}$ j. $-13 - 6\sqrt{5}$ k. 26

4. a. $\dfrac{1}{27}$ b. $\dfrac{1}{x^2}$ c. $\dfrac{4}{x}$ d. $\dfrac{3}{2x}$

 e. $\dfrac{4}{9}$ f. $\dfrac{1}{8x^9}$

5. a. 5.32×10^6 b. 6.4×10^{-3} c. 4.26×10^5 d. 6.0×10

6. a. 2.4×10^6 b. 2.0×10^{-3}

7. a. 9 b. $2a^2$ c. 125 d. 9

 e. $\dfrac{1}{2}$ f. $\dfrac{1}{1{,}000}$ g. $\dfrac{\sqrt{y}}{x}$ h. $x + x^2$

 i. $\dfrac{y^3}{x^3}$ j. x^2

8. **a.** $\sqrt{x}$ **b.** $3x\sqrt{x}$ **c.** $\sqrt[6]{108}$ **d.** $x\sqrt[6]{x}$

9. **a.** all of them **b.** $0, \sqrt{2}, 5 + \sqrt{2}$

c. $2 + 3i, \sqrt{5}i, (2 + \sqrt{3})i$ **d.** $\sqrt{5}i, (2 + \sqrt{3})i$

10. **a.** (1) 2 (2) 7 **b.** (1) $2 + \sqrt{5}$ (2) 0

c. (1) 0 (2) $\sqrt{5}$ **d.** (1) 5 (2) 0

11. **a.** $5 + 6i$ **b.** $-1 - 2i$ **c.** -32 **d.** $-12 + 8i$

e. $11 + 10i$ **f.** 25 **g.** $2 - \frac{4}{3}i$ **h.** $2 + 4i$

Chapter 5

Quadratic Equations and Inequalities

The properties developed for solving linear equations and inequalities will be used when solving quadratic equations and inequalities. Solving a quadratic equation in x means converting it to the form $x = a$ or $x = b$, where a and b are real numbers.

5.1 Solving Quadratic Equations by Factoring

1

A *quadratic equation*, or a *second-degree equation* in x, is one that can be written in the *standard quadratic form*, $ax^2 + bx + c = 0$, where a, b, and c are real numbers and $a \neq 0$. For example, $-3x^2 + 2x - 7 = 0$ is written in standard form with $a = -3$, $b = 2$, and $c = -7$. When a quadratic equation is written in standard form it is "equated to 0."

Example: Write the equation $3x^2 = 5x - 2$ in standard form.

Solution

$3x^2 = 5x - 2$

Adding $-5x$ and 2 to both sides of the equation yields

$3x^2 - 5x + 2 = 0$

which is in standard form with $a = 3$, $b = -5$, and $c = 2$.

Q1 Determine a, b, and c for the equations:

a. $\sqrt{3}x^2 - 2x + 5 = 0$

$a =$ ____ $b =$ ____ $c =$ ____

b. $x^2 - x + 1 = 0$

$a =$ ____ $b =$ ____ $c =$ ____

STOP • STOP • STOP • STOP • STOP • STOP • STOP • STOP • STOP

A1 a. $\sqrt{3}, -2, 5$

b. $1, -1, 1$: recall that x^2 can be written $1 \cdot x^2$

Q2 Write the equations in standard form:

a. $7x - 2 = x^2$

b. $-\sqrt{7} = 9x + 2x^2$

STOP • STOP • STOP • STOP • STOP • STOP • STOP • STOP • STOP

A2

a. $-x^2 + 7x - 2 = 0$ **b.** $0 = 2x^2 + 9x + \sqrt{7}$

It would be equally correct to write:

a. $0 = x^2 - 7x + 2$ **b.** $-2x^2 - 9x - \sqrt{7} = 0$

2

Quadratic equations of the form $ax^2 + c = 0$ can be solved by solving the equation for x^2 and then taking the square root of both sides.

$$ax^2 + c = 0$$
$$ax^2 = -c$$
$$x^2 = \frac{-c}{a}$$
$$x = \pm\sqrt{\frac{-c}{a}}$$

The symbolism $\pm$ is read "positive or negative."

The quadratic equation $ax^2 + c = 0$ has two solutions, $x = \sqrt{\frac{-c}{a}}$ or $x = -\sqrt{\frac{-c}{a}}$. If the solutions are to be real, $\frac{-c}{a}$ must be nonnegative.

Example 1: Solve $4x^2 - 100 = 0$.

Solution

$$4x^2 - 100 = 0$$
$$4x^2 = 100$$
$$x^2 = 25$$
$$x = \pm\sqrt{25}$$
$$= \pm 5$$

Check: Let $x = -5$: $4(-5)^2 - 100 \stackrel{?}{=} 0$, $4(25) - 100 \stackrel{?}{=} 0$, $0 = 0$

Let $x = 5$: $4(5)^2 - 100 \stackrel{?}{=} 0$, $4(25) - 100 \stackrel{?}{=} 0$, $0 = 0$

Example 2: Solve $3x^2 = 21$.

Solution

$$3x^2 = 21$$
$$x^2 = 7$$
$$x = \pm\sqrt{7}$$

Q3

Solve:

a. $4x^2 - 16 = 0$ **b.** $\frac{x^2}{4} = 5$

STOP • STOP • STOP • STOP • STOP • STOP • STOP • STOP • STOP

A3

a. $x = \pm 2$: $4x^2 - 16 = 0$
$$4x^2 = 16$$
$$x^2 = 4$$
$$x = \pm\sqrt{4}$$
$$x = \pm 2$$

b. $x = \pm 2\sqrt{5}$: $\frac{x^2}{4} = 5$
$$x^2 = 20$$
$$x = \pm\sqrt{20}$$
$$x = \pm\sqrt{4}\cdot\sqrt{5}$$
$$x = \pm 2\sqrt{5}$$

Q4

Solve for x:

a. $9x^2 = a^2$

b. $x^2 + 20 = 0$

STOP • STOP • STOP • STOP • STOP • STOP • STOP • STOP • STOP

A4

a. $x = \pm\frac{a}{3}$: $9x^2 = a^2$
$$x^2 = \frac{a^2}{9}$$
$$x = \pm\sqrt{\frac{a^2}{9}}$$
$$x = \pm\frac{a}{3}$$

b. $x = \pm 2i\sqrt{5}$: $x^2 + 20 = 0$
$$x^2 = -20$$
$$x = \pm\sqrt{-20}$$
$$x = \pm 2i\sqrt{5}$$

(*Note:* In an imaginary number it is common practice to write the i before the radical. This procedure will be followed in the remainder of the text.)

Q5

Solve for r in the equation $A = \pi r^2$.

STOP • STOP • STOP • STOP • STOP • STOP • STOP • STOP • STOP

A5

$r = \sqrt{\frac{A}{\pi}}$: the radius of a circle is always nonnegative, so the negative solution is rejected

---

3

Many quadratic equations can be solved using the following property: The product of two real numbers is zero if and only if one of the factors is zero. That is, for all real numbers a and b, $ab = 0$ if and only if $a = 0$ or $b = 0$. Hence

$(x - 2)(x + 5) = 0$ if and only if $x - 2 = 0$ or $x + 5 = 0$

Thus $x = 2$ or $x = -5$.

Example: Solve $x(x + 3) = 0$.

Solution

The equation is expressed as the product of two factors equated to zero; hence $x(x + 3) = 0$ if and only if $x = 0$ or $x + 3 = 0$. Thus $x = 0$ or $x = -3$.

Q6 Solve:

a. $(x + 7)\left(x - \frac{3}{2}\right) = 0$

b. $x(x + 2) = 0$

STOP • STOP • STOP • STOP • STOP • STOP • STOP • STOP • STOP

A6 a. $x = -7$ or $x = \frac{3}{2}$:

$x + 7 = 0$ or $x - \frac{3}{2} = 0$

$x = -7$ or $x = \frac{3}{2}$

b. $x = 0$ or $x = -2$:

$x = 0$ or $x + 2 = 0$

$x = 0$ or $x = -2$

---

4 Many quadratic equations can be solved by factoring and utilizing the property developed in Frame 3. A quadratic equation lacking the constant term, c, can be solved by first removing the common factor, x.

$ax^2 + bx = 0$

$x(ax + b) = 0$

$x = 0$ or $ax + b = 0$

$x = 0$ or $x = \frac{-b}{a}$

Example: Solve $x^2 - 4x = 0$.

Solution

$x^2 - 4x = 0$

$x(x - 4) = 0$

$x = 0$ or $x - 4 = 0$

$x = 0$ or $x = 4$

Q7 Solve:

a. $x^2 + 5x = 0$

b. $2x^2 - 7x = 0$

STOP • STOP • STOP • STOP • STOP • STOP • STOP • STOP • STOP

A7 a. $x = 0$ or $x = -5$:

$x^2 + 5x = 0$

$x(x + 5) = 0$

$x = 0$ or $x + 5 = 0$

$x = 0$ or $x = -5$

b. $x = 0$ or $x = \frac{7}{2}$:

$2x^2 - 7x = 0$

$x(2x - 7) = 0$

$x = 0$ or $2x - 7 = 0$

$x = 0$ or $x = \frac{7}{2}$

Q8 Solve:

a. $x^2 = -x$

b. $2x^2 = 8x$

STOP • STOP • STOP • STOP • STOP • STOP • STOP • STOP • STOP

A8

a. $x = 0$ or $x = -1$:

$$x^2 = -x$$
$$x^2 + x = 0$$
$$x(x + 1) = 0$$
$$x = 0 \quad \text{or} \quad x = -1$$

b. $x = 0$ or $x = 4$:

$$2x^2 = 8x$$
$$2x^2 - 8x = 0$$
$$2x(x - 4) = 0$$
$$2x = 0 \quad \text{or} \quad x - 4 = 0$$
$$x = 0 \quad \text{or} \quad x = 4$$

5

Quadratic equations of the form $ax^2 + bx + c = 0$, where a, b, and $c \neq 0$, can often be factored by use of the techniques developed in Section 2.5.

Example 1: Solve $x^2 - 5x + 4 = 0$.

Solution

$$x^2 - 5x + 4 = 0$$
$$(x - 4)(x - 1) = 0$$
$$x - 4 = 0 \quad \text{or} \quad x - 1 = 0$$
$$x = 4 \quad \text{or} \quad x = 1$$

Check: Let $x = 4$:

$$4^2 - 5(4) + 4 \stackrel{?}{=} 0$$
$$16 - 20 + 4 \stackrel{?}{=} 0$$
$$0 = 0$$

Let $x = 1$:

$$1^2 - 5(1) + 4 \stackrel{?}{=} 0$$
$$1 - 5 + 4 \stackrel{?}{=} 0$$
$$0 = 0$$

Example 2: Solve $y^2 + 10 = 7y$.

Solution

Recall from Frame 3 that the equation must first be equated to 0 before factoring.

$$y^2 + 10 = 7y$$
$$y^2 - 7y + 10 = 0$$
$$(y - 5)(y - 2) = 0$$
$$y - 5 = 0 \quad \text{or} \quad y - 2 = 0$$
$$y = 5 \quad \text{or} \quad y = 2$$

Q9 Solve:

a. $x^2 + 6x + 8 = 0$

b. $m^2 - 5m - 6 = 0$

STOP • STOP • STOP • STOP • STOP • STOP • STOP • STOP • STOP

A9

a. $x = -4$ or $x = -2$:

$$(x + 4)(x + 2) = 0$$
$$x + 4 = 0 \quad \text{or} \quad x + 2 = 0$$

b. $m = 6$ or $m = -1$:

$$(m - 6)(m + 1) = 0$$
$$m - 6 = 0 \quad \text{or} \quad m + 1 = 0$$

Q10 Solve:

a. $x(x + 4) = 21$

b. $3x^2 + x - 2 = 0$

STOP • STOP • STOP • STOP • STOP • STOP • STOP • STOP • STOP

A10 **a.** $x = -7$ or $x = 3$:

$x^2 + 4x = 21$

$x^2 + 4x - 21 = 0$

$(x + 7)(x - 3) = 0$

$x + 7 = 0$ or $x - 3 = 0$

b. $x = \frac{2}{3}$ or $x = -1$:

$(3x - 2)(x + 1) = 0$

$3x - 2 = 0$ or $x + 1 = 0$

Q11 Solve:

a. $x^2 + 6x + 9 = 0$

b. $x^2 - 10x + 25 = 0$

STOP • STOP • STOP • STOP • STOP • STOP • STOP • STOP • STOP

A11 **a.** $x = -3$:

$(x + 3)(x + 3) = 0$

$x + 3 = 0$ or $x + 3 = 0$

b. $x = 5$

Q12 Solve:

a. $\frac{9}{4}x^2 = \frac{1}{2}x^2 + 1$

b. $x^2 + 9x + 12 = 0$

STOP • STOP • STOP • STOP • STOP • STOP • STOP • STOP • STOP

A12 **a.** $x = \frac{\pm 2\sqrt{7}}{7}$:

$9x^2 = 2x^2 + 4$ (multiplying both sides by 4)

$7x^2 = 4$

$x^2 = \frac{4}{7}$

$x = \pm\sqrt{\frac{4}{7}}$

$x = \frac{2}{\pm\sqrt{7}}$

$x = \frac{\pm 2\sqrt{7}}{7}$ (the $\pm$ is usually placed with the radical)

b. The polynomial is prime and the equation cannot be solved using the techniques developed thus far.

6

Equations that contain rational expressions (*rational equations*) may be solved using the technique of "clearing the equation of fractions." This technique involves multiplying both sides of the equation by a polynomial that is the LCD of the denominators.

Example: Solve $\frac{2}{x-1} = \frac{x}{x+2}$.

Solution

LCD $= (x-1)(x+2)$.

$$\overset{1}{\cancel{(x-1)}}(x+2)\cdot\frac{2}{\underset{1}{\cancel{x-1}}} = (x-1)\overset{1}{\cancel{(x+2)}}\cdot\frac{x}{\underset{1}{\cancel{x+2}}}$$

$$\begin{aligned} (x+2)2 &= (x-1)x \\ 2x+4 &= x^2 - x \\ 0 &= x^2 - 3x - 4 \\ 0 &= (x-4)(x+1) \\ x-4 = 0 \quad &\text{or} \quad x+1 = 0 \\ x = 4 \quad &\text{or} \quad x = -1 \end{aligned}$$

Check: Let $x = 4$: $\frac{2}{4-1} \stackrel{?}{=} \frac{4}{4+2}$

$$\frac{2}{3} = \frac{4}{6}$$

Let $x = -1$: $\frac{2}{-1-1} \stackrel{?}{=} \frac{-1}{-1+2}$

$$-1 = -1$$

Q13 Solve and check the equation $\frac{1}{y} = \frac{1}{3y^2}$.

STOP • STOP • STOP • STOP • STOP • STOP • STOP • STOP • STOP

A13 $y = \frac{1}{3}$: LCD $= 3y^2$

$$\begin{aligned} 3y^2\left(\frac{1}{y}\right) &= 3y^2\left(\frac{1}{3y^2}\right) \\ 3y &= 1 \\ y &= \frac{1}{3} \end{aligned}$$

Check: $\frac{1}{\frac{1}{3}} \stackrel{?}{=} \frac{1}{3\left(\frac{1}{3}\right)^2}$

$$3 \stackrel{?}{=} \frac{1}{3\left(\frac{1}{9}\right)}$$

$$3 = 3$$

Q14 Solve:

a. $\frac{3}{2x} - \frac{13}{2(x+2)} = 6$

b. $x + 3 = \frac{4}{x}$

STOP • STOP • STOP • STOP • STOP • STOP • STOP • STOP • STOP

A14 a. $x = -3$ or $x = \frac{1}{6}$:

$\text{LCD} = 2x(x+2)$

$2x(x+2)\frac{3}{2x} - 2x(x+2)\frac{13}{2(x+2)} = 2x(x+2)6$

$(x+2)3 - x(13) = 12x(x+2)$

b. $x = -4$ or $x = 1$

7 Multiplying both sides of an equation by the LCD of the denominators sometimes results in an equation not equivalent to the original equation. For this reason it is especially important to check the solution when solving rational equations.

Example: Solve $\frac{x}{x^2-4} = \frac{2}{x^2-4} + \frac{x}{x+2}$.

Solution

Factor the denominators to aid in finding the LCD.

$$\frac{x}{(x-2)(x+2)} = \frac{2}{(x-2)(x+2)} + \frac{x}{x+2}$$

$$\text{LCD} = (x-2)(x+2)$$

$$(x-2)(x+2)\frac{x}{(x-2)(x+2)} = (x-2)(x+2)\frac{2}{(x-2)(x+2)} + (x-2)(x+2)\frac{x}{x+2}$$

$$x = 2 + (x-2)x$$

$$x = 2 + x^2 - 2x$$

$$0 = x^2 - 3x + 2$$

$$0 = (x-2)(x-1)$$

$$x - 2 = 0 \quad \text{or} \quad x - 1 = 0$$

$$x = 2 \quad \text{or} \quad x = 1$$

Check: Let $x = 2$: $\frac{2}{2^2-4} \stackrel{?}{=} \frac{2}{2^2-4} + \frac{2}{2+2}$

Since division by zero is impossible, 2 is not a solution to the original equation.

Let $x = 1$:

$$\frac{1}{1^2 - 4} \stackrel{?}{=} \frac{2}{1^2 - 4} + \frac{1}{1 + 2}$$

$$\frac{-1}{3} \stackrel{?}{=} \frac{-2}{3} + \frac{1}{3}$$

$$\frac{-1}{3} = \frac{-1}{3}$$

1 is a solution of the original equation; hence $x = 1$.

Q15 Solve and check the equation $\frac{x}{x - 2} - \frac{x + 4}{3(x - 2)} = \frac{1}{3}$.

STOP • STOP • STOP • STOP • STOP • STOP • STOP • STOP • STOP

A15 $\varnothing$: LCD $= 3(x - 2)$

$$3(x - 2)\frac{x}{x - 2} - 3(x - 2)\frac{x + 4}{3(x - 2)} = 3(x - 2)\frac{1}{3}$$

$$3x - (x + 4) = x - 2$$

$$3x - x - 4 = x - 2$$

$$x = 2$$

Check: Substituting 2 in the original equation would result in division by zero; hence 2 is a solution of the derived equation but not of the original equation. The equation $\frac{x}{x - 2} - \frac{x + 4}{3(x - 2)} = \frac{1}{3}$ has no solution.

Q16 Solve:

a. $\frac{x + 2}{x - 3} = \frac{x - 6}{x - 1}$

b. $\frac{3x}{x^2 - 4} = \frac{x - 4}{x + 2}$

STOP • STOP • STOP • STOP • STOP • STOP • STOP • STOP • STOP

A16 **a.** $x = 2$
LCD $= (x - 3)(x - 1)$

b. $x = 1$ or $x = 8$
LCD $= (x - 2)(x + 2)$

Q17 Solve:

a. $\dfrac{x}{x + 2} + \dfrac{x}{2x + 1} = \dfrac{2}{3}$

b. $\dfrac{2}{x + 2} - \dfrac{4}{x + 1} + \dfrac{x}{x + 2} = 0$

STOP • STOP • STOP • STOP • STOP • STOP • STOP • STOP • STOP

A17 **a.** $x = \dfrac{-4}{5}$ or $x = 1$

b. $x = 3$

---

8 Quadratic equations can be used to solve certain verbal problems. Recall the steps for solving verbal problems developed in Frame 12, Section 3.1.

Step 1: List the parts of the problem mathematically; use a variable for the unknown quantity. Give a pictorial representation, if possible.

Step 2: Write a mathematical sentence.

Step 3: Solve the resulting equation or inequality.

Step 4: List and check the solutions.

Example: Find three positive consecutive odd integers such that twice the square of the first is one less than the product of the second and third.

Solution

Step 1: Let

$x =$ first consecutive odd integer
$x + 2 =$ second consecutive odd integer
$x + 4 =$ third consecutive odd integer

Step 2: Twice the square of the first is one less than the product of the second and third; hence

$2x^2 = (x + 2)(x + 4) - 1$

Step 3: $2x^2 = x^2 + 6x + 8 - 1$
$x^2 - 6x - 7 = 0$
$(x - 7)(x + 1) = 0$
$x = 7$ or $x = -1$

Step 4: The solution -1 is rejected because the problem asks for consecutive positive integers. Therefore, $x = 7$ is the solution. $x + 2 = 9$ and $x + 4 = 11$.

Check: $2 \cdot 7^2 \stackrel{?}{=} 9 \cdot 11 - 1$
$$98 = 99 - 1$$

Q18 If 4 is subtracted from 9 times a certain number, the result is equal to 2 times the square of the number. Find the number.

STOP • STOP • STOP • STOP • STOP • STOP • STOP • STOP • STOP

A18 $x = \frac{1}{2}$ or $x = 4$:

Step 1:
$x =$ the number
$9x - 4 =$ 4 subtracted from 9 times the number
$2x^2 =$ two times the square of the number

Step 2: $9x - 4 = 2x^2$

Step 3: $0 = 4 - 9x + 2x^2$

$$0 = (2x - 1)(x - 4)$$

$$x = \frac{1}{2} \quad \text{or} \quad x = 4$$

Step 4: Check: Let $x = \frac{1}{2}$:

$$9\left(\frac{1}{2}\right) - 4 \stackrel{?}{=} 2\left(\frac{1}{2}\right)^2$$

$$\frac{9}{2} - \frac{8}{2} \stackrel{?}{=} 2\left(\frac{1}{4}\right)$$

$$\frac{1}{2} = \frac{1}{2}$$

Let $x = 4$:

$$9(4) - 4 \stackrel{?}{=} 2(4)^2$$

$$36 - 4 \stackrel{?}{=} 2(16)$$

$$32 = 32$$

Hence the solution is $x = \frac{1}{2}$ or $x = 4$.

Q19 The perimeter of a rectangle is 16 centimeters and its area is 15 square centimeters. Find the dimensions of the rectangle. (*Hint:* $A = lw$, $p = 2l + 2w$; solve for one variable and substitute into the other equation.)

STOP • STOP • STOP • STOP • STOP • STOP • STOP • STOP • STOP

A19 $l = 5$ cm or $l = 3$ cm
$w = 3$ cm $\quad w = 5$ cm:

Step 1: $A = lw, \quad 15 = lw$
$p = 2l + 2w, \quad 16 = 2l + 2w$

Step 2: Solve for one variable and substitute into the other equation.

$$2l + 2w = 16$$
$$w = 8 - l$$
$$15 = l \cdot w$$
$$15 = l(8 - l)$$

Step 3:

$$15 = 8l - l^2$$
$$l^2 - 8l + 15 = 0$$
$$(l - 5)(l - 3) = 0$$
$$l = 5 \quad \text{or} \quad l = 3$$

Step 4: If $l = 5$, $w = 3$, and if $l = 3$, $w = 5$, either solution satisfies the conditions of the problem.

This completes the instruction for this section.

5.1 Exercises

1. Give the standard form of the quadratic equation.
2. Solve the equations:

a. $3x^2 - 6x = 0$

b. $2x^2 - 5x = 0$

c. $3y = 5y(y - 2)$

d. $\frac{x^2}{3} = 12$

e. $x + 7 = \frac{1}{x - 7}$

f. $x^2 + 10x + 21 = 0$

g. $y^2 - 12y + 35 = 0$

h. $-18k - 40 = -k^2$

i. $m^2 = 14m - 24$

j. $x^2 + 4x + 4 = 0$

k. $3x^2 - 11x + 6 = 0$

l. $2x^2 - 5x - 12 = 0$

m. $-2x^2 + 3x + 20 = 0$

n. $\dfrac{4x + 3}{x + 2} = x$

o. $\dfrac{y + 3}{2y - 1} = 3y + 1$

p. $\dfrac{x^2 - 8}{x^2 - 4} = \dfrac{1}{x + 2} + \dfrac{1}{x - 2}$

q. $\dfrac{6}{x - 1} + \dfrac{16}{x^2 - 1} = 5$

*r. $\dfrac{x + 2}{x + 1} - \dfrac{32}{x^2 + 4x + 3} = \dfrac{-4}{5}$

3. Solve for x:
   a. $x^2 + a^2 = c^2$ b. $nx^2 - r^2 = 0$
4. The larger of two integers exceeds the smaller by 3. The sum of the squares of the two integers is 89. Find the integers.
5. The base of a triangle is 4 centimeters shorter than its altitude, and the area of the triangle is 126 square centimeters. Find the measures of the base and altitude.
6. Find two consecutive positive even integers whose product is 168.

5.1 Exercise Answers

1. $ax^2 + bx + c$, $a,b,c \in R$ with $a \neq 0$

2. a. $x = 0$ or $x = 2$

b. $x = 0$ or $x = \dfrac{5}{2}$

c. $y = 0$ or $y = \dfrac{13}{5}$

d. $x = \pm 6$

e. $x = \pm 5\sqrt{2}$

f. $x = -3$ or $x = -7$

g. $y = 5$ or $y = 7$

h. $k = 20$ or $k = -2$

i. $m = 12$ or $m = 2$

j. $x = -2$

k. $x = \dfrac{2}{3}$ or $x = 3$

l. $x = \dfrac{-3}{2}$ or $x = 4$

m. $x = \dfrac{-5}{2}$ or $x = 4$

n. $x = -1$ or $x = 3$

o. $y = \dfrac{-2}{3}$ or $y = 1$

p. $x = 4$

q. $x = \dfrac{-9}{5}$ or $x = 3$

*r. $x = \dfrac{-59}{9}$ or $x = 2$

3. a. $x = \pm\sqrt{c^2 - a^2}$ b. $x = \dfrac{\pm r\sqrt{n}}{n}$

4. $x = -8$ and $x + 3 = -5$ or $x = 5$ and $x + 3 = 8$
5. base $= 14$ cm, altitude $= 18$ cm
6. 12 and 14

5.2 Solving Quadratic Equations by Completing the Square and the Quadratic Formula

1 The technique called "*completing the square*" can be used to solve quadratic equations. In Section 2.6 perfect square trinomials were discussed; however, in this section trinomials will be formed into perfect squares only when the coefficient of x^2 is 1. Examples of perfect square trinomials:

$(x - 2)^2 = x^2 - 4x + 4$
$(x + 3)^2 = x^2 + 6x + 9$
$(x + 5)^2 = x^2 + 10x + 25$

A trinomial is a perfect square if the constant term is the square of one-half the coefficient of x and the coefficient of x^2 is 1. That is, from the examples above, $4 = \left(\frac{1}{2} \cdot {}^-4\right)^2$, $9 = \left(\frac{1}{2} \cdot 6\right)^2$, and $25 = \left(\frac{1}{2} \cdot 10\right)^2$. In general, the polynomial $x^2 + bx + c$ is a perfect square if

$$c = \left(\frac{1}{2} \cdot b\right)^2$$

Q1 The trinomial $x^2 - 6x + k$ is a perfect square if $k = \left(\frac{1}{2} \cdot \_\_\_\_\right)^2 = \_\_\_\_$.

STOP • STOP • STOP • STOP • STOP • STOP • STOP • STOP • STOP

A1 $-6, 9$: $x^2 - 6x + 9 = (x - 3)^2$

Q2 The trinomial $x^2 + 2x + k$ is a perfect square if $k = \left(\frac{1}{2} \cdot \_\_\_\_\right)^2 = \_\_\_\_$.

STOP • STOP • STOP • STOP • STOP • STOP • STOP • STOP • STOP

A2 $2, 1$: $x^2 + 2x + 1 = (x + 1)^2$

Q3 Determine the value of k that makes the trinomial a perfect square and factor the result:

a. $x^2 + 16x + k = x^2 + 16x + \_\_\_\_ = (\_\_\_\_)^2$

b. $x^2 - 12x + k = x^2 - 12x + \_\_\_\_ = (\_\_\_\_)^2$

c. $x^2 + 7x + k = x^2 + 7x + \_\_\_\_ = (\_\_\_\_)^2$

d. $x^2 + 2ax + k = x^2 + 2ax + \_\_\_\_ = (\_\_\_\_)^2$

STOP • STOP • STOP • STOP • STOP • STOP • STOP • STOP • STOP

A3 **a.** $64, x + 8$ **b.** $36, x - 6$ **c.** $\frac{49}{4}, x + \frac{7}{2}$ **d.** $a^2, x + a$

2 Quadratic equations can now be solved by completing the square.

Example: Solve $x^2 - 6x + 8 = 0$ by completing the square.

Solution

Place the constant on the right side, leaving space for a constant to be added, to make the left side a perfect square.

$x^2 - 6x + \_\_\_\_ = -8 + \_\_\_\_$

constant to be added

Note that -8 was added to both sides of the equation and that the constant to be added must be added to both sides of the equation. Properties of equality must be used if the resulting equation is to be equivalent to the original equation.

Complete the square: $x^2 - 6x + \underline{9} = -8 + \underline{9}$

Factor: $(x - 3)^2 = 1$

Square root of both sides: $x - 3 = \pm\sqrt{1}$

$x - 3 = \pm 1$

Hence $x - 3 = 1$ or $x - 3 = -1$. Thus $x = 4$ or $x = 2$.

Q4 Solve $x^2 - 6x - 7 = 0$ by completing the steps illustrated in Frame 2.

$x^2 - 6x - 7 = 0$

$x^2 - 6x + \_\_\_\_ = 7 + \_\_\_\_$

STOP • STOP • STOP • STOP • STOP • STOP • STOP • STOP • STOP

A4

$$x^2 - 6x + \underline{9} = 7 + \underline{9}$$
$$(x - 3)^2 = 16$$
$$x - 3 = \pm\sqrt{16}$$
$$x - 3 = \pm 4$$
$$x = 3 - 4 \quad \text{or} \quad x = 3 + 4$$
$$x = -1 \quad \text{or} \quad x = 7$$

Q5 Solve by completing the square:

a. $x^2 + 2x - 3 = 0$

b. $y^2 + 4y - 21 = 0$

STOP • STOP • STOP • STOP • STOP • STOP • STOP • STOP • STOP

A5

a. $x = -3$ or $x = 1$:

$$x^2 + 2x + \_\_\_\_ = 3 + \_\_\_\_$$
$$x^2 + 2x + 1 = 3 + 1$$
$$(x + 1)^2 = 4$$
$$x + 1 = \pm\sqrt{4}$$
$$x + 1 = -2 \quad \text{or} \quad x + 1 = 2$$
$$x = -3 \quad \text{or} \quad x = 1$$

b. $y = -7$ or $y = 3$:

$$y^2 + 4y + 4 = 21 + 4$$
$$(y + 2)^2 = 25$$
$$y + 2 = \pm\sqrt{25}$$
$$y + 2 = 5 \quad \text{or} \quad y + 2 = -5$$
$$y = 3 \quad \text{or} \quad y = -7$$

3

The technique of completing the square can be used to solve quadratic equations that contain prime trinomials.

Solve $x^2 - 6x - 1 = 0$. $x^2 - 6x - 1$ is prime and cannot be factored using the techniques developed in Chapter 2. The equation can be solved by completing the square.

$$x^2 - 6x - 1 = 0$$
$$x^2 - 6x + \_\_\_\_ = 1 + \_\_\_\_$$
$$x^2 - 6x + 9 = 1 + 9$$
$$(x - 3)^2 = 10$$
$$x - 3 = \pm\sqrt{10}$$
$$x - 3 = -\sqrt{10} \quad \text{or} \quad x - 3 = \sqrt{10}$$
$$x = -\sqrt{10} + 3 \quad \text{or} \quad x = \sqrt{10} + 3$$

The solution is usually written

$$x = 3 - \sqrt{10} \quad \text{or} \quad x = 3 + \sqrt{10}$$

which can be combined as

$$x = 3 \pm \sqrt{10}$$

Q6

Solve $x^2 - 2x - 7 = 0$.

STOP • STOP • STOP • STOP • STOP • STOP • STOP • STOP • STOP

A6

$x = 1 \pm 2\sqrt{2}$:

$$x^2 - 2x + 1 = 7 + 1$$
$$(x - 1)^2 = 8$$
$$x - 1 = \pm\sqrt{8}$$
$$x - 1 = \pm 2\sqrt{2}$$
$$x = 1 \pm 2\sqrt{2}$$

Q7 Solve:

a. $x^2 - 4x - 13 = 0$

b. $x^2 + 3x - 2 = 0$

STOP • STOP • STOP • STOP • STOP • STOP • STOP • STOP • STOP

A7

a. $x = 2 \pm \sqrt{17}$:

$$x^2 - 4x + 4 = 13 + 4$$
$$(x - 2)^2 = 17$$
$$x - 2 = \pm\sqrt{17}$$
$$x = 2 \pm \sqrt{17}$$

b. $x = \dfrac{-3 \pm \sqrt{17}}{2}$:

$$x^2 + 3x + \frac{9}{4} = 2 + \frac{9}{4}$$
$$\left(x + \frac{3}{2}\right)^2 = \frac{17}{4}$$
$$x + \frac{3}{2} = \pm\sqrt{\frac{17}{4}}$$
$$x + \frac{3}{2} = \frac{\pm\sqrt{17}}{2}$$
$$x = \frac{-3}{2} \pm \frac{\sqrt{17}}{2}$$
$$x = \frac{-3 \pm \sqrt{17}}{2}$$

4

Completing the square can be used to solve the quadratic equation $ax^2 + bx + c = 0$, where $a \neq 1$, by multiplying both sides by $\frac{1}{a}$, which makes 1 the coefficient of x^2.

Example: Solve $2x^2 - 8x - 6 = 0$ by completing the square.

Solution

Multiply both sides by $\frac{1}{2}$, which will make 1 the coefficient of x^2.

$$\frac{1}{2}(2x^2 - 8x - 6) = \frac{1}{2}(0)$$
$$x^2 - 4x - 3 = 0$$

Now complete the square and solve.

$$x^2 - 4x + \underline{4} = 3 + \underline{4}$$
$$(x - 2)^2 = 7$$
$$x = 2 \pm \sqrt{7}$$

Q8 Solve $2x^2 - 4x - 5 = 0$ by completing the square.

STOP • STOP • STOP • STOP • STOP • STOP • STOP • STOP • STOP

A8 $x = \frac{2 \pm \sqrt{14}}{2}$:

$$\frac{1}{2}(2x^2 - 4x - 5) = \frac{1}{2}(0)$$
$$x^2 - 2x - \frac{5}{2} = 0$$
$$x^2 - 2x + 1 = \frac{5}{2} + 1$$
$$(x - 1)^2 = \frac{7}{2}$$
$$x - 1 = \pm\sqrt{\frac{7}{2}}$$
$$x = 1 \pm \frac{\sqrt{14}}{2}$$
$$x = \frac{2 \pm \sqrt{14}}{2}$$

5 In actual practice quadratic equations are usually solved by factoring or the quadratic formula. The *quadratic formula* can be developed using the technique of completing the square to solve $ax^2 + bx + c = 0$, where $a,b,c \in R$ and $a \neq 0$. Multiply both sides of $ax^2 + bx + c = 0$ by $\frac{1}{a}$.

$$\frac{1}{a}(ax^2 + bx + c) = \frac{1}{a}(0)$$
$$x^2 + \frac{b}{a}x + \frac{c}{a} = 0$$
$$x^2 + \frac{b}{a}x + \_\_\_\_ = \frac{-c}{a} + \_\_\_\_$$

Complete the square. The constant term is $\left(\frac{1}{2} \cdot \frac{b}{a}\right)^2$.

$$x^2 + \frac{b}{a}x + \underline{\frac{b^2}{4a^2}} = \frac{-c}{a} + \underline{\frac{b^2}{4a^2}}$$

Factor the left side and add the fractions on the right side.

$$\left(x + \frac{b}{2a}\right)^2 = \frac{-4ac + b^2}{4a^2}$$

Take the square root of both sides.

$$x + \frac{b}{2a} = \pm\sqrt{\frac{b^2 - 4ac}{4a^2}}$$

Simplify the right side.

$$x + \frac{b}{2a} = \frac{\pm\sqrt{b^2 - 4ac}}{\sqrt{4a^2}}$$

$$x + \frac{b}{2a} = \frac{\pm\sqrt{b^2 - 4ac}}{2a}$$

Add $\frac{-b}{2a}$ to both sides.

$$x = \frac{-b}{2a} \pm \frac{\sqrt{b^2 - 4ac}}{2a}$$

Add the fractions.

$$x = \frac{-b \pm \sqrt{b^2 - 4ac}}{2a}$$

This equation is called the *quadratic formula* and gives the solutions

$$x = \frac{-b - \sqrt{b^2 - 4ac}}{2a} \qquad \text{or} \qquad x = \frac{-b + \sqrt{b^2 - 4ac}}{2a}$$

for the quadratic equation $ax^2 + bx + c = 0$.

Example: Solve $3x^2 - 8x + 5 = 0$ by use of the quadratic formula.

Solution

$a = 3 \qquad b = -8 \qquad c = 5$

$$x = \frac{-b \pm \sqrt{b^2 - 4ac}}{2a}$$

$$x = \frac{-(-8) \pm \sqrt{(-8)^2 - 4(3)(5)}}{2(3)}$$

$$x = \frac{8 \pm \sqrt{4}}{6}$$

$$x = \frac{8 \pm 2}{6}$$

$$x = \frac{8 - 2}{6} \quad \text{or} \quad x = \frac{8 + 2}{6}$$

$$x = 1 \qquad \text{or} \quad x = \frac{5}{3}$$

Q9 Solve $2x^2 - 8x + 3 = 0$ by completing the steps illustrated in Frame 5.

$a = 2 \qquad b = -8 \qquad c = 3$

$$x = \frac{-b \pm \sqrt{b^2 - 4ac}}{2a}$$

STOP • STOP • STOP • STOP • STOP • STOP • STOP • STOP • STOP

A9 $x = \frac{4 \pm \sqrt{10}}{2}$: $x = \frac{-(-8) \pm \sqrt{(-8)^2 - 4(2)(3)}}{2(2)}$

$$x = \frac{8 \pm \sqrt{40}}{4}$$

$$x = \frac{8 \pm 2\sqrt{10}}{4}$$

$$x = \frac{2(4 \pm \sqrt{10})}{4}$$

$$x = \frac{4 \pm \sqrt{10}}{2}$$

Q10 Before the quadratic formula can be used, the equation must be written in standard form as ________________.

STOP • STOP • STOP • STOP • STOP • STOP • STOP • STOP • STOP

A10 $ax^2 + bx + c = 0$

Q11 Give the solution to $mx^2 + kx + d = 0$, $m \neq 0$; use the quadratic formula.

STOP • STOP • STOP • STOP • STOP • STOP • STOP • STOP • STOP

A11 $$x = \frac{-k \pm \sqrt{k^2 - 4md}}{2m}$$

Q12 Solve $x^2 - 2x = -2$; use the quadratic formula.

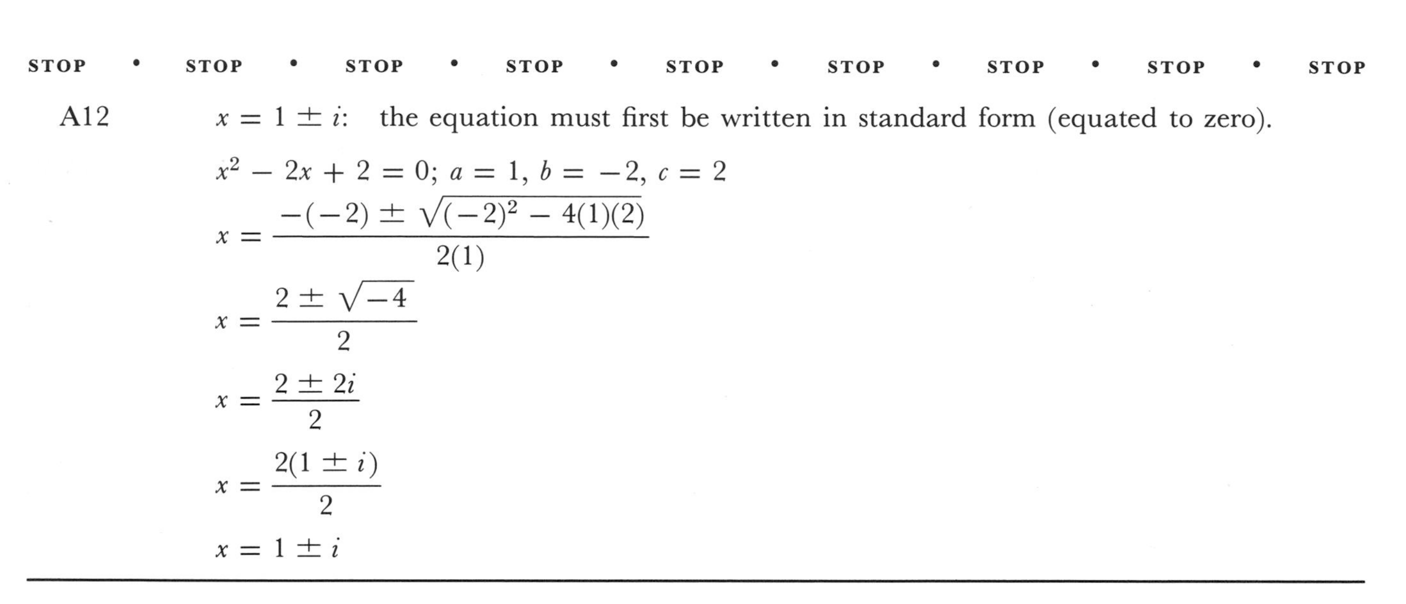

STOP • STOP • STOP • STOP • STOP • STOP • STOP • STOP • STOP

A12 $x = 1 \pm i$: the equation must first be written in standard form (equated to zero).

$x^2 - 2x + 2 = 0$; $a = 1$, $b = -2$, $c = 2$

$$x = \frac{-(-2) \pm \sqrt{(-2)^2 - 4(1)(2)}}{2(1)}$$

$$x = \frac{2 \pm \sqrt{-4}}{2}$$

$$x = \frac{2 \pm 2i}{2}$$

$$x = \frac{2(1 \pm i)}{2}$$

$$x = 1 \pm i$$

Q13 Solve by use of the quadratic formula:

a. $3x^2 - 4x + 1 = 0$ **b.** $2x^2 + 3x - 1 = 0$

STOP • STOP • STOP • STOP • STOP • STOP • STOP • STOP • STOP

A13 **a.** $x = \frac{1}{3}$ or $x = 1$ **b.** $x = \frac{-3 \pm \sqrt{17}}{4}$

Q14 Solve:

a. $17y = -6y^2 + 14$ **b.** $4x^2 = 8x - 9$

STOP • STOP • STOP • STOP • STOP • STOP • STOP • STOP • STOP

A14 **a.** $y = \frac{-7}{2}$ or $y = \frac{2}{3}$ **b.** $x = \frac{2 \pm i\sqrt{5}}{2}$

Q15 Solve:

a. $x^2 - 6x - 3 = 0$ **b.** $y^2 = 1 - 3y$

c. $2a^2 + 3 = 6a$ **d.** $x^2 = 2x + 3$

STOP • STOP • STOP • STOP • STOP • STOP • STOP • STOP • STOP

A15 **a.** $x = 3 \pm 2\sqrt{3}$ **b.** $y = \frac{-3 \pm \sqrt{13}}{2}$ **c.** $a = \frac{3 \pm \sqrt{3}}{2}$

d. $x = -1$ or $x = 3$

6	Some quadratic equations are solved more simply by factoring rather than using the quadratic formula. If factoring does not produce a quick solution, use the quadratic formula.

Q16 Solve:

a. $2x^2 + 3x = 0$

b. $x^2 - 4x + 13 = 0$

c. $6x = \frac{1 - x}{x}$

d. $y^2 = 3y - \frac{3}{2}$

STOP • STOP • STOP • STOP • STOP • STOP • STOP • STOP • STOP

A16

a. $x = 0$ or $x = \frac{-3}{2}$

b. $x = 2 \pm 3i$

c. $x = \frac{-1}{2}$ or $x = \frac{1}{3}$

d. $y = \frac{3 \pm \sqrt{3}}{2}$

7

Radical equations were solved in Section 4.2 by squaring both sides of the equation. The same technique can be used to solve radical equations involving quadratic equations.

Example: Solve $\sqrt{5x - 6} = x$.

Solution

Square both sides of the equation to eliminate the radical.

$$(\sqrt{5x - 6})^2 = x^2$$
$$5x - 6 = x^2$$
$$0 = x^2 - 5x + 6$$
$$(x - 3)(x - 2) = 0$$
$$x = 3 \quad \text{or} \quad x = 2$$

Recall that squaring both sides of an equation may produce a solution to the resulting equation that is not a solution to the original equation; hence it is important that the solutions are checked.

Check: Let $x = 3$: $\sqrt{5(3) - 6} \stackrel{?}{=} 3$, $3 = 3$ Let $x = 2$: $\sqrt{5(2) - 6} \stackrel{?}{=} 2$, $2 = 2$

Hence $x = 3$ or $x = 2$ is the solution.

Q17 Solve $\sqrt{3y} = \sqrt{2 - y}$.

STOP • STOP • STOP • STOP • STOP • STOP • STOP • STOP • STOP

A17 $y = \frac{2}{3}$: $(\sqrt{3y})^2 = (\sqrt{2-y})^2$

$$3y^2 = 2 - y$$
$$3y^2 + y - 2 = 0$$
$$y = \frac{-1 \pm \sqrt{1^2 - 4(3)(-2)}}{2(3)}$$
$$y = \frac{-1 \pm 5}{6}$$
$$y = -1 \quad \text{or} \quad y = \frac{2}{3}$$

Check: Let $y = -1$:

$\sqrt{3(-1)} \stackrel{?}{=} \sqrt{2 - (-1)}$

$-\sqrt{3} = \sqrt{3}$ (false)

Let $y = \frac{2}{3}$:

$\sqrt{3\left(\frac{2}{3}\right)} \stackrel{?}{=} \sqrt{2 - \frac{2}{3}}$

$\frac{2\sqrt{3}}{3} = \frac{2\sqrt{3}}{3}$ (true)

Q18 Solve:

a. $\sqrt{2x} = \sqrt{5x - 2}$

b. $\sqrt{x^2 + x + 2} = x$

STOP • STOP • STOP • STOP • STOP • STOP • STOP • STOP • STOP

A18 **a.** $x = \frac{1}{2}$ or $x = 2$

b. $\varnothing$

8 When solving an equation that involves radicals it may be necessary to square the equation twice.

Example: Solve $\sqrt{2x + 3} = 2 + \sqrt{x - 2}$.

Solution

$(\sqrt{2x + 3})^2 = (2 + \sqrt{x - 2})^2$

[*Hint:* Use the special product $(a + b)^2 = a^2 + 2ab + b^2$ to simplify the right member.]

$2x + 3 = 4 + 4\sqrt{x - 2} + x - 2$

Isolate the remaining radical on one side of the equation and square again.

$$\begin{aligned} x + 1 &= 4\sqrt{x - 2} \\ (x + 1)^2 &= (4\sqrt{x - 2})^2 \\ x^2 + 2x + 1 &= 16(x - 2) \\ x^2 - 14x + 33 &= 0 \\ (x - 11)(x - 3) &= 0 \\ x = 11 \quad &\text{or} \quad x = 3 \end{aligned}$$

Both solutions check in the original equation.

Q19 Solve $\sqrt{2x + 5} + \sqrt{x + 3} = 2$. (*Hint:* First place the radicals on opposite sides of the equation.)

STOP • STOP • STOP • STOP • STOP • STOP • STOP • STOP • STOP

A19 $x = -2$:

$$\begin{aligned} \sqrt{2x + 5} &= 2 - \sqrt{x + 3} \\ (\sqrt{2x + 5})^2 &= (2 - \sqrt{x + 3})^2 \\ 2x + 5 &= 4 - 4\sqrt{x + 3} + x + 3 \\ x - 2 &= -4\sqrt{x + 3} \\ x^2 - 4x + 4 &= 16(x + 3) \\ x^2 - 20x - 44 &= 0 \\ (x - 22)(x + 2) &= 0 \\ x = 22 \quad &\text{or} \quad x = -2 \end{aligned}$$

The solution 22 does not check in the original equation; hence $x = -2$ is the solution.

Q20 Solve $\sqrt{y + 5} = \sqrt{y} + 1$.

STOP • STOP • STOP • STOP • STOP • STOP • STOP • STOP • STOP

A20 $y = 4$

Q21 Find two consecutive integers whose product is 72.

STOP • STOP • STOP • STOP • STOP • STOP • STOP • STOP • STOP

A21 -9 and -8 or 8 and 9:

Step 1: Let $x =$ first consecutive integer
$x + 1 =$ second consecutive integer

Step 2: $x(x + 1) = 72$

Step 3:
$$x^2 + x - 72 = 0$$
$$(x + 9)(x - 8) = 0$$
$$x = -9 \quad \text{or} \quad x = 8$$

Step 4: If $x = -9$, $x + 1 = -8$: $(-9)(-8) = 72$
If $x = 8$, $x + 1 = 9$: $(8)(9) = 72$

Q22 The length of a rectangle is two more than twice the width. The area is 60 square inches. Find the dimensions of the rectangle.

STOP • STOP • STOP • STOP • STOP • STOP • STOP • STOP • STOP

A22 $w = 5$, $l = 12$:

Step 1: Let w = width
$2w + 2$ = length

Step 2: $w(2w + 2) = 60$

Step 3: $2w^2 + 2w - 60 = 0$
$w^2 + w - 30 = 0$
$w = -6$ or $w = 5$

Step 4: Since the width cannot be negative, $w = 5$ and $l = 2w + 2 = 12$.

Q23 Frances will not tell her age, because she is 3 years older than her husband, Larry. In 1 year she will be twice as old as he was when they were married 18 years ago. How old is she?

STOP • STOP • STOP • STOP • STOP • STOP • STOP • STOP • STOP

A23 43 years old:

	Age 18 years ago:
Step 1: Let x = Frances' age now	$x - 18$
$x - 3$ = Larry's age now	$x - 21$

Step 2: $x + 1 = 2(x - 21)$

Step 3: $x + 1 = 2x - 42$
$43 = x$

Step 4: Frances 43, Larry 40: $44 = 2(40 - 18)$

This completes the instruction for this section.

5.2 Exercises

1. What must k equal if $x^2 + mx + k = 0$ is to be a perfect square?
2. Solve the equations by completing the square:
   a. $x^2 - 2x - 63 = 0$
   b. $x^2 + 4x - 12 = 0$
   c. $y^2 + 9y + 20 = 0$
   d. $m^2 - 2m - 1 = 0$
   e. $2x^2 + 5x - 2 = 0$
   f. $3x^2 - 1 = -6x$
3. Solve $px^2 + qx + r = 0$, $p \neq 0$; use the quadratic formula.
4. Solve:
   a. $x^2 + 5x + 5 = 0$
   b. $x^2 = 3 - x$
   c. $y^2 - 3y + 1 = 0$
   d. $x^2 + 2x + 10 = 0$
   e. $6x^2 - 7x - 3 = 0$
   f. $2y^2 - 3y + 2 = 0$
   g. $x^2 + 8x = -15$
   h. $4x^2 - 3x - 5 = 0$
   i. $\sqrt{2x + 3} = 2 + \sqrt{x - 2}$
   j. $\sqrt{3x + 3} + 2 = \sqrt{x + 5}$
   k. $x^2 - 5x - 14 = 0$
   l. $3k^2 - 18k + 6 = 0$
   m. $5x^2 - 6x = -5$
   n. $x^2 + 3x - 40 = 0$
   o. $2\sqrt{x + 2} = 3x - 2$
   p. $\sqrt{x + 4} = 2 - \sqrt{x + 16}$
5. Helen travels 280 miles at a certain speed. If she had increased her speed by 5 miles per hour, she could have made the trip in 1 hour less. Find her speed.
6. Sue is twice as old as her brother Gayle, but in 4 years Gayle will be two-thirds as old as Sue is then. How old is Sue now?
7. The length of a rectangle is 2 meters greater than its width. If the perimeter is 14 meters, find the width.

5.2 Exercise Answers

1. $k = \left(\frac{1}{2}m\right)^2$ or $\frac{1}{4}m^2$ or $\frac{m^2}{4}$
2. a. $x = -7$ or $x = 9$
   b. $x = 2$ or $x = -6$
   c. $y = -5$ or $y = -4$
   d. $m = 1 \pm \sqrt{2}$
   e. $x = \dfrac{-5 \pm \sqrt{41}}{4}$
   f. $x = \dfrac{-3 \pm 2\sqrt{3}}{3}$
3. $x = \dfrac{-q \pm \sqrt{q^2 - 4pr}}{2p}$
4. a. $x = \dfrac{-5 \pm \sqrt{5}}{2}$
   b. $x = \dfrac{-1 \pm \sqrt{13}}{2}$
   c. $y = \dfrac{3 \pm \sqrt{5}}{2}$
   d. $x = -1 \pm 3i$
   e. $x = \dfrac{-1}{3}$ or $x = \dfrac{3}{2}$
   f. $y = \dfrac{3 \pm i\sqrt{7}}{4}$
   g. $x = -3$ or $x = -5$
   h. $x = \dfrac{3 \pm \sqrt{89}}{8}$
   i. $x = 3$ or $x = 11$
   j. $x = -1$
   k. $x = -2$ or $x = 7$
   l. $k = 3 \pm \sqrt{7}$
   m. $x = \dfrac{3 \pm 4i}{5}$
   n. $x = -8$ or $x = 5$
   o. $x = 2$
   p. $\varnothing$
5. 35 miles per hour
6. 8 years old
7. 2.5 meters

5.3 The Discriminant and Writing Quadratic Equations from Solutions

1

The solutions of the quadratic equation $ax^2 + bx + c = 0$, $a,b,c \in R$ and $a \neq 0$, are $x = \frac{-b \pm \sqrt{b^2 - 4ac}}{2a}$. The number $b^2 - 4ac$ is called the *discriminant* of the quadratic equation and can be used to examine the solutions. Consider the following three examples:

$2x^2 + 3x + 2 = 0$

$$x = \frac{-3 \pm \sqrt{3^2 - 4(2)(2)}}{2(2)}$$

$$x = \frac{-3 \pm \sqrt{-7}}{4}$$

$x^2 - 4x + 4 = 0$

$$x = \frac{-(-4) \pm \sqrt{(-4)^2 - 4(1)(4)}}{2(1)}$$

$$x = \frac{4 \pm \sqrt{0}}{2}$$

$$x = 2$$

$x^2 - 2x - 3 = 0$

$$x = \frac{-(-2) \pm \sqrt{(-2)^2 - 4(1)(-3)}}{2(1)}$$

$$x = \frac{2 \pm \sqrt{16}}{2}$$

$$x = -1 \text{ or } x = 3$$

Examining the discriminant for these examples shows that in the first, $b^2 - 4ac = -7$ and the solutions are imaginary; in the second, $b^2 - 4ac = 0$ and the solutions are real and equal; and in the third, $b^2 - 4ac = 16$ and the solutions are real and unequal. These results are generalized in the following table:

Discriminant	Two solutions for $ax^2 + bx + c = 0$
$b^2 - 4ac > 0$	Real and unequal
$b^2 - 4ac = 0$	Real and equal (one solution)
$b^2 - 4ac < 0$	Imaginary

Hence the discriminant can be used to determine the type of solution.

Q1 (1) Find the discriminant and (2) describe the solution:

a. $x^2 + 6x + 1 = 0$

(1) $b^2 - 4ac =$ _____

(2) ____________

b. $3x^2 + 4x + 2 = 0$

(1) $b^2 - 4ac =$ _____

(2) ____________

STOP • STOP • STOP • STOP • STOP • STOP • STOP • STOP • STOP

A1

a. (1) 32
(2) real and unequal

b. (1) −8
(2) imaginary

Q2

(1) Find the discriminant and (2) describe the solution:

a. $9x^2 - 12x + 4 = 0$

(1) ____

(2) ____________

b. $x^2 + 5x + 8 = 0$

(1) ____

(2) ____________

STOP • STOP • STOP • STOP • STOP • STOP • STOP • STOP • STOP

A2

a. (1) 0
(2) real and equal

b. (1) −7
(2) imaginary

Q3

(1) Find the discriminant and (2) describe the solution:

a. $4x^2 - 4x - 3 = 0$

(1) ____

(2) ____________

b. $2x^2 + 3x + 1 = 0$

(1) ____

(2) ____________

STOP • STOP • STOP • STOP • STOP • STOP • STOP • STOP • STOP

A3

a. (1) 64
(2) real and unequal

b. (1) 1
(2) real and unequal

2

In Q3 the discriminant was greater than zero ($b^2 - 4ac > 0$) and the solutions were real and unequal; however, the discriminant was also a perfect square. When the discriminant is a perfect square and a, b, and c are integers, the solutions will be rational.

$4x^2 - 4x - 3 = 0$

$$x = \frac{-(-4) \pm \sqrt{(-4)^2 - 4(4)(-3)}}{2(4)}$$

$$x = \frac{4 \pm \sqrt{64}}{8}: \qquad b^2 - 4ac = 64$$

$$x = \frac{4 \pm 8}{8}$$

$$x = \frac{-1}{2} \text{ or } x = \frac{3}{2}$$

Q4

Describe the solutions as rational, irrational, or imaginary:

a. $25x^2 + 20x = 2$

b. $4x^2 - 4x - 3 = 0$

STOP • STOP • STOP • STOP • STOP • STOP • STOP • STOP • STOP

A4

a. irrational: $b^2 - 4ac = 600$ (the equation must first be placed in standard form)

b. rational: $b^2 - 4ac = 64$ and $64 = 8^2$

Q5 Find b in the equation $x^2 + bx + 25 = 0$ so that the resulting equation will have exactly one rational solution.

STOP • STOP • STOP • STOP • STOP • STOP • STOP • STOP • STOP

A5 $b = \pm 10$: $b^2 - 4ac = 0$ (1 rational solution)

$$b^2 - (4)(1)(25) = 0$$
$$b^2 - 100 = 0$$
$$b^2 = 100$$
$$b = \pm 10$$

Q6 Find b in the equation $9x^2 + bx + 4 = 0$ so that the resulting equation will have exactly one rational solution.

STOP • STOP • STOP • STOP • STOP • STOP • STOP • STOP • STOP

A6 $b = \pm 12$

3 A quadratic equation can be formed from its solutions by utilizing the property that if $a = 0$ or $b = 0$, then $ab = 0$. For example, if $x - 2 = 0$ or $x + 3 = 0$, then $(x - 2)(x + 3) = 0$, which can be written in standard quadratic form as $x^2 + x - 6 = 0$ by multiplying the factors.

Example: Write a quadratic equation whose solutions are $\frac{-1}{2}$ and 2.

Solution

$$x = \frac{-1}{2} \quad \text{or} \quad x = 2$$

$$x + \frac{1}{2} = 0 \quad \text{or} \quad x - 2 = 0$$

$$\left(x + \frac{1}{2}\right)(x - 2) = 0$$

$$x^2 - \frac{3}{2}x - 1 = 0$$

The equation could be cleared of fractions by multiplying both sides by 2.

$$2x^2 - 3x - 2 = 0$$

Q7 Write a quadratic equation with the given solutions:

a. $x = -1$ or $x = 8$ **b.** $x = \pm 3$

STOP • STOP • STOP • STOP • STOP • STOP • STOP • STOP • STOP

A7

a. $x^2 - 7x - 8 = 0$:

$x = -1$ or $x = 8$

$x + 1 = 0$ or $x - 8 = 0$

$(x + 1)(x - 8) = 0$

$x^2 - 7x - 8 = 0$

b. $x^2 - 9 = 0$:

$x = -3$ or $x = 3$

$x + 3 = 0$ or $x - 3 = 0$

$(x + 3)(x - 3) = 0$

$x^2 - 9 = 0$

Q8

Write a quadratic equation with the given solutions:

a. $x = \frac{-3}{2}$ or $x = 2$

b. $x = 2$ or $x = -\sqrt{3}$

STOP • STOP • STOP • STOP • STOP • STOP • STOP • STOP • STOP

A8

a. $2x^2 - x - 6 = 0$:

$$\left(x + \frac{3}{2}\right)(x - 2) = 0$$

$$x^2 - \frac{1}{2}x - 3 = 0$$

$$2x^2 - x - 6 = 0$$

b. $x^2 + (\sqrt{3} - 2)x - 2\sqrt{3} = 0$:

$(x - 2)(x + \sqrt{3}) = 0$

Q9

Write a quadratic equation with the given solutions:

a. $x = \pm\sqrt{5}$

b. $x = -2 \pm \sqrt{3}$

STOP • STOP • STOP • STOP • STOP • STOP • STOP • STOP • STOP

A9

a. $x^2 - 5 = 0$:

$x = -\sqrt{5}$ or $x = \sqrt{5}$

$(x + \sqrt{5})(x - \sqrt{5}) = 0$

b. $x^2 + 4x + 1 = 0$:

$x = -2 - \sqrt{3}$ or $x = -2 + \sqrt{3}$

$(x + 2 + \sqrt{3})(x + 2 - \sqrt{3}) = 0$

Q10

Check the results of Q9b by solving $x^2 + 4x + 1 = 0$.

STOP • STOP • STOP • STOP • STOP • STOP • STOP • STOP • STOP

A10 $x = -2 \pm \sqrt{3}$: $x = \dfrac{-4 \pm \sqrt{4^2 - 4(1)(1)}}{2(1)}$

$$x = \frac{-4 \pm \sqrt{12}}{2}$$

$$x = \frac{-4 \pm 2\sqrt{3}}{2}$$

$$x = -2 \pm \sqrt{3}$$

Q11 Write a quadratic equation with the given solutions:

a. $x = 2$ **b.** $x = r$ or $x = s$

STOP • STOP • STOP • STOP • STOP • STOP • STOP • STOP • STOP

A11 **a.** $x^2 - 4x + 4 = 0$:
$x = 2$ or $x = 2$
$(x - 2)(x - 2) = 0$

b. $x^2 - (r + s)x + rs = 0$:
$x = r$ or $x = s$
$x - r = 0$ or $x - s = 0$
$(x - r)(x - s) = 0$
$x^2 - rx - sx + rs = 0$

This completes the instruction for this section.

5.3 Exercises

1. Write the expression for the discriminant for the equation $ax^2 + bx + c = 0$.
2. Give the value of the discriminant that yields the following solutions for the equation $ax^2 + bx + c = 0$:
   a. real and equal **b.** rational and unequal **c.** imaginary
   d. real and unequal
3. Describe the solutions by examining the discriminant (be as specific as possible):
   a. $x^2 - 4x = x - 2$ **b.** $5x^2 - 3x + 7 = 0$ **c.** $5x^2 = -3x - 1$
   d. $x^2 - 6x + 9 = 0$ **e.** $9x^2 - 6x = 0$ **f.** $a^2 = 2a - 10$
4. Determine the value of the coefficient a that will make the solutions of $ax^2 - 2x + 7 = 0$:
   a. real and equal **b.** imaginary **c.** real and unequal
5. Write a quadratic equation with the given solutions:
   a. $x = -1$ or $x = 4$ **b.** $y = 7$ or $y = -7$ **c.** $\frac{-1}{2}$ or $\frac{3}{4}$
   d. $\sqrt{13}$ or $-\sqrt{13}$ **e.** π or $\sqrt{2}$
   f. $x = -4 + \sqrt{3}$ or $x = -4 - \sqrt{3}$ **g.** $x = -6 \pm \sqrt{2}$
   h. $y = -2$ **i.** $x = \pm i$ **j.** $x = 0$ or $x = 3$

5.3 Exercise Answers

1. $b^2 - 4ac$
2. **a.** $b^2 - 4ac = 0$ **b.** $b^2 - 4ac =$ perfect square, not zero
   c. $b^2 - 4ac < 0$ **d.** $b^2 - 4ac > 0$

3. **a.** real and unequal **b.** imaginary **c.** imaginary
   d. rational and equal (one solution) **e.** rational and unequal **f.** imaginary

4. **a.** $a = \frac{1}{7}$ **b.** $a > \frac{1}{7}$ **c.** $a < \frac{1}{7}$

5. **a.** $x^2 - 3x - 4 = 0$ **b.** $y^2 - 49 = 0$
   c. $8x^2 - 2x - 3 = 0$ **d.** $x^2 - 13 = 0$
   e. $x^2 - (\pi + \sqrt{2})x + \pi\sqrt{2} = 0$ **f.** $x^2 + 8x + 13 = 0$
   g. $x^2 + 12x + 34 = 0$ **h.** $y^2 + 4y + 4 = 0$
   i. $x^2 + 1 = 0$ **j.** $x^2 - 3x = 0$

5.4 Solution of Quadratic Inequalities

1

Quadratic inequalities can be solved algebraically or geometrically. The first technique presented will be a geometric solution. For example, to solve $x^2 + x - 6 < 0$ it is first factored as $(x - 2)(x + 3) < 0$. The *equality points* (the numbers making the left side equal to zero) are then marked on the number line.

-3 0 2

The equality points, $x = -3$ or $x = 2$, divide the number line into three intervals: $x < -3$, $-3 < x < 2$, and $x > 2$. Now, if the product of two factors is negative, one factor is negative and the other positive. That is, if $ab < 0$, then $a < 0$ and $b > 0$ or $a > 0$ and $b < 0$. This property can be used to find the solution for $(x - 2)(x + 3) < 0$ by finding the intervals in which the product $(x - 2)(x + 3)$ is negative. First consider the factor $x - 2$.

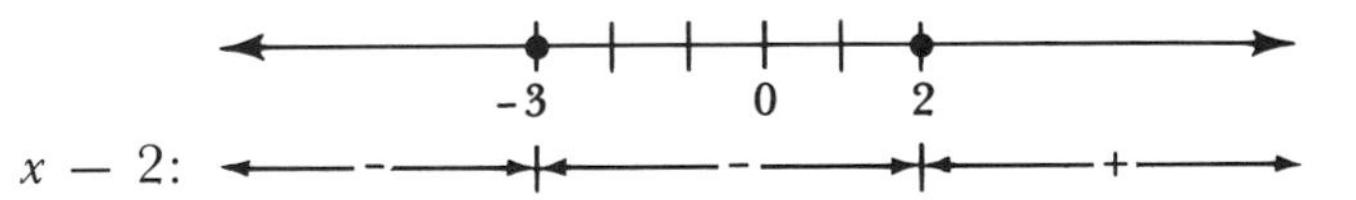

First interval $(x < -3)$: $x - 2$ is negative
Second interval $(-3 < x < 2)$: $x - 2$ is negative
Third interval $(x > 2)$: $x - 2$ is positive

Consider the factor $x + 3$.

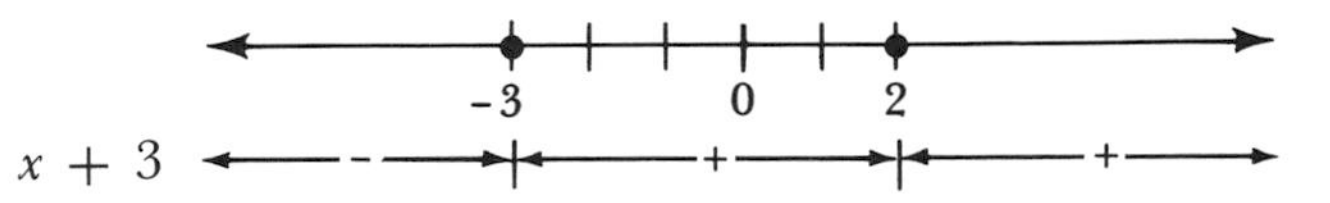

First interval: $x + 3$ is negative
Second interval: $x + 3$ is positive
Third interval: $x + 3$ is positive
Now, consider the product of the signs of the factors in each interval.

-3 0 2

(−)(−) = + (−)(+) = − (+)(+) = +

Hence the product $(x - 2)(x + 3)$ is negative only in the second interval and the solution is $-3 < x < 2$.

Q1 Determine the equality points for the quadratic inequality $x^2 - x - 2 < 0$.

STOP • STOP • STOP • STOP • STOP • STOP • STOP • STOP • STOP

A1 $x = -1$ or $x = 2$: $(x + 1)(x - 2) < 0$

Q2 List the three intervals into which the equality points of Q1 divide the number line.

____

STOP • STOP • STOP • STOP • STOP • STOP • STOP • STOP • STOP

A2 $x < -1$, $-1 < x < 2$, $x > 2$

Q3 Find and mark the equality points for $x^2 - 5x - 6 < 0$.

0

STOP • STOP • STOP • STOP • STOP • STOP • STOP • STOP • STOP

A3 $x = -1$ or $x = 6$:

-1 0 6

---

2 Since a factor is either positive or negative for all points of an interval, quadratic inequalities can be solved by testing only one point from each interval. If one point satisfies the inequality, all points of that interval will satisfy the inequality; conversely, if one point fails to satisfy the inequality, all points of that interval fail.

Example: Solve $x^2 - 5x - 6 < 0$.

Solution

$(x - 6)(x + 1) < 0$; hence the equality points are $x = 6$ or $x = -1$.

-1 0 6

Test one point from each interval:

1. First interval ($x < -1$): Let $x = -2$ (other choices of x could have been made).

$$(-2)^2 - 5(-2) - 6 \overset{?}{<} 0$$
$$4 + 10 - 6 \overset{?}{<} 0$$
$$8 < 0 \quad \text{(false)}$$

Hence $x < -1$ is not part of the solution.

2. Second interval ($-1 < x < 6$): Let $x = 0$.

$$(0)^2 - 5(0) - 6 \overset{?}{<} 0$$
$$0 - 0 - 6 \overset{?}{<} 0$$
$$-6 < 0 \quad \text{(true)}$$

Hence $-1 < x < 6$ is part of the solution.

3. Third interval ($x > 6$): Let $x = 7$.

$$(7)^2 - 5(7) - 6 \overset{?}{<} 0$$
$$49 - 35 - 6 \overset{?}{<} 0$$
$$8 < 0 \quad \text{(false)}$$

Hence $x > 6$ is not part of the solution.

Therefore, the solution for $x^2 - 5x - 6 < 0$ is $-1 < x < 6$.

Q4 Solve $x^2 - x - 6 < 0$ by testing one point from each interval determined by the equality points.

STOP • STOP • STOP • STOP • STOP • STOP • STOP • STOP • STOP

A4 $-2 < x < 3$:

-2 0 3

First interval: Let $x = -3$: $(-3)^2 - (-3) - 6 \overset{?}{<} 0$

$$9 + 3 - 6 \overset{?}{<} 0$$
$$6 < 0 \quad \text{(false)}$$

Hence $x < -2$ is not part of the solution.

Second interval: Let $x = 0$: $(0)^2 - (0) - 6 \overset{?}{<} 0$

$$-6 < 0 \quad \text{(true)}$$

Hence $-2 < x < 3$ is part of the solution.

Third interval: Let $x = 4$: $(4)^2 - (4) - 6 \overset{?}{<} 0$

$$6 < 0 \quad \text{(false)}$$

Hence $x > 3$ is not part of the solution.

Q5 Solve $x^2 + 4x - 5 < 0$.

STOP • STOP • STOP • STOP • STOP • STOP • STOP • STOP • STOP

A5 $-5 < x < 1$:

First: Let $x = -6$: $(-6)^2 + 4(-6) - 5 < 0$ (false)
Second: Let $x = 0$: $(0)^2 + 4(0) - 5 < 0$ (true)
Third: Let $x = 2$: $(2)^2 + 4(2) - 5 < 0$ (false)

Q6 Solve $2x^2 + 5x - 3 < 0$.

STOP • STOP • STOP • STOP • STOP • STOP • STOP • STOP • STOP

A6 $-3 < x < \frac{1}{2}$:

First: Let $x = -4$: $2(-4)^2 + 5(-4) - 3 < 0$ (false)
Second: Let $x = 0$: $2(0)^2 + 5(0) - 3 < 0$ (true)
Third: Let $x = 1$: $2(1)^2 + 5(1) - 3 < 0$ (false)

3 Inequalities that involve the greater-than symbol can also be solved by testing intervals. It will sometimes be necessary to test fractional numbers if the equality points are 1 unit apart.

Example: Solve $x^2 - 11x + 30 > 0$.

Solution

$(x - 5)(x - 6) > 0$

The equality points are $x = 5$ and $x = 6$. They divide the number line into three intervals: $x < 5$, $5 < x < 6$, and $x > 6$. Test each interval.

1. First interval ($x < 5$): Let $x = 0$: $(0)^2 - 11(0) + 30 > 0$ (true)
2. Second interval ($5 < x < 6$): Let $x = \frac{11}{2}$: $\left(\frac{11}{2}\right)^2 - 11\left(\frac{11}{2}\right) + 30 \overset{?}{>} 0$

$$\frac{121}{4} - \frac{242}{4} + \frac{120}{4} > 0 \quad \text{(false)}$$

3. Third interval ($x > 6$): Let $x = 7$: $7^2 - 11(7) + 30 > 0$ (true)

Hence the solution is $x < 5$ or $x > 6$.

Q7 Solve $x^2 - 5x + 6 > 0$.

STOP • STOP • STOP • STOP • STOP • STOP • STOP • STOP • STOP

A7 $x < 2$ or $x > 3$:

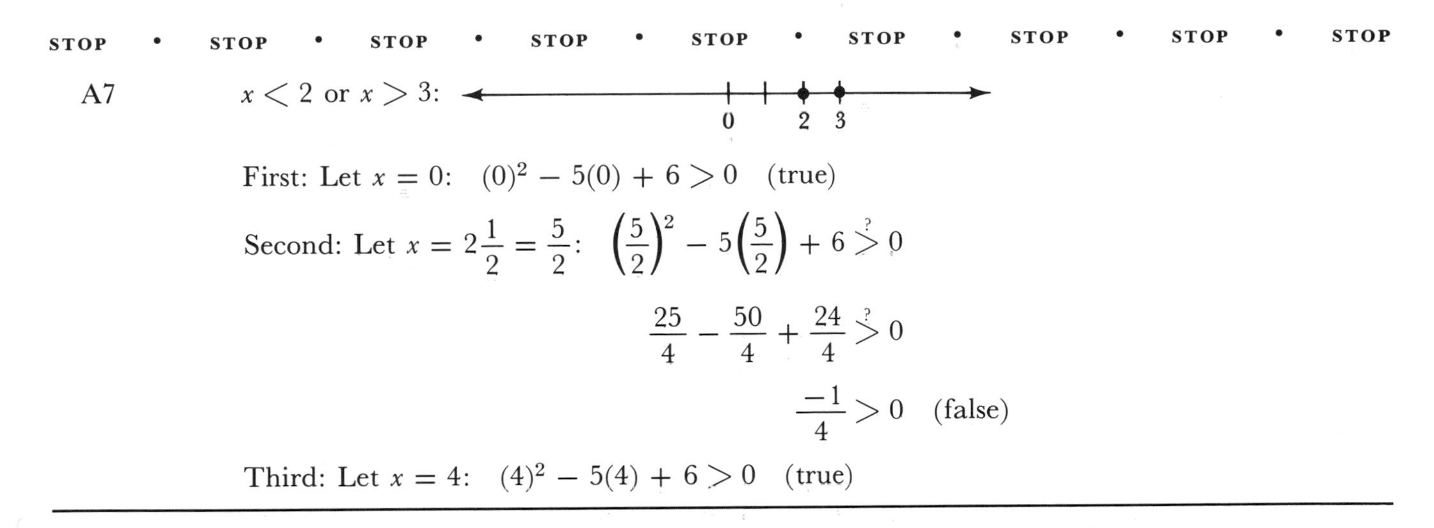

First: Let $x = 0$: $(0)^2 - 5(0) + 6 > 0$ (true)

Second: Let $x = 2\frac{1}{2} = \frac{5}{2}$: $\left(\frac{5}{2}\right)^2 - 5\left(\frac{5}{2}\right) + 6 \overset{?}{>} 0$

$$\frac{25}{4} - \frac{50}{4} + \frac{24}{4} \overset{?}{>} 0$$

$$\frac{-1}{4} > 0 \quad \text{(false)}$$

Third: Let $x = 4$: $(4)^2 - 5(4) + 6 > 0$ (true)

Q8 Solve $x^2 - 7x + 6 \geqslant 0$ (the equality points are included in the solution).

STOP • STOP • STOP • STOP • STOP • STOP • STOP • STOP • STOP

A8 $x \leqslant 1$ or $x \geqslant 6$:

0 1 6

First: Let $x = 0$: $(0)^2 - 7(0) + 6 \geqslant 0$ (true)
Second: Let $x = 2$: $(2)^2 - 7(2) + 6 \geqslant 0$ (false)
Third: Let $x = 7$: $(7)^2 - 7(7) + 6 \geqslant 0$ (true)

Q9 Solve $x^2 - 4x + 1 > 0$. (*Hint:* The equality points are irrational.)

STOP • STOP • STOP • STOP • STOP • STOP • STOP • STOP • STOP

A9 $x < 2 - \sqrt{3}$ or $x > 2 + \sqrt{3}$

0 2 5
$2 - \sqrt{3}$ $2 + \sqrt{3}$

First: Let $x = 0$: $(0)^2 - 4(0) + 1 > 0$ (true)
Second: Let $x = 1$: $(1)^2 - 4(1) + 1 > 0$ (false)
Third: Let $x = 4$: $(4)^2 - 4(4) + 1 > 0$ (true)

Q10 Solve $x^2 - 4x + 4 < 0$.

STOP • STOP • STOP • STOP • STOP • STOP • STOP • STOP • STOP

A10 $\varnothing$: (two intervals)

0 2

First: Let $x = 0$: $(0)^2 - 4(0) + 4 < 0$ (false)
Second: Let $x = 3$: $(3)^2 - 4(3) + 4 < 0$ (false)

4 It is possible for an inequality to have no solution or for the solution to be all real numbers.

Example 1: Solve $x^2 + 8 < 0$.

Solution

Subtracting 8 from both sides:

$x^2 < -8$

Since all real numbers squared are clearly nonnegative, x^2 cannot be less than -8. Hence there is no solution to the inequality $x^2 + 8 < 0$.

Example 2: Solve $x^2 + 2x + 3 > 0$.

Solution

The discriminant $b^2 - 4ac = 2^2 - 4(1)(3) = -8$; hence there are no equality points. When there are no equality points, it is necessary to test only one point on the number line. If the resulting statement is true, the solution is all real numbers. If the statement is false, there is no solution. Let $x = 0$:

$(0)^2 + 2(0) + 3 > 0$ (true)

Hence the solution is all real numbers.

The first example could have been solved in this manner. That is, in the first example, let $x = 0$.

$(0)^2 + 8 < 0$ (false)

Hence there is no solution.

Q11 Solve:

a. $x^2 + 4 < 0$ **b.** $x^2 + 4x + 5 < 0$

STOP • STOP • STOP • STOP • STOP • STOP • STOP • STOP • STOP

A11 **a.** $\varnothing$ **b.** $\varnothing$

(*Note:* There are no equality points to either problem because the solutions to the points of equality are imaginary.)

Q12 Solve:

a. $x^2 + 4 > 0$ **b.** $x^2 < 9$

STOP • STOP • STOP • STOP • STOP • STOP • STOP • STOP • STOP

A12 **a.** R **b.** $-3 < x < 3$: $x^2 - 9 < 0$

$(x - 3)(x + 3) < 0$

---

5 Quadratic inequalities can also be solved using the technique called *solution by cases.* Recall the property that if $ab < 0$, then (case 1) $a < 0$ and $b > 0$ or (case 2) $a > 0$ and $b < 0$.

Example: Solve $x^2 - x - 2 < 0$ by cases.

Solution

$$x^2 - x - 2 < 0$$
$$(x + 1)(x - 2) < 0$$

Case 1: or Case 2:

Case 1			or	Case 2		
$x + 1 < 0$	and	$x - 2 > 0$		$x + 1 > 0$	and	$x - 2 < 0$
$x < -1$	and	$x > 2$		$x > -1$	and	$x < 2$
	no solution				$-1 < x < 2$	

There is no solution for Case 1 because x cannot be less than -1 and greater than 2 simultaneously. Hence the solution for $x^2 - x - 2 < 0$ is $-1 < x < 2$.

Q13 **a.** Solve $x^2 - 8x - 20 < 0$ by cases.

b. Graph the solution to Q13a.

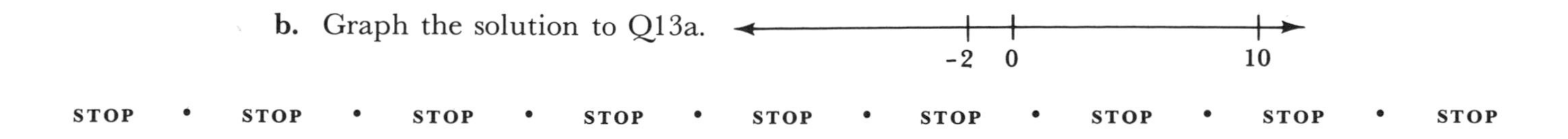

STOP • STOP • STOP • STOP • STOP • STOP • STOP • STOP • STOP

A13

a. $-2 < x < 10$:

Case 1: $x + 2 < 0$ and $x - 10 > 0$

$x < -2$ and $x > 10$

no solution

or

Case 2: $x + 2 > 0$ and $x - 10 < 0$

$x > -2$ and $x < 10$

$-2 < x < 10$

b.

6

The property that $ab > 0$ means that either (case 1) $a > 0$ and $b > 0$ or (case 2) $a < 0$ and $b < 0$.

Example: Solve $x^2 - 5x + 6 > 0$ by cases.

Solution

$$x^2 - 5x + 6 > 0$$
$$(x - 2)(x - 3) > 0$$

Case 1:

$x - 2 > 0$ and $x - 3 > 0$

$x > 2$ and $x > 3$

$x > 3$

or

Case 2:

$x - 2 < 0$ and $x - 3 < 0$

$x < 2$ and $x < 3$

$x < 2$

Hence the solution for $x^2 - 5x + 6 > 0$ is $x < 2$ or $x > 3$.

Q14

a. Solve $x^2 + 3x - 4 > 0$ by cases.

b. Graph the solution to Q14a.

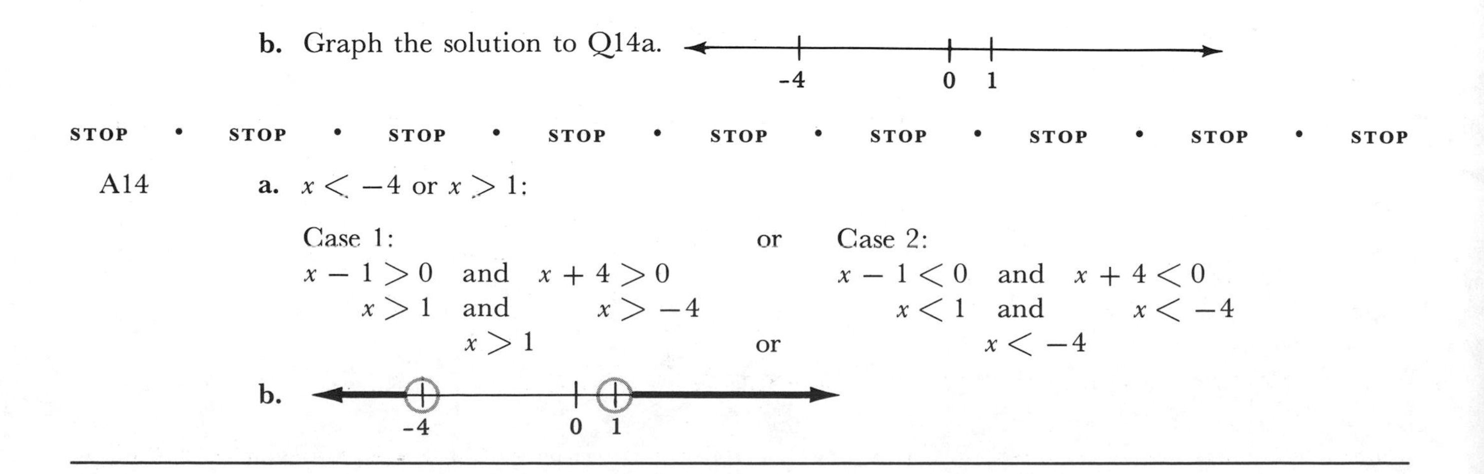

STOP • STOP • STOP • STOP • STOP • STOP • STOP • STOP • STOP

A14

a. $x < -4$ or $x > 1$:

Case 1: $x - 1 > 0$ and $x + 4 > 0$

$x > 1$ and $x > -4$

$x > 1$

or

Case 2: $x - 1 < 0$ and $x + 4 < 0$

$x < 1$ and $x < -4$

$x < -4$

b.

Q15 Solve by cases:

a. $x^2 - x - 2 > 0$ **b.** $3x^2 - 7x + 2 > 0$

STOP • STOP • STOP • STOP • STOP • STOP • STOP • STOP • STOP

A15 **a.** $x < -1$ or $x > 2$ **b.** $x < \frac{1}{3}$ or $x > 2$

This completes the instruction for this section.

5.4 Exercises

1. Given that $x, y \in R$:
   a. $xy < 0$ means x ____ 0 and y ____ 0 or x ____ 0 and y ____ 0.
   b. $xy > 0$ means x ____ 0 and y ____ 0 or x ____ 0 and y ____ 0.
2. Solve:
   a. $x^2 - 5 < 0$ **b.** $x^2 + 8x + 7 > 0$ **c.** $x^2 - 4x + 3 < 0$
   d. $x^2 + 2x - 15 > 0$ **e.** $-x^2 + 5x - 4 \geqslant 0$ **f.** $x^2 - 9 < 0$
   g. $3x^2 - 7x + 2 \geqslant 0$ **h.** $x^2 < 16$ **i.** $2x^2 + 3x - 9 < 0$
   j. $4x^2 - 8x + 3 > 0$ **k.** $x(x - 3) \leqslant 0$ **l.** $x^2 + 1 < 0$
   m. $x(x + 2) > 2$ **n.** $x^2 + 2x - 5 \leqslant 0$ **o.** $x^2 + 1 > 0$
   p. $\frac{x^2}{2} + x + 1 < 0$ **q.** $x^2 - 6x + 9 \leqslant 0$ **r.** $4x^2 + 7x - 15 \geqslant 0$
3. Solve by cases:
   a. $x^2 - 3x + 2 > 0$ **b.** $4x^2 - 9 \leqslant -9x$ **c.** $x^2 > 3$
   d. $7x < 6x^2 + 2$
4. Graph the solutions to problems 3a and 3b.

5.4 Exercise Answers

1. **a.** $x < 0$ and $y > 0$ or $x > 0$ and $y < 0$.
   b. $x > 0$ and $y > 0$ or $x < 0$ and $y < 0$.
2. **a.** $-\sqrt{5} < x < \sqrt{5}$ **b.** $x < -7$ or $x > -1$
   c. $1 < x < 3$ **d.** $x < -5$ or $x > 3$
   e. $1 \leqslant x \leqslant 4$ **f.** $-3 < x < 3$
   g. $x \leqslant \frac{1}{3}$ or $x \geqslant 2$ **h.** $-4 < x < 4$
   i. $-3 < x < \frac{3}{2}$ **j.** $x < \frac{1}{2}$ or $x > \frac{3}{2}$
   k. $0 \leqslant x \leqslant 3$ **l.** $\varnothing$
   m. $x < -1 - \sqrt{3}$ or $x > -1 + \sqrt{3}$ **n.** $-1 - \sqrt{6} \leqslant x \leqslant -1 + \sqrt{6}$

o. R p. $\varnothing$

q. $x = 3$ r. $x \leqslant -3$ or $x \geqslant \frac{5}{4}$

3. a. $x < 1$ or $x > 2$ b. $-3 \leqslant x \leqslant \frac{3}{4}$

c. $x < -\sqrt{3}$ or $x > \sqrt{3}$ d. $x < \frac{1}{2}$ or $x > \frac{2}{3}$

4. (3a) (3b)

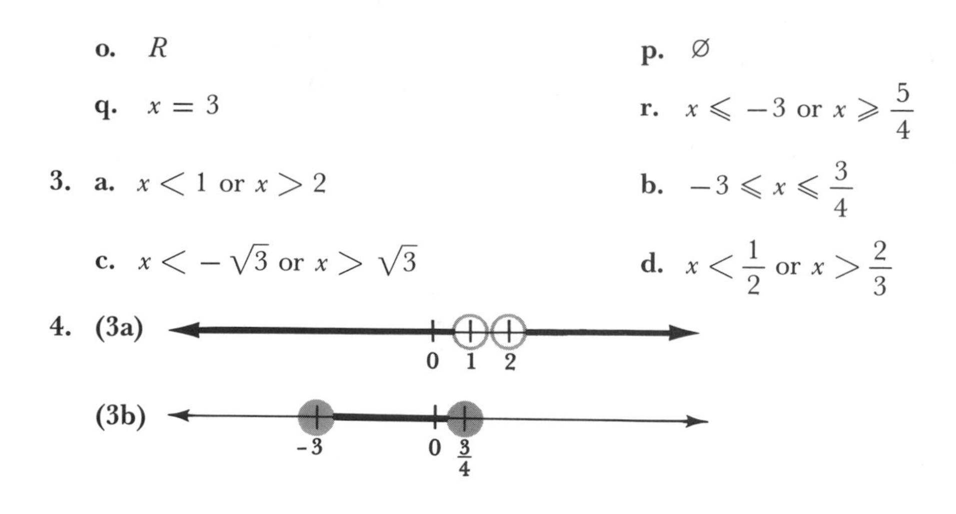

Chapter 5 Sample Test

At the completion of Chapter 5 it is expected that you will be able to work the following problems.

5.1 Solving Quadratic Equations by Factoring

1. Solve the quadratic equations:
   a. $x^2 - 5x + 6 = 0$ b. $x^2 + 4x = 0$
   c. $x^2 + 10x + 25 = 0$ d. $2x^2 + 5x - 3 = 0$
   e. $x^2 - 5x = 24$ f. $x^2 + 72 = 17x$

5.2 Solving Quadratic Equations by Completing the Square and the Quadratic Formula

2. Solve by completing the square:
   a. $x^2 + x = 2$ b. $x^2 - 4x - 7 = 0$
3. Solve by use of the quadratic formula:
   a. $x^2 - 2x + 5 = 0$ b. $x^2 - 1 = 2x$

5.3 The Discriminant and Writing Quadratic Equations from Solutions

4. Describe the solutions by analyzing the discriminant:
   a. $x^2 + 3x - 2 = 0$ b. $x^2 - 2x + 3 = 0$
5. Write the equation with the given solutions:
   a. $x = -2$ or $x = 1$ b. $x = -1 \pm \sqrt{3}$

5.4 Solution of Quadratic Inequalities

6. Solve the quadratic inequalities:
   a. $x^2 + x - 20 < 0$ b. $2x^2 + 5x - 3 \geqslant 0$

Chapter 5 Sample Test Answers

1. **a.** $x = 2$ or $x = 3$ **b.** $x = 0$ or $x = -4$ **c.** $x = -5$
   d. $x = \frac{1}{2}$ or $x = -3$ **e.** $x = 8$ or $x = -3$ **f.** $x = 8$ or $x = 9$
2. **a.** $x = -2$ or $x = 1$ **b.** $x = 2 \pm \sqrt{11}$
3. **a.** $x = 1 \pm 2i$ **b.** $x = 1 \pm \sqrt{2}$
4. **a.** real and unequal **b.** imaginary
5. **a.** $x^2 + x - 2 = 0$ **b.** $x^2 + 2x - 2 = 0$
6. **a.** $-5 < x < 4$ **b.** $x \leqslant -3$ or $x \geqslant \frac{1}{2}$

Chapter 6

Functions and Their Graphs

In Chapter 6 many types of sets will be graphed. Most students will probably have had exposure to graphing in previous mathematics courses. Circle, bar, and line graphs as presented in newspapers and magazines are usually pictorial representations of a certain set of data. Graphs make it possible to quickly draw conclusions from tables of data. The graphs presented in this chapter will be pictorial representations of mathematical sets.

6.1 Relations and the Coordinate System

1 If the elements in one set are matched with the elements of another set, a *correspondence* is formed. In the sets A and B below, the elements of A are matched with the elements of B such that each element of A is 1 greater than the corresponding element of B. The arrow is read "corresponds to."

Set A		*Set B*
1	$\longrightarrow$	0
2	$\longrightarrow$	1
3	$\longrightarrow$	2
4	$\longrightarrow$	3

Q1 Complete the correspondence so that each element of the first set is 2 less than the corresponding element of the second set.

a. 0 $\longrightarrow$ ____

b. 2 $\longrightarrow$ ____

c. ____ $\longrightarrow$ -3

STOP • STOP • STOP • STOP • STOP • STOP • STOP • STOP • STOP

A1 **a.** 2 **b.** 4 **c.** -5

2 When a correspondence is formed, the matched elements are often expressed as *ordered pairs*. This set of ordered pairs is called a *relation*. For example, the correspondence

$1 \longrightarrow -1$
$2 \longrightarrow -2$
$3 \longrightarrow -3$

expressed as a relation is $\{(1, -1), (2, -2), (3, -3)\}$. The symbolism $(1, -1)$ is called an *ordered pair* because the order of the numbers is important. The number 1 is called the *first member* and the number -1 is called the *second member* of the ordered pair. $(1, -1)$ means $1 \longrightarrow -1$ and hence is *not* equivalent to $(-1, 1)$.

Q2 Write the correspondence as a relation (set of ordered pairs):

a. $0 \rightarrow 1$, $1 \rightarrow 2$, $2 \rightarrow 3$, $3 \rightarrow 4$

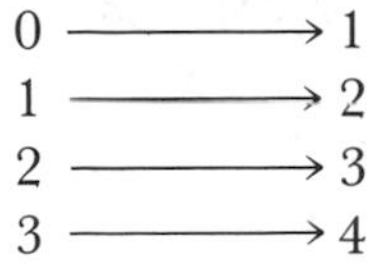

b. $-1 \rightarrow 1$, $1 \rightarrow 1$, $-2 \rightarrow 2$, $2 \rightarrow 2$

________________ ________________

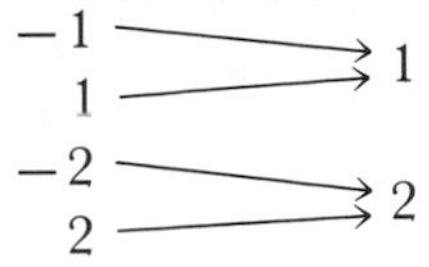

STOP • STOP • STOP • STOP • STOP • STOP • STOP • STOP • STOP

A2 **a.** $\{(0, 1), (1, 2), (2, 3), (3, 4)\}$ **b.** $\{(-1, 1), (1, 1), (-2, 2), (2, 2)\}$

Q3 Write the relation as a correspondence using arrows:

a. $\{(1, 2), (1, 3), (0, 8), (-1, 4)\}$ **b.** $\{(3, 2), (4, 2), (6, 2)\}$

STOP • STOP • STOP • STOP • STOP • STOP • STOP • STOP • STOP

A3 **a.** $1 \rightarrow 2$, $1 \rightarrow 3$, $0 \rightarrow 8$, $-1 \rightarrow 4$

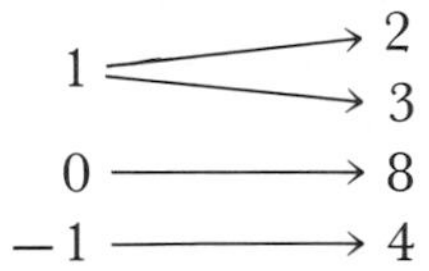

b. $3 \rightarrow 2$, $4 \rightarrow 2$, $6 \rightarrow 2$

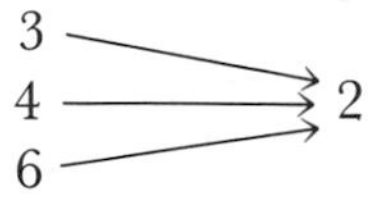

3 Given two sets of numbers A and B, many relations can be formed between the sets by matching the elements of A with corresponding elements of B.

Example: Find the relation such that each element of A is greater than the corresponding element of B.

$A = \{-3, 0, 3, 6\}$ and $B = \{-1, 0, 1, 4\}$

Solution

Consider each element of A separately.

-3: -3 does not match any element of B because -3 is less than all elements of B
0: $(0, -1)$ because $0 > -1$
3: $(3, -1), (3, 0), (3, 1)$ because $3 > -1$, $3 > 0$, and $3 > 1$
6: $(6, -1), (6, 0), (6, 1), (6, 4)$ because 6 is greater than all elements of B

Hence the relation is $\{(0, -1), (3, -1), (3, 0), (3, 1), (6, -1), (6, 0), (6, 1), (6, 4)\}$.

Q4 Find the relation matching elements of A with its opposite element in B.
$A = \{-3, -2, -1, 0, 1, 2, 3\}$ and $B = \{0, 1, 2, 3\}$

STOP • STOP • STOP • STOP • STOP • STOP • STOP • STOP • STOP

A4 $\{(-3, 3), (-2, 2), (-1, 1), (0, 0)\}$

Q5 Find the relation such that each element of A is less than the corresponding element of B.
$A = \{-4, -3, 2\}$ and $B = \{-5, -3, 2\}$

STOP • STOP • STOP • STOP • STOP • STOP • STOP • STOP • STOP

A5 $\{(-4, -3), (-4, 2), (-3, 2)\}$

4

Recall that a set of ordered pairs is a relation. The set consisting of the first members of the ordered pairs is called the *domain* of the relation and the set consisting of the second members is called the *range*. If the relation is represented as a correspondence (using arrows), the domain is the first set and the range is the second set.

Example: Find the domain and range for the relation $\{(-3, -1), (-2, 0), (-1, 1), (0, 2)\}$.

Solution

domain $= \{-3, -2, -1, 0\}$
range $= \{-1, 0, 1, 2\}$

$$\begin{array}{ccc} -3 & \longrightarrow & -1 \\ -2 & \longrightarrow & 0 \\ -1 & \longrightarrow & 1 \\ 0 & \longrightarrow & 2 \\ \uparrow & & \uparrow \\ \text{domain} & & \text{range} \end{array}$$

Q6

Find the (1) domain and (2) range:

a. $\{(1, 3), (3, 7), (5, 11), (7, 15)\}$

(1) ____________

(2) ____________

b. $\{(3, 4), (4, 3), (3, -4), (4, -3)\}$

(1) ____________

(2) ____________

STOP • STOP • STOP • STOP • STOP • STOP • STOP • STOP • STOP

A6

a. (1) $\{1, 3, 5, 7\}$
(2) $\{3, 7, 11, 15\}$

b. (1) $\{3, 4\}$
(2) $\{4, 3, -4, -3\}$

Q7

Find the (1) domain and (2) range:

a. $\{(2, 5), (4, 6), (-2, 5), (3, 4)\}$

(1) ____________

(2) ____________

b. $\{(0, 1), (0, 3), (0, -1), (0, -2)\}$

(1) ____________

(2) ____________

STOP • STOP • STOP • STOP • STOP • STOP • STOP • STOP • STOP

A7

a. (1) $\{2, 4, -2, 3\}$
(2) $\{5, 6, 4\}$

b. (1) $\{0\}$
(2) $\{1, 3, -1, -2\}$

5

Previously, each relation (and its domain and range) has been finite; however, relations can and often do have infinite domains and ranges.

Example: List the domain and range for the relation $\{(1, 2), (2, 3), (3, 4), (4, 5), \ldots\}$.

Solution

domain $= \{1, 2, 3, 4, \ldots\}$ or domain $= N$
range $= \{2, 3, 4, 5, \ldots\}$

Q8

Find the domain and range.
$\{\ldots, (-3, 1), (-2, 1), (-1, 1), (0, 1), (1, 1), (2, 1), \ldots\}$

domain = ____________________ range = ____________________

STOP • STOP • STOP • STOP • STOP • STOP • STOP • STOP • STOP

A8 domain $= \{\ldots, -3, -2, -1, 0, 1, 2, \ldots\}$ or domain $= I$
range $= \{1\}$

6 A relation can be graphed on the coordinate system. The *coordinate system* consists of two perpendicular number lines whose intersection is called the *origin.*

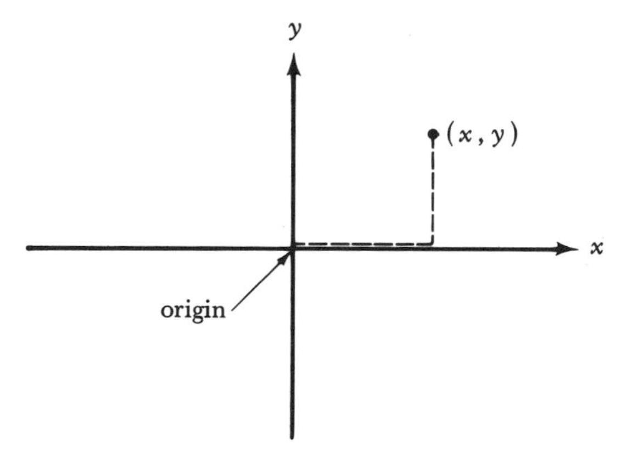

The horizontal and vertical lines, called the *coordinate axes,* are usually labeled the x axis and y axis, respectively. The x axis is positive to the right of the origin and the y axis is positive above the origin.

An ordered pair of real numbers, (x, y), is graphed by first determining the horizontal distance from the y axis, x. From this position the vertical distance from the x axis, y, is determined. The first member, x, is called the *abscissa* and the second member, y, is called the *ordinate.* The members x and y are called the *coordinates* of the point (x, y).

Q9 Graph and label the ordered pair $(3, -2)$ on the coordinate system. (When a scale is not indicated on a coordinate system, each space represents 1 unit.)

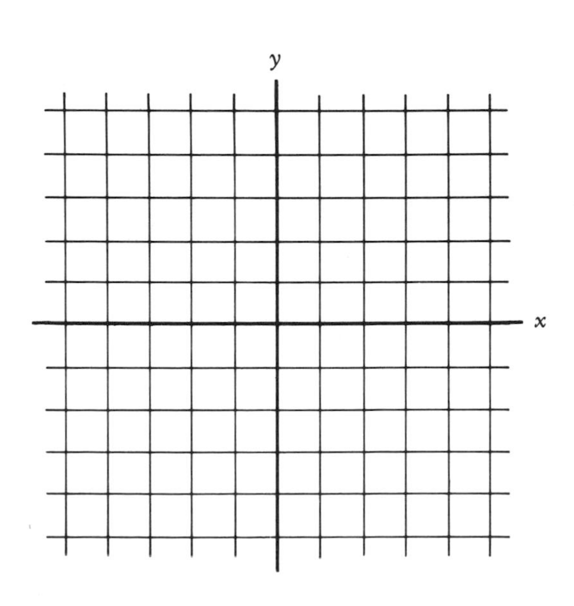

STOP • STOP • STOP • STOP • STOP • STOP • STOP • STOP • STOP

A9

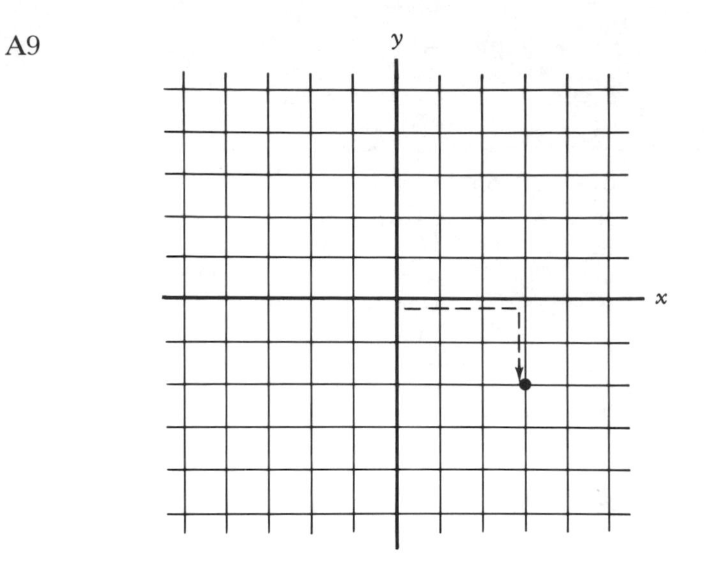

The ordered pair $(3, -2)$ is graphed by moving, from the origin, to the right 3 units and down 2.

Q10 Graph and label the ordered pairs:

$A(0, 3)$, $B(2, 3)$, $C(3, 2)$, $D(-1, -3)$, $E(3, 0)$, $F\left(\frac{-5}{2}, \sqrt{3}\right)$

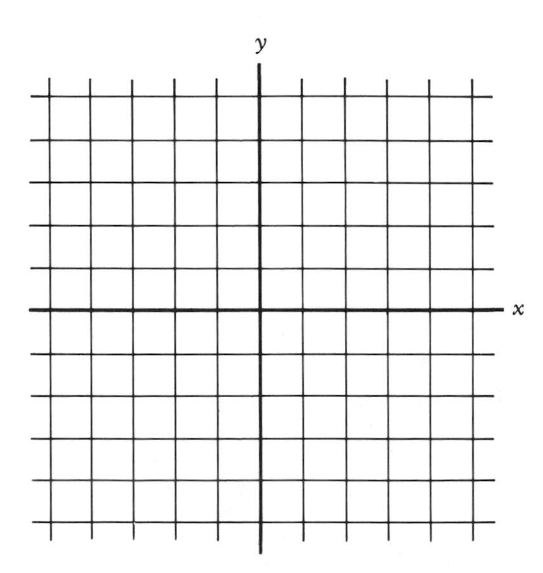

STOP • STOP • STOP • STOP • STOP • STOP • STOP • STOP • STOP

A10

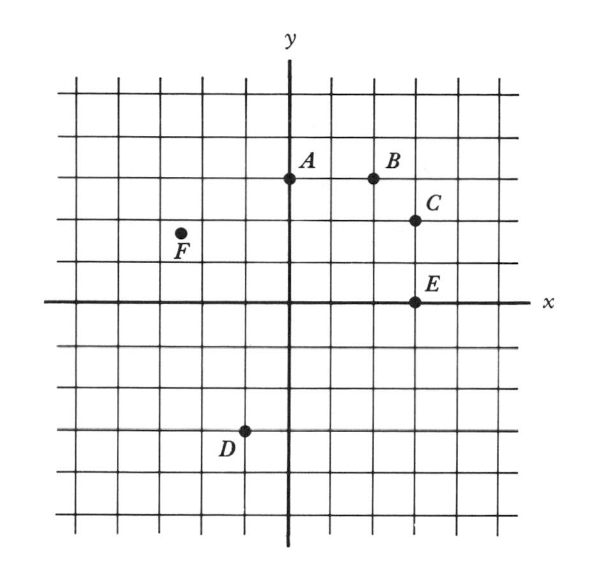

Q11 Write the ordered pair that corresponds to each point:

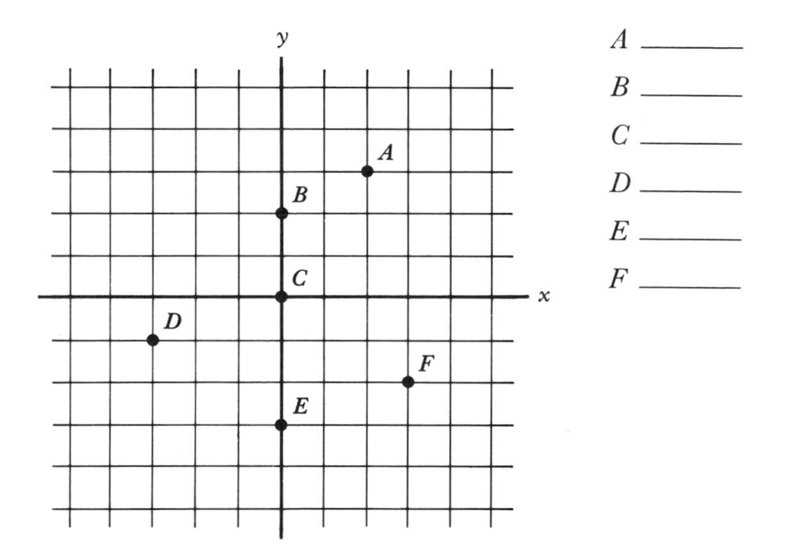

STOP • STOP • STOP • STOP • STOP • STOP • STOP • STOP • STOP

A11 $A(2, 3)$, $B(0, 2)$, $C(0, 0)$, $D(-3, -1)$, $E(0, -3)$, $F(3, -2)$

7 Since a relation is defined as a set of ordered pairs, a relation can be graphed on the coordinate system.

Example: Graph the relation $\{\ldots, (-3, -2), (-2, -1), (-1, 0), (0, 1), (1, 2), \ldots\}$.

Solution

The infinite set is shown graphically by placing arrowheads to indicate the position and direction of the infinite set.

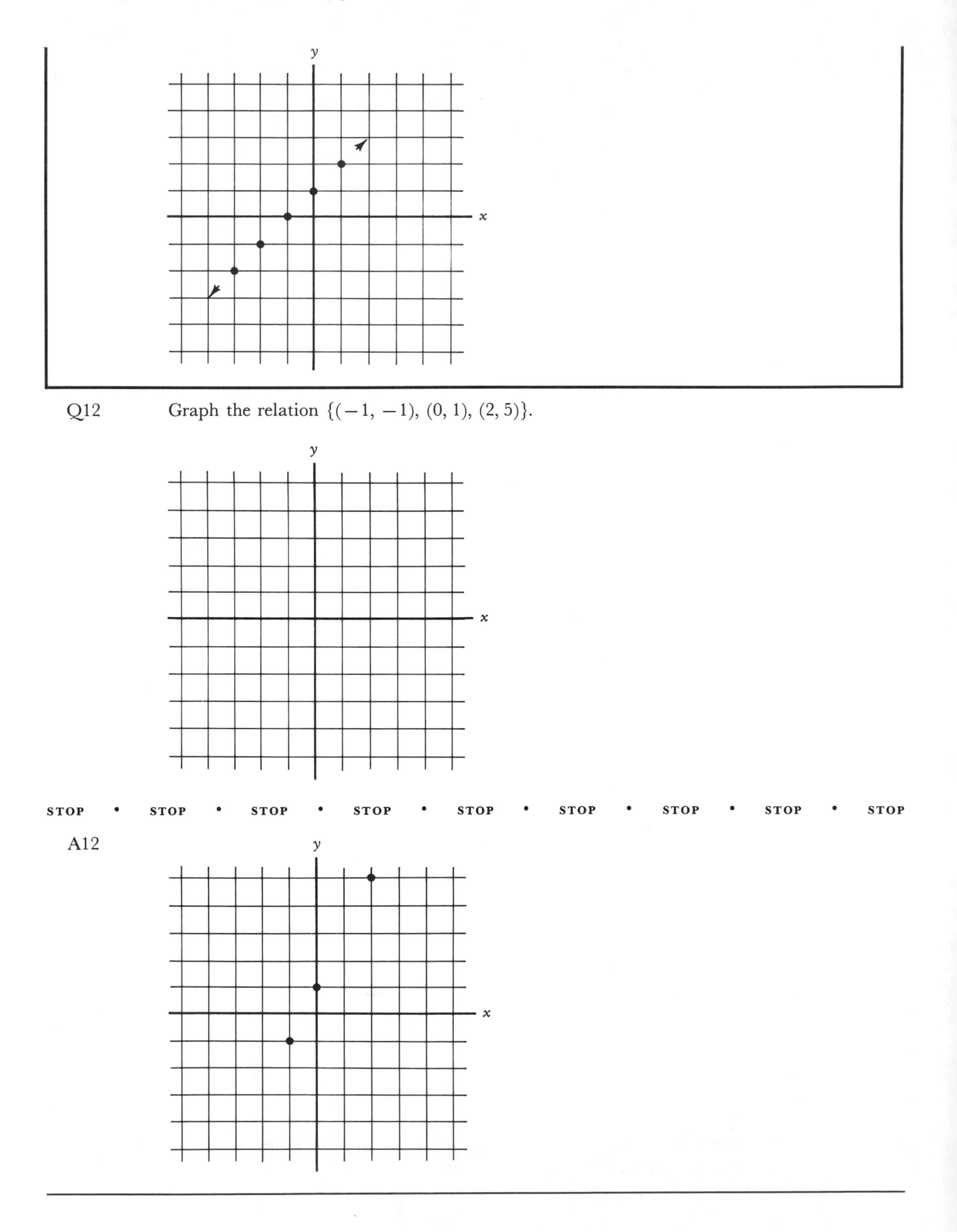
y
x
Q12
Graph the relation {(−1, −1), (0, 1), (2, 5)}.
y
x
STOP • STOP • STOP • STOP • STOP • STOP • STOP • STOP • STOP
A12
y
x

Q13 Graph:

a. $\{(-2, 4), (-1, 1), (0, 0), (1, 1), (2, 4)\}$

b. $\{\ldots, (-2, 3), (-1, 3), (0, 3), (1, 3), (2, 3), \ldots\}$

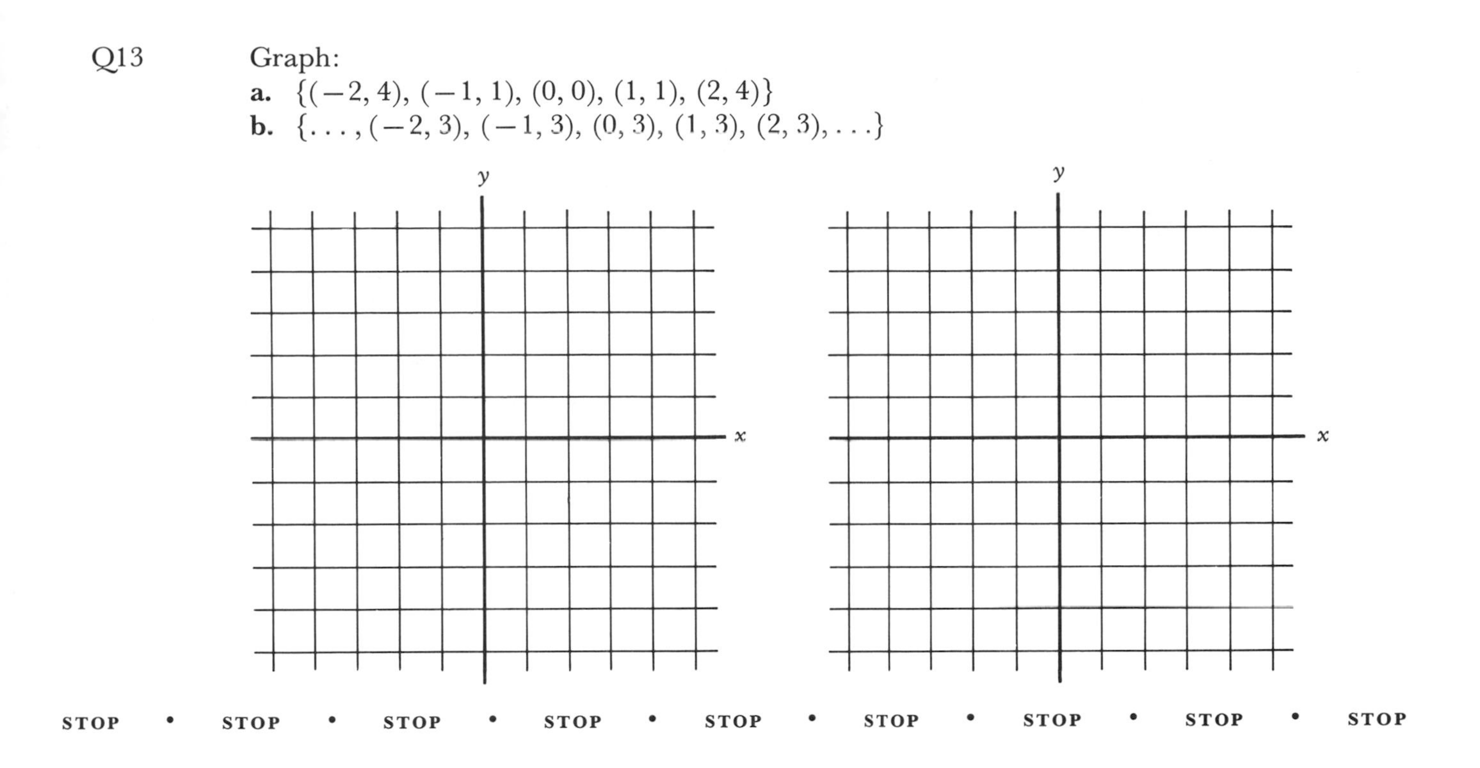

STOP • STOP • STOP • STOP • STOP • STOP • STOP • STOP • STOP

A13

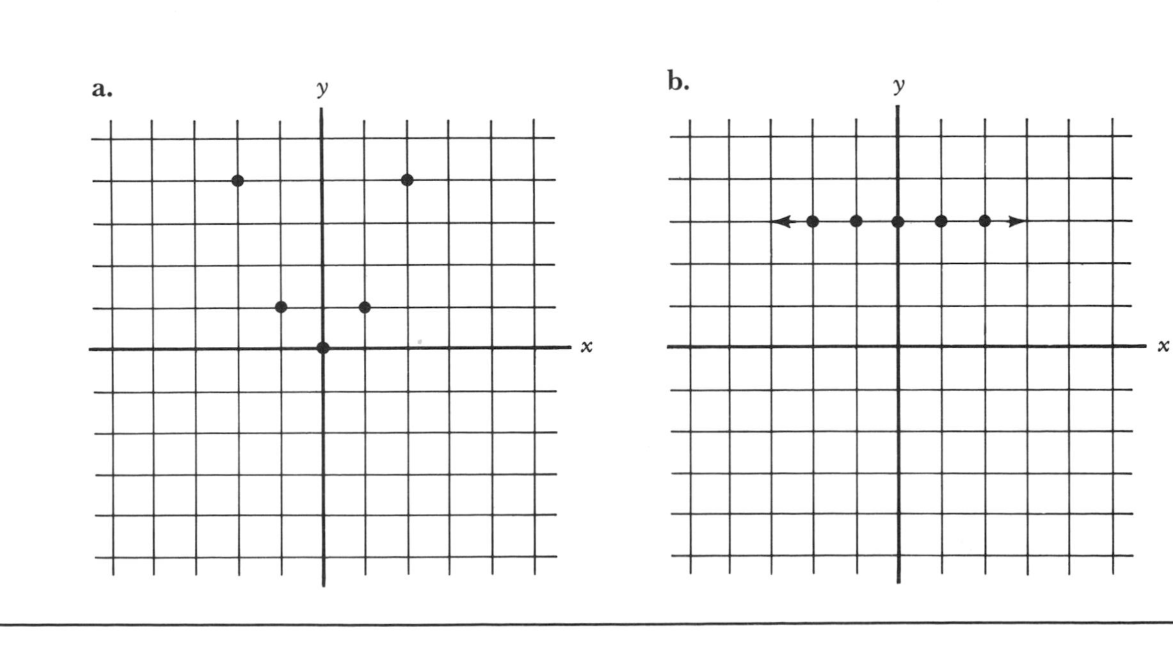

Q14 List the relation represented by the graph:

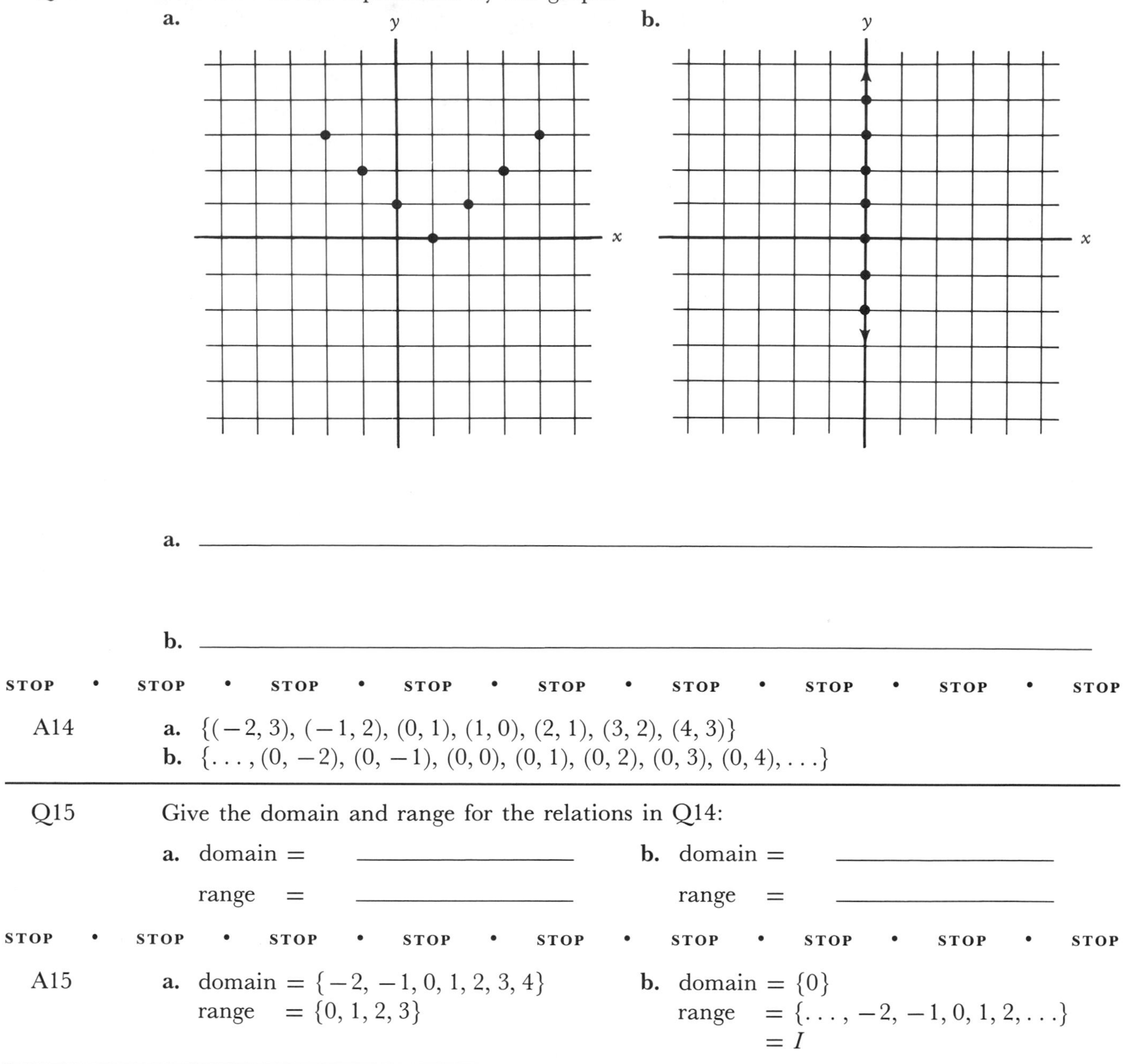

a. ______________________________

b. ______________________________

STOP • STOP • STOP • STOP • STOP • STOP • STOP • STOP • STOP

A14 **a.** $\{(-2, 3), (-1, 2), (0, 1), (1, 0), (2, 1), (3, 2), (4, 3)\}$
b. $\{\ldots, (0, -2), (0, -1), (0, 0), (0, 1), (0, 2), (0, 3), (0, 4), \ldots\}$

Q15 Give the domain and range for the relations in Q14:

a. domain = ______________ **b.** domain = ______________

range = ______________ range = ______________

STOP • STOP • STOP • STOP • STOP • STOP • STOP • STOP • STOP

A15 **a.** domain $= \{-2, -1, 0, 1, 2, 3, 4\}$
range $= \{0, 1, 2, 3\}$

b. domain $= \{0\}$
range $= \{\ldots, -2, -1, 0, 1, 2, \ldots\}$
$= I$

This completes the instruction for this section.

6.1 Exercises

1. Define:
   a. relation **b.** domain of a relation
   c. range of a relation
2. Find the relation between A and B such that each element of A is the absolute value of the elements in B.
   $A = \{-2, -1, 0, 1, 2, 3\}$ and $B = \{-2, 0, 2\}$
3. Find the (1) domain and (2) range:
   a. $\{(-5, 3), (-4, 0), (2, 3), (0, 1)\}$
   b. $\{\ldots, (-2, 2), (-1, 1), (0, 0), (1, 1), (2, 2), \ldots\}$

c. $\{(-2, 2), (-1, 2), (0, 2), (1, 2)\}$
d. $\{\ldots, (4, -3), (4, -2), (4, -1), (4, 0), (4, 1), (4, 2), \ldots\}$

4. Graph the relations in Q3.
5. List the relation represented by the graph:

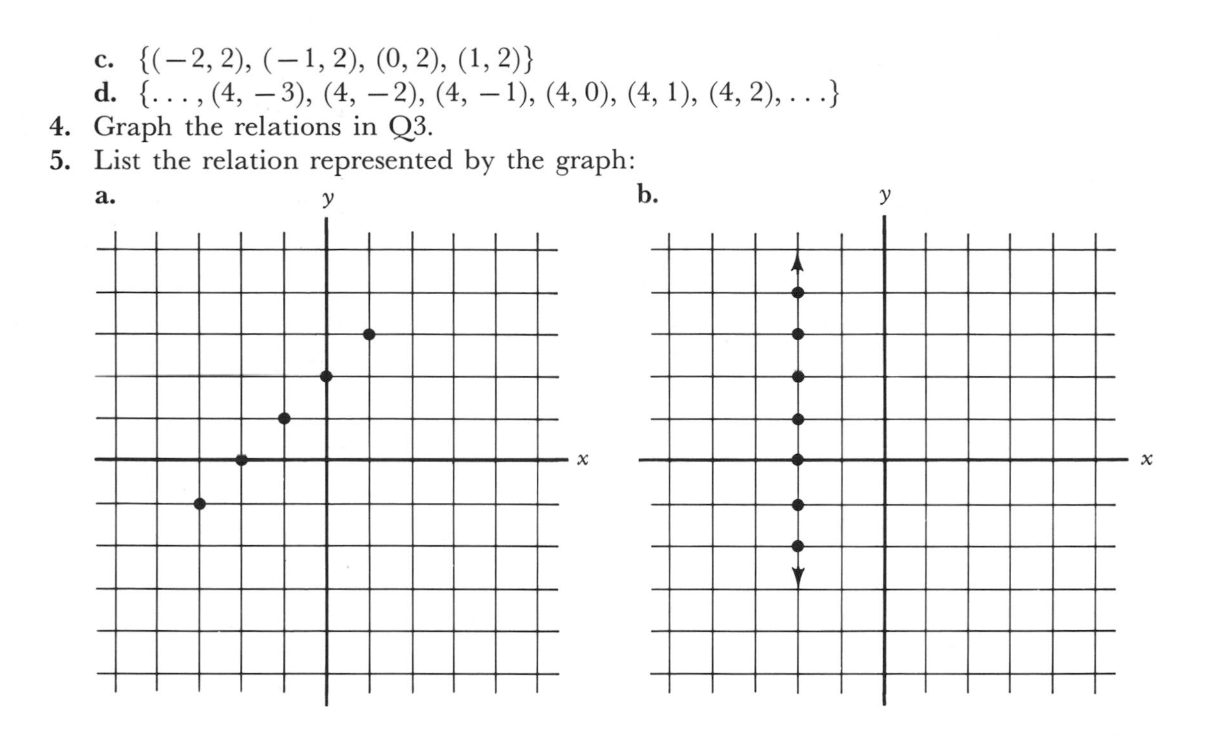

6. Give the domain and range for the relations in Q5.

6.1 Exercise Answers

1. a. a set of ordered pairs
   b. the set of first coordinates of each ordered pair
   c. the set of second coordinates of each ordered pair
2. $\{(0, 0), (2, -2), (2, 2)\}$
3. a. (1) $\{-5, -4, 2, 0\}$ (2) $\{0, 1, 3\}$
   b. (1) $\{\ldots, -2, -1, 0, 1, 2, \ldots\} = I$ (2) $\{0, 1, 2, \ldots\} = W$
   c. (1) $\{-2, -1, 0, 1\}$ (2) $\{2\}$
   d. (1) $\{4\}$ (2) I
4. a.

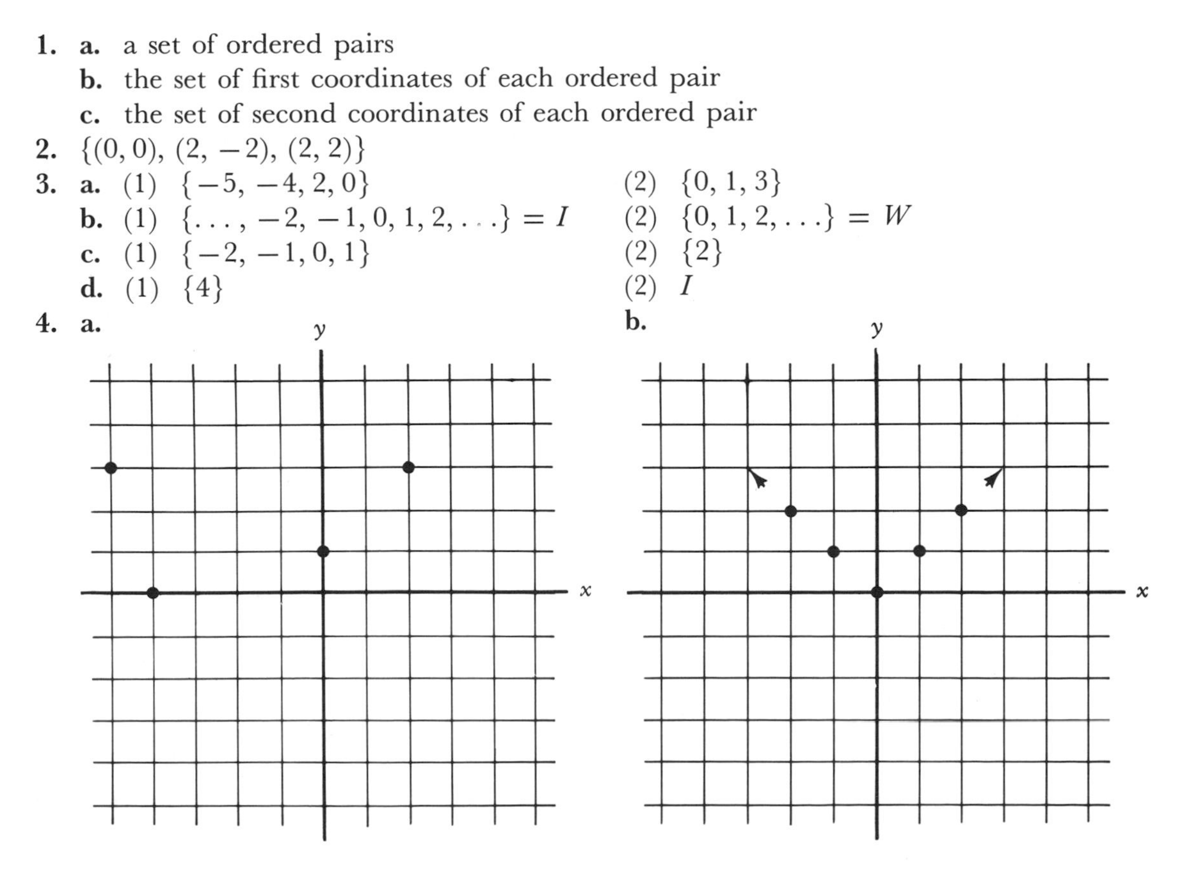

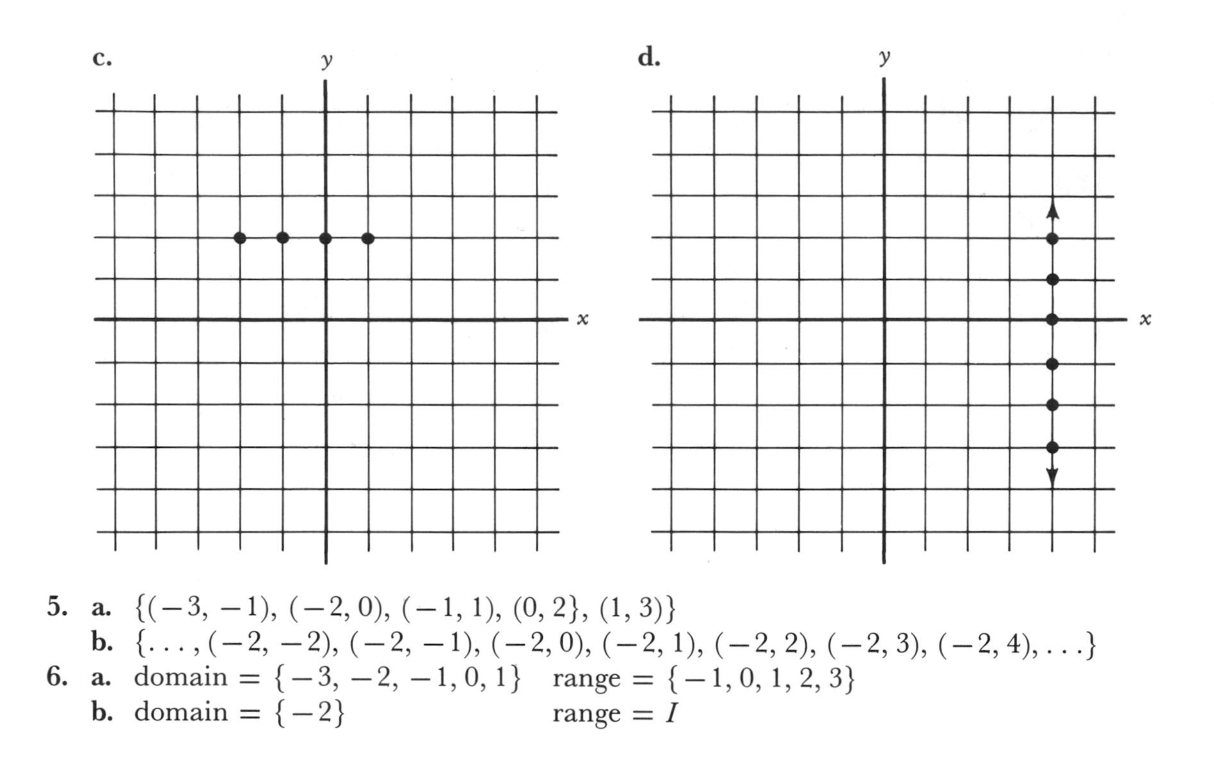

5. a. $\{(-3, -1), (-2, 0), (-1, 1), (0, 2\}, (1, 3)\}$
   b. $\{\ldots, (-2, -2), (-2, -1), (-2, 0), (-2, 1), (-2, 2), (-2, 3), (-2, 4), \ldots\}$
6. a. domain = $\{-3, -2, -1, 0, 1\}$ range = $\{-1, 0, 1, 2, 3\}$
   b. domain = $\{-2\}$ range = I

6.2 Functions and Function Notation

1

A relation in which no two ordered pairs have the same first coordinate is called a *function*. That is, if each domain element of the relation corresponds to exactly one range element, the relation is a function.

Examples: In the following four examples, is the relation also a function?

1. 1 ⟶ 4; 2 ⟶ 4; 3 ⟶ 5
2. 0 ⟶ 6; 1 ⟶ −1; 1 ⟶ 1
3. $\{(-1, 1), (0, 0), (1, 1), (2, 2)\}$
4. $\{(-1, 1), (2, 0), (-1, -1), (3, 0)\}$

Solutions

Your reasoning should proceed as follows:

1. Yes; the correspondencc is a function because each domain element corresponds to exactly one range element. 1 and 2 each corresponds to exactly one element, even though it is the same element for each.
2. No; the domain element, 1, corresponds to two different range elements, -1 and 1.
3. Yes; no two ordered pairs have the same first coordinate.
4. No; the ordered pairs $(-1, 1)$ and $(-1, -1)$ have the same first coordinate.

Q1 Is the relation a function? Explain.

a. $0 \longrightarrow 0$, $1 \longrightarrow 3$, $2 \longrightarrow 5$

b. $-2 \longrightarrow 3$, $-1 \longrightarrow 2$, $0 \longrightarrow 2$, $1 \longrightarrow 2$, $2 \longrightarrow 3$

c. $1 \longrightarrow -4$, $1 \longrightarrow 4$, $2 \longrightarrow 8$, $7 \longrightarrow 8$, $6 \longrightarrow 12$

a. ______; ______________________

b. ______; ______________________

c. ______; ______________________

STOP • STOP • STOP • STOP • STOP • STOP • STOP • STOP • STOP

A1 **a.** yes; each domain element corresponds to exactly one range element
b. yes; same as part a
c. no; the domain element, 1, corresponds to two different range elements, −4 and 4.

Q2 Is the relation a function? Explain.

a.* $x_1 \longrightarrow y$, $x_2 \longrightarrow y$, $x_3 \longrightarrow y$

b. $x \longrightarrow y_1$, $x \longrightarrow y_2$

a. ______; ______________________

b. ______; ______________________

STOP • STOP • STOP • STOP • STOP • STOP • STOP • STOP • STOP

A2 **a.** yes; the domain elements, x_1, x_2, and x_3, correspond to exactly one range element, y
b. no; the domain element, x, corresponds to two different range elements, y_1 and y_2

Q3 Is the relation a function? Explain.

a. $\{(4, 3), (3, 3), (2, 7), (2, -1)\}$ **b.** $\{(1, 1), (2, 2), (3, 3)\}$

a. ______; ______________________

b. ______; ______________________

STOP • STOP • STOP • STOP • STOP • STOP • STOP • STOP • STOP

A3 **a.** no; the ordered pairs (2, 7) and (2, −1) have the same first coordinate
b. yes; no two ordered pairs have the same first coordinate

Q4 Is the relation a function? Explain.

a. $\{\ldots, (-2, 3), (-1, 3), (0, 3), \ldots\}$ **b.** $\{\ldots, (2, -1), (2, 0), (2, 1), \ldots\}$

a. ______; ______________________

b. ______; ______________________

STOP • STOP • STOP • STOP • STOP • STOP • STOP • STOP • STOP

A4 **a.** yes; no two ordered pairs have the same first coordinate
b. no; each ordered pair has the same first coordinate

---

*The numbers written slightly below and to the right of a variable are called *subscripts* (x_1 ← subscript). Variables with subscripts represent different numbers and are usually used to represent a relationship between the variables. For example, y_1 and y_2 are used to represent different range elements.

Q5 Which of the relations are also functions? ________

a. $\{(1, 3), (4, 5), (7, 7), (8, 3)\}$

b. $\{(0, 1), (2, 0), (-2, 0), (0, 0)\}$

c. $\{(0, 0), (1, 3), (2, 6), (3, 9), \ldots\}$

d. $\{(a, b)\}, (d, e), (a, c)\}$

STOP • STOP • STOP • STOP • STOP • STOP • STOP • STOP • STOP

A5 a and c: no two ordered pairs have the same first coordinate

---

Q6 Do these graphs of relations represent functions? Explain.

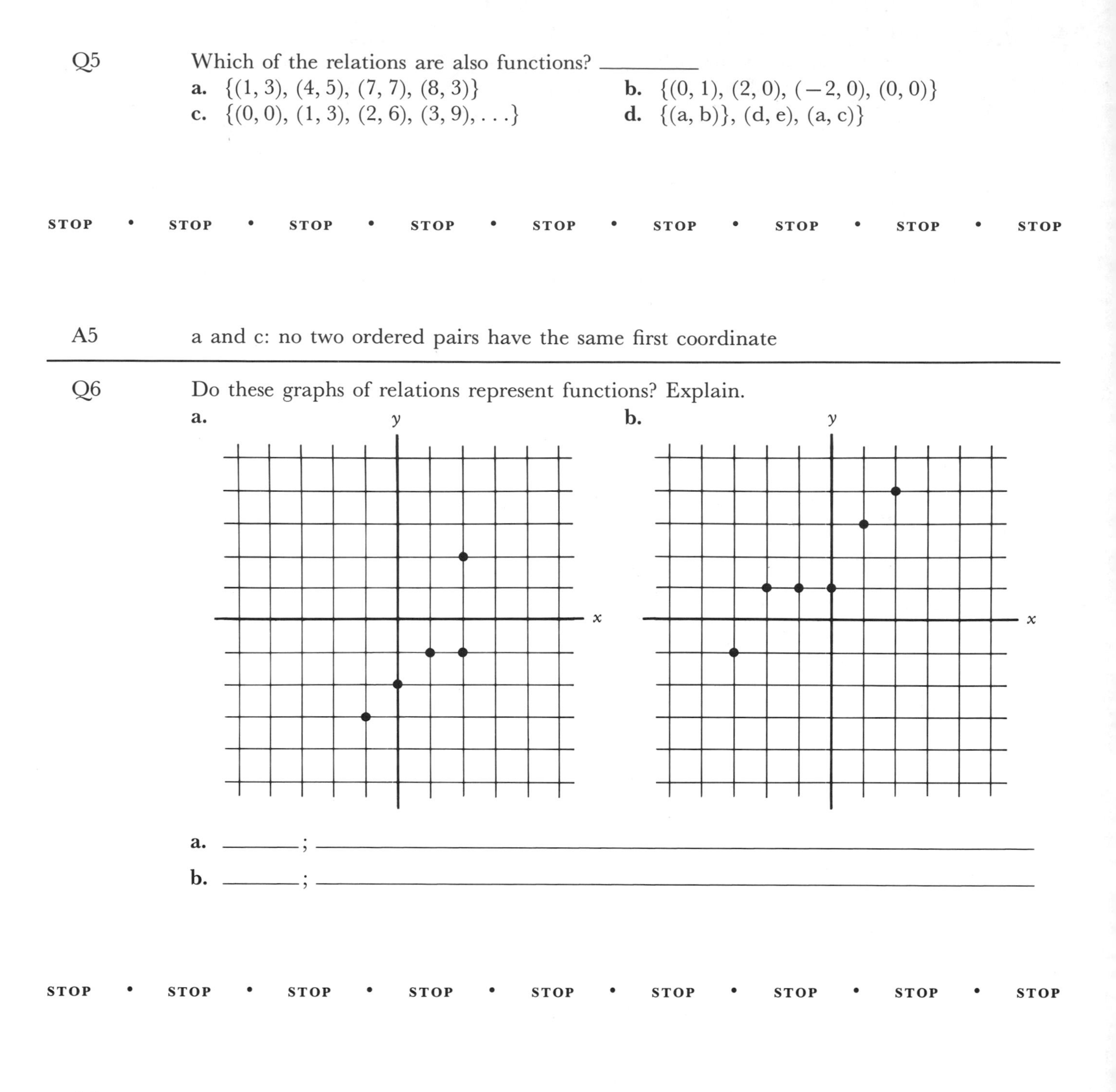

a. ________; ____________________________________

b. ________; ____________________________________

STOP • STOP • STOP • STOP • STOP • STOP • STOP • STOP • STOP

A6 **a.** no; the ordered pairs (2, 2) and $(2, -1)$ have the same first coordinate

b. yes; no two ordered pairs have the same first coordinate

2 If two ordered pairs have the same first coordinate, one point is directly above the other on the coordinate system. That is, some vertical line would contain the two points. Hence, if any vertical line contains two points of the graph of a relation, the relation is not a function. If no vertical line contains two points of the graph, the relation is a function. This technique for determining whether the graph of a relation is a function is called the *vertical-line test.*

Examples: Does the graph of the relation in the following two examples represent a function?

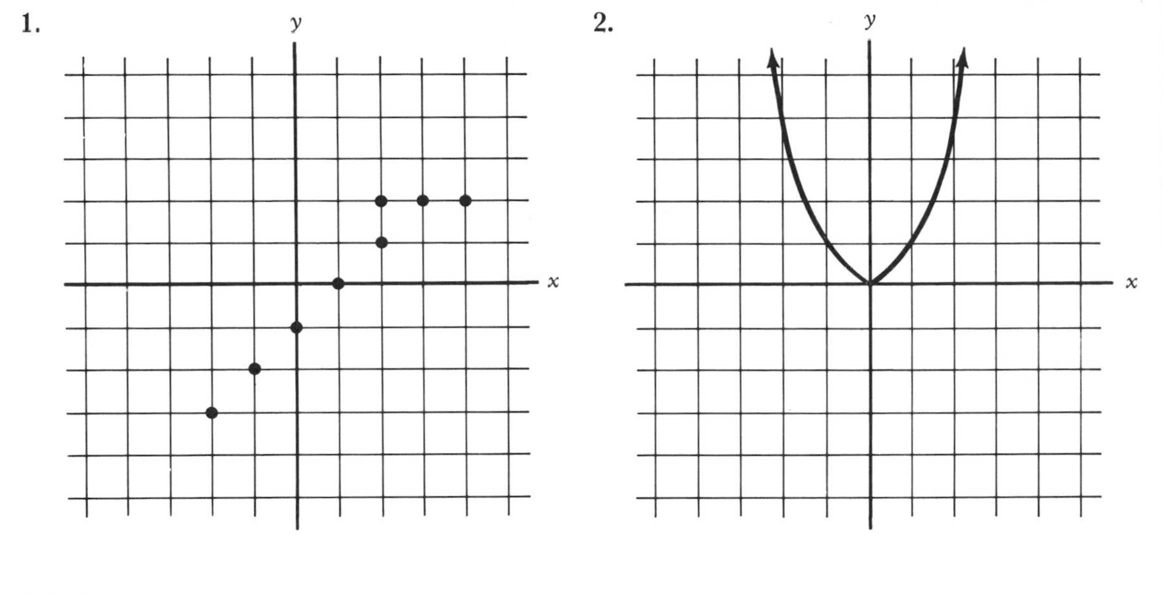

Solutions:

1. No; a vertical line contains two points of the graph. That is, the ordered pairs (2, 1) and (2, 2) have the same first coordinate.
2. Yes; no vertical line contains two points of the graph.

Q7 Is the relation a function? Explain.

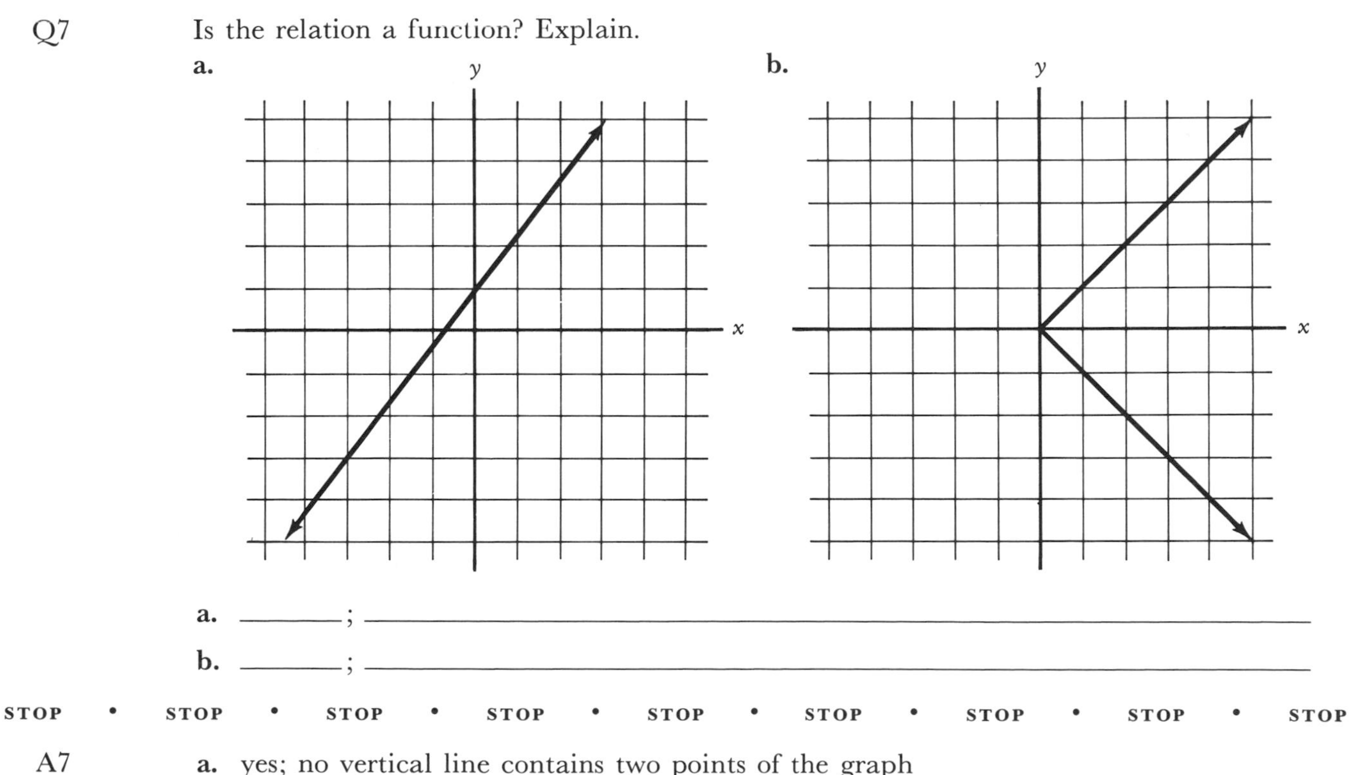

a. ________; ________________________________

b. ________; ________________________________

STOP • STOP • STOP • STOP • STOP • STOP • STOP • STOP • STOP

A7
a. yes; no vertical line contains two points of the graph
b. no; a vertical line does contain two points of the graph, as illustrated (in fact, an infinite number of vertical lines contain two points of the graph)

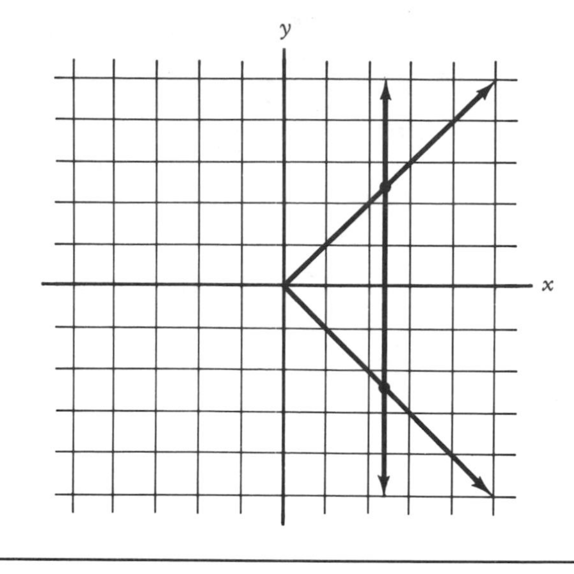

Q8 Which graphs of relations represent functions? __________

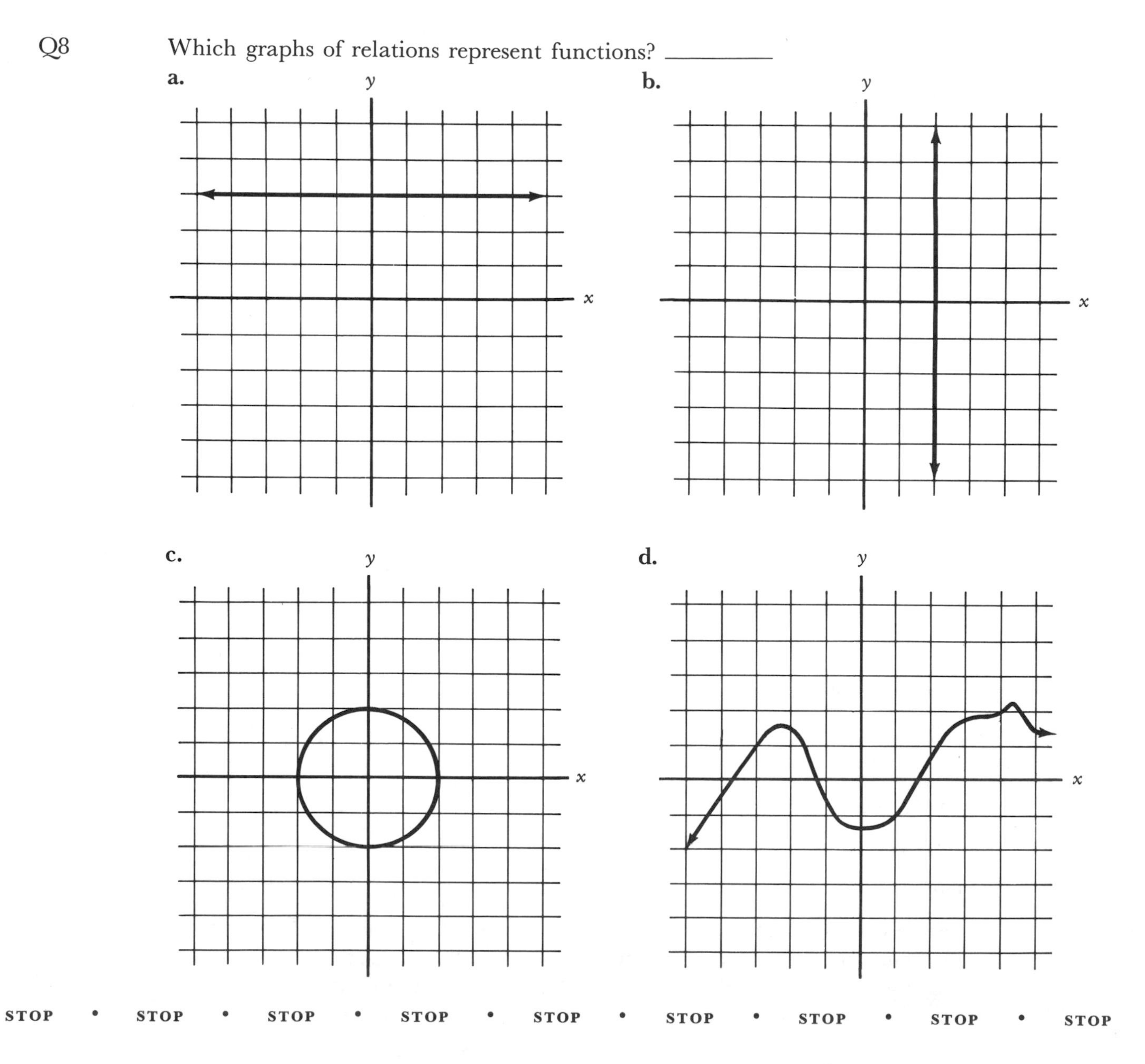

STOP • STOP • STOP • STOP • STOP • STOP • STOP • STOP • STOP

A8 a and **d**

3

A function is usually represented by a rule that describes how the ordered pairs are formed. The notion of a "function machine" can be used to represent this function rule. Consider the following function machine:

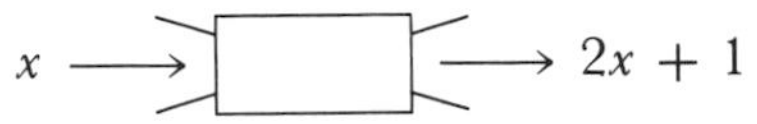

$x \longrightarrow \boxed{} \longrightarrow 2x + 1$

The function machine processes the number, x, and produces exactly one number, $2x + 1$. The function process is: Multiply by 2 and add 1. For the above function multiply and add are the function "operations."

Q9 Write the function process represented by the function machine:

a. $a \longrightarrow \boxed{} \longrightarrow -3a + 7$

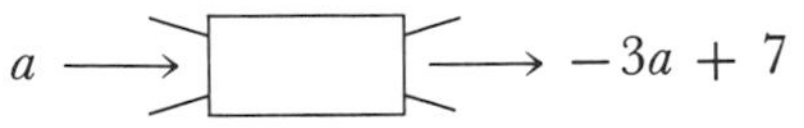

b. $x \longrightarrow \boxed{} \longrightarrow \frac{x}{2} - 10$

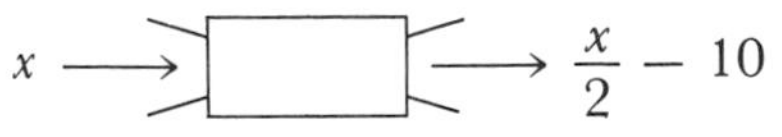

a. ____________________

b. ____________________

STOP • STOP • STOP • STOP • STOP • STOP • STOP • STOP • STOP

A9 **a.** multiply by -3 and add 7
b. divide by 2 and subtract 10

Q10 Determine the expression produced from the function process:

a. multiply by 7 and subtract 4

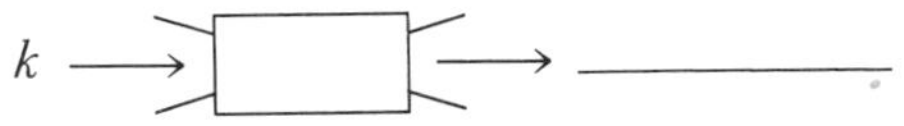

$k \longrightarrow \boxed{} \longrightarrow$ ____________

b. divide by 6 and add $\frac{2}{3}$

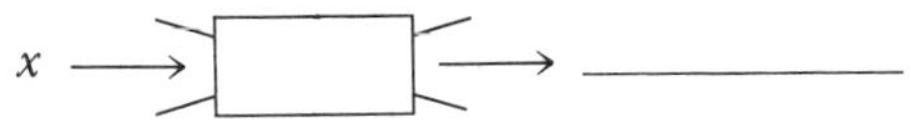

$x \longrightarrow \boxed{} \longrightarrow$ ____________

STOP • STOP • STOP • STOP • STOP • STOP • STOP • STOP • STOP

A10 **a.** $7k - 4$ **b.** $\frac{x}{6} + \frac{2}{3}$

4

The function machine aids in the understanding of the function concept; however, it would not be practical to always describe functions in this manner. The following notation is developed to replace the function machine.

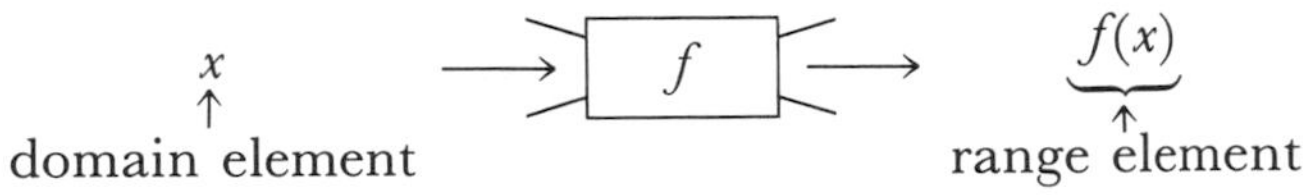

x (↑ domain element) $\longrightarrow \boxed{f} \longrightarrow$ $f(x)$ (↑ range element)

The function, f, processes the domain elements, x, and produces the range elements, $f(x)$. The symbolism $f(x)$ is called *function notation* and is read "f of x." The function

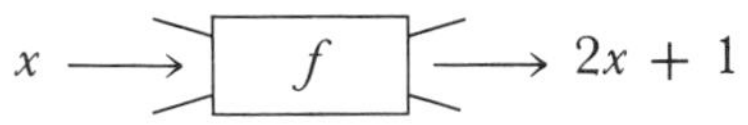

$x \longrightarrow \boxed{f} \longrightarrow 2x + 1$

can be expressed as

$f(x) = 2x + 1$

The notation $f(x) = 2x + 1$ should be interpreted as follows. When x is placed into the function, f, the function multiplies by 2 and adds 1. Note that $f(x)$ does *not* mean f times x.

Q11 Express using function notation:

a. $x \longrightarrow$ [f] $\longrightarrow 7x - 3$

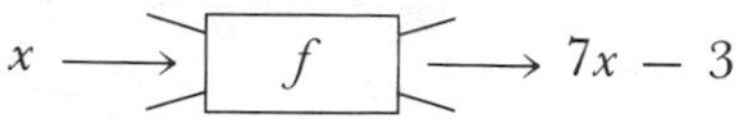

b. $x \longrightarrow$ [g] $\longrightarrow \frac{-2}{3}x + 2$

________________ ________________

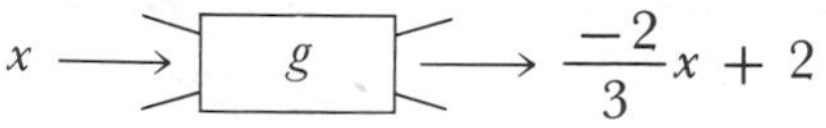

STOP • STOP • STOP • STOP • STOP • STOP • STOP • STOP • STOP

A11 a. $f(x) = 7x - 3$ b. $g(x) = \frac{-2}{3}x + 2$

Q12 $g(x)$ is read "__________."

STOP • STOP • STOP • STOP • STOP • STOP • STOP • STOP • STOP

A12 g of x

Q13 Use function machines to represent the function:

a. $f(x) = -8x + 3$

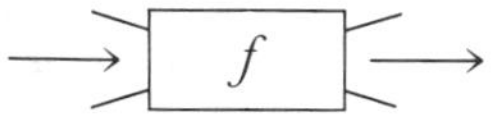

______ $\longrightarrow$ [f] $\longrightarrow$ __________

b. $g(x) = 0x + 1$

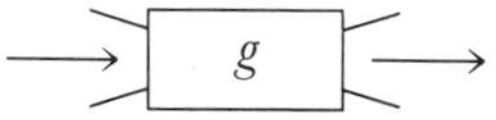

______ $\longrightarrow$ [g] $\longrightarrow$ __________

STOP • STOP • STOP • STOP • STOP • STOP • STOP • STOP • STOP

A13 a. $x, -8x + 3$ b. $x, 0x + 1$

5 Function values can be determined using function notation. For example, for the function $f(x) = 2x + 1$, $f(-3)$ means

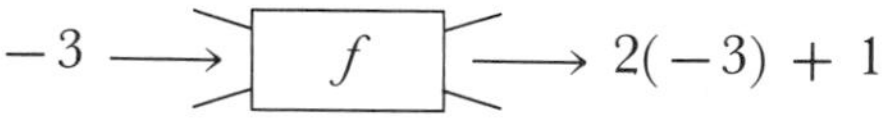

$-3 \longrightarrow$ [f] $\longrightarrow 2(-3) + 1$

That is,

$$f(-3) = 2(-3) + 1$$
$$= -5$$

Example: If $f(x) = 2x + 1$, find $f(0)$, $f(10)$, $f\left(\frac{5}{2}\right)$, $f(-2)$, and $f(a)$.

Solution

The function process is: Multiply by 2 and add 1.

$f(0) = 2(0) + 1$ $= 1$ $f(10) = 2(10) + 1$ $= 21$

$f\left(\frac{5}{2}\right) = 2\left(\frac{5}{2}\right) + 1$ $= 6$ $f(-2) = 2(-2) + 1$ $= -3$ $f(a) = 2(a) + 1$ $= 2a + 1$

Q14 Write the function process for $f(x) = 7x - 2$.

____

STOP • STOP • STOP • STOP • STOP • STOP • STOP • STOP • STOP

A14 multiply by 7 and subtract 2

Q15 If $f(x) = 7x - 2$, find:

a. $f(1)$ **b.** $f(0)$

c. $f(-1)$ **d.** $f(k)$

STOP • STOP • STOP • STOP • STOP • STOP • STOP • STOP • STOP

A15 **a.** 5: $f(1) = 7(1) - 2$ **b.** -2: $f(0) = 7(0) - 2$
c. -9: $f(-1) = 7(-1) - 2$ **d.** $7k - 2$: $f(k) = 7(k) - 2$

Q16 If $f(x) = \frac{-2}{3}x + 1$, find:

a. $f(0)$ **b.** $f(1)$

c. $f\left(\frac{3}{2}\right)$ **d.** $f(a)$

STOP • STOP • STOP • STOP • STOP • STOP • STOP • STOP • STOP

A16 **a.** 1: $f(0) = \frac{-2}{3}(0) + 1$ **b.** $\frac{1}{3}$: $f(1) = \frac{-2}{3}(1) + 1$

c. 0: $f\left(\frac{3}{2}\right) = \frac{-2}{3}\left(\frac{3}{2}\right) + 1$ **d.** $\frac{-2}{3}a + 1$: $f(a) = \frac{-2}{3}(a) + 1$

6 More complicated functional values can be determined by remembering the function process.

Example 1: If $f(x) = -2x + 3$, find $f(a + b)$.

Solution

The function process is: Multiply by -2 and add 3.

$$\begin{aligned} f(a + b) &= -2(a + b) + 3 \\ &= -2a - 2b + 3 \end{aligned}$$

Example 2: If $f(x) = x^2 - 2x + 3$, find $f(3a)$.

Solution

The function process is: Square the domain element, subtract twice the domain element, and add 3.

$$f(3a) = (3a)^2 - 2(3a) + 3$$
$$= 9a^2 - 6a + 3$$

Q17 If $f(x) = -3x + 5$, find:

a. $f(-4x)$

b. $f(a + b)$

STOP • STOP • STOP • STOP • STOP • STOP • STOP • STOP • STOP

A17 a. $12x + 5$: $f(-4x) = -3(-4x) + 5$

b. $-3a - 3b + 5$: $f(a + b) = -3(a + b) + 5$

Q18 If $f(x) = x^2 - x$, find:

a. $f(4)$

b. $f(a + b)$

STOP • STOP • STOP • STOP • STOP • STOP • STOP • STOP • STOP

A18 a. 12

b. $a^2 + 2ab + b^2 - a - b$:
$f(a + b) = (a + b)^2 - (a + b)$

Q19 If $f(x) = 3x - 1$, find:

a. $f(x + h)$

b. $f(3x^2)$

STOP • STOP • STOP • STOP • STOP • STOP • STOP • STOP • STOP

A19 a. $3x + 3h - 1$:
$f(x + h) = 3(x + h) - 1$

b. $9x^2 - 1$:
$f(3x^2) = 3(3x^2) - 1$

Q20 If $f(x) = 2x^2 - x + 5$, find:

a. $f(0)$

b. $f(2)$

c. $f(-2)$

d. $f(a + b)$

STOP • STOP • STOP • STOP • STOP • STOP • STOP • STOP • STOP

A20

a. 5:
$f(0) = 2(0)^2 - (0) + 5$

b. 11:
$f(2) = 2(2)^2 - (2) + 5$

c. 15:
$f(-2) = 2(-2)^2 - (-2) + 5$

d. $2a^2 + 4ab + 2b^2 - a - b + 5$:
$f(a + b) = 2(a + b)^2 - (a + b) + 5$

This completes the instruction for this section.

6.2 Exercises

1. Define a function.
2. Which of the relations also represent functions?
   a. $\{(1, 2), (2, 3), (3, 4), (4, 5)\}$
   b. $\{(0, 5), (1, 5), (2, 5), (3, 5)\}$
   c. $\{(5, 0), (5, 1), (5, 2), (5, 3)\}$
   d. $\{(2, 4), (1, 1), (2, -4), (1, -1)\}$
3. a. For what is the vertical-line test used?
   b. How is the vertical-line test used?
4. Which of the graphs of relations represent functions?

a.

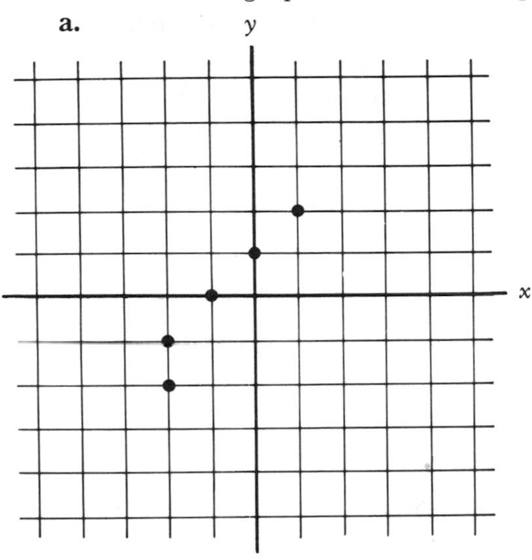

b.

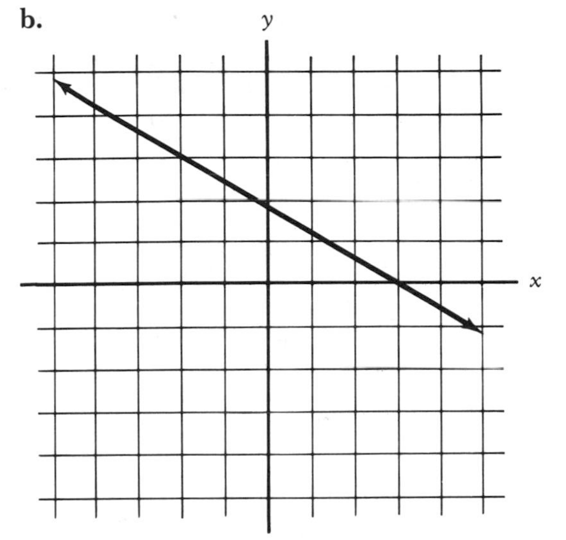

c.

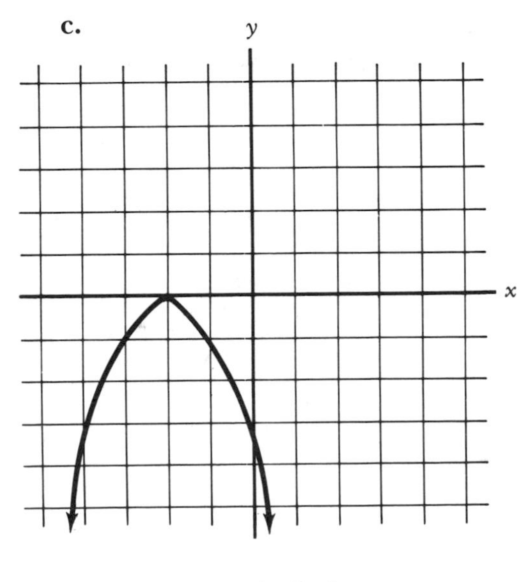

d.

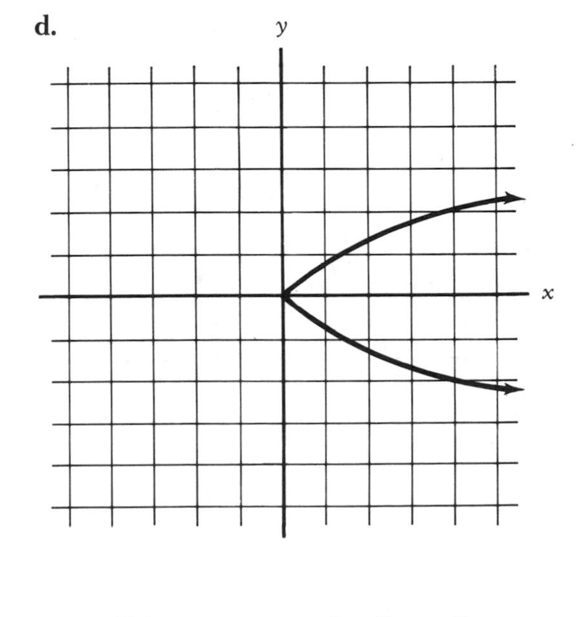

5. If $f(x) = 3x - 1$, find:
   a. $f(-4)$
   b. $f(0)$
   c. $f(a)$
   d. $f(a + b)$

6. If $f(x) = -2x + 5$, find:

   a. $f(0)$ b. $f\left(\frac{1}{2}\right)$ c. $f(2)$ d. $f(2x^2)$

7. If $g(x) = x^2 + 1$, find:
   a. $g(0)$ b. $g(-1)$ c. $g(1)$ d. $g(x + 2)$
8. If $f(x) = x^2 + 4x - 1$, find:
   a. $f(3)$ b. $f(x + h)$

9. If $f(x) = \frac{x + 1}{x - 1}$, find:

   a. $f(0)$ b. $f(1)$
10. If $g(x) = mx + b$, find:
    a. $g(0)$ b. $g(3)$

6.2 Exercise Answers

1. a relation in which no two ordered pairs have the same first coordinate
2. a and b
3. a. to determine whether the graph of a relation also represents a function
   b. if no vertical line intersects two points of the graph, the graph represents a function
4. b and c
5. a. -13 b. -1 c. $3a - 1$ d. $3a + 3b - 1$
6. a. 5 b. 4 c. 1 d. $-4x^2 + 5$
7. a. 1 b. 2 c. 2 d. $x^2 + 4x + 5$
8. a. 20 b. $x^2 + 2hx + h^2 + 4x + 4h - 1$
9. a. -1 b. undefined, because division by zero is impossible
10. a. b b. $3m + b$

6.3 Constant and Linear Functions: Slope

1

The function $f(x) = b$, where b is a constant, is called the *constant function* because each domain element "produces" the same range element, b. The function machine can be used to illustrate the constant function:

x (domain) $\longrightarrow$ [f] $\longrightarrow$ b (range)

If $f(x) = 4$, then

$f(1) = 4, \quad f(0) = 4, \quad f(-3) = 4, \quad f(\sqrt{7}) = 4, \quad$ etc.

Q1 If $f(x) = 5$, then:

a. $f(0) =$ ____ b. $f\left(\frac{5}{2}\right) =$ ____ c. $f(-3) =$ ____

STOP • STOP • STOP • STOP • STOP • STOP • STOP • STOP • STOP

A1 a. 5 b. 5 c. 5

Q2 If $f(x) = -8$, then:

a. $f(-2) =$ ____ b. $f(-1) =$ ____ c. $f(0) =$ ____

STOP • STOP • STOP • STOP • STOP • STOP • STOP • STOP • STOP

A2 a. -8 b. -8 c. -8

---

2 Ordered pairs can be generated from function notation by use of x to represent an element of the domain and y an element of the range. Consider the following function machine:

$$x \longrightarrow \boxed{f} \longrightarrow y$$

$$f = \{(x, y) | y = f(x)\}$$

The function f is equal to the set of ordered pairs (x, y) such that y is a function of x, which means that x is placed into the function f and y is "produced."

The statements on the left are represented as ordered pairs on the right.

$f(2) = 6$	$(2, 6)$
$f(-1) = 7$	$(-1, 7)$
$f(0) = 6$	$(0, 6)$
$f(a) = b$	(a, b)
$f(x) = y$	(x, y)

Q3 Determine the function value and express the function notation as an ordered pair. If $f(x) = 3$:

	Ordered pair
a. $f(-1) =$ ____	____
b. $f(0) =$ ____	____
c. $f(1) =$ ____	____
d. $f\left(\frac{1}{2}\right) =$ ____	____
e. $f(\sqrt{3}) =$ ____	____
f. $f(a) =$ ____, $a \in R$	____

STOP • STOP • STOP • STOP • STOP • STOP • STOP • STOP • STOP

A3 a. 3, $(-1, 3)$ b. 3, $(0, 3)$ c. 3, $(1, 3)$ d. 3, $\left(\frac{1}{2}, 3\right)$

e. 3, $(\sqrt{3}, 3)$ f. 3, $(a, 3)$

---

3 A function can be "pictured" by graphing all possible ordered pairs generated from the domain. The domain will be assumed to be the real numbers unless otherwise specified.

Example: Graph $f(x) = 3$.

Solution

Since the domain is the set of real numbers, an infinite number of ordered pairs are generated. A few ordered pairs (in table form on the left) are graphed below.

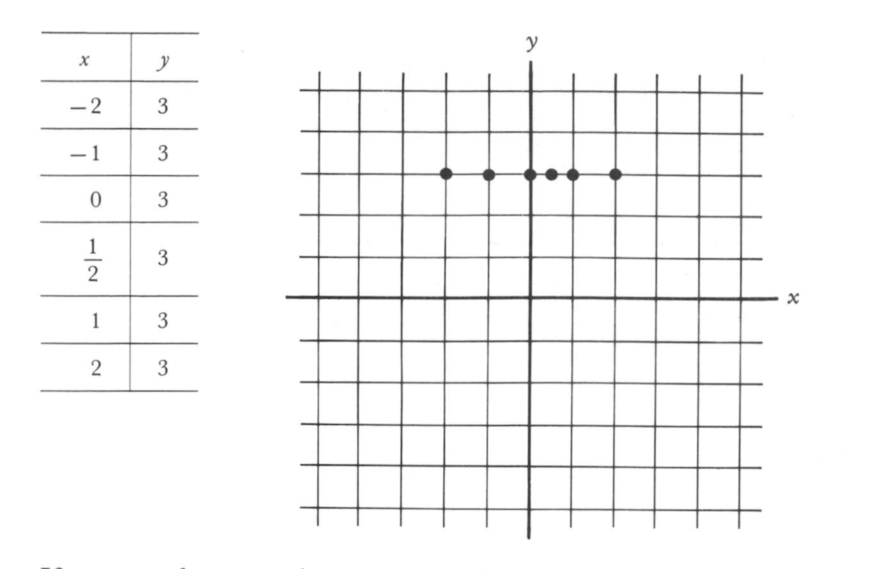

x	y
-2	3
-1	3
0	3
$\frac{1}{2}$	3
1	3
2	3

If more and more points were graphed, the dotted line would appear to form a solid line. The graph of $f(x) = 3$ is formed by connecting the graphed points with a solid line, as shown. In general, the graph of $f(x) = b$ is a straight line parallel to the x axis.

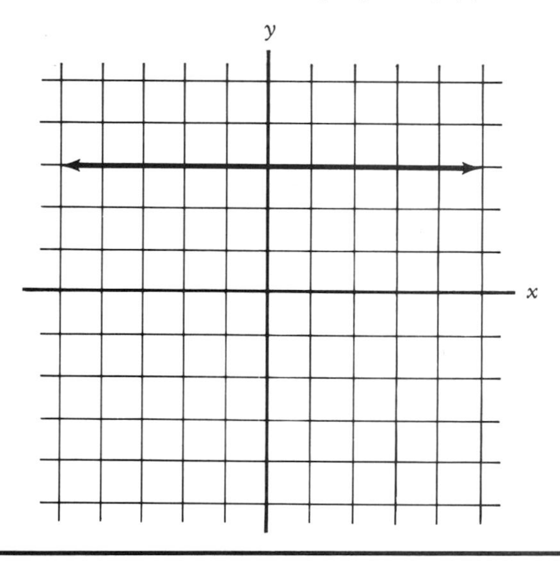

Q4 (1) Complete the table of values, plotting each ordered pair and (2) graph each function:

a. $f(x) = -3$

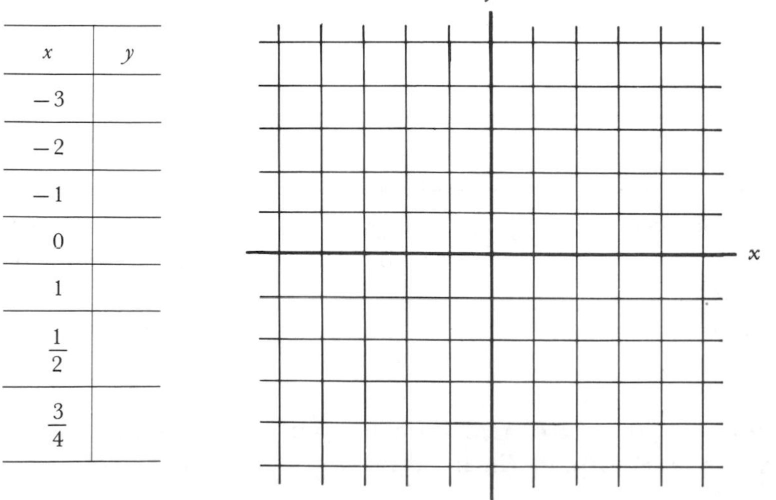

x	y
-3	
-2	
-1	
0	
1	
$\frac{1}{2}$	
$\frac{3}{4}$	

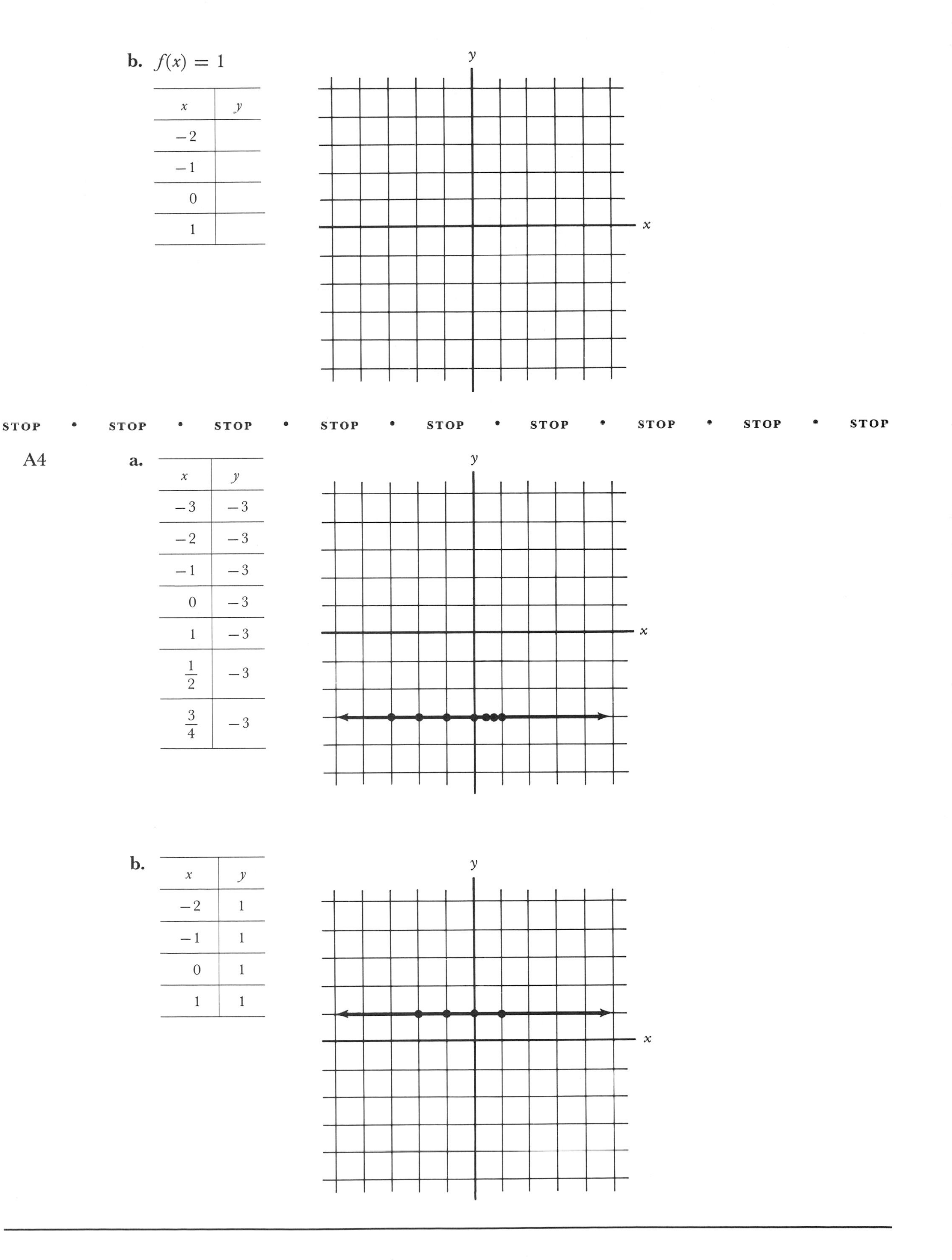

b. $f(x) = 1$

x	y
-2	
-1	
0	
1	

STOP • STOP • STOP • STOP • STOP • STOP • STOP • STOP • STOP

A4 **a.**

x	y
-3	-3
-2	-3
-1	-3
0	-3
1	-3
$\frac{1}{2}$	-3
$\frac{3}{4}$	-3

b.

x	y
-2	1
-1	1
0	1
1	1

Q5 Graph:

a. $f(x) = \frac{5}{2}$ **b.** $f(x) = -2$

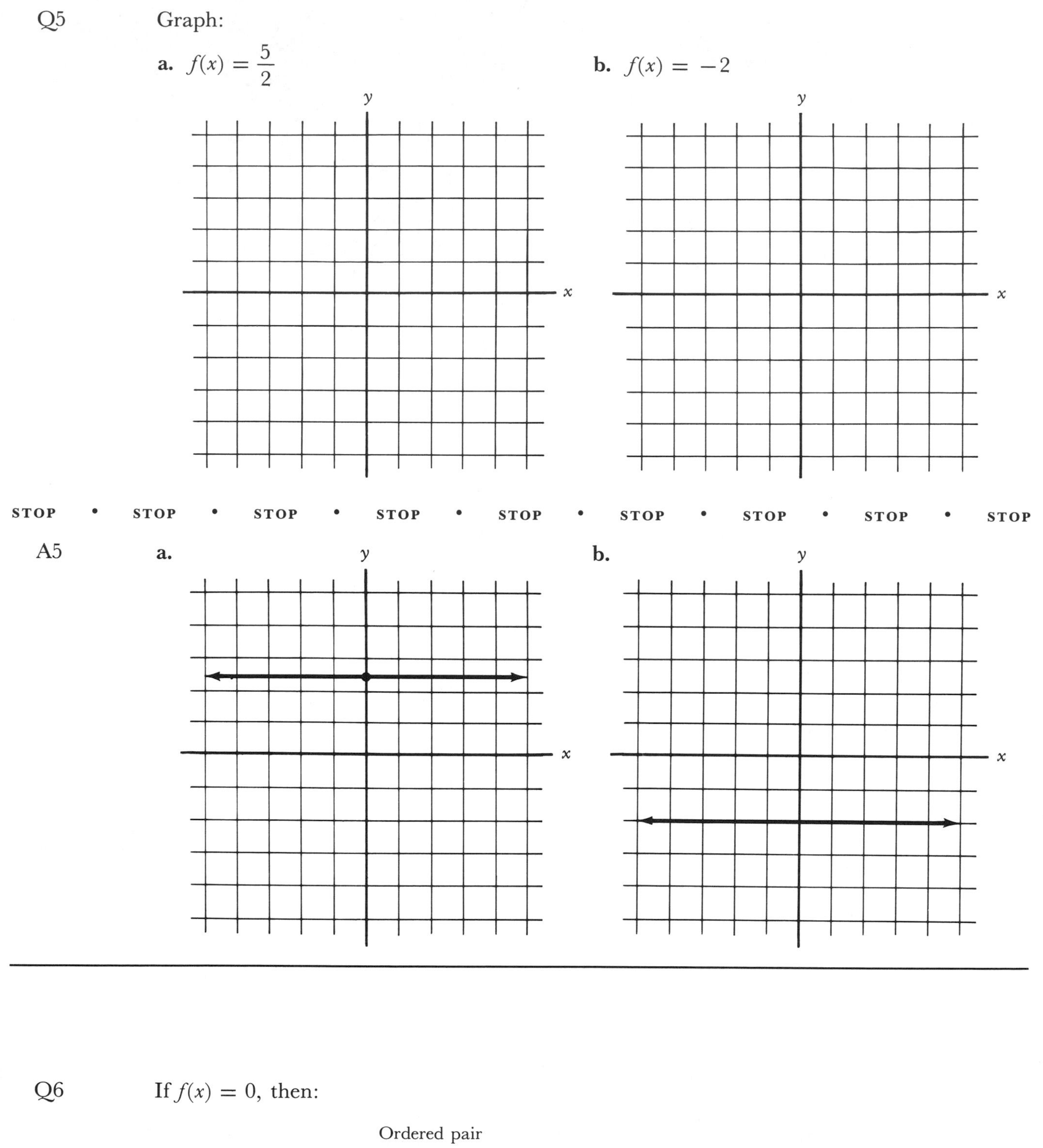

STOP • STOP • STOP • STOP • STOP • STOP • STOP • STOP • STOP

A5 **a.** **b.**

Q6 If $f(x) = 0$, then:

		Ordered pair
a. $f(-1)$	= _____	________
b. $f(0)$	= _____	________
c. $f(2)$	= _____	________

STOP • STOP • STOP • STOP • STOP • STOP • STOP • STOP • STOP

A6 **a.** 0, $(-1, 0)$ **b.** 0, $(0, 0)$ **c.** 0, $(2, 0)$

Q7 Graph $f(x) = 0$.

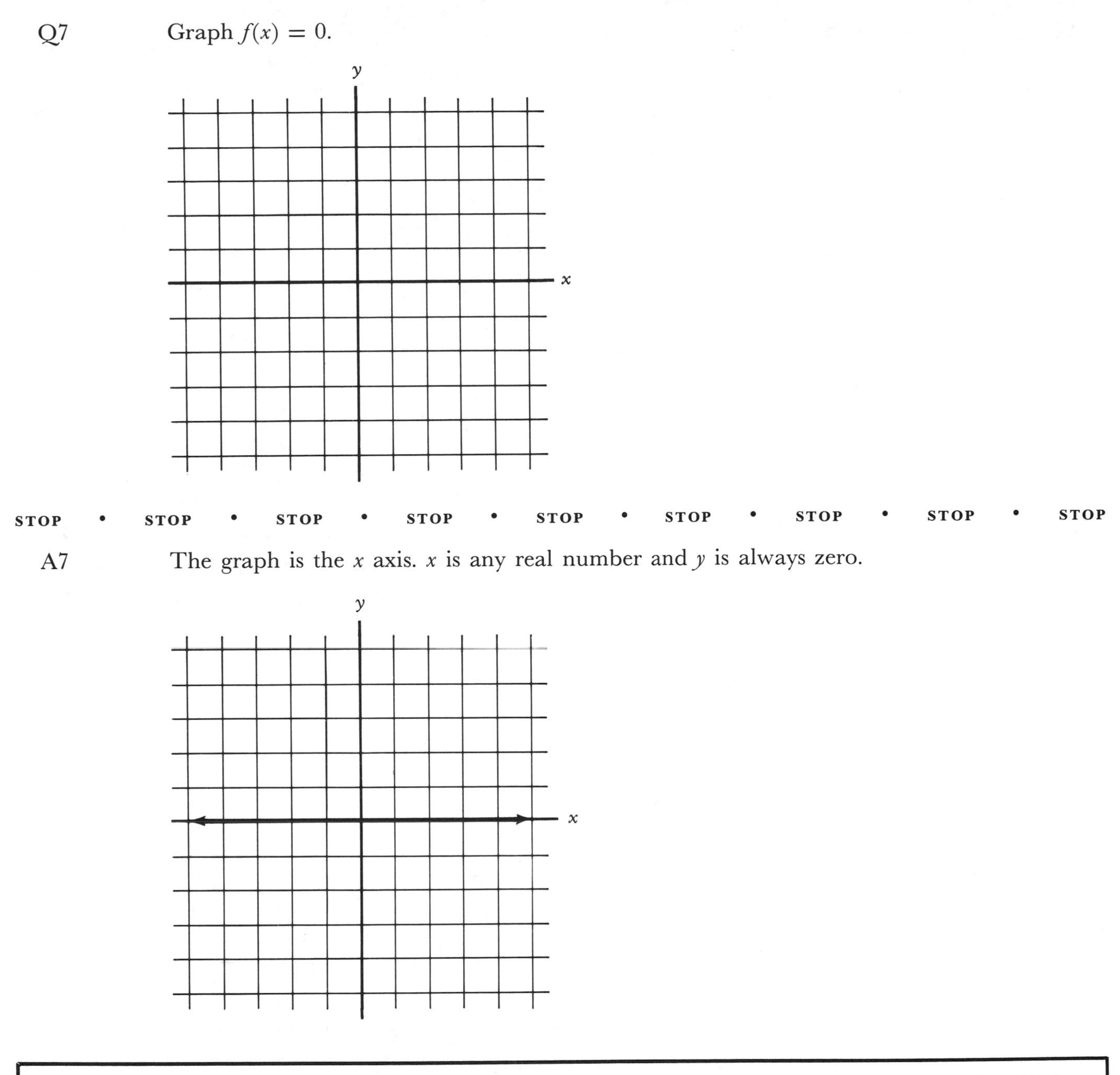

STOP • STOP • STOP • STOP • STOP • STOP • STOP • STOP • STOP

A7 The graph is the x axis. x is any real number and y is always zero.

---

4 A function of the form $f(x) = mx + b$, $m \neq 0$, where m and b are constants, is called a *linear function*. It is called that because its graph is always a straight line.

Example: Given $f(x) = -2x + 5$, determine the function values $f(0)$ and $f(-1)$ and represent the function notation as ordered pairs.

Solution

	Ordered pair
$f(0) = -2(0) + 5$ $= 5$	$(0, 5)$
$f(-1) = -2(-1) + 5$ $= 7$	$(-1, 7)$

Q8 If $f(x) = 3x - 2$, then:

		Ordered pair
a. $f(-1)$	= _____	________
b. $f(0)$	= _____	________
c. $f\left(\frac{1}{2}\right)$	= _____	________
d. $f(1)$	= _____	________
e. $f(2)$	= _____	________

STOP • STOP • STOP • STOP • STOP • STOP • STOP • STOP • STOP

A8 **a.** $-5, (-1, -5)$ **b.** $-2, (0, -2)$ **c.** $\frac{-1}{2}, \left(\frac{1}{2}, \frac{-1}{2}\right)$ **d.** $1, (1, 1)$

e. $4, (2, 4)$

---

5 The linear function $f(x) = mx + b$ is graphed in the same manner that the constant function $f(x) = b$ was graphed. That is, all possible ordered pairs generated from the function are graphed on the coordinate system.

Example: Graph $f(x) = -2x + 1$.

Solution

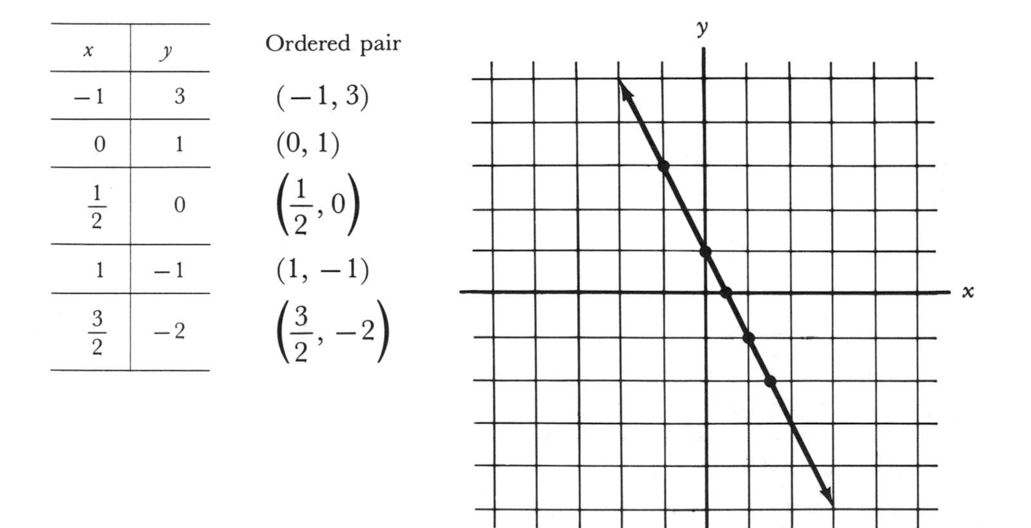

x	y	Ordered pair
-1	3	$(-1, 3)$
0	1	$(0, 1)$
$\frac{1}{2}$	0	$\left(\frac{1}{2}, 0\right)$
1	-1	$(1, -1)$
$\frac{3}{2}$	-2	$\left(\frac{3}{2}, -2\right)$

The function is graphed by connecting the plotted points with a straight line. Even though a straight line is determined by only two points, it will be useful to plot three points when graphing linear functions (as a check).

Q9 Complete the table and graph $f(x) = 3x - 2$.

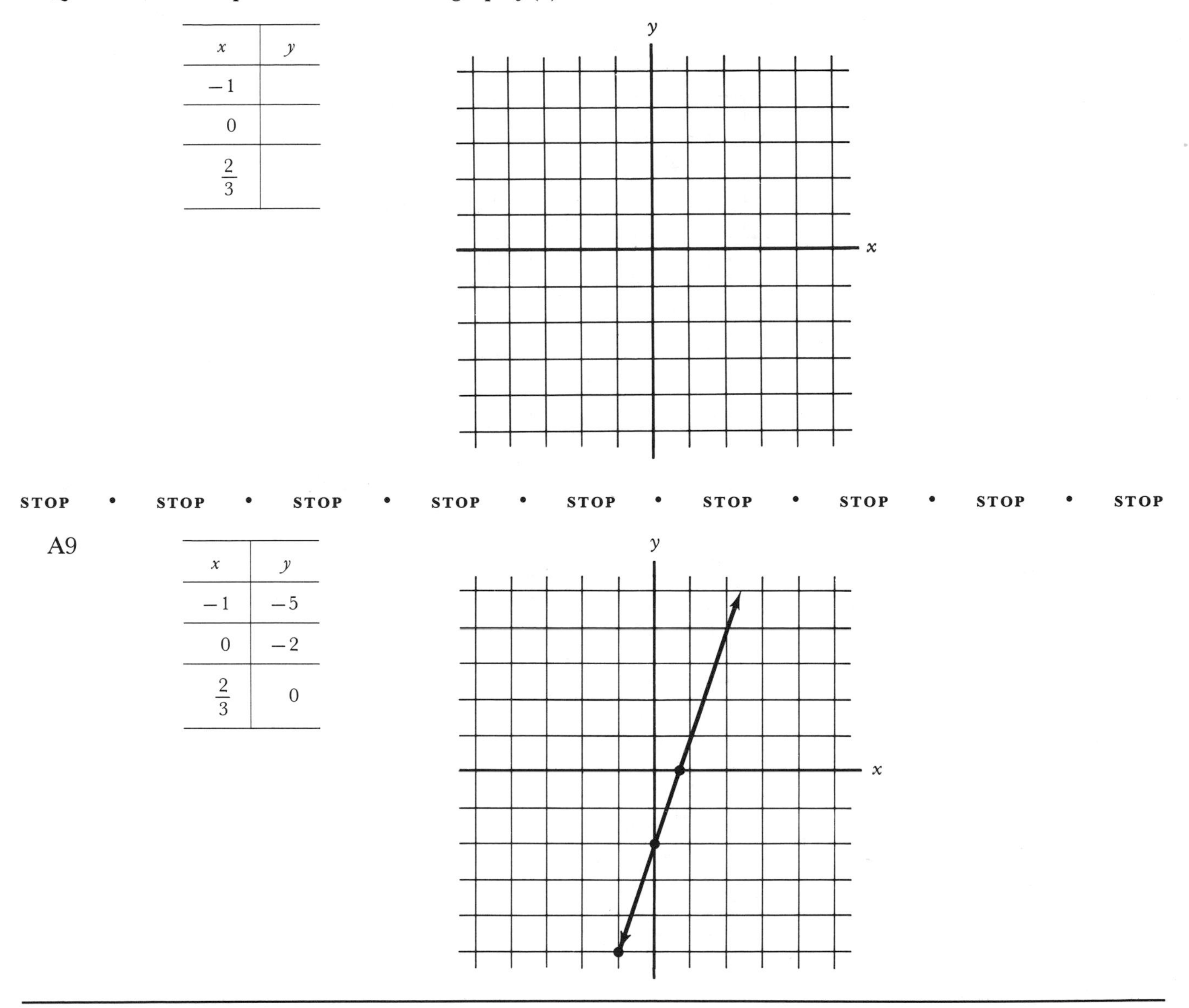

x	y
-1	
0	
$\frac{2}{3}$	

STOP • STOP • STOP • STOP • STOP • STOP • STOP • STOP • STOP

A9

x	y
-1	-5
0	-2
$\frac{2}{3}$	0

Q10 Complete the table and graph:

a. $f(x) = \frac{2}{3}x - 1$

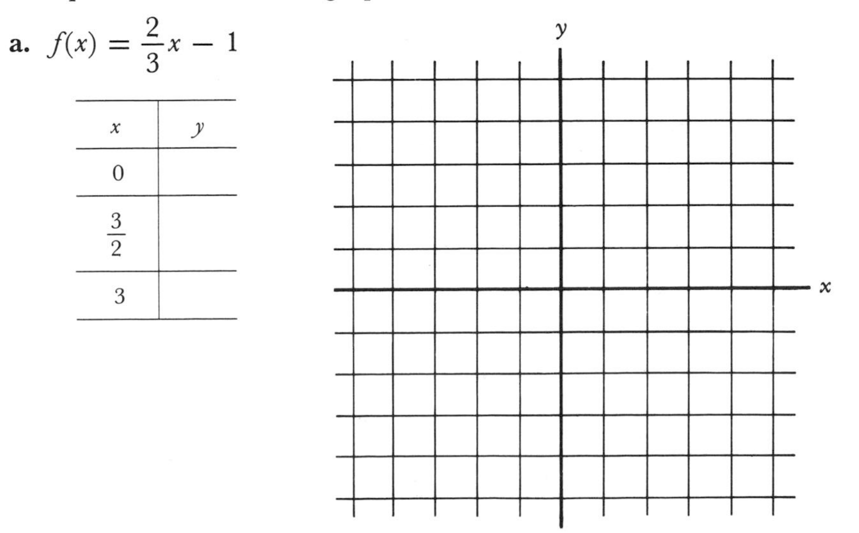

x	y
0	
$\frac{3}{2}$	
3	

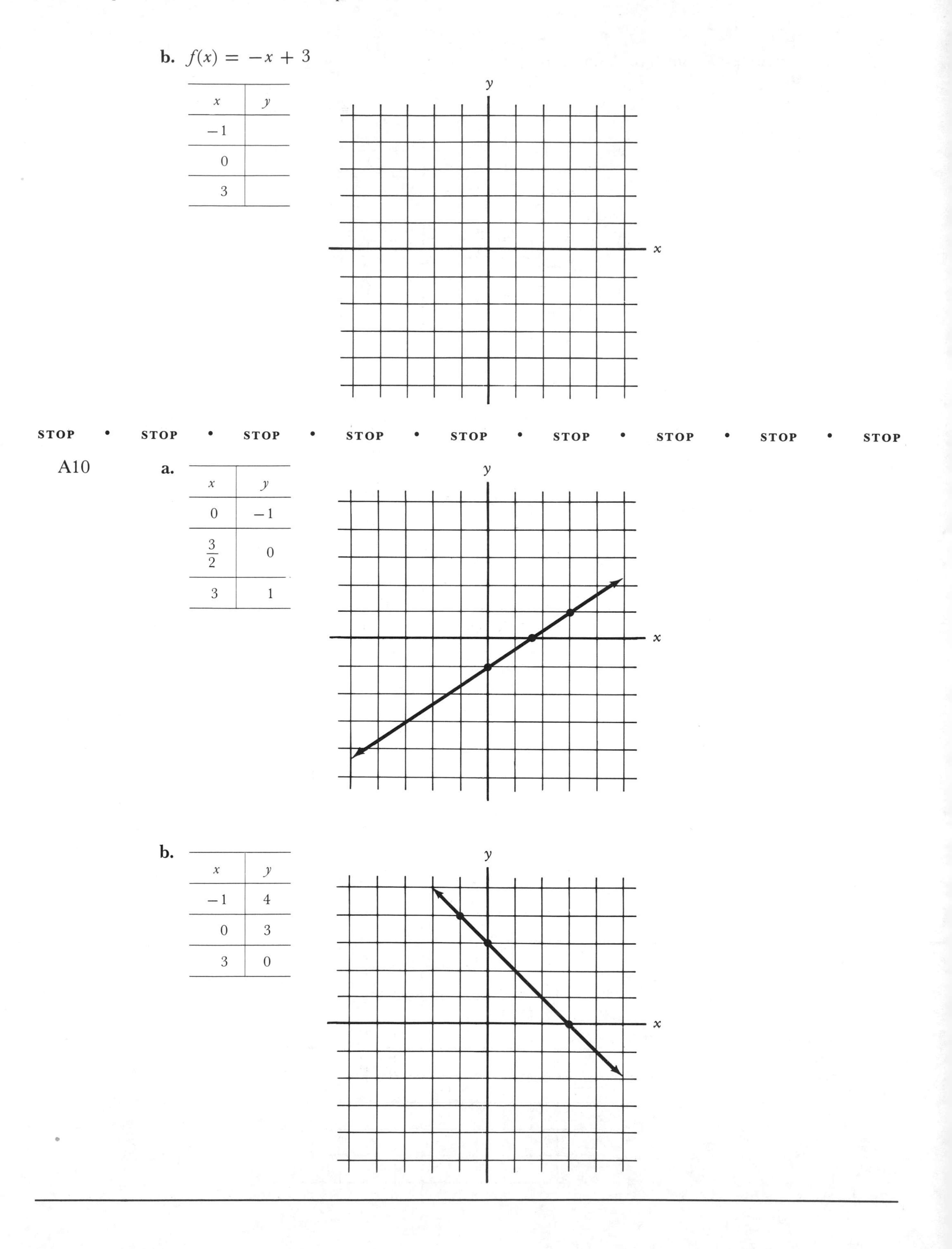

b. $f(x) = -x + 3$

x	y
-1	
0	
3	

STOP • STOP • STOP • STOP • STOP • STOP • STOP • STOP • STOP

A10 **a.**

x	y
0	-1
$\frac{3}{2}$	0
3	1

b.

x	y
-1	4
0	3
3	0

Q11 Complete the table and graph:

a. $f(x) = \frac{1}{2}x - 2$

x	y
-2	
0	
2	

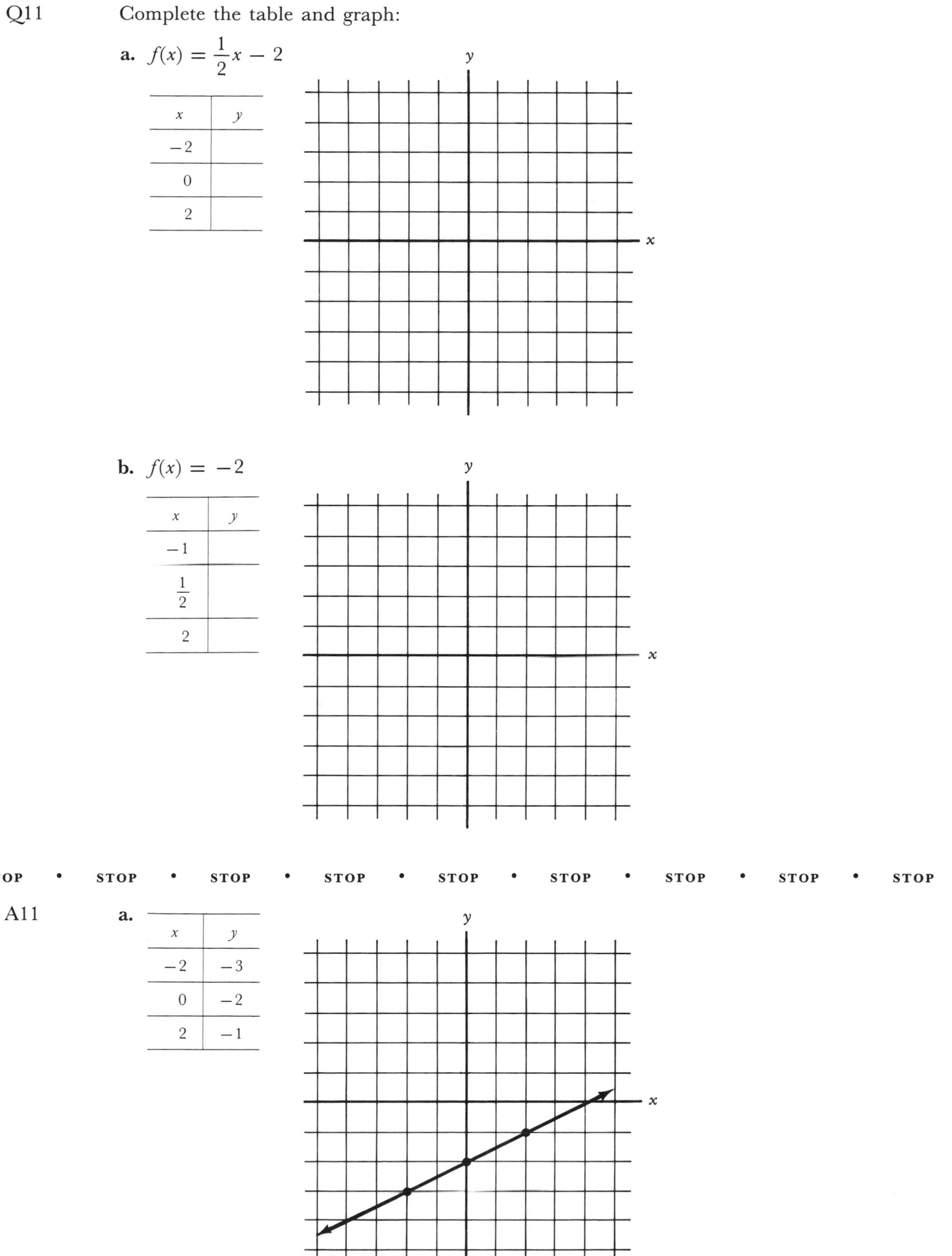

b. $f(x) = -2$

x	y
-1	
$\frac{1}{2}$	
2	

STOP • STOP • STOP • STOP • STOP • STOP • STOP • STOP • STOP

A11 **a.**

x	y
-2	-3
0	-2
2	-1

b.

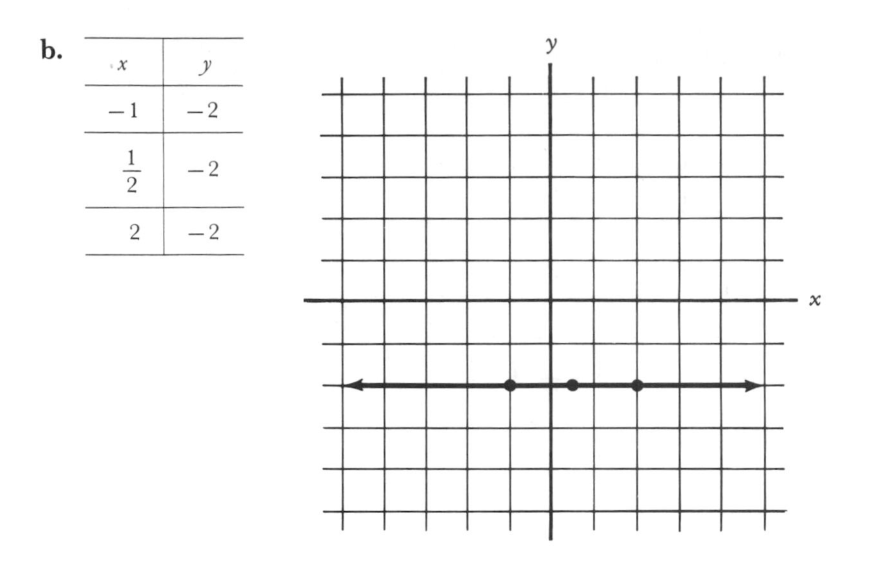

x	y
-1	-2
$\frac{1}{2}$	-2
2	-2

6 The value of x chosen to produce the corresponding y value is completely arbitrary. The value zero was often chosen because it is easy to compute with zero. Special techniques will be developed in Section 6.4 to aid in graphing linear functions.

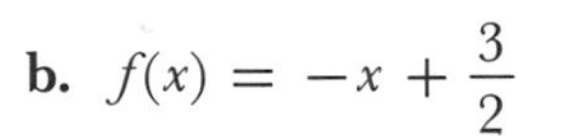

Q12 Graph, plotting at least three ordered pairs:

a. $f(x) = 3x - 1$ **b.** $f(x) = -x + \frac{3}{2}$

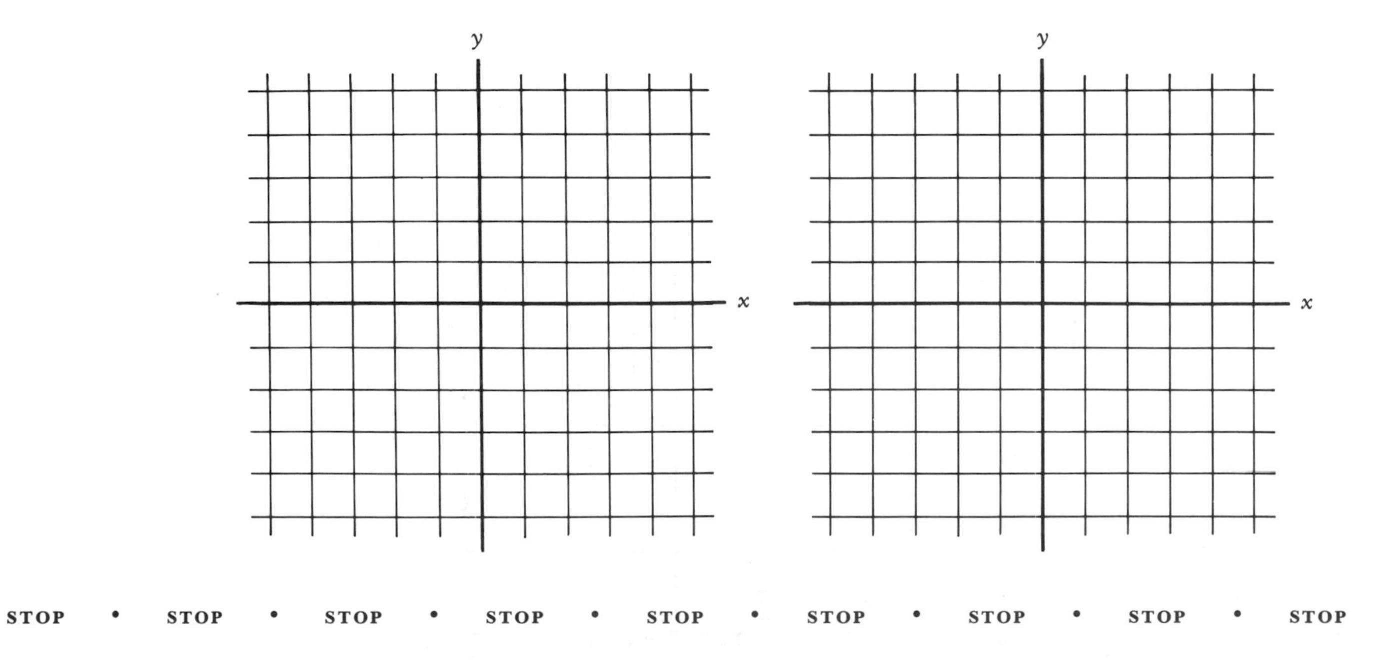

STOP • STOP • STOP • STOP • STOP • STOP • STOP • STOP • STOP

A12 **a.** **b.**

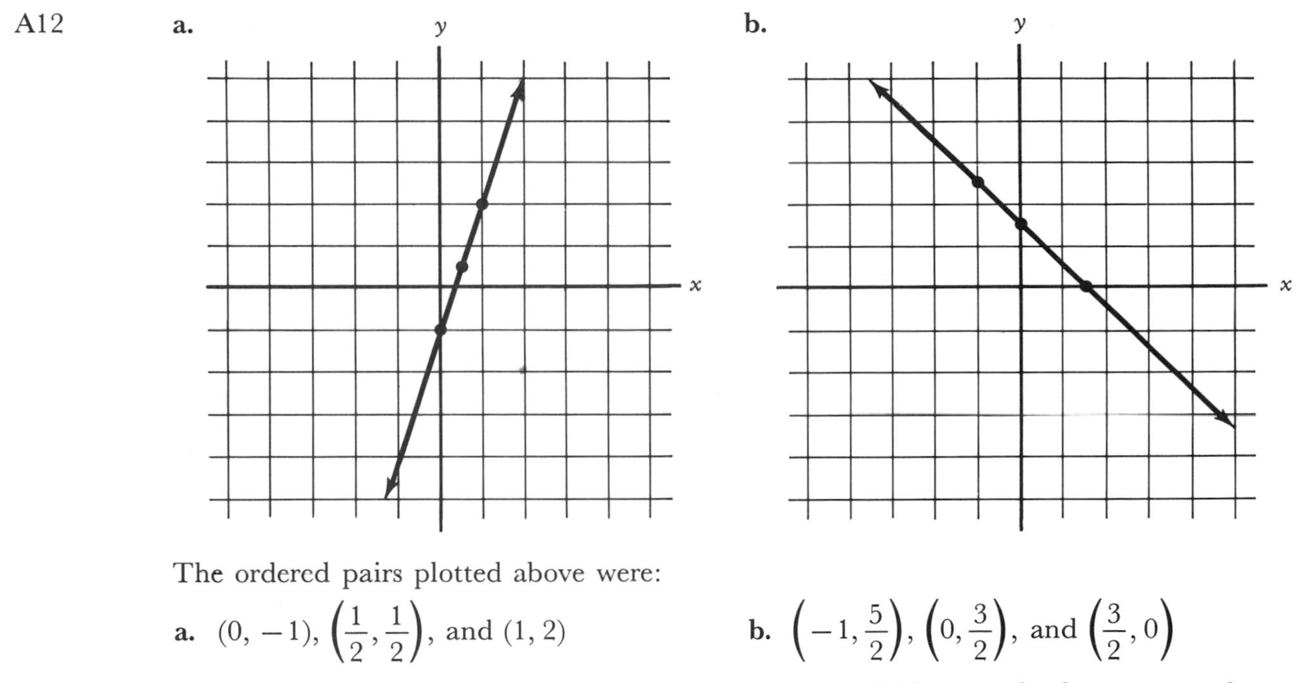

The ordered pairs plotted above were:

a. $(0, -1)$, $\left(\frac{1}{2}, \frac{1}{2}\right)$, and $(1, 2)$ **b.** $\left(-1, \frac{5}{2}\right)$, $\left(0, \frac{3}{2}\right)$, and $\left(\frac{3}{2}, 0\right)$

Regardless of the ordered pairs chosen, the graphs should be exactly the same as above.

7 The *slope* of a line will be discussed in this section and used in Section 6.4 as an aid in graphing lines. The slope of a line can be visualized in the same manner as one would visualize the slope of the surrounding landscape. That is, a hill is said to have a steep slope while flatland has little slope. The slope of a line is computed between any two points of the line and is defined as the "rise" divided by the "run" between the two points.

Example: Find the slope of the line containing the points $(3, 3)$ and $(1, -1)$.

Solution

Plot the points and compute:

$$\text{slope} = \frac{\text{rise}}{\text{run}}$$
$$= \frac{4}{2}$$
$$= 2$$

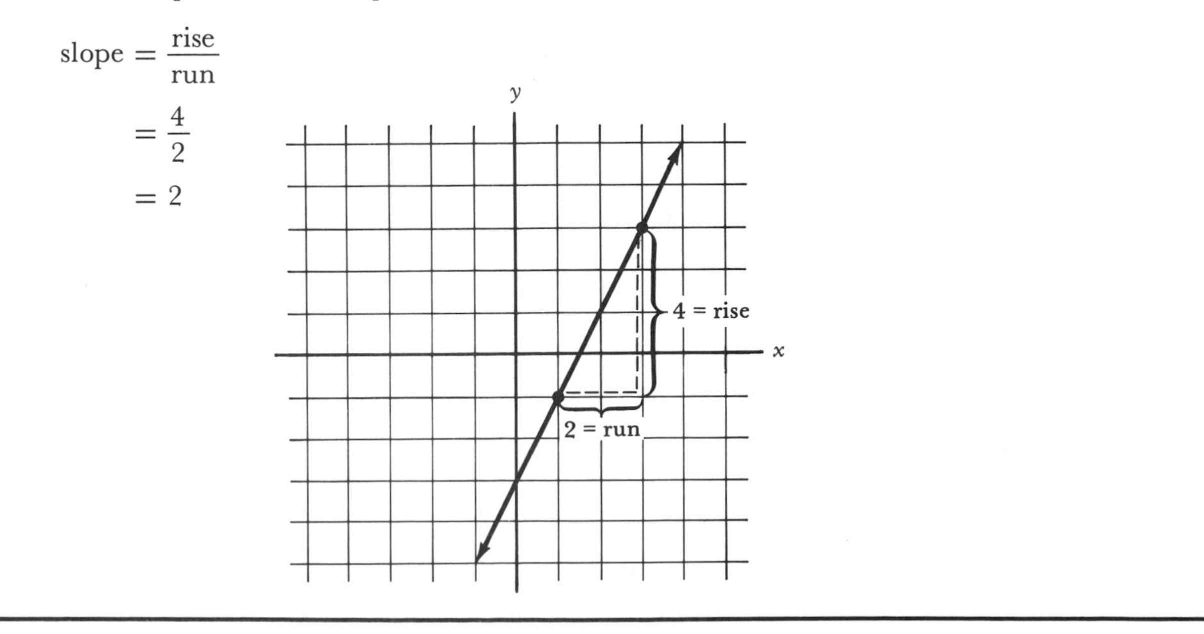

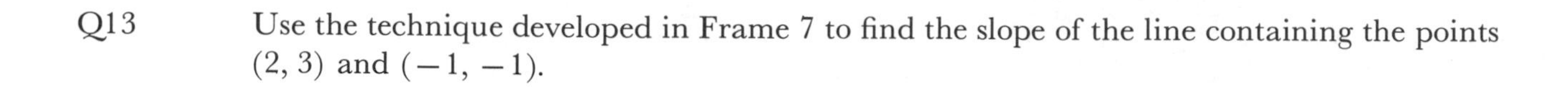

Q13 Use the technique developed in Frame 7 to find the slope of the line containing the points $(2, 3)$ and $(-1, -1)$.

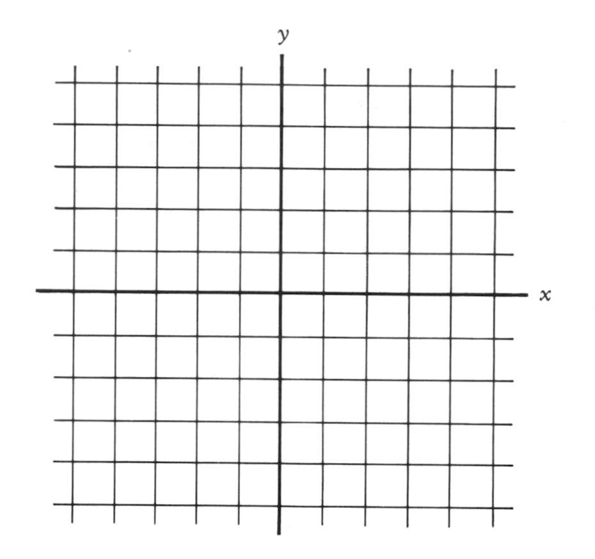

STOP • STOP • STOP • STOP • STOP • STOP • STOP • STOP • STOP

A13 $\frac{4}{3}$:

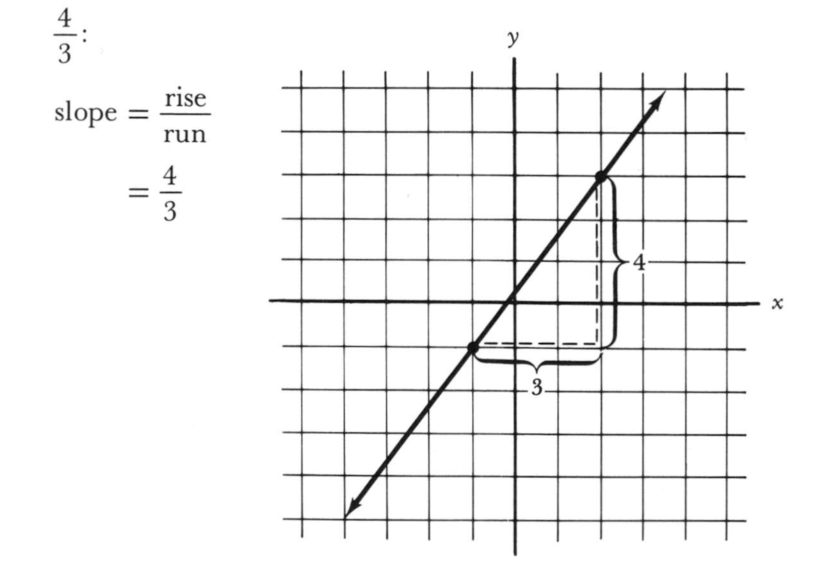

$$\text{slope} = \frac{\text{rise}}{\text{run}}$$
$$= \frac{4}{3}$$

8	A line can have a negative slope if either the rise or the run is negative. Consider the slope of the line containing the points $(-3, 2)$ and $(1, -3)$.

$$\text{slope} = \frac{\text{rise}}{\text{run}} = \frac{5}{-4} = \frac{-5}{4}$$

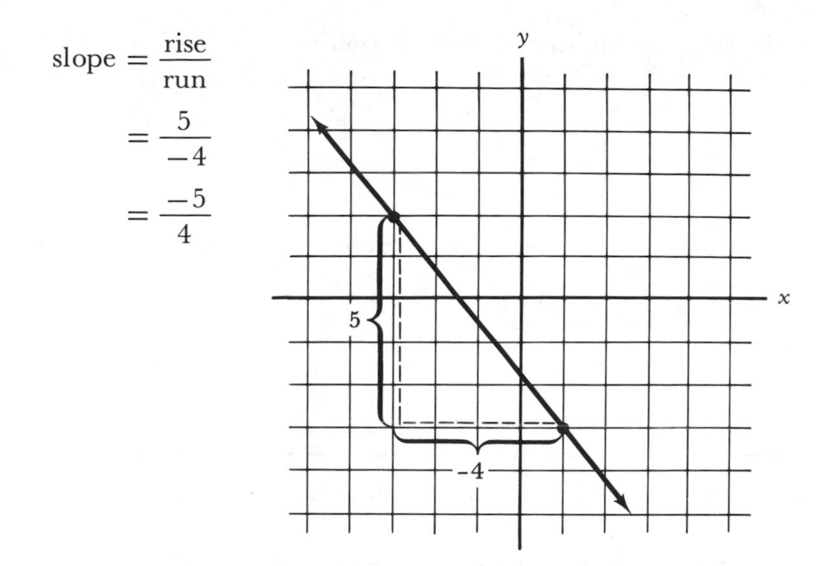

The rise is usually the difference between the y coordinates, and the run is the difference between the x coordinates. That is, to find the slope of the line containing the points (x_1,y_1) and (x_2,y_2), find:

$$\text{slope} = \frac{\text{rise}}{\text{run}} = \frac{y_1 - y_2}{x_1 - x_2} \quad \text{or} \quad \frac{y_2 - y_1}{x_2 - x_1}$$

Example: Find the slope of the line containing the points $(-3, 2)$ and $(1, -3)$.

Solution

$$\text{slope} = \frac{y_1 - y_2}{x_1 - x_2} = \frac{2 - (-3)}{-3 - 1} = \frac{5}{-4} = \frac{-5}{4} \quad \text{or} \quad \text{slope} = \frac{y_2 - y_1}{x_2 - x_1} = \frac{-3 - 2}{1 - (-3)} = \frac{-5}{4}$$

The slope will be the same regardless of which ordered pair is used first in the subtraction; however, the order of the subtraction must be the same for both x and y.

Q14 Find the slope of the line containing the points $(-3, 2)$ and $(5, -3)$ by completing these steps:

$$\text{slope} = \frac{y_1 - y_2}{x_1 - x_2} = \frac{(\quad) - (\quad)}{(\quad) - (\quad)} =$$

STOP • STOP • STOP • STOP • STOP • STOP • STOP • STOP • STOP

A14 $\frac{-5}{8}$:

$$\text{slope} = \frac{(2) - (-3)}{(-3) - (5)} = \frac{5}{-8} = \frac{-5}{8} \quad \text{or} \quad \text{slope} = \frac{(-3) - (2)}{(5) - (-3)} = \frac{-5}{8}$$

Q15 Find the slope of the line containing the pairs of points:
a. (3, 2) and (1, −8) **b.** (−2, 3) and (2, −4)

STOP • STOP • STOP • STOP • STOP • STOP • STOP • STOP • STOP

A15 **a.** 5: slope $= \dfrac{y_1 - y_2}{x_1 - x_2} = \dfrac{2 - (-8)}{3 - 1} = \dfrac{10}{2}$

b. $\dfrac{-7}{4}$

9 A *horizontal* line has a slope of zero and a *vertical* line has no slope.

Example 1: Find the slope of line l containing (−2, 3) and (1, 3).

Solution

$$\text{slope} = \frac{\text{rise}}{\text{run}} = \frac{y_1 - y_2}{x_1 - x_2} = \frac{3 - 3}{-2 - 1} = \frac{0}{-3} = 0$$

Hence l has zero slope.

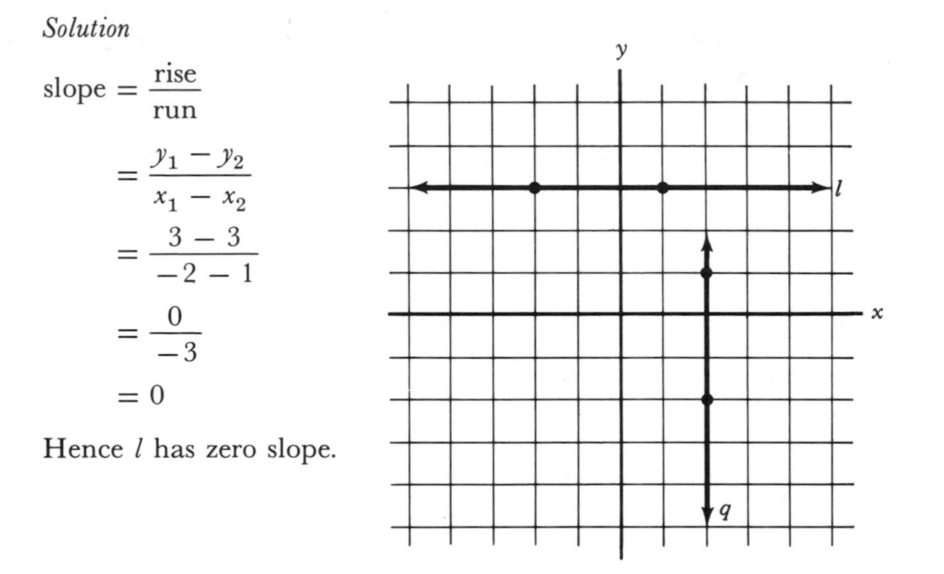

Example 2: Find the slope of line q containing (2, 1) and (2, −2).

Solution

$$\text{slope} = \frac{1 - (-2)}{2 - 2} = \frac{3}{0} \text{ (undefined)}$$

Hence q has no slope.

Q16 Find the slope of the line containing the pairs of points:
a. (−5, 4) and (1, 4) **b.** (−1, 3) and (−1, −2)

STOP • STOP • STOP • STOP • STOP • STOP • STOP • STOP • STOP

A16 **a.** 0: slope $= \dfrac{4 - 4}{-5 - 1} = \dfrac{0}{-6}$ **b.** no slope: slope $= \dfrac{3 - (-2)}{-1 - (-1)} = \dfrac{5}{0}$

Q17 Find the slope of the line containing the pairs of points:

a. $(-3, 3)$ and $(3, -2)$ **b.** $(-4, 8)$ and $(-4, 0)$ **c.** $(-3, -2)$ and $(5, -2)$

STOP • STOP • STOP • STOP • STOP • STOP • STOP • STOP • STOP

A17 **a.** $\frac{-5}{6}$ **b.** no slope **c.** 0

This completes the instruction for this section.

6.3 Exercises

1. Define:
   a. constant function **b.** linear function
2. If $f(x) = 17$, find (also write as ordered pairs):
   a. $f(-2)$ **b.** $f(1)$ **c.** $f\left(\frac{1}{2}\right)$ **d.** $f(\sqrt{3})$
   e. $f(a)$, $a \in R$
3. If $f(x) = -2$, find (also write as ordered pairs):
   a. $f(-4)$ **b.** $f(15)$ **c.** $f(x)$, $x \in R$ **d.** $f(0)$
   e. $f(100)$
4. If $f(x) = \frac{2}{3}x - 4$, find (also write as ordered pairs):
   a. $f(-1)$ **b.** $f(0)$ **c.** $f\left(\frac{3}{2}\right)$ **d.** $f(6)$
5. If $f(x) = -2x + 7$, find (also write as ordered pairs):
   a. $f(-2)$ **b.** $f\left(\frac{-1}{2}\right)$ **c.** $f(0)$ **d.** $f\left(\frac{7}{2}\right)$
6. Graph:
   a. $f(x) = -2$ **b.** $f(x) = 3$ **c.** $f(x) = \frac{5}{2}$ **d.** $f(x) = 0$
7. Graph:
   a. $f(x) = \frac{2}{3}x - 4$ **b.** $f(x) = -2x + 3$
   c. $f(x) = -x + 2$ **d.** $f(x) = -x - 3$
8. Find the slope of the line containing the pairs of points:
   a. $(3, 2)$ and $(1, -8)$ **b.** $(2, 5)$ and $(-4, 3)$
   c. $(5, 2)$ and $(-3, 7)$ **d.** $(-3, 6)$ and $(9, 11)$
   e. $(-3, 2)$ and $(-1, 2)$ **f.** $(-3, 5)$ and $(-3, 1)$

9. Do the following graphs of lines have positive slope, negative slope, zero slope, or no slope?

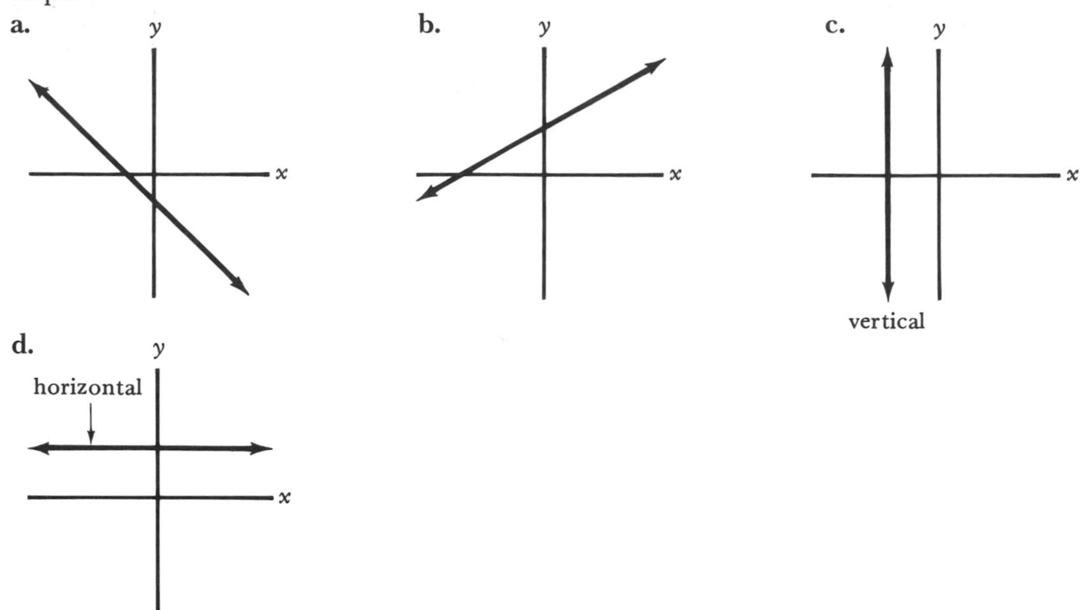

6.3 Exercise Answers

1. a. $f(x) = b$, where b is a constant.
   b. $f(x) = mx + b$, $m \neq 0$, m and b constants.
2. a. 17, $(-2, 17)$ b. 17, $(1, 17)$ c. 17, $\left(\frac{1}{2}, 17\right)$ d. 17, $(\sqrt{3}, 17)$
   e. 17, (a, 17)
3. a. -2, $(-4, -2)$ b. -2, $(15, -2)$ c. -2, $(x, -2)$ d. -2, $(0, -2)$
   e. -2, $(100, -2)$
4. a. $\frac{-14}{3}$, $\left(-1, \frac{-14}{3}\right)$ b. -4, $(0, -4)$ c. -3, $\left(\frac{3}{2}, -3\right)$
   d. 0, $(6, 0)$
5. a. 11, $(-2, 11)$ b. 8, $\left(\frac{-1}{2}, 8\right)$ c. 7, $(0, 7)$ d. 0, $\left(\frac{7}{2}, 0\right)$
6. a. b.

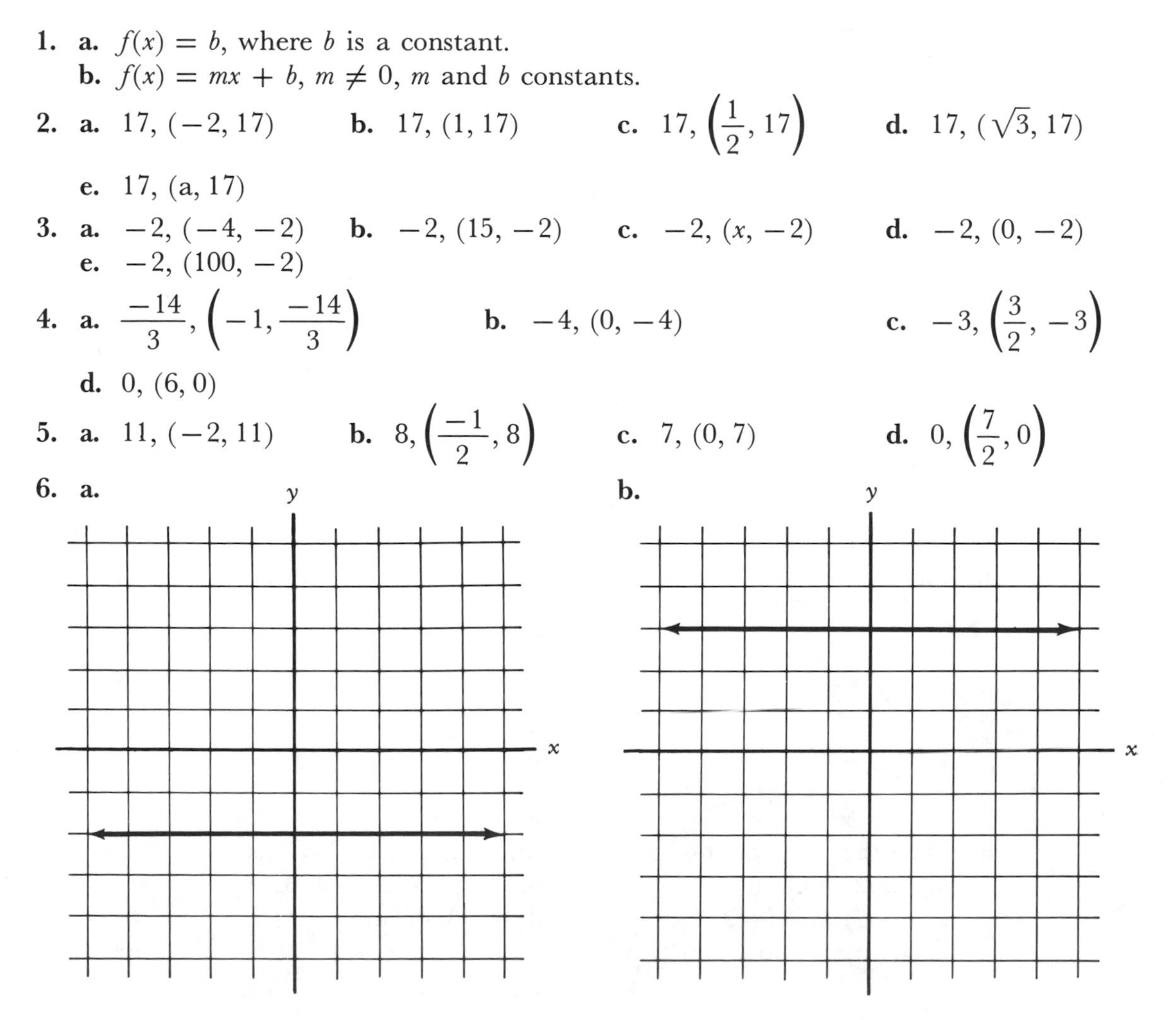

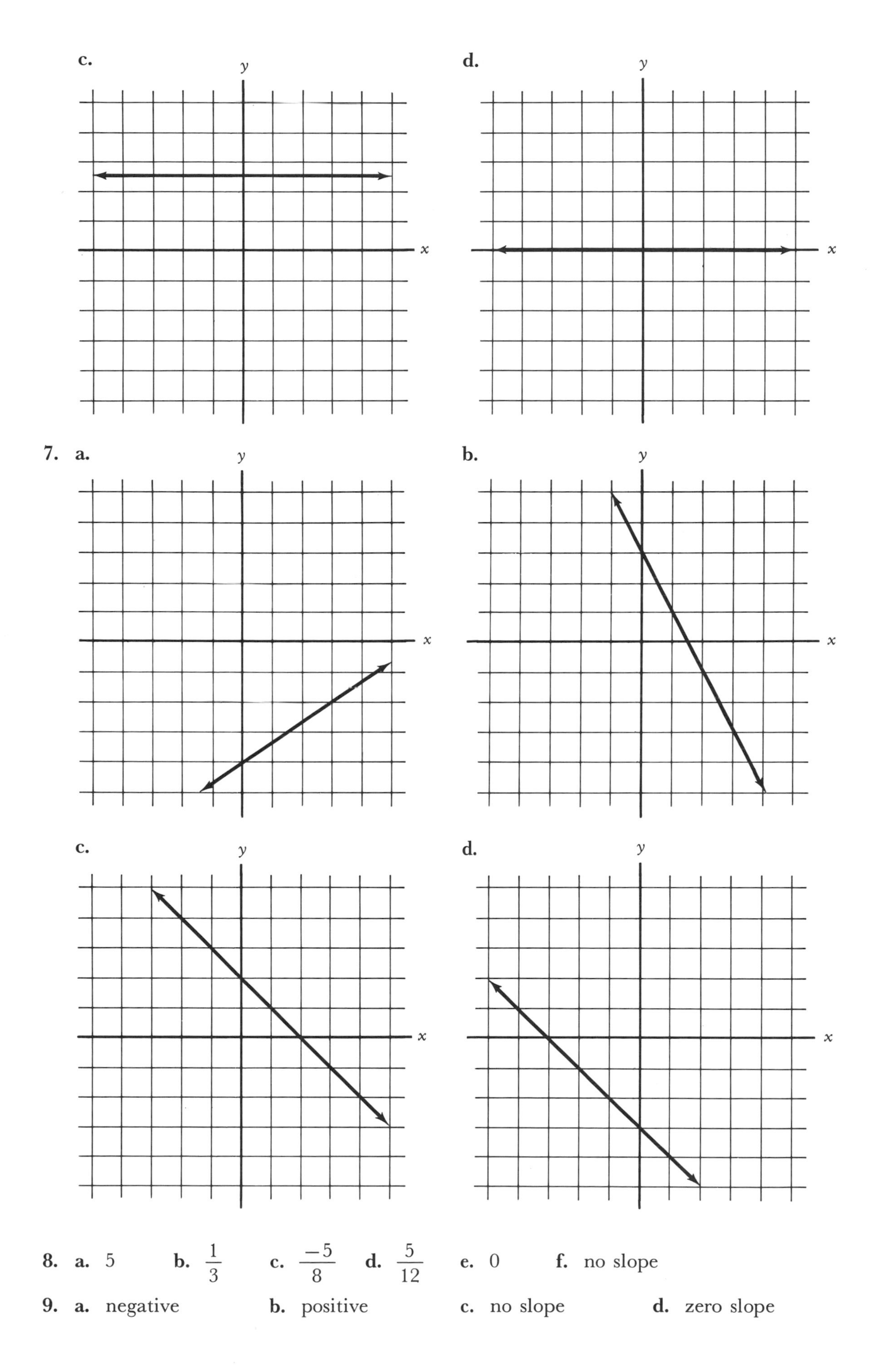

8. **a.** 5 **b.** $\frac{1}{3}$ **c.** $\frac{-5}{8}$ **d.** $\frac{5}{12}$ **e.** 0 **f.** no slope

9. **a.** negative **b.** positive **c.** no slope **d.** zero slope

6.4 Equations of the Line

1

The function $f(x) = mx + b$ was called a linear function because its graph was a straight line. Equations whose graphs are straight lines are called *linear equations* and are often expressed as $\{(x, y)|Ax + By + C = 0$, where A, B, and C are constants$\}$. The set notation is usually assumed; hence linear equations in *standard form* are written $Ax + By + C = 0$. Since the polynomial $Ax + By + C$ is of degree 1, linear equations are also called *first-degree equations;* conversely, all first-degree equations are linear equations.

Example 1: Which of the following represent linear equations (first-degree equations)?

a. $7x - 3y = 10$ **b.** $2x^2 - 1 = y$ **c.** $\frac{1}{x} + y = 3x$

Solution

a does, because each of its terms is of degree 1 or 0. In **b,** $2x^2$ is of second degree; in **c,** $\frac{1}{x}$ would be written x^{-1}, hence is not of degree 1.

Example 2: Write $-2y = 7x + 3$ in standard form.

Solution

$$-2y = 7x + 3$$
$$0 = 7x + 2y + 3$$
$$7x + 2y + 3 = 0$$

or

$$-2y = 7x + 3$$
$$-7x - 2y - 3 = 0$$

Each equation is of the form $Ax + By + C = 0$. It can be shown that the equations are equivalent by multiplying both sides of either by -1. That is, $-1(-7x - 2y - 3) = -1(0)$ is the same as $7x + 2y + 3 = 0$. It is common practice to write the standard form with a positive x coefficient.

Q1 Which of the following represent linear (first-degree) equations? ________

a. $y = 2x + 5$ **b.** $x^2y - 3x = 2$

c. $y = \frac{-2}{x}$ **d.** $\sqrt{2}x - \frac{2}{3}y = \sqrt{13}$

STOP • STOP • STOP • STOP • STOP • STOP • STOP • STOP • STOP

A1 a and d

Q2 Write in standard form:

a. $2x - 3y = 6$ **b.** $x = 2y - 3$

STOP • STOP • STOP • STOP • STOP • STOP • STOP • STOP • STOP

A2 **a.** $2x - 3y - 6 = 0$ **b.** $x - 2y + 3 = 0$

Q3 Write $y = \frac{2}{3}x - 7$ in standard form.

STOP • STOP • STOP • STOP • STOP • STOP • STOP • STOP • STOP

A3 $2x - 3y - 21 = 0$: $y = \frac{2}{3}x - 7$

$$3y = 2x - 21 \qquad \text{(multiplying both sides by 3)}$$
$$0 = 2x - 3y - 21$$

The standard form could have been $\frac{2}{3}x - y - 7 = 0$; however, it is often convenient to express the standard form without fractional coefficients.

Q4 Write in standard form:

a. $y = \frac{3}{5}x - 1$ **b.** $\frac{x}{9} - \frac{y}{4} = 1$

STOP • STOP • STOP • STOP • STOP • STOP • STOP • STOP • STOP

A4 **a.** $3x - 5y - 5 = 0$ **b.** $4x - 9y - 36 = 0$

Q5 Write in standard form:

a. $2x = 3y$ **b.** $Mx = Ny + Q$

STOP • STOP • STOP • STOP • STOP • STOP • STOP • STOP • STOP

A5 **a.** $2x - 3y = 0$: $C = 0$ **b.** $Mx - Ny - Q = 0$

---

2 When graphing linear equations, *x and y intercepts* are the points of intersection of a line with the x axis and y axis, respectively. Since every ordered pair on the x axis has a second coordinate of zero, the x intercept is of the form $(a, 0)$, $a \in R$. Every ordered pair on the y axis will have a first coordinate of zero; hence the y intercept is of the form $(0, b)$, $b \in R$.

Example: Determine the x and y intercepts for the graph of the line shown here.

Solution

x intercept = $(-2, 0)$
y intercept = $(0, 3)$

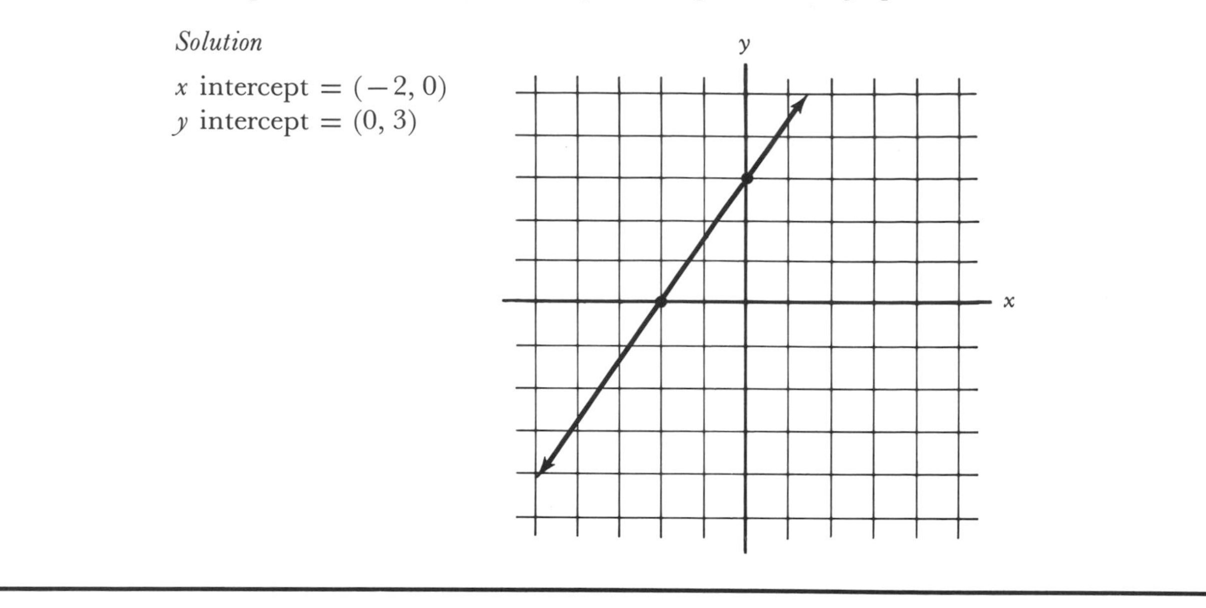

Q6 Determine the x and y intercepts:

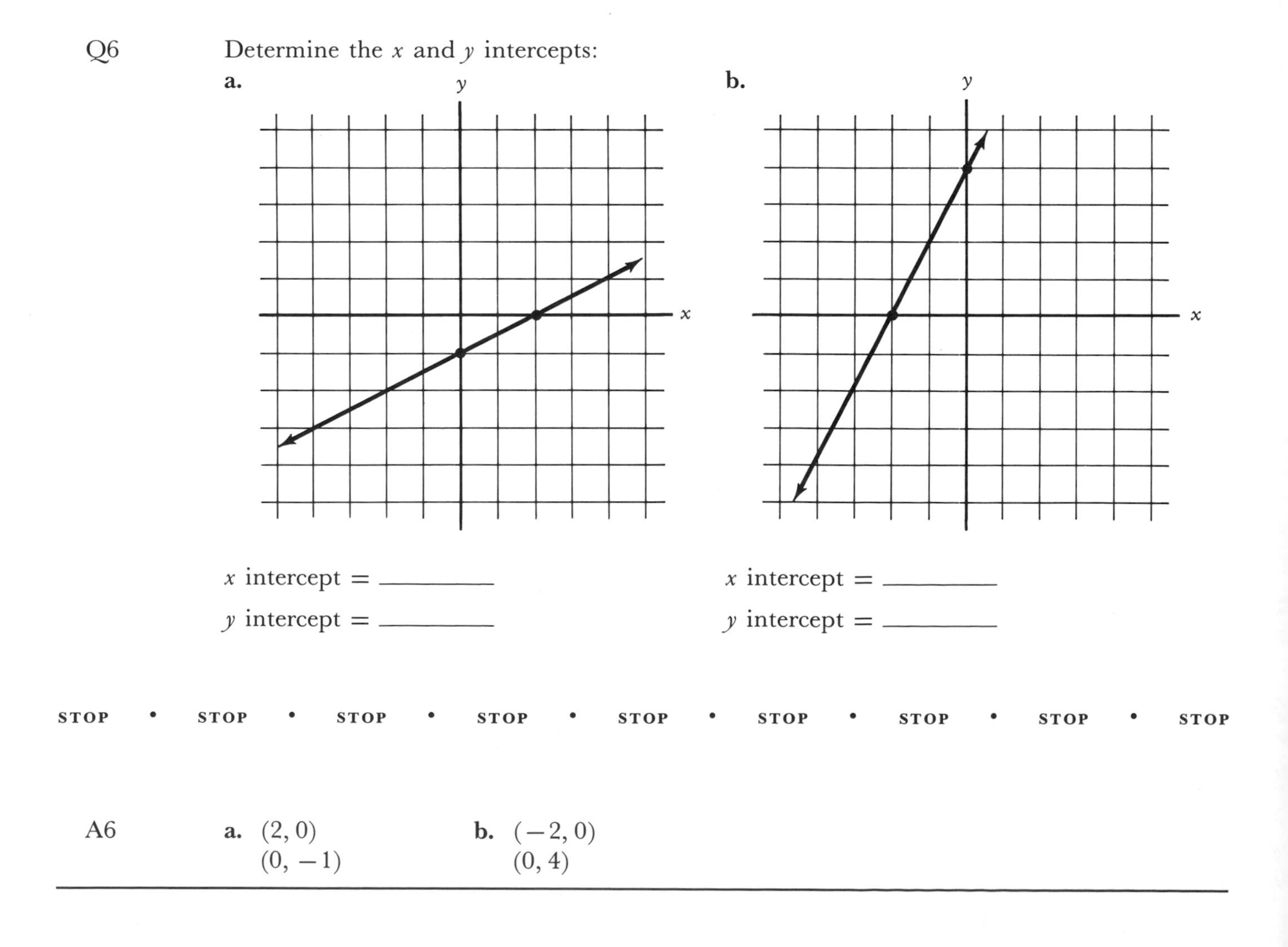

a.

x intercept = ________

y intercept = ________

b.

x intercept = ________

y intercept = ________

STOP • STOP • STOP • STOP • STOP • STOP • STOP • STOP • STOP

A6 **a.** $(2, 0)$
$(0, -1)$

b. $(-2, 0)$
$(0, 4)$

Q7 Graph the line containing x intercept $(3, 0)$ and y intercept $\left(0, \frac{3}{2}\right)$.

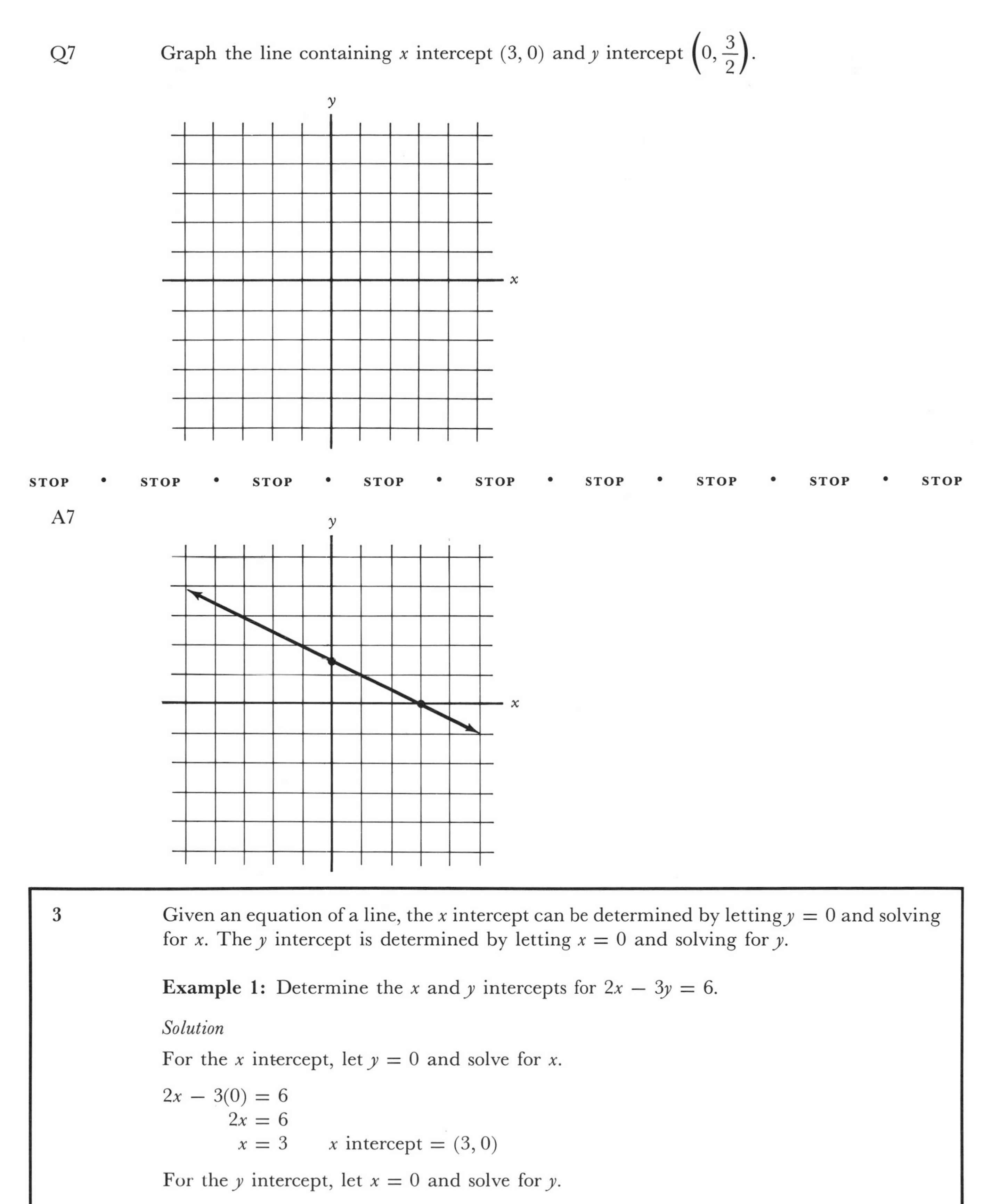

STOP • STOP • STOP • STOP • STOP • STOP • STOP • STOP • STOP

A7

3 Given an equation of a line, the x intercept can be determined by letting $y = 0$ and solving for x. The y intercept is determined by letting $x = 0$ and solving for y.

Example 1: Determine the x and y intercepts for $2x - 3y = 6$.

Solution

For the x intercept, let $y = 0$ and solve for x.

$$
\begin{aligned}
2x - 3(0) &= 6 \\
2x &= 6 \\
x &= 3 \qquad x \text{ intercept} = (3, 0)
\end{aligned}
$$

For the y intercept, let $x = 0$ and solve for y.

$$
\begin{aligned}
2(0) - 3y &= 6 \\
-3y &= 6 \\
y &= -2 \qquad y \text{ intercept} = (0, -2)
\end{aligned}
$$

Example 2: Determine the x and y intercepts for $y = 2x - 3$.

Solution

The solution is found more easily if the x and y terms are on one side of the equation with the constant term on the other side.

$$y = 2x - 3$$
$$-2x + y = -3$$

Let $y = 0$: $-2x + 0 = -3$

$$x = \frac{3}{2}$$

$$x \text{ intercept} = \left(\frac{3}{2}, 0\right)$$

Let $x = 0$: $-2(0) + y = -3$

$$y = -3$$

$$y \text{ intercept} = (0, -3)$$

Q8 Find the (1) x intercept and (2) y intercept:

a. $2x + 3y = 6$

b. $3x - 7y - 21 = 0$

STOP • STOP • STOP • STOP • STOP • STOP • STOP • STOP • STOP

A8 **a.** (1) (3, 0) (2) (0, 2):

Let $y = 0$: $2x + 3(0) = 6$

$x = 3$

Let $x = 0$: $2(0) + 3y = 6$

$y = 2$

b. (1) (7, 0) (2) (0, −3):

Let $y = 0$: $3x = 21$

$x = 7$

Let $x = 0$: $-7y = 21$

$y = -3$

4 If the equation of a line is written in the form $Ax + By = C$, the x and y intercepts are usually found mentally. For example, if $-2x + 5y = 10$, the x intercept is found by letting $y = 0$; however, when $y = 0$, the y term is equal to zero. Hence the x intercept can be found by "covering up" the y term and solving for x.

$-2x +$ [finger] $= 10$

$x = -5$

Thus the x intercept is $(-5, 0)$. The same technique can be used to find the y intercept.

[finger] $+ 5y = 10$

$y = 2$

Therefore, the y intercept is (0, 2).

Q9 Find the (1) x intercept and (2) y intercept:

a. $x + y = \sqrt{7}$

b. $6x - 5y - 15 = 0$

STOP • STOP • STOP • STOP • STOP • STOP • STOP • STOP • STOP

A9 **a.** (1) $(\sqrt{7}, 0)$ (2) $(0, \sqrt{7})$

b. (1) $\left(\frac{5}{2}, 0\right)$ (2) $(0, -3)$: $6x - 5y = 15$

Q10 Graph by finding the x and y intercepts:

a. $-2x + 3y = 6$

b. $x + 2y = 4$

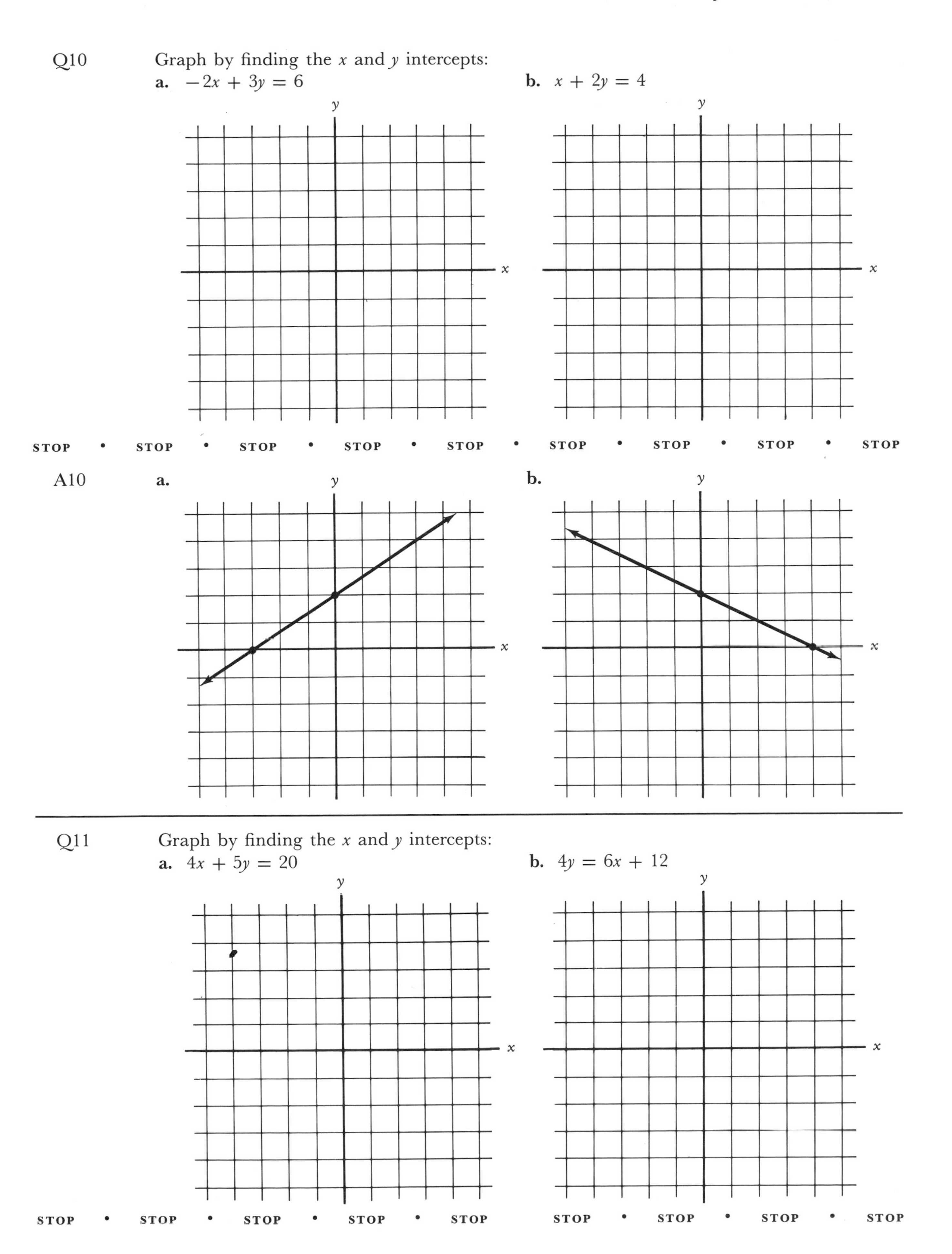

STOP • STOP • STOP • STOP • STOP • STOP • STOP • STOP • STOP

A10 a.

b.

Q11 Graph by finding the x and y intercepts:

a. $4x + 5y = 20$

b. $4y = 6x + 12$

STOP • STOP • STOP • STOP • STOP • STOP • STOP • STOP • STOP

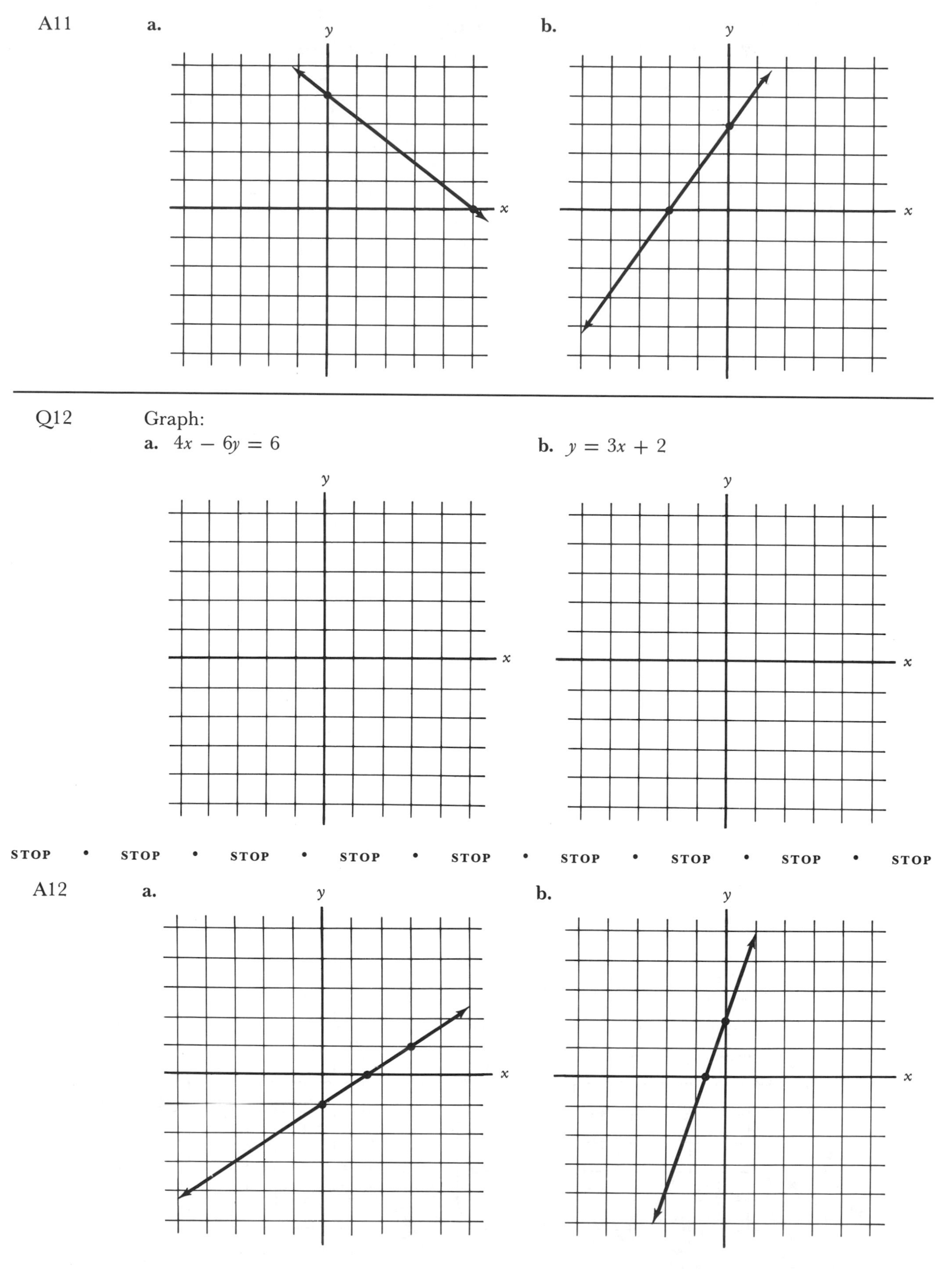
A11
a.
y
x
b.
y
x
Q12
Graph:
a. $4x - 6y = 6$
b. $y = 3x + 2$
y
x
y
x
STOP • STOP • STOP • STOP • STOP • STOP • STOP • STOP • STOP
A12
a.
y
x
b.
y
x

5

If the constant term in a linear equation is zero, the graph will pass through the origin; therefore, the x intercept and the y intercept are both at the point (0, 0). Thus, to determine the graph, another point of the line will have to be plotted.

Example: Graph $2x - y = 0$.

Solution

If $x = 0$, $y = 0$; hence the line contains the origin. Choose an arbitrary value of x and solve for y. Let $x = 1$:

$$2(1) - y = 0$$
$$-y = -2$$
$$y = 2$$

Hence the line also contains the point (1, 2).

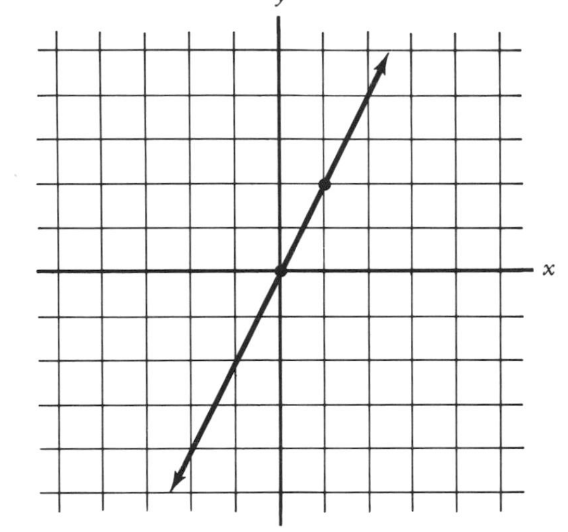

Q13 Complete the table and graph:

a. $x - y = 0$

x	y
0	
	0
1	

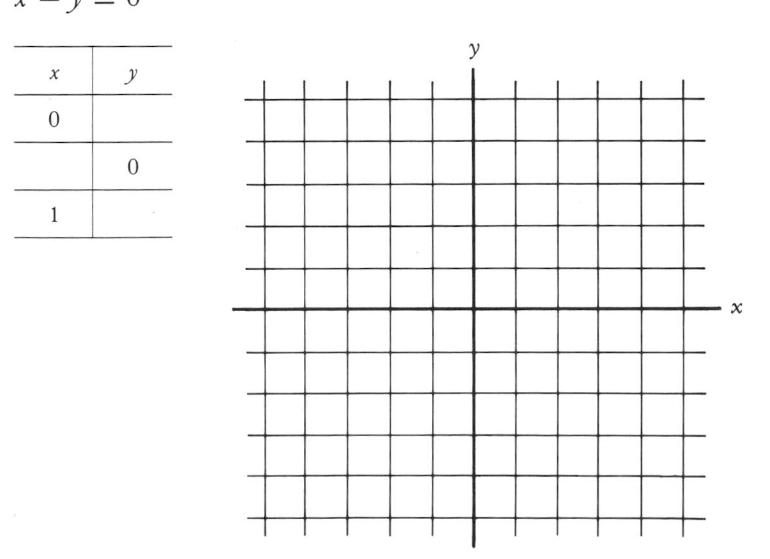

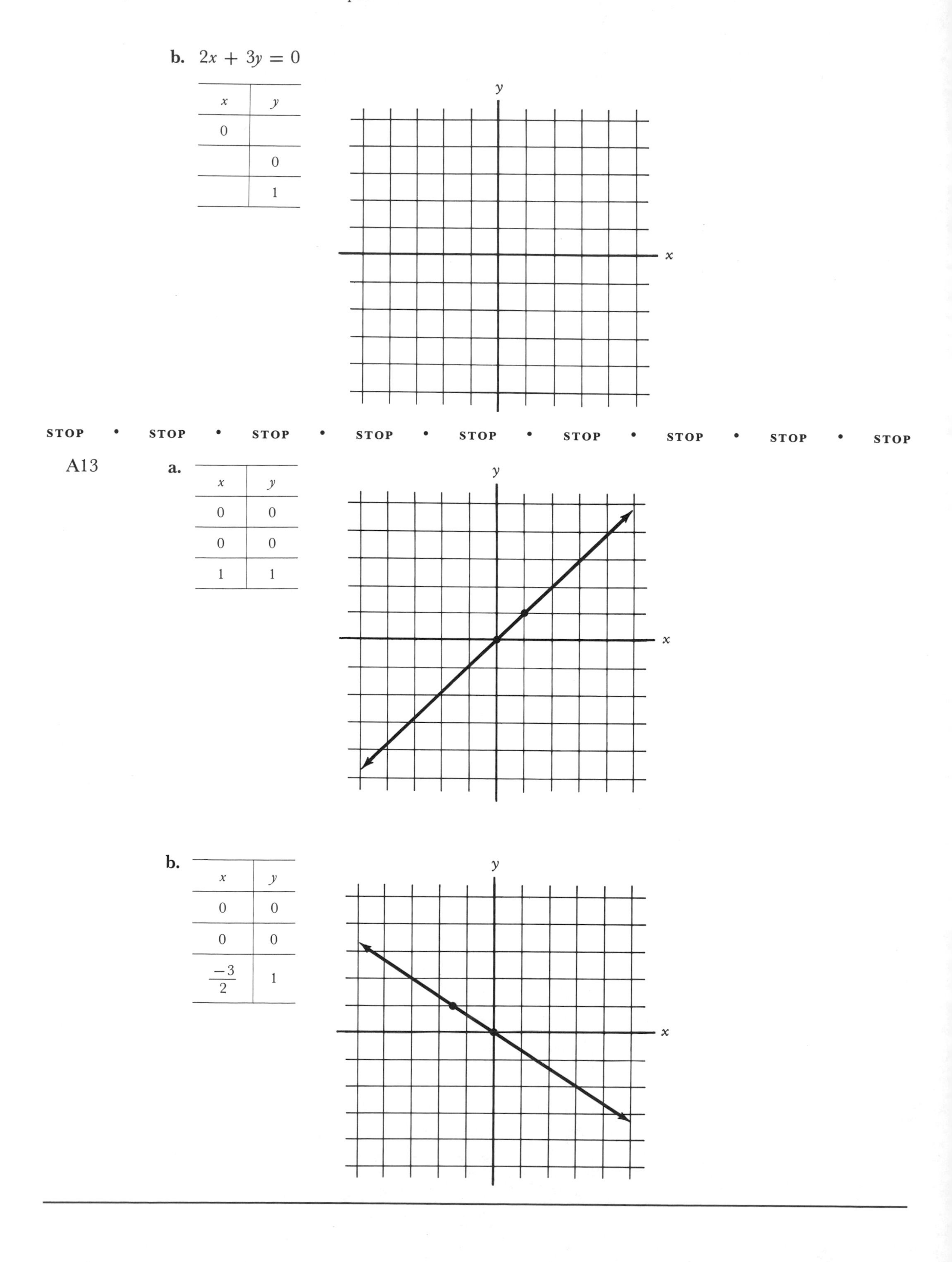

b. $2x + 3y = 0$

x	y
0	
	0
	1

STOP • STOP • STOP • STOP • STOP • STOP • STOP • STOP • STOP

A13 **a.**

x	y
0	0
0	0
1	1

b.

x	y
0	0
0	0
$\frac{-3}{2}$	1

Q14 Graph:

a. $y = -x$

b. $y = \frac{2}{3}x$

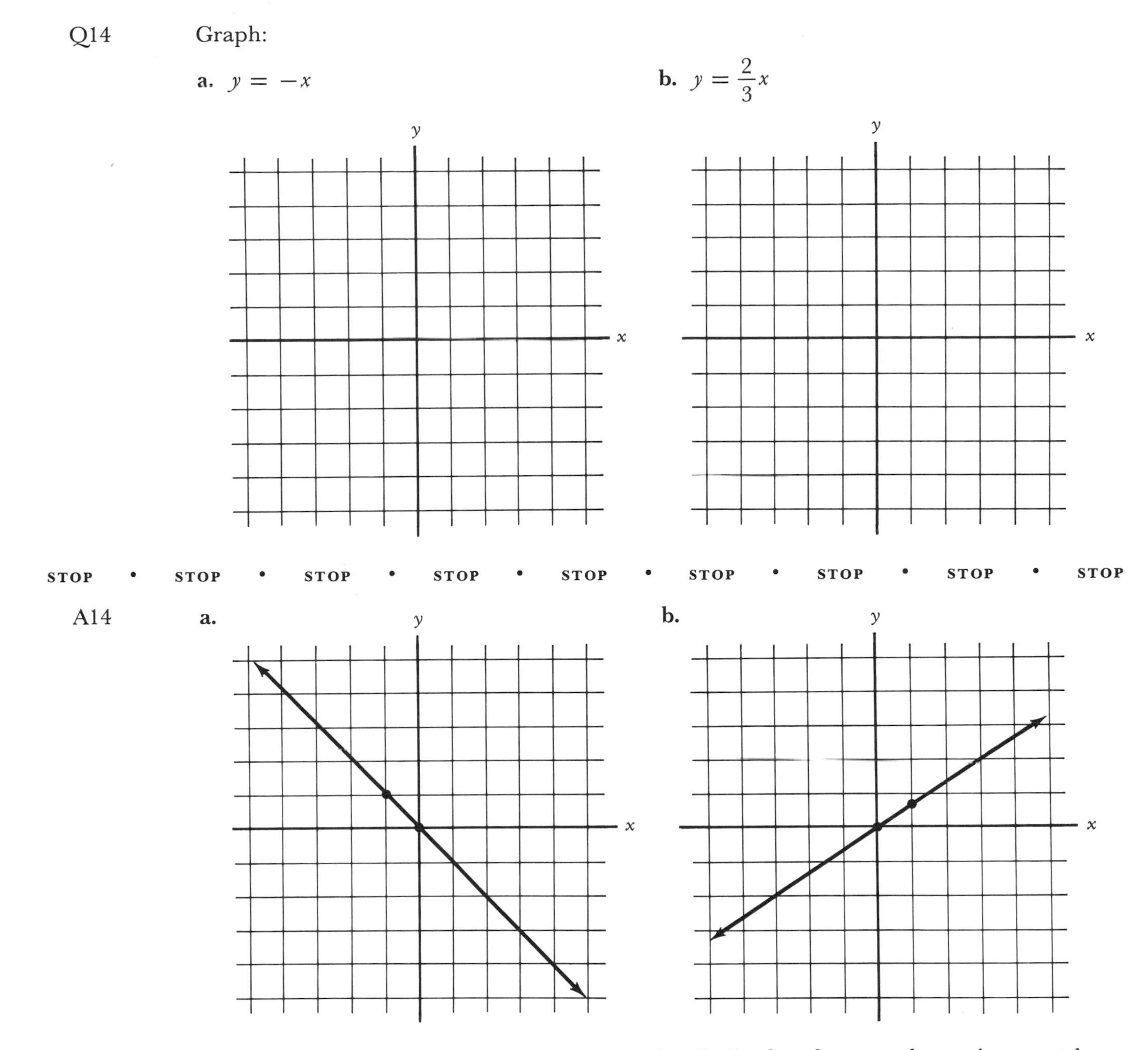

In each case the x and y intercept is the point $(0, 0)$; therefore, another point must be determined to graph the line. The choice of points is arbitrary; however, there is exactly one graph representing each equation.

---

6 Equations of the form $\{(x, y) | x = a\}$ and $\{(x, y) | y = b\}$ (one variable missing) are graphed as lines parallel to the y axis and x axis, respectively.

Example 1: Graph $\{(x, y) | x = 2\}$.

Solution

The equation is graphed by finding all ordered pairs such that the first coordinate is always 2. Since there is no restriction on y, the second coordinate can be any real number. Possible ordered pairs would be: $(2, 0)$, $(2, -1)$, $\left(2, \frac{-5}{2}\right)$, $(2, 10)$, $(2, \sqrt{3})$, and so on.

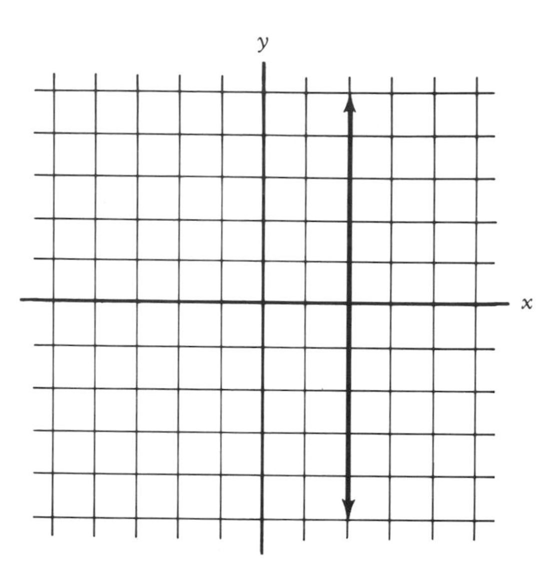

Example 2: Graph $\{(x, y) | y = -1\}$.

Solution

The second coordinate is always -1, whereas the first coordinate is any real number. Possible ordered pairs are: $(-3, -1)$, $(-2, -1)$, $(-\sqrt{2}, -1)$, $(0, -1)$, $(5, -1)$, and so on. The statement $\{(x, y) | y = -1\}$ is usually written $y = -1$, with the set builder notation understood.

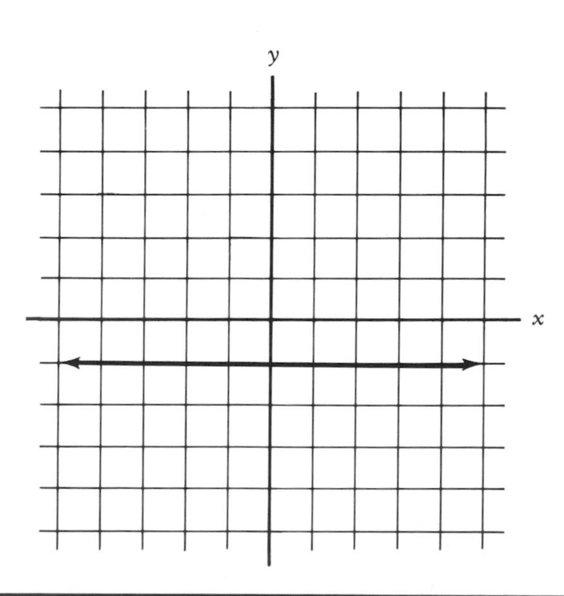

Q15 Complete the table and graph:

a. $x = -1$

x	y
	-1
	0
	2

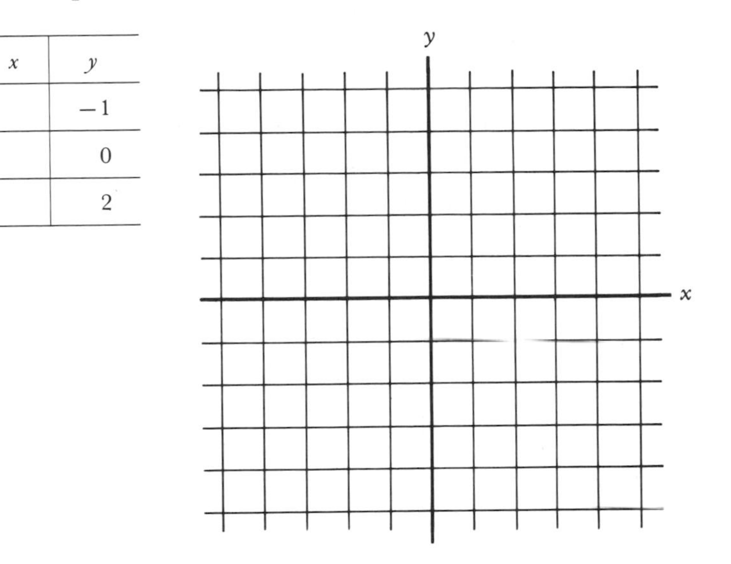

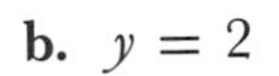

b. $y = 2$

x	y
-1	
0	
1	

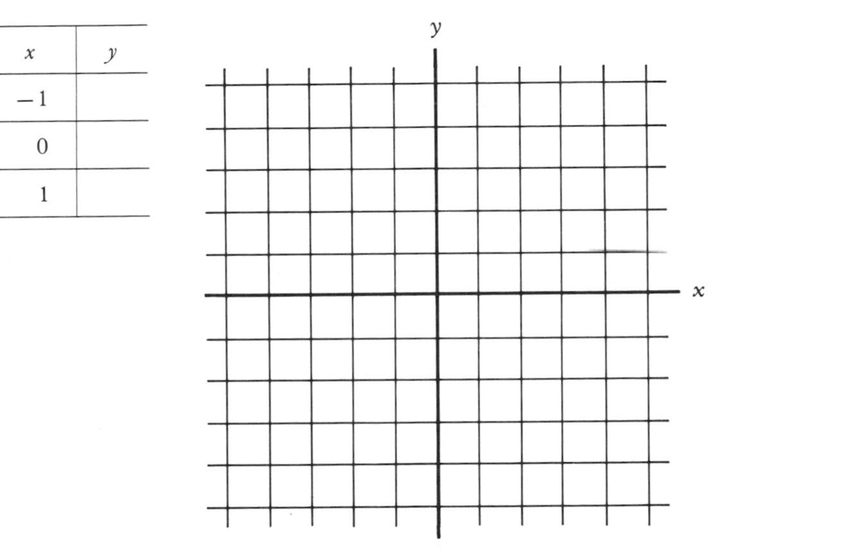

STOP • STOP • STOP • STOP • STOP • STOP • STOP • STOP • STOP

A15 **a.**

x	y
-1	-1
-1	0
-1	2

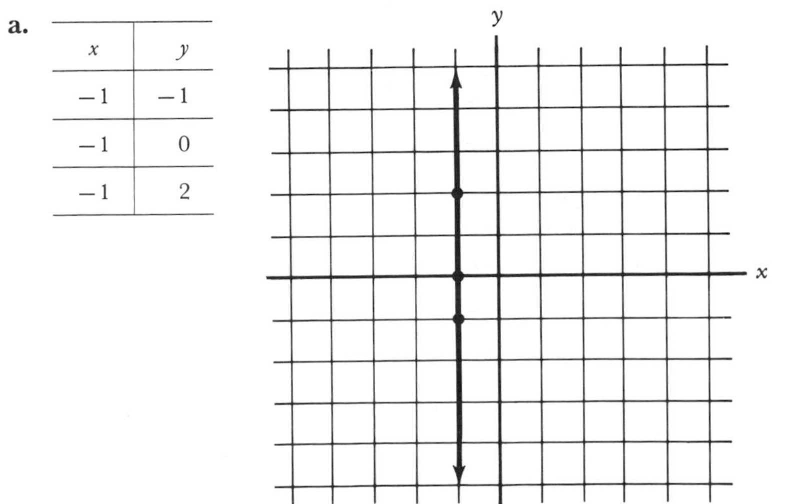

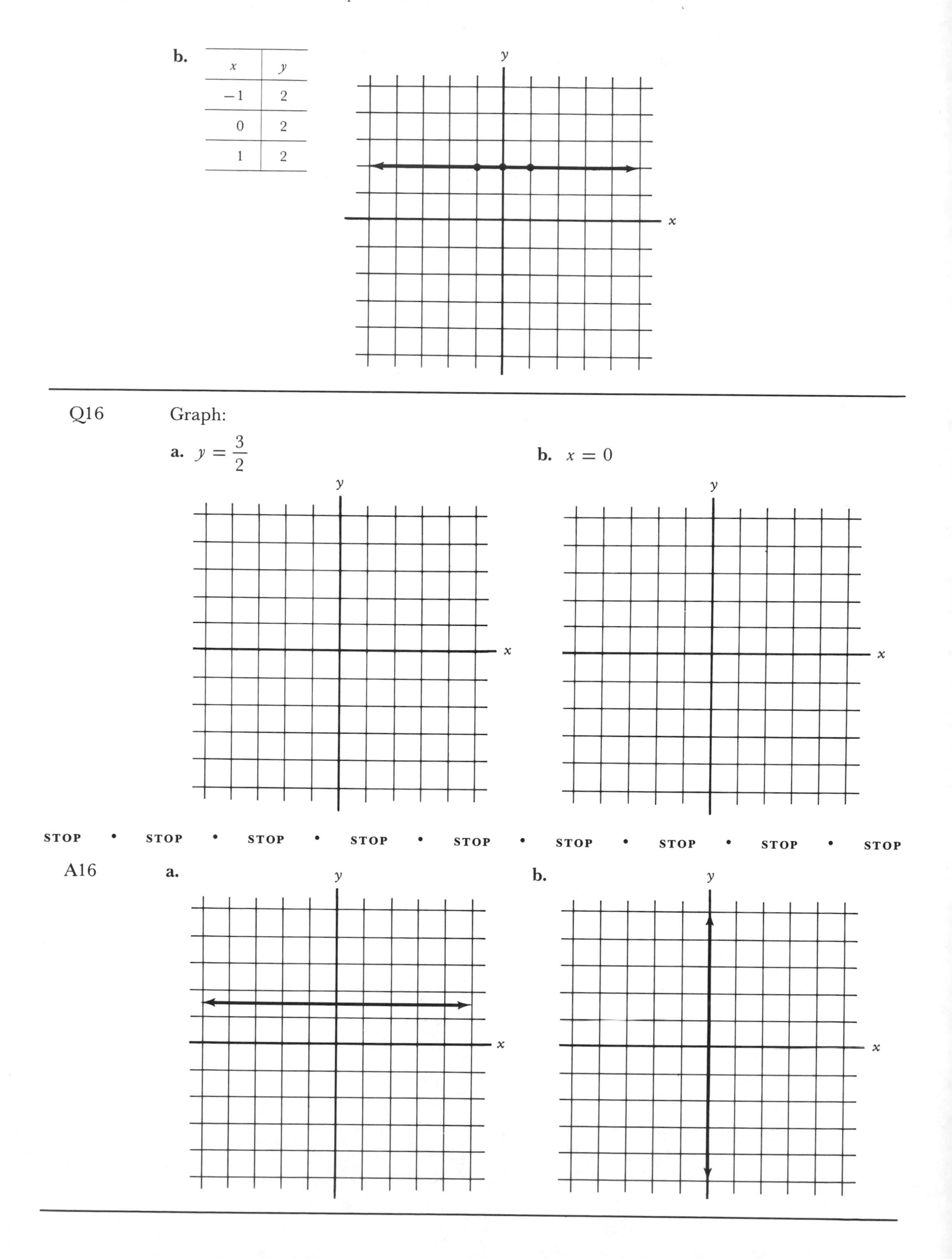

b.

x	y
-1	2
0	2
1	2

Q16 Graph:

a. $y = \frac{3}{2}$

b. $x = 0$

STOP • STOP • STOP • STOP • STOP • STOP • STOP • STOP • STOP

A16 **a.**

b.

Q17 What can be said about all ordered pairs on the y axis?

____

STOP • STOP • STOP • STOP • STOP • STOP • STOP • STOP • STOP

A17 The first coordinate is zero.

Q18 The graph of $y = 0$ is the ______ axis.

STOP • STOP • STOP • STOP • STOP • STOP • STOP • STOP • STOP

A18 x

---

7 The slope of a line containing the points (x_1, y_1) and (x_2, y_2) was defined in Section 6.3 as $\dfrac{y_1 - y_2}{x_1 - x_2}$ or $\dfrac{y_2 - y_1}{x_2 - x_1}$. The letter m is often used to represent slope; hence

$$m = \frac{y_2 - y_1}{x_2 - x_1}$$

The definition of slope can be used to find the equation of a line if the slope, m, and one point, (x_1, y_1), of the line are known. Let (x, y) represent an arbitrary unknown point of the line. Substituting (x, y) for (x_2, y_2) in the slope expression above yields

$$m = \frac{y - y_1}{x - x_1} \qquad x - x_1 \neq 0$$

Now, multiply both sides by $x - x_1$.

$$m(x - x_1) = y - y_1$$

or

$$y - y_1 = m(x - x_1)$$

The latter equation is called the *point-slope form* of a linear equation and can be used to find the equation of a line if the *slope and one point are known.*

Example: Find the equation of the line containing $(1, -2)$ with $m = 2$.

Solution

$(x_1, y_1) = (1, -2) \qquad m = 2$

$$\begin{aligned} y - y_1 &= m(x - x_1) \\ y - (-2) &= 2(x - 1) \\ y + 2 &= 2x - 2 \\ -2x + y + 4 &= 0 \\ 2x - y - 4 &= 0 \qquad \text{(multiplying both sides by } -1) \end{aligned}$$

Q19 Find the equation of the line containing $(1, 2)$ with $m = 3$ (complete the steps).

$$y - y_1 = m(x - x_1)$$

$y - (\qquad) =$ ______$(x -$ ______$)$

STOP • STOP • STOP • STOP • STOP • STOP • STOP • STOP • STOP

A19

$y - 2 = 3(x - 1)$
$y - 2 = 3x - 3$
$3x - y - 1 = 0$

Q20 Find the equation of the line containing the given point with slope m:

a. $(1, 4)$, $m = 2$ **b.** $(-1, 5)$, $m = 3$

STOP • STOP • STOP • STOP • STOP • STOP • STOP • STOP • STOP

A20

a. $2x - y + 2 = 0$:
$y - 4 = 2(x - 1)$
$y - 4 = 2x - 2$

b. $3x - y + 8 = 0$:
$y - 5 = 3(x + 1)$
$y - 5 = 3x + 3$

Q21 Find the equation of the line containing the given point with slope m:

a. $(-2, 3)$, $m = \frac{1}{2}$ **b.** $(0, 0)$, $m = \frac{-3}{4}$

STOP • STOP • STOP • STOP • STOP • STOP • STOP • STOP • STOP

A21

a. $x - 2y + 8 = 0$:

$$y - 3 = \frac{1}{2}(x + 2)$$

$$y - 3 = \frac{1}{2}x + 1$$

$$2y - 6 = x + 2$$

b. $3x + 4y = 0$:

$$y - 0 = \frac{-3}{4}(x - 0)$$

$$y = \frac{-3}{4}x$$

$$4y = -3x$$

---

8 The equation of a line containing two points can be determined by first finding the slope and then using one of the given points with the point-slope form of the equation of a line.

Example: Find the equation of the line containing the points $(2, 4)$ and $(4, 8)$.

Solution

$$m = \frac{4 - 8}{2 - 4} = \frac{-4}{-2} = 2$$

$y - y_1 = m(x - x_1)$
$y - 4 = 2(x - 2)$ or $y - 8 = 2(x - 4)$
$y - 4 = 2x - 4$ $y - 8 = 2x - 8$
$2x - y = 0$ $2x - y = 0$

Q22 Find the equation of the line containing the given points:

a. (2, 1) and (5, 4) **b.** (−2, 7) and (2, 10)

STOP • STOP • STOP • STOP • STOP • STOP • STOP • STOP • STOP

A22 **a.** $x - y - 1 = 0$:

$m = 1, y - 1 = 1(x - 2)$

or $y - 4 = 1(x - 5)$

b. $3x - 4y + 34 = 0$:

$m = \frac{3}{4}, y - 7 = \frac{3}{4}(x + 2)$

or $y - 10 = \frac{3}{4}(x - 2)$

9 The equation of a line whose graph is parallel to the x axis can be found by observation or by using the point-slope method. The equation of a line parallel to the y axis must be found by observation.

Example 1: Find the equation of the line containing the points (2, 3) and (−1, 3).

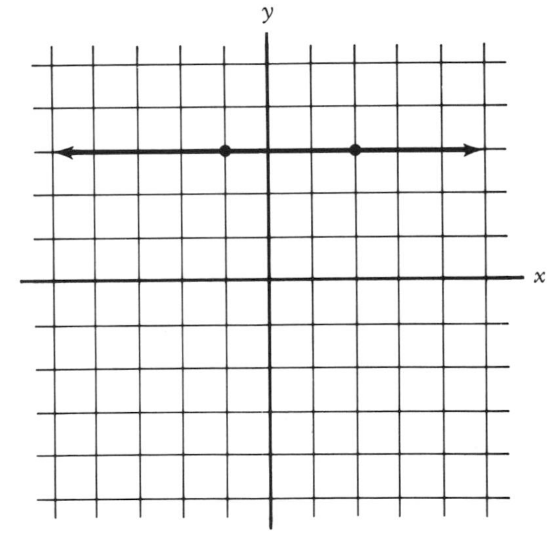

Solution

The line is parallel to the x axis and 3 units above; hence the second coordinate of all ordered pairs is 3. Thus the equation of the line is $y = 3$.

Since the slope of a horizontal line is zero, the point-slope method could also be used.

$y - y_1 = m(x - x_1)$ $m = 0, (x_1, y_1) = (2, 3)$

$y - 3 = 0(x - 2)$

$y - 3 = 0$

$y = 3$

Example 2: Find the equation of the line containing the points (2, 3) and (2, −1).

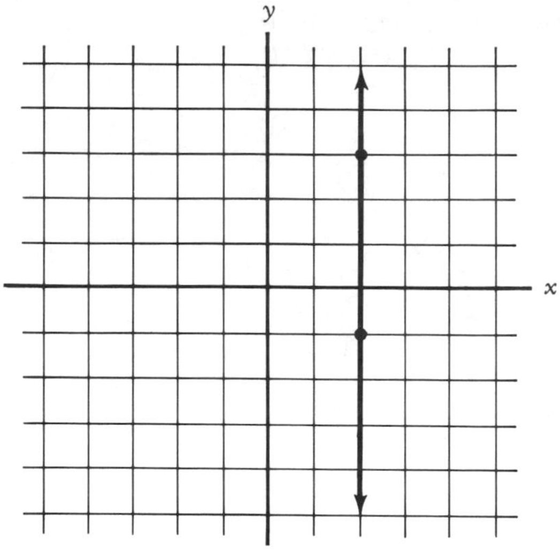

Solution

The line is parallel to the y axis and 2 units to the right; hence the first coordinate of all ordered pairs is 2. Thus the equation of the line is $x = 2$. Since a vertical line has no slope, the point-slope method cannot be used.

Q23 **a.** Find the equation of the line parallel to the x axis and containing $(0, -4)$. ________

b. Find the equation of the line parallel to the y axis and containing $(-1, -3)$. ________

STOP • STOP • STOP • STOP • STOP • STOP • STOP • STOP • STOP

A23 **a.** $y = -4$ **b.** $x = -1$

Q24 Find the equation of the line containing the given points:

a. $(2, 4)$ and $(2, 8)$ **b.** $(-3, -2)$ and $(3, -2)$

STOP • STOP • STOP • STOP • STOP • STOP • STOP • STOP • STOP

A24 **a.** $x = 2$ **b.** $y = -2$

Q25 Find the equation of the line containing the given points:

a. $(0, 5)$ and $(2, 0)$ **b.** $(-3, -2)$ and $(5, 6)$

c. $(2, -5)$ and $(2, 3)$ **d.** $(1, 3)$ and $(-5, 3)$

STOP • STOP • STOP • STOP • STOP • STOP • STOP • STOP • STOP

A25 **a.** $5x + 2y - 10 = 0$ **b.** $x - y + 1 = 0$ **c.** $x = 2$
d. $y = 3$

10

A special form of a linear equation is determined by examining the slope and y intercept of the line. Consider the equation of the line with slope m and y intercept $(0, b)$.

$$y - y_1 = m(x - x_1) \qquad (x_1, y_1) = (0, b)$$
$$y - b = m(x - 0)$$
$$y - b = mx$$
$$y = mx + b$$

An equation in this form is said to be written in *slope-intercept form,* where m, the coefficient of x, is the slope; b, the constant term, is the second coordinate of the y intercept.

Example: Find the slope and y intercept for $2x - 3y = 6$.

Solution

The equation is written in the slope-intercept form by solving for y.

$$2x - 3y = 6$$
$$-3y = -2x + 6$$
$$y = \frac{2}{3}x - 2$$
$$y = mx + b$$

Hence $m = \frac{2}{3}$ and the y intercept $= (0, -2)$.

Q26 Find the (1) slope and (2) y intercept:
a. $y = 2x + 5$ **b.** $y = -x + 1$
(1) ____ (2) ________ (1) ____ (2) ________

STOP • STOP • STOP • STOP • STOP • STOP • STOP • STOP • STOP

A26 **a.** (1) 2 (2) (0, 5) **b.** (1) -1 (2) (0, 1)

Q27 Find the slope and y intercept:
a. $2y = 3x + 2$ **b.** $6x + 2y = 8$

STOP • STOP • STOP • STOP • STOP • STOP • STOP • STOP • STOP

A27 **a.** $m = \frac{3}{2}$, (0, 1):

$$y = \frac{3}{2}x + 1$$

b. $m = -3$, (0, 4):

$$2y = -6x + 8$$
$$y = -3x + 4$$

Q28 Find the slope and y intercept:

a. $2x = 3y + 8$ **b.** $5x + 2y = 10$

STOP • STOP • STOP • STOP • STOP • STOP • STOP • STOP • STOP

A28 **a.** $m = \frac{2}{3}, \left(0, \frac{-8}{3}\right)$ **b.** $m = \frac{-5}{2}$, (0, 5)

11 The slope and y intercept can be used to graph a line by first plotting the y intercept and then using the slope to find another point of the line.

Example: Graph $y = \frac{-2}{3}x - 1$ by use of the slope-intercept method.

Solution

$$m = \frac{-2}{3}$$

y intercept = (0, −1)

Plot (0, −1). Now, since $m = -2/3$ = rise/run, a second point of the line can be found by moving down 2 and to the right 3, obtaining (3, −3). Since $-2/3 = 2/-3$, a second point could have been obtained by moving up 2 and to the left 3, obtaining (−3, 1). Other points of the line could be determined by continual application of this method to a point known to be on the line.

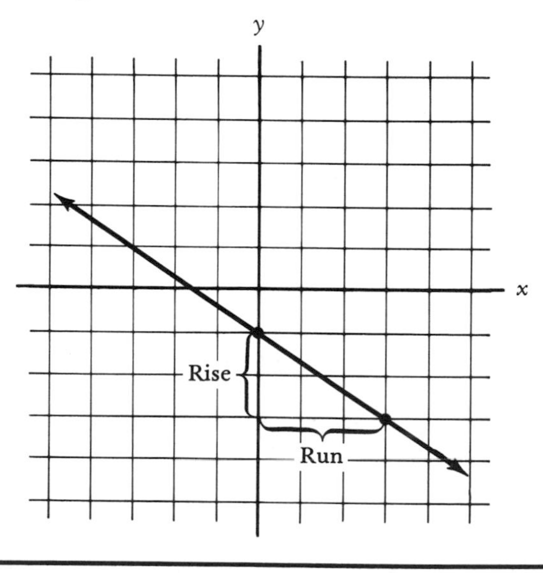

Q29 Graph $y = \frac{1}{2}x - 2$ by use of the slope-intercept method. Besides the y intercept find two points of the line using $m = \frac{1}{2}$ and $m = \frac{-1}{-2}$.

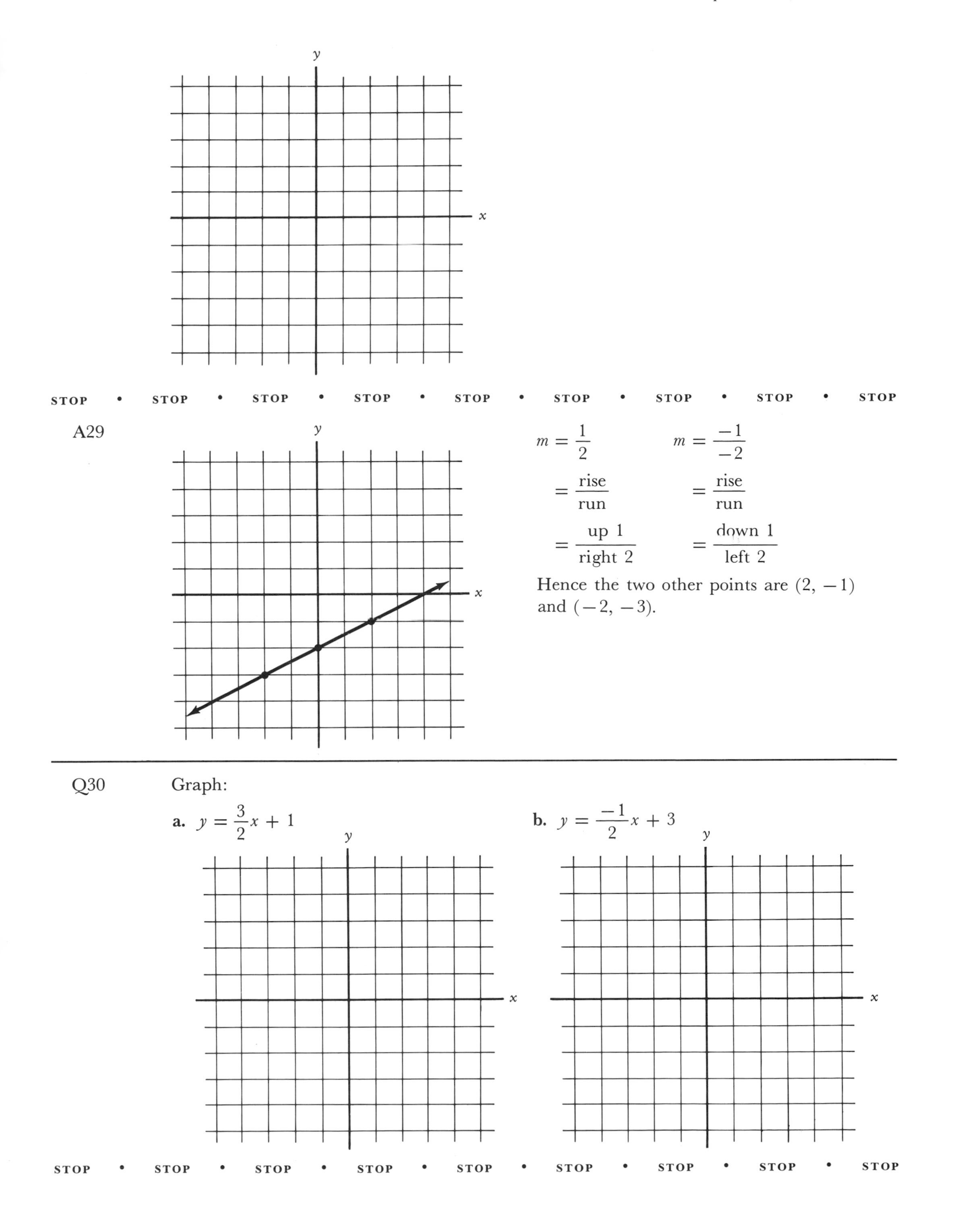

STOP • STOP • STOP • STOP • STOP • STOP • STOP • STOP • STOP

A29

$$m = \frac{1}{2} = \frac{\text{rise}}{\text{run}} = \frac{\text{up } 1}{\text{right } 2} \qquad m = \frac{-1}{-2} = \frac{\text{rise}}{\text{run}} = \frac{\text{down } 1}{\text{left } 2}$$

Hence the two other points are $(2, -1)$ and $(-2, -3)$.

Q30 Graph:

a. $y = \frac{3}{2}x + 1$

b. $y = \frac{-1}{2}x + 3$

STOP • STOP • STOP • STOP • STOP • STOP • STOP • STOP • STOP

A30

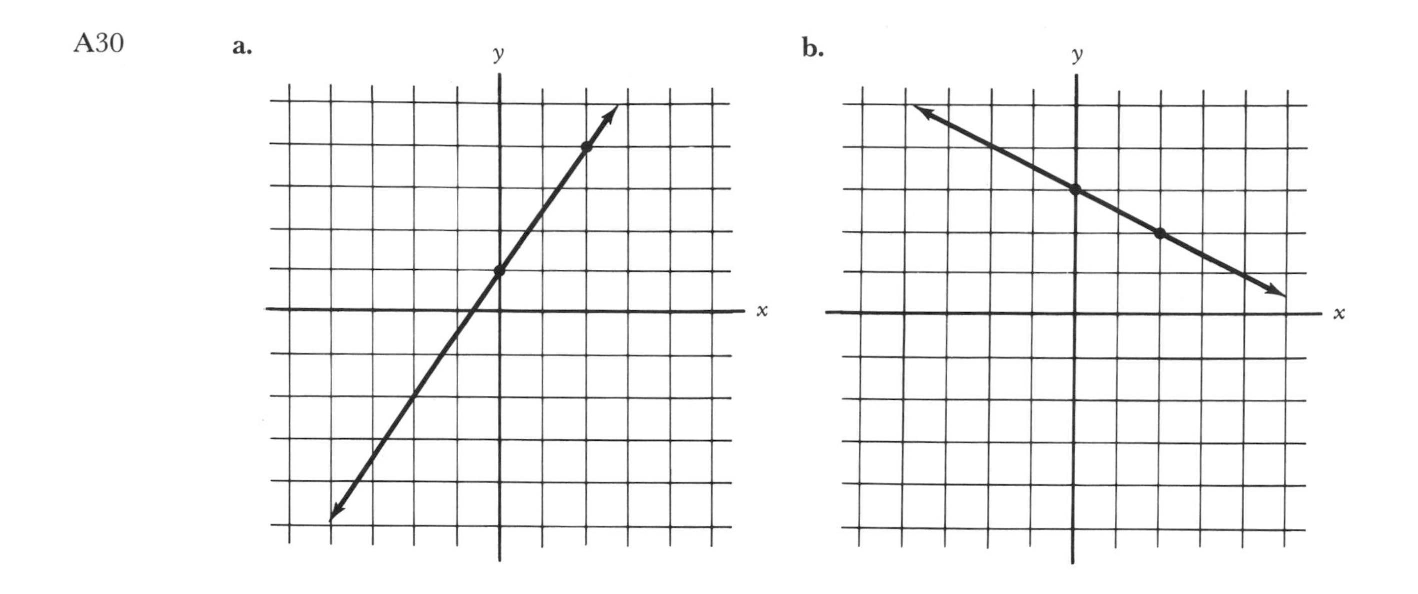

12 Since $2 = \frac{2}{1}$, the slope-intercept form can be used to graph equations of the form $y = 2x - 1$.

Example: Graph $y = 2x - 1$.

Solution

y intercept $= (0, -1)$

$m = 2 = \frac{2}{1} = \frac{\text{rise}}{\text{run}}$

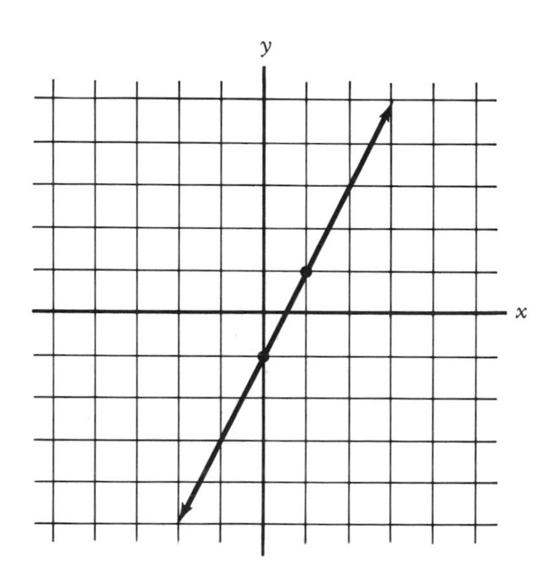

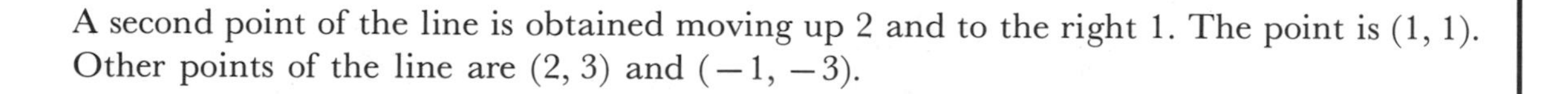
A second point of the line is obtained moving up 2 and to the right 1. The point is (1, 1). Other points of the line are (2, 3) and (−1, −3).

Q31 Graph:

a. $y = -2x + 1$ **b.** $y = -3x + \frac{1}{2}$

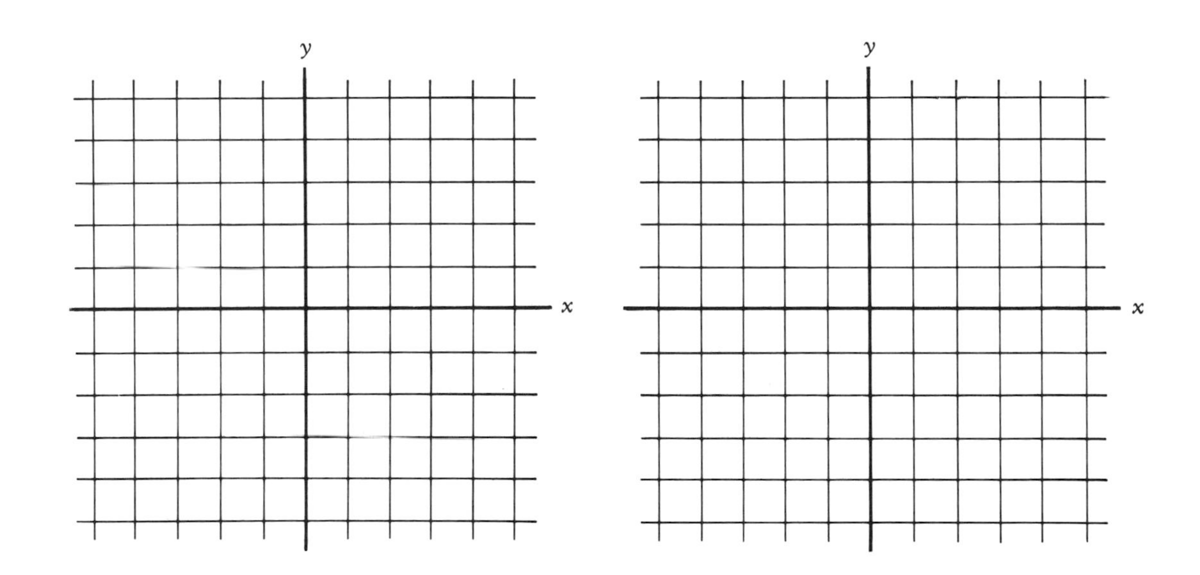

STOP • STOP • STOP • STOP • STOP • STOP • STOP • STOP • STOP

A31 **a.** **b.**

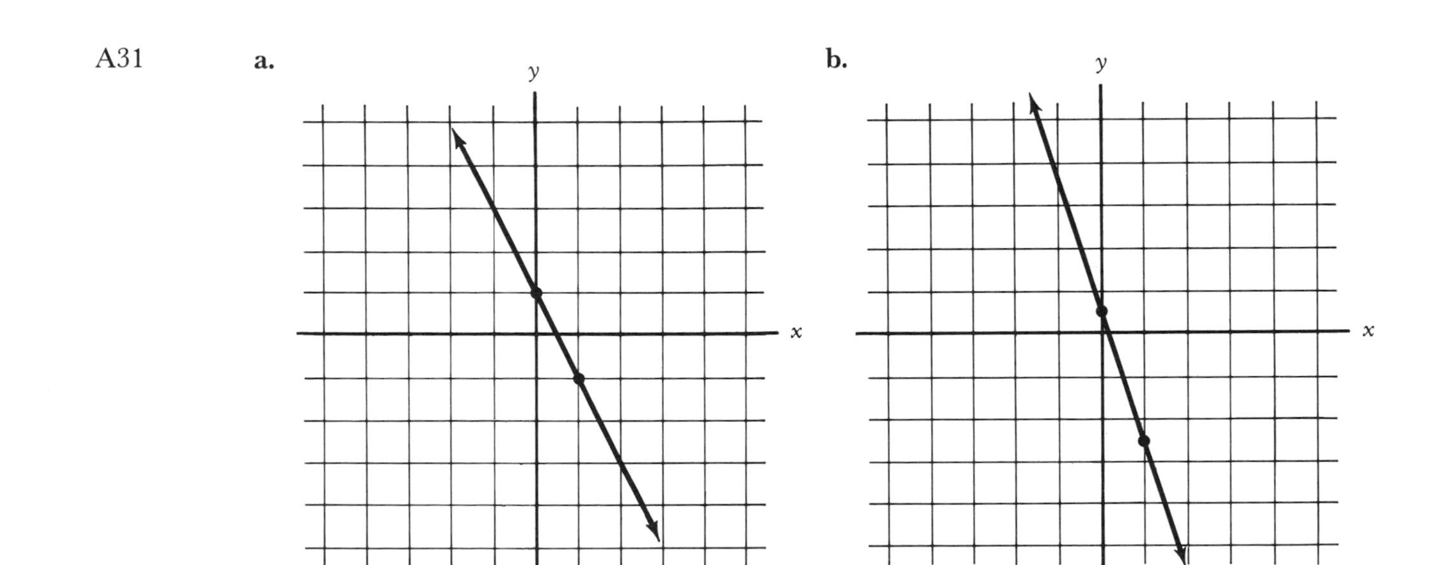

Q32 Graph:

a. $y = \frac{2}{3}x$

b. $2y = 3x - 8$

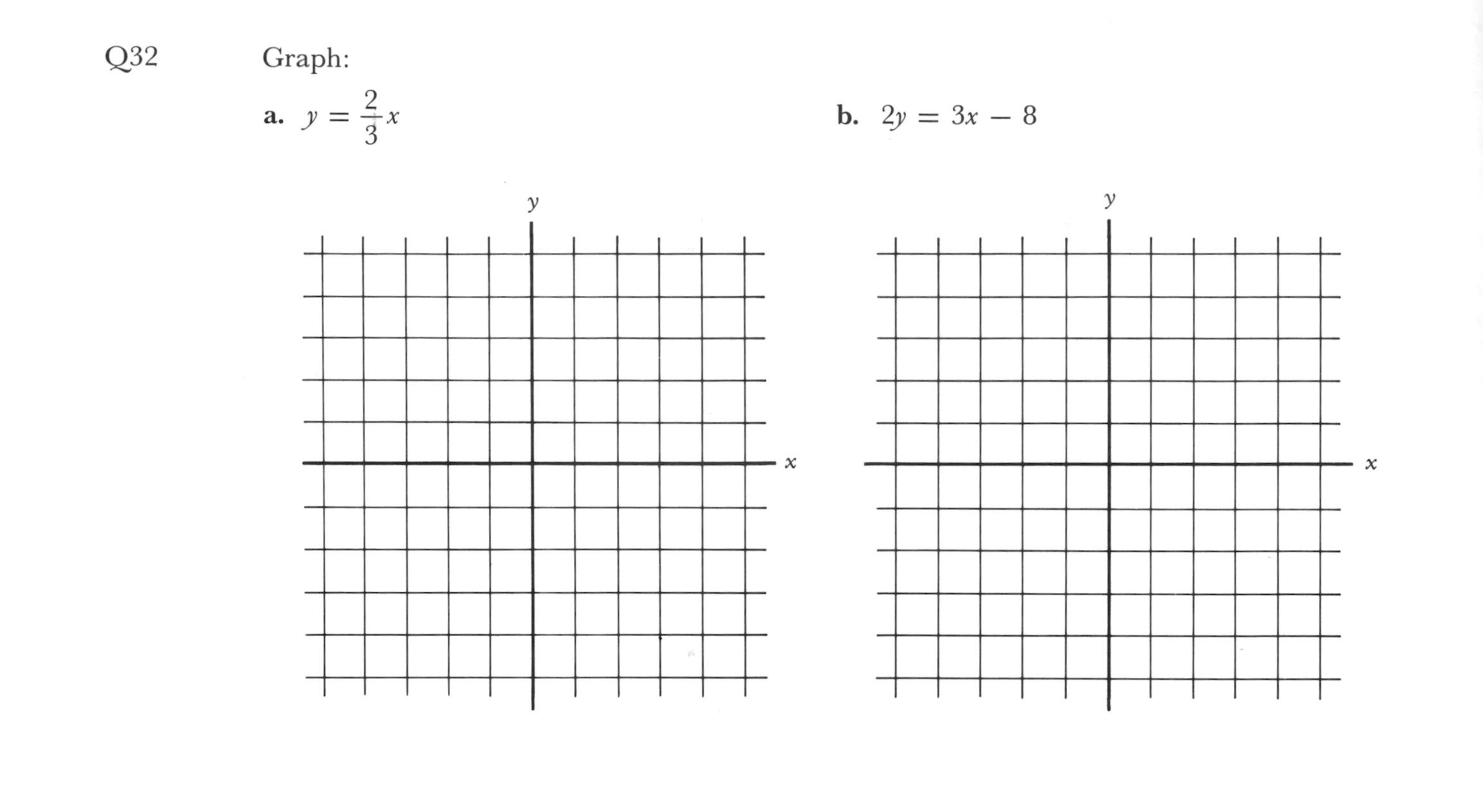

STOP • STOP • STOP • STOP • STOP • STOP • STOP • STOP • STOP

A32 **a.**

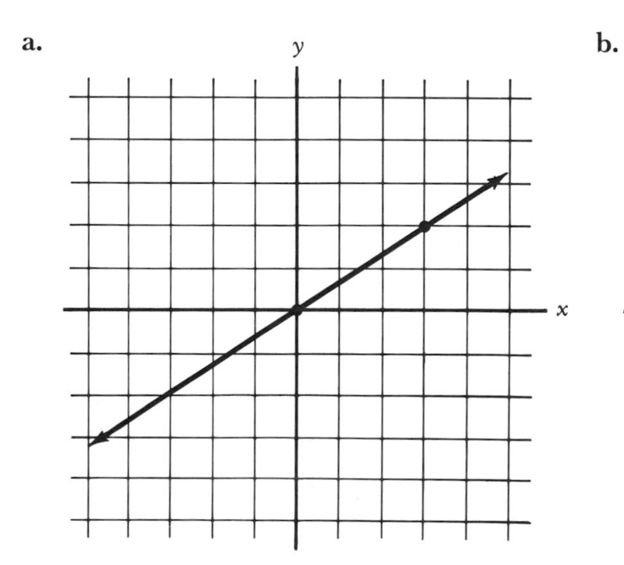

b.

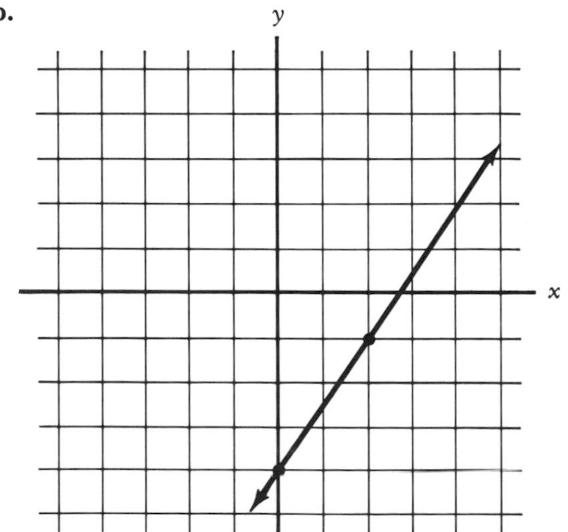

Q33 Graph:

a. $y = 3 - x$

b. $3x + 5y - 6 = 0$

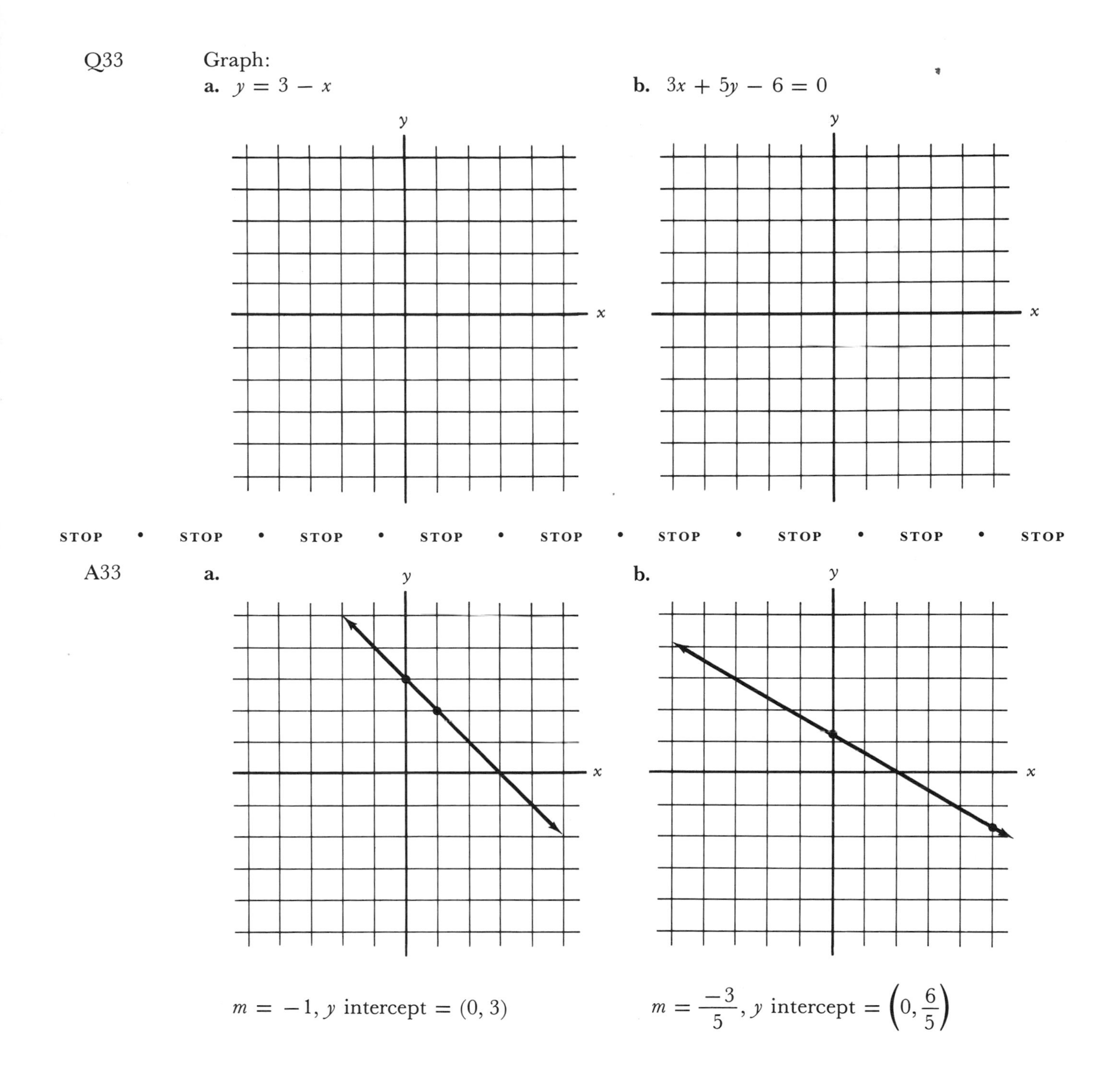

STOP • STOP • STOP • STOP • STOP • STOP • STOP • STOP • STOP

A33 a. $m = -1$, y intercept $= (0, 3)$

b. $m = \dfrac{-3}{5}$, y intercept $= \left(0, \dfrac{6}{5}\right)$

---

13 Linear inequalities in two variables can be graphed by graphing all points above or below the line $y = mx + b$.

Example: Graph $y < 2x + 1$.

Solution

Since $y < 2x + 1$ is strictly less than, $y = 2x + 1$ is not part of the solution. To show this, graph $y = 2x + 1$ as a dashed line. The dashed line represents all points where y is equal to $2x + 1$. Since y is less than $2x + 1$, the solution is all points below the dashed line. This is shown by shading (or drawing lines) from the dashed line $y = 2x + 1$.

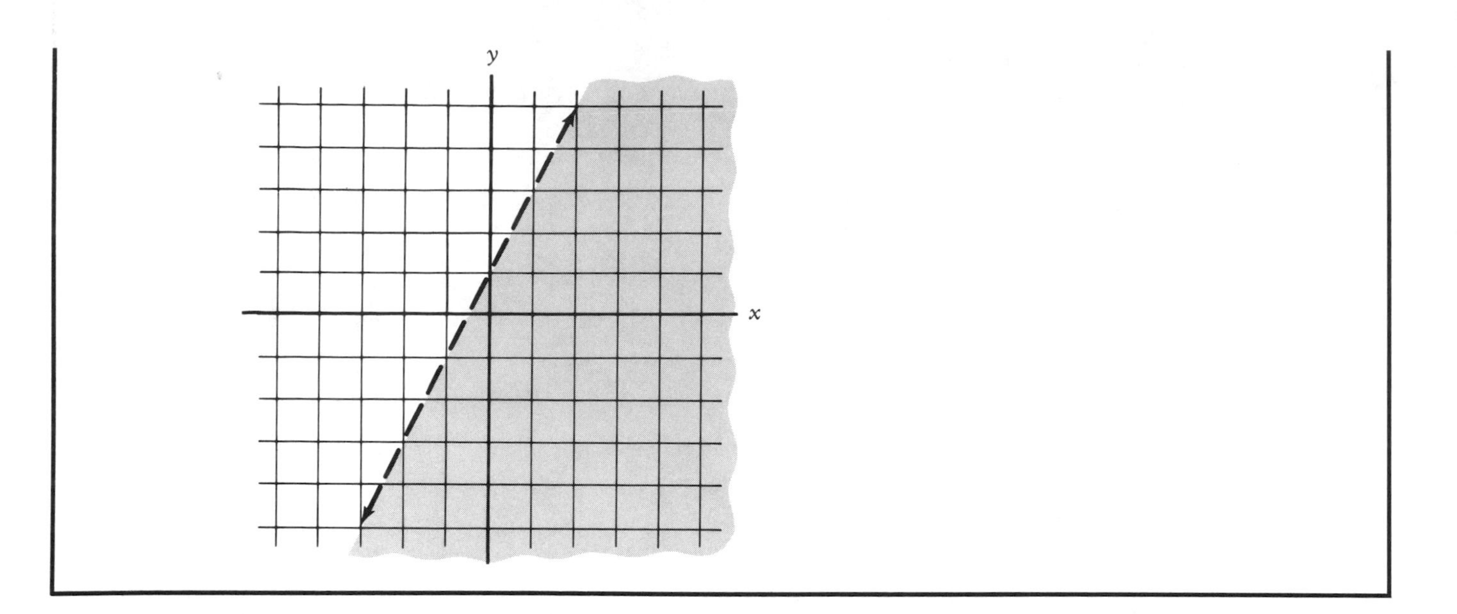

Q34 Graph:

a. $y > 2x + 1$ **b.** $y \leqslant \frac{2}{3}x - 2$

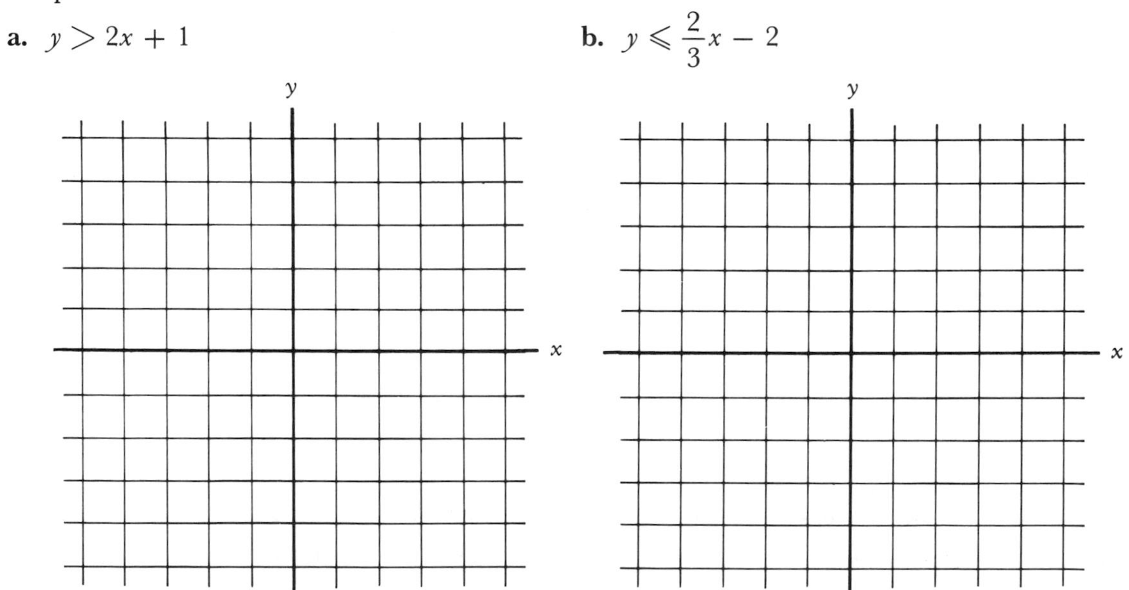

STOP • STOP • STOP • STOP • STOP • STOP • STOP • STOP • STOP

A34

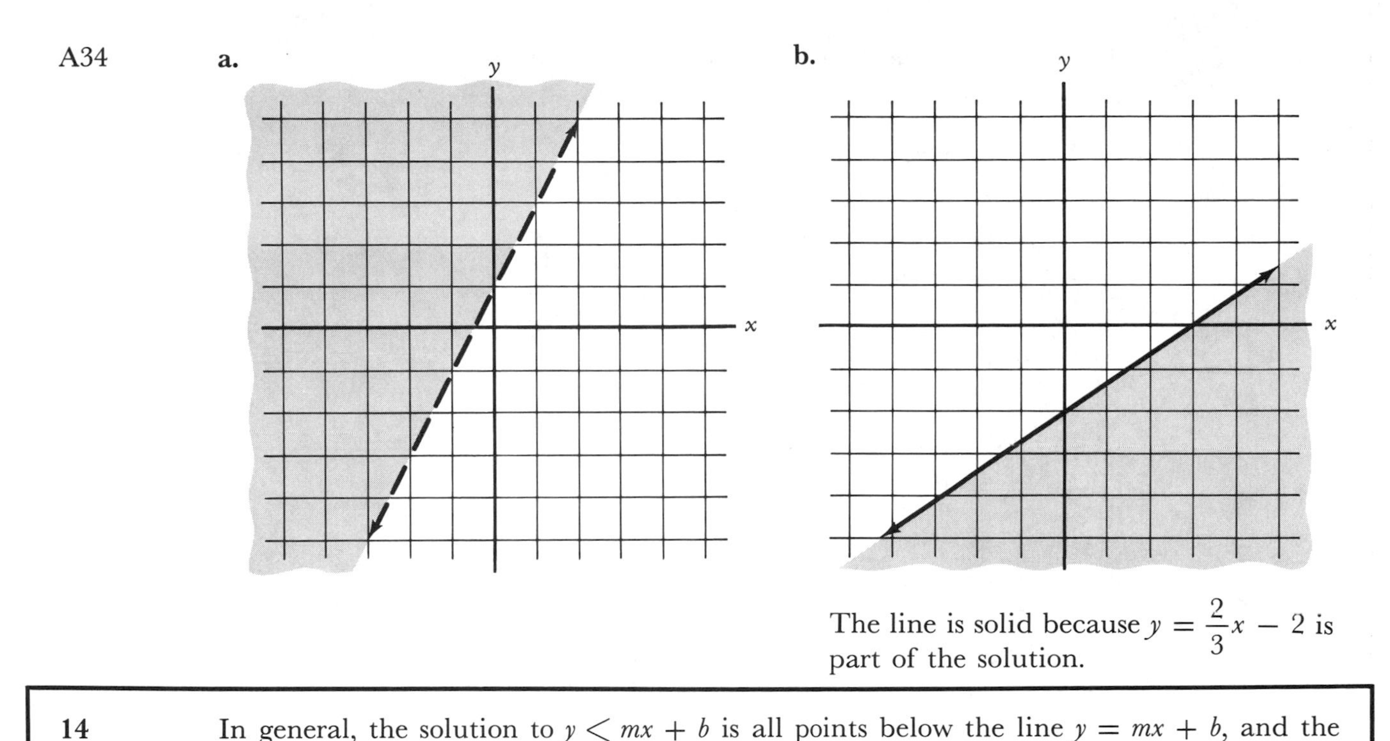

The line is solid because $y = \frac{2}{3}x - 2$ is part of the solution.

14

In general, the solution to $y < mx + b$ is all points below the line $y = mx + b$, and the solution to $y > mx + b$ is all points above the line $y = mx + b$. The solution can be checked by choosing a point from the graph of the solution and checking in the inequality.

Example: Graph $3x - 2y > 6$.

Solution

The inequality should first be written in the form $y < mx + b$ or $y > mx + b$.

$$3x - 2y > 6$$
$$-2y > -3x + 6$$
$$y < \frac{3}{2}x - 3$$

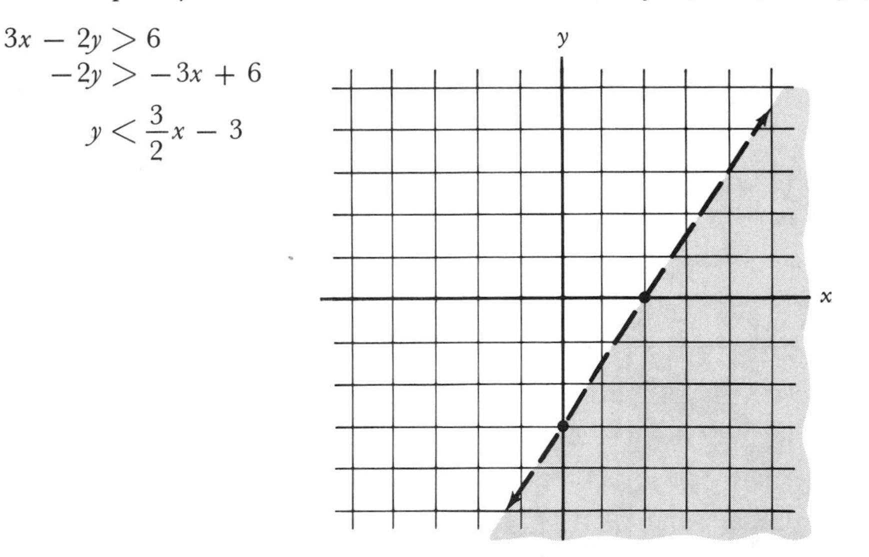

Recall that the direction of the inequality changes when both sides are multiplied by a negative number.

Check: Choosing the point (3, 0):

$$3(3) - 2(0) \overset{?}{>} 6$$
$$9 - 0 > 6 \quad \text{(true)}$$

Hence (3, 0) is in the solution.

Q35 Graph (check by choosing a point from the graph):

a. $2x + y > -2$ b. $x - 2y \geqslant 0$

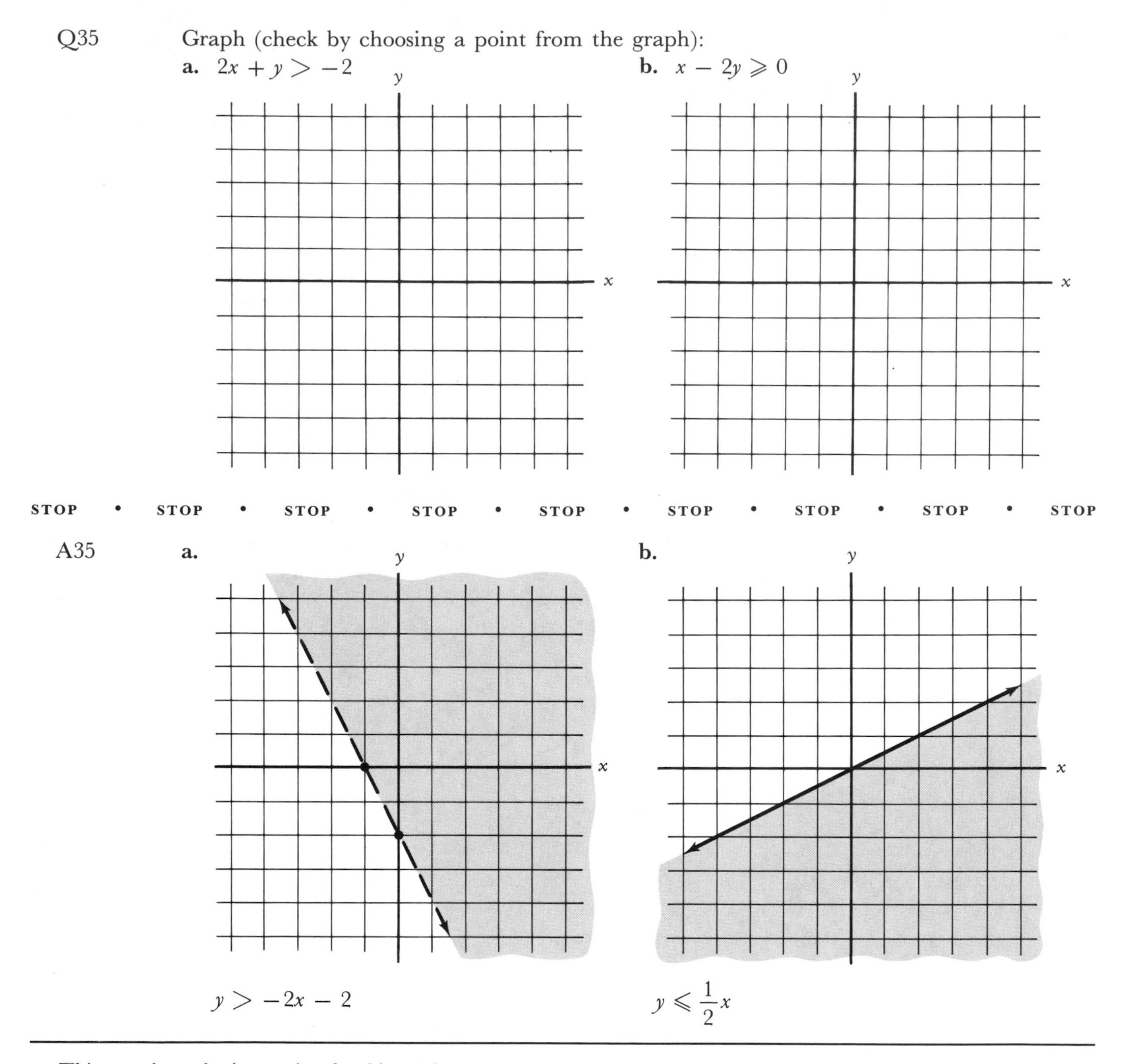

This completes the instruction for this section.

6.4 Exercises

1. Define a linear equation.
2. Which are linear equations?

   a. $2x - 3y = \sqrt{7}$ b. $\frac{2}{3}x - y = 12$ c. $\frac{1}{x} - y = 0$ d. $\sqrt{x} - 2y = 17$

3. Find the equation of the line (standard form) containing the given point with the given slope:

   a. $(2, 3)$, $m = 1$ b. $(4, -1)$, $m = \frac{2}{3}$ c. $(7, 0)$, $m = \frac{-1}{2}$

   d. $(-1, 2)$, $m = 0$ e. $(3, -5)$, no slope f. $(0, 2)$, $m = \frac{2}{3}$

4. Find the equation of the line (slope-intercept form) containing the given points:
   a. $(4, 2)$ and $(5, -1)$ **b.** $(-1, 2)$ and $(3, -1)$
   c. $\left(\frac{-7}{3}, 0\right)$ and $\left(\frac{-5}{2}, 0\right)$ **d.** $\left(-2, \frac{2}{3}\right)$ and $(-2, 4)$
5. Graph by finding the x and y intercepts (if possible):
   a. $x + y = 2$ **b.** $x - 4y = 4$ **c.** $y = 4x$
   d. $3x - 5y = 15$ **e.** $3x + 4y - 6 = 0$ **f.** $y = -3$
   g. $y = 3x + 2$ **h.** $2x + 5y - 10 = 0$ **i.** $x - 2y = 0$
6. Find the slope and y intercept:
   a. $y = 2x - 5$ **b.** $y = \frac{2}{3}x - \sqrt{5}$ **c.** $2x - 3y = 6$ **d.** $3x + 5y = 15$
7. Graph using the slope-intercept method:
   a. $y = 2x - 4$ **b.** $2x - 3y = 6$ **c.** $3x + 5y = 15$ **d.** $y = -4x + 2$
   e. $y = -x - 2$ **f.** $y = 3x$ **g.** $y = \frac{-3}{2}x + 2$ **h.** $y - 2x = 2$
   i. $y = -3$ **j.** $x = 0$
8. Graph:
   a. $y \leqslant x + 1$ **b.** $2x + y - 3 > 0$
9. Write the indicated form of the equation of a line:
   a. standard form
   b. slope-intercept form
   c. point-slope form
10. Find the slope of the line $Ax + By + C = 0$, $B \neq 0$.
11. Find the y intercept for $Ax + By + C = 0$, $B \neq 0$.

6.4 Exercise Answers

1. An equation whose graph is a straight line of the form $Ax + By + C = 0$ in standard form or $y = mx + b$ in slope-intercept form.
2. a and b
3. **a.** $x - y + 1 = 0$ **b.** $2x - 3y - 11 = 0$ **c.** $x + 2y - 7 = 0$
   d. $y - 2 = 0$ **e.** $x - 3 = 0$ **f.** $2x - 3y + 6 = 0$
4. **a.** $y = -3x + 14$ **b.** $y = \frac{-3}{4}x + \frac{5}{4}$ **c.** $y = 0$
   d. $x = -2$
5. **a.** **b.**

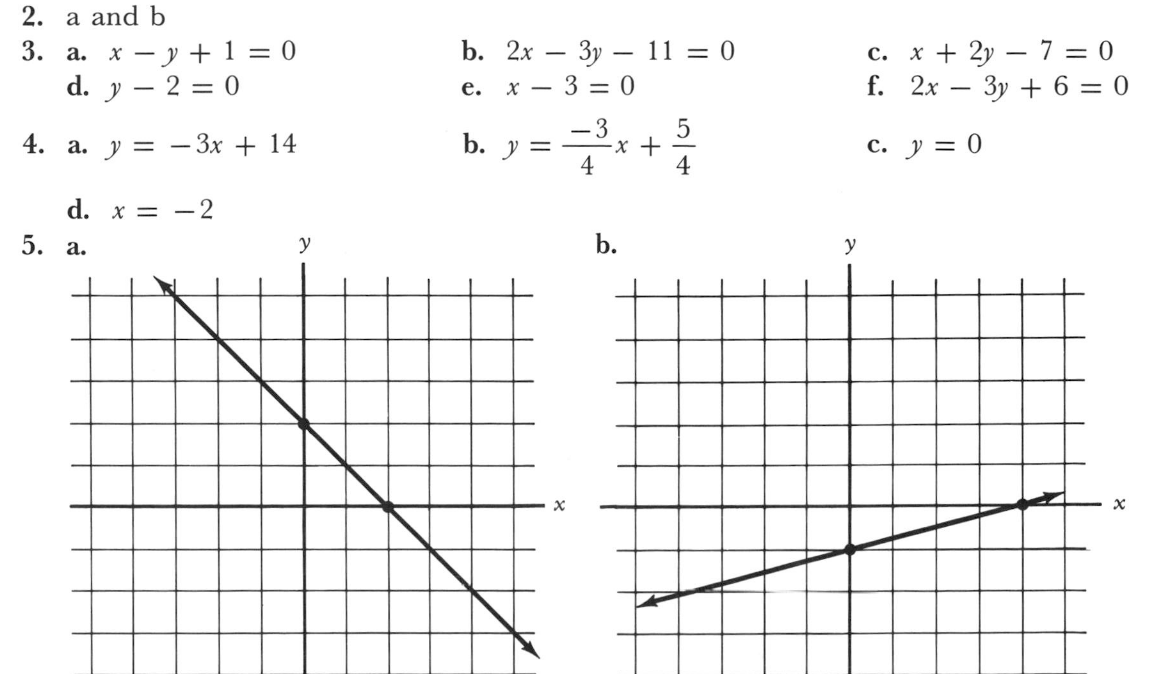

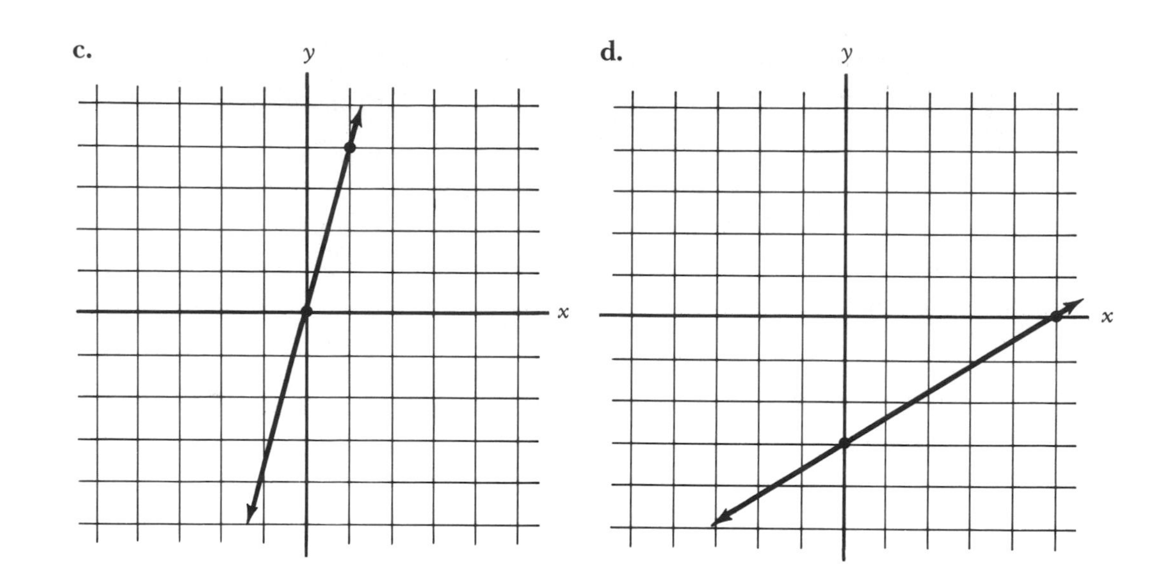
c.
y
x
d.
y
x

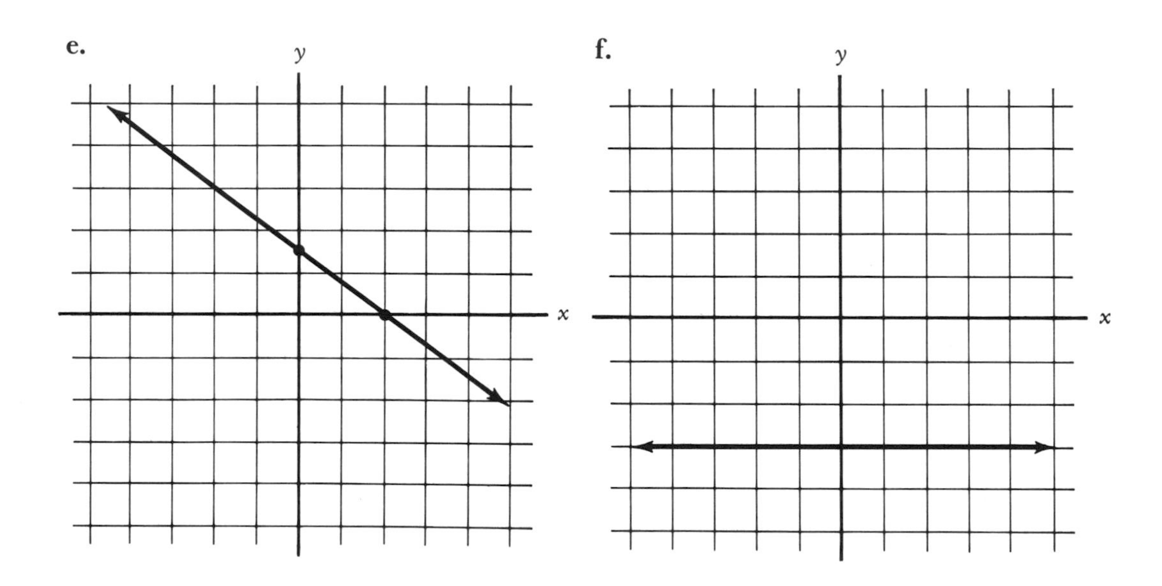
e.
y
x
f.
y
x

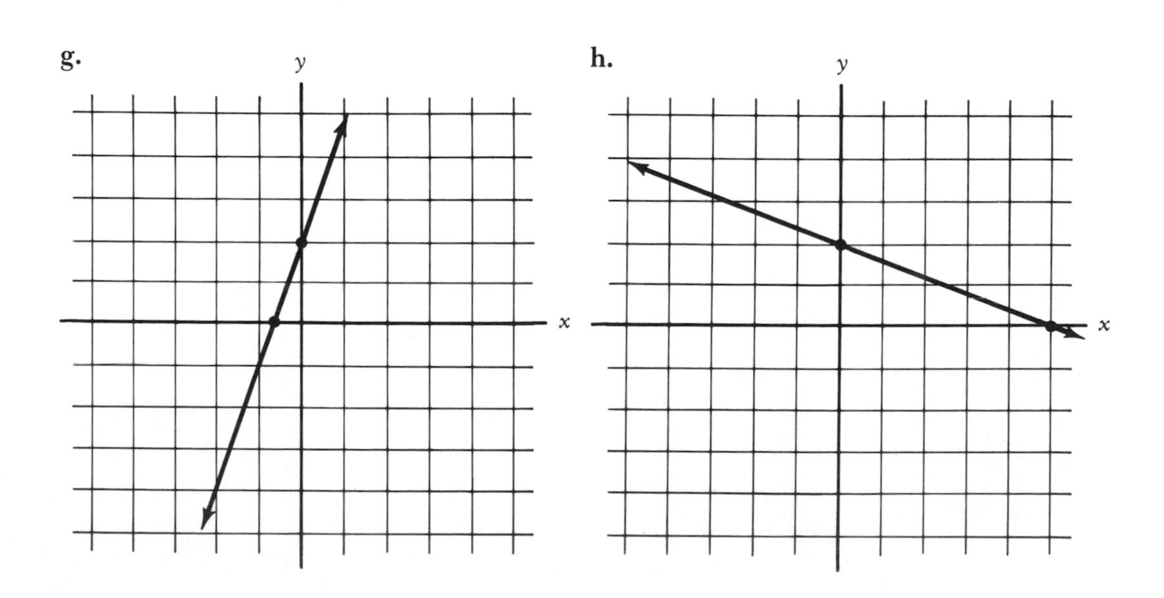
g.
y
x
h.
y
x

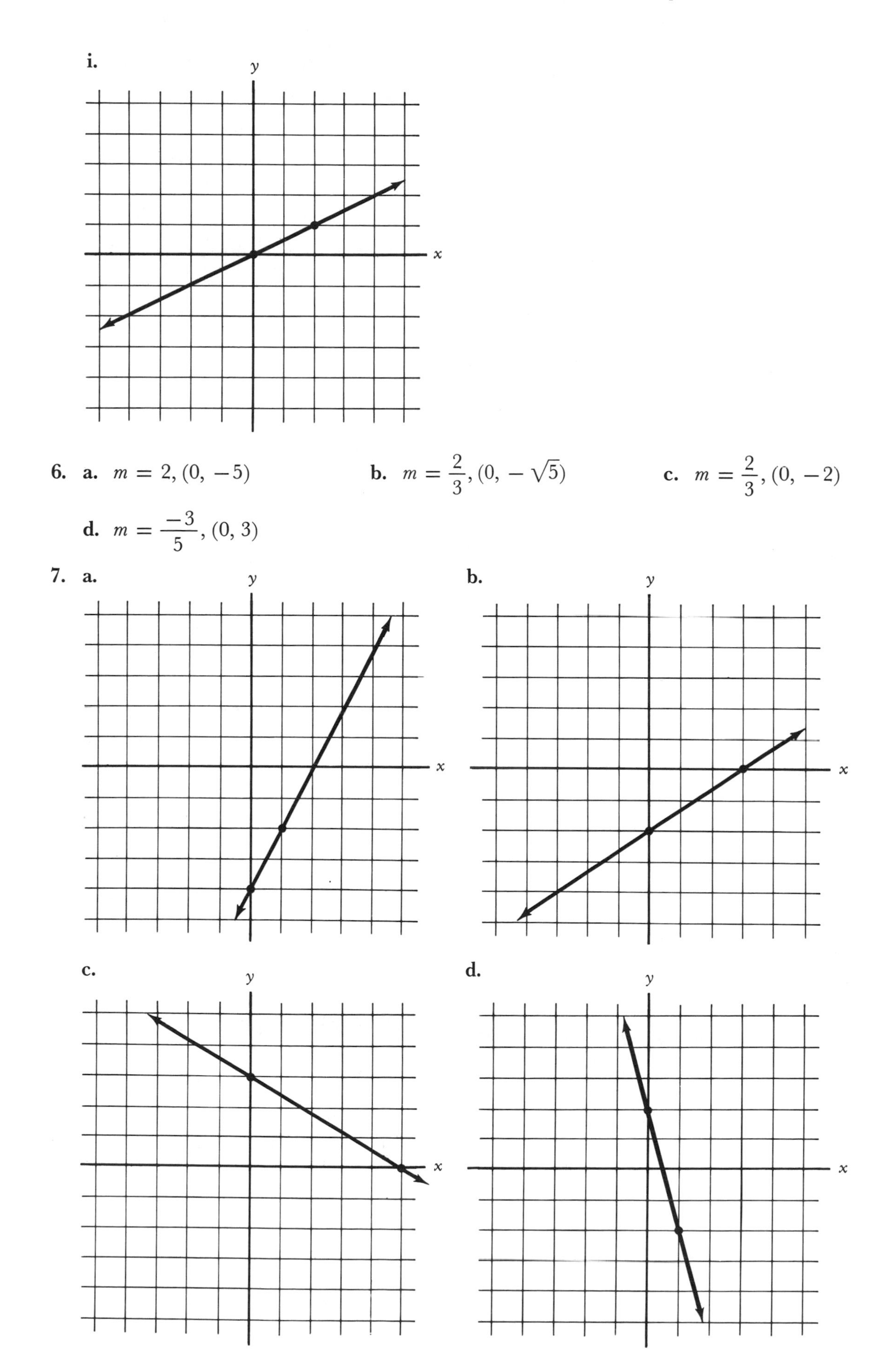

6. **a.** $m = 2, (0, -5)$ **b.** $m = \frac{2}{3}, (0, -\sqrt{5})$ **c.** $m = \frac{2}{3}, (0, -2)$

d. $m = \frac{-3}{5}, (0, 3)$

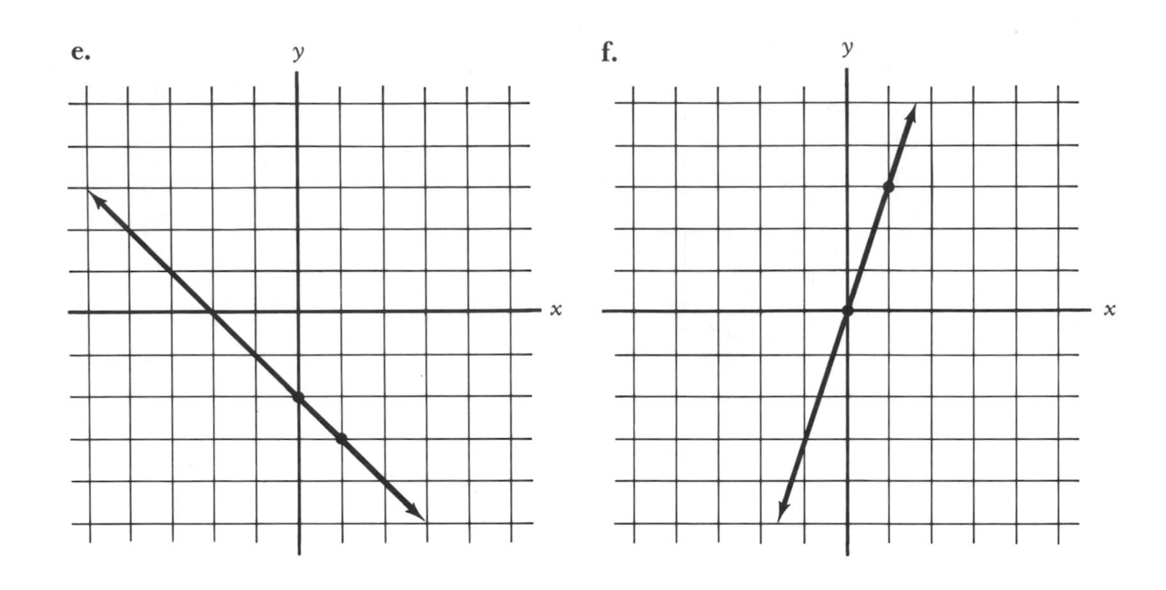
e.
y
x
f.
y
x

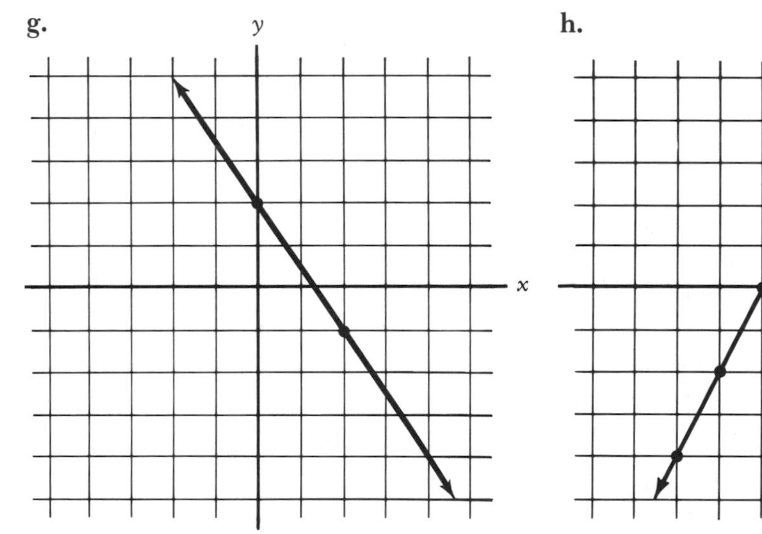
g.
y
x

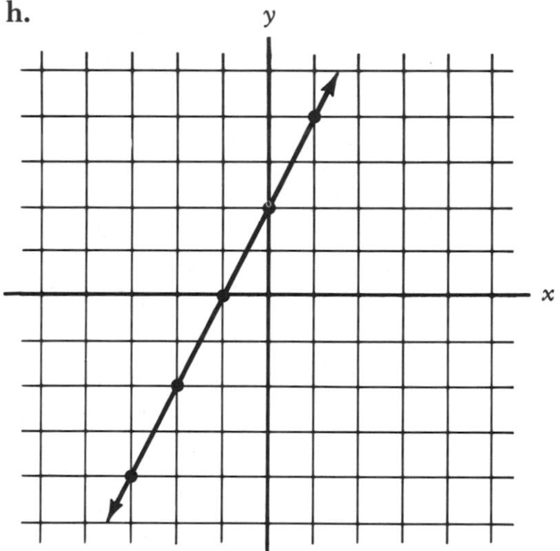
h.
y
x

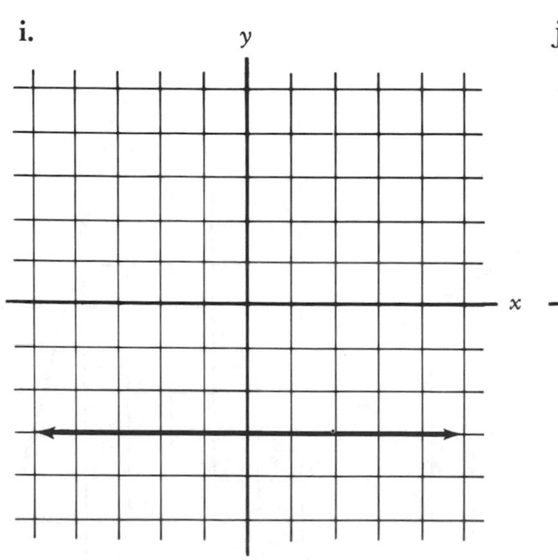
i.
y
x

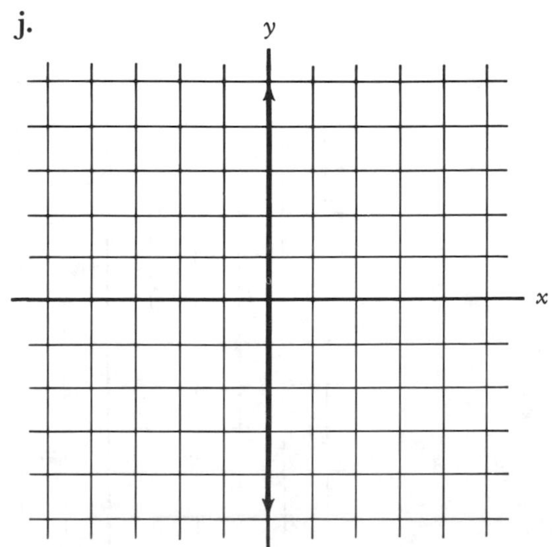
j.
y
x

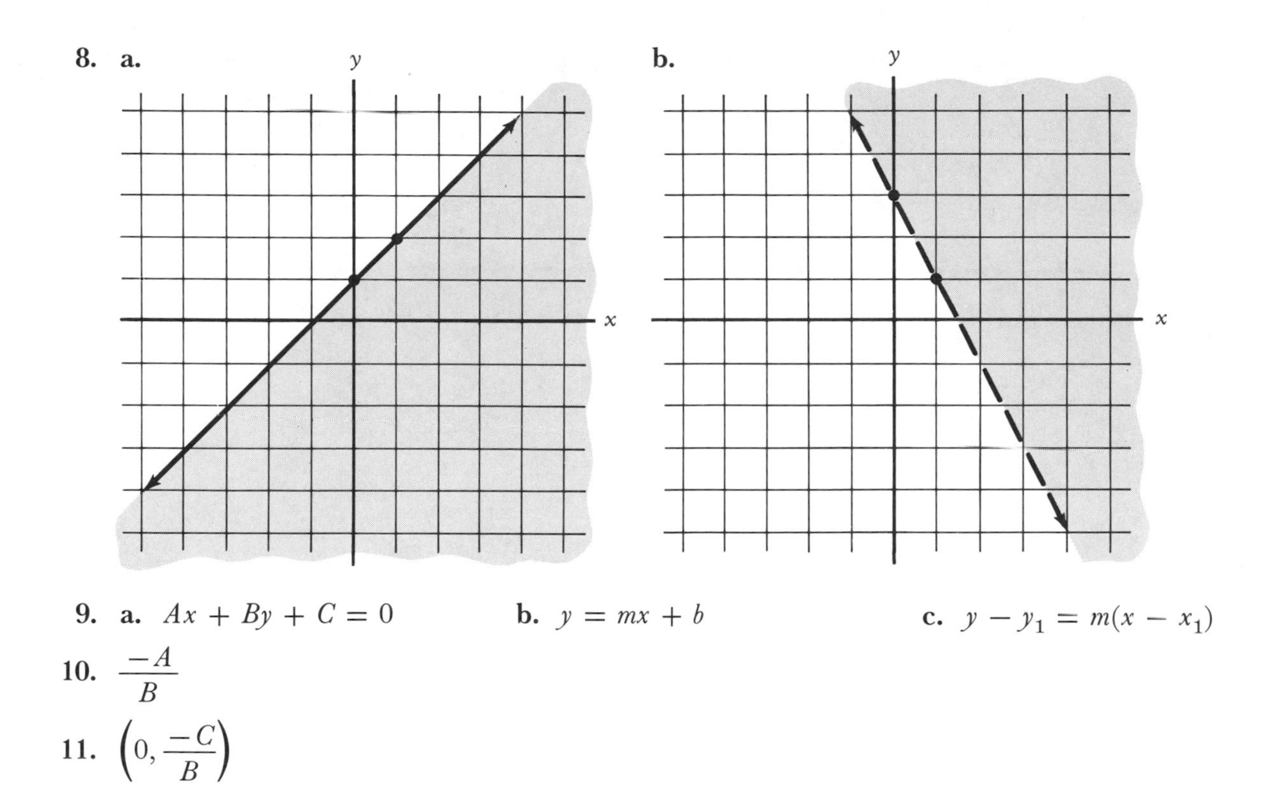

9. **a.** $Ax + By + C = 0$ **b.** $y = mx + b$ **c.** $y - y_1 = m(x - x_1)$

10. $\dfrac{-A}{B}$

11. $\left(0, \dfrac{-C}{B}\right)$

6.5 Quadratic Functions

1 A quadratic function is defined as

$$f(x) = ax^2 + bx + c$$

where a, b, and c are constants and $a \neq 0$. The condition that $a \neq 0$ is necessary because if $a = 0$, $f(x) = ax^2 + bx + c$ becomes $f(x) = bx + c$, which is a linear function. A quadratic function can be represented by a second-degree equation, $y = ax^2 + bx + c$.

Example: Graph $f(x) = x^2$.

Solution

x	y
-2	4
$\frac{-3}{2}$	$\frac{9}{4}$
-1	1
$\frac{-1}{2}$	$\frac{1}{4}$
0	0
$\frac{1}{2}$	$\frac{1}{4}$
1	1
$\frac{3}{2}$	$\frac{9}{4}$
2	4

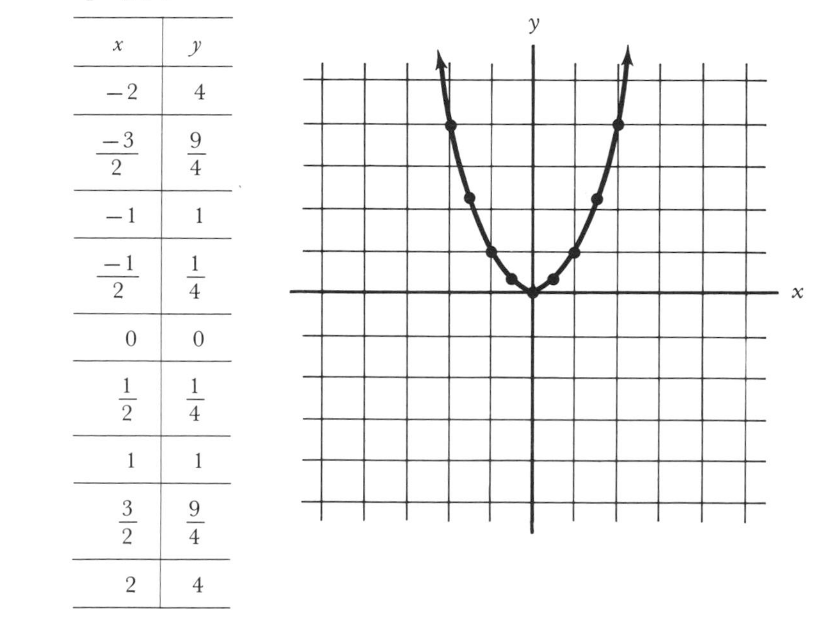

Connect the plotted points with a smooth curve. The assumption is made that all the points on this smooth curve have coordinates that satisfy the equation of the graph. The graph of the quadratic function is called a *parabola.* The above graph is symmetric* to the y axis.

*Symmetry to a line means that if the paper were folded on the line, the two halves of the graph would coincide.

Q1 Complete the table and graph $f(x) = 2x^2$ (it is necessary to lower the x axis to plot a sufficient number of points).

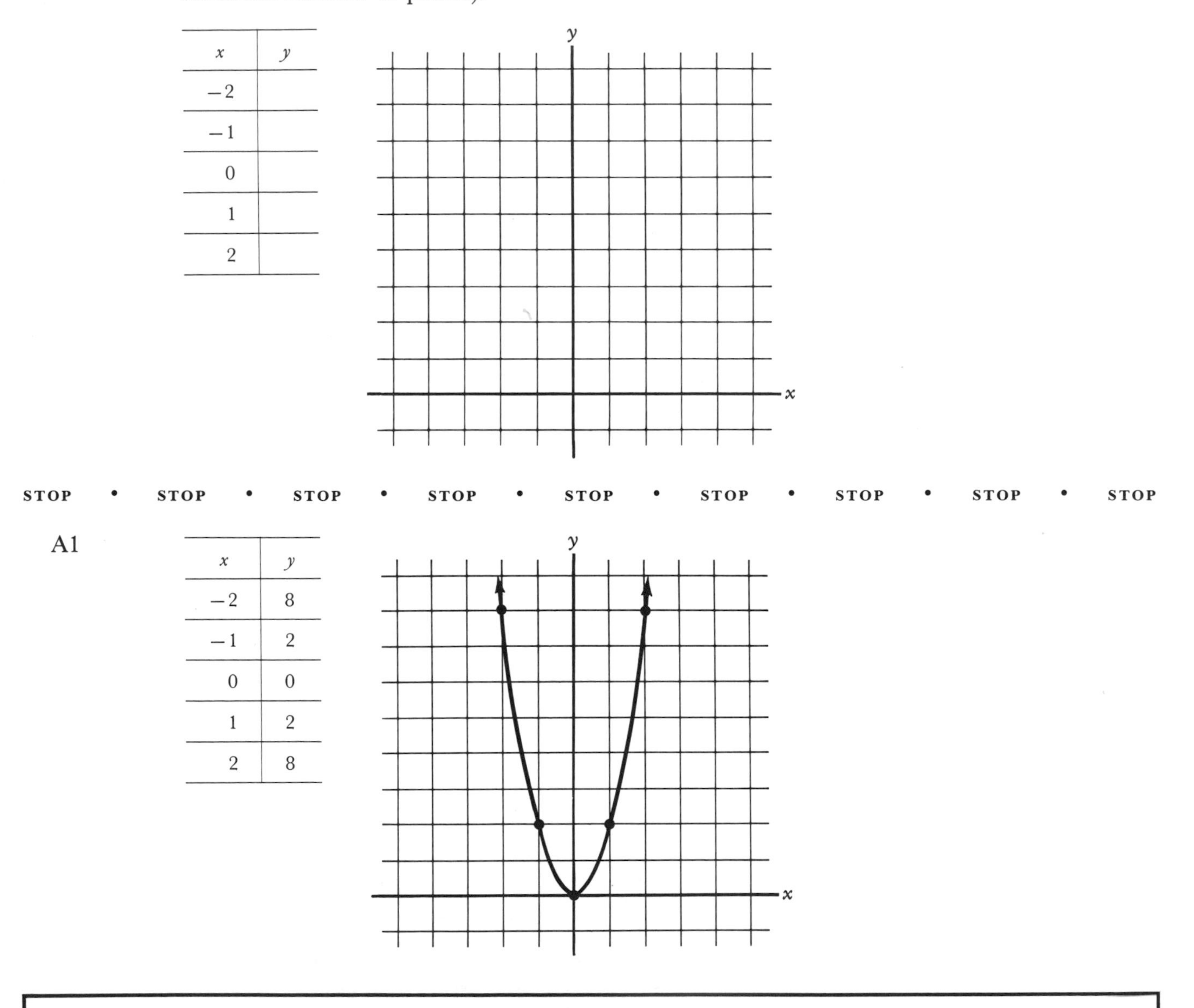

x	y
-2	
-1	
0	
1	
2	

STOP • STOP • STOP • STOP • STOP • STOP • STOP • STOP • STOP

A1

x	y
-2	8
-1	2
0	0
1	2
2	8

2 The highest or lowest point on the graph of a quadratic function is called the *vertex* of the parabola. The vertex of the graph of $f(x) = x^2$ in Frame 1 is (0, 0). Since the parabola opens upward, the vertex is the lowest point on the graph.

To graph $f(x) = -x^2$, it should be noted that $-x^2$ does *not* equal $-x \cdot -x$. The expression $-x^2$ means $-(x)(x)$. $f(x) = -x^2$ could be written as $f(x) = -1 \cdot x^2$; that is, $a = -1$. Examples are:

$$-2^2 = -(2)(2) = -4$$
$$-3^2 = -(3)(3) = -9$$

$-a^2 = -a \cdot a$
$(-2)^2 = (-2)(-2) = 4$
$(-3)^2 = (-3)(-3) = 9$
$(-a)^2 = (-a)(-a) = a^2$

Q2 Evaluate:

a. -7^2 ______ **b.** $(-2x)^2$ ______ **c.** $-x^2,\ x = 3$ ______

d. $-x^2,\ x = -3$ ______

STOP • STOP • STOP • STOP • STOP • STOP • STOP • STOP • STOP

A2 **a.** -49 **b.** $4x^2$ **c.** -9

d. -9

Q3 Complete the table and graph:

a. $f(x) = -x^2$

x	y
-2	
-1	
0	
1	
2	

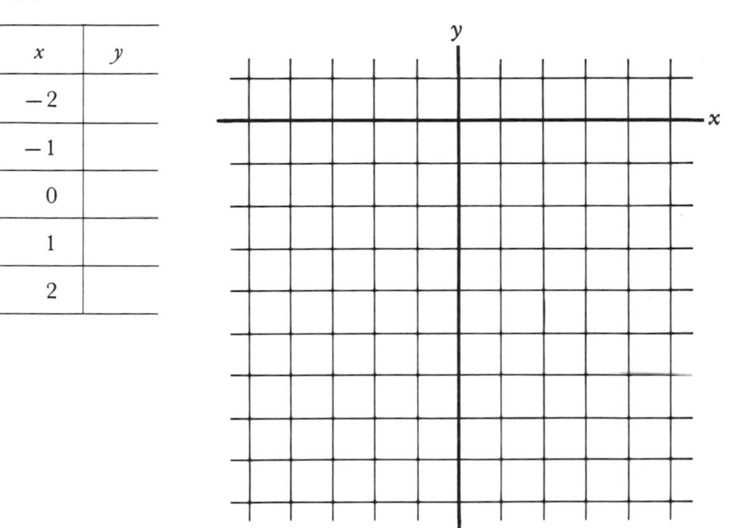

b. $g(x) = -2x^2$

x	y
-2	
-1	
0	
1	
2	

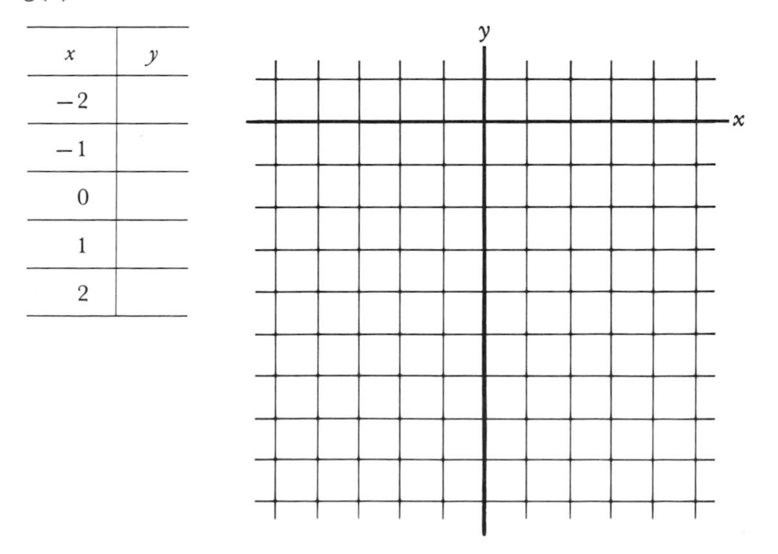

STOP • STOP • STOP • STOP • STOP • STOP • STOP • STOP • STOP

A3

a.

x	y
-2	-4
-1	-1
0	0
1	-1
2	-4

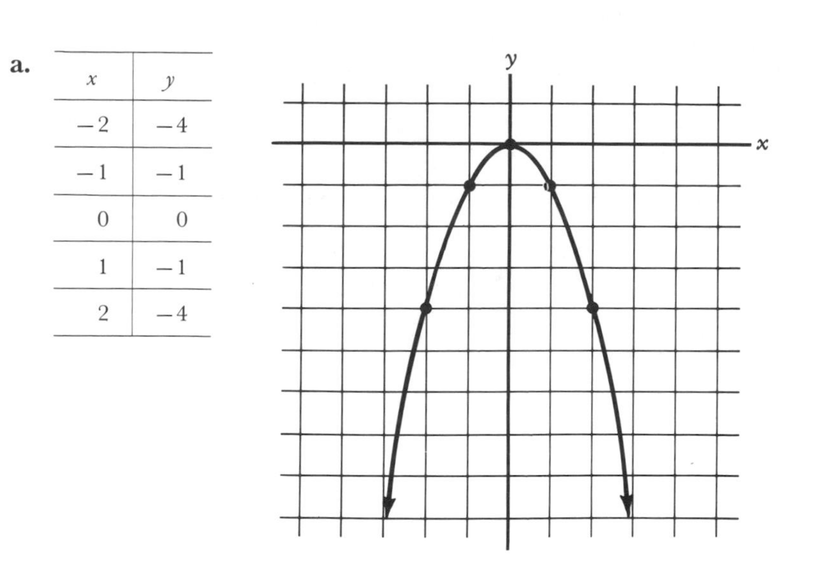

b.

x	y
-2	-8
-1	-2
0	0
1	-2
2	-8

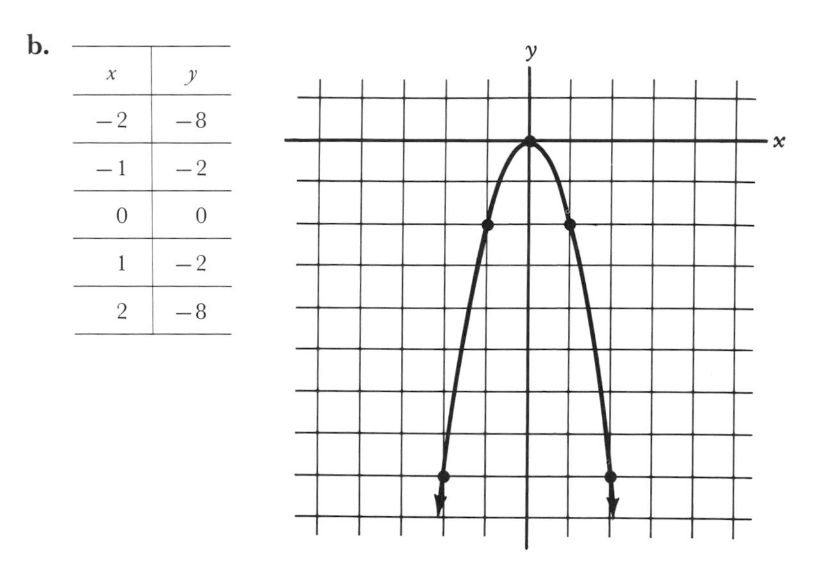

Q4

a. If $f(x) = -x^2$ is considered to be in the form $f(x) = ax^2 + bx + c$, then $a =$ _____ $b =$ _____ $c =$ _____.

b. For $g(x) = -2x^2$, $a =$ _____ $b =$ _____ $c =$ _____.

c. The vertex of both f and g is __________.

d. In Q3, is the vertex the highest or lowest point of the parabola? (Circle answer.)

STOP • STOP • STOP • STOP • STOP • STOP • STOP • STOP • STOP

A4 **a.** $-1, 0, 0$ **b.** $-2, 0, 0$ **c.** $(0, 0)$ **d.** highest

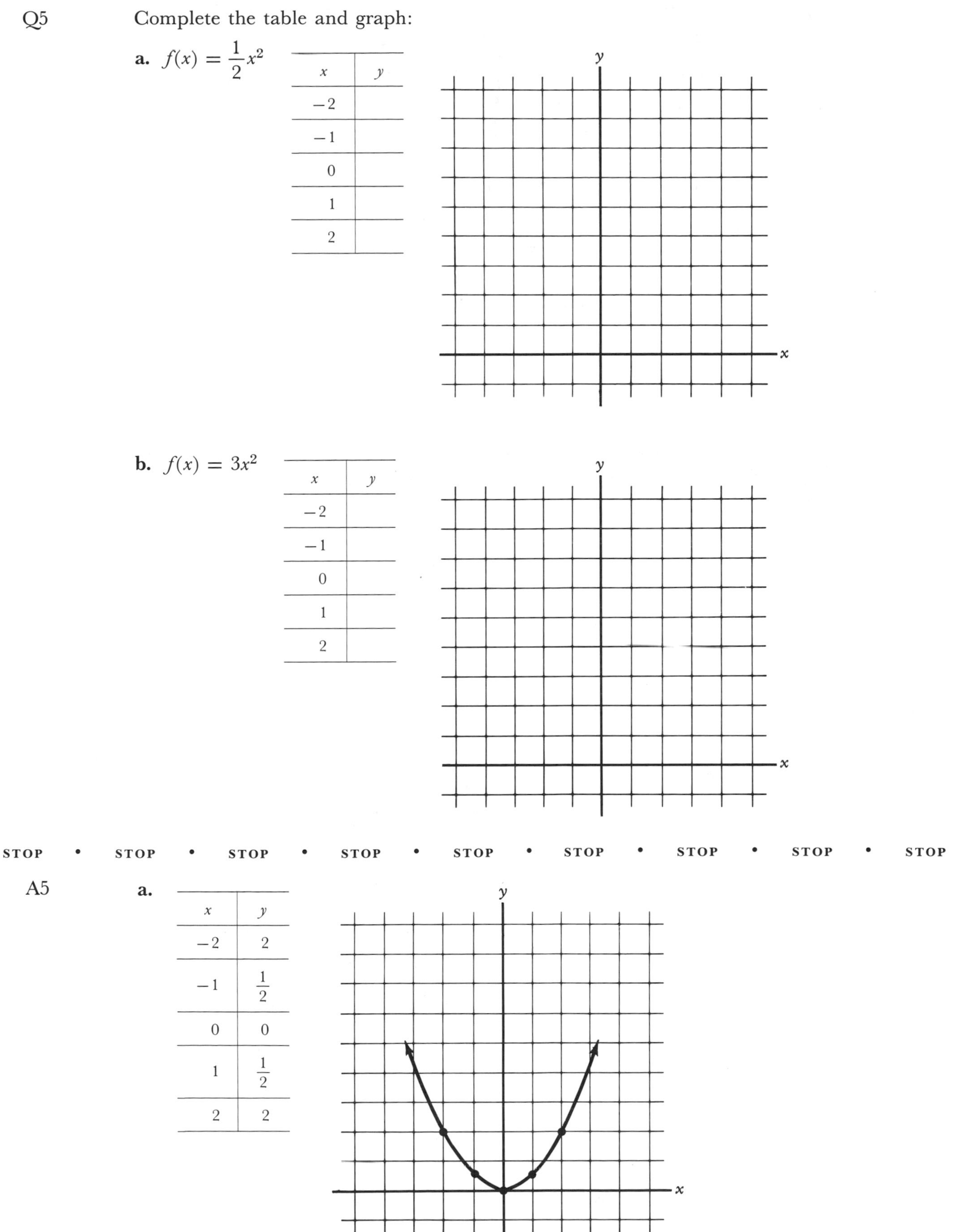

Q5 Complete the table and graph:

a. $f(x) = \frac{1}{2}x^2$

x	y
-2	
-1	
0	
1	
2	

b. $f(x) = 3x^2$

x	y
-2	
-1	
0	
1	
2	

STOP • STOP • STOP • STOP • STOP • STOP • STOP • STOP • STOP

A5 **a.**

x	y
-2	2
-1	$\frac{1}{2}$
0	0
1	$\frac{1}{2}$
2	2

b.

x	y
-2	12
-1	3
0	0
1	3
2	12

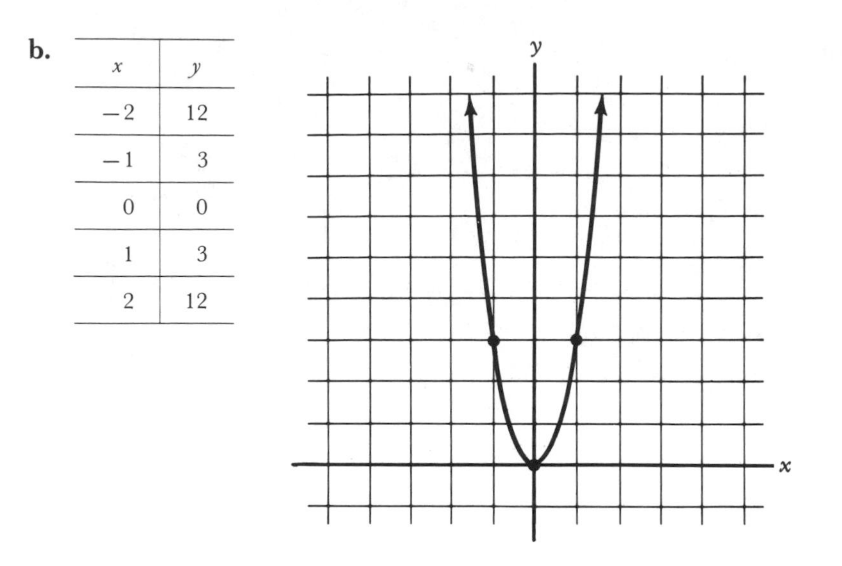

> **3** The results of the previous problems can be summarized as follows. When graphing $f(x) = ax^2$:
>
> 1. If $a > 0$, the parabola opens upward.
> 2. If $a < 0$, the parabola opens downward.
> 3. The larger $|a|$, the "narrower" the parabola.

Q6 Match the graph with its function:

a. $f(x) = \dfrac{1}{16}x^2$ ______

b. $g(x) = \dfrac{-1}{4}x^2$ ______

c. $h(x) = 3x^2$ ______

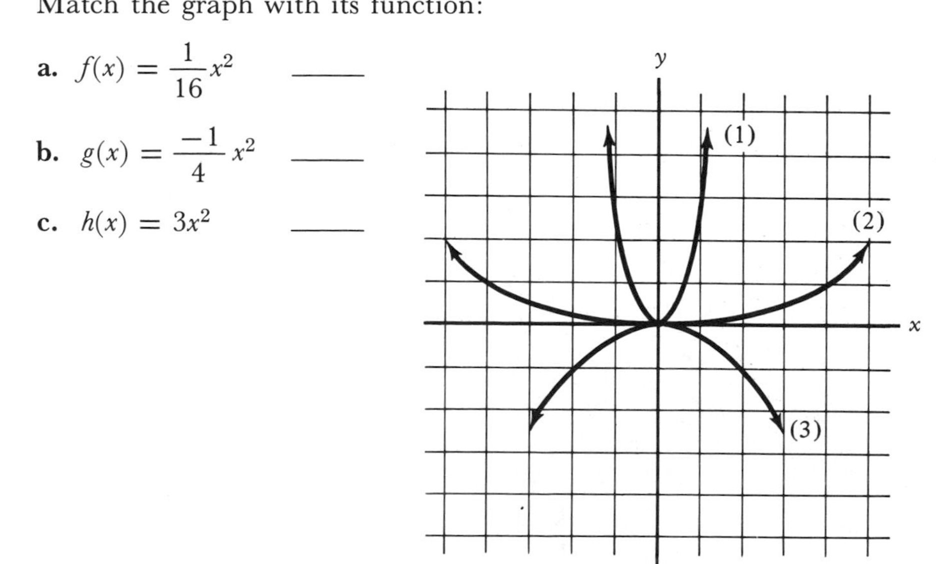

STOP • STOP • STOP • STOP • STOP • STOP • STOP • STOP • STOP

A6 **a.** 2 **b.** 3 **c.** 1

Q7 Complete the table and graph:

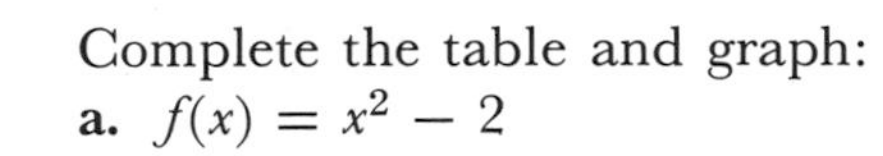

a. $f(x) = x^2 - 2$

x	y
-2	
-1	
0	
1	
2	

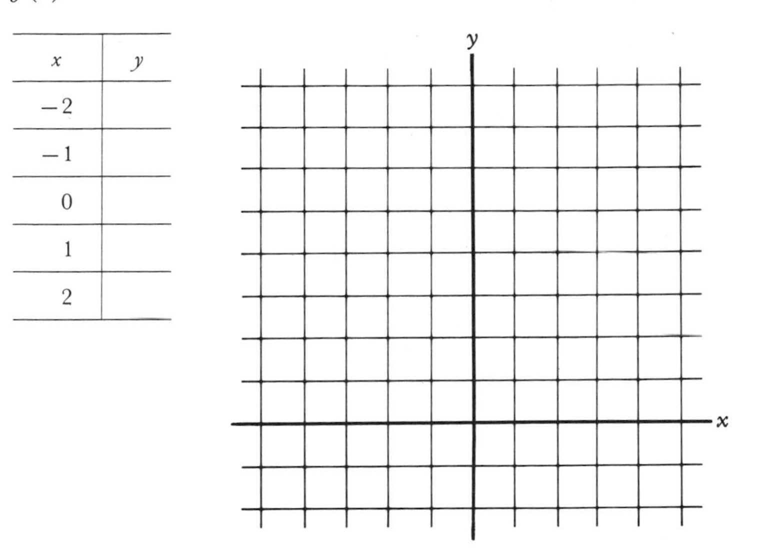

b. $f(x) = x^2 + 3$

x	y
-2	
-1	
0	
1	
2	

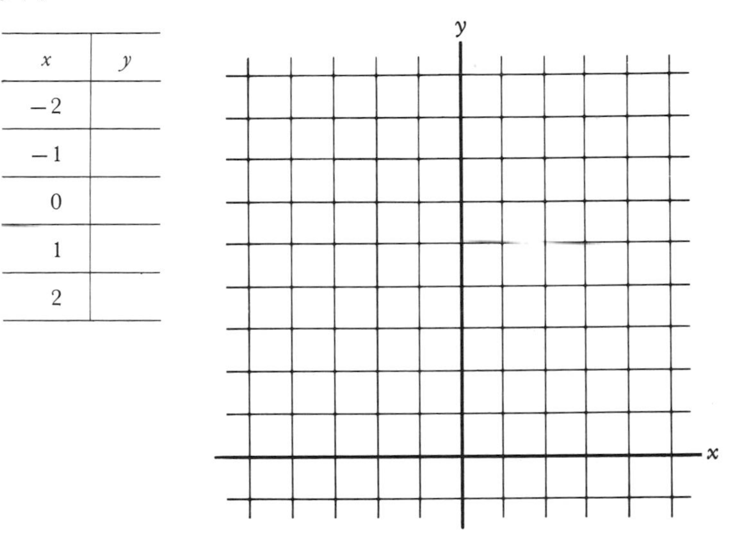

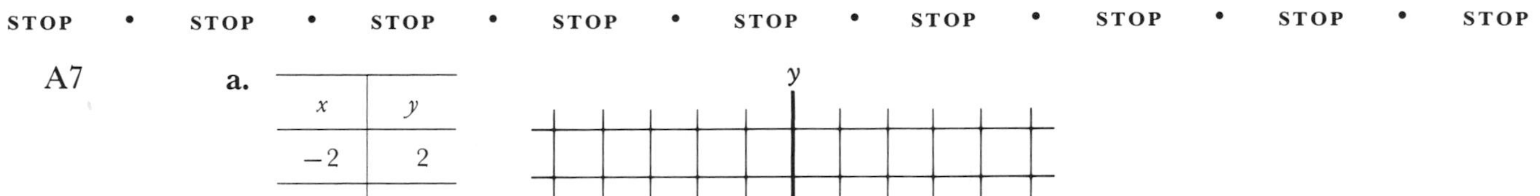

STOP • STOP • STOP • STOP • STOP • STOP • STOP • STOP • STOP

A7 **a.**

x	y
-2	2
-1	-1
0	-2
1	-1
2	2

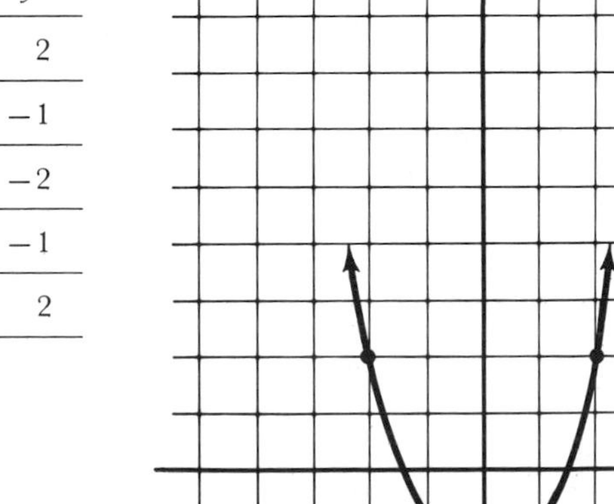

b.

x	y
-2	7
-1	4
0	3
1	4
2	7

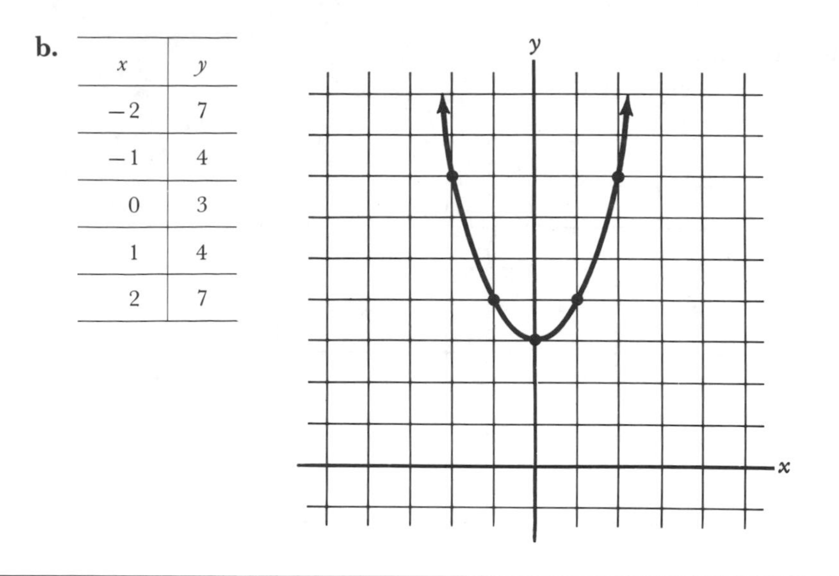

Q8 Graph (if necessary complete a table):

a. $f(x) = -x^2 - 2$ **b.** $f(x) = -2x^2 + 4$

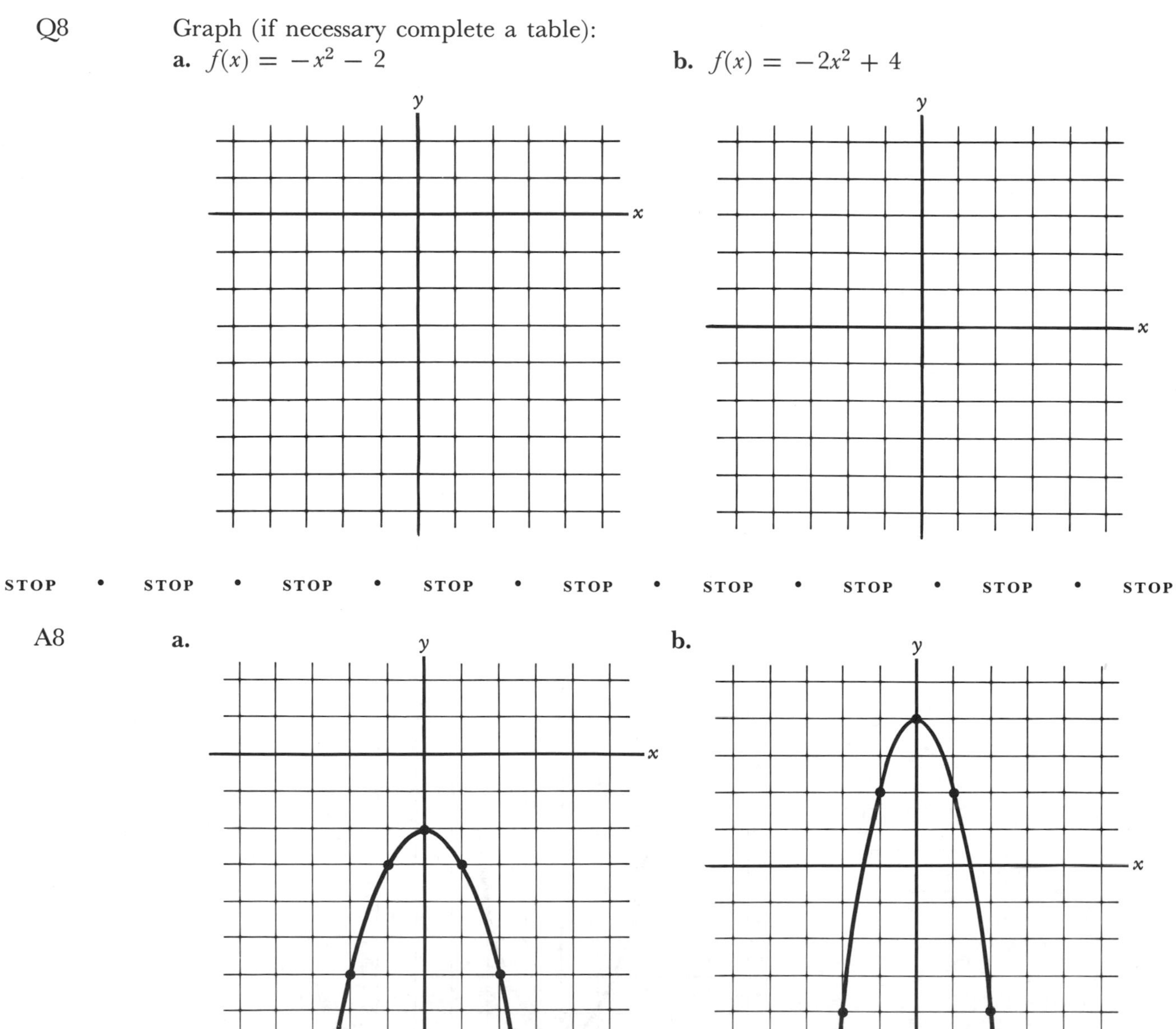

STOP • STOP • STOP • STOP • STOP • STOP • STOP • STOP • STOP

A8 **a.** **b.**

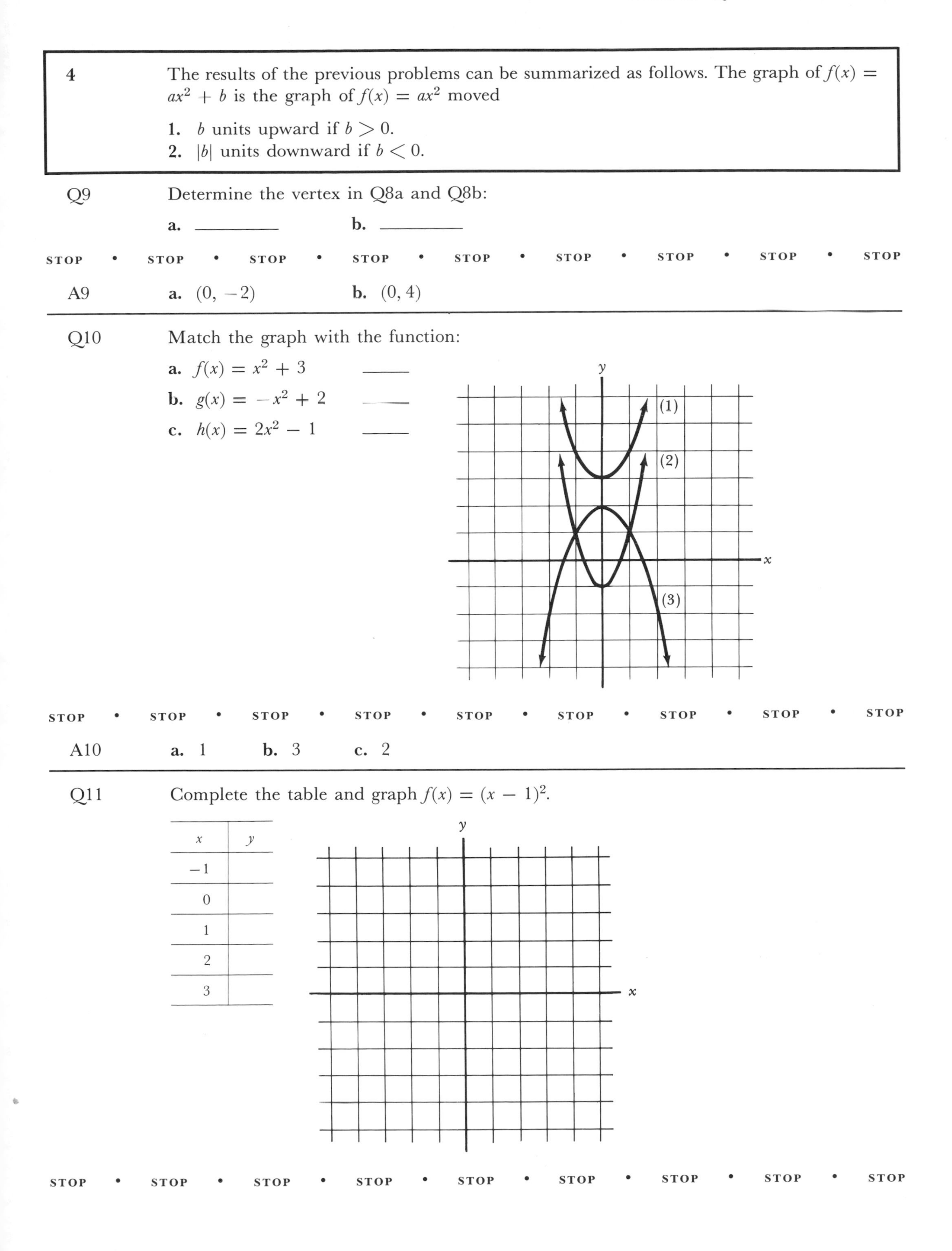

4 The results of the previous problems can be summarized as follows. The graph of $f(x) = ax^2 + b$ is the graph of $f(x) = ax^2$ moved

1. b units upward if $b > 0$.
2. $|b|$ units downward if $b < 0$.

Q9 Determine the vertex in Q8a and Q8b:

a. ________ **b.** ________

STOP • STOP • STOP • STOP • STOP • STOP • STOP • STOP • STOP

A9 **a.** $(0, -2)$ **b.** $(0, 4)$

Q10 Match the graph with the function:

a. $f(x) = x^2 + 3$ ____

b. $g(x) = -x^2 + 2$ ____

c. $h(x) = 2x^2 - 1$ ____

STOP • STOP • STOP • STOP • STOP • STOP • STOP • STOP • STOP

A10 **a.** 1 **b.** 3 **c.** 2

Q11 Complete the table and graph $f(x) = (x - 1)^2$.

x	y
−1	
0	
1	
2	
3	

STOP • STOP • STOP • STOP • STOP • STOP • STOP • STOP • STOP

A11

x	y
-1	4
0	1
1	0
2	1
3	4

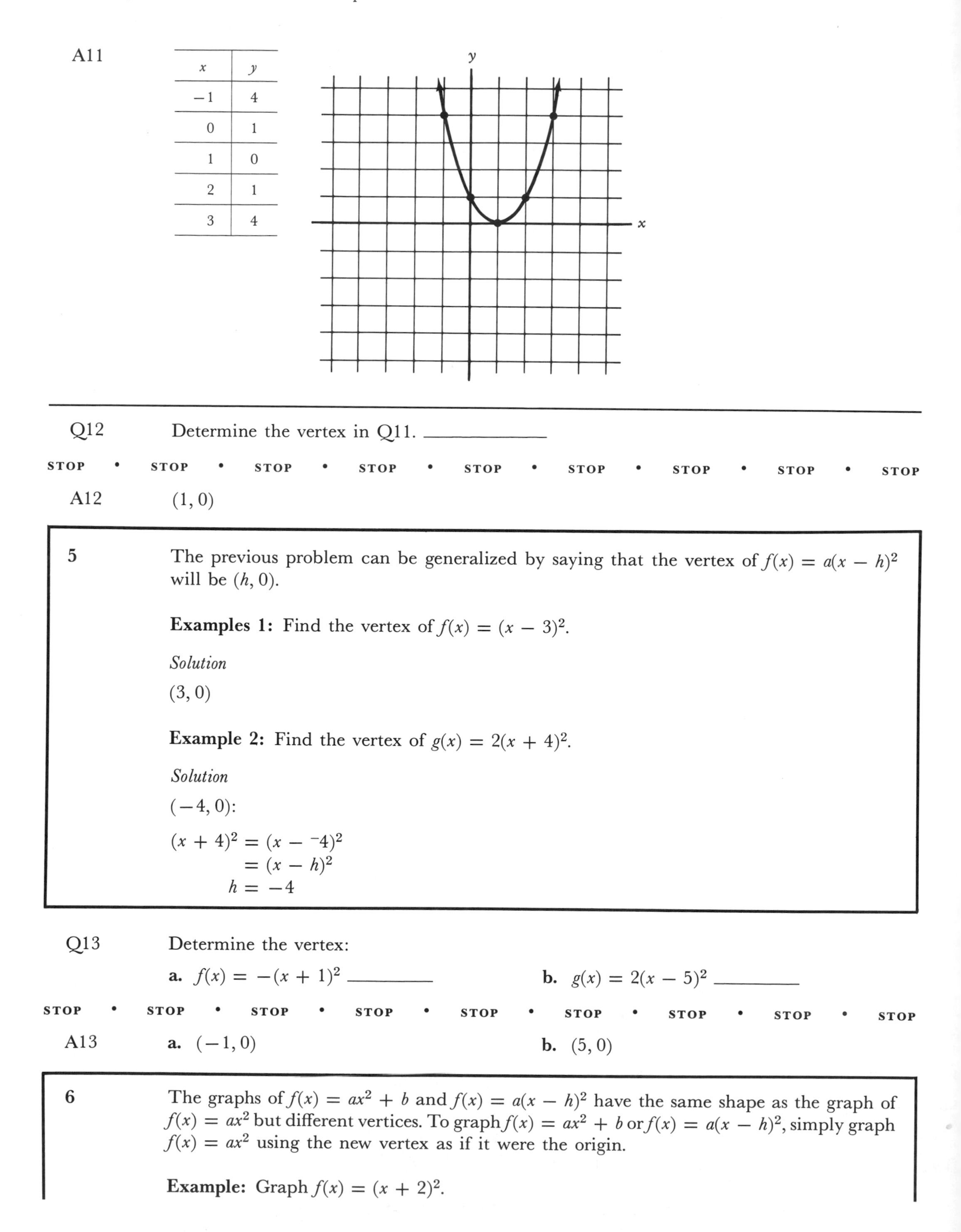

Q12 Determine the vertex in Q11. ________

STOP • STOP • STOP • STOP • STOP • STOP • STOP • STOP • STOP

A12 $(1, 0)$

5 The previous problem can be generalized by saying that the vertex of $f(x) = a(x - h)^2$ will be $(h, 0)$.

Examples 1: Find the vertex of $f(x) = (x - 3)^2$.

Solution

$(3, 0)$

Example 2: Find the vertex of $g(x) = 2(x + 4)^2$.

Solution

$(-4, 0)$:

$$\begin{aligned}(x + 4)^2 &= (x - {}^-4)^2 \\ &= (x - h)^2 \\ h &= -4\end{aligned}$$

Q13 Determine the vertex:

a. $f(x) = -(x + 1)^2$ ________ b. $g(x) = 2(x - 5)^2$ ________

STOP • STOP • STOP • STOP • STOP • STOP • STOP • STOP • STOP

A13 a. $(-1, 0)$ b. $(5, 0)$

6 The graphs of $f(x) = ax^2 + b$ and $f(x) = a(x - h)^2$ have the same shape as the graph of $f(x) = ax^2$ but different vertices. To graph $f(x) = ax^2 + b$ or $f(x) = a(x - h)^2$, simply graph $f(x) = ax^2$ using the new vertex as if it were the origin.

Example: Graph $f(x) = (x + 2)^2$.

Solution

From the vertex $(-2, 0)$, graph $f(x) = x^2$.

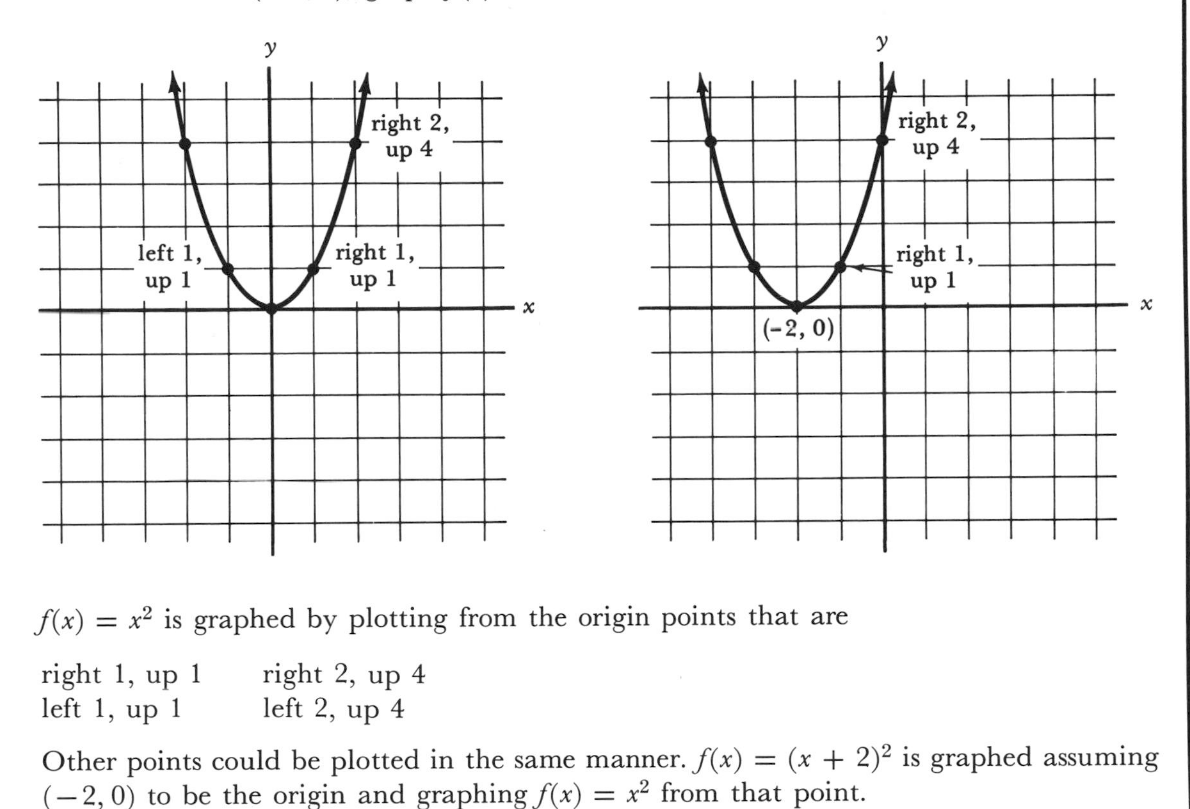

$f(x) = x^2$ is graphed by plotting from the origin points that are

right 1, up 1 right 2, up 4
left 1, up 1 left 2, up 4

Other points could be plotted in the same manner. $f(x) = (x + 2)^2$ is graphed assuming $(-2, 0)$ to be the origin and graphing $f(x) = x^2$ from that point.

Q14 Graph by graphing $f(x) = x^2$ from the proper vertex:

a. $f(x) = (x + 1)^2$ **b.** $f(x) = (x - 2)^2$

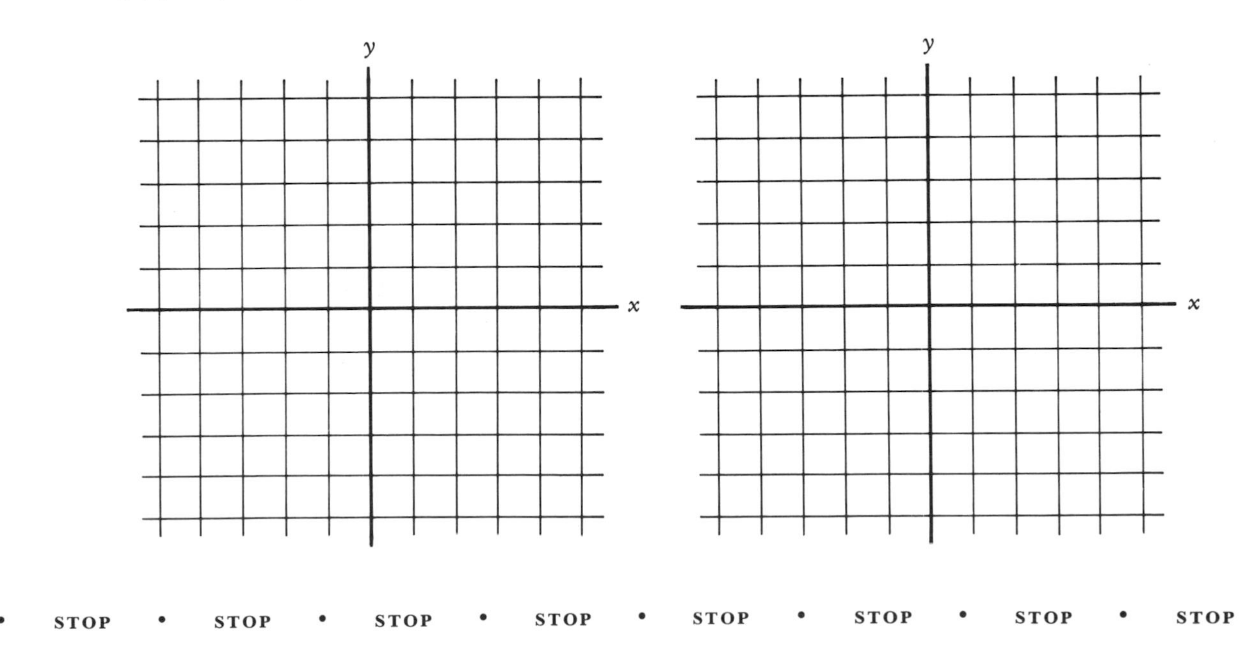

STOP • STOP • STOP • STOP • STOP • STOP • STOP • STOP • STOP

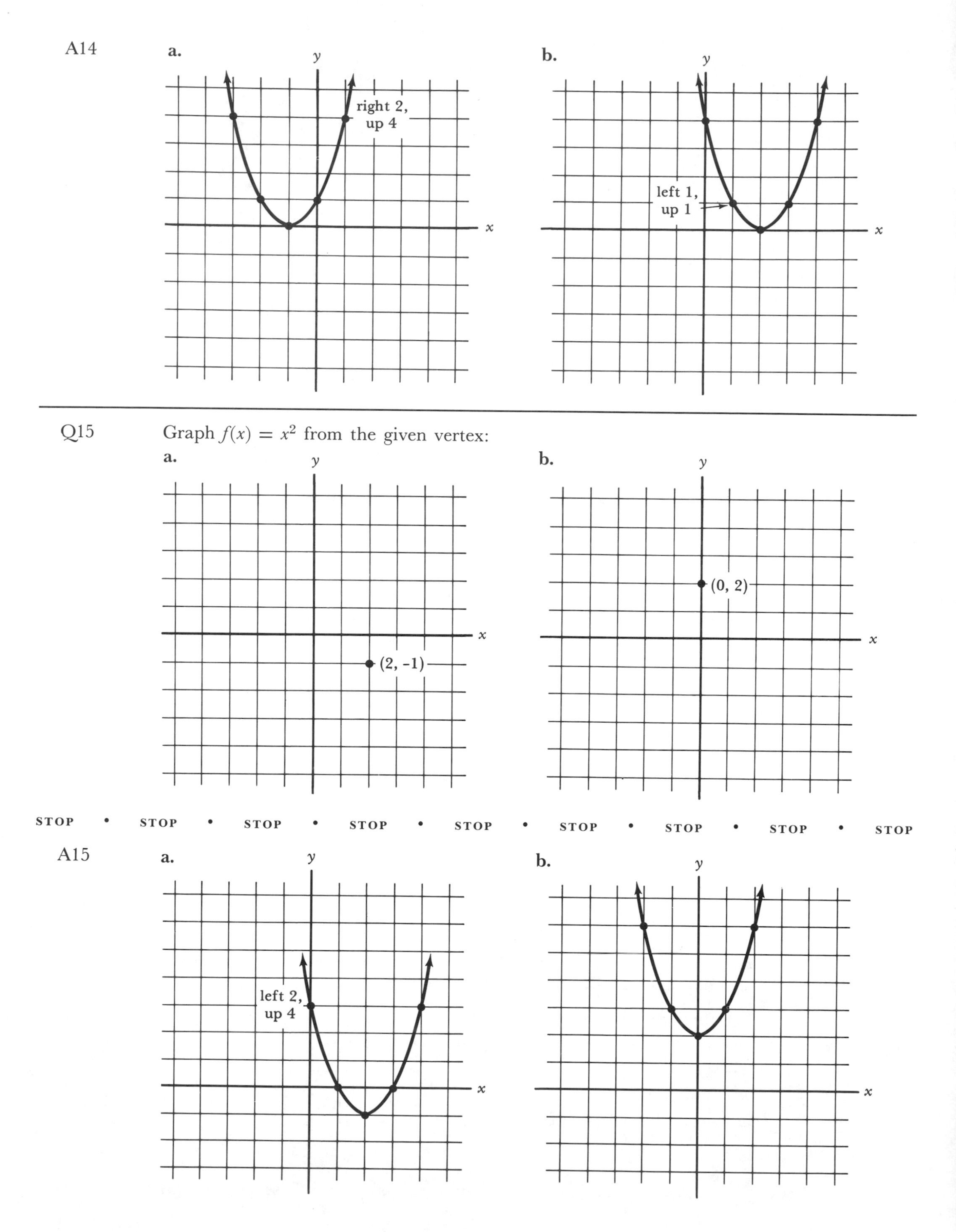
A14
a.
right 2,
up 4
b.
left 1,
up 1
Q15
Graph $f(x) = x^2$ from the given vertex:
a.
(2, -1)
b.
(0, 2)
STOP • STOP • STOP • STOP • STOP • STOP • STOP • STOP • STOP
A15
a.
left 2,
up 4
b.

7 The quadratic function $f(x) = a(x - h)^2 + k$ has its vertex at (h, k) and can be graphed by graphing $f(x) = ax^2$ from the vertex (h, k).

Example: Graph $f(x) = (x + 1)^2 + 2$.

Solution

$h = -1$ and $k = 2$; hence the vertex is $(-1, 2)$. Graph $f(x) = x^2$ from the vertex $(-1, 2)$.

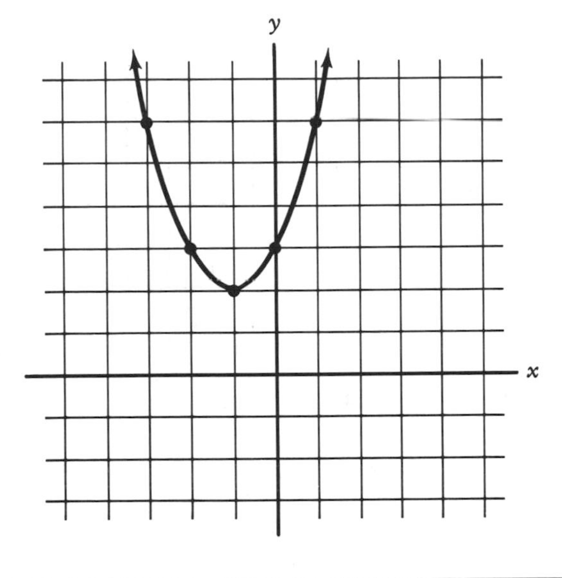

Q16 Determine the vertex:

a. $f(x) = (x - 2)^2 + 3$

b. $f(x) = 2(x + 3)^2 - 4$

c. $f(x) = -2(x + 2)^2 + 2$

d. $f(x) = m(x + n)^2 - t$

STOP • STOP • STOP • STOP • STOP • STOP • STOP • STOP • STOP

A16

a. $(2, 3)$

b. $(-3, -4)$

c. $(-2, 2)$

d. $(-n, -t)$:

$f(x) = m(x + n)^2 - t$

$= m(x - {}^{-}n)^2 + {}^{-}t$

Compare to:

$f(x) = a(x - h)^2 + k$

Hence $h = -n$ and $k = -t$

Q17 Graph:

a. $f(x) = (x - 2)^2 - 1$

b. $f(x) = (x + 1)^2 - 3$

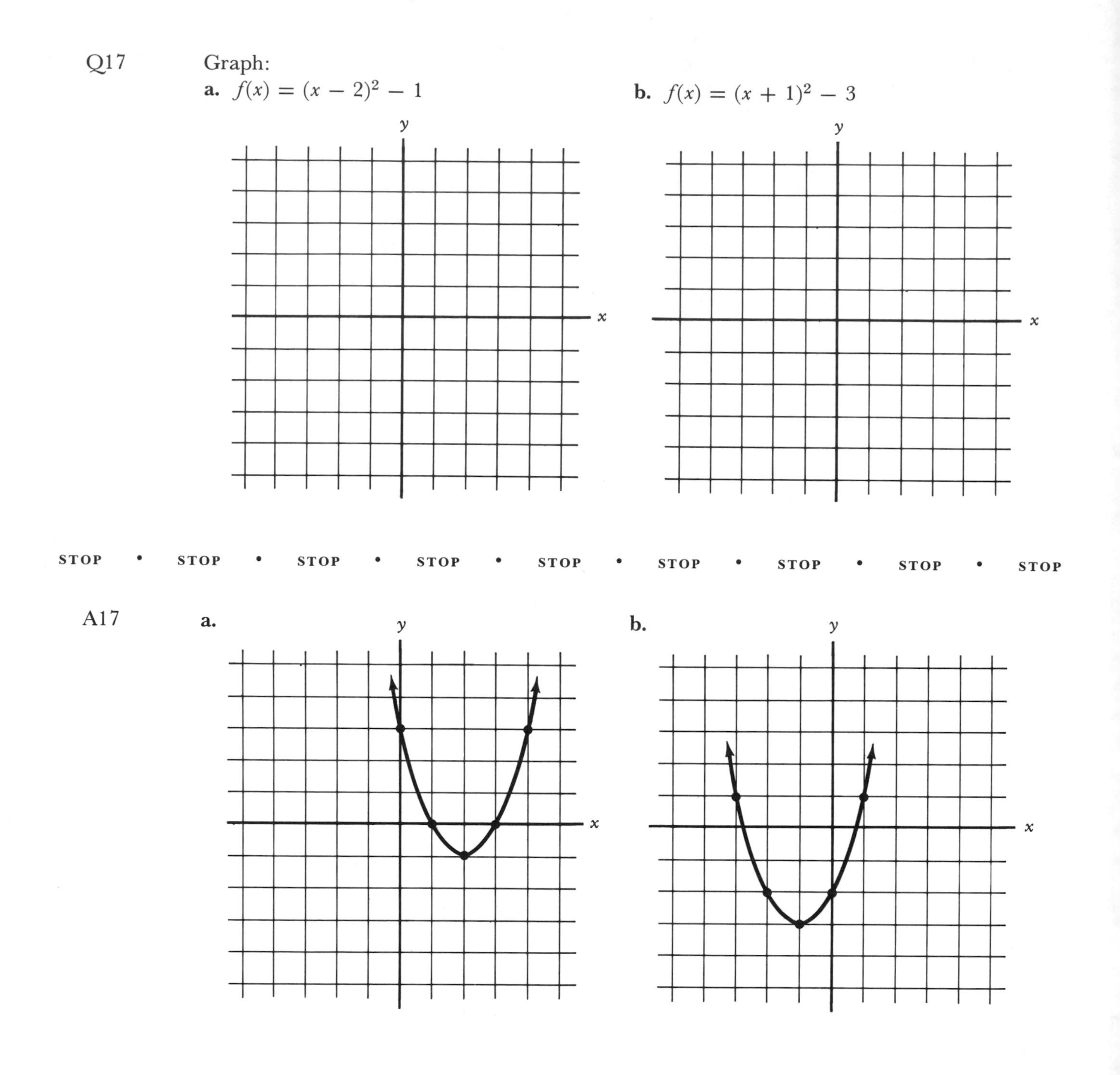

---

8 The function $f(x) = -(x + 1)^2 + 1$ can be graphed by graphing $f(x) = -x^2$ from the vertex $(-1, 1)$. From $(-1, 1)$ plot:

right 1, down 1
right 2, down 4
left 1, down 1
left 2, down 4

Other points could be plotted in the same manner.

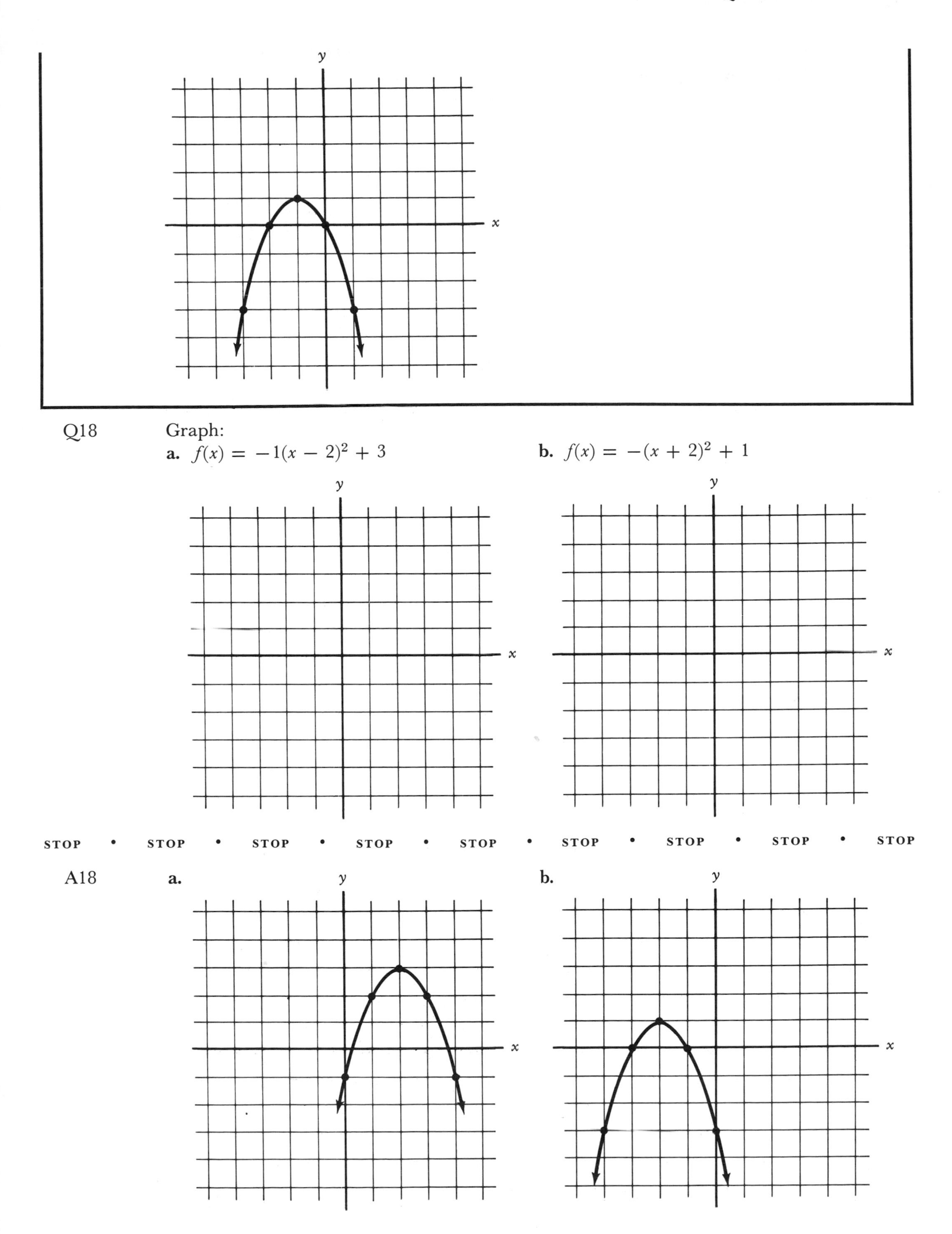

Q18 Graph:

a. $f(x) = -1(x - 2)^2 + 3$

b. $f(x) = -(x + 2)^2 + 1$

STOP • **STOP** • **STOP** • **STOP** • **STOP** • **STOP** • **STOP** • **STOP** • **STOP**

A18 **a.** **b.**

9 The quadratic function $f(x) = ax^2 + bx + c$ can be graphed by converting to the form $f(x) = a(x - h)^2 + k$ by completing the square.

Example: Write $f(x) = x^2 + 6x + 5$ in the form $f(x) = (x - h)^2 + k$, where $a = 1$.

Solution

Group the x^2 and x terms inside parentheses, leaving a space for a constant to be added. Leave the given constant outside the parentheses.

$f(x) = (x^2 + 6x + \_\_\_\_) + 5 - \_\_\_\_$

add constant ↗ subtract constant ↑

Now determine the constant that makes the trinomial inside the parentheses a perfect square. Since the converted equation must be equivalent to the original, the constant added (inside the parentheses) must be subtracted (outside the parentheses) so that the total added is zero.

$f(x) = (x^2 + 6x + \underline{9}) + 5 - \underline{9}$

$f(x) = (x + 3)^2 - 4$

Q19 Write in the form $f(x) = a(x - h)^2 + k$ by completing the steps.

$f(x) = x^2 + 4x + 5$

$f(x) = (x^2 + 4x + \_\_\_\_) + 5 - \_\_\_\_$

$f(x) = \_\_\_\_\_\_\_\_$

STOP • STOP • STOP • STOP • STOP • STOP • STOP • STOP • STOP

A19 $f(x) = (x^2 + 4x + \underline{4}) + 5 - \underline{4}$

$f(x) = (x + 2)^2 + 1$

Q20 Write in the form $f(x) = a(x - h)^2 + k$:

a. $f(x) = x^2 - 4x - 4$

b. $f(x) = x^2 - 8x + 10$

STOP • STOP • STOP • STOP • STOP • STOP • STOP • STOP • STOP

A20 **a.** $f(x) = (x - 2)^2 - 8$:

$f(x) = (x^2 - 4x + \_\_\_\_) - 4 - \_\_\_\_$

$f(x) = (x - 2)^2 - 8$

b. $f(x) = (x - 4)^2 - 6$:

$f(x) = (x^2 - 8x + \_\_\_\_) + 10 - \_\_\_\_$

$f(x) = (x^2 - 8x + 16) + 10 - 16$

$f(x) = (x - 4)^2 - 6$

Q21 Determine the vertex:

a. $f(x) = x^2 + 3x + 1$

b. $f(x) = x^2 - 6x + 2$

STOP • STOP • STOP • STOP • STOP • STOP • STOP • STOP • STOP

A21 **a.** $\left(\frac{-3}{2}, \frac{-5}{4}\right)$:

$f(x) = \left(x + \frac{3}{2}\right)^2 - \frac{5}{4}$

b. $(3, ^-7)$:

$f(x) = (x - 3)^2 - 7$

Q22 Graph:

a. $f(x) = x^2 - 4x + 6$

b. $f(x) = x^2 + 6x + 7$

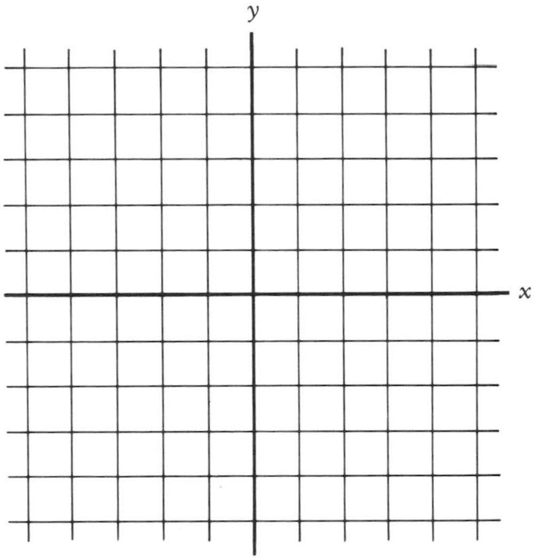

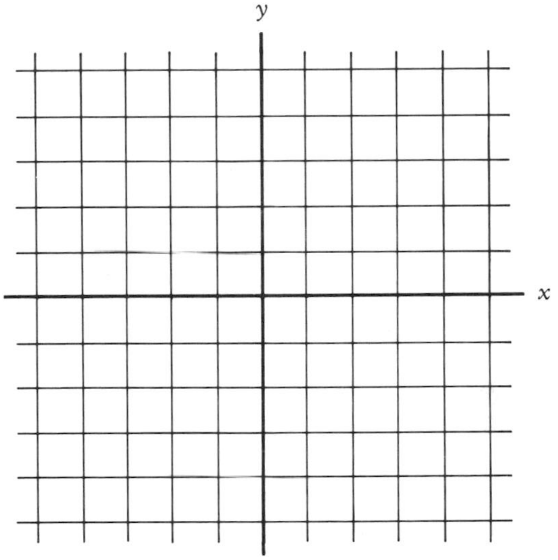

STOP • STOP • STOP • STOP • STOP • STOP • STOP • STOP • STOP

A22 **a.**

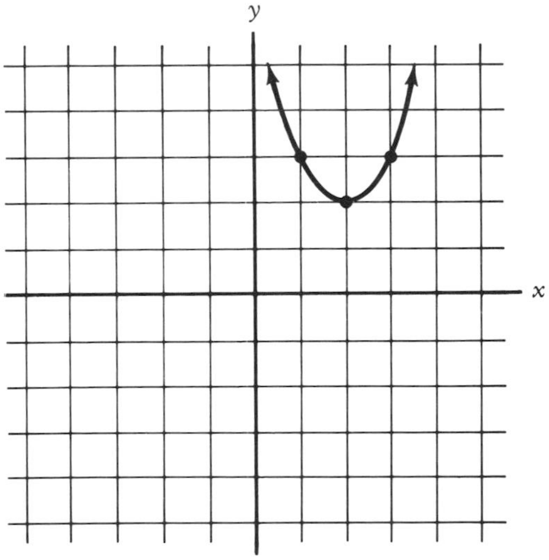

b.

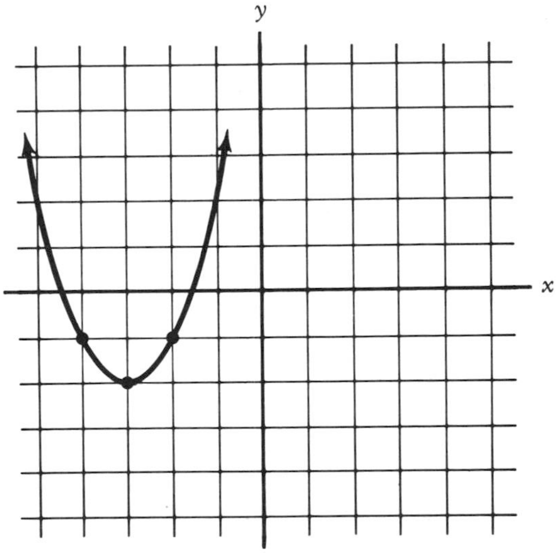

A parabola is fixed by plotting only three points.

---

10 The function $f(x) = ax^2 + bx + c$ can be converted to the form $f(x) = a(x - h)^2 + k$ when $a \neq 1$ by also completing the square.

Example: Write $f(x) = -x^2 + 4x - 2$ in the form $f(x) = a(x - h)^2 + k$.

Solution

Factor out the coefficient of x^2 from the first two terms.

$f(x) = -1(x^2 - 4x + \underline{\quad\quad}) - 2 + \underline{\quad\quad}$

constant subtracted (↑ first blank) add constant (↖ second blank)

The constant 4 makes the trinomial a perfect square; however, 4 is actually being multiplied by -1 and is being subtracted inside the parentheses. Hence 4 must be added outside the parentheses.

$f(x) = -1(x^2 - 4x + \underline{4}) - 2 + \underline{4}$
$f(x) = -1(x - 2)^2 + 2$

Q23 Write in the form $f(x) = a(x - h)^2 + k$ by completing the steps:
$f(x) = -x^2 + 2x + 3$

$f(x) = -1(x^2 - 2x + \underline{\quad\quad}) + 3 + \underline{\quad\quad}$

$f(x) = \underline{\quad\quad\quad\quad\quad\quad}$

STOP • STOP • STOP • STOP • STOP • STOP • STOP • STOP • STOP

A23 $f(x) = -1(x^2 - 2x + 1) + 3 + 1$
$f(x) = -1(x - 1)^2 + 4$

Q24 Write in the form $f(x) = a(x - h)^2 + k$:
a. $f(x) = -x^2 + 2x + 2$ **b.** $f(x) = -x^2 + 6x - 7$

STOP • STOP • STOP • STOP • STOP • STOP • STOP • STOP • STOP

A24 **a.** $f(x) = -(x - 1)^2 + 3$:
$f(x) = -(x^2 - 2x + 1) + 2 + 1$
$f(x) = -(x - 1)^2 + 3$

b. $f(x) = -(x - 3)^2 + 2$:
$f(x) = -(x^2 - 6x + 9) - 7 + 9$
$f(x) = -(x - 3)^2 + 2$

Q25 Write in the form $f(x) = a(x - h)^2 + k$ by completing the steps:
$f(x) = 2x^2 + 8x + 3$

$f(x) = 2(x^2 + 4x + \underline{\quad\quad}) + 3 - \underline{\quad\quad}$

remember the multiplication by 2

$f(x) = \underline{\quad\quad\quad\quad\quad\quad}$

STOP • STOP • STOP • STOP • STOP • STOP • STOP • STOP • STOP

A25 $f(x) = 2(x^2 + 4x + 4) + 3 - 8$
$f(x) = 2(x + 2)^2 - 5$

Q26 Write in the form $f(x) = a(x - h)^2 + k$:

a. $f(x) = 2x^2 - 20x + 70$ **b.** $f(x) = 3x^2 - 12x + 30$

STOP • STOP • STOP • STOP • STOP • STOP • STOP • STOP • STOP

A26 **a.** $f(x) = 2(x - 5)^2 + 20$ **b.** $f(x) = 3(x - 2)^2 + 18$

Q27 What is the vertex of the functions in Q26?

a. __________ **b.** __________

STOP • STOP • STOP • STOP • STOP • STOP • STOP • STOP • STOP

A27 **a.** (5, 20) **b.** (2, 18)

Q28 Graph:

a. $f(x) = -x^2 + 4x - 2$ **b.** $f(x) = -x^2 + 2x - 2$

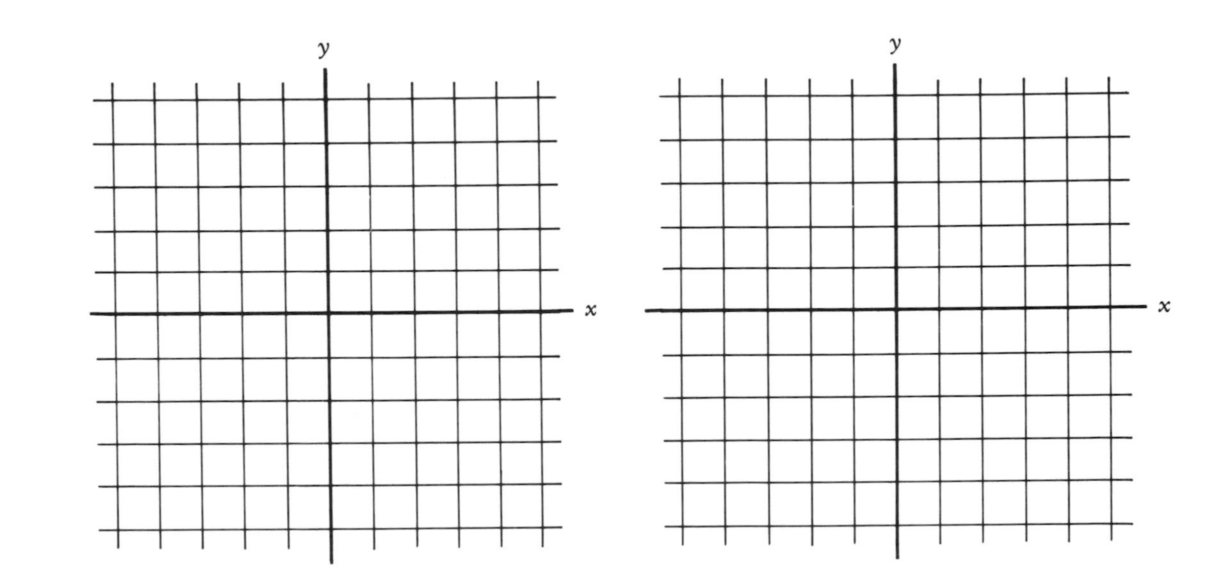

STOP • STOP • STOP • STOP • STOP • STOP • STOP • STOP • STOP

A28

a. $f(x) = -x^2 + 4x - 2$
$f(x) = -(x - 2)^2 + 2$

b. $f(x) = -x^2 + 2x - 2$
$f(x) = -(x - 1)^2 - 1$

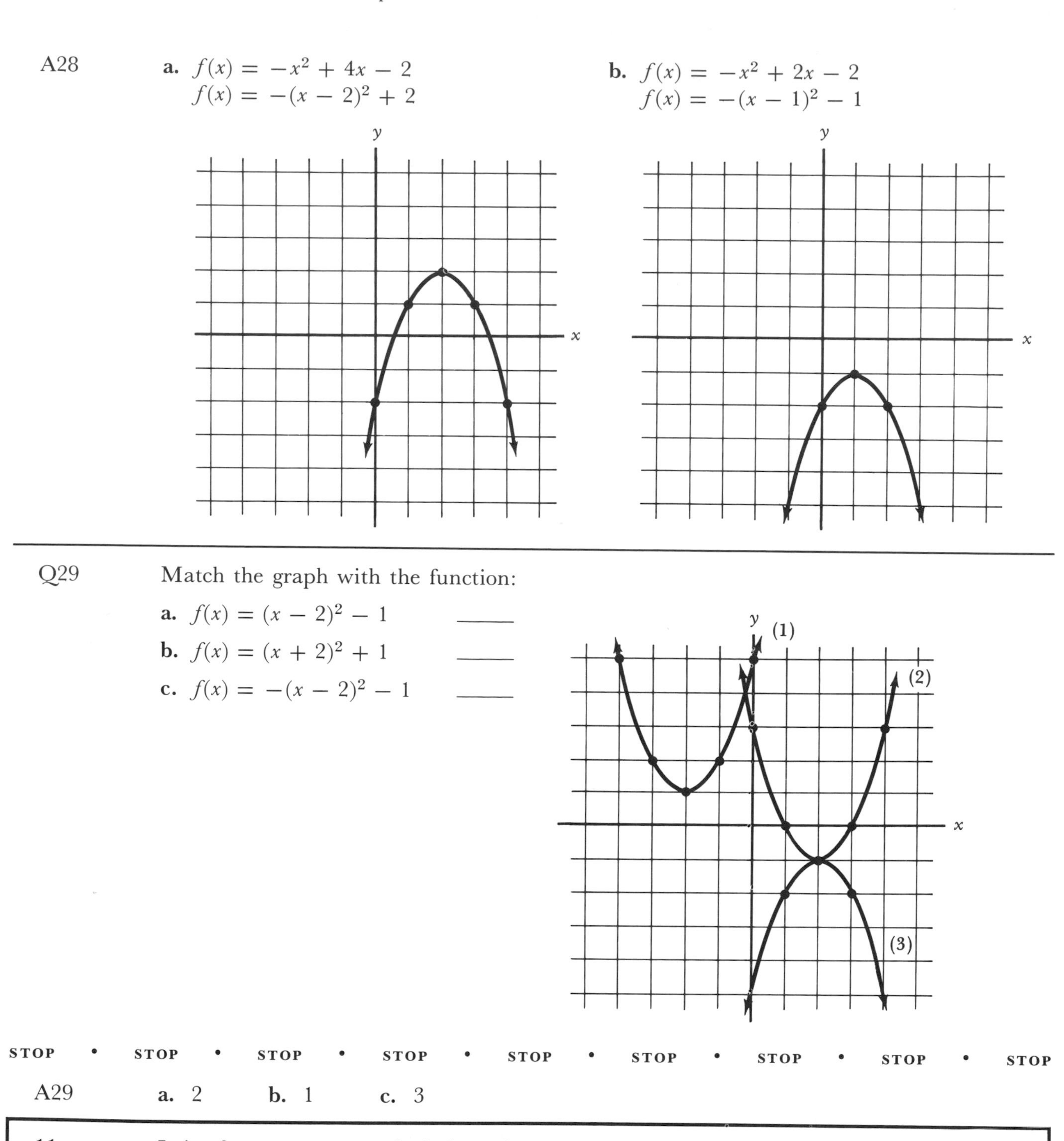

Q29 Match the graph with the function:

a. $f(x) = (x - 2)^2 - 1$ _____

b. $f(x) = (x + 2)^2 + 1$ _____

c. $f(x) = -(x - 2)^2 - 1$ _____

STOP • STOP • STOP • STOP • STOP • STOP • STOP • STOP • STOP

A29 **a.** 2 **b.** 1 **c.** 3

11 It is often necessary to find the points where the graph of a quadratic function crosses the x axis. This is called finding the x intercepts. Since every point on the x axis has a second coordinate of zero, the x intercepts for $f(x) = ax^2 + bx + c$ are where $f(x) = 0$. The x coordinates of the intercepts can be found by solving the quadratic equation $ax^2 + bx + c = 0$.

Example: (1) Graph $f(x) = x^2 - 4x + 3$ and (2) find the x intercepts algebraically.

Solution

(1) $f(x) = x^2 - 4x + 3$
$f(x) = (x - 2)^2 - 1$
vertex $= (2, -1)$

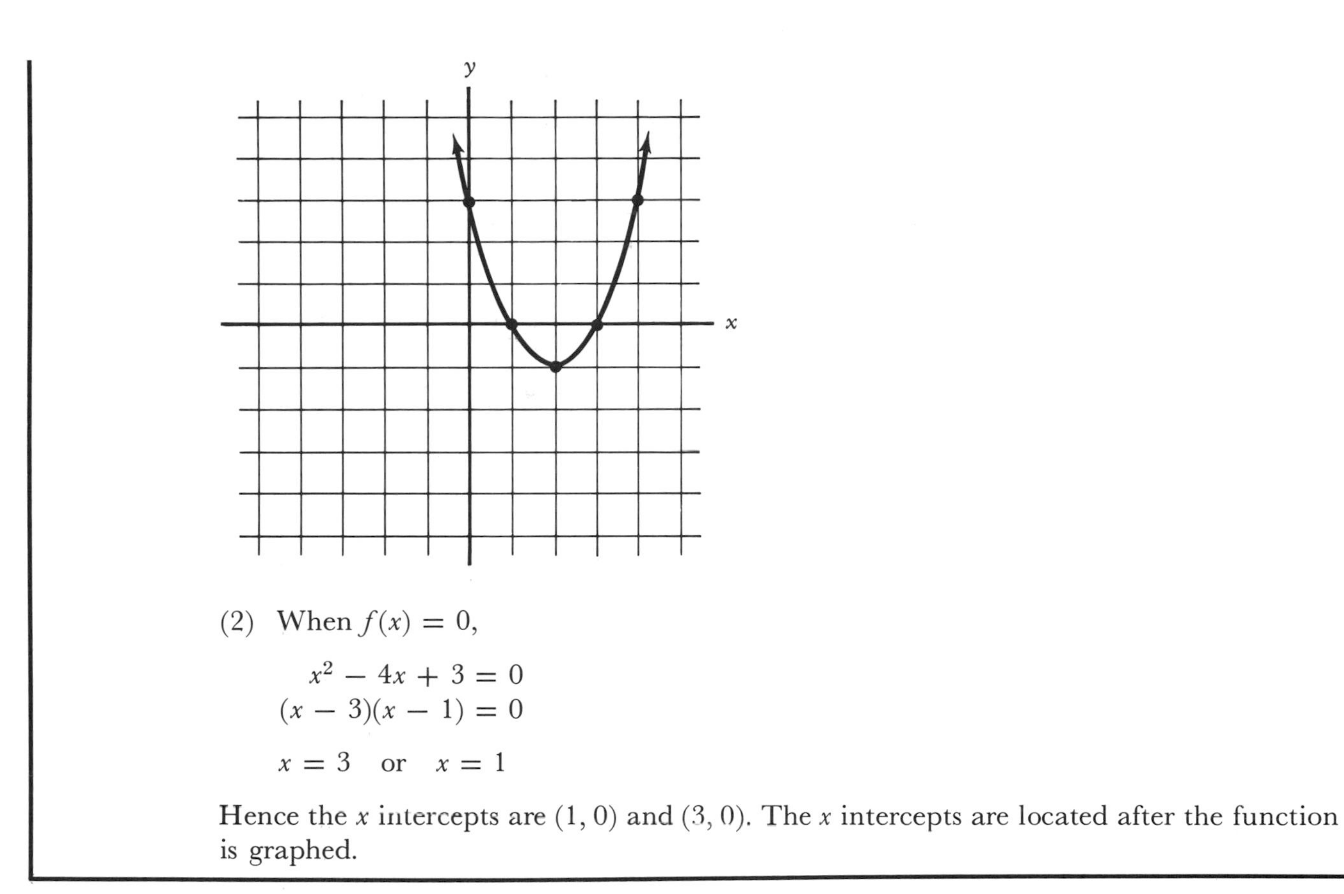

(2) When $f(x) = 0$,

$$x^2 - 4x + 3 = 0$$
$$(x - 3)(x - 1) = 0$$
$$x = 3 \quad \text{or} \quad x = 1$$

Hence the x intercepts are (1, 0) and (3, 0). The x intercepts are located after the function is graphed.

Q30 (1) Graph and (2) find the x intercepts algebraically:

a. $f(x) = -x^2 + 6x - 8$ **b.** $f(x) = x^2 + 2x - 2$

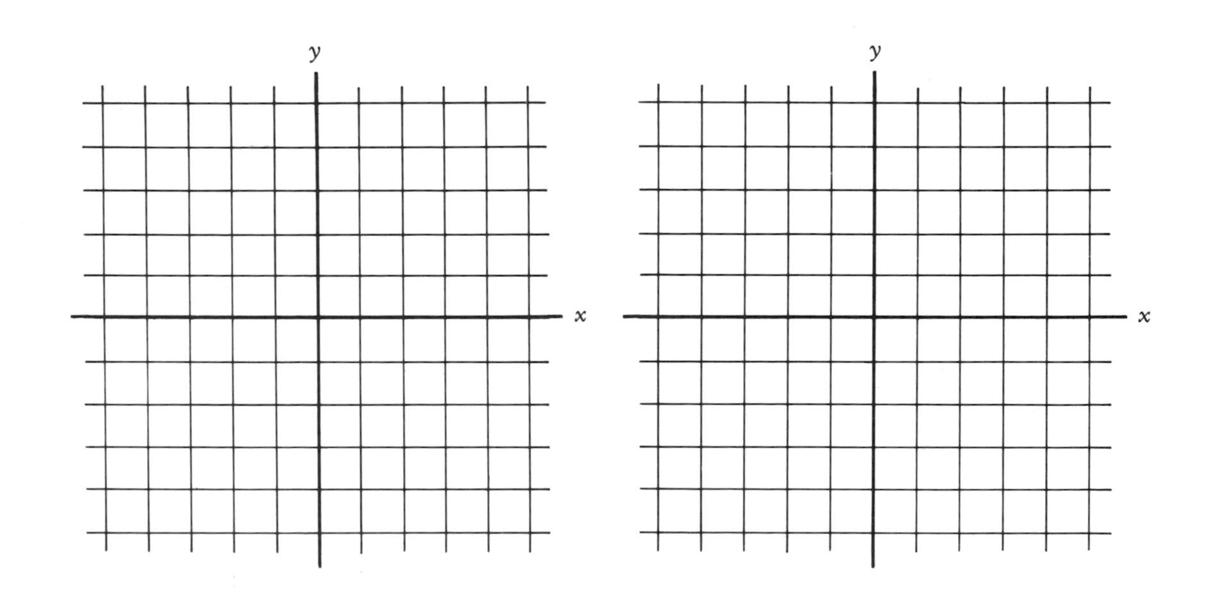

STOP • STOP • STOP • STOP • STOP • STOP • STOP • STOP • STOP

A30

a. (1) $f(x) = -(x - 3)^2 + 1$

(2)
$$-x^2 + 6x - 8 = 0$$
$$x^2 - 6x + 8 = 0$$
$$(x - 2)(x - 4) = 0$$
$$x = 2 \quad \text{or} \quad x = 4$$
$$x \text{ intercepts} = (2, 0), (4, 0)$$

b. (1) $f(x) = (x + 1)^2 - 3$

(2)
$$x^2 + 2x - 2 = 0$$
$$x = \frac{-2 \pm \sqrt{(2)^2 - 4(1)(-2)}}{2}$$
$$x = \frac{-2 \pm \sqrt{12}}{2}$$
$$x = -1 \pm \sqrt{3}$$
$$x \text{ intercepts} = (-1 - \sqrt{3}, 0), (-1 + \sqrt{3}, 0)$$

Q31 Graph, locating the x intercepts:

a. $f(x) = x^2 - 4x + 5$

b. $f(x) = x^2 - 2x$

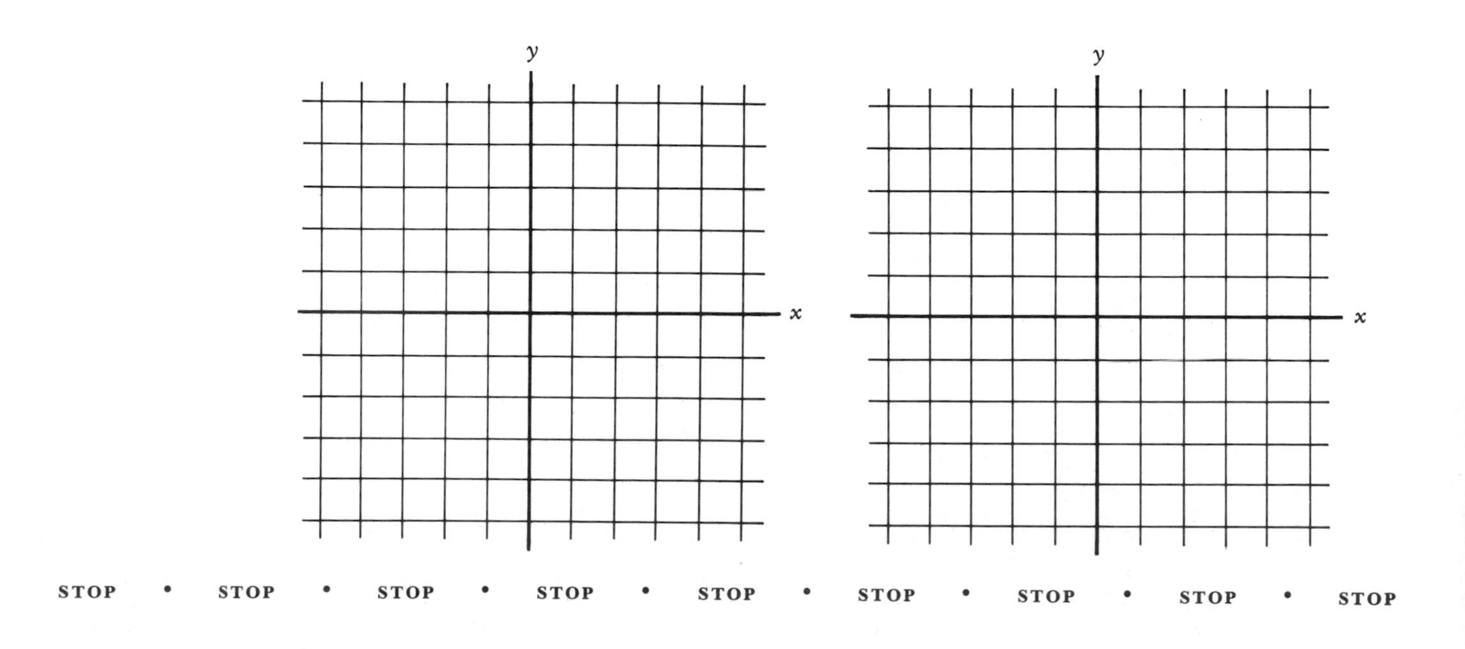

STOP • STOP • STOP • STOP • STOP • STOP • STOP • STOP • STOP

A31 **a.** $f(x) = (x - 2)^2 + 1$ **b.**

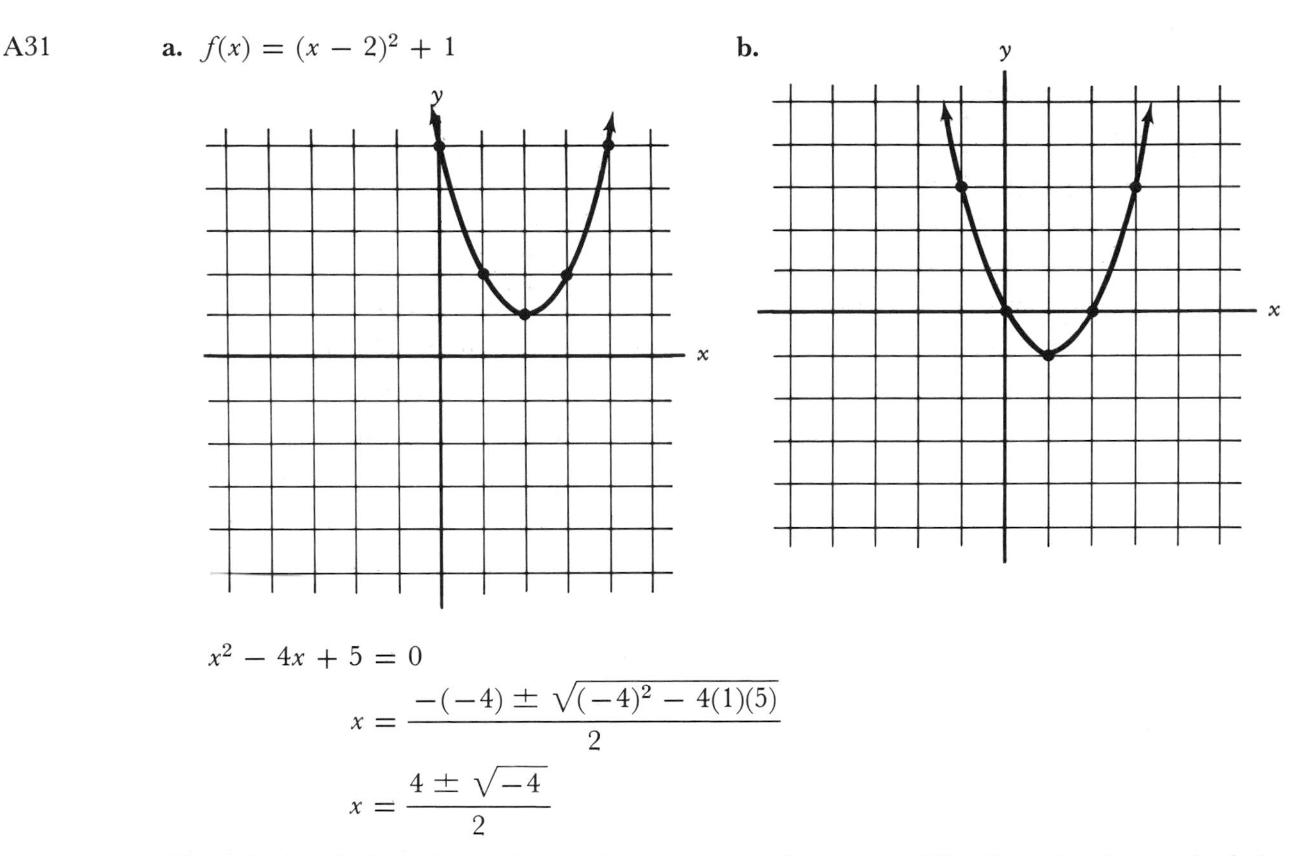

$$x^2 - 4x + 5 = 0$$

$$x = \frac{-(-4) \pm \sqrt{(-4)^2 - 4(1)(5)}}{2}$$

$$x = \frac{4 \pm \sqrt{-4}}{2}$$

The solution is imaginary; hence there are no x intercepts. The function is graphed, in the normal manner, without need of locating the x intercepts.

This completes the instruction for this section.

6.5 Exercises

1. **a.** Define a quadratic function.
   b. The graph of a quadratic function is called a __________.
   c. The highest or lowest point of the graph of a quadratic function is called the __________.
   d. A quadratic function is represented by a __________-degree equation.
2. Graph:
   a. $f(x) = x^2$ **b.** $f(x) = -x^2$ **c.** $f(x) = 2x^2 - 4$
   d. $f(x) = (x - 2)^2$ **e.** $f(x) = (x + 2)^2 - 3$ **f.** $f(x) = x^2 - 2x$
   g. $f(x) = x^2 + 6x + 11$ **h.** $f(x) = x^2 - 4x + 5$
   i. $f(x) = -x^2 + 2x + 3$ **j.** $f(x) = -x^2 - 4x - 4$
3. Graph locating the x intercepts:
   a. $f(x) = x^2 + 6x + 8$ **b.** $f(x) = x^2 - 4x + 6$

*4. Determine the vertex for $f(x) = ax^2 + bx + c$ by completing the square.

*5. From the graph of the function, determine the equation:

a. b.

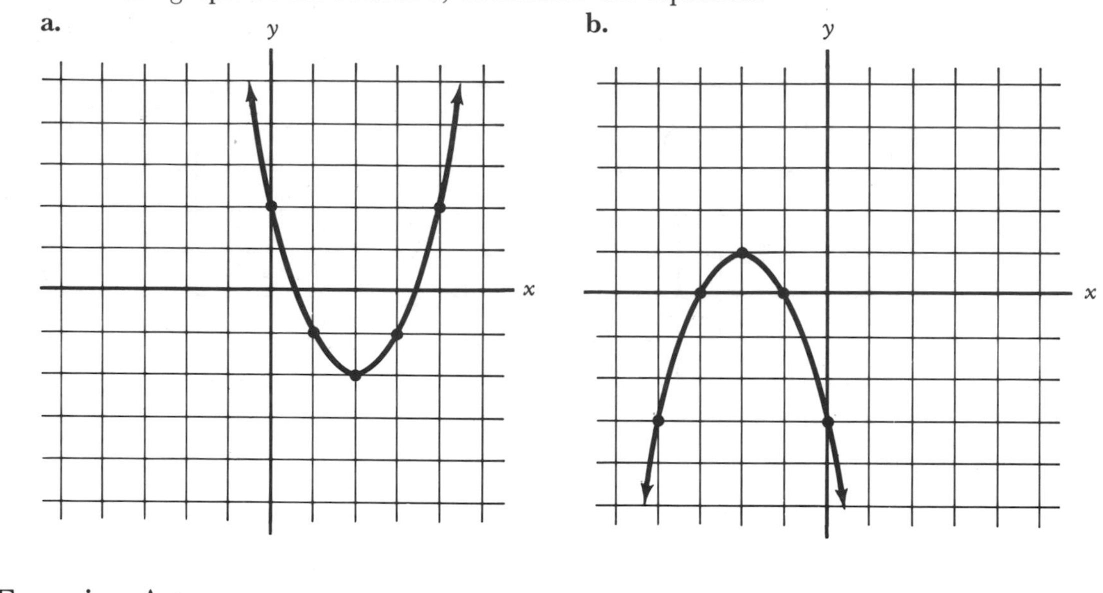

6.5 Exercise Answers

1. a. $f(x) = ax^2 + bx + c$, where a, b, and c are constants with $a \neq 0$.
   b. parabola c. vertex d. second
2. a. b. c. d.

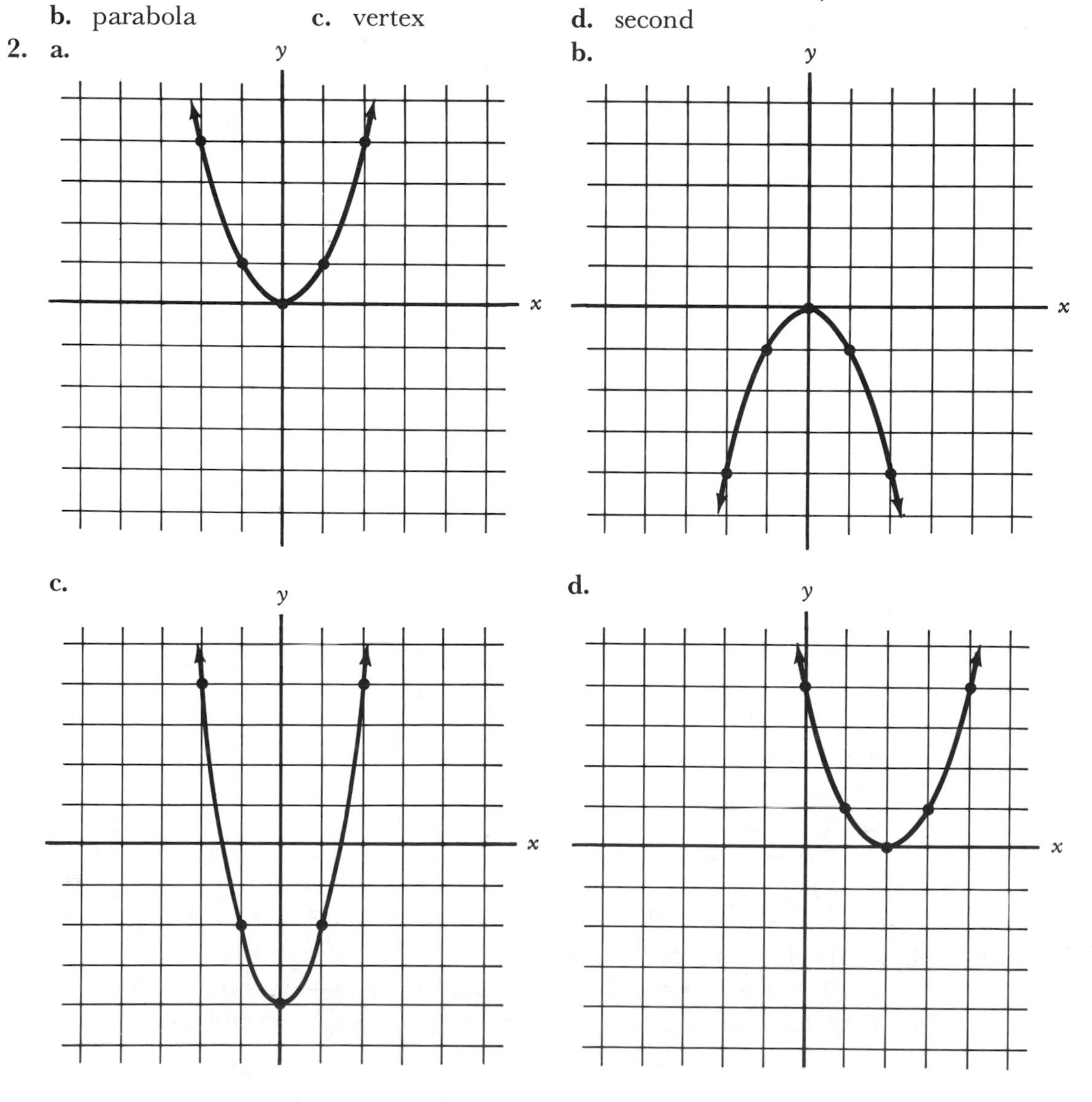

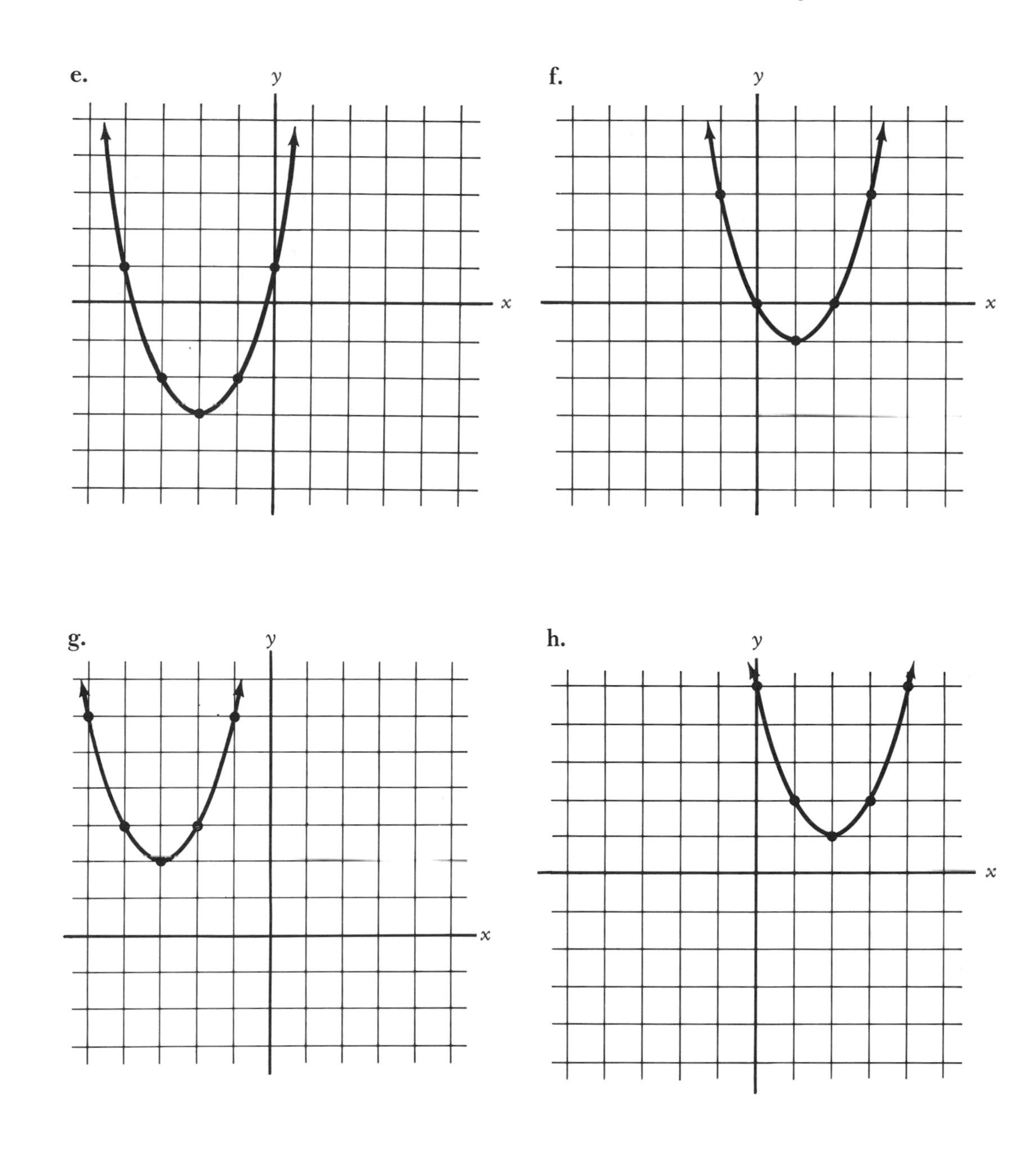
e.
y
x
f.
y
x
g.
y
x
h.
y
x

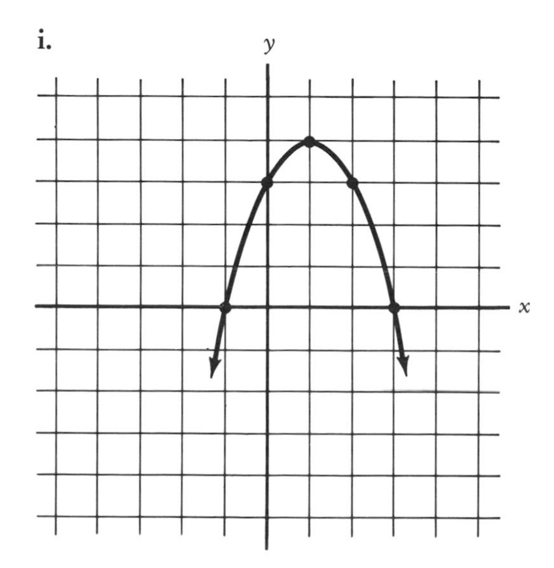
i.
y
x

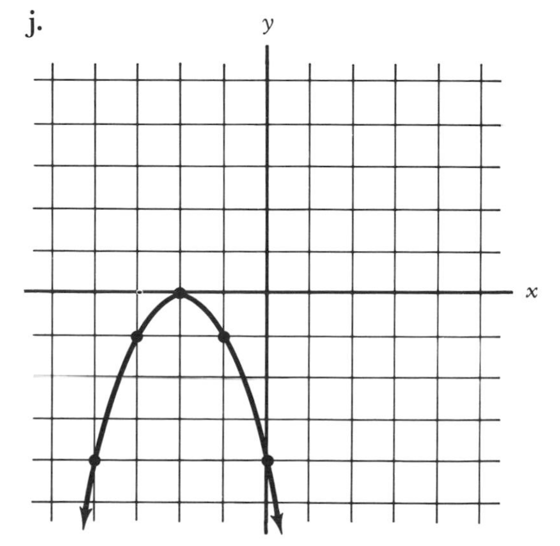
j.
y
x

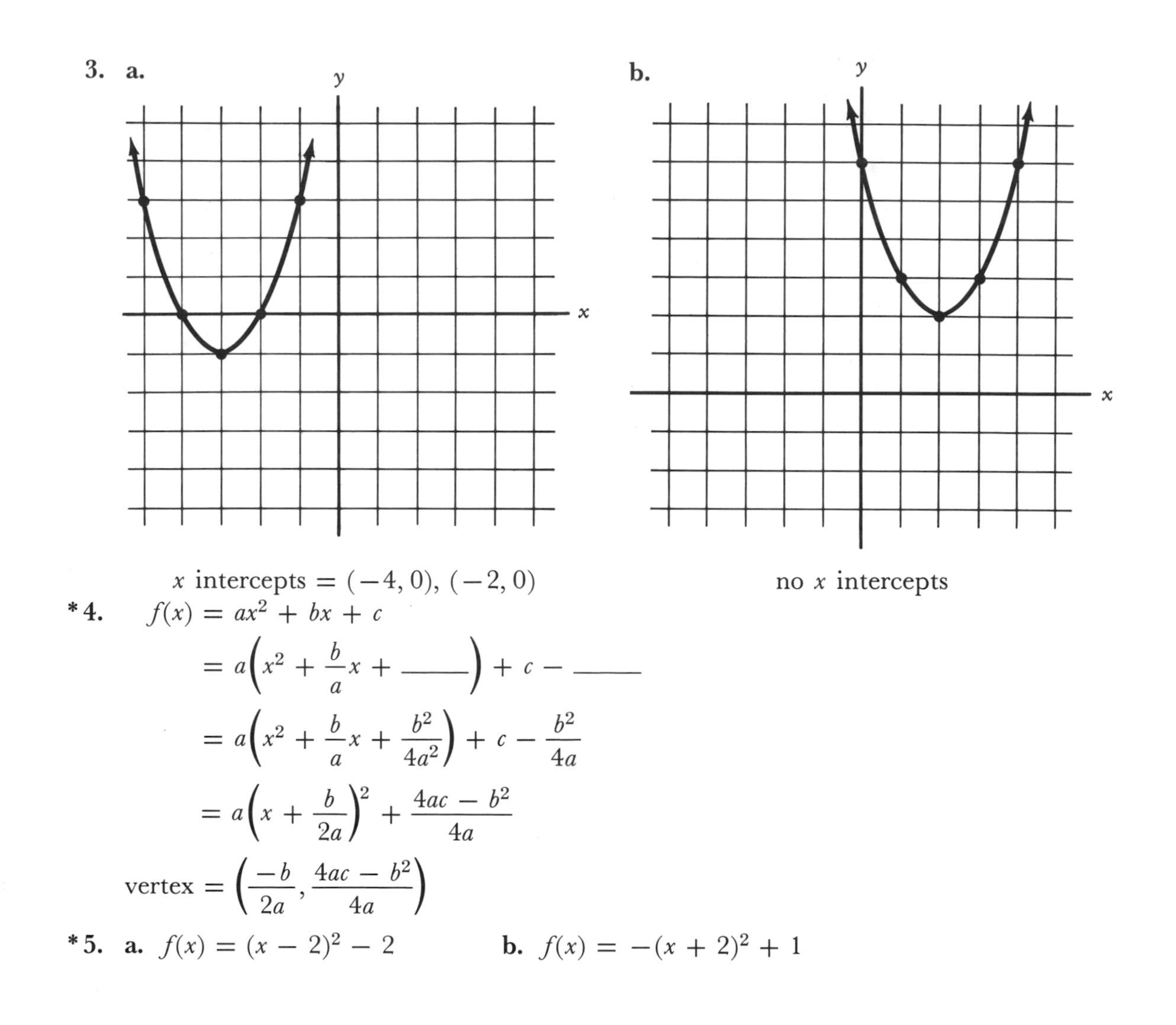

3. a. x intercepts $= (-4, 0), (-2, 0)$

b. no x intercepts

***4.**
$$f(x) = ax^2 + bx + c$$
$$= a\left(x^2 + \frac{b}{a}x + \_\_\_\_\right) + c - \_\_\_\_$$
$$= a\left(x^2 + \frac{b}{a}x + \frac{b^2}{4a^2}\right) + c - \frac{b^2}{4a}$$
$$= a\left(x + \frac{b}{2a}\right)^2 + \frac{4ac - b^2}{4a}$$
$$\text{vertex} = \left(\frac{-b}{2a}, \frac{4ac - b^2}{4a}\right)$$

***5. a.** $f(x) = (x - 2)^2 - 2$ **b.** $f(x) = -(x + 2)^2 + 1$

6.6 Inverse Functions

1 The *inverse* of a function can be thought of as "undoing" the operations of the function. If the function "produces" the ordered pair (2, 8), the inverse would produce the ordered pair (8, 2). That is, the inverse relation is formed by interchanging the coordinates of each ordered pair of the function.

Example: Given $f = \{(0, 0), (1, 1), (2, 8), (3, 27)\}$, find the inverse.

Solution

f inverse $= \{(0, 0), (1, 1), (8, 2), (27, 3)\}$

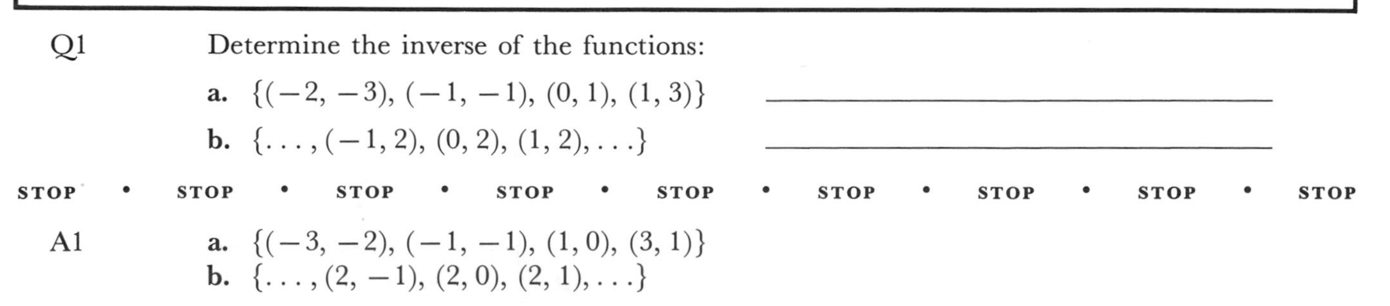

Q1 Determine the inverse of the functions:

a. $\{(-2, -3), (-1, -1), (0, 1), (1, 3)\}$ ____________

b. $\{\ldots, (-1, 2), (0, 2), (1, 2), \ldots\}$ ____________

STOP • STOP • STOP • STOP • STOP • STOP • STOP • STOP • STOP

A1 **a.** $\{(-3, -2), (-1, -1), (1, 0), (3, 1)\}$

b. $\{\ldots, (2, -1), (2, 0), (2, 1), \ldots\}$

2

The inverse of a function, f, is denoted by f^{-1}. The notation f^{-1} is read "f inverse." Recall that in Chapter 4, a^{-1} meant $\frac{1}{a}$. However, in this section f^{-1} will mean the inverse of f and does *not* mean $\frac{1}{f}$. In general, the meaning of the symbol will be clear from the context in which it is found.

If a function, f, has two ordered pairs with the same second coordinate, the inverse, f^{-1}, will have two ordered pairs with the same first coordinate; hence f^{-1} is not a function.

Example: Does the function have an inverse that is a function? Explain.

$f = \{\ldots, (-2, 2), (-1, 1), (0, 0), (1, 1), (2, 2), \ldots\}$
$g = \{(4, 8), (5, 10), (6, 12), (7, 14)\}$

Solutions

f does not; since two ordered pairs of f havc the same second coordinate, two ordered pairs of f^{-1} will have the same first coordinate. g^{-1} is a function; since no two ordered pairs of f have the same second coordinate, no two ordered pairs of f^{-1} will have the same first coordinate.

Q2

(1) Does the function have an inverse that is a function? Explain. (2) Determine the inverse.

a. $f = \{(4, 6), (10, 13), (11, 15)\}$ **b.** $g = \{(2, -1), (4, 3), (8, 3), (1, 7)\}$

a. (1) _____; ____________________

(2) ____________________

b. (1) _____; ____________________

(2) ____________________

STOP • STOP • STOP • STOP • STOP • STOP • STOP • STOP • STOP

A2

a. (1) yes; no two ordered pairs have the same second coordinate
(2) $f^{-1} = \{(6, 4), (13, 10), (15, 11)\}$

b. (1) no; two ordered pairs have the same second coordinate
(2) $g^{-1} = \{(-1, 2), (3, 4), (3, 8), (7, 1)\}$

3

The inverse of a function described using function notation will be determined by writing the "function process." The function process for $f(x) = 2x + 1$ is: Multiply by 2 and add 1. Multiply and add are called the "function operators." They are the verbs of the function process.

Example: Determine the function process and underline the function operators.

$f(x) = -3x + 5$

Solution

<u>Multiply</u> by -3 and <u>add</u> 5.

Q3 Determine the function process and underline the function operators:

a. $f(x) = \frac{2}{3}x - 7$ ______

b. $g(x) = \frac{x-3}{2}$ ______

STOP • STOP • STOP • STOP • STOP • STOP • STOP • STOP • STOP

A3 a. Multiply by $\frac{2}{3}$ and subtract 7.

b. Subtract 3 and divide by 2.

Q4 Determine the function process and underline the operators:

a. $f(x) = -5x - \sqrt{3}$ ______

b. $g(x) = \frac{x+2}{-3}$ ______

STOP • STOP • STOP • STOP • STOP • STOP • STOP • STOP • STOP

A4 a. Multiply by -5 and subtract $\sqrt{3}$.

b. Add 2 and divide by -3.

Q5 Write the function process by use of function notation:

a. Multiply by $\frac{1}{2}$ and add 4. ______

b. Multiply by -7 and subtract 2. ______

STOP • STOP • STOP • STOP • STOP • STOP • STOP • STOP • STOP

A5 a. $f(x) = \frac{1}{2}x + 4$ b. $f(x) = -7x - 2$

Q6 Write the function process by use of function notation:

a. Add 1 and divide by 7. ______

b. Subtract 4 and divide by 2. ______

STOP • STOP • STOP • STOP • STOP • STOP • STOP • STOP • STOP

A6 a. $f(x) = \frac{x+1}{7}$ b. $f(x) = \frac{x-4}{2}$

---

4 The inverse function can now be determined by "reversing" the function process. A function and its inverse can be pictured by use of function machines as:

$x \longrightarrow$ [f] $\longrightarrow y$ $x \longleftarrow$ [f^{-1}] $\longleftarrow y$

The inverse function is formed using the opposite* function operators in the opposite order.

Example: Find the inverse of $f(x) = 2x + 1$.

*Addition and subtraction are opposite operations; multiplication and division are opposite operations.

Solution

Write f as a function process. f: multiply by 2 and add 1. f^{-1} is formed by using the opposite function operators from *right* to *left*. f^{-1}: subtract 1 and divide by 2. The inverse can now be written using function notation.

$$f^{-1}(y) = \frac{y-1}{2}$$

Q7 Find the inverse of $f(x) = -2x + 5$ by first writing f and f^{-1} as function processes.

a. f: ________________

b. f^{-1}: ________________

c. $f^{-1}(y) =$ ________________

STOP • STOP • STOP • STOP • STOP • STOP • STOP • STOP • STOP

A7 **a.** f: multiply by -2 and add 5.
b. f^{-1}: subtract 5 and divide by -2.
c. $f^{-1}(y) = \frac{y-5}{-2}$

Q8 Find f^{-1} if $f(x) = \frac{2}{3}x - 7$.

a. f: ________________

b. f^{-1}: ________________

c. $f^{-1}(y) =$ ________________

STOP • STOP • STOP • STOP • STOP • STOP • STOP • STOP • STOP

A8 **a.** f: multiply by $\frac{2}{3}$ and subtract 7.

b. f^{-1}: add 7 and divide by $\frac{2}{3}$.

c. $f^{-1}(y) = \frac{3y+21}{2}$: $f^{-1}(y) = \frac{y+7}{\frac{2}{3}} = (y+7)\frac{3}{2}$

Q9 Find f^{-1} if $f(x) = -3x$.

a. f: ________________

b. f^{-1}: ________________

c. $f^{-1}(y) =$ ________________

STOP • STOP • STOP • STOP • STOP • STOP • STOP • STOP • STOP

A9 **a.** f: multiply by -3.
b. f^{-1}: divide by -3.
c. $f^{-1}(y) = \frac{y}{-3}$

Q10 Find f^{-1} if $f(x) = x - 5$.

a. f: ______________________

b. f^{-1}: ______________________

c. $f^{-1}(y) =$ ______________________

STOP • STOP • STOP • STOP • STOP • STOP • STOP • STOP • STOP

A10 **a.** f: subtract 5.
b. f^{-1}: add 5.
c. $f^{-1}(y) = y + 5$

Q11 For the function in Q10, find:

a. $f^{-1}(2)$ ______ **b.** $f^{-1}(0)$ ______

c. $f^{-1}(k)$ __________ **d.** $f^{-1}(x)$ __________

STOP • STOP • STOP • STOP • STOP • STOP • STOP • STOP • STOP

A11 **a.** 7 **b.** 5 **c.** $k + 5$ **d.** $x + 5$

Q12 If $f(x) = -6x + 2$, find:
a. $f^{-1}(0)$ **b.** $f^{-1}(2)$ **c.** $f^{-1}(x)$

STOP • STOP • STOP • STOP • STOP • STOP • STOP • STOP • STOP • STOP

A12 **a.** $\frac{1}{3}$ **b.** 0 **c.** $\frac{x - 2}{-6}$

f^{-1}: subtract 2 and divide by -6.

Q13 If $f(x) = 4x + 1$, find:
a. $f^{-1}(9)$ **b.** $f^{-1}(13)$ **c.** $f^{-1}(a)$ **d.** $f^{-1}(x)$

STOP • STOP • STOP • STOP • STOP • STOP • STOP • STOP • STOP

A13 **a.** 2 **b.** 3 **c.** $\frac{a - 1}{4}$ **d.** $\frac{x - 1}{4}$

Q14 If $f(x) = mx + b$, find f^{-1}:

a. f: ____

b. f^{-1}: ____

c. $f^{-1}(y) =$ ____

STOP • STOP • STOP • STOP • STOP • STOP • STOP • STOP • STOP

A14

a. f: multiply by m and add b.
b. f^{-1}: subtract b and divide by m, $m \neq 0$.
c. $f^{-1}(y) = \dfrac{y - b}{m}$

5

The inverse of a quadratic function can be found using the technique used for finding the inverse of linear functions; however, the inverse of a quadratic function will not be a function.

Example: If $f(x) = 2x^2 + 1$, find f^{-1}.

Solution

f: square, multiply by 2, and add 1. f^{-1}: subtract 1, divide by 2, and find the square roots. Because the last operation is "find the square roots," f^{-1} does not represent a function. Two different values are produced as a result of the last operation.

Q15

If $f(x) = x^2$, find:

a. f: ____________________

b. f^{-1}: ____________________

c. Is f^{-1} a function? ____________________

STOP • STOP • STOP • STOP • STOP • STOP • STOP • STOP • STOP

A15

a. square **b.** find the square roots
c. no: two different values result from the last operation.

Q16

If $f(x) = x^2 - 3$, find:

a. f: ____________________

b. f^{-1}: ____________________
c. Is f^{-1} a function? Explain.

________; ______________________________

STOP • STOP • STOP • STOP • STOP • STOP • STOP • STOP • STOP

A16

a. square and subtract 3
b. add 3 and find the square roots
c. no; two different values result from the last operation.

Q17

For f^{-1} in Q16, what second coordinate corresponds to a first coordinate of:

a. 1 ______ **b.** 6 ______

STOP • STOP • STOP • STOP • STOP • STOP • STOP • STOP • STOP

A17

a. ± 2 **b.** ± 3

6

In Frame 1 the inverse of a function was formed by interchanging the coordinates in each ordered pair. If an equation is written in terms of x and y, the coordinates of each ordered pair can be interchanged by interchanging x and y; thus forming the inverse.

Example 1: If $f = \{(0, 2), (1, 3), (2, 4)\}$, find f^{-1}.

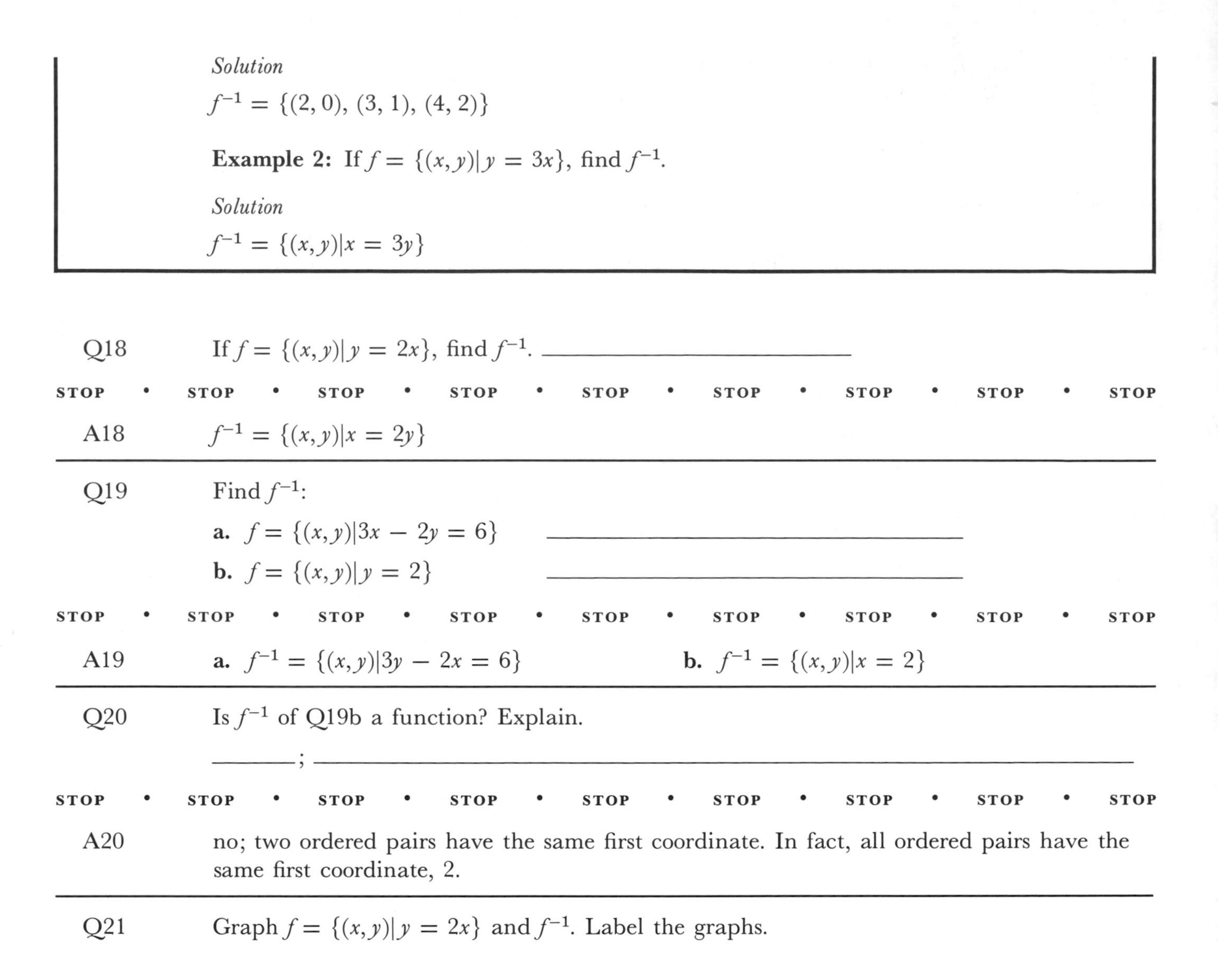

Solution

$f^{-1} = \{(2, 0), (3, 1), (4, 2)\}$

Example 2: If $f = \{(x, y) | y = 3x\}$, find f^{-1}.

Solution

$f^{-1} = \{(x, y) | x = 3y\}$

Q18 If $f = \{(x, y) | y = 2x\}$, find f^{-1}. ________________

STOP • STOP • STOP • STOP • STOP • STOP • STOP • STOP • STOP

A18 $f^{-1} = \{(x, y) | x = 2y\}$

Q19 Find f^{-1}:

a. $f = \{(x, y) | 3x - 2y = 6\}$ ________________

b. $f = \{(x, y) | y = 2\}$ ________________

STOP • STOP • STOP • STOP • STOP • STOP • STOP • STOP • STOP

A19 **a.** $f^{-1} = \{(x, y) | 3y - 2x = 6\}$ **b.** $f^{-1} = \{(x, y) | x = 2\}$

Q20 Is f^{-1} of Q19b a function? Explain.

________; ________________________

STOP • STOP • STOP • STOP • STOP • STOP • STOP • STOP • STOP

A20 no; two ordered pairs have the same first coordinate. In fact, all ordered pairs have the same first coordinate, 2.

Q21 Graph $f = \{(x, y) | y = 2x\}$ and f^{-1}. Label the graphs.

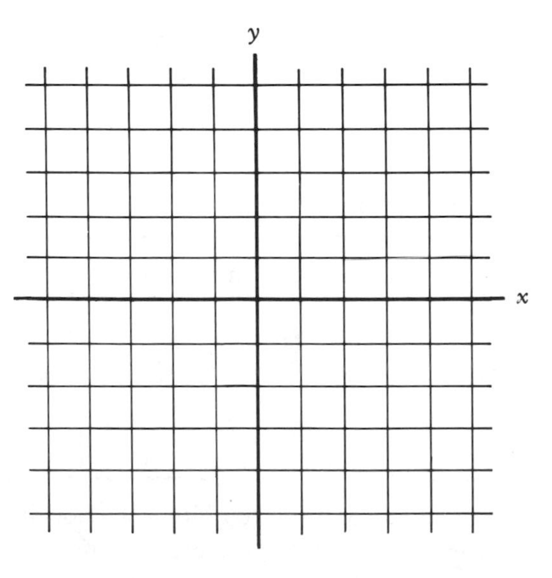

STOP • STOP • STOP • STOP • STOP • STOP • STOP • STOP • STOP

A21

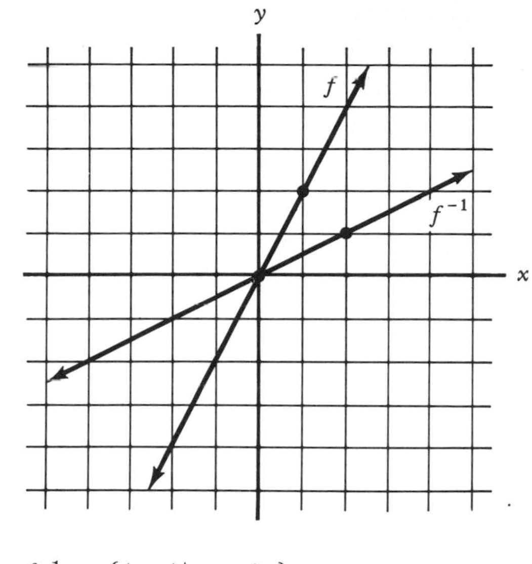

$f^{-1} = \{(x, y) | x = 2y\}$

7

The graph of a function and its inverse will be symmetric to the line $y = x$.

Example: Graph $f = \{(x, y) | 3x - 2y = 6\}$ and f^{-1}. Graph $y = x$ as a dashed line.

Solution

$f^{-1} = \{(x, y) | 3y - 2x = 6\}$

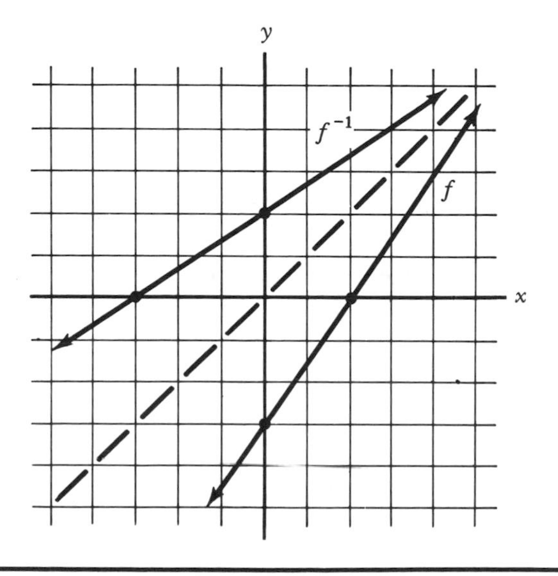

Q22 Graph $f = \{(x, y) | x + 2y = 4\}$ and f^{-1}. Graph $y = x$ dashed.

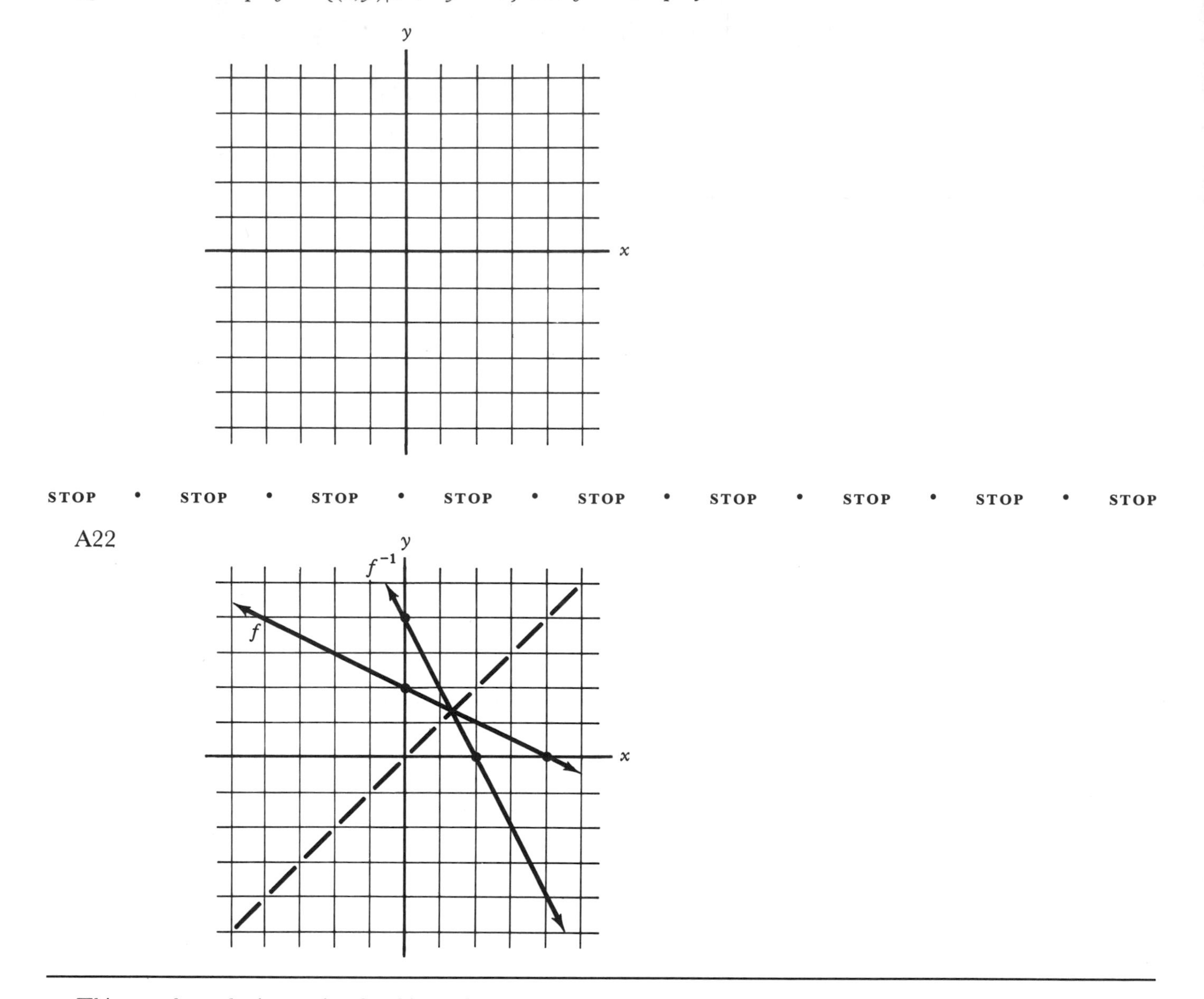

This completes the instruction for this section.

6.6 Exercises

1. Find the inverse of each function:
   a. $\{(0, 4), (2, 6), (4, 8)\}$ b. $\{\ldots, (-1, 2), (0, 2), (1, 2), \ldots\}$
   c. $\{\ldots, (-2, -2), (-1, -1), (0, 0), \ldots\}$
   d. $\{(0, 0), (1, 0), (2, 0), (3, 0), \ldots\}$
2. Which of the inverses in problem 1 are also functions?
3. Find f^{-1}:
   a. $f(x) = 2x$ b. $f(x) = 2x - 4$ c. $f(x) = -8x + 10$
   d. $f(x) = \dfrac{x - 3}{2}$ e. $f(x) = \dfrac{2}{3}x - 1$ f. $f(x) = 2x^3$
4. If $f(x) = -7x + 1$, find:
   a. $f^{-1}(0)$ b. $f^{-1}(x)$

5. Find f^{-1}:
   a. $f = \{(x, y) | y = -3x\}$ b. $f = \{(x, y) | 2x + 3y = 6\}$
   c. $f = \{(x, y) | y = -1\}$ d. $f = \{(x, y) | 2x - y = 2\}$
6. Graph f and f^{-1}. Graph $y = x$ dashed.
   a. $f = \{(x, y) | y = -1\}$ b. $f = \{(x, y) | 2x - y = 2\}$

6.6 Exercise Answers

1. a. $\{(4, 0), (6, 2), (8, 4)\}$ b. $\{\ldots, (2, -1), (2, 0), (2, 1), \ldots\}$
   c. $\{\ldots, (-2, -2), (-1, -1), (0, 0), \ldots\}$
   d. $\{(0, 0), (0, 1), (0, 2), (0, 3), \ldots\}$
2. a and c
3. a. $f^{-1}(y) = \dfrac{y}{2}$ b. $f^{-1}(y) = \dfrac{y + 4}{2}$ c. $f^{-1}(y) = \dfrac{y - 10}{-8}$
   d. $f^{-1}(y) = 2y + 3$ e. $f^{-1}(y) = \dfrac{3y + 3}{2}$ f. $f^{-1}(y) = \sqrt[3]{\dfrac{y}{2}}$
4. a. $\dfrac{1}{7}$ b. $f^{-1}(x) = \dfrac{x - 1}{-7}$
5. a. $f^{-1} = \{(x, y) | x = -3y\}$ b. $f^{-1} = \{(x, y) | 2y + 3x = 6\}$
   c. $f^{-1} = \{(x, y) | x = -1\}$ d. $f^{-1} = \{(x, y) | 2y - x = 2\}$
6. a. b.

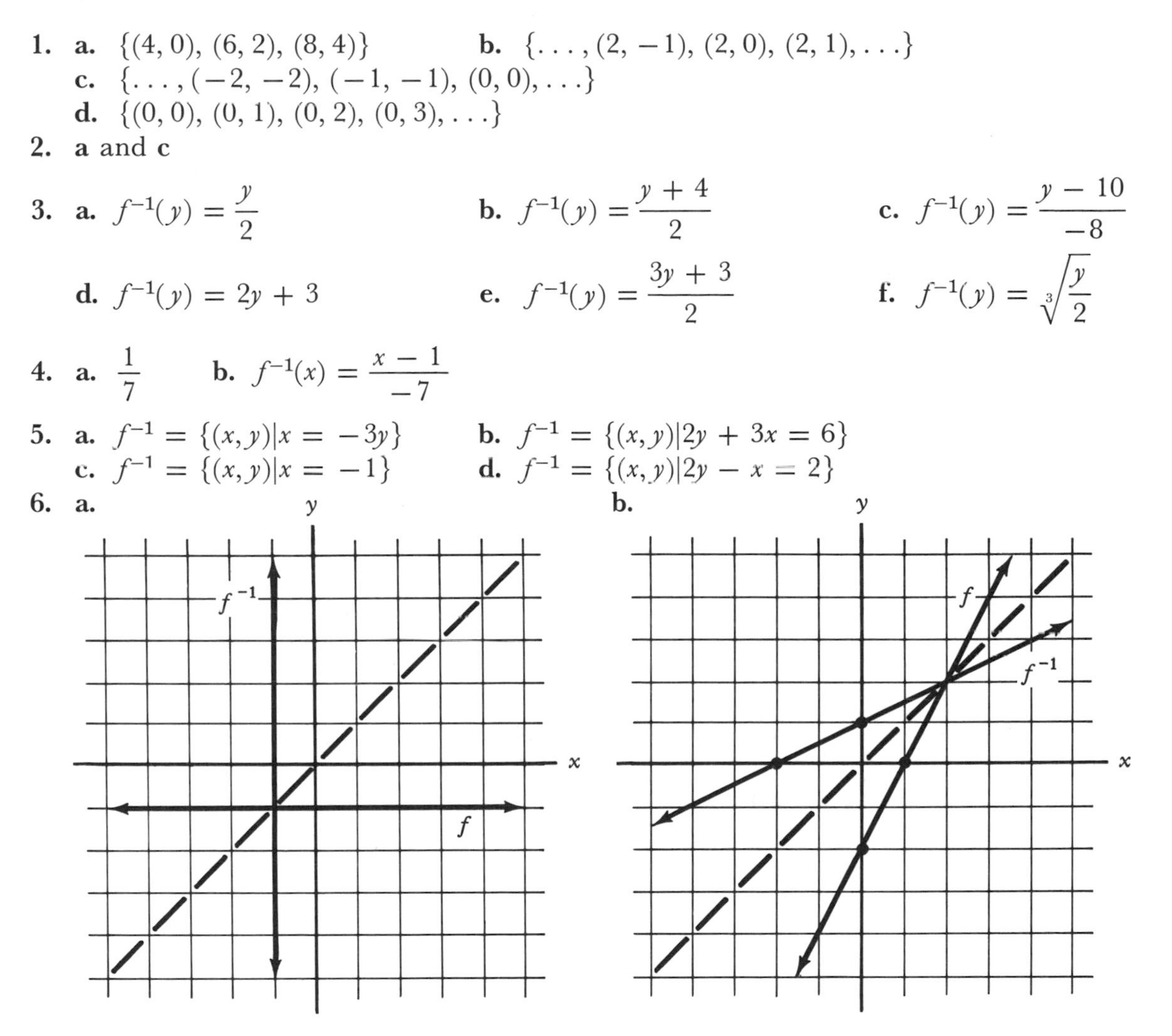

Chapter 6 Sample Test

At the completion of Chapter 6 it is expected that you will be able to work the following problems.

6.1 Relations and the Coordinate System

1. Graph each relation listing the domain and range:
   a. $\{(0, 1), (2, 3), (4, -2)\}$ b. $\{\ldots, (-1, 2), (0, 2), (1, 2), \ldots\}$

2. List the relation represented by the graph:

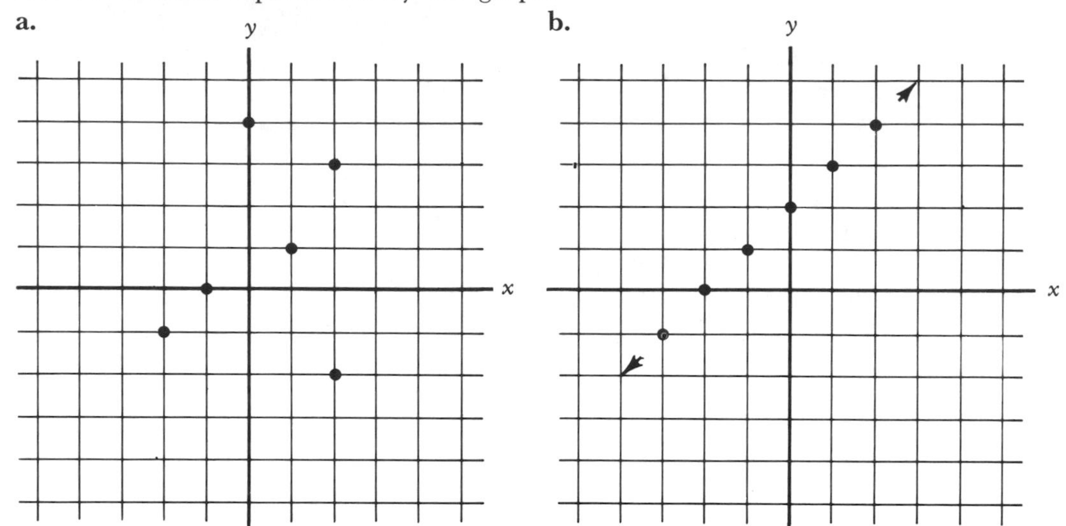

6.2 Functions and Function Notation

3. Are the relations in problem 1 functions? Explain.
4. Do the graphs of relations in problem 2 represent functions? Explain.
5. If $f(x) = 3x + 5$, find:
   a. $f(0)$ **b.** $f(-1)$ **c.** $f(3a)$ **d.** $f(x + h)$
6. If $f(x) = -3x^2$, find:
   a. $f(-1)$ **b.** $f(1)$ **c.** $f(0)$ **d.** $f(a + b)$

6.3 Constant and Linear Functions: Slope

7. If $f(x) = 3$, find:
   a. $f(0)$ **b.** $f(-10)$ **c.** $f\left(\frac{1}{2}\right)$ **d.** $f(a)$
8. Graph:
   a. $f(x) = -2$ **b.** $f(x) = 2x + 1$
9. Find the slope of the line containing the pairs of points:
   a. $(2, -1)$ and $(2, 3)$ **b.** $(4, -2)$ and $(-2, 4)$

6.4 Equations of the Line

10. Find the equation of the line (slope-intercept form) that contains the given point with the given slope:
    a. $(4, -2)$, $m = \frac{1}{2}$ **b.** $(-2, 3)$, $m = 0$
11. Graph by finding the x and y intercepts:
    a. $x - 2y = 4$ **b.** $x = -1$
12. Find the slope and y intercept:
    a. $y = \frac{-2}{3}x + 3$ **b.** $2x - 6y = 8$

13. Graph by use of the slope-intercept method:
  a. $y = \frac{-2}{3}x + 3$ b. $2x - 6y = 8$

6.5 Quadratic Functions

14. Write in the form $f(x) = a(x - h)^2 + k$:
  a. $f(x) = x^2 + 4x + 5$ b. $f(x) = x^2 + 2x$
15. Graph:
  a. $f(x) = -2x^2 + 4$ b. $f(x) = x^2 + 4x + 5$

6.6 Inverse Functions

16. Find the inverse:
  a. $f(x) = 7x + 1$ b. $f = \{(x, y) | 2x + y = 1\}$
  c. $f(x) = \frac{2x - 1}{6}$ d. $f(x) = 7x^3$

Chapter 6 Sample Test Answers

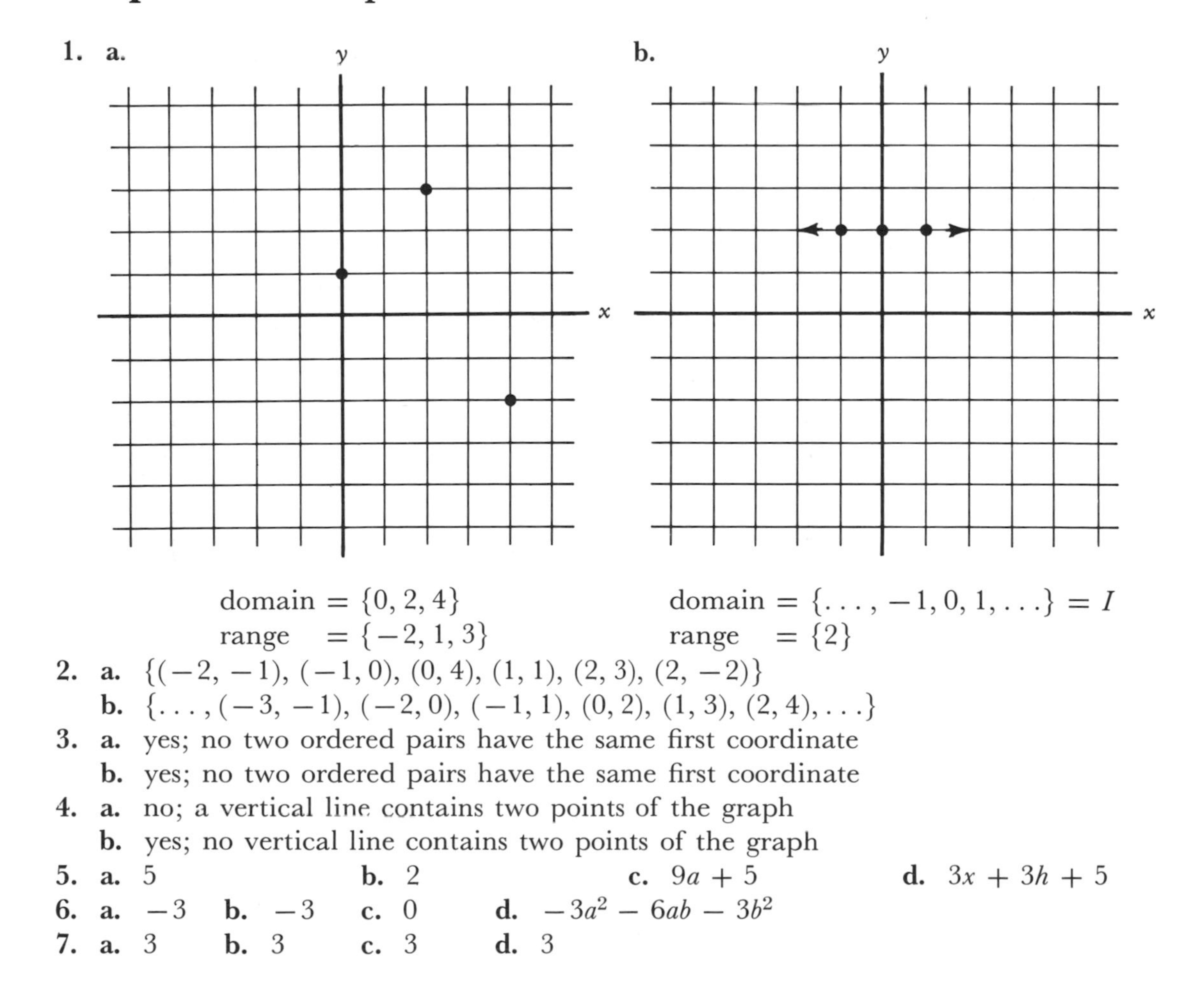

domain = $\{0, 2, 4\}$ domain = $\{\ldots, -1, 0, 1, \ldots\} = I$
range = $\{-2, 1, 3\}$ range = $\{2\}$

2. a. $\{(-2, -1), (-1, 0), (0, 4), (1, 1), (2, 3), (2, -2)\}$
  b. $\{\ldots, (-3, -1), (-2, 0), (-1, 1), (0, 2), (1, 3), (2, 4), \ldots\}$
3. a. yes; no two ordered pairs have the same first coordinate
  b. yes; no two ordered pairs have the same first coordinate
4. a. no; a vertical line contains two points of the graph
  b. yes; no vertical line contains two points of the graph
5. a. 5 b. 2 c. $9a + 5$ d. $3x + 3h + 5$
6. a. -3 b. -3 c. 0 d. $-3a^2 - 6ab - 3b^2$
7. a. 3 b. 3 c. 3 d. 3

8. a. b.

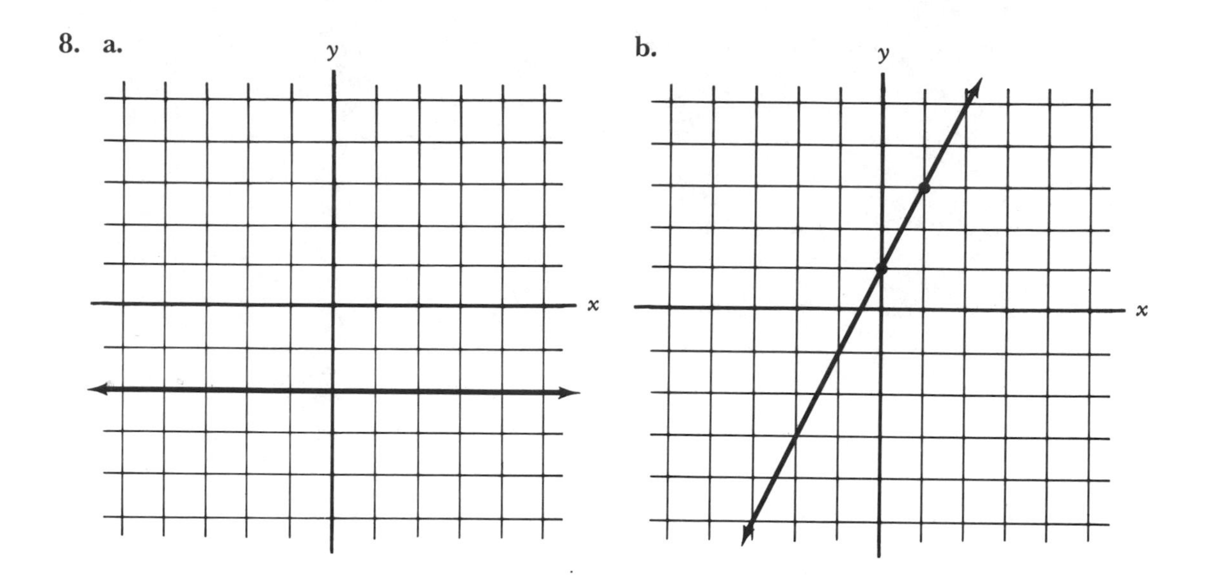

9. a. no slope b. -1

10. a. $y = \frac{1}{2}x - 4$ b. $y = 3$

11. a. b.

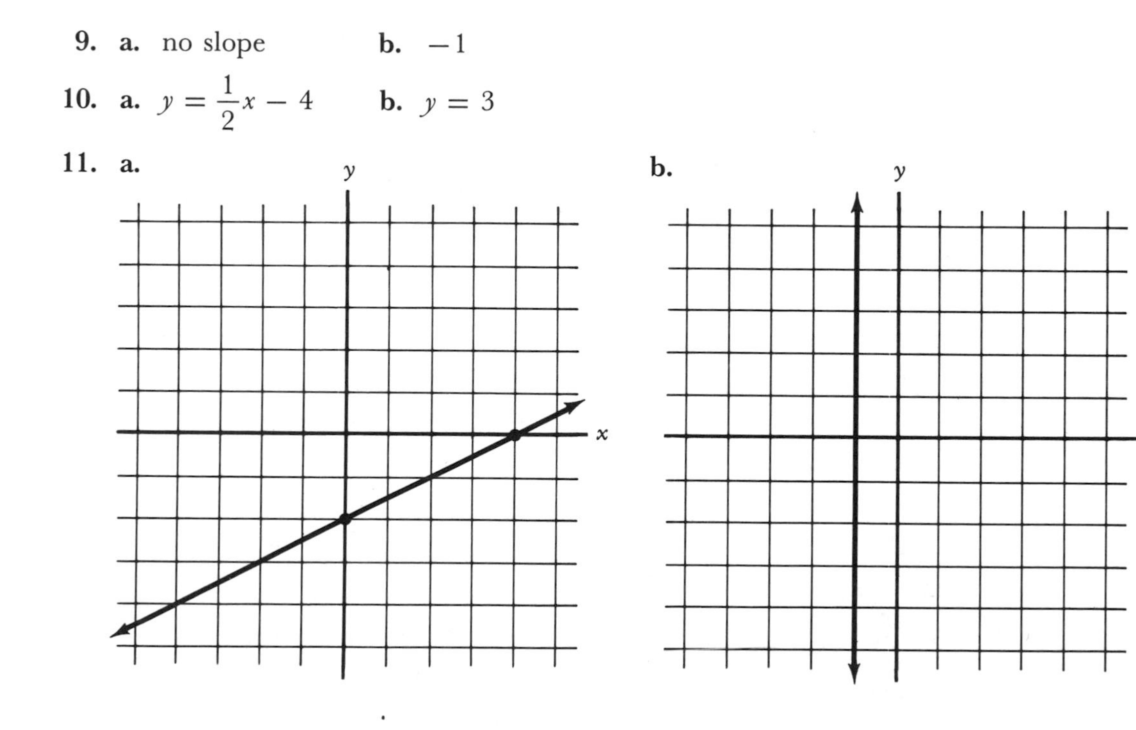

12. a. $m = \frac{-2}{3}$, y intercept $= (0, 3)$

b. $m = \frac{1}{3}$, y intercept $= \left(0, \frac{-4}{3}\right)$

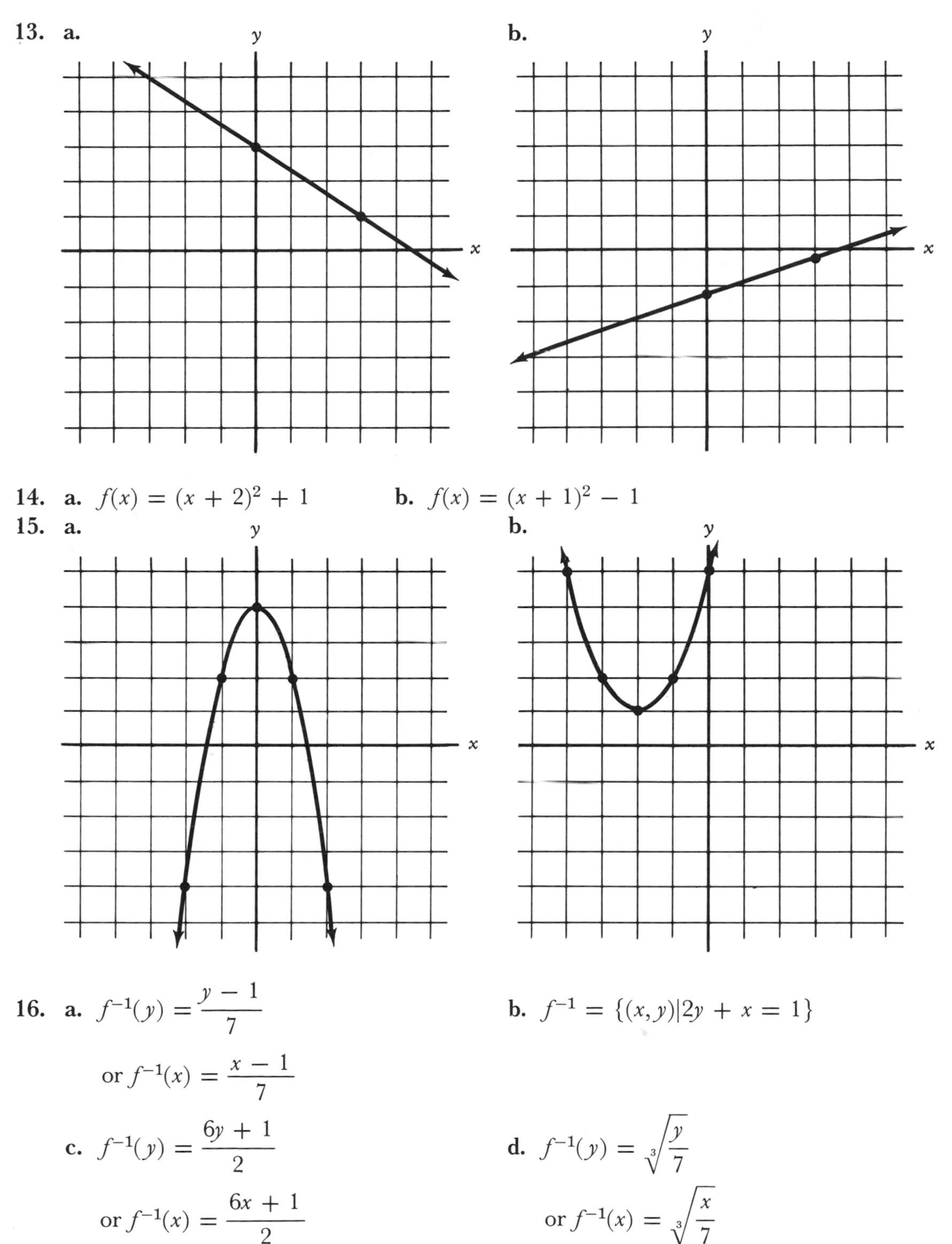

13. **a.**

b.

14. **a.** $f(x) = (x + 2)^2 + 1$ **b.** $f(x) = (x + 1)^2 - 1$

15. **a.**

b.

16. **a.** $f^{-1}(y) = \dfrac{y - 1}{7}$

or $f^{-1}(x) = \dfrac{x - 1}{7}$

b. $f^{-1} = \{(x, y) | 2y + x = 1\}$

c. $f^{-1}(y) = \dfrac{6y + 1}{2}$

or $f^{-1}(x) = \dfrac{6x + 1}{2}$

d. $f^{-1}(y) = \sqrt[3]{\dfrac{y}{7}}$

or $f^{-1}(x) = \sqrt[3]{\dfrac{x}{7}}$

Chapter 7

Systems of Equations

Solving a system of two equations or solving two equations means to find the point of intersection of two graphs. In this chapter that point of intersection will be found graphically and algebraically.

7.1 Parallel, Perpendicular, and Intersecting Lines

1

If two linear equations are graphed, on the same coordinate system, three possibilities exist: (1) The lines intersect in exactly one point, (2) the lines are parallel, or (3) they coincide (contain the same points). In (1) the intersection point is a point on both lines; therefore, its coordinates must satisfy the conditions of both linear equations.

Example: Graph $x - y + 3 = 0$ and $2x + y = 6$, finding the point of intersection and check the ordered pair in both equations.

Solution

The lines intersect at $(1, 4)$.
Check: Substitute $(1, 4)$ in both equations.

$$x - y + 3 = 0$$
$$1 - 4 + 3 \stackrel{?}{=} 0$$
$$0 = 0 \text{ (true)}$$
$$2x + y = 6$$
$$2(1) + 4 \stackrel{?}{=} 6$$
$$6 = 6 \text{ (true)}$$

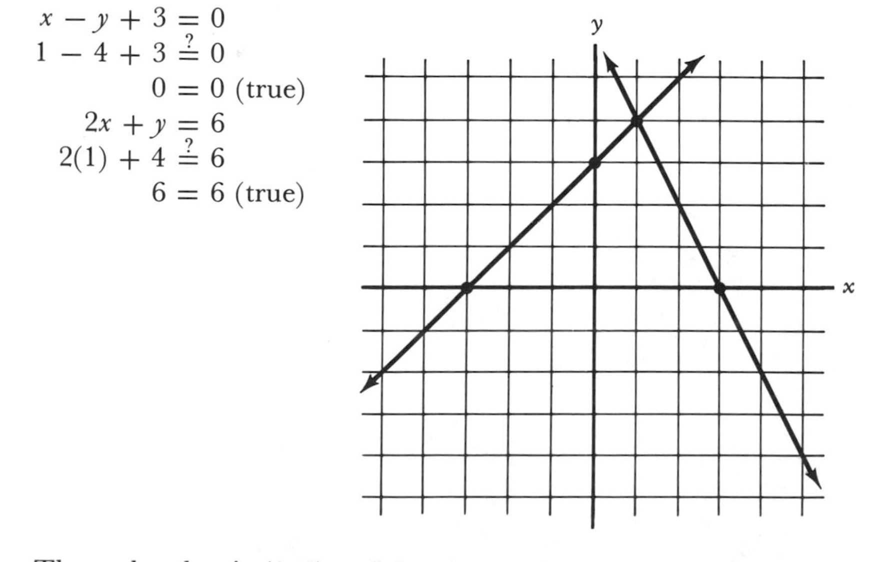

The ordered pair $(1, 4)$ satisfies the conditions of both equations.

Q1 Solve the system of equations graphically. Check.

$y = -x + 3$ and $y = \frac{1}{2}x$

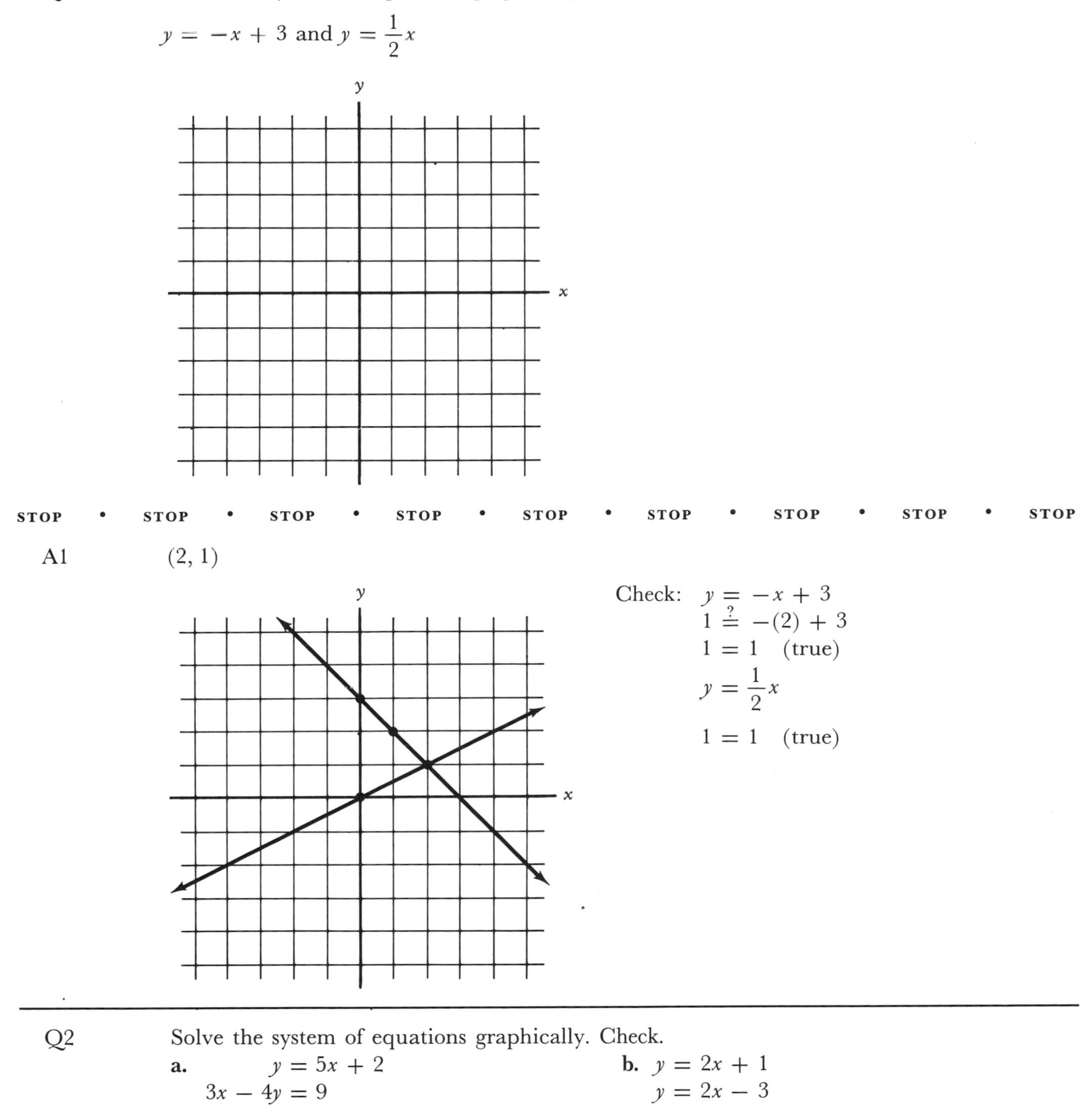

STOP • STOP • STOP • STOP • STOP • STOP • STOP • STOP • STOP

A1 (2, 1)

Check:

$$y = -x + 3$$
$$1 \stackrel{?}{=} -(2) + 3$$
$$1 = 1 \quad \text{(true)}$$
$$y = \frac{1}{2}x$$
$$1 = 1 \quad \text{(true)}$$

Q2 Solve the system of equations graphically. Check.

a. $y = 5x + 2$
$3x - 4y = 9$

b. $y = 2x + 1$
$y = 2x - 3$

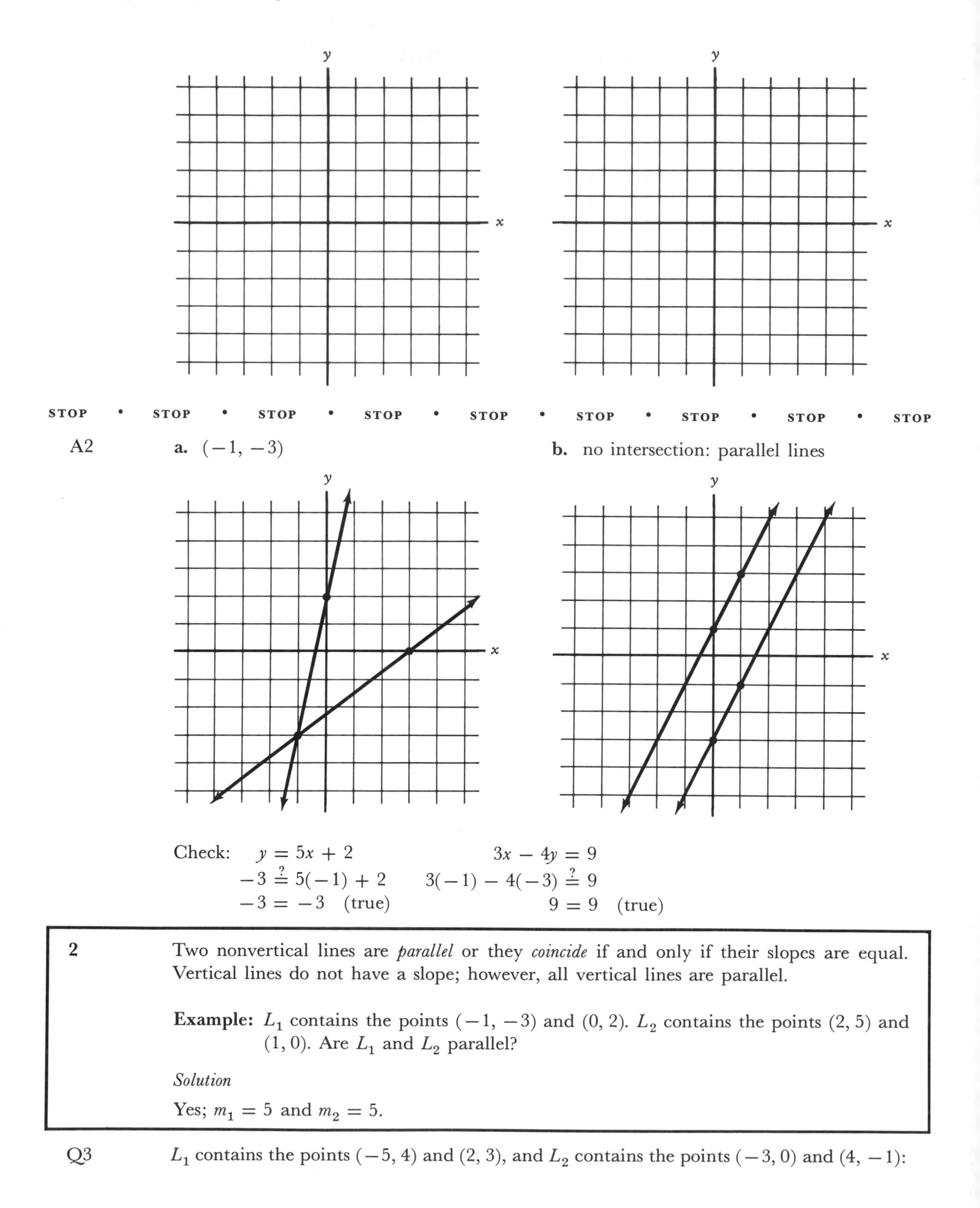

STOP • STOP • STOP • STOP • STOP • STOP • STOP • STOP • STOP

A2 **a.** $(-1, -3)$ **b.** no intersection: parallel lines

Check:

$$y = 5x + 2 \qquad 3x - 4y = 9$$

$$-3 \stackrel{?}{=} 5(-1) + 2 \qquad 3(-1) - 4(-3) \stackrel{?}{=} 9$$

$$-3 = -3 \text{ (true)} \qquad 9 = 9 \text{ (true)}$$

2 Two nonvertical lines are *parallel* or they *coincide* if and only if their slopes are equal. Vertical lines do not have a slope; however, all vertical lines are parallel.

Example: L_1 contains the points $(-1, -3)$ and $(0, 2)$. L_2 contains the points $(2, 5)$ and $(1, 0)$. Are L_1 and L_2 parallel?

Solution

Yes; $m_1 = 5$ and $m_2 = 5$.

Q3 L_1 contains the points $(-5, 4)$ and $(2, 3)$, and L_2 contains the points $(-3, 0)$ and $(4, -1)$:

a. $m_1 =$ ____ **b.** $m_2 =$ ____ **c.** Hence L_1 and L_2 are ________.

STOP • STOP • STOP • STOP • STOP • STOP • STOP • STOP • STOP

A3 **a.** $\frac{-1}{7}$ **b.** $\frac{-1}{7}$ **c.** parallel

---

Q4 L_1 contains the points $(2, 5)$ and $(2, -1)$, and L_2 contains the points $(-4, 0)$ and $(-4, 8)$:

a. $m_1 =$ ________ **b.** $m_2 =$ ________

c. Hence L_1 and L_2 are ________.

STOP • STOP • STOP • STOP • STOP • STOP • STOP • STOP • STOP

A4 **a.** no slope **b.** no slope

c. parallel (vertical lines)

---

3 To determine if two lines are parallel it will be convenient to write the equations in the slope-intercept form.

1. If the slopes are *not* equal, the lines intersect.
2. If the slopes are equal and the y intercepts are not equal, the lines are parallel.
3. If the slopes are equal and the y intercepts are also equal, the lines coincide.

Example: Describe the system as (1) intersecting lines, (2) parallel lines, or (3) lines that coincide.

$$2x + 5y - 10 = 0$$
$$4x + 10y = 30$$

Solution

$$2x + 5y - 10 = 0 \qquad 4x + 10y = 30$$
$$5y = -2x + 10 \qquad 10y = -4x + 30$$
$$y = \frac{-2}{5}x + 2 \qquad y = \frac{-2}{5}x + 3$$

Since the slopes are equal and the y intercepts different, the lines are parallel.

Q5 Describe the system as intersecting lines, parallel lines, or lines that coincide:

a. $2x - 3y = 6$
$4x - 6y = 1$

b. $2x - y = 3$
$x + 2y + 6 = 0$

STOP • STOP • STOP • STOP • STOP • STOP • STOP • STOP • STOP

A5 **a.** parallel lines:

$$2x - 3y = 6;\ y = \frac{2}{3}x - 2$$
$$4x - 6y = 1;\ y = \frac{2}{3}x - \frac{1}{6}$$

b. intersecting lines:

$$2x - y = 3;\ y = 2x - 3$$
$$x + 2y + 6 = 0;\ y = \frac{-1}{2}x - 3$$

Q6 Describe the system:

a. $y = 2x + 1$
$4x - 2y = 0$

b. $2x - y = 1$
$4x - 2y = 2$

STOP • STOP • STOP • STOP • STOP • STOP • STOP • STOP • STOP

A6 **a.** parallel lines:
$4x - 2y = 0; y = 2x$

b. lines that coincide:
$2x - y = 1; y = 2x - 1$
$4x - 2y = 2; y = 2x - 1$

Q7 Solve the system of equations graphically:

a. $2x - y + 1 = 0$
$4x - 2y = 0$

b. $2x - y = 1$
$4x - 2y = 2$

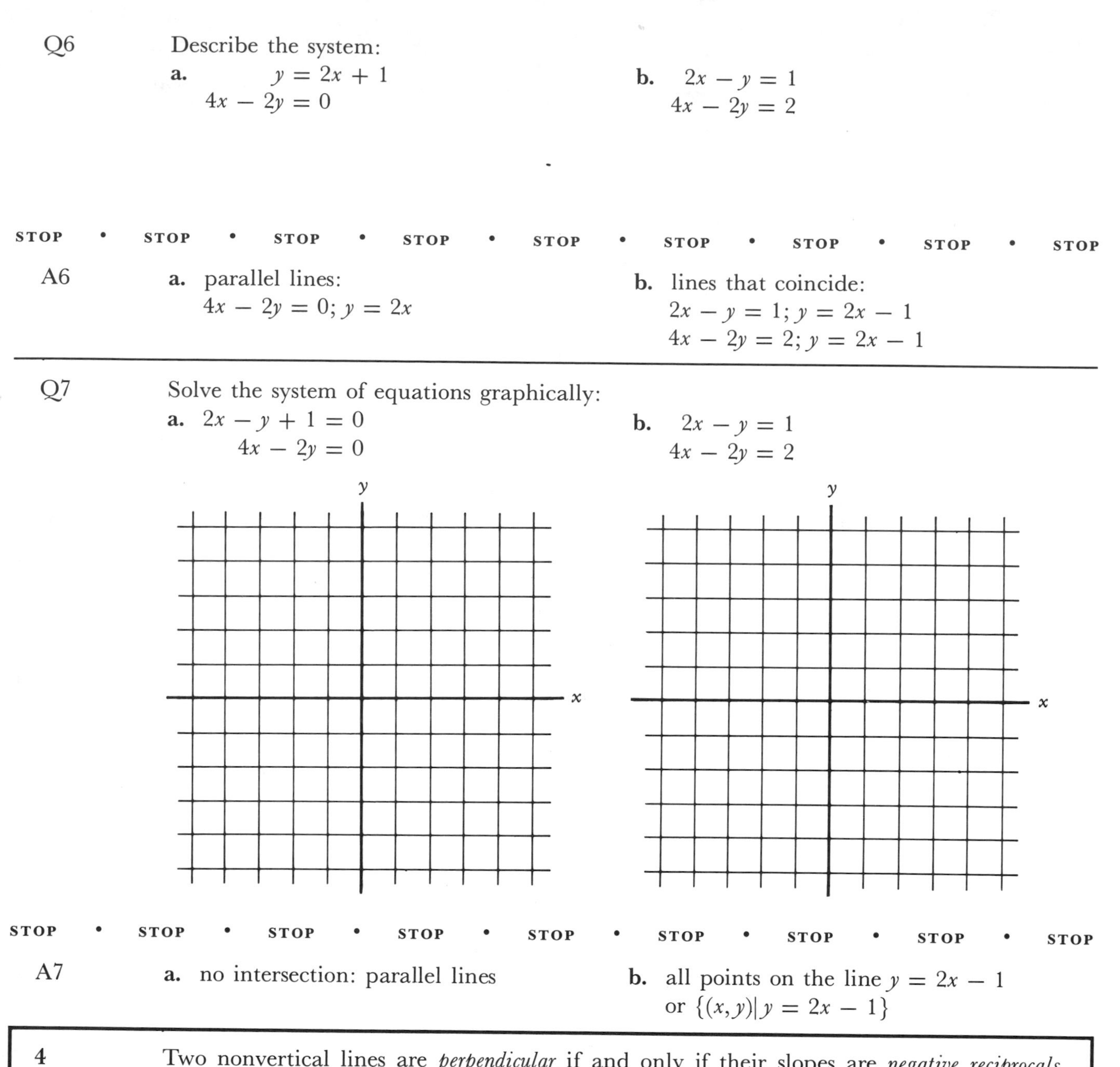

STOP • STOP • STOP • STOP • STOP • STOP • STOP • STOP • STOP

A7 **a.** no intersection: parallel lines

b. all points on the line $y = 2x - 1$ or $\{(x, y) | y = 2x - 1\}$

---

4 Two nonvertical lines are *perpendicular* if and only if their slopes are *negative reciprocals.* Horizontal and vertical lines are also perpendicular. Negative reciprocal means the opposite of the reciprocal.

Number	Negative reciprocal
$\frac{2}{3}$	$\frac{-3}{2}$
-4	$\frac{1}{4}$
$\frac{1}{2}$	-2

Example: Determine if the system represents perpendicular lines.

$2x - y = 4$
$x + 2y = 0$

Solution

Yes.

$2x - y = 4$ $\qquad x + 2y = 0$

$y = 2x - 4$ $\qquad y = \frac{-1}{2}x$

$m = 2$ $\qquad m = \frac{-1}{2}$

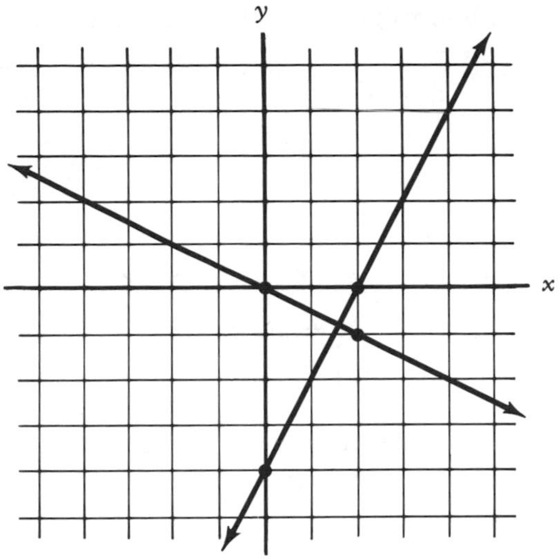

The slopes are negative reciprocals. The lines are shown in the graph.

Q8 Determine if the system represents perpendicular lines:

a. $y = \frac{2}{3}x - 1$

$y = \frac{-3}{2}x + 2$

b. $2x - 5y = 10$

$10x + 4y = 6$

STOP • STOP • STOP • STOP • STOP • STOP • STOP • STOP • STOP

A8 **a.** yes

b. yes: $2x - 5y = 10; y = \frac{2}{5}x - 2$

$10x + 4y = 6; y = \frac{-5}{2}x + \frac{3}{2}$

The slopes are negative reciprocals.

Q9 Determine if the system represents perpendicular lines:

a. $2x - 6y = 7$

$3x + y = 2$

b. $x - 2y = 3$

$2x - y = 5$

STOP • STOP • STOP • STOP • STOP • STOP • STOP • STOP • STOP

A9 **a.** yes: $m_1 = \frac{1}{3}$

$m_2 = -3$

b. no: $m_1 = \frac{1}{2}$

$m_2 = 2$

In part b the slopes are reciprocals but not negative reciprocals.

Q10 Determine if the system represents perpendicular lines:

a. $x = 2$
$y = -1$

b. $y = 4$
$x = 1$

STOP • STOP • STOP • STOP • STOP • STOP • STOP • STOP • STOP

A10 **a.** yes: $m_1 =$ no slope
$m_2 = 0$

b. yes: $m_1 = 0$
$m_2 =$ no slope

In part a, $x = 2$ represents a vertical line and $y = -1$ represents a horizontal line. Vertical and horizontal lines are perpendicular.

Q11 Solve the system graphically:

a. $x = 2$
$y = -1$

b. $y = 4$
$x = 1$

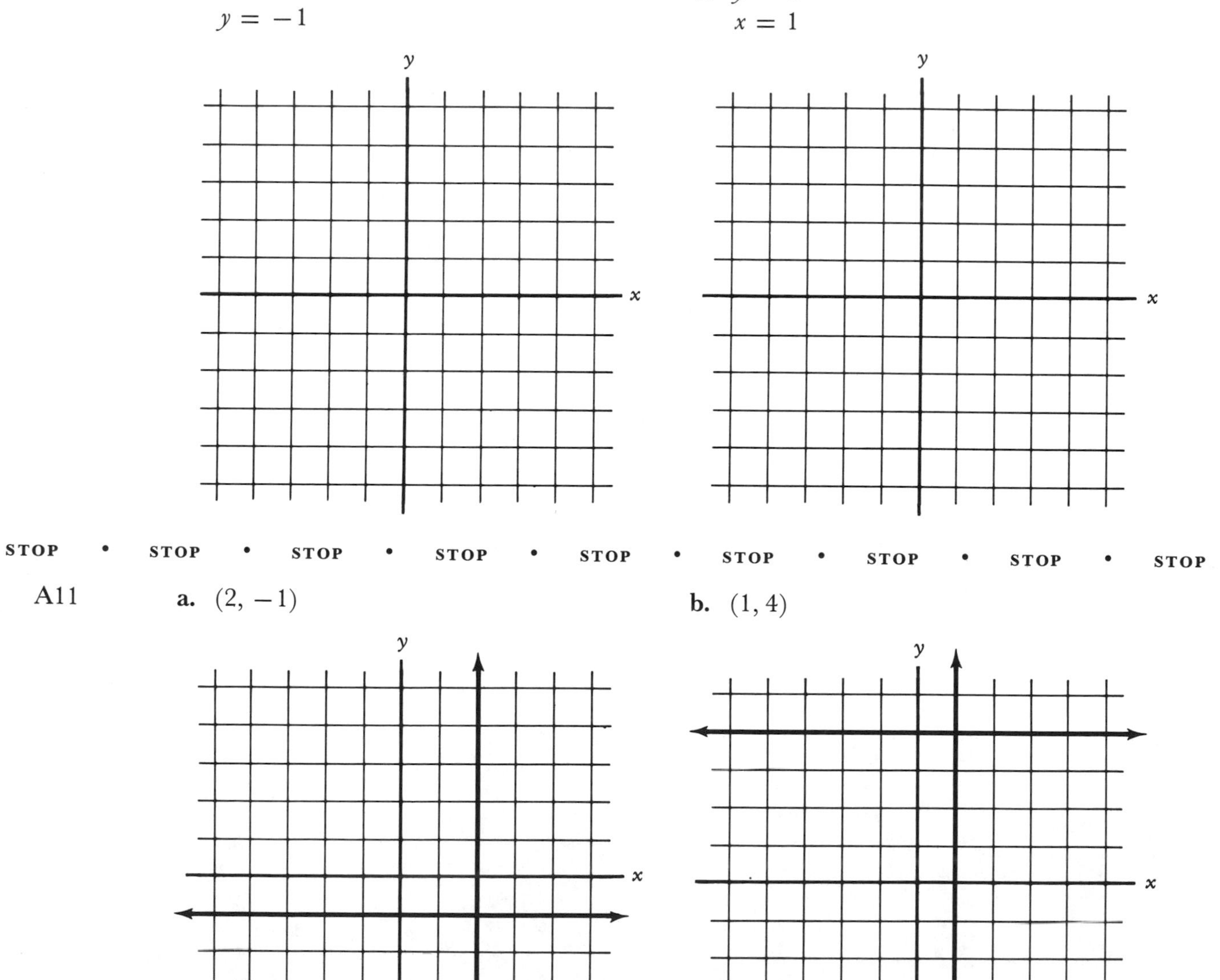

STOP • STOP • STOP • STOP • STOP • STOP • STOP • STOP • STOP

A11 **a.** $(2, -1)$

b. $(1, 4)$

5 The equation of a line, containing a known point and parallel (or perpendicular) to a given line, can now be determined.

Example: Find the equation of the line containing $(1, -3)$ and perpendicular to the line $2x - y = 4$.

Solution

Since the lines are perpendicular, their slopes are negative reciprocals.

$$2x - y = 4$$
$$y = 2x - 4 \qquad m = 2$$

Hence the slope of the unknown line is $\frac{-1}{2}$. Now, use the point-slope form of the equation of a line with $(x_1, y_1) = (1, -3)$ and $m = \frac{-1}{2}$.

$$y - y_1 = m(x - x_1)$$
$$y + 3 = \frac{-1}{2}(x - 1)$$
$$y + 3 = \frac{-1}{2}x + \frac{1}{2}$$
$$y = \frac{-1}{2}x - \frac{5}{2}$$

The equations are shown in the graph.

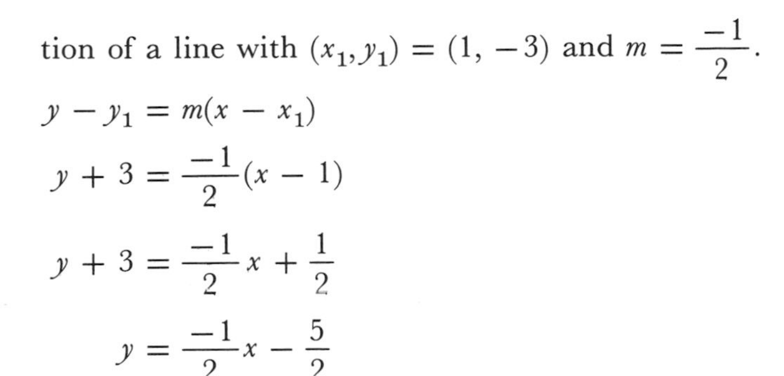

Q12 Find the equation of the line parallel to the given line and containing the given point. Graph both lines.

a. $2x + y - 3 = 0$; $(-2, 1)$ **b.** $x = 2$; $(-3, 1)$

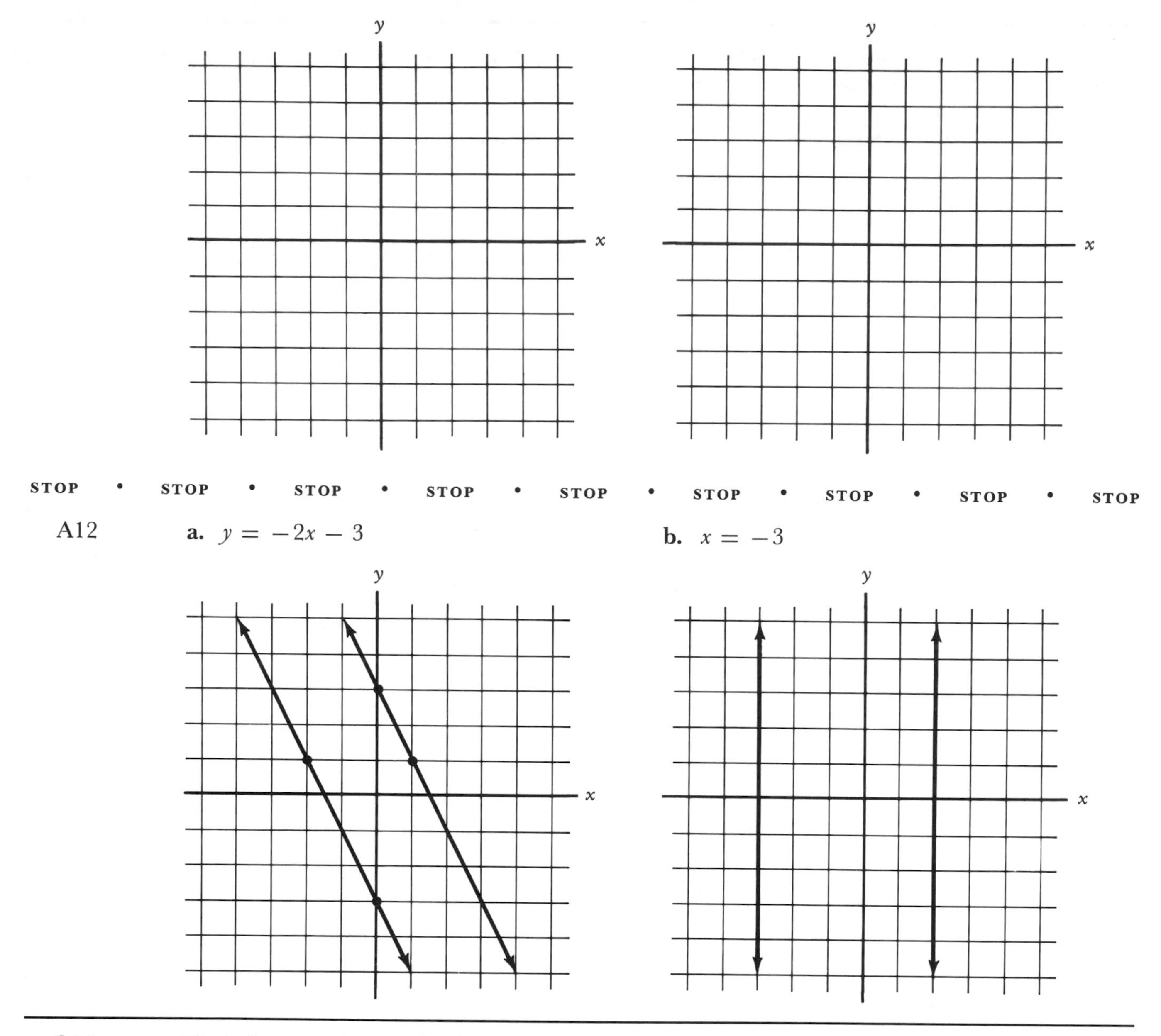

STOP • **STOP** • **STOP** • **STOP** • **STOP** • **STOP** • **STOP** • **STOP** • **STOP**

A12 **a.** $y = -2x - 3$ **b.** $x = -3$

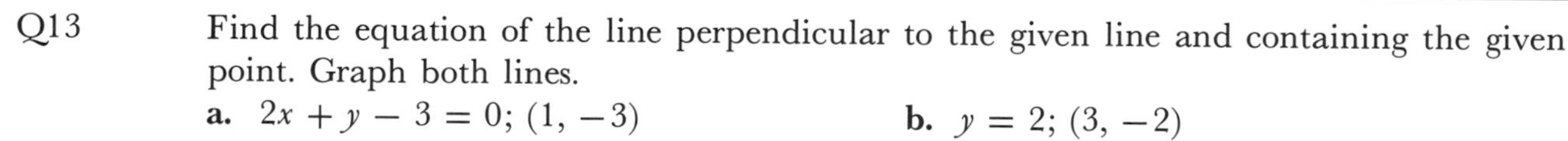

Q13 Find the equation of the line perpendicular to the given line and containing the given point. Graph both lines.

a. $2x + y - 3 = 0$; $(1, -3)$ **b.** $y = 2$; $(3, -2)$

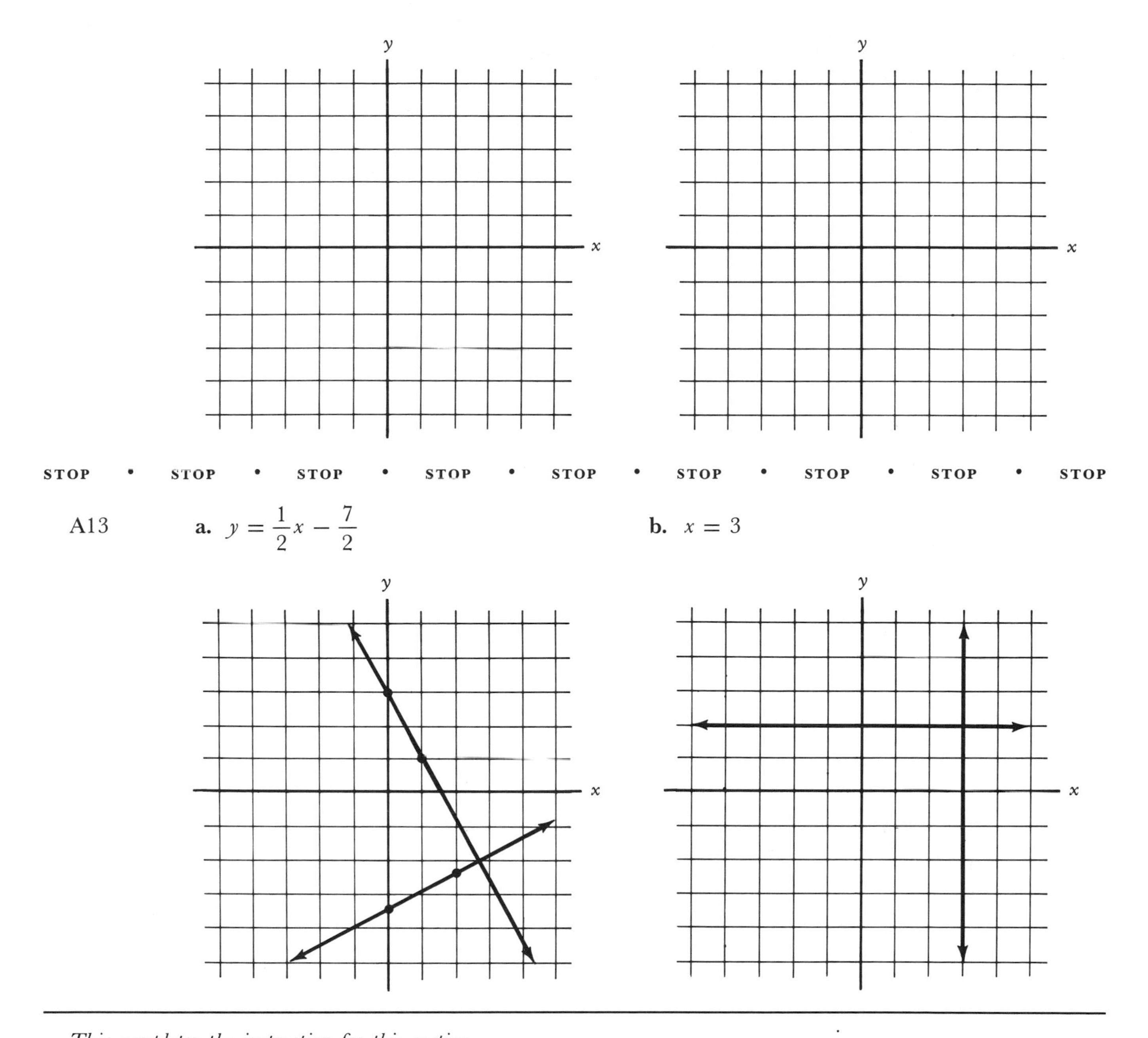

STOP • STOP • STOP • STOP • STOP • STOP • STOP • STOP • STOP

A13 **a.** $y = \frac{1}{2}x - \frac{7}{2}$ **b.** $x = 3$

This completes the instruction for this section.

7.1 Exercises

1. Solve the systems of equations graphically:

a. $x + y = 3$
$x + y = -1$

b. $x - y = 2$
$x + y = -2$

c. $3x + 2y = 5$
$x + 2y = 3$

d. $y = -2x$
$y = \frac{1}{3}x$

e. $x = 2$
$y = -1$

f. $x - 2y = 2$
$2x - 4y = 4$

2. Describe the system as intersecting lines, parallel lines, or lines that coincide:

a. $2x - y = 7$
$x + y = 5$

b. $x - 2y = 12$
$-2y = -x + 6$

c. $3x - 5y = 9$
$x + 2y = 4$

d. $2x - y = 4$
$x - \frac{1}{2}y = 2$

3. Describe the system as parallel lines or perpendicular lines:

a. $2x - y = 4$
$x + 2y = 7$

b. $y = \frac{2}{3}x - 1$
$-2x + 3y + 5 = 0$

c. $x = 2$
$y = 1$

d. $y = -8$
$y = 0$

4. Find the slope of the line parallel to the given line:
a. $y = -2x + 3$ **b.** $x = -1$ **c.** $y = 4$

5. Find the slope of the line perpendicular to the given line:
a. $y = -2x + 3$ **b.** $x = -1$ **c.** $y = 4$

6. Find the equation of the line satisfying the given conditions. Graph each line.
a. Contains $(2, -3)$ parallel to $2x - y = 0$.
b. Contains $(2, -3)$ perpendicular to $2x - y = 0$.
c. Contains $(2, -3)$ parallel to $y = 1$.
d. Contains $(2, -3)$ perpendicular to $x = -1$.

*7. Prove that the triangle with vertices $A(-4, -2)$, $B(2, -8)$, and $C(4, 6)$ is a right triangle.

*8. Determine the real number k so that the two lines $5x - 3y = 12$ and $kx - y = 2$ will be:
a. parallel **b.** perpendicular

*9. When are the lines $Ax + By + C = 0$ and $Dx + Ey + F = 0$ (neither represents vertical or horizontal lines)
a. parallel? **b.** perpendicular?

7.1 Exercise Answers

1. **a.** no intersection

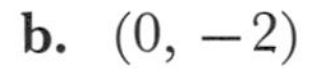
b. $(0, -2)$

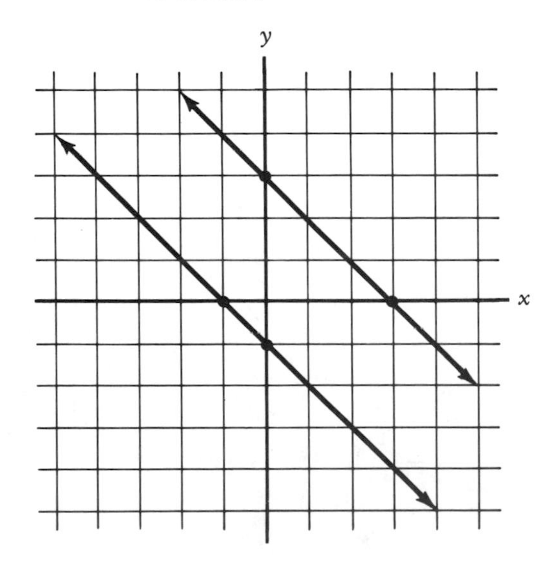

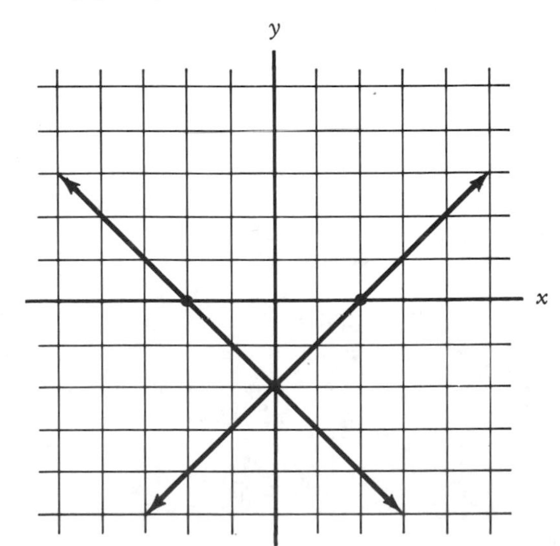

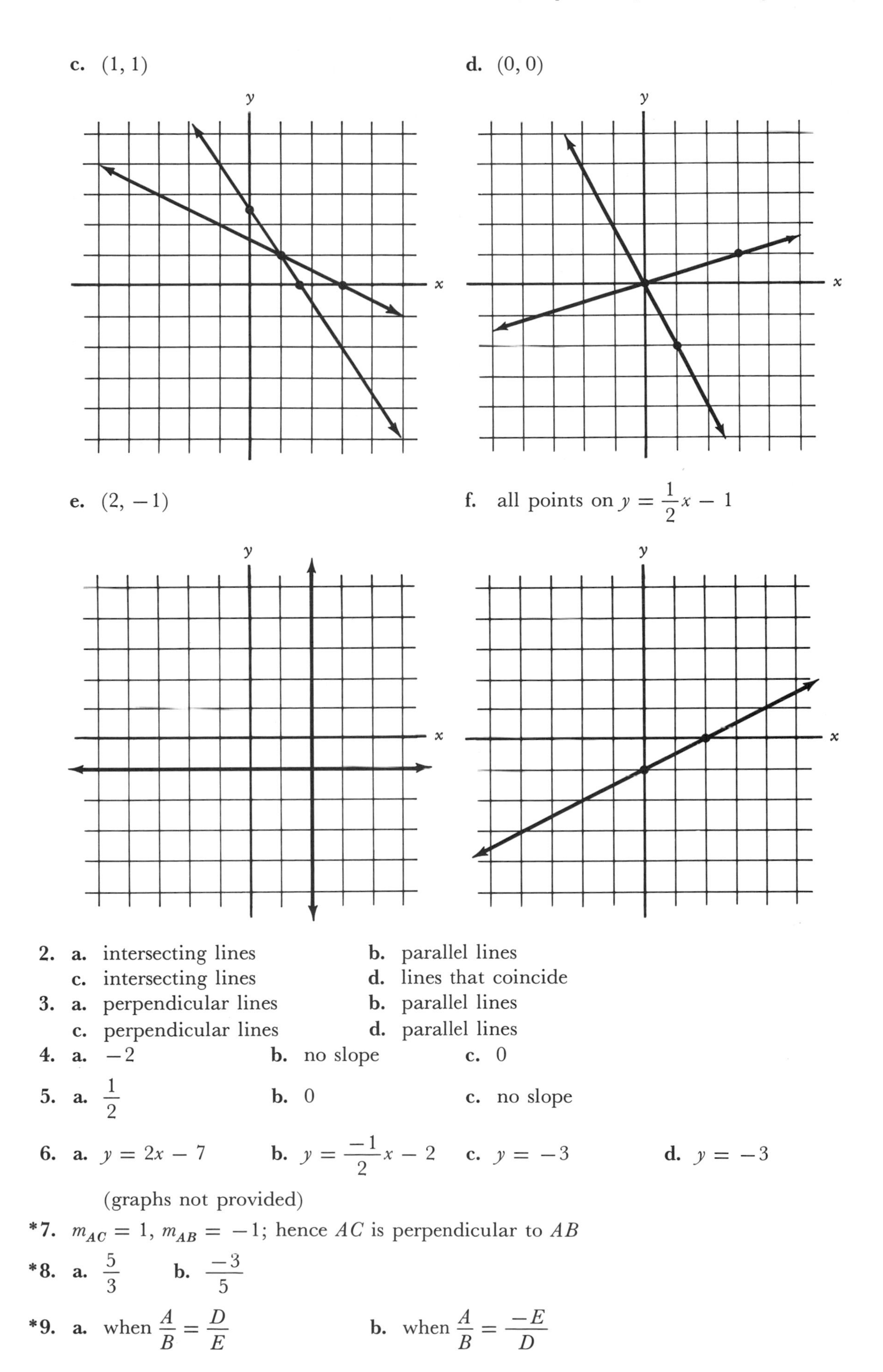

c. (1, 1)

d. (0, 0)

e. (2, −1)

f. all points on $y = \frac{1}{2}x - 1$

2. **a.** intersecting lines **b.** parallel lines
c. intersecting lines **d.** lines that coincide

3. **a.** perpendicular lines **b.** parallel lines
c. perpendicular lines **d.** parallel lines

4. **a.** -2 **b.** no slope **c.** 0

5. **a.** $\frac{1}{2}$ **b.** 0 **c.** no slope

6. **a.** $y = 2x - 7$ **b.** $y = \frac{-1}{2}x - 2$ **c.** $y = -3$ **d.** $y = -3$

(graphs not provided)

***7.** $m_{AC} = 1$, $m_{AB} = -1$; hence AC is perpendicular to AB

***8.** **a.** $\frac{5}{3}$ **b.** $\frac{-3}{5}$

***9.** **a.** when $\frac{A}{B} = \frac{D}{E}$ **b.** when $\frac{A}{B} = \frac{-E}{D}$

7.2 Addition and Substitution Methods

1

Because graphing equations is a time-consuming process, and, more important, because graphical results are not always precise, solutions to systems of linear equations are usually sought by analytic methods. One such method is the addition method, which relies on the following theorem: If (x, y) satisfies the equations

(1) $Ax + By + C = 0$
(2) $Dx + Ey + F = 0$

then (x, y) will satisfy the sum of the two equations,

(3) $Ax + By + C + Dx + Ey + F = 0$

Example: Show that the ordered pair (5, 7) satisfies equations (1), (2), and (3). Equation (3) is the sum of (1) and (2).

(1) $-2x + 3y = 11$
(2) $x - y = -2$
(3) $-x + 2y = 9$

Solution

Substitute (5, 7) into each equation.

$-2x + 3y = 11$ $\quad$ $x - y = -2$ $\quad$ $-x + 2y = 9$

$-2(5) + 3(7) = 11$ (true) $\quad$ $5 - 7 = -2$ (true) $\quad$ $-5 + 2(7) = 9$ (true)

This result illustrates that if an ordered pair satisfies a system of linear equations, the ordered pair will satisfy the sum of the equations.

Q1 Determine the sum of equations (1) and (2).

(1) $-3x - 2y = 7$
(2) $2x + y = -2$

STOP • STOP • STOP • STOP • STOP • STOP • STOP • STOP • STOP

A1 $-x - y = 5$

Q2 Show that the ordered pair $(3, -8)$ satisfies equations (1), (2), and (3).

(1) $-3x - 2y = 7$ $\quad$ (2) $2x + y = -2$ $\quad$ (3) $-x - y = 5$

STOP • STOP • STOP • STOP • STOP • STOP • STOP • STOP • STOP

A2 (1) $-3(3) - 2(-8) \stackrel{?}{=} 7$
$-9 + 16 = 7$ (true)

(2) $2(3) + (-8) \stackrel{?}{=} -2$
$6 - 8 = -2$ (true)

(3) $-3 - (-8) \stackrel{?}{=} 5$
$-3 + 8 = 5$ (true)

2

Since the solution to a system of linear equations satisfies the sum of the equations, the system of equations can often be solved by adding the two equations. The intent is to eliminate one variable in the sum and solve the resulting equation in terms of the other variable.

Example: Solve the system by adding the two equations.

$$\begin{array}{l} x + y = 8 \\ \underline{x - y = 2} \end{array}$$

Solution

$2x = 10$

The variable y was eliminated in the sum. Now solve the resulting equation obtaining the x coordinate of the solution (intersection).

$x = 5$

This value can be substituted into either of the original equations to obtain the y coordinate of the solution.

$$\begin{array}{rl} x + y = 8 & \\ 5 + y = 8 & \\ y = 3 & \end{array} \quad \text{or} \quad \begin{array}{r} x - y = 2 \\ 5 - y = 2 \\ -y = -3 \\ y = 3 \end{array}$$

The solution (intersection) is (5, 3).

Check:

$$\begin{array}{ll} x + y = 8 & \\ 5 + 3 = 8 & \text{(true)} \end{array} \qquad \begin{array}{ll} x - y = 2 & \\ 5 - 3 = 2 & \text{(true)} \end{array}$$

Q3 Solve the system for the variable x.

$$\begin{array}{r} x - 2y = 3 \\ \underline{2x + 2y = 6} \end{array}$$

STOP • STOP • STOP • STOP • STOP • STOP • STOP • STOP • STOP

A3 $x = 3$: $3x = 9$

Q4 Solve the system in Q3 for y by substituting $x = 3$ into both equations.

$x - 2y = 3 \qquad 2x + 2y = 6$

STOP • STOP • STOP • STOP • STOP • STOP • STOP • STOP • STOP

A4 $y = 0$:
$$\begin{aligned} 3 - 2y &= 3 \\ -2y &= 0 \\ y &= 0 \end{aligned} \qquad \begin{aligned} 2(3) + 2y &= 6 \\ 6 + 2y &= 6 \\ 2y &= 0 \end{aligned}$$

Q5 The solution to the system in Q3 is the ordered pair ________.

STOP • STOP • STOP • STOP • STOP • STOP • STOP • STOP • STOP

A5 $(3, 0)$

Q6 Check the solution to Q3 in both equations.
$x - 2y = 3 \qquad 2x + 2y = 6$

STOP • STOP • STOP • STOP • STOP • STOP • STOP • STOP • STOP

A6 $3 - 2(0) = 3$ (true) $\qquad 2(3) + 2(0) = 6$ (true)

Q7 Solve by the addition method.
$$\begin{aligned} x + 7y &= 2 \\ 5x - 7y &= -32 \end{aligned}$$

STOP • STOP • STOP • STOP • STOP • STOP • STOP • STOP • STOP

A7 $(-5, 1)$:
$$\begin{aligned} x + 7y &= 2 \\ 5x - 7y &= -32 \\ \hline 6x \qquad &= -30 \\ x &= -5 \end{aligned} \qquad \begin{aligned} x + 7y &= 2 \\ -5 + 7y &= 2 \\ y &= 1 \end{aligned}$$

The solution for y could have been obtained by substituting $x = -5$ into $5x - 7y = -32$.

Q8 Solve, eliminating x when adding.
$$\begin{aligned} -2x + y &= 7 \\ 2x + 3y &= 5 \end{aligned}$$

STOP • STOP • STOP • STOP • STOP • STOP • STOP • STOP • STOP

A8 $(-2, 3)$:
$$\begin{aligned} -2x + y &= 7 \\ 2x + 3y &= 5 \\ \hline 4y &= 12 \\ y &= 3 \end{aligned} \qquad \begin{aligned} -2x + y &= 7 \\ -2x + 3 &= 7 \\ -2x &= 4 \\ x &= -2 \end{aligned}$$

Q9 Solve:

a. $\begin{aligned} x - 3y &= -4 \\ -x - 2y &= -1 \end{aligned}$

b. $\begin{aligned} 2m + 3n &= -8 \\ -5m - 3n &= 2 \end{aligned}$

STOP • STOP • STOP • STOP • STOP • STOP • STOP • STOP • STOP

A9 **a.** $(-1, 1)$ **b.** $(2, -4)$

Q10 Solve:

a. $\begin{aligned} x - y &= 12 \\ -x - y &= 6 \end{aligned}$

b. $\begin{aligned} 3x - 3y &= -12 \\ x + 3y &= 12 \end{aligned}$

STOP • STOP • STOP • STOP • STOP • STOP • STOP • STOP • STOP

A10 **a.** $(3, -9)$ **b.** $(0, 4)$

3 If the sum of the equations of a system will not result in the elimination of one of the variables, it is often possible to multiply both sides of one of the equations by the same integer in order that one of the variables will be eliminated in the sum.

$$\begin{aligned} -2x + 3y = 11 &\longrightarrow \quad -2x + 3y = 11 \\ x - y = -2 &\xrightarrow{\text{multiply by 3}} \quad 3x - 3y = -6 \\ & \qquad\qquad\quad x \qquad\; = 5 \end{aligned}$$

The y value is found by substituting $x = 5$ into either of the original equations.

$$\begin{aligned} x - y &= -2 \\ 5 - y &= -2 \\ -y &= -7 \\ y &= 7 \end{aligned}$$

Hence the solution for the system is $(5, 7)$.

Q11 Solve:

$$\begin{aligned} 2x - 5y = 10 &\longrightarrow \_\_\_\_\_\_\_\_\_\_ \\ 3x + y = 15 &\xrightarrow{\text{multiply by 5}} \_\_\_\_\_\_\_\_\_\_ \end{aligned}$$

STOP • STOP • STOP • STOP • STOP • STOP • STOP • STOP • STOP

A11 $(5, 0)$:

$$\begin{aligned} 2x - 5y &= 10 \\ 15x + 5y &= 75 \\ 17x \qquad &= 85 \\ x &= 5 \end{aligned} \qquad \begin{aligned} 2x - 5y &= 10 \\ 2(5) - 5y &= 10 \\ -5y &= 0 \\ y &= 0 \end{aligned}$$

Q12 Solve:

$$6x + y = 5 \longrightarrow \underline{\qquad\qquad}$$

$$5x + y = 3 \xrightarrow{\text{multiply by } -1} \underline{\qquad\qquad}$$

STOP • STOP • STOP • STOP • STOP • STOP • STOP • STOP • STOP

A12 (2, −7):

$$\begin{array}{r} 6x + y = 5 \\ -5x - y = -3 \\ \hline x \qquad = 2 \end{array} \qquad \begin{array}{r} 6x + y = 5 \\ 6(2) + y = 5 \\ y = -7 \end{array}$$

Q13 Solve:

$$4x - y = 3 \longrightarrow \underline{\qquad\qquad}$$

$$-2x + 3y = 1 \xrightarrow{\text{multiply by } 2} \underline{\qquad\qquad}$$

STOP • STOP • STOP • STOP • STOP • STOP • STOP • STOP • STOP

A13 (1, 1):

$$\begin{array}{r} 4x - y = 3 \\ -4x + 6y = 2 \\ \hline 5y = 5 \\ y = 1 \end{array} \qquad \begin{array}{r} 4x - y = 3 \\ 4x - 1 = 3 \\ 4x = 4 \\ x = 1 \end{array}$$

Q14 Solve:

$$4x - y = 3 \xrightarrow{\text{multiply by } 3} \underline{\qquad\qquad}$$

$$-2x + 3y = 1 \longrightarrow \underline{\qquad\qquad}$$

STOP • STOP • STOP • STOP • STOP • STOP • STOP • STOP • STOP

A14 (1, 1):

$$\begin{array}{r} 12x - 3y = 9 \\ -2x + 3y = 1 \\ \hline 10x \qquad = 10 \\ x = 1 \end{array} \qquad \begin{array}{r} -2x + 3y = 1 \\ -2(1) + 3y = 1 \\ 3y = 3 \\ y = 1 \end{array}$$

4 Note that in Q13 and Q14 the same system was solved in two different ways. There are usually different options that can be followed when solving systems. As long as both sides of an equation are multiplied by the same nonzero number, an equivalent equation is obtained.

Q15 Solve:

$$\begin{aligned} 2x + 3y &= 12 \\ 5x + 3y &= 14 \end{aligned}$$

STOP • STOP • STOP • STOP • STOP • STOP • STOP • STOP • STOP

A15 $\left(\frac{2}{3}, \frac{32}{9}\right)$:

$$\begin{aligned} 2x + 3y &= 12 \xrightarrow{\qquad\qquad} & 2x + 3y &= 12 \\ 5x + 3y &= 14 \xrightarrow{\text{multiply by } -1} & -5x - 3y &= -14 \\ & & \hline -3x \qquad &= -2 \\ & & x &= \frac{2}{3} \end{aligned}$$

$$\begin{aligned} 2\left(\frac{2}{3}\right) + 3y &= 12 \\ 3y &= \frac{36}{3} - \frac{4}{3} \\ 3y &= \frac{32}{3} \\ y &= \frac{32}{9} \end{aligned}$$

Q16 Solve:

$$\begin{aligned} -3x - 5y &= 7 \\ 2x - 5y &= -3 \end{aligned}$$

STOP • STOP • STOP • STOP • STOP • STOP • STOP • STOP • STOP

A16 $\left(-2, \frac{-1}{5}\right)$:

$$\begin{aligned} -3x - 5y &= 7 \xrightarrow{\text{multiply by } -1} & 3x + 5y &= -7 \\ 2x - 5y &= -3 \xrightarrow{\qquad\qquad} & 2x - 5y &= -3 \\ & & \hline 5x \qquad &= -10 \\ & & x \qquad &= -2 \end{aligned}$$

$$\begin{aligned} 2x - 5y &= -3 \\ 2(-2) - 5y &= -3 \\ -5y &= 1 \\ y &= \frac{-1}{5} \end{aligned}$$

Q17 Solve:

$$\begin{aligned} x - 7y &= 4 \\ 3x + 5y &= 12 \end{aligned}$$

STOP • STOP • STOP • STOP • STOP • STOP • STOP • STOP • STOP

A17 $(4, 0)$:

$$\begin{aligned} x - 7y &= 4 \xrightarrow{\text{multiply by } -3} & -3x + 21y &= -12 \\ 3x + 5y &= 12 \xrightarrow{\qquad\qquad} & 3x + 5y &= 12 \\ & & \hline \end{aligned}$$

Q18 Solve:

a. $\begin{aligned} x - y &= 3 \\ 2x + 3y &= 16 \end{aligned}$

b. $\begin{aligned} 2x + 3y &= 7 \\ 3x + y &= 7 \end{aligned}$

STOP • STOP • STOP • STOP • STOP • STOP • STOP • STOP • STOP

A18 **a.** (5, 2) **b.** (2, 1)

5 A system may require that both equations be multiplied by an integer in order that one variable is eliminated in the sum.

$$2x - 7y = 2 \xrightarrow{\text{multiply by 2}} 4x - 14y = 4$$
$$3x + 2y = 3 \xrightarrow{\text{multiply by 7}} 21x + 14y = 21$$
$$25x = 25$$
$$x = 1$$

$$2x - 7y = 2$$
$$2(1) - 7y = 2$$
$$-7y = 0$$
$$y = 0$$

Hence the solution to the system is (1, 0).

Q19 Solve:

$$5x - 3y = 4 \xrightarrow{\text{multiply by 3}} \_\_\_\_\_\_\_\_\_\_$$
$$3x - 4y = -2 \xrightarrow{\text{multiply by } -5} \_\_\_\_\_\_\_\_\_\_$$

STOP • STOP • STOP • STOP • STOP • STOP • STOP • STOP • STOP

A19 (2, 2): $15x - 9y = 12$
$-15x + 20y = 10$

Q20 Solve eliminating the y term first.
$3x - 2y = 7$
$4x + 5y = -6$

STOP • STOP • STOP • STOP • STOP • STOP • STOP • STOP • STOP

A20 (1, −2): $3x - 2y = 7 \xrightarrow{\text{multiply by 5}} 15x - 10y = 35$
$4x + 5y = -6 \xrightarrow{\text{multiply by 2}} 8x + 10y = -12$

Q21 Determine an equivalent system so that the x terms would be eliminated first (do not solve).
$11x + 6y = 1$
$-7x + 5y = 17$

STOP • STOP • STOP • STOP • STOP • STOP • STOP • STOP • STOP

A21 $77x + 42y = 7$
$-77x + 55y = 187$

Q22 Determine an equivalent system so that the y terms would be eliminated first (do not solve).

$$11x + 6y = 1$$
$$-7x + 5y = 17$$

STOP • STOP • STOP • STOP • STOP • STOP • STOP • STOP • STOP

A22

$$55x + 30y = 5$$
$$42x - 30y = -102$$

Q23 Solve:

a. $3x - 5y = 22$
$7x - 4y = 13$

b. $13m - 5n = -40$
$9m - 3n = -24$

STOP • STOP • STOP • STOP • STOP • STOP • STOP • STOP • STOP

A23 a. $(-1, -5)$ b. $(0, 8)$

---

6 The solution to a system of equivalent linear equations will be all ordered pairs satisfying either equation. If the sum of two linear equations produces the equation $0 = 0$, the equations are equivalent.

$$\begin{array}{rcl} x - 2y = 5 & \xrightarrow{\text{multiply by } 2} & 2x - 4y = 10 \\ -2x + 4y = -10 & \longrightarrow & -2x + 4y = -10 \\ & & \overline{0 + 0 = 0} \end{array}$$

The equations are equivalent and the solution is all ordered pairs satisfying either equation. The second equation can be obtained from the first by multiplying the first by -2.

Q24 Solve:

$$2x - 3y = 6$$
$$-4x + 6y = -12$$

STOP • STOP • STOP • STOP • STOP • STOP • STOP • STOP • STOP

A24 All ordered pairs satisfying $2x - 3y = 6$:

$$\begin{array}{rcl} 2x - 3y = 6 & \xrightarrow{\text{multiply by } 2} & 4x - 6y = 12 \\ -4x + 6y = -12 & \longrightarrow & -4x + 6y = -12 \\ & & \overline{0 + 0 = 0} \end{array}$$

---

7 The system of equations represented by parallel lines will have no solution. If the sum of two linear equations produces a false statement, the system has no solution.

$$\begin{array}{rcl} x + 3y = 6 & \xrightarrow{\text{multiply by } -2} & -2x - 6y = -12 \\ 2x + 6y = 1 & \longrightarrow & 2x + 6y = 1 \\ & & \overline{0 + 0 = -11} \end{array}$$

The sum of the two equations produces the false statement $0 = -11$; hence the system has no solution. In this case, the solution set is $\emptyset$. When the equations are written in the slope-intercept form, they can be seen to represent parallel lines.

$$x + 3y = 6 \longrightarrow y = \frac{-1}{3}x + 2$$

$$2x + 6y = 1 \longrightarrow y = \frac{-1}{3}x + \frac{1}{6}$$

Q25 Solve:

$$\begin{aligned} 5x - 2y &= 7 \\ -10x + 4y &= 4 \end{aligned}$$

STOP • STOP • STOP • STOP • STOP • STOP • STOP • STOP • STOP

A25 $\emptyset$:

$$\begin{aligned} 5x + 2y = 7 &\xrightarrow{\text{multiply by 2}} 10x - 4y = 14 \\ -10x + 4y = 4 &\longrightarrow -10x + 4y = 4 \\ & \qquad\qquad\quad\; 0 = 18 \end{aligned}$$

Q26 Solve:

a. $\begin{aligned} -H + 3K &= -3 \\ 2H - 6K &= 5 \end{aligned}$

b. $\begin{aligned} 4x - 2y &= 10 \\ 2x - y &= 5 \end{aligned}$

STOP • STOP • STOP • STOP • STOP • STOP • STOP • STOP • STOP

A26 a. $\emptyset$ b. All ordered pairs satisfying $2x - y = 5$

8 A system of linear equations can also be solved by use of the *substitution method.* This is done by writing each equation in terms of y, assuming that y represents the same number in both equations, to solve for x.

$$2x - y = -1 \longrightarrow y = 2x + 1$$
$$3x - y = 0 \longrightarrow y = 3x$$

Since y represents the same number in both equations:

$$\begin{aligned} 2x + 1 &= 3x \\ 1 &= x \end{aligned}$$

The y value is found substituting $x = 1$ into either of the equations solved in terms of y.

$$\begin{aligned} y &= 3x \\ y &= 3(1) \\ y &= 3 \end{aligned}$$

Hence the solution to the system is (1, 3). Each equation could have been written in terms of x and solved for y.

Q27 Solve by the substitution method by solving both equations for y.

$$2x + y = -4$$
$$-2y - 3 = 3x$$

STOP • STOP • STOP • STOP • STOP • STOP • STOP • STOP • STOP

A27 $(-5, 6)$:

$$2x + y = -4 \longrightarrow y = -2x - 4$$
$$-2y - 3 = 3x \longrightarrow y = \frac{-3}{2}x - \frac{3}{2}$$
$$-2x - 4 = \frac{-3}{2}x - \frac{3}{2}$$
$$-4x - 8 = -3x - 3$$
$$-5 = x$$
$$y = -2x - 4$$
$$y = -2(-5) - 4$$
$$y = 6$$

Q28 Solve by the substitution method.

$$4x + 3y - 15 = 0$$
$$3x + 5y - 14 = 0$$

STOP • STOP • STOP • STOP • STOP • STOP • STOP • STOP • STOP

A28 $(3, 1)$:

$$y = \frac{-4}{3}x + 5$$
$$y = \frac{-3}{5}x + \frac{14}{5}$$
$$\frac{-4}{3}x + 5 = \frac{-3}{5}x + \frac{14}{5}$$
$$-20x + 75 = -9x + 42$$
$$x = 3$$

Q29 Solve by the substitution method:

a. $4x - 3y = 1$
$3x - 4y = 6$

b. $x - 2y = 5$
$3x - 6y = 4$

STOP • STOP • STOP • STOP • STOP • STOP • STOP • STOP • STOP

A29 **a.** $(-2, -3)$ **b.** $\varnothing$

9

Linear systems as presented in the above problems are more easily solved by use of the addition method. However, some systems may be solved more conveniently by use of the substitution method. This will be especially true when solving a system made up of a linear and a nonlinear equation (not presented here).

The technique of substitution can also be used by substituting an equivalent expression into one of the equations to solve for x or y.

Example: Solve the system.

$$x + 2y = 10$$
$$y = 2x$$

Solution

Since $y = 2x$, $2x$ can be substituted for y in the first equation.

$$x + 2(2x) = 10$$
$$x + 4x = 10$$
$$5x = 10$$
$$x = 2$$

$$y = 2x$$
$$y = 2(2)$$
$$y = 4$$

Hence the solution is $(2, 4)$.

Q30 Solve:

a. $x - 2y = 4$
$y = -x$

b. $2x + y = 6$
$x = 3$

STOP • STOP • STOP • STOP • STOP • STOP • STOP • STOP • STOP

A30

a. $\left(\frac{4}{3}, \frac{-4}{3}\right)$:

$$x - 2y = 4$$
$$x - 2(-x) = 4$$
$$x + 2x = 4$$
$$3x = 4$$
$$x = \frac{4}{3}$$

$$y = -x$$
$$y = \frac{-4}{3}$$

b. (3, 0):

$$2x + y = 6$$
$$2(3) + y = 6$$
$$y = 0$$

10

A system of equations can often be used to solve a verbal problem by the following steps:

Step 1: Use different variables to represent the unknown quantities in the problem.

Step 2: Translate the given relationships in the problem into a system of equations.

Step 3: Solve the system.

Step 4: Check the solution.

Example: The sum of two numbers is 21 and their difference is 9. Find the numbers.

Solution

Step 1: Let x = larger number
y = smaller number

Step 2: The sum of two numbers is 21:

$x + y = 21$

Their difference is 9:

$x - y = 9$

Step 3:

$$\begin{aligned} x + y &= 21 \\ x - y &= 9 \\ \hline 2x &= 30 \\ x &= 15 \end{aligned}$$

$$\begin{aligned} x + y &= 21 \\ 15 + y &= 21 \\ y &= 6 \end{aligned}$$

Step 4 (check):

$15 + 6 = 21$ (true)
$15 - 6 = 9$ (true)

Q31

The sum of two numbers is 56. Twice the smaller number exceeds one-half the larger number by 22. Find the numbers, completing each step.

Step 1: Let x = smaller number
y = larger number

Step 2: ____________

$$2x - \frac{1}{2}y = 22$$

Step 3:

Step 4:

STOP • STOP • STOP • STOP • STOP • STOP • STOP • STOP • STOP

A31 20, 36:

Step 2:

$$x + y = 56$$
$$2x - \frac{1}{2}y = 22$$

Step 3:

$$\begin{aligned} x + y &= 56 \\ 4x - y &= 44 \\ \hline 5x \quad &= 100 \\ x &= 20 \\ x + y &= 56 \\ 20 + y &= 56 \\ y &= 36 \end{aligned}$$

Step 4: Check:

$$20 + 36 = 56 \quad \text{(true)}$$
$$2(20) - \frac{1}{2}(36) \stackrel{?}{=} 22$$
$$40 - 18 = 22 \quad \text{(true)}$$

Q32 The sum of two numbers is 50. If twice the larger is subtracted from 4 times the smaller, the result is 8. Find the numbers.

STOP • STOP • STOP • STOP • STOP • STOP • STOP • STOP • STOP

A32 18, 32:

Step 1: Let x = smaller number
Let y = larger number

Step 2:

$$x + y = 50$$
$$4x - 2y = 8$$

---

11 Mixture problems are often solved by writing the conditions of the problems as a system of linear equations.

Example: One bar of tin alloy is 25 percent pure tin and another bar is 10 percent pure tin. How many kilograms of each alloy must be used to make 75 kilograms of a new alloy that is 20 percent pure?

Solution

Step 1: Let x = kilograms of 25% alloy
y = kilograms of 10% alloy

Step 2: There must be 75 kilograms of new alloy; hence $x + y = 75$.

The actual amount of tin in each alloy must total the amount of tin in the new alloy.

Tin in 25% alloy: 25% of x kilograms $= 0.25x$
Tin in 10% alloy: 10% of y kilograms $= 0.1y$
Tin in 20% alloy: 20% of 75 kilograms $= 0.2(75)$

$$0.25x + 0.1y = 0.2(75)$$

Step 3:

$$\begin{array}{rcl} x + y = 75 & \longrightarrow x + y = 75 \longrightarrow & -10x - 10y = -750 \\ 0.25x + 0.1y = 15 & \longrightarrow 25x + 10y = 1500 \longrightarrow & 25x + 10y = 1500 \\ \hline & & 15x = 750 \\ & & x = 50 \text{ kilograms} \end{array}$$

Step 4:

$$\begin{array}{ccc} x + y = 75 & 0.25x + 0.1y = 0.2(75) & x + y = 75 \\ 50 + 25 = 75 & 0.25(50) + 0.1(25) \stackrel{?}{=} 0.2(75) & y = 25 \text{ kilograms} \\ & 12.5 + 2.5 = 15 \quad \text{(true)} & \end{array}$$

Hence the solution is 50 kilograms of the 25 percent alloy and 25 kilograms of the 10 percent alloy.

Q33 A chemist has a solution that is 18 percent pure salt and a second solution that is 45 percent pure salt. How many cubic centimeters of each solution should the chemist use to make 24 cubic centimeters of a solution 36 percent pure salt?

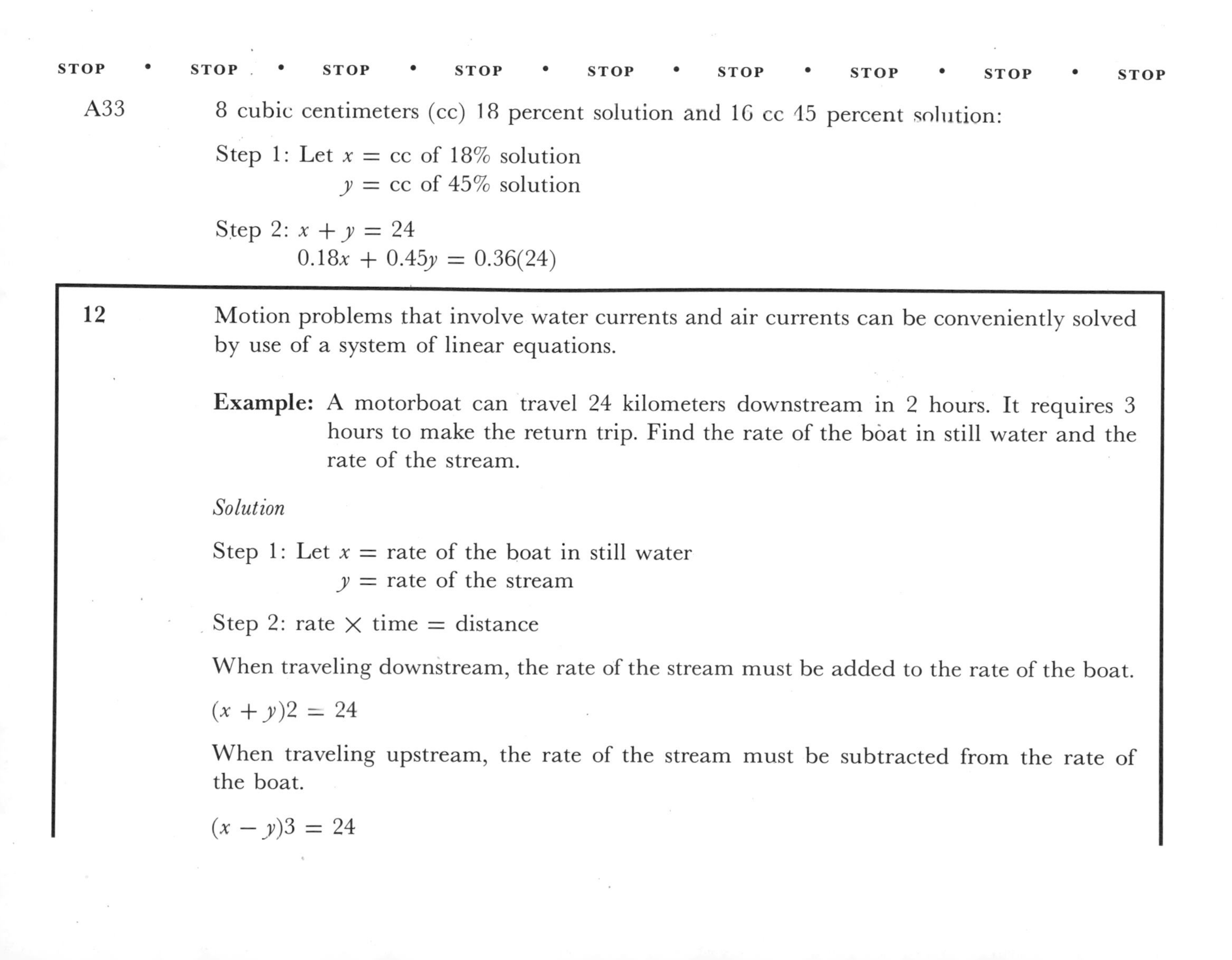

STOP • STOP • STOP • STOP • STOP • STOP • STOP • STOP • STOP

A33 8 cubic centimeters (cc) 18 percent solution and 16 cc 45 percent solution:

Step 1: Let x = cc of 18% solution
y = cc of 45% solution

Step 2: $x + y = 24$
$0.18x + 0.45y = 0.36(24)$

12 Motion problems that involve water currents and air currents can be conveniently solved by use of a system of linear equations.

Example: A motorboat can travel 24 kilometers downstream in 2 hours. It requires 3 hours to make the return trip. Find the rate of the boat in still water and the rate of the stream.

Solution

Step 1: Let x = rate of the boat in still water
y = rate of the stream

Step 2: rate × time = distance

When traveling downstream, the rate of the stream must be added to the rate of the boat.

$(x + y)2 = 24$

When traveling upstream, the rate of the stream must be subtracted from the rate of the boat.

$(x - y)3 = 24$

Step 3: $2(x + y) = 24 \longrightarrow x + y = 12$
$3(x - y) = 24 \longrightarrow x - y = 8$

$$\begin{aligned} 2x \quad &= 20 \\ x &= 10 \\ x + y &= 12 \\ y &= 2 \end{aligned}$$

Step 4: $2(10 + 2) = 24$ (true)
$3(10 - 2) = 24$ (true)

Hence the solution is 10 kilometers per hour for the rate of the boat and 2 kilometers per hour for the rate of the stream.

Q34 William canoed 20 kilometers downstream in 2 hours. The trip upstream took 5 hours. Find his rate in still water and the rate of the current.

STOP • STOP • STOP • STOP • STOP • STOP • STOP • STOP • STOP

A34 7 kilometers per hour (William's rate)
3 kilometers per hour (rate of current):

Step 1: Let x = rate of the canoe in still water
y = rate of the current

Step 2: $2(x + y) = 20$
$5(x - y) = 20$

This completes the instruction for this section.

7.2 Exercises

1. Solve:
   a. $x - y = 5$
   $x + y = 7$
   b. $2x - 3y = 4$
   $-2x - 3y = 8$
   c. $2x + y = 3$
   $-3x + 3y = 9$
   d. $3x - 4y = 5$
   $-2x + 5y = 6$
2. Solve:
   a. $2x - 3y = 7$
   $-2x - y - 9 = 0$
   b. $m - n = 2$
   $m - n + 5 = 0$
   c. $5x - 2y = 0$
   $-x - y = 7$

d. $12x - 3y = 6$
$-4x + y = -2$

3. Solve:

a. $-3x + 5y = 2$
$2x - 3y = 1$

b. $5x + 3y = 1$
$-3x - 4y = 6$

c. $5x + 2y = 3$
$-10x - 4y = -1$

d. $\frac{x}{2} + \frac{y}{3} = \frac{-1}{3}$

$\frac{x}{2} + 2y = -7$

4. Solve using the substitution method:

a. $2x = y + 5$
$x = -3$

b. $\frac{x}{2} + \frac{y}{3} = 3$
$y = 3x$

c. $-2x + y = 3$
$y = x + 2$

5. Given two grades of zinc ore, the first containing 45 percent zinc and the second 25 percent zinc, how many kilograms of each grade must be used to make a mixture of 2,000 kilograms containing 40 percent zinc?

6. The larger of two numbers exceeds the smaller by 15. One-half the smaller number tripled is equal to the larger number. Find the numbers.

7. The sum of two numbers is 84. If twice the smaller number is three more than the larger number, find the numbers.

8. Diana invests part of $8,000 at 5 percent interest and the rest at $5\frac{1}{2}$ percent interest. Her return on the investment at the end of 1 year is $430. How much is invested at each rate?

9. Two hikers start at the same time from points that are 18 kilometers apart. One walks 1 kilometer per hour faster than the other. They meet in 2 hours. What is the average rate of each hiker?

10. In his boat, Ronald travels a distance of 18 kilometers upstream in 1 hour. Returning downstream, he makes the trip in $\frac{3}{4}$ hour. What is the speed of the boat and current?

11. How many cubic centimeters each of 5 percent hydrochloric acid and 20 percent hydrochloric acid must be added together to get 10 cubic centimeters of solution that is 12.5 percent hydrochloric acid?

12. An automobile radiator holds 12 liters. How much pure antifreeze must be added to a mixture that is 4 percent antifreeze to make enough mixture that is 20 percent antifreeze to fill the radiator?

13. Two cars start out together and travel in opposite directions. At the end of 3 hours, they are 345 kilometers apart. If one car travels 15 kilometers per hour faster than the other, what are their speeds?

***14.** Solve the system

$Ax + By = C$
$Dx + Ey = F$

7.2 Exercise Answers

1. **a.** $(6, 1)$ **b.** $(-1, -2)$ **c.** $(0, 3)$ **d.** $(7, 4)$

2. **a.** $\left(\frac{-5}{2}, -4\right)$ **b.** $\varnothing$ **c.** $(-2, -5)$

d. all ordered pairs satisfying either equation

3. a. (11, 7) b. (2, −3) c. ∅ d. (2, −4)
4. a. (−3, −11) b. (2, 6) c. (−1, 1)
5. 1500 kilograms of 45 percent ore, 500 kilograms of 25 percent ore
6. 30 and 45
7. 29 and 55
8. \$6,000 at $5\frac{1}{2}$ percent and \$2,000 at 5 percent
9. 4 kilometers per hour and 5 kilometers per hour
10. 21 kilometers per hour (speed of the boat)
    3 kilometers per hour (speed of the current)
11. 5 cc of each solution is required
12. 2 liters
13. 50 kilometers per hour and 65 kilometers per hour

*14. $x = \frac{CE - BF}{AE - BD}; y = \frac{AF - CD}{AE - BD}$

7.3 Cramer's Rule

1 A *matrix* is a rectangular array of numbers. Matrices are named according to the number of rows and columns they contain. These matrices contain two rows and three columns. They are called 2 × 3 matrices (read "two by three").

$$\begin{bmatrix} 1 & 2 & 0 \\ -1 & 8 & 4 \end{bmatrix} \quad \begin{bmatrix} 0 & 0 & 1 \\ 2 & -1 & 0 \end{bmatrix}$$

A matrix in which the number of rows equals the number of columns is referred to as a *square matrix.*

Q1 Label the matrices according to their rows and columns:

a. $\begin{bmatrix} 1 & 2 \\ -1 & 0 \end{bmatrix}$ ________ b. $\begin{bmatrix} 1 & 2 \\ 3 & 4 \\ 5 & 6 \end{bmatrix}$ ________ c. $\begin{bmatrix} 1 \\ 2 \end{bmatrix}$ ________

STOP • STOP • STOP • STOP • STOP • STOP • STOP • STOP • STOP

A1 a. 2 × 2 b. 3 × 2 c. 2 × 1

Q2 Which of the matrices in Q1 are square matrices? ________

STOP • STOP • STOP • STOP • STOP • STOP • STOP • STOP • STOP

A2 a

2 Associated with each square matrix is a real number called the *determinant* of the matrix. For a 2 × 2 matrix,

$$\begin{bmatrix} a & b \\ c & d \end{bmatrix}$$

the determinant is written

$$\begin{vmatrix} a & b \\ c & d \end{vmatrix}$$

and is defined as

$$\begin{vmatrix} a & b \\ c & d \end{vmatrix} = ad - bc$$

Example:

$$\begin{vmatrix} 8 & -1 \\ 3 & 2 \end{vmatrix} = 8 \cdot 2 - 3(-1)$$
$$= 16 - {}^{-}3 = 19$$

Q3 Evaluate the determinant: $\begin{vmatrix} -5 & 6 \\ 2 & 1 \end{vmatrix}$

STOP • STOP • STOP • STOP • STOP • STOP • STOP • STOP • STOP

A3 -17: $-5 \cdot 1 - 2 \cdot 6 = -17$

Q4 Evaluate the determinants:

a. $\begin{vmatrix} 3 & 4 \\ -2 & -6 \end{vmatrix}$ b. $\begin{vmatrix} A & 2 \\ 3A & 4 \end{vmatrix}$

STOP • STOP • STOP • STOP • STOP • STOP • STOP • STOP • STOP

A4 a. -10 b. $-2A$: $A \cdot 4 - 3A \cdot 2 = 4A - 6A = -2A$

Q5 Evaluate:

a. $\begin{vmatrix} 1 & 0 \\ 0 & 1 \end{vmatrix}$ b. $\begin{vmatrix} -2 & 5 \\ 3 & -7 \end{vmatrix}$ c. $\begin{vmatrix} 5 & 5 \\ 5 & 5 \end{vmatrix}$

STOP • STOP • STOP • STOP • STOP • STOP • STOP • STOP • STOP

A5 a. 1 b. -1 c. 0

Q6 Evaluate the ratios:

a. $\dfrac{\begin{vmatrix} 3 & 4 \\ 3 & 3 \end{vmatrix}}{\begin{vmatrix} 0 & 1 \\ 1 & 0 \end{vmatrix}}$ b. $\dfrac{\begin{vmatrix} -1 & 3 \\ 2 & 6 \end{vmatrix}}{\begin{vmatrix} 0 & 2 \\ -2 & 6 \end{vmatrix}}$

STOP • STOP • STOP • STOP • STOP • STOP • STOP • STOP • STOP

A6 a. 3 b. -3

Q7 Evaluate:

a. $\dfrac{\begin{vmatrix} 0 & 4 \\ -3 & 1 \end{vmatrix}}{\begin{vmatrix} 3 & 1 \\ 4 & 0 \end{vmatrix}}$ b. $\dfrac{\begin{vmatrix} C & B \\ F & E \end{vmatrix}}{\begin{vmatrix} A & B \\ D & E \end{vmatrix}}$

STOP • STOP • STOP • STOP • STOP • STOP • STOP • STOP • STOP

A7 a. -3 b. $\dfrac{CE - BF}{AE - BD}$

3 The following equations can be solved simultaneously by the addition method:

(1) $Ax + By = C$
(2) $Dx + Ey = F$

Multiplying equation (1) by E and (2) by $-B$,

(3) $AEx + BEy = CE$
(4) $-BDx - BEy = -BF$

Adding equations (3) and (4) gives

$AEx - BDx = CE - BF$

Factoring,

$x(AE - BD) = CE - BF$

Therefore,

$$x = \frac{CE - BF}{AE - BD}$$

This sentence is equivalent to

$$x = \frac{\begin{vmatrix} C & B \\ F & E \end{vmatrix}}{\begin{vmatrix} A & B \\ D & E \end{vmatrix}}$$

When solving for x, the denominator, D, is the determinant of the matrix consisting of the coefficients of x and y. The numerator, D_x, is the determinant of the matrix consisting of the coefficients of y with the constant terms replacing the coefficients of x.

$Ax + By = C$
$Dx + Ey = F$

$$x = \frac{D_x}{D} = \frac{\begin{vmatrix} C & B \\ F & E \end{vmatrix}}{\begin{vmatrix} A & B \\ D & E \end{vmatrix}}$$

Q8

$3x - 5y = 4$
$7x + 4y = 25$ Complete: $x = \frac{D_x}{D} = \frac{\begin{vmatrix} 4 & -5 \\ 25 & 4 \end{vmatrix}}{\begin{vmatrix} 3 & -5 \\ 7 & 4 \end{vmatrix}} = \frac{(\quad)}{(\quad)} = (\quad)$

STOP • STOP • STOP • STOP • STOP • STOP • STOP • STOP • STOP

A8 $\frac{141}{47} = 3$

Q9 Using the same procedure, solve for x.
$x + 2y = 8$
$2x - 3y = 2$

STOP • STOP • STOP • STOP • STOP • STOP • STOP • STOP • STOP

A9 4: $x = \frac{D_x}{D} = \frac{\begin{vmatrix} 8 & 2 \\ 2 & -3 \end{vmatrix}}{\begin{vmatrix} 1 & 2 \\ 2 & -3 \end{vmatrix}} = \frac{-28}{-7} = 4$

4 When solving for y, the denominator, D, is the determinant of the matrix consisting of the coefficients of x and y, the same as when solving for x. The numerator, D_y, is the determinant of the matrix consisting of the coefficients of x with the constant terms replacing the coefficients of y.

$Ax + By = C$
$Dx + Ey = F$ $y = \frac{D_y}{D} = \frac{\begin{vmatrix} A & C \\ D & F \end{vmatrix}}{\begin{vmatrix} A & B \\ D & E \end{vmatrix}}$

Q10

$3x - 5y = 4$
$7x + 4y = 25$ Complete: $y = \frac{D_y}{D} = \frac{\begin{vmatrix} 3 & 4 \\ 7 & 25 \end{vmatrix}}{\begin{vmatrix} 3 & -5 \\ 7 & 4 \end{vmatrix}} = \frac{(\quad)}{(\quad)} = (\quad)$

STOP • STOP • STOP • STOP • STOP • STOP • STOP • STOP • STOP

A10 $\frac{47}{47} = 1$

Q11 Using the same procedure, solve for y.

$x + 2y = 8$
$2x - 3y = 2$

STOP • STOP • STOP • STOP • STOP • STOP • STOP • STOP • STOP

A11 2: $y = \frac{D_y}{D} = \frac{\begin{vmatrix} 1 & 8 \\ 2 & 2 \end{vmatrix}}{\begin{vmatrix} 1 & 2 \\ 2 & -3 \end{vmatrix}} = \frac{-14}{-7} = 2$

5 This method for solving simultaneous equations is credited to Gabriel Cramer (1750). It is called *Cramer's rule.* To solve

$Ax + By = C$
$Dx + Ey = F$

$$x = \frac{D_x}{D} = \frac{\begin{vmatrix} C & B \\ F & E \end{vmatrix}}{\begin{vmatrix} A & B \\ D & E \end{vmatrix}} \quad \text{and} \quad y = \frac{D_y}{D} = \frac{\begin{vmatrix} A & C \\ D & F \end{vmatrix}}{\begin{vmatrix} A & B \\ D & E \end{vmatrix}}$$

Since it involves evaluating only three determinants, work is usually shown as follows.

Example: Solve

$x + y = 5$
$x - y = -1$

by use of Cramer's rule.

Solution

$$D_x = \begin{vmatrix} 5 & 1 \\ -1 & -1 \end{vmatrix} = -5 + 1 = -4 \qquad D_y = \begin{vmatrix} 1 & 5 \\ 1 & -1 \end{vmatrix} = -1 - 5 = -6$$

$$D = \begin{vmatrix} 1 & 1 \\ 1 & -1 \end{vmatrix} = -1 - 1 = -2 \qquad x = \frac{D_x}{D} = \frac{-4}{-2} = 2$$

$$y = \frac{D_y}{D} = \frac{-6}{-2} = 3$$

Therefore, the solution is (2, 3).

Q12 Use Cramer's rule to solve $\begin{aligned} 3x + 4y &= 5 \\ -2x - 3y &= -4 \end{aligned}$

STOP • STOP • STOP • STOP • STOP • STOP • STOP • STOP • STOP

A12 $(-1, 2)$: $D_x = \begin{vmatrix} 5 & 4 \\ -4 & -3 \end{vmatrix} = 1$, $D_y = \begin{vmatrix} 3 & 5 \\ -2 & -4 \end{vmatrix} = -2$, $D = \begin{vmatrix} 3 & 4 \\ -2 & -3 \end{vmatrix} = -1$

$$x = \frac{D_x}{D} = \frac{1}{-1} = -1 \qquad y = \frac{D_y}{D} = \frac{-2}{-1} = 2$$

---

6 Division by zero is undefined. If $D = 0$ and D_x and D_y are nonzero, the system represents two parallel lines, and hence there is no common solution.

Example 1: Solve

$$\begin{aligned} -x + 2y &= -6 \\ -x + 2y &= 8 \end{aligned}$$

by use of Cramer's rule.

Solution

$$D = \begin{vmatrix} -1 & 2 \\ -1 & 2 \end{vmatrix} = 0 \qquad D_x = \begin{vmatrix} -6 & 2 \\ 8 & 2 \end{vmatrix} = -28 \qquad D_y = \begin{vmatrix} -1 & -6 \\ -1 & 8 \end{vmatrix} = -14$$

Therefore, the system represents two parallel lines and hence there is no common solution. The solution set is $\varnothing$. If $D = 0$, $D_x = 0$, and $D_y = 0$, the system represents two equivalent equations and hence the solution is any ordered pair satisfying either equation.

Example 2: Solve

$$\begin{aligned} 2x + y &= -3 \\ -4x - 2y &= 6 \end{aligned}$$

by use of Cramer's rule.

Solution

$$D = \begin{vmatrix} 2 & 1 \\ -4 & -2 \end{vmatrix} = 0 \qquad D_x = \begin{vmatrix} -3 & 1 \\ 6 & -2 \end{vmatrix} = 0 \qquad D_y = \begin{vmatrix} 2 & -3 \\ -4 & 6 \end{vmatrix} = 0$$

Therefore, the system represents two equivalent equations. The solution is any ordered pair that satisfies either equation.

Q13 Determine the nature of the solution to the following systems:

a. $2x + 2y = 7$
$-x - y = 3$

b. $3x - 4y = 5$
$-6x + 8y = -10$

STOP • STOP • STOP • STOP • STOP • STOP • STOP • STOP • STOP

A13 **a.** $\varnothing$
$D = 0, D_x = -13, D_y = 13$

b. any ordered pair satisfying either equation: $D = 0, D_x = 0, D_y = 0$

7 Cramer's rule does not necessarily save time when solving a system of two equations in two variables.

Example: Solve

$$x = 4y$$
$$x + y = 175$$

by use of Cramer's rule.

Solution

First, transform the original system to

$$x - 4y = 0$$
$$x + y = 175$$

Then

$$D = \begin{vmatrix} 1 & -4 \\ 1 & 1 \end{vmatrix} = 5 \qquad D_x = \begin{vmatrix} 0 & -4 \\ 175 & 1 \end{vmatrix} = 700 \qquad D_y = \begin{vmatrix} 1 & 0 \\ 1 & 175 \end{vmatrix} = 175$$

$$x = \frac{D_x}{D} = \frac{700}{5} = 140 \qquad y = \frac{D_y}{D} = \frac{175}{5} = 35$$

Therefore, the solution is (140, 35).

Normally, when solving a system of two equations in two variables, use the method of solution that appears to be easiest.

Q14 Use x and y to write a system of equations for the following problem. The sum of two numbers is 561. One number is twice the other.

STOP • STOP • STOP • STOP • STOP • STOP • STOP • STOP • STOP

A14 Step 1: Let x = the larger number and y = the smaller number

Step 2: $x + y = 561$
$x = 2y$

Q15 Use Cramer's rule to solve the above system. (Cramer's rule is not the easiest method.)

STOP • STOP • STOP • STOP • STOP • STOP • STOP • STOP • STOP

A15 (374, 187):

$$D = \begin{vmatrix} 1 & 1 \\ 1 & -2 \end{vmatrix} = -3 \qquad D_x = \begin{vmatrix} 561 & 1 \\ 0 & -2 \end{vmatrix} = -1{,}122 \qquad D_y = \begin{vmatrix} 1 & 561 \\ 1 & 0 \end{vmatrix} = -561$$

$$x = \frac{D_x}{D} = \frac{-1{,}122}{-3} = 374 \qquad y = \frac{D_y}{D} = \frac{-561}{-3} = 187$$

---

8 Cramer's rule can be applied to solving systems of three equations in *three variables.* First you must be able to find the determinant of a given (3×3) matrix such as A.

$$A = \begin{bmatrix} A & B & C \\ D & E & F \\ G & H & I \end{bmatrix}$$

It can be shown algebraically that the determinant of A is

$$\begin{vmatrix} A & B & C \\ D & E & F \\ G & H & I \end{vmatrix} = AEI + BFG + CDH - GEC - HFA - IDB$$

To obtain these values quickly, repeat the first and second columns on the right-hand side of the given (3×3) determinant in the following manner:

$$\left|\begin{array}{ccc|cc} A & B & C & A & B \\ D & E & F & D & E \\ G & H & I & G & H \end{array}\right.$$

Then draw diagonals as indicated.

$$\begin{array}{ccccc} 1 & 2 & 3 & & \\ A & B & C & A & B \\ D & E & F & D & E \\ G & H & I & G & H \\ 4 & 5 & 6 & & \end{array}$$

1. Diagonals 1, 2, and 3: multiply and add. That is,

   $AEI + BFG + CDH$

2. Diagonals 4, 5, and 6: multiply and subtract. That is,

$$-GEC - HFA - IDB$$

Therefore, the determinant of the matrix is

$$AEI + BFG + CDH - GEC - HFA - IDB$$

Example: Find the determinant

$$\begin{vmatrix} 3 & -2 & -6 \\ 1 & 2 & -2 \\ 2 & -3 & -3 \end{vmatrix}$$

Solution

$$\begin{array}{ccc|cc} 1 & 2 & 3 & & \\ 3 & -2 & -6 & 3 & -2 \\ 1 & 2 & -2 & 1 & 2 \\ 2 & -3 & -3 & 2 & -3 \\ 4 & 5 & 6 & & \end{array}$$

$$= (3)(2)(-3) + (-2)(-2)(2) + (-6)(1)(-3) - (2)(2)(-6) - (-3)(-2)(3) - (-3)(1)(-2)$$
$$= -18 + 8 + 18 + 24 - 18 - 6 = 8$$

Q16 Solve using the procedure outlined in Frame 8:

a. $\begin{vmatrix} 3 & -2 & 1 \\ 1 & 2 & -1 \\ 2 & -3 & 2 \end{vmatrix}$

b. $\begin{vmatrix} 3 & 4 & 0 \\ 2 & -2 & 1 \\ 1 & 0 & -1 \end{vmatrix}$

STOP • STOP • STOP • STOP • STOP • STOP • STOP • STOP • STOP

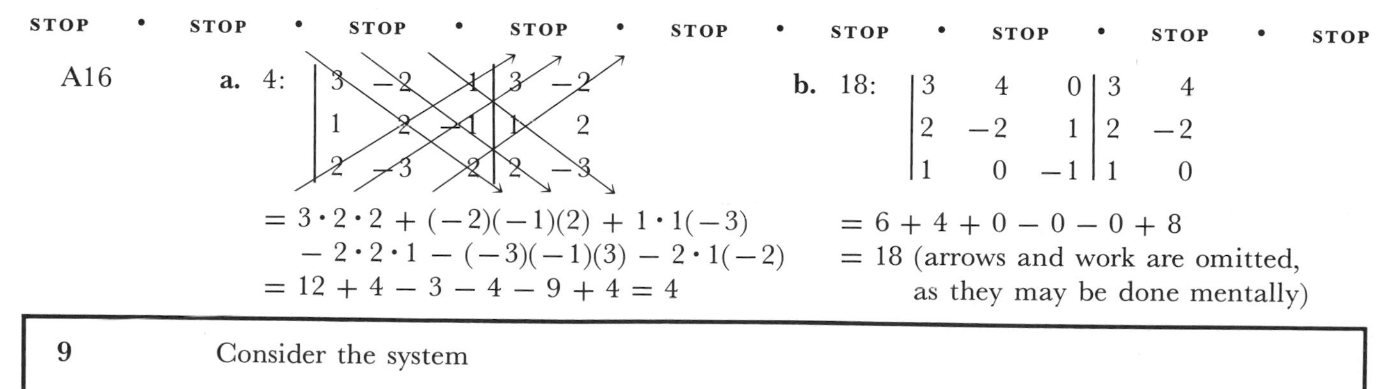

A16 a. 4: $\begin{array}{ccc|cc} 3 & -2 & 1 & 3 & -2 \\ 1 & 2 & -1 & 1 & 2 \\ 2 & -3 & 2 & 2 & -3 \end{array}$

$$= 3\cdot 2\cdot 2 + (-2)(-1)(2) + 1\cdot 1(-3) - 2\cdot 2\cdot 1 - (-3)(-1)(3) - 2\cdot 1(-2)$$
$$= 12 + 4 - 3 - 4 - 9 + 4 = 4$$

b. 18: $\begin{array}{ccc|cc} 3 & 4 & 0 & 3 & 4 \\ 2 & -2 & 1 & 2 & -2 \\ 1 & 0 & -1 & 1 & 0 \end{array}$

$$= 6 + 4 + 0 - 0 - 0 + 8$$
$= 18$ (arrows and work are omitted, as they may be done mentally)

9 Consider the system

$$\begin{aligned} Ax + By + Cz &= D \\ Ex + Fy + Gz &= H \\ Ix + Jy + Kz &= L \end{aligned}$$

By extending the method developed for (2 × 2) systems of equations, determinants can be used to solve systems of three equations in three variables.

$$D = \begin{vmatrix} A & B & C \\ E & F & G \\ I & J & K \end{vmatrix} \quad D_x = \begin{vmatrix} D & B & C \\ H & F & G \\ L & J & K \end{vmatrix} \quad D_y = \begin{vmatrix} A & D & C \\ E & H & G \\ I & L & K \end{vmatrix} \quad D_z = \begin{vmatrix} A & B & D \\ E & F & H \\ I & J & L \end{vmatrix}$$

$$x = \frac{D_x}{D} \quad y = \frac{D_y}{D} \quad z = \frac{D_z}{D}$$

When solving for D_x, D_y, and D_z, the constant terms replace the coefficients of x, y, or z, respectively. The solution to the system is the ordered triplet (x, y, z). This procedure is also called Cramer's rule.

Q17 Use Cramer's rule to complete the solution for the given system.

$$5x - y + 7z = -2$$
$$9x - y + 5z = 4$$
$$4x + y + 3z = 1$$

$$D = \begin{vmatrix} 5 & -1 & 7 \\ 9 & -1 & 5 \\ 4 & 1 & 3 \end{vmatrix} = \_\_\_\_$$

$$D_x = \begin{vmatrix} -2 & -1 & 7 \\ 4 & -1 & 5 \\ 1 & 1 & 3 \end{vmatrix} = \_\_\_\_$$

$$D_y = \begin{vmatrix} 5 & -2 & 7 \\ 9 & 4 & 5 \\ 4 & 1 & 3 \end{vmatrix} = \_\_\_\_$$

$$D_z = \begin{vmatrix} 5 & -1 & -2 \\ 9 & -1 & 4 \\ 4 & 1 & 1 \end{vmatrix} = \_\_\_\_$$

$$x = \frac{D_x}{D} = \_\_\_\_ \quad y = \frac{D_y}{D} = \_\_\_\_ \quad z = \frac{D_z}{D} = \_\_\_\_$$

Therefore, the solution is (____, ____, ____).

STOP • STOP • STOP • STOP • STOP • STOP • STOP • STOP • STOP

A17 $D = 58, D_x = 58, D_y = 0, D_z = -58, x = 1, y = 0, z = -1, (1, 0, -1)$

Q18 Use Cramer's rule to solve

$$\begin{aligned} 3x + 4y - 2z &= 5 \\ -x + 2y - 3z &= -6 \\ x - 3y + z &= -2 \end{aligned}$$

STOP • STOP • STOP • STOP • STOP • STOP • STOP • STOP • STOP

A18 (1, 2, 3): $D = -31, D_x = -31, D_y = -62, D_z = -93$

Q19 Use Cramer's rule to solve

$$\begin{aligned} 4x + 3y - 2z &= 6 \\ -x + 5y + 3z &= 10 \\ 3x - 2y + z &= -4 \end{aligned}$$

STOP • STOP • STOP • STOP • STOP • STOP • STOP • STOP • STOP

Q19 (0, 2, 0)

10 When variables are missing, their coefficients must be zero. When using Cramer's rule, zeros must be placed accordingly in the proper position in the determinant.

Example: Solve

$$\begin{aligned} 2x + y &= 7 \\ -3x + 5z &= -1 \\ -y - 4z &= 5 \end{aligned}$$

This system is equivalent to

$$\begin{aligned} 2x + y + 0z &= 7 \\ -3x + 0y + 5z &= -1 \\ 0x - y - 4z &= 5 \end{aligned}$$

Partial solution

$$D = \begin{vmatrix} 2 & 1 & 0 \\ -3 & 0 & 5 \\ 0 & -1 & -4 \end{vmatrix} \qquad D_x = \begin{vmatrix} 7 & 1 & 0 \\ -1 & 0 & 5 \\ 5 & -1 & -4 \end{vmatrix}$$

$$D_y = \begin{vmatrix} 2 & 7 & 0 \\ -3 & -1 & 5 \\ 0 & 5 & -4 \end{vmatrix} \qquad D_z = \begin{vmatrix} 2 & 1 & 7 \\ -3 & 0 & -1 \\ 0 & -1 & 5 \end{vmatrix}$$

Q20 Write D, D_x, D_y, and D_z for the following system of equations:

$$\begin{aligned} x + y &= 2 \\ 2x - z &= 1 \\ 2y - 3z &= -1 \end{aligned}$$

STOP • STOP • STOP • STOP • STOP • STOP • STOP • STOP • STOP

A20

$$D = \begin{vmatrix} 1 & 1 & 0 \\ 2 & 0 & -1 \\ 0 & 2 & -3 \end{vmatrix} \quad D_x = \begin{vmatrix} 2 & 1 & 0 \\ 1 & 0 & -1 \\ -1 & 2 & -3 \end{vmatrix} \quad D_y = \begin{vmatrix} 1 & 2 & 0 \\ 2 & 1 & -1 \\ 0 & -1 & -3 \end{vmatrix} \quad D_z = \begin{vmatrix} 1 & 1 & 2 \\ 2 & 0 & 1 \\ 0 & 2 & -1 \end{vmatrix}$$

11 When $D \neq 0$, the system has exactly one solution. When $D = 0$, the system has no solution or an infinite number of solutions: If $D = 0$ and either $D_x \neq 0$ or $D_y \neq 0$ or $D_z \neq 0$, the solution set is empty; if $D = 0$ and $D_x = D_y = D_z = 0$, the solution set is infinite.

Q21 Indicate the nature of the solution to the following system.

$$\begin{aligned} 2x + 2z &= 7 \\ 3x + y + 2z &= -3 \\ -x - z &= -2 \end{aligned}$$

STOP • STOP • STOP • STOP • STOP • STOP • STOP • STOP • STOP

A21 $\varnothing$: $D = 0$, $D_x = -3$, D_y and D_z are not needed.

12 Many applied problems may be solved using three variables and writing a system of three equations. In such cases, the use of Cramer's rule to solve the system usually saves time.

Example: The sum of three numbers is 3. The first number minus twice the second minus the third is 4. Twice the first number plus the other two is 5. Find the numbers.

Solution

Let x = first number, y = second number, and z = the third number. Translating each statement into an equation:

$$\begin{aligned} x + y + z &= 3 \\ x - 2y - z &= 4 \\ 2x + y + z &= 5 \end{aligned}$$

The system can now be solved.

Q22 Use Cramer's rule to solve the system in Frame 12.

STOP • STOP • STOP • STOP • STOP • STOP • STOP • STOP • STOP

A22 $(2, -3, 4)$

Q23 Use x, y, and z to write the system of equations needed for the following problem. Find three numbers such that the sum of the first and third is 1. The difference between the second and third is three. Three times the first minus the second is -28.

STOP • STOP • STOP • STOP • STOP • STOP • STOP • STOP • STOP

A23

$$\begin{aligned} x + z &= 1 \\ y - z &= 3 \\ 3x - y &= -28 \end{aligned}$$

Q24 Solve the system above by use of Cramer's rule.

STOP • STOP • STOP • STOP • STOP • STOP • STOP • STOP • STOP

A24 $(-6, 10, 7)$

13 Although the above examples are not very practical, they illustrate a technique that may be applied in real situations. Solving systems of equations in three variables is essential in electrical applications and in numerous areas relating to business and engineering. You should now have the necessary skills to solve such problems when you are confronted with them.

This completes the instruction for this section.

7.3 Exercises

1. Evaluate:

a. $\begin{vmatrix} 2 & 5 \\ -1 & 4 \end{vmatrix}$ **b.** $\begin{vmatrix} 1 & 3 \\ 7 & 2 \end{vmatrix}$ **c.** $\begin{vmatrix} -2 & 2 \\ 2 & -2 \end{vmatrix}$ **d.** $\begin{vmatrix} 0 & 1 \\ 0 & 1 \end{vmatrix}$

2. Evaluate:

a. $\begin{vmatrix} 1 & 2 & 3 \\ -3 & -2 & -1 \\ 2 & -1 & 1 \end{vmatrix}$ **b.** $\begin{vmatrix} 1 & 0 & 2 \\ 0 & 2 & 3 \\ 1 & 0 & 5 \end{vmatrix}$

3. Using Cramer's rule, find the solutions to the given system:

a. $2x - y = 3$
$-x + 2y = -3$

b. $4x + 3y = 7$
$6y = -8x - 12$

c. $2x - y = 5$
$4y = 8x - 20$

d. $4x - 3y = 1$
$3x - 4y = 6$

e. $3x + y = 1$
$9x + 3y = -4$

f. $2x - y = 0$
$x + y - 1 = 0$

4. Using Cramer's rule, find the solution to the given systems:

a. $x + y + 4z = 0$
$3x - y + z = -1$
$x + 3y - 2z = 12$

b. $x + y + z = 6$
$x - y + 2z = 12$
$2x + y - z = 1$

c. $2x + y + z = 1$
$-2x - y + z = 2$
$4x + 2y + 3z = 1$

d. $2x + z = 7$
$y - z = -2$
$x + y = 2$

5. Solve by use of Cramer's rule:

a. The length of a lot exceeds its width by 6 meters. The perimeter is 628 meters. Find the dimensions.

b. Find three numbers whose sum is 105 if the third is 11 less than 10 times the second, and twice the first is 7 more than 3 times the second.

7.3 Exercise Answers

1. **a.** 13 **b.** -19 **c.** 0 **d.** 0

2. **a.** 20 **b.** 6

3. **a.** $(1, -1)$ **b.** $\varnothing$ **c.** any ordered pair satisfying either equation
d. $(-2, -3)$ **e.** $\varnothing$ **f.** $\left(\frac{1}{3}, \frac{2}{3}\right)$

4. **a.** $(1, 3, -1)$ **b.** $(3, -1, 4)$ **c.** $\varnothing$ **d.** $(3, -1, 1)$

5. **a.** 154 meters by 160 meters **b.** 17, 9, and 79

7.4 Nonlinear Equations and Systems of Inequalities

1 The intersection of the graph of a first-degree equation with the graph of a second-degree equation representing a parabola will be one of the following:

one point two points

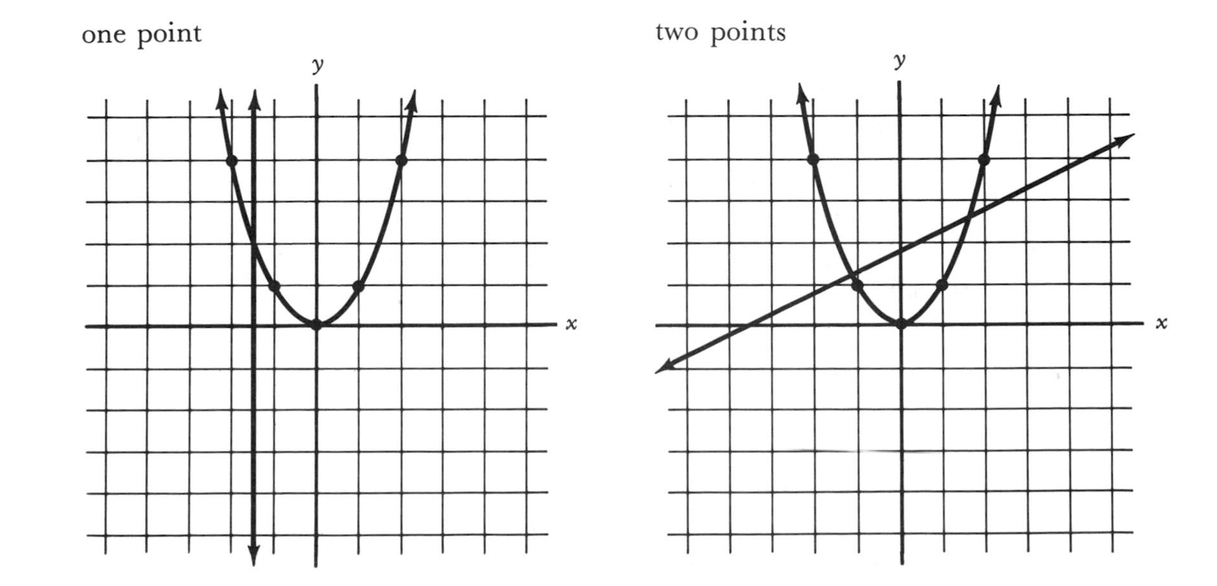

no intersection

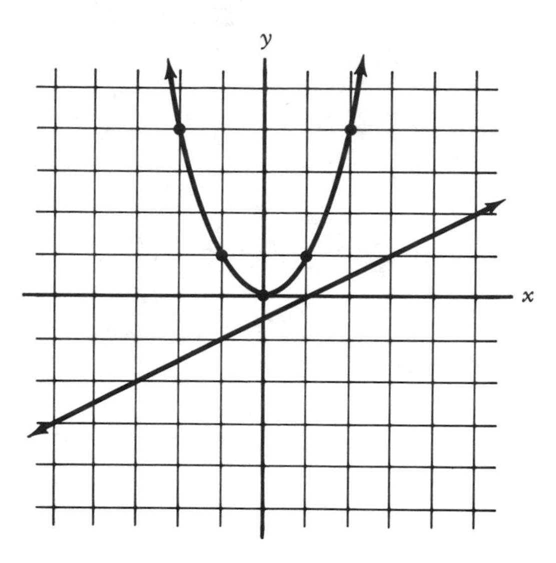

The points of intersection are found by use of the substitution method.

Example: Solve the system

$y = x$
$y = x^2$

Solution

First assume that y represents the same number in both equations. Since $y = x^2$, x^2 is substituted for y in the first equation.

$x^2 = x$

Solve for x:

$$x^2 - x = 0$$
$$x(x - 1) = 0$$
$$x = 0 \quad \text{or} \quad x = 1$$

Obtain the y coordinates of the intersection by substituting each x value into either of the original equations.

Let $x = 0$: $y = x$, $y = 0$ Let $x = 1$: $y = x$, $y = 1$

Hence the points of intersection are $(0, 0)$ and $(1, 1)$.

Q1 Solve the system and graph each equation on the coordinate system.

$y = x^2$
$y = -x$

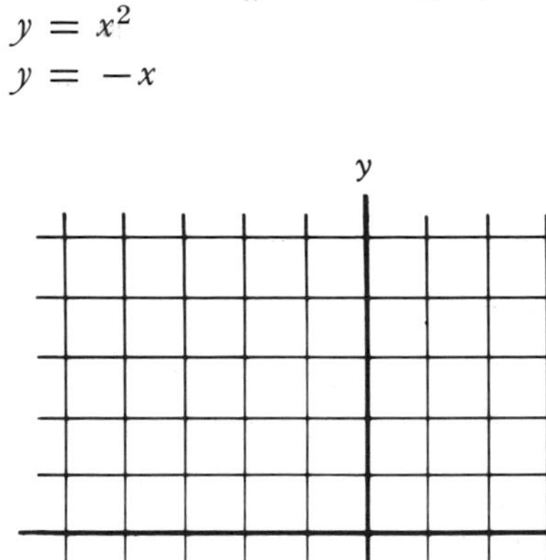

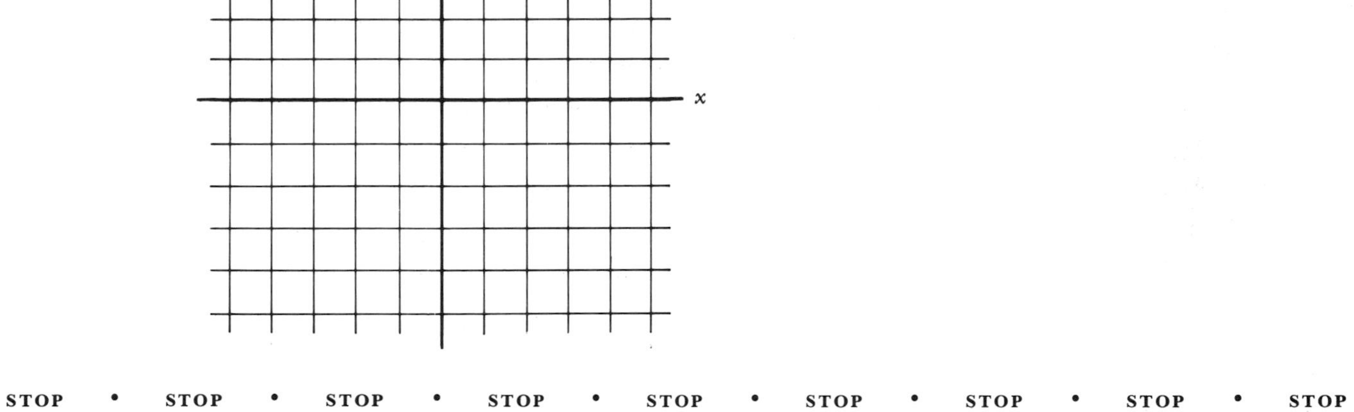

STOP • **STOP** • **STOP** • **STOP** • **STOP** • **STOP** • **STOP** • **STOP** • **STOP**

A1 $(0, 0), (-1, 1)$: $y = -x$ and $y = x^2$. Therefore,

$$x^2 = -x$$
$$x^2 + x = 0$$
$$x(x + 1) = 0$$
$$x = 0 \quad \text{or} \quad x = -1$$

If $x = 0, y = 0$. If $x = -1, y = 1$.

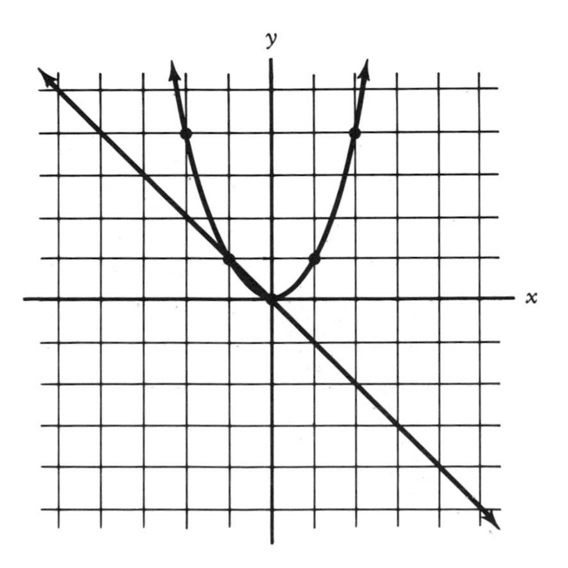

Q2 Solve and graph:
$y = x - 2$
$y = x^2 + 1$

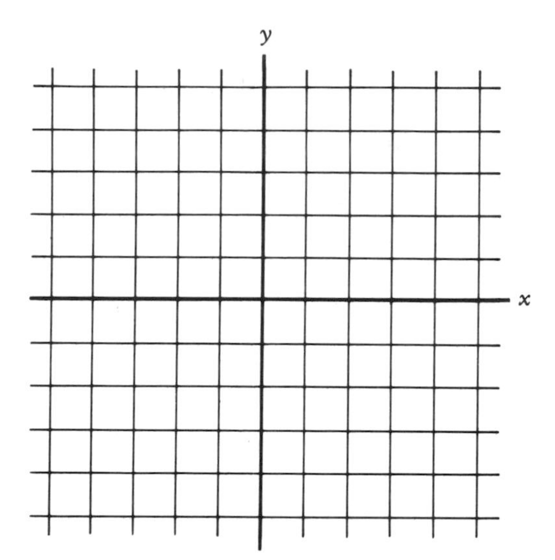

STOP • STOP • STOP • STOP • STOP • STOP • STOP • STOP • STOP

A2 $\varnothing$:

$$x^2 + 1 = x - 2$$
$$x^2 - x + 3 = 0$$
$$x = \frac{-(-1) \pm \sqrt{(-1)^2 - 4(1)(3)}}{2(1)}$$
$$x = \frac{1 \pm \sqrt{-11}}{2}$$

$\sqrt{-11}$ is imaginary; therefore, there is no intersection. Hence the solution set is $\varnothing$.

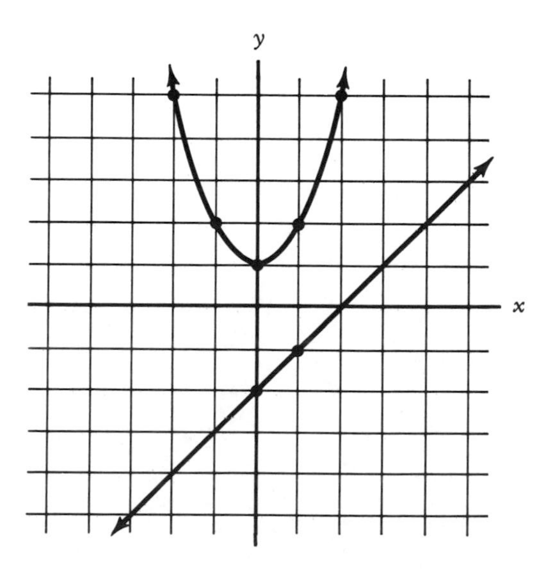

Q3 Solve:

a. $y = 2x^2 - 1$
$y = -1$

b. $y = x^2$
$x = 2$

STOP • STOP • STOP • STOP • STOP • STOP • STOP • STOP • STOP

A3 **a.** $(0, -1)$:

$2x^2 - 1 = -1$
$2x^2 = 0$
$x^2 = 0$
$x = 0$

If $x = 0, y = -1$.

b. $(2, 4)$: substitute 2 for x into the first equation

$y = (2)^2$
$y = 4$

Q4 Solve:

a. $y = x^2 + 2x$
$y = x$

b. $y = 2x^2 + 1$
$x - y = 2$

STOP • STOP • STOP • STOP • STOP • STOP • STOP • STOP • STOP

A4 **a.** $(0, 0), (-1, -1)$ **b.** $\varnothing$

Q5 Solve:

a. $y = x^2 - 4x + 4$
$x + y = 2$

b. $x^2 = y$
$2x - y - 2 = 0$

STOP • STOP • STOP • STOP • STOP • STOP • STOP • STOP • STOP

A5 **a.** $(1, 1), (2, 0)$ **b.** $\varnothing$

---

2 The intersection of the graphs for two quadratic functions can be found using the substitution method.

Example: Solve the system

$y = x^2$
$y = x^2 + 4x + 4$

Solution

Substitute $x^2 + 4x + 4$ for y in the first equation.

$x^2 + 4x + 4 = x^2$
$4x + 4 = 0$
$x = -1$

Now substitute $x = -1$ into either of the original equations.

$y = x^2$
$y = (-1)^2$
$y = 1$

Hence the solution is $(-1, 1)$.

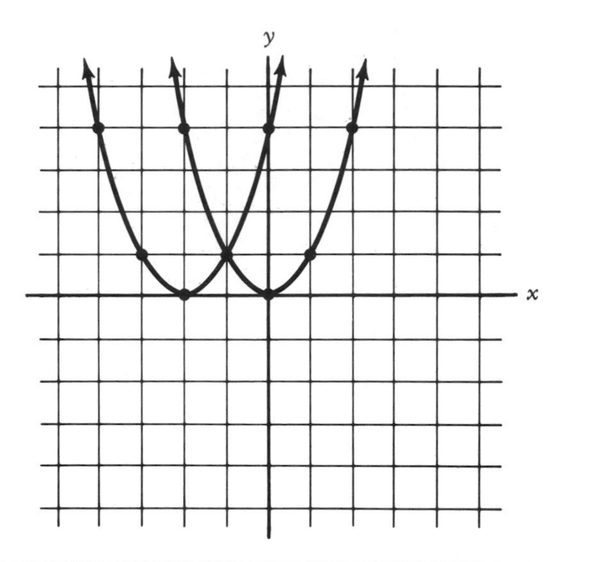

Q6 Solve and graph:
$y = x^2$
$y = (x - 1)^2$

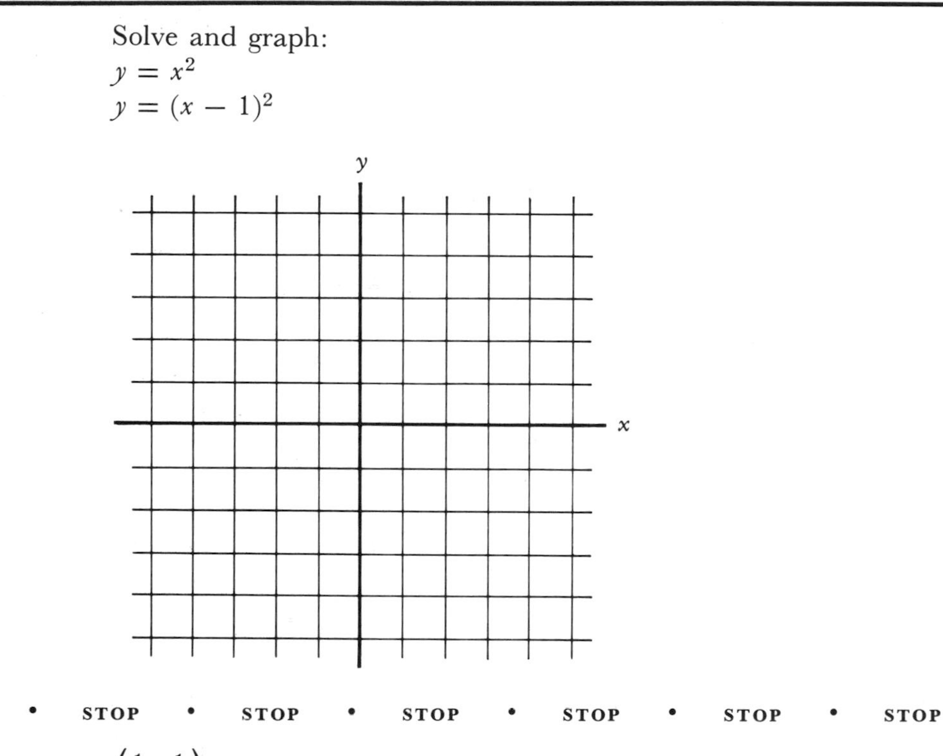

STOP • STOP • STOP • STOP • STOP • STOP • STOP • STOP • STOP

A6 $\left(\frac{1}{2}, \frac{1}{4}\right)$:

$$(x - 1)^2 = x^2$$
$$x^2 - 2x + 1 = x^2$$
$$-2x + 1 = 0$$
$$x = \frac{1}{2}$$
$$y = x^2$$
$$y = \left(\frac{1}{2}\right)^2$$

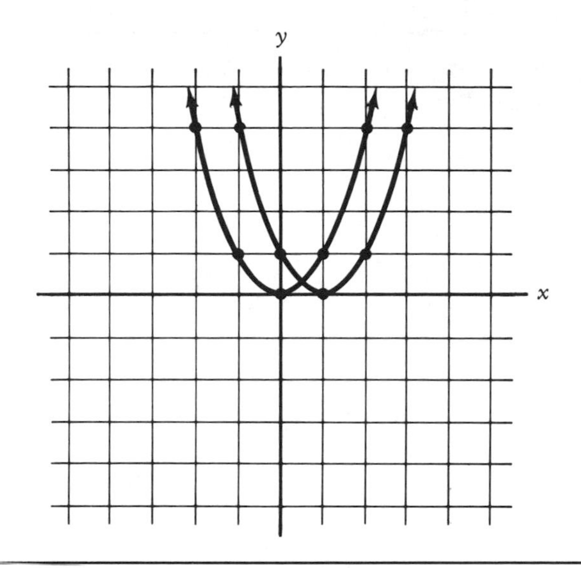

Q7 Solve:

a. $y = x^2 + 4x + 3$
$y = x^2 + 8x + 19$

b. $y = 2x^2 + 4$
$y = -x^2 + 1$

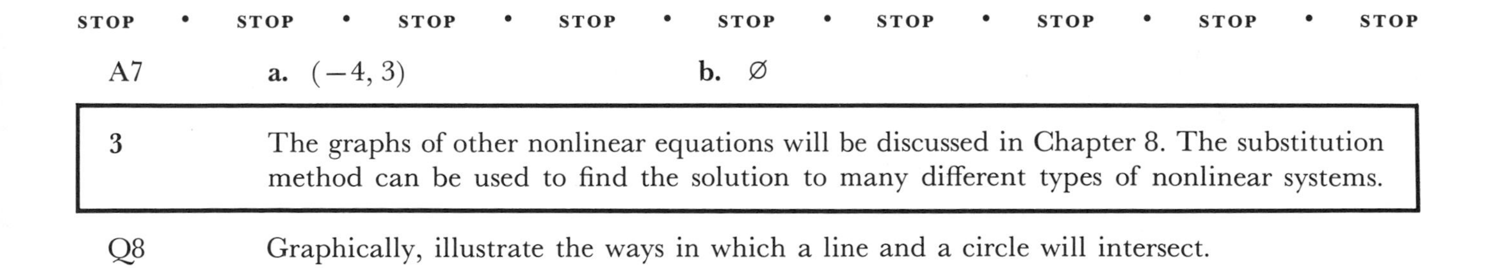

STOP • STOP • STOP • STOP • STOP • STOP • STOP • STOP • STOP

A7 **a.** $(-4, 3)$ **b.** $\varnothing$

3	The graphs of other nonlinear equations will be discussed in Chapter 8. The substitution method can be used to find the solution to many different types of nonlinear systems.

Q8 Graphically, illustrate the ways in which a line and a circle will intersect.

STOP • STOP • STOP • STOP • STOP • STOP • STOP • STOP • STOP

A8

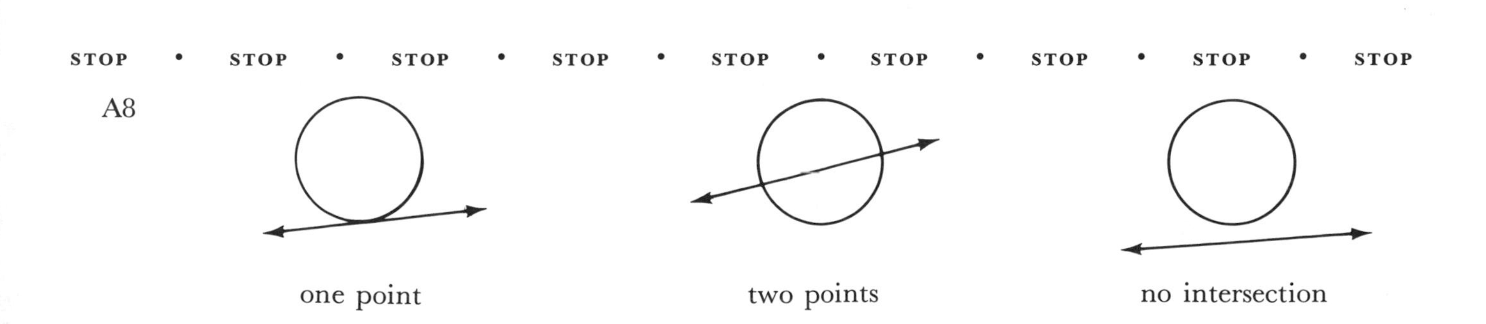

4 Systems of inequalities can be graphed by graphing each inequality on the same coordinate system. The common intersection is usually cross-hatched or more heavily shaded.

Example: Solve the system

$y < 2x - 1$
$y \leqslant -x + 1$

graphically.

Solution

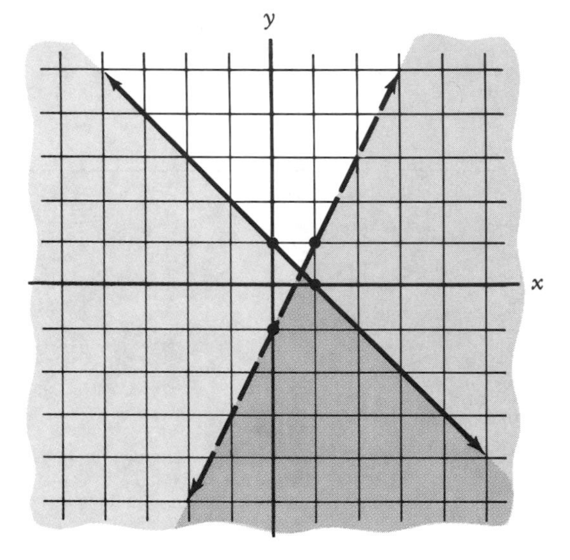

Graph $y < 2x - 1$. The line is dashed because the points on the line are not part of the solution. Graph $y \leqslant -x + 1$. The line is solid because the points on the line are part of the solution. The solution to the system is the intersection of the two graphs, the more heavily shaded portion of the graph.

Q9 Solve graphically.
$y < 2x - 1$
$y \geqslant -x + 1$

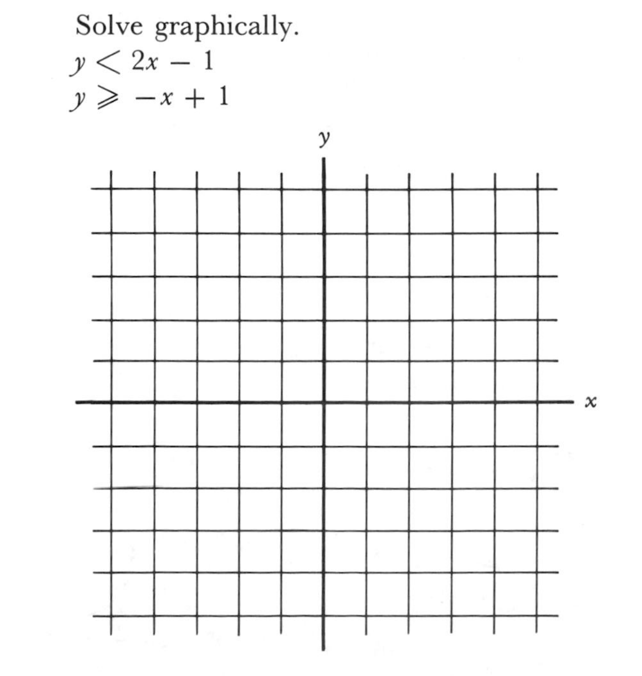

STOP • STOP • STOP • STOP • STOP • STOP • STOP • STOP • STOP

A9

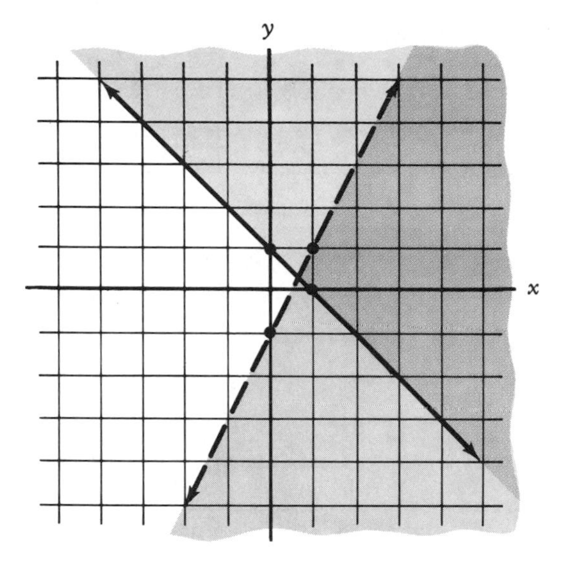

Q10 Solve graphically:

a. $y > \frac{1}{2}x - 2$

$y \leqslant -2x + 3$

b. $2x + y > -3$

$y \leqslant 2$

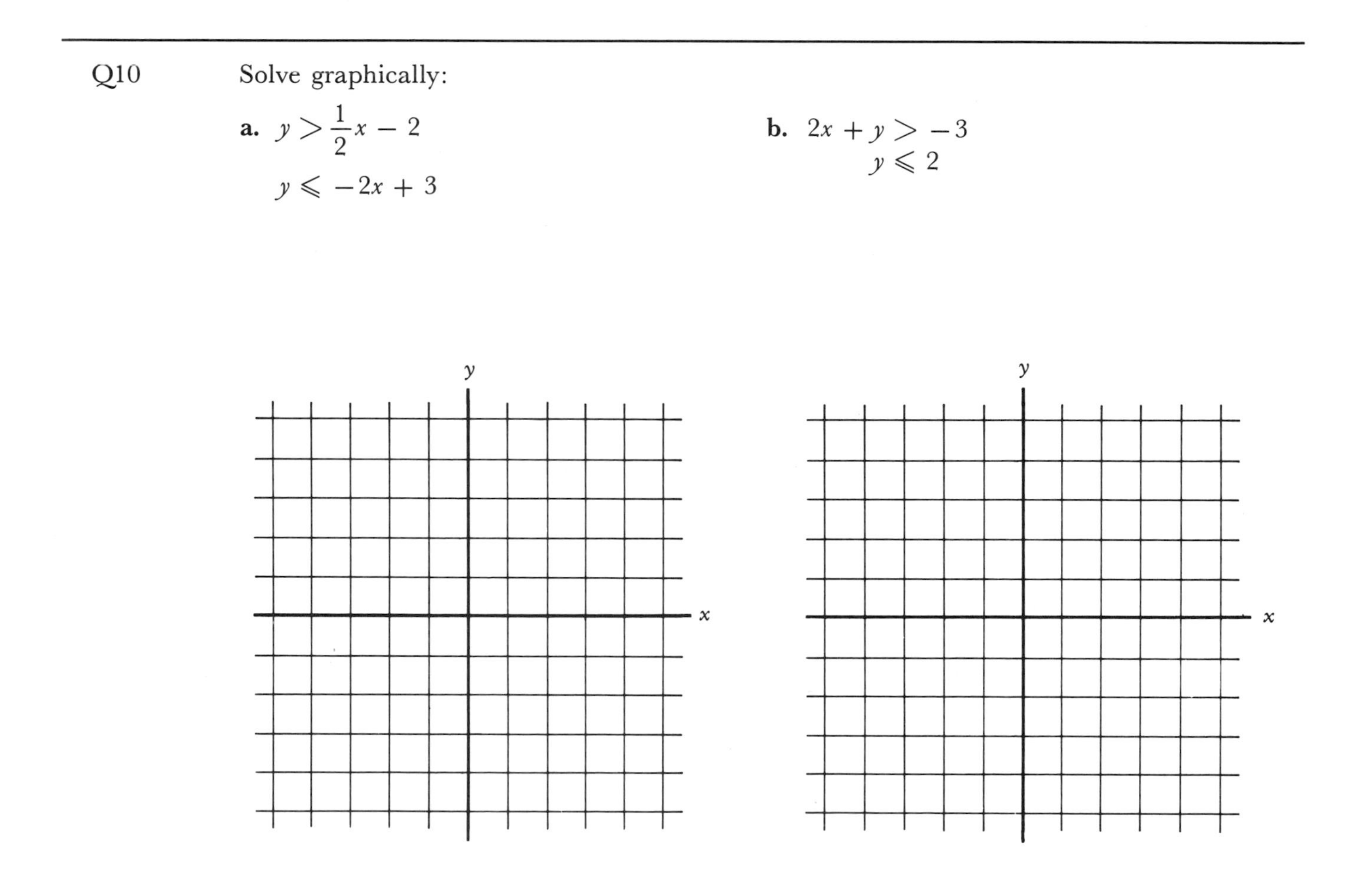

STOP • STOP • STOP • STOP • STOP • STOP • STOP • STOP • STOP

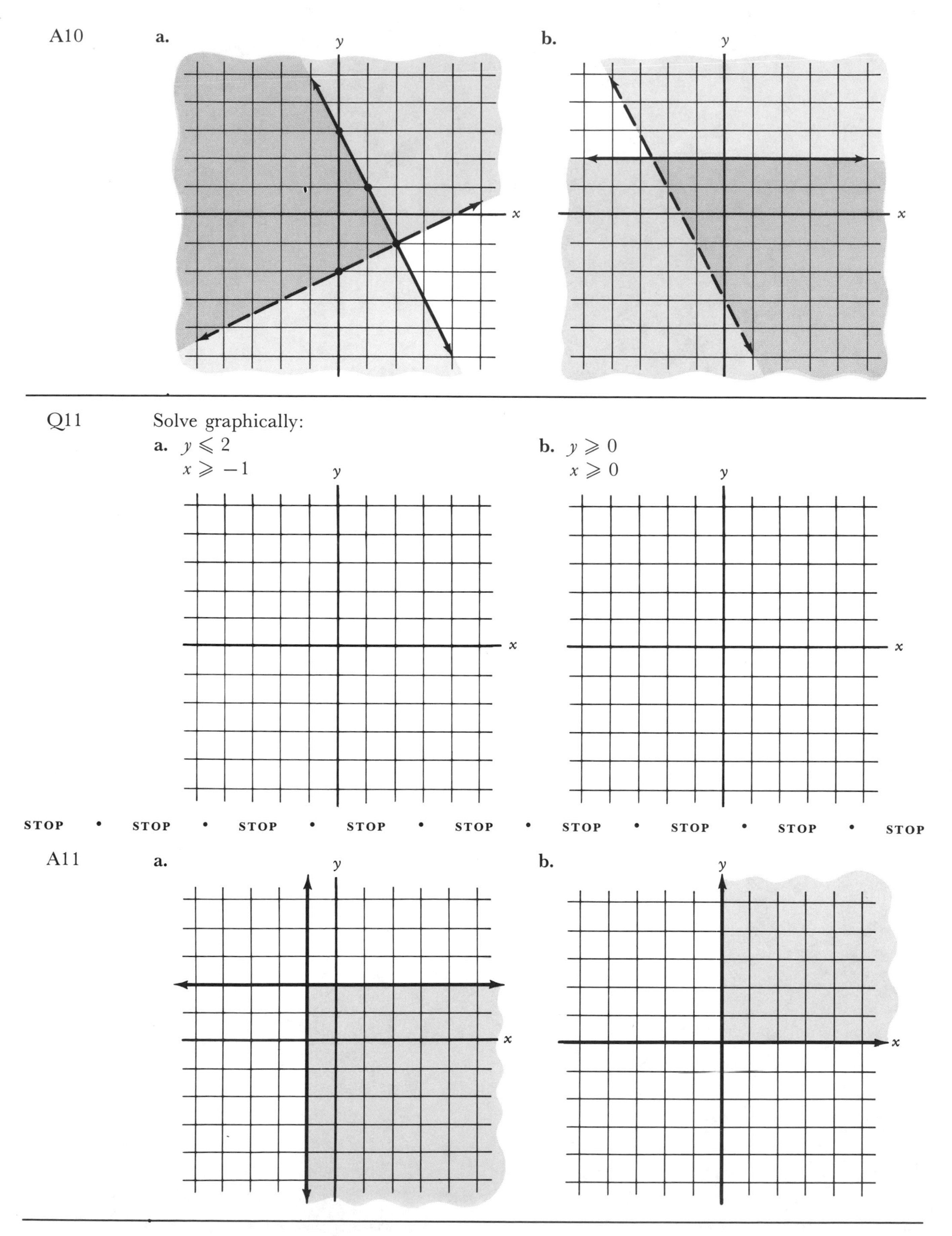

This completes the instruction for this section.

7.4 Exercises

1. Solve the systems and graph the pairs of equations on the same coordinate system:
   a. $y = -3(x + 1)^2 + 2$, $y = x$ **b.** $y = -x$, $y = x^2 + 1$ **c.** $y = x^2 + 4x + 4$, $y = x^2 + 2$
2. Solve:
   a. $y = 2x + 4$, $y = x^2 + 1$ **b.** $y = 2x^2 - x$, $4x - 2y = 0$
   c. $y = x^2 - 5$, $4x - y = 0$ **d.** $y = x^2 + 2x + 1$, $x - y + 3 = 0$
   e. $y = x^2$, $y = (x + 2)^2$ **f.** $y = x^2 + 6x + 9$, $y = -x^2 - 2$

*3. Solve:
   a. $x^2 + y^2 = 1$, $x + 2y = 1$ **b.** $x - y = 2$, $x^2 - y^2 = 16$

4. Solve graphically:
   a. $y \leqslant x - 2$, $y \geqslant 1$ **b.** $y < x$, $x > -2$

7.4 Exercise Answers

1. **a.** $\left(\frac{-7 - \sqrt{37}}{6}, \frac{-7 - \sqrt{37}}{6}\right)$ $\left(\frac{-7 + \sqrt{37}}{6}, \frac{-7 + \sqrt{37}}{6}\right)$ **b.** $\varnothing$

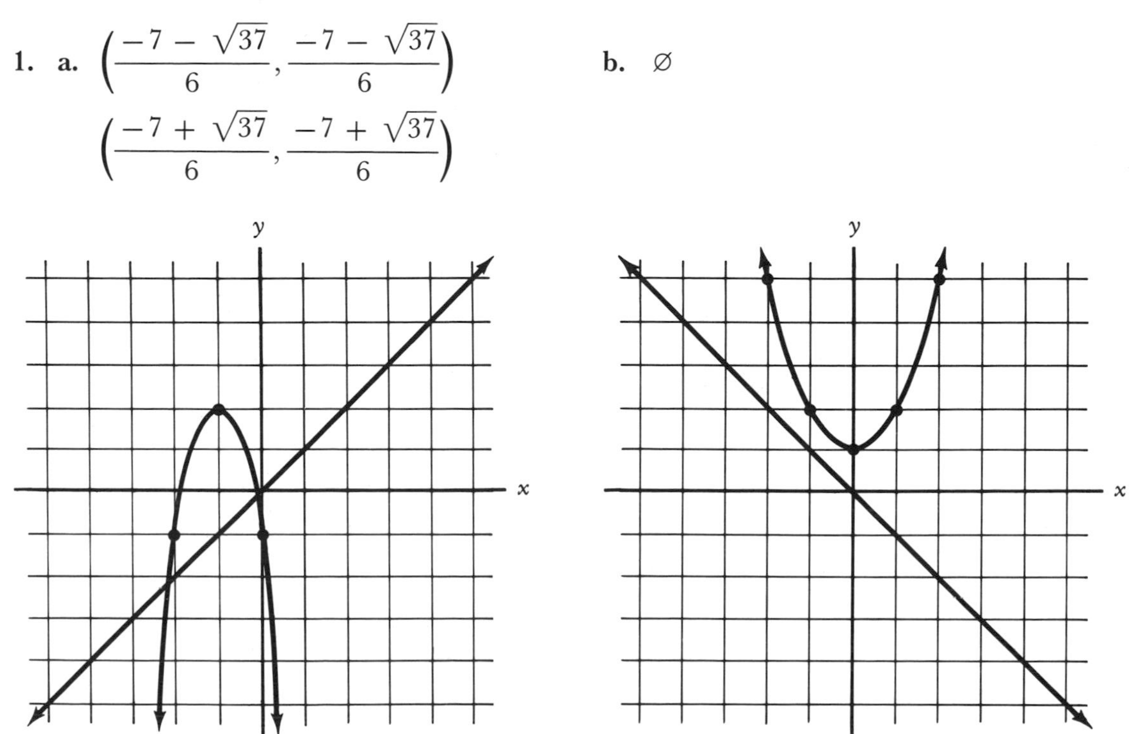

c. $\left(\frac{-1}{2}, \frac{9}{4}\right)$

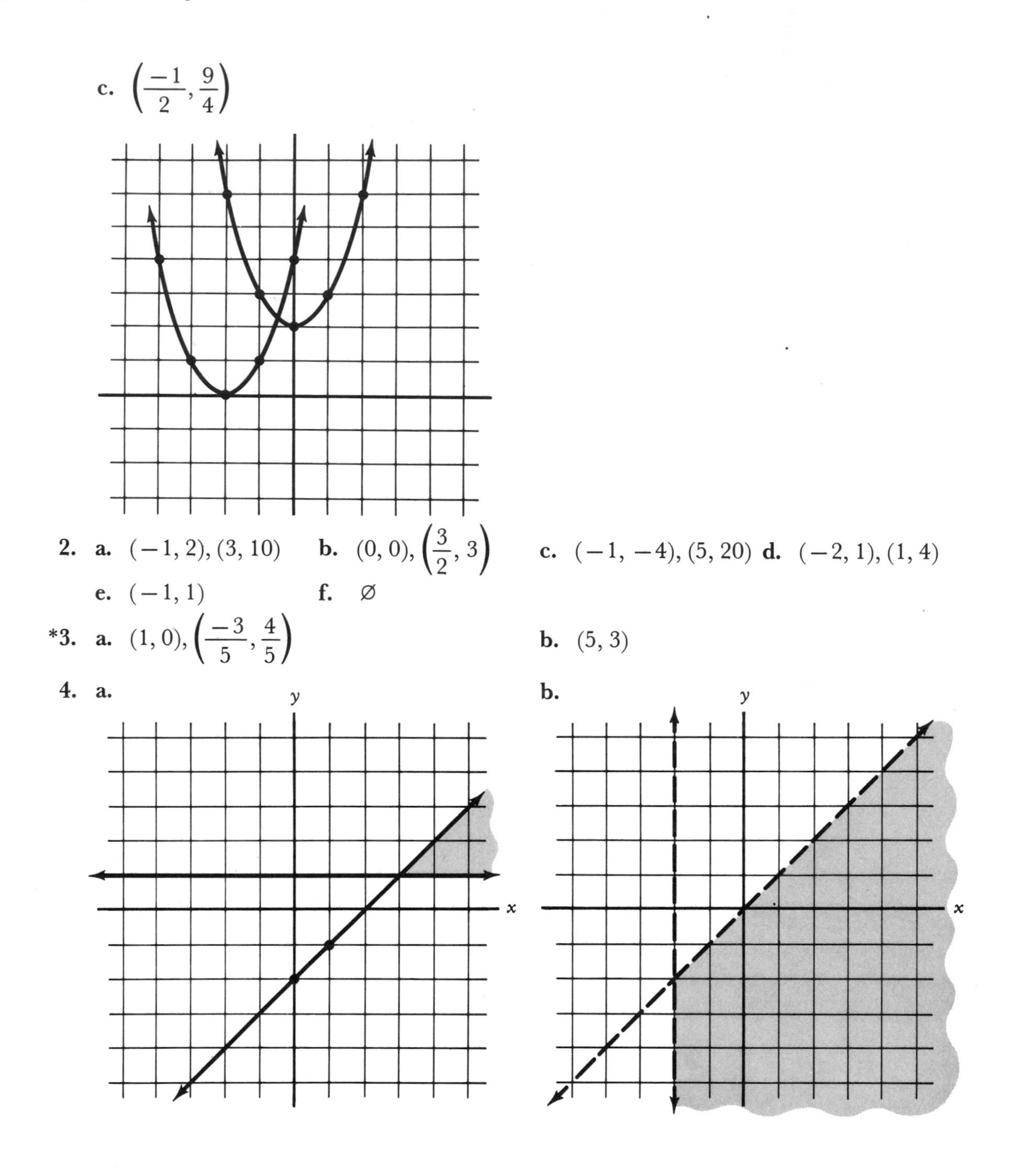

2. a. $(-1, 2), (3, 10)$ b. $(0, 0), \left(\frac{3}{2}, 3\right)$ c. $(-1, -4), (5, 20)$ d. $(-2, 1), (1, 4)$
   e. $(-1, 1)$ f. $\varnothing$

*3. a. $(1, 0), \left(\frac{-3}{5}, \frac{4}{5}\right)$ b. $(5, 3)$

4. a. b.

Chapter 7 Sample Test

At the completion of Chapter 7 it is expected that you will be able to work the following problems.

7.1 Parallel, Perpendicular, and Intersecting Lines

1. Solve the system graphically:
   a. $x + y = 2$
      $y = \frac{1}{2}x + 2$
   b. $y = 2$
      $x = -1$

2. Find the slope of a line parallel to the line:
   a. $y = -2x + 1$ b. $x - 4y = 10$
3. Find the slope of a line perpendicular to the line:
   a. $y = -2x + 1$ b. $x - 4y = 10$
4. Find the equation of the line:
   a. parallel to $x + 2y = 4$ containing $(0, 0)$
   b. perpendicular to $x + 2y = 4$ containing $(0, 0)$
   c. Graph the solution to parts (a) and (b) on the same coordinate system.

7.2 Addition and Substitution Methods

5. Solve:
   a. $\begin{aligned} 2x - 3y &= 10 \\ 2x + 2y &= 5 \end{aligned}$ b. $\begin{aligned} 2x + 3y &= 10 \\ -3x + 2y &= 11 \end{aligned}$
6. Solve:
   a. $\begin{aligned} 2x &= y + 5 \\ x &= -3 \end{aligned}$ b. $\begin{aligned} y &= -x \\ x + y &= 6 \end{aligned}$
7. a. The sum of two numbers is 62. Their difference is 16. Find the numbers.
   b. A chemist needs 100 liters of 50 percent alcohol solution. He has a 30 percent alcohol solution that he can mix with a 80 percent alcohol solution. How many liters of each will be required to make 100 liters of 50 percent solution?

7.3 Cramer's Rule

8. Evaluate:
   a. $\begin{vmatrix} -3 & 2 \\ -1 & 4 \end{vmatrix}$ b. $\begin{vmatrix} 2 & -1 & 0 \\ 0 & 3 & 1 \\ -2 & 0 & 1 \end{vmatrix}$
9. Solve using Cramer's rule:
   a. $\begin{aligned} 4x + 3y &= -5 \\ 3x - 4y &= 7 \end{aligned}$ b. $\begin{aligned} 2x + 2y + z &= 6 \\ 4x - y + z &= 12 \\ -x + y - 2z &= -3 \end{aligned}$

7.4 Nonlinear Equations and Systems of Inequalities

10. Solve:
    a. $\begin{aligned} y &= x^2 + 2x + 1 \\ x - y &= -7 \end{aligned}$ b. $\begin{aligned} y &= 2x^2 + 4x \\ y &= x^2 - 3x - 10 \end{aligned}$
11. Solve graphically:
    a. $\begin{aligned} y &\leqslant x \\ x + y &> 2 \end{aligned}$ b. $\begin{aligned} y &\leqslant 1 \\ x &> 0 \end{aligned}$

Chapter 7 Sample Test Answers

1. **a.** (0, 2) **b.** (−1, 2)

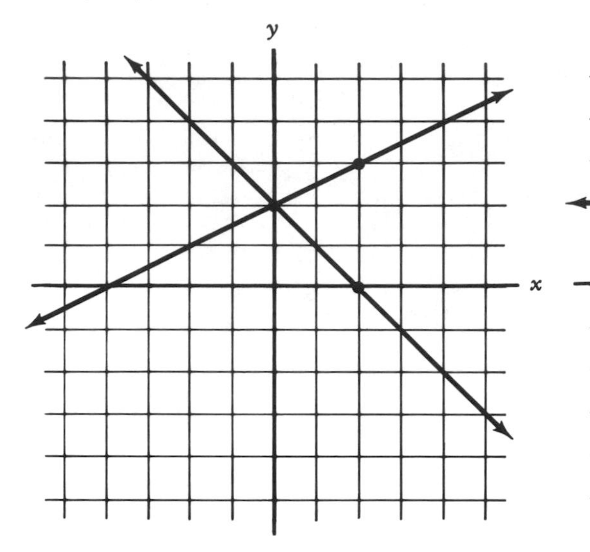

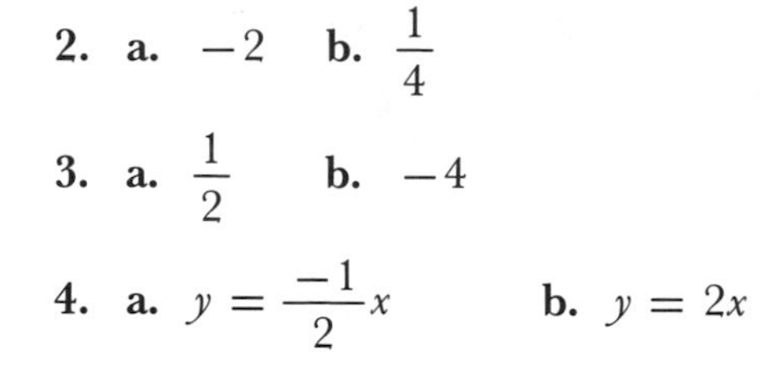

2. **a.** -2 **b.** $\frac{1}{4}$

3. **a.** $\frac{1}{2}$ **b.** -4

4. **a.** $y = \frac{-1}{2}x$ **b.** $y = 2x$ **c.**

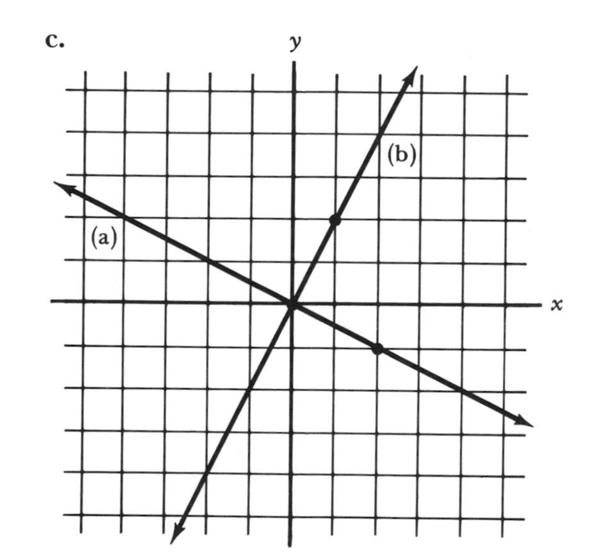

5. **a.** $\left(\frac{7}{2}, -1\right)$ **b.** $(-1, 4)$
6. **a.** $(-3, -11)$ **b.** $\varnothing$
7. **a.** 39 and 23
   b. 60 liters of 30 percent solution, 40 liters of 80 percent solution
8. **a.** -10 **b.** 8
9. **a.** $\left(\frac{1}{25}, \frac{-43}{25}\right)$ **b.** $(3, 0, 0)$
10. **a.** $(-3, 4)$, $(2, 9)$ **b.** $(-2, 0)$, $(-5, 30)$

11. a. b.

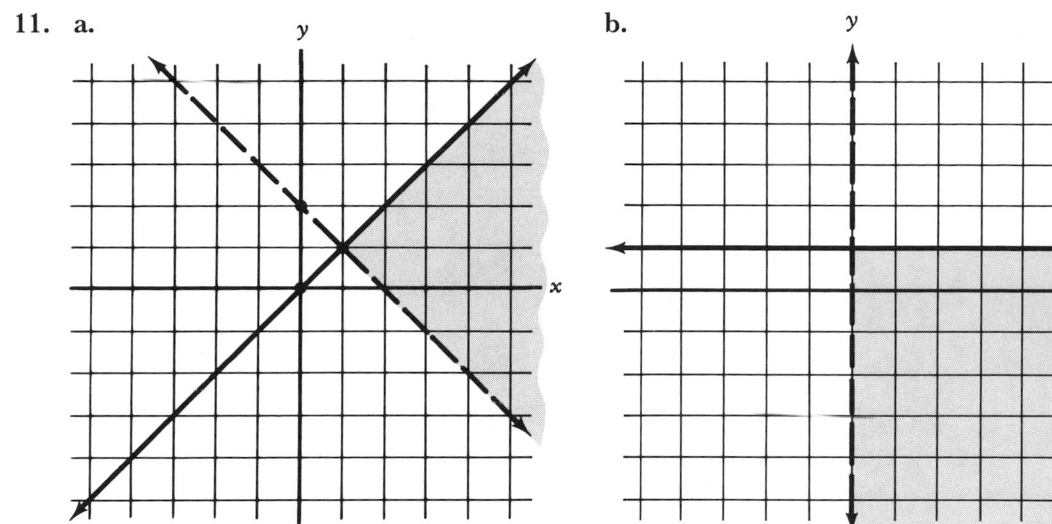

Chapter 8

Special, Exponential, and Logarithmic Functions

Many different types of relations and functions will be presented in Chapter 8. A critical analysis of some of these functions is beyond the scope of this textbook. Exponential and logarithmic functions have many applications to different scientific fields. One often hears mention of bacteria, population, and other organisms growing at an exponential rate. These growth rates can be represented by exponential functions.

8.1 Special Functions and the Conics

1 A *polynomial function* is defined by

$$P(x) = a_n x^n + a_{n-1} x^{n-1} + \cdots + a_0$$

where $n \in W$ and the a's are constants. The polynomial function where $n = 2$ was graphed in Chapter 6. If $n = 3$, the polynomial function is called a *cubic function.* The simplest cubic function is $f(x) = x^3$ and is graphed by plotting a few well-chosen points.

x	y
-3	-27
-2	-8
-1	-1
0	0
1	1
2	8
3	27

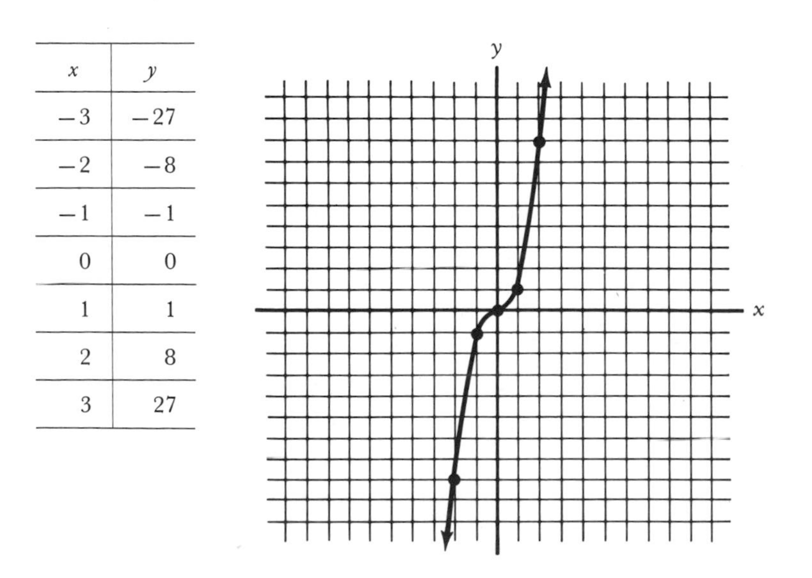

The ordered pairs $(-3, -27)$ and $(3, 27)$ are not plotted because of their large y values.

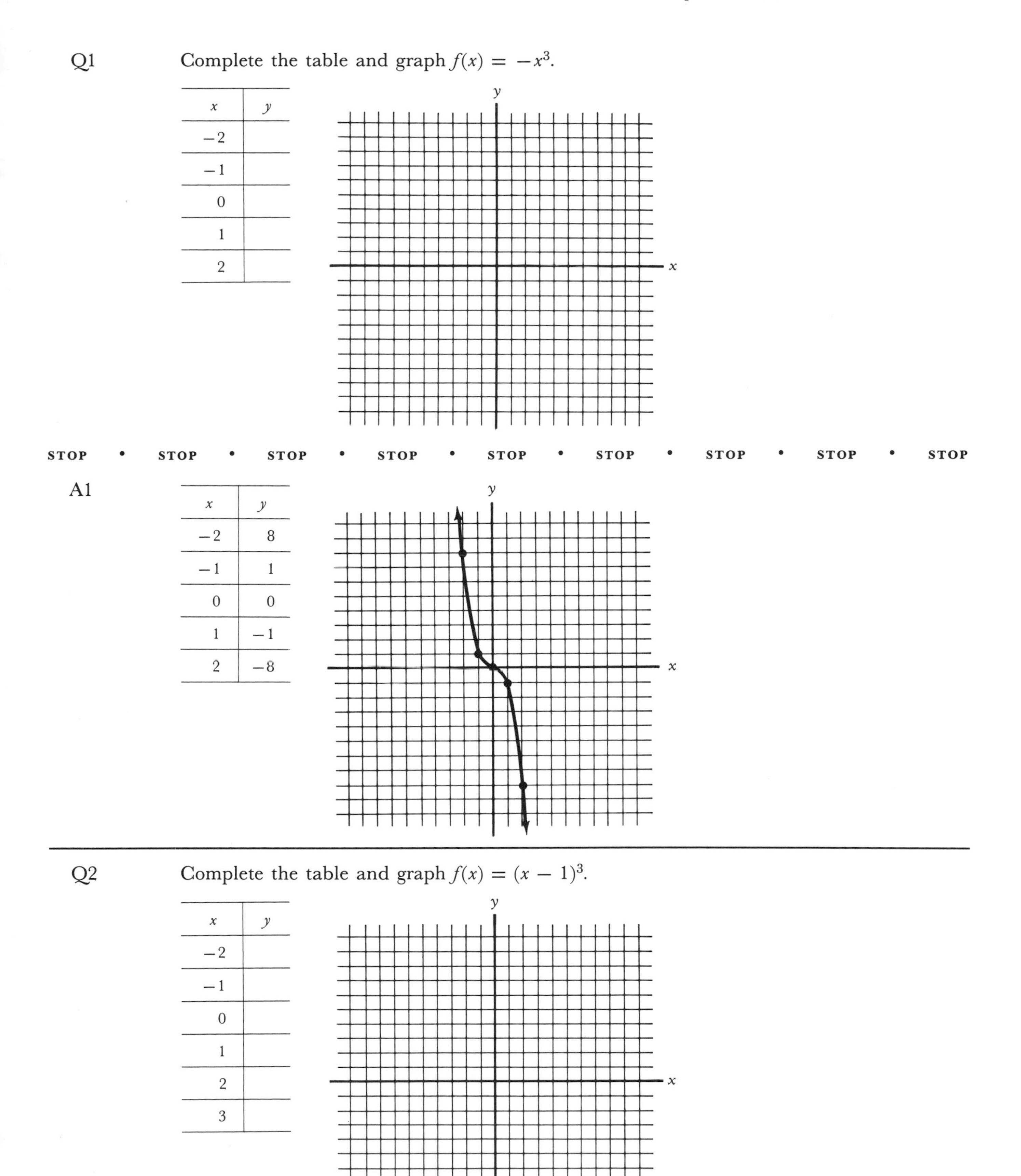

Q1 Complete the table and graph $f(x) = -x^3$.

x	y
-2	
-1	
0	
1	
2	

STOP • STOP • STOP • STOP • STOP • STOP • STOP • STOP • STOP

A1

x	y
-2	8
-1	1
0	0
1	-1
2	-8

Q2 Complete the table and graph $f(x) = (x - 1)^3$.

x	y
-2	
-1	
0	
1	
2	
3	

STOP • STOP • STOP • STOP • STOP • STOP • STOP • STOP • STOP

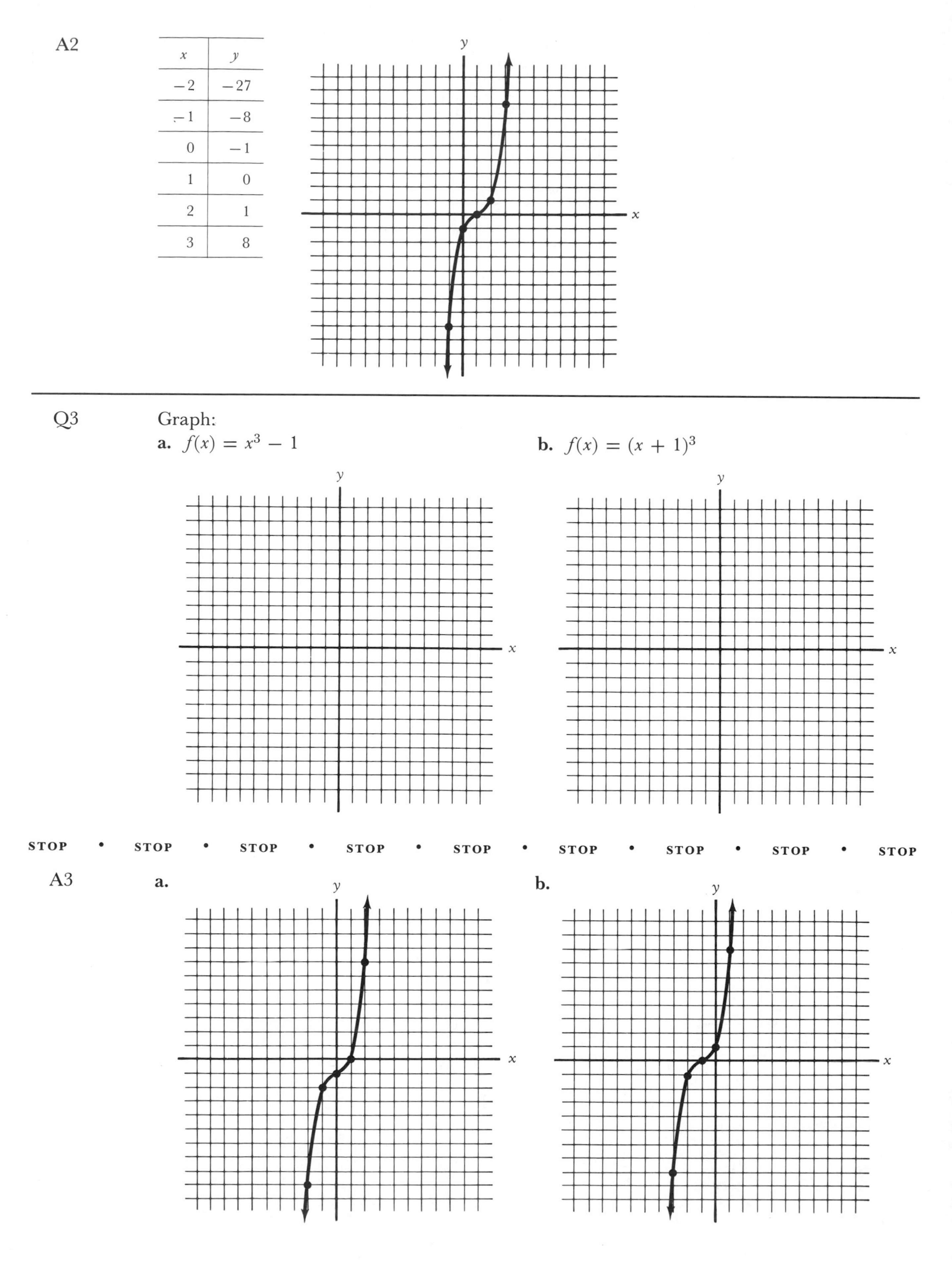

A2

x	y
-2	-27
-1	-8
0	-1
1	0
2	1
3	8

Q3 Graph:

a. $f(x) = x^3 - 1$

b. $f(x) = (x + 1)^3$

STOP • STOP • STOP • STOP • STOP • STOP • STOP • STOP • STOP

A3 **a.**

b.

2

A *rational function* is defined by

$$R(x) = \frac{P(x)}{Q(x)} \qquad Q(x) \neq 0$$

where $P(x)$ and $Q(x)$ are polynomials. One of the simplest rational functions is $f(x) = \frac{1}{x}$, $x \neq 0$. The function is often called the *reciprocal function,* because $\frac{1}{x}$ is the reciprocal of x.

Examples:

$$f(8) = \frac{1}{8}$$
$$f(-1) = -1$$
$$f\left(\frac{1}{10}\right) = 10$$
$f(0)$ is undefined

Q4 Determine the function values of $f(x) = \frac{1}{x}$:

a. $f\left(\frac{1}{4}\right) =$ _____ b. $f\left(\frac{1}{2}\right) =$ _____

c. $f(1) =$ _____ d. $f(2) =$ _____

e. $f(10) =$ _____ f. $f(50) =$ _____

STOP • STOP • STOP • STOP • STOP • STOP • STOP • STOP • STOP

A4 a. 4 b. 2 c. 1 d. $\frac{1}{2}$ e. $\frac{1}{10}$ f. $\frac{1}{50}$

Q5 Determine the function values of $f(x) = \frac{1}{x}$:

a. $f\left(\frac{-1}{4}\right) =$ _____ b. $f\left(\frac{-1}{2}\right) =$ _____

c. $f(-1) =$ _____ d. $f(-2) =$ _____

e. $f(-10) =$ _____ f. $f(-50) =$ _____

STOP • STOP • STOP • STOP • STOP • STOP • STOP • STOP • STOP

A5 a. -4 b. -2 c. -1 d. $\frac{-1}{2}$ e. $\frac{-1}{10}$ f. $\frac{-1}{50}$

3 The function $f(x) = \frac{1}{x}$ can be graphed by observing:

1. As $|x|$ gets larger and larger, y gets close to zero.
2. As $|x|$ gets close to zero, $|y|$ gets larger and larger.
3. Since $x \neq 0$ and $y \neq 0$, the graph will not intersect the x or y axis.

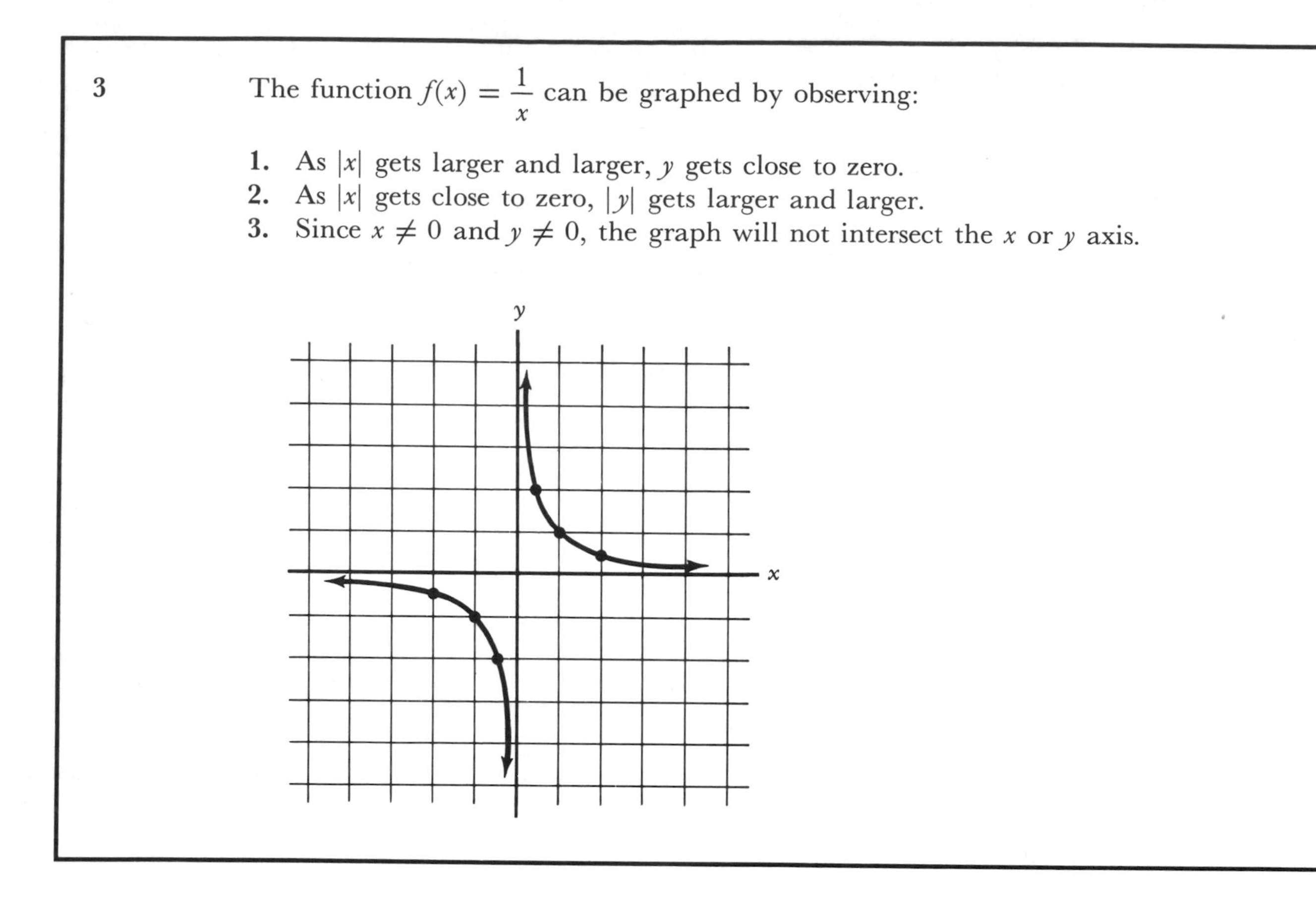

Q6 Complete the table and graph $f(x) = \frac{2}{x}$.

x	y
-2	
-1	
$\frac{-1}{2}$	
0	
$\frac{1}{2}$	
1	
2	
4	

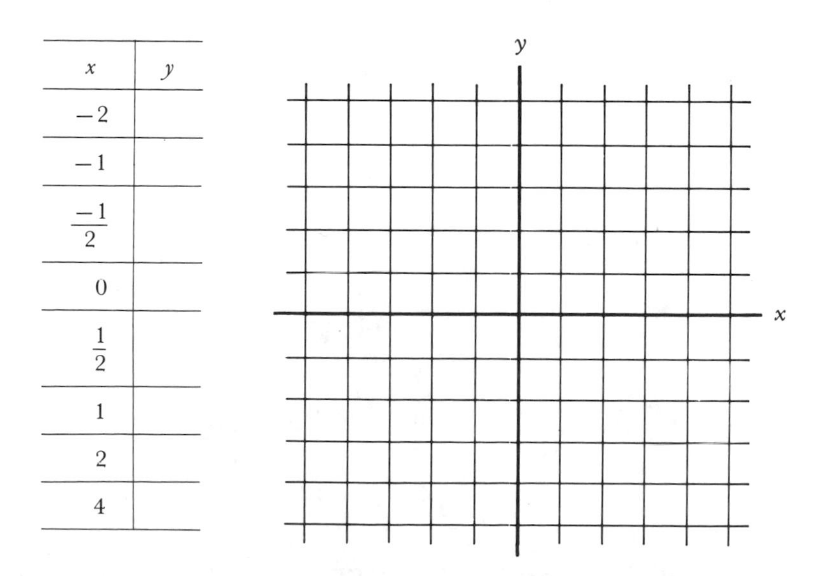

STOP • STOP • STOP • STOP • STOP • STOP • STOP • STOP • STOP

A6

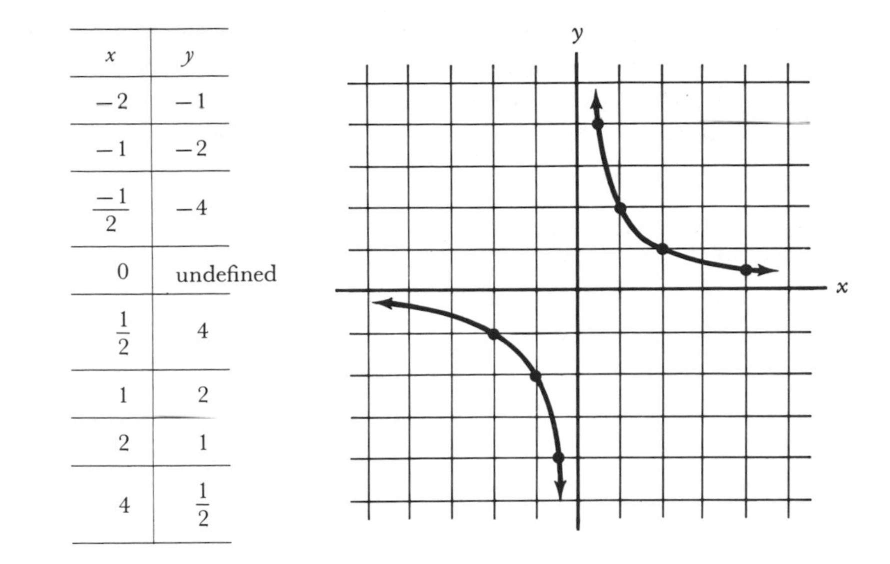

x	y
-2	-1
-1	-2
$\frac{-1}{2}$	-4
0	undefined
$\frac{1}{2}$	4
1	2
2	1
4	$\frac{1}{2}$

Q7 Graph:

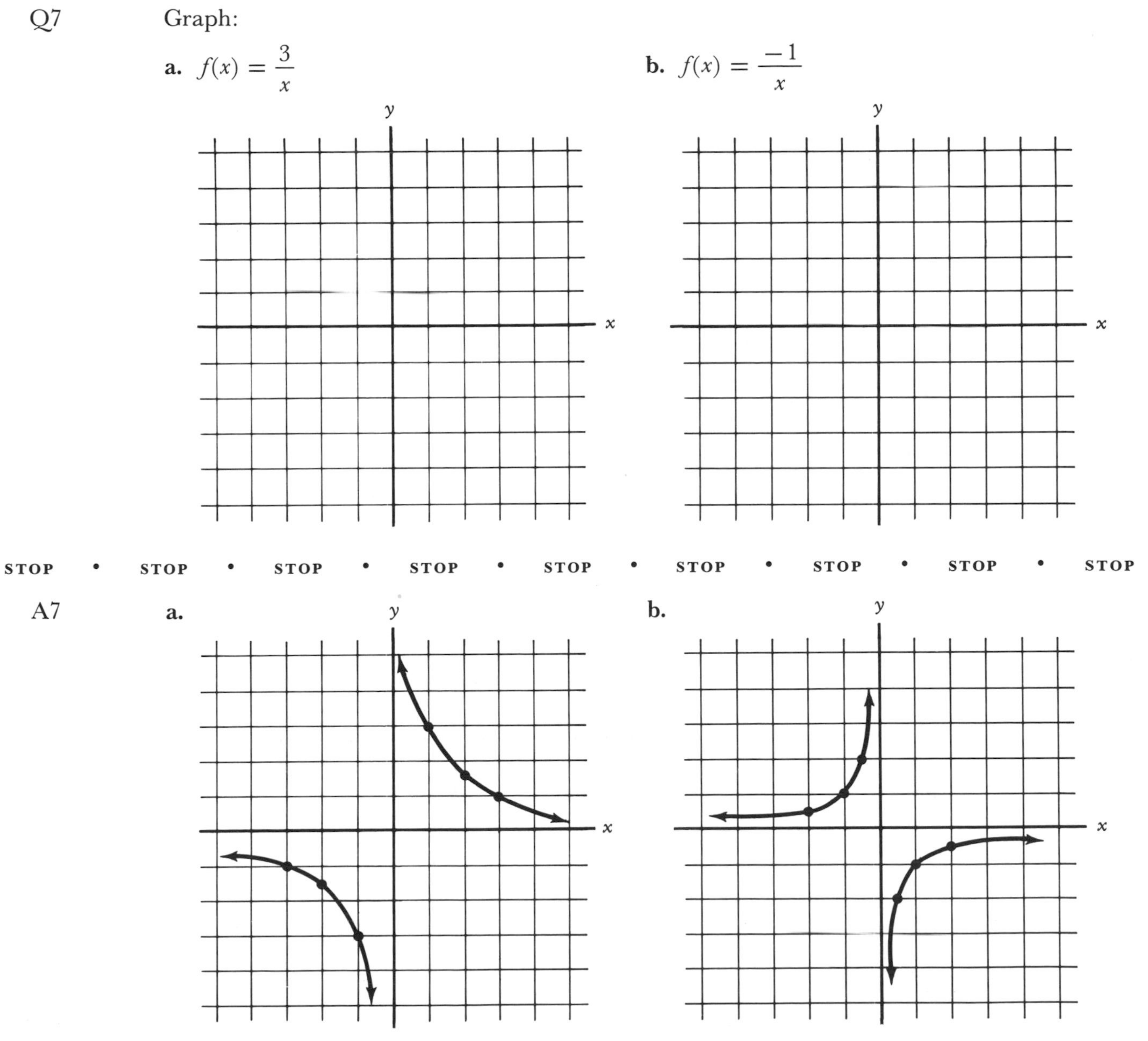

a. $f(x) = \frac{3}{x}$

b. $f(x) = \frac{-1}{x}$

STOP • STOP • STOP • STOP • STOP • STOP • STOP • STOP • STOP

A7 a.

b.

4

If x and y *vary directly,* then $y = kx$, where k is a constant called the constant of variation. If two variables vary directly, an increase in one causes a corresponding increase in the other. For example, the faster one drives, the greater the distance traveled in a given time; hence speed and distance traveled vary directly.

Example: Hooke's law states that the distance a spring stretches varies directly as the force applied. If a force of 15 kilograms stretches a certain spring 8 centimeters, how much will a force of 20 kilograms stretch the spring?

Solution

Step 1: Let

$d =$ distance stretched
$f =$ force applied

Step 2: Since d and f vary directly,

$$d = kf$$

where k is some constant. The constant k can be determined from the information that a force of 15 kilograms stretches the spring 8 centimeters. That is, when $f = 15$, $d = 8$.

$$d = kf$$
$$8 = k \cdot 15$$
$$k = \frac{8}{15}$$

Step 3:

$$d = \frac{8}{15}f$$
$$d = \frac{8}{15}(20)$$
$$d = \frac{32}{3} \text{ centimeters}$$

Step 4 (check):

$$d = \frac{8}{15}f$$
$$\frac{32}{3} = \frac{8}{15}(20) \quad \text{(true)}$$

Q8

The distance that a body falls from rest varies directly as the square of the time it falls (disregarding air resistance). If an object falls 19.6 meters in 2 seconds, how far will it fall in 5 seconds?

STOP • STOP • STOP • STOP • STOP • STOP • STOP • STOP • STOP

A8 122.5 m:

Step 2: $d = kt^2$
$19.6 = k(2)^2$
$4.9 = k$

Step 3: $d = 4.9t^2$
$d = 4.9(5)^2$
$d = 122.5$ meters

Q9 The pressure exerted by a certain liquid varies directly as the depth of a point beneath the surface of the liquid. The pressure at 20 meters is 50 kilograms per square centimeter. What pressure is exerted at 4 meters?

STOP • STOP • STOP • STOP • STOP • STOP • STOP • STOP • STOP

A9 10 kg/cm^2:

Step 2: $p = kd$
$50 = k \cdot 20$
$k = \frac{5}{2}$

Step 3: $p = \frac{2}{5}d$
$p = \frac{5}{2}(4)$
$p = 10$ kg/cm^2

---

5 If x and y *vary indirectly*, $y = \frac{k}{x}$, where k is a constant. If two variables *vary indirectly* (inversely), an increase in one causes a corresponding decrease in the other. For example, an increase in speed will cause a decrease in the time required to travel from one point to another. Hence speed and time are indirectly related.

Example: The current (measured in amperes) in a simple electrical circuit varies indirectly with the resistance (measured in ohms). If the current is 20 amperes when the resistance is 5 ohms, find the current if the resistance is 25 ohms.

Solution

Step 1: Let

$C =$ amperes of current
$R =$ ohms of resistance

Step 2:

$C = \frac{k}{R}$

$20 = \frac{k}{5}$

$100 = k$

Step 3:

$C = \frac{100}{R}$

$C = \frac{100}{25}$

$C = 4$ amperes

Step 4: (check):

$C = \frac{100}{R}$

$4 = \frac{100}{25}$ (true)

Q10 The illumination produced by a light source varies indirectly as the square of the distance from the source. If the illumination, I, produced 4 feet from a light source is 50 footcandles, find the illumination produced 5 feet from the same source.

STOP • STOP • STOP • STOP • STOP • STOP • STOP • STOP • STOP

A10 32 footcandles:

Step 2: $I = \frac{k}{d^2}$

$50 = \frac{k}{(4)^2}$

$800 = k$

Step 3: $I = \frac{800}{d^2}$

$I = \frac{800}{(5)^2}$

$I = 32$ footcandles

6 Often one variable varies directly with one variable and indirectly with another. Such variation is called *combined variation.*

Example: Write an equation of variation if y varies directly as x and the square of z and indirectly as m.

Solution

y varies directly as x and the square of z:

$y = kxz^2$

and indirectly as m:

$y = \frac{kxz^2}{m}$

Q11 Write an equation of variation if z varies directly as x and y. ______

STOP • STOP • STOP • STOP • STOP • STOP • STOP • STOP • STOP

A11 $z = kxy$

Q12 Write an equation of variation as y varies directly as the square of x and inversely as the cube of p. ______

STOP • STOP • STOP • STOP • STOP • STOP • STOP • STOP • STOP

A12 $y = \frac{kx^2}{p^3}$

Q13 The volume (V) of a gas varies directly as its temperature (T) and inversely as its pressure (P). A gas occupies 20 cubic centimeters at a temperature of 300 Kelvin and a pressure of 30 grams per square centimeter. What will the volume be if the temperature is raised to 360 K and the pressure decreased to 20 grams per square centimeter?

STOP • STOP • STOP • STOP • STOP • STOP • STOP • STOP • STOP

A13 36 cubic centimeters:

$$V = \frac{kT}{P} \qquad V = \frac{2T}{P}$$

$$20 = \frac{k(300)}{30} \qquad V = \frac{2(360)}{20}$$

$$2 = k \qquad V = 36$$

7 Graphs of second-degree equations are called the *conic sections* because they can be formed by the intersection of a cone and a plane.

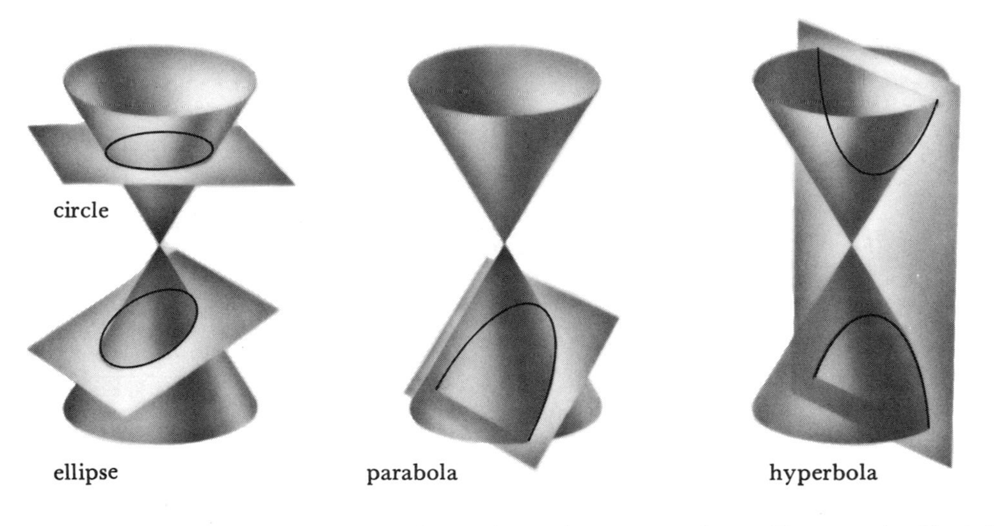

A circle is the set of all points in a plane that are a given distance (radius) from a given point (center). The general form of a circle with center at the origin and radius r is

$x^2 + y^2 = r^2$

Example: Graph the circle $x^2 + y^2 = 9$.

Solution

The x intercepts are found by letting $y = 0$.

$$x^2 + y^2 = 9$$
$$x^2 + 0 = 9$$
$$x = \pm 3$$

x intercepts $= (-3, 0)$ and $(3, 0)$

The y intercepts are found by letting $x = 0$.

$x^2 + y^2 = 9$
$0 + y^2 = 9$
$y = \pm 3$
y intercepts $= (0, -3)$ and $(0, 3)$

A circle is now graphed connecting the x and y intercepts with a circle of radius 3.

Q14 Graph:

a. $x^2 + y^2 = 4$

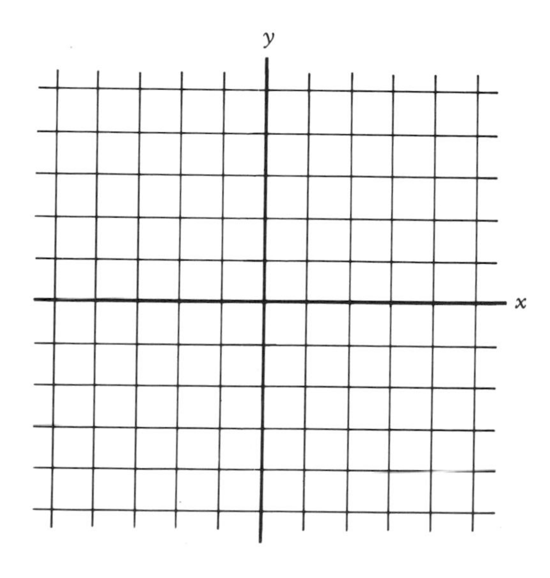

b. $x^2 + y^2 = 6$

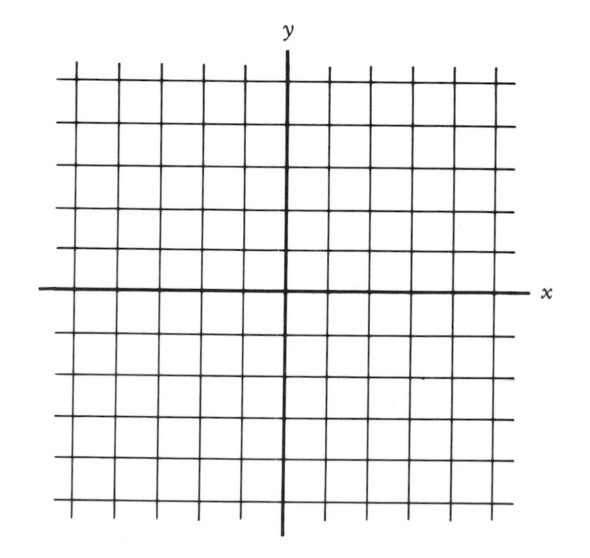

STOP • STOP • STOP • STOP • STOP • STOP • STOP • STOP • STOP

A14

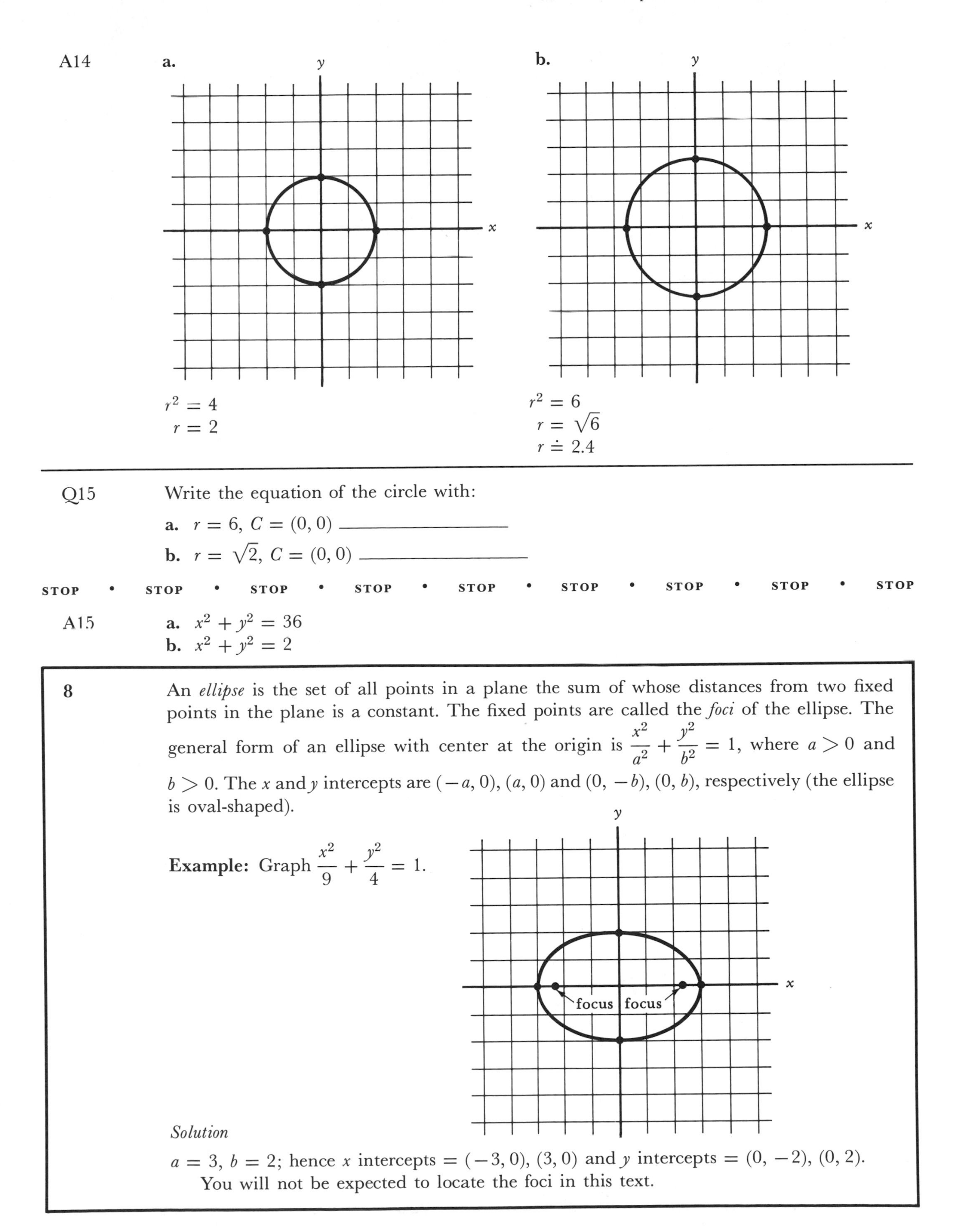

a.

$r^2 = 4$
$r = 2$

b.

$r^2 = 6$
$r = \sqrt{6}$
$r \doteq 2.4$

Q15 Write the equation of the circle with:

a. $r = 6$, $C = (0, 0)$ ______________

b. $r = \sqrt{2}$, $C = (0, 0)$ ______________

STOP • STOP • STOP • STOP • STOP • STOP • STOP • STOP • STOP

A15

a. $x^2 + y^2 = 36$
b. $x^2 + y^2 = 2$

8 An *ellipse* is the set of all points in a plane the sum of whose distances from two fixed points in the plane is a constant. The fixed points are called the *foci* of the ellipse. The general form of an ellipse with center at the origin is $\frac{x^2}{a^2} + \frac{y^2}{b^2} = 1$, where $a > 0$ and $b > 0$. The x and y intercepts are $(-a, 0)$, $(a, 0)$ and $(0, -b)$, $(0, b)$, respectively (the ellipse is oval-shaped).

Example: Graph $\frac{x^2}{9} + \frac{y^2}{4} = 1$.

Solution

$a = 3$, $b = 2$; hence x intercepts $= (-3, 0)$, $(3, 0)$ and y intercepts $= (0, -2)$, $(0, 2)$. You will not be expected to locate the foci in this text.

Q16 Find the x intercepts for $\frac{x^2}{9} + \frac{y^2}{4} = 1$ by letting $y = 0$ and solving for x.

STOP • STOP • STOP • STOP • STOP • STOP • STOP • STOP • STOP

A16 $(-3, 0), (3, 0)$: $\frac{x^2}{9} + \frac{(0)^2}{4} = 1$

$$\frac{x^2}{9} = 1$$

$$x^2 = 9$$

$$x = \pm 3$$

Q17 Find the y intercepts for $\frac{x^2}{9} + \frac{y^2}{4} = 1$ by letting $x = 0$ and solving for y.

STOP • STOP • STOP • STOP • STOP • STOP • STOP • STOP • STOP

A17 $(0, -2), (0, 2)$: $\frac{(0)^2}{9} + \frac{y^2}{4} = 1$

$$\frac{y^2}{4} = 1$$

$$y^2 = 4$$

$$y = \pm 2$$

Q18 Find a and b for $\frac{x^2}{25} + \frac{y^2}{7} = 1$.

STOP • STOP • STOP • STOP • STOP • STOP • STOP • STOP • STOP

A18 $a = 5$, $b = \sqrt{7}$: $\frac{x^2}{25} + \frac{y^2}{7} = \frac{x^2}{(5)^2} + \frac{y^2}{(\sqrt{7})^2}$

Q19 Determine the x and y intercepts for the ellipse $\frac{x^2}{4} + \frac{y^2}{16} = 1$. x intercepts = ______________, y intercepts = ______________

STOP • STOP • STOP • STOP • STOP • STOP • STOP • STOP • STOP

A19 $(-2, 0), (2, 0)$; $(0, -4), (0, 4)$; $\frac{x^2}{4} + \frac{y^2}{16} = \frac{x^2}{(2)^2} + \frac{y^2}{(4)^2}$ $(a = 2, b = 4)$

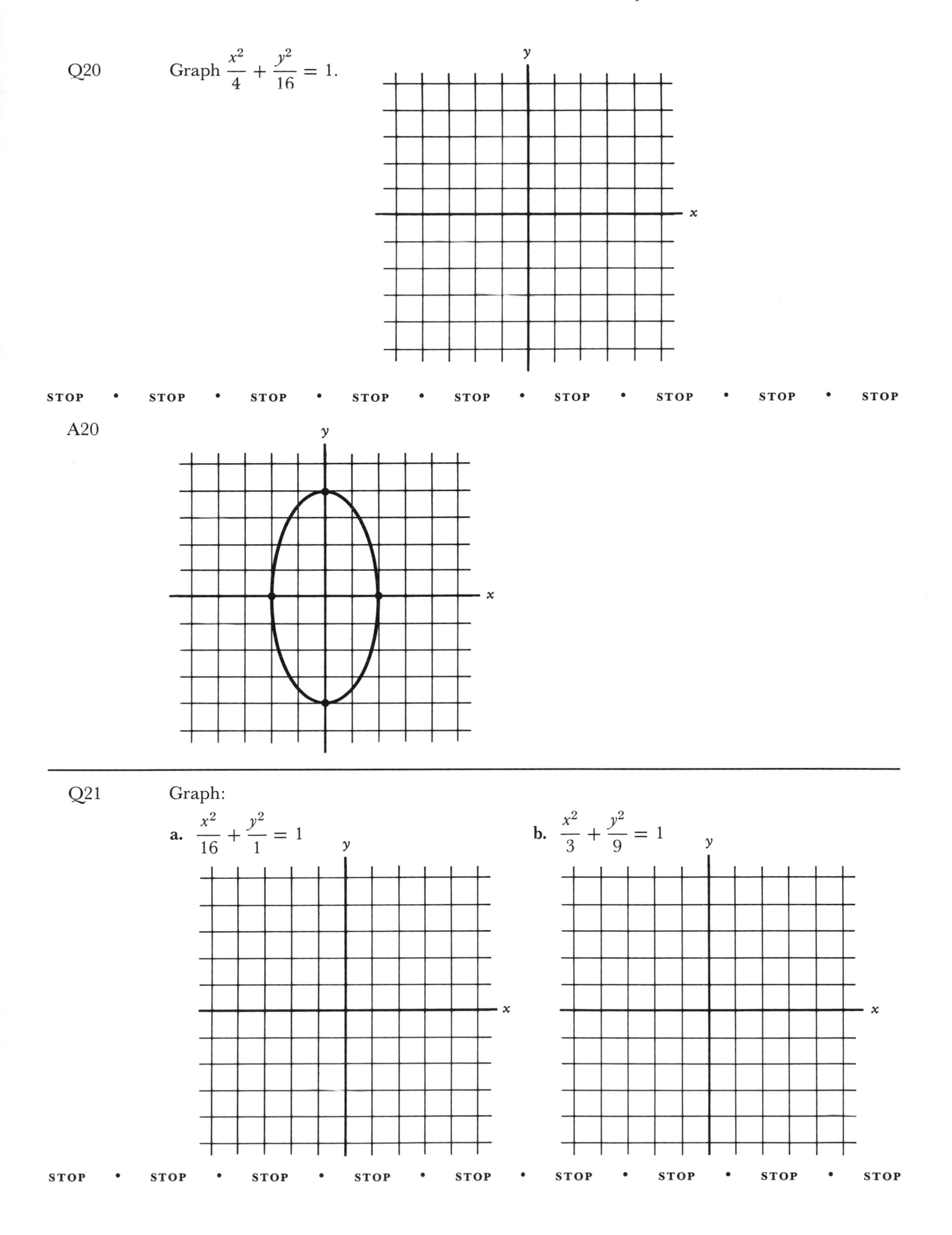

Q20 Graph $\frac{x^2}{4} + \frac{y^2}{16} = 1.$

STOP • STOP • STOP • STOP • STOP • STOP • STOP • STOP • STOP

A20

Q21 Graph:

a. $\frac{x^2}{16} + \frac{y^2}{1} = 1$

b. $\frac{x^2}{3} + \frac{y^2}{9} = 1$

STOP • STOP • STOP • STOP • STOP • STOP • STOP • STOP • STOP

A21

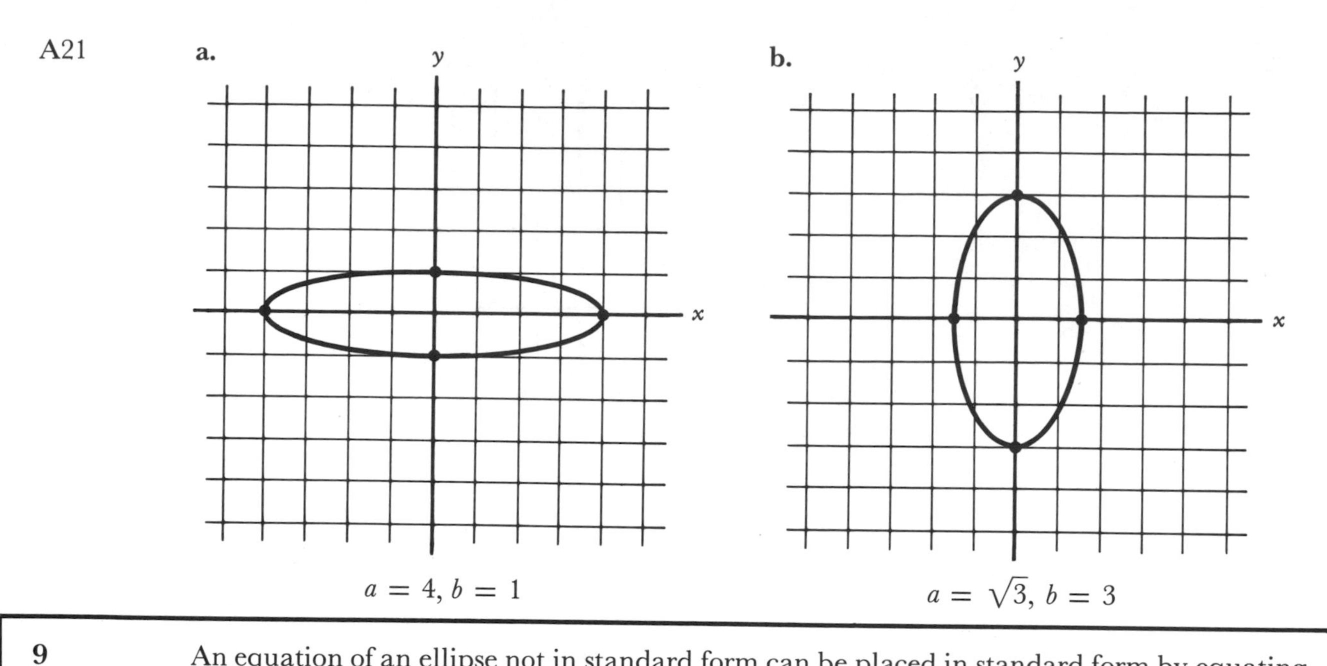

a. $a = 4, b = 1$

b. $a = \sqrt{3}, b = 3$

9

An equation of an ellipse not in standard form can be placed in standard form by equating the left member of the equation to 1.

Example 1: Write $x^2 + 4y^2 = 4$ in standard form.

Solution

Divide both members of the equation by 4.

$$x^2 + 4y^2 = 4$$

$$\frac{x^2}{4} + \frac{4y^2}{4} = \frac{4}{4}$$

$$\frac{x^2}{4} + \frac{y^2}{1} = 1 \qquad a = 2, \quad b = 1$$

Example 2: Write $9x^2 + y^2 = 4$ in standard form.

Solution

Divide both members of the equation by 4.

$$\frac{9x^2}{4} + \frac{y^2}{4} = 1$$

The equation is not in standard form because the first term is not of the form $\frac{x^2}{a^2}$; however,

$$\frac{9x^2}{4} = \frac{x^2}{\frac{4}{9}}$$

Therefore, the equation in standard form is

$$\frac{x^2}{\frac{4}{9}} + \frac{y^2}{4} = 1 \qquad a = \frac{2}{3}, \quad b = 2$$

Q22 Write in standard form:

a. $x^2 + 4y^2 = 36$ **b.** $25x^2 + 3y^2 = 9$

STOP • STOP • STOP • STOP • STOP • STOP • STOP • STOP • STOP

A22 **a.** $\dfrac{x^2}{36} + \dfrac{y^2}{9} = 1$ **b.** $\dfrac{x^2}{\frac{9}{25}} + \dfrac{y^2}{3} = 1$

Q23 Graph:

a. $x^2 + 4y^2 = 4$ **b.** $4x^2 + y^2 = 16$

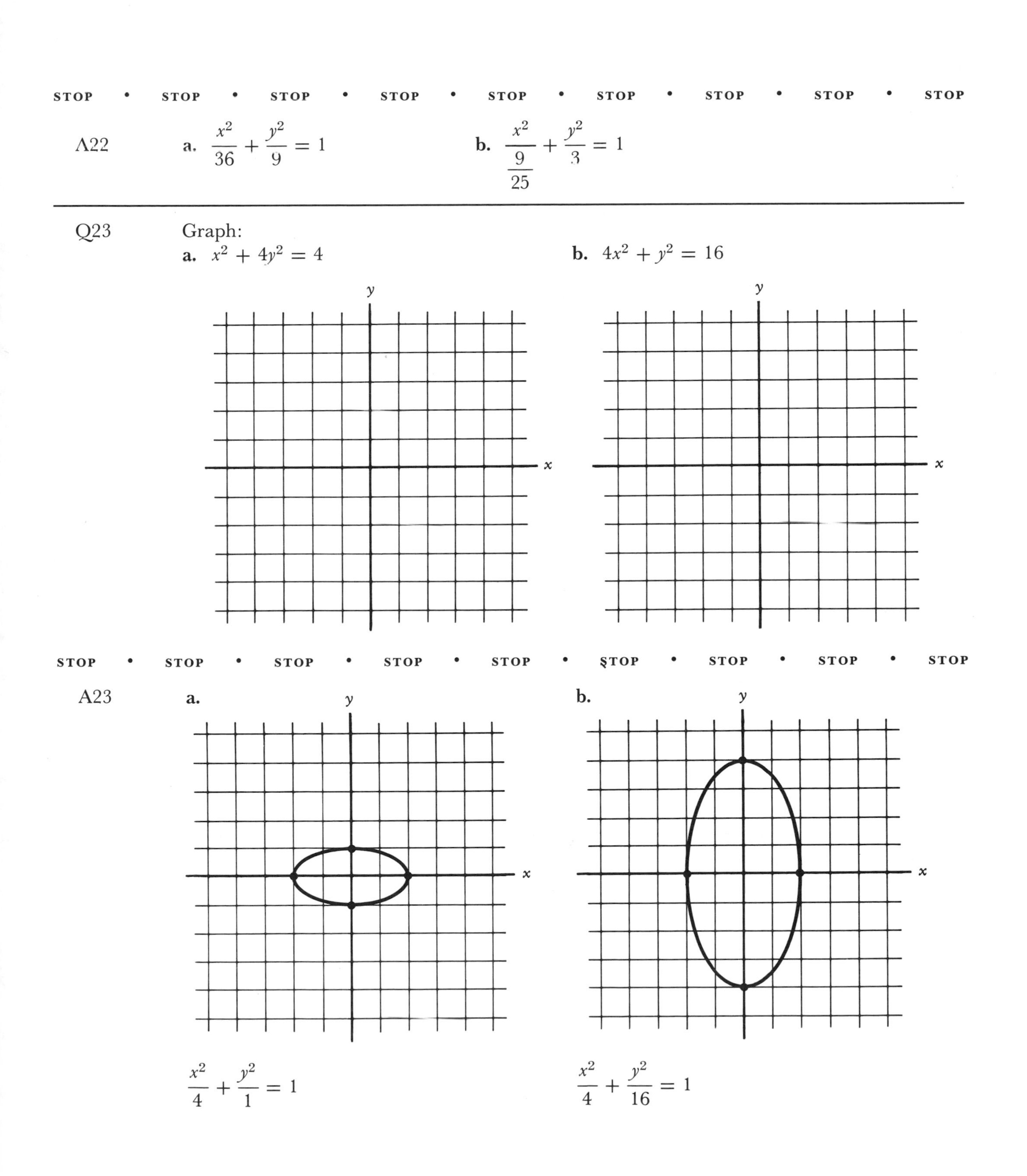

STOP • STOP • STOP • STOP • STOP • STOP • STOP • STOP • STOP

A23 **a.** **b.**

$\dfrac{x^2}{4} + \dfrac{y^2}{1} = 1$ $\dfrac{x^2}{4} + \dfrac{y^2}{16} = 1$

10

A *hyperbola* is the set of all points in a plane such that the difference of their distances from two fixed points (foci) in the plane is a constant. Two standard forms of the hyperbola are

$$\frac{x^2}{a^2} - \frac{y^2}{b^2} = 1$$

$$\frac{y^2}{b^2} - \frac{x^2}{a^2} = 1, \qquad \text{where } a > 0, \quad b > 0$$

The hyperbola can be graphed quickly if the lines $y = \frac{b}{a}x$ and $y = \frac{-b}{a}x$ are first graphed. The farther away the graph is from the origin, the closer it gets to these lines. They are called *asymptotes*. Then find the intercepts, if they exist, and draw curves from the intercepts toward the asymptotes.

Example: Determine the x and y intercepts for $\frac{x^2}{4} - \frac{y^2}{9} = 1$.

Solution

If $y = 0$:

$$\frac{x^2}{4} - \frac{(0)^2}{9} = 1$$

$$x^2 = 4$$

$$x = \pm 2$$

Hence the x intercepts are $(-2, 0)$ and $(2, 0)$. If $x = 0$:

$$\frac{(0)^2}{4} - \frac{y^2}{9} = 1$$

$$-y^2 = 9$$

$$y^2 = -9$$

$$y = \pm\sqrt{-9}$$

$$y = \pm i\sqrt{3}$$

The imaginary result indicates that there are no y intercepts. In general, for $\frac{x^2}{a^2} - \frac{y^2}{b^2} = 1$, the x intercepts $= (-a, 0), (a, 0)$ and there are no y intercepts.

Q24 Find the x and y intercepts:

a. $\frac{x^2}{4} - \frac{y^2}{25} = 1$

x intercepts = __________

y intercepts = __________

b. $\frac{x^2}{6} - \frac{y^2}{9} = 1$

x intercepts = __________

y intercepts = __________

STOP • STOP • STOP • STOP • STOP • STOP • STOP • STOP • STOP

A24 a. $(-2, 0), (2, 0)$
no y intercepts

b. $(-\sqrt{6}, 0), (\sqrt{6}, 0)$
no y intercepts

11 The hyperbola is graphed by first graphing the asymptotes $y = \frac{b}{a}x$ and $y = \frac{-b}{a}x$, and then drawing curves from the intercepts toward the asymptotes.

Example: Graph $\frac{x^2}{4} - \frac{y^2}{9} = 1$.

Solution

$a = 2$, $b = 3$. First graph $y = \frac{3}{2}x$ and $y = \frac{-3}{2}x$. The asymptotes are graphed as dashed lines because they are merely an aid in graphing the hyperbola. The x intercepts are $(-2, 0)$ and $(2, 0)$.

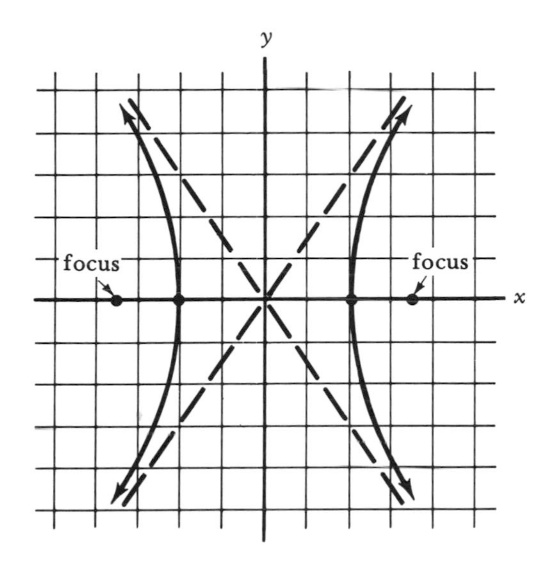

Q25 Graph the asymptotes for $\frac{x^2}{9} - \frac{y^2}{4} = 1$.

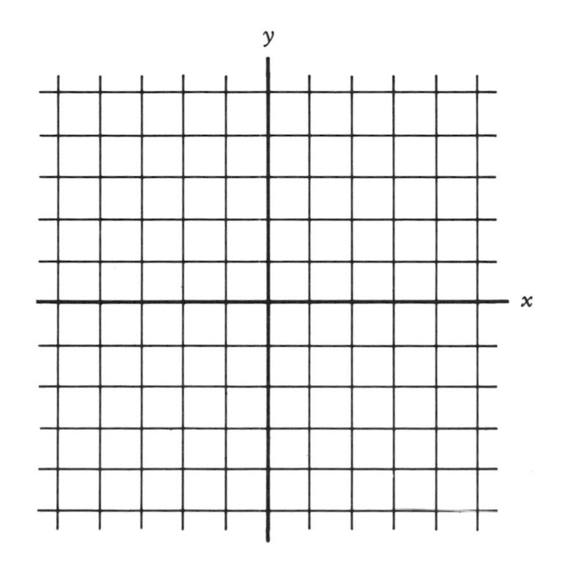

STOP • STOP • STOP • STOP • STOP • STOP • STOP • STOP • STOP

A25 $a = 3,\ b = 2$

$$y = \frac{-b}{a}x \quad \text{and} \quad y = \frac{b}{a}x$$

$$y = \frac{-2}{3}x \quad \text{and} \quad y = \frac{2}{3}x$$

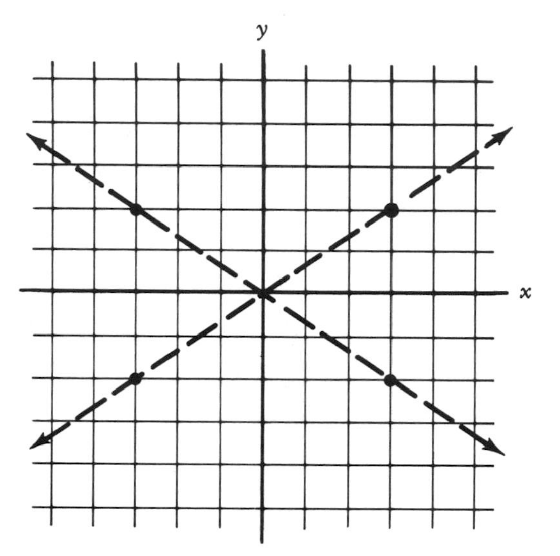

Q26 Graph $\frac{x^2}{9} - \frac{y^2}{4} = 1$.

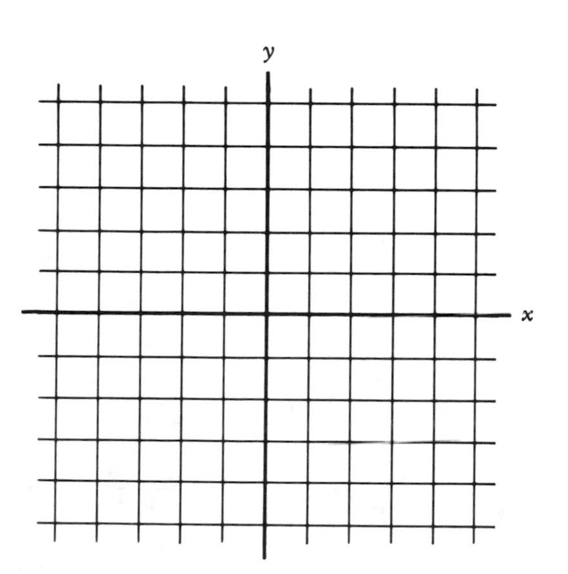

STOP • STOP • STOP • STOP • STOP • STOP • STOP • STOP • STOP

A26 x intercepts $= (-3, 0)$ and $(3, 0)$

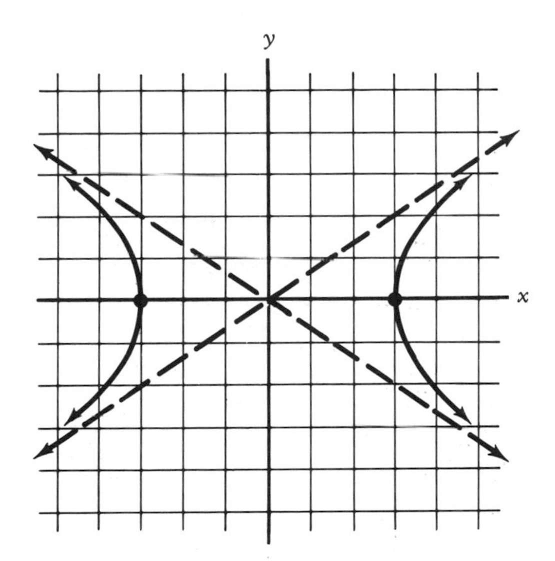

Q27 Graph:

a. $\dfrac{x^2}{4} - \dfrac{y^2}{1} = 1$

b. $4x^2 - y^2 = 4$

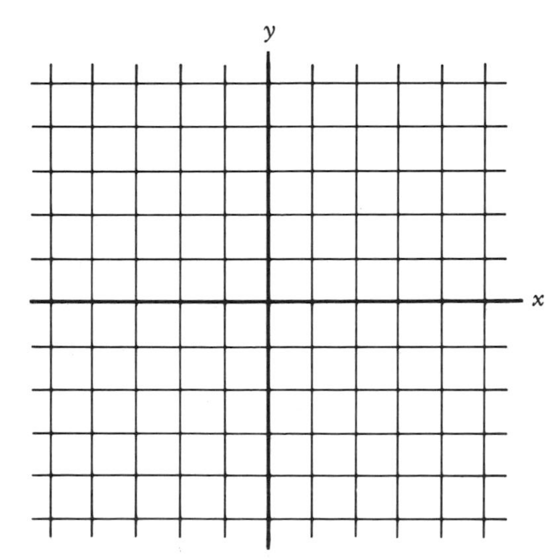

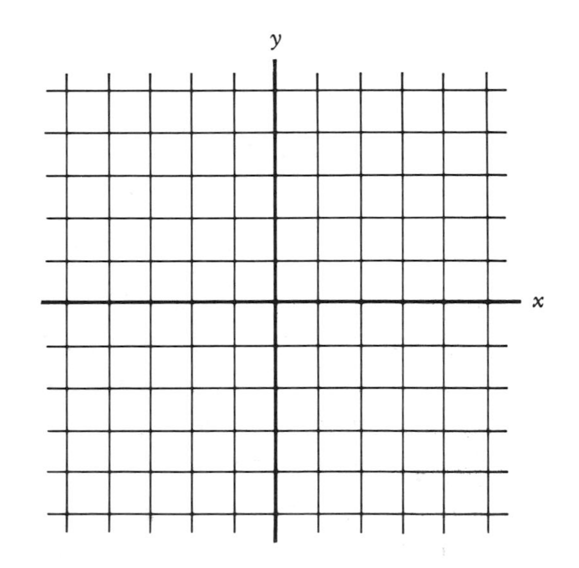

STOP • STOP • STOP • STOP • STOP • STOP • STOP • STOP • STOP

A27

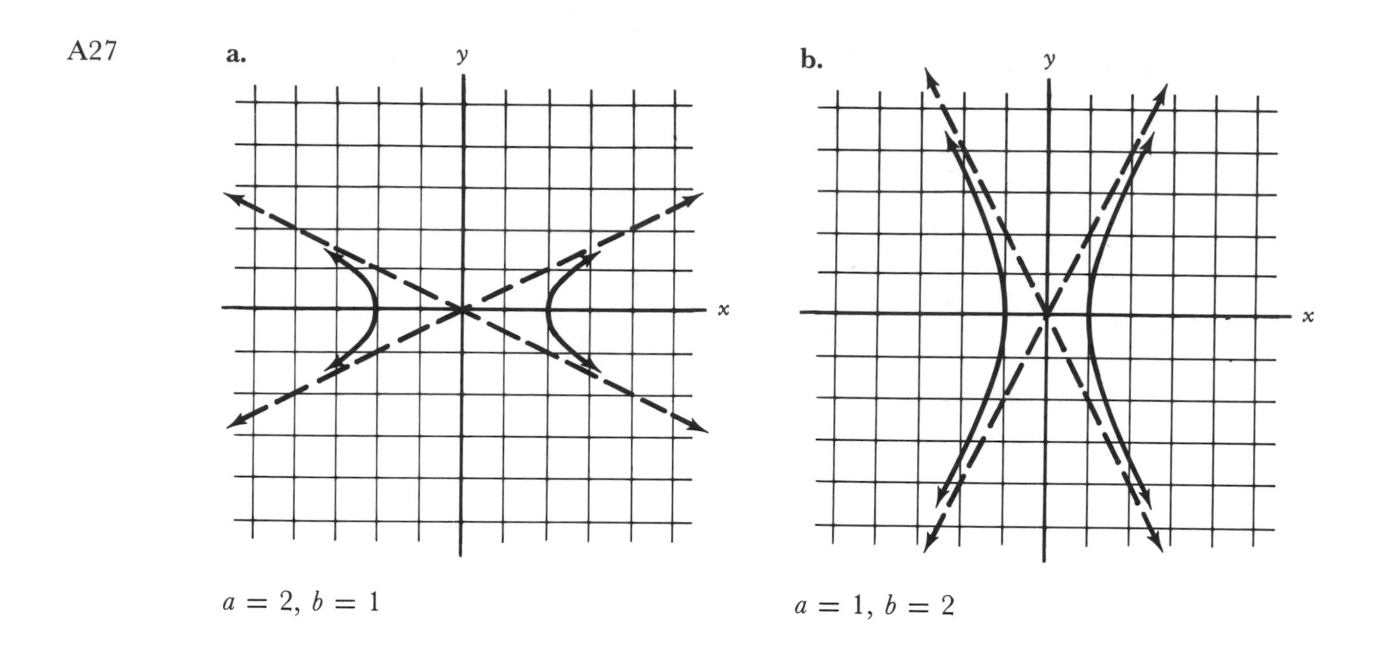

12 Hyperbolas of the form $\frac{y^2}{b^2} - \frac{x^2}{a^2} = 1$ also have asymptotes $y = \frac{-b}{a}x$ and $y = \frac{b}{a}x$; however, the y intercepts are $(0, -b)$ and $(0, b)$. There are no x intercepts.

Example: Graph

$$\frac{y^2}{4} - \frac{x^2}{9} = 1.$$

Solution

$a = 3 \qquad b = 2$

$y = \frac{-2}{3}x$ and $y = \frac{2}{3}x$ (asymptotes)

y intercepts $= (0, -2)$ and $(0, 2)$

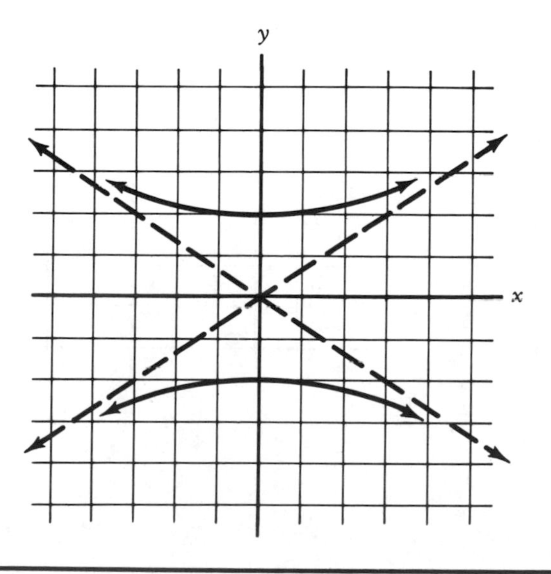

Q28 Graph:

a. $\frac{y^2}{4} - x^2 = 1$

b. $y^2 - x^2 = 1$

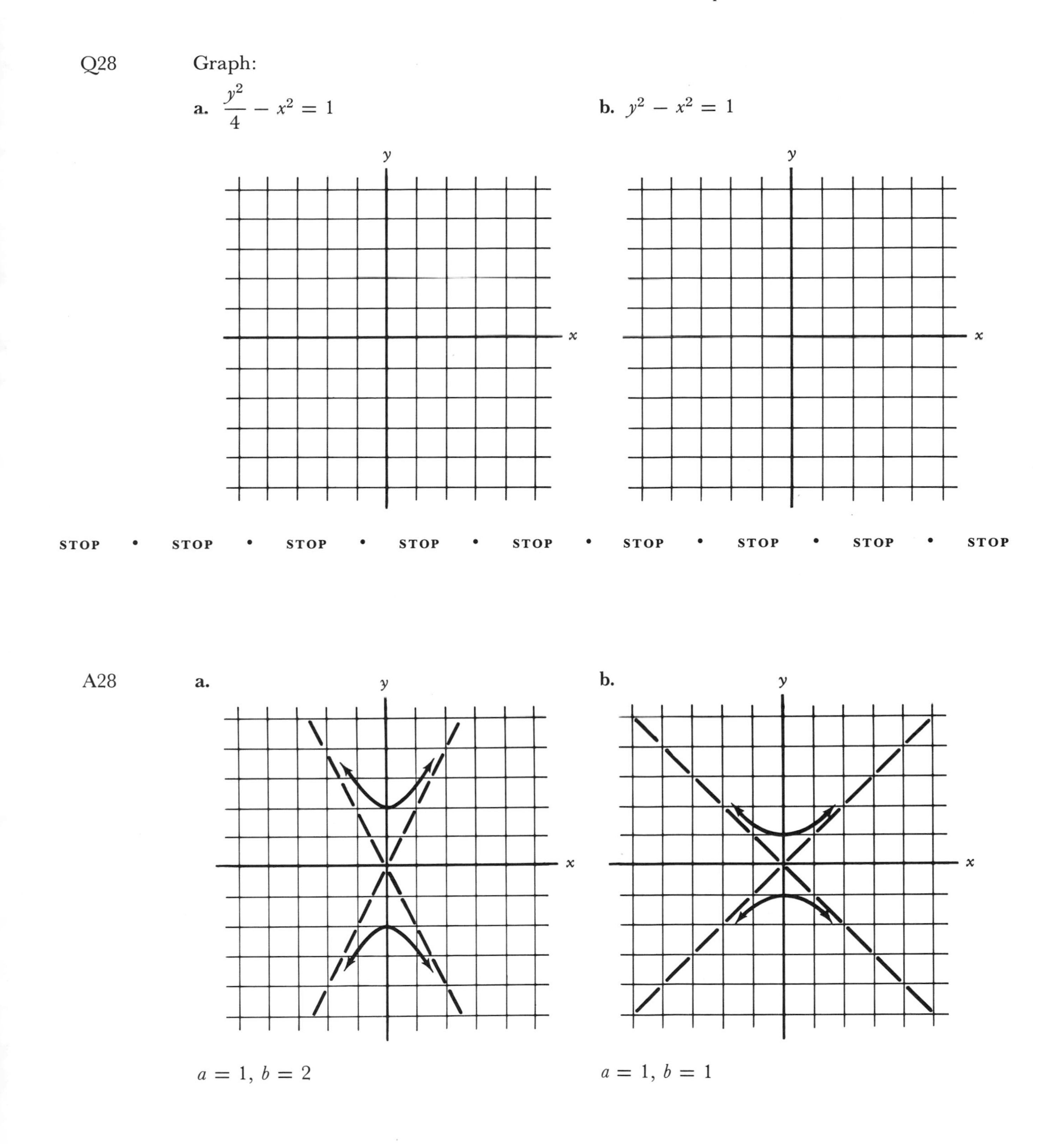

STOP • STOP • STOP • STOP • STOP • STOP • STOP • STOP • STOP

A28 a.

$a = 1,\ b = 2$

b.

$a = 1,\ b = 1$

---

13 The *parabola* is the set of all points in a plane that are equidistant from a fixed point (*focus*) and a fixed line (*directrix*) in the plane. The parabola was presented in detail in Section 6.5 as the graph of a quadratic function. However, all parabolas do not represent functions.

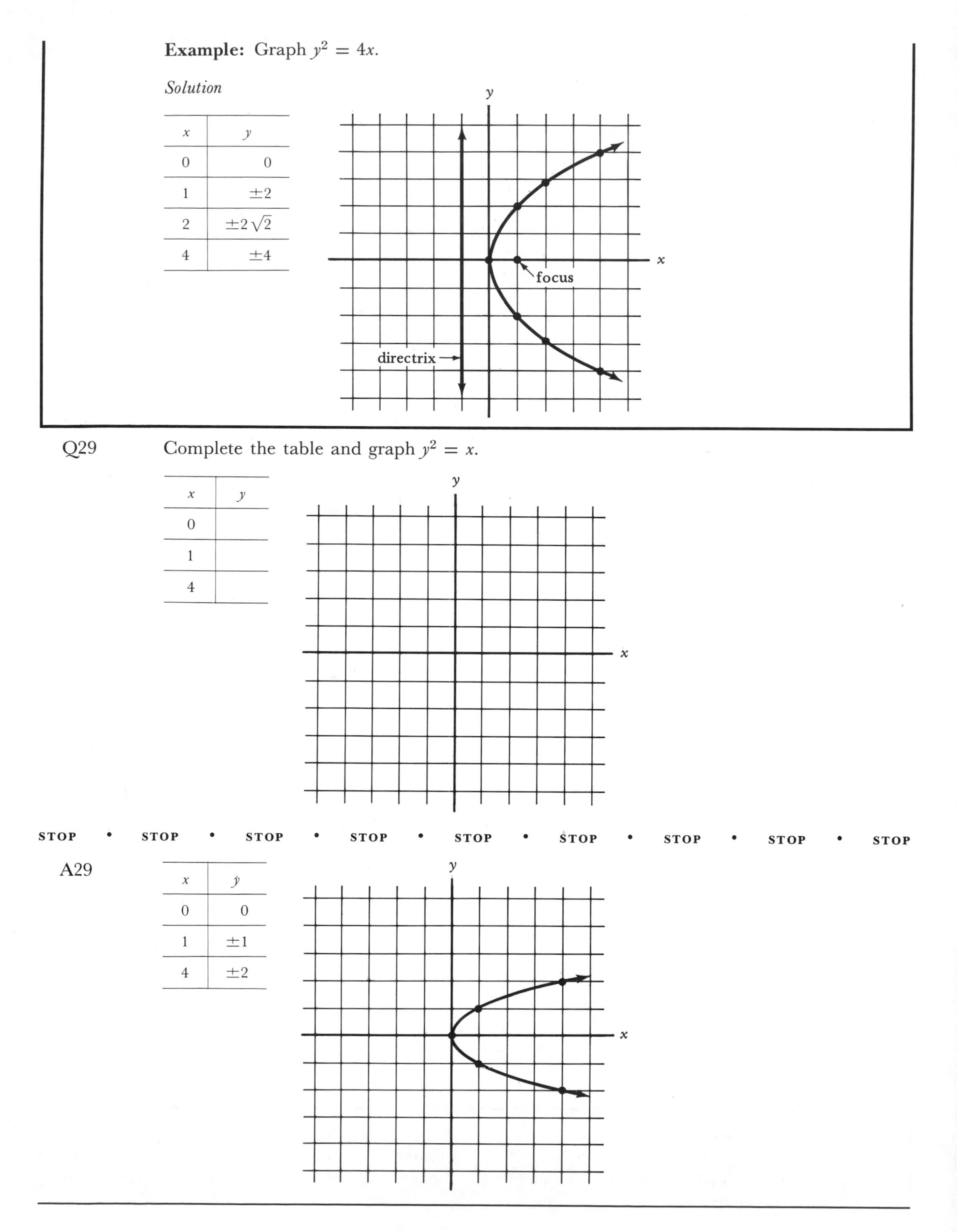

Example: Graph $y^2 = 4x$.

Solution

x	y
0	0
1	± 2
2	$\pm 2\sqrt{2}$
4	± 4

Q29 Complete the table and graph $y^2 = x$.

x	y
0	
1	
4	

STOP • STOP • STOP • STOP • STOP • STOP • STOP • STOP • STOP

A29

x	y
0	0
1	± 1
4	± 2

Q30 Graph:

a. $y^2 = -x$ **b.** $y^2 = -2x$

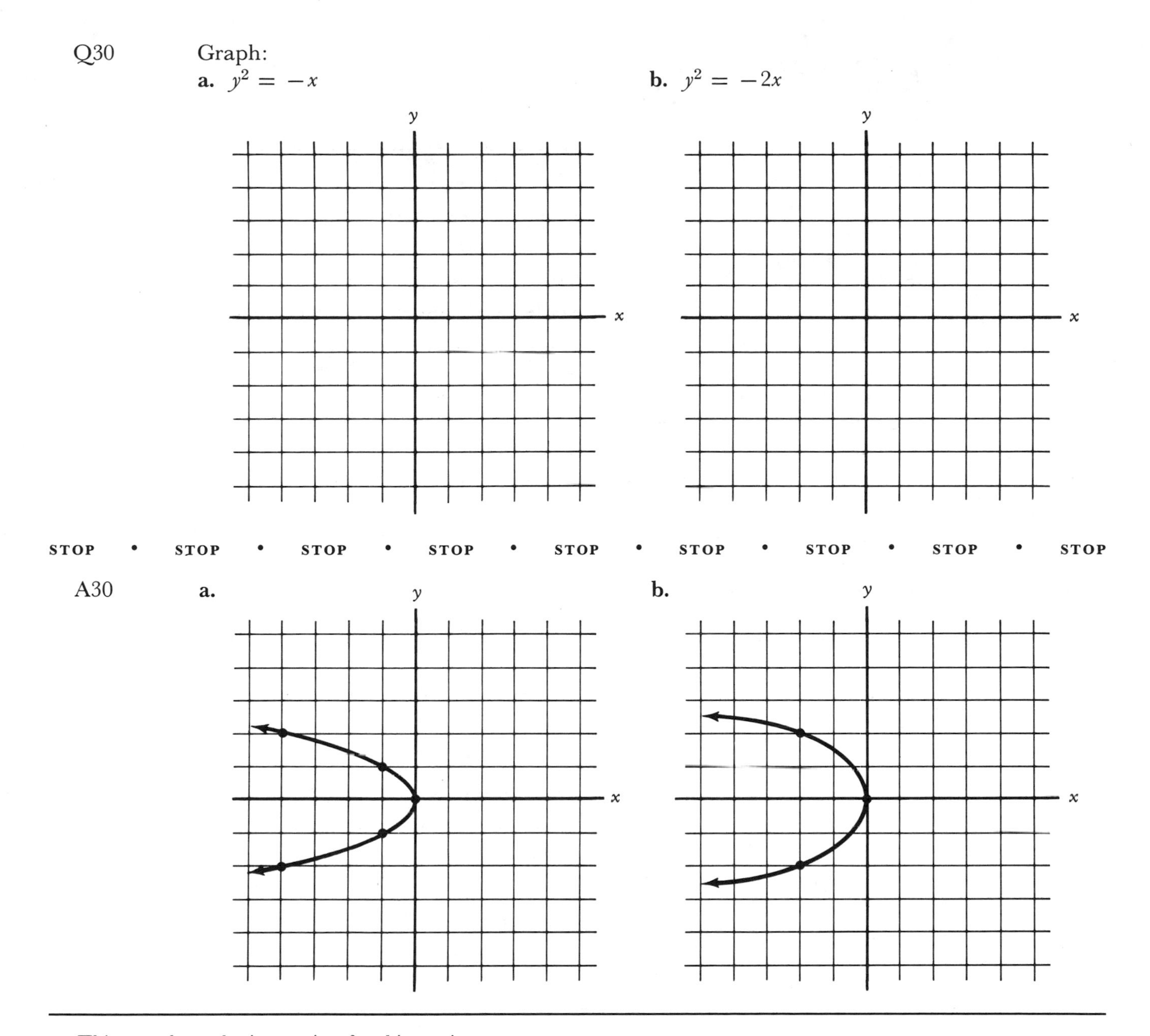

STOP • STOP • STOP • STOP • STOP • STOP • STOP • STOP • STOP

A30 **a.** **b.**

This completes the instruction for this section.

8.1 Exercises

1. Graph:
   a. $f(x) = -x^3 + 1$ **b.** $f(x) = x^3 - 1$
2. Graph:
   a. $f(x) = \frac{-1}{x}$ **b.** $f(x) = \frac{2}{x}$
3. Graph:
   a. $x^2 + y^2 = 9$ **b.** $x^2 + y^2 = 5$ **c.** $x^2 - y^2 = 9$
   d. $xy = 4$ **e.** $\frac{x^2}{16} + \frac{y^2}{4} = 1$ **f.** $x^2 + 2y^2 = 8$

g. $\frac{x^2}{16} - \frac{y^2}{4} = 1$ h. $\frac{y^2}{4} - \frac{x^2}{16} = 1$ i. $y^2 = -4x$

j. $y = 2x^2$

4. Translate each equation of variation into an appropriate equation that uses k as the constant of variation:
   a. n varies directly as v.
   b. G varies directly as the square of t.
   c. I varies inversely as t.
   d. Geologists have found in studies of earth erosion that the erosive force P of a swiftly flowing stream varies directly as the sixth power of the velocity v of the water.
5. The intensity of illumination E at a given point varies inversely as the square of the distance d that the point is from the light source. If the illumination is 50 footcandles at a point 2 feet from a light, what will be the illumination at a point 5 feet from the same light? At what distance will the illumination be 100 footcandles?
6. The distance that a particle falls in a certain medium is directly proportional to the square of the length of time it falls. If the particle falls 16 meters in 2 seconds, how far will it fall in 10 seconds?
7. The pressure exerted by a liquid at a given point varies directly as the depth of the point beneath the surface of the liquid. If a certain liquid exerts a pressure of 40 kilograms per square meter at a depth of 10 meters, what would be the pressure at 40 meters?
8. The volume of a gas varies inversely as the pressure and directly as the temperature. If a certain gas occupies a volume of 1.3 liters at 300 K and a pressure of 18 kilograms per square centimeter, find the volume at 340 K and a pressure of 24 kilograms per square centimeter.

8.1 Exercise Answers

1. a. b.

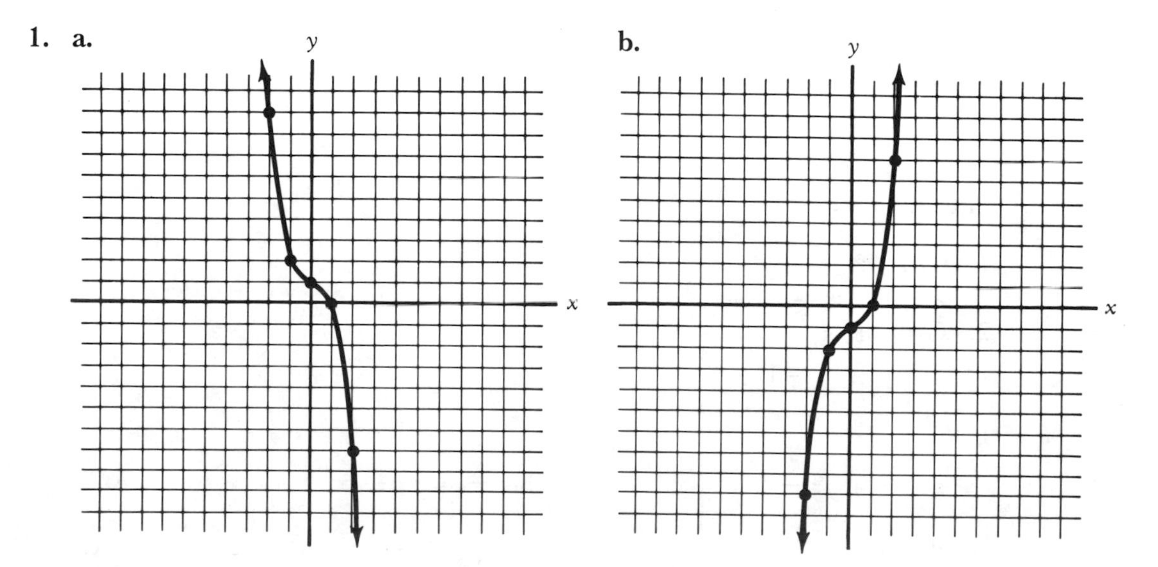

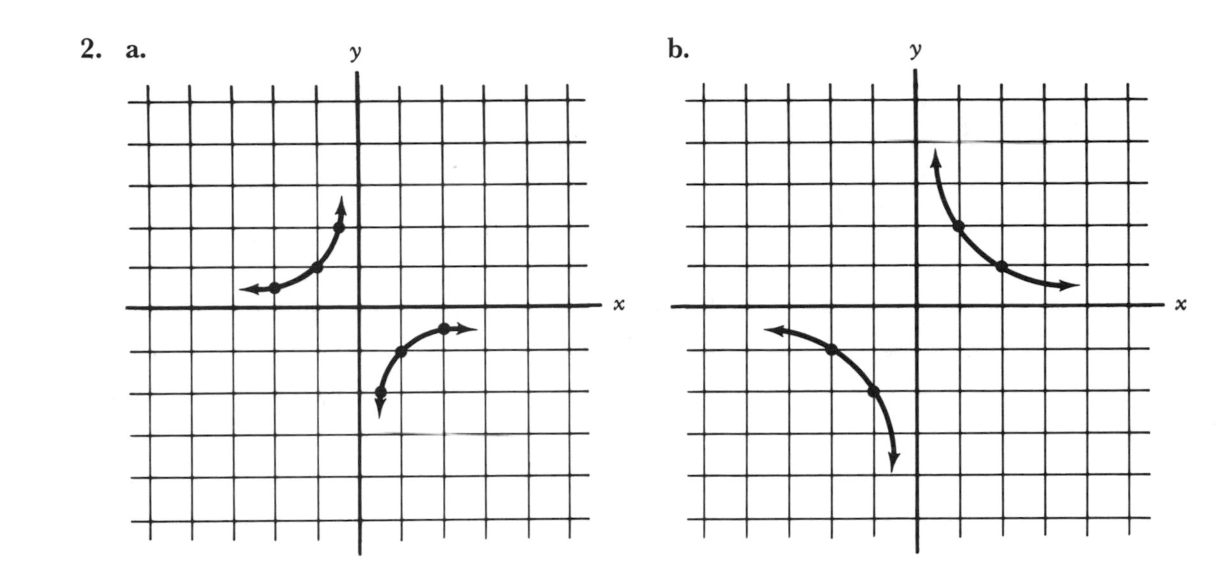
2. a.
y
x
b.
y
x

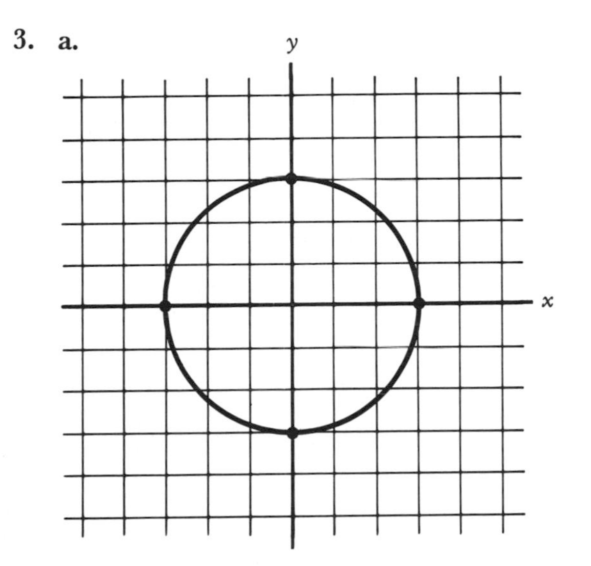
3. a.
y
x

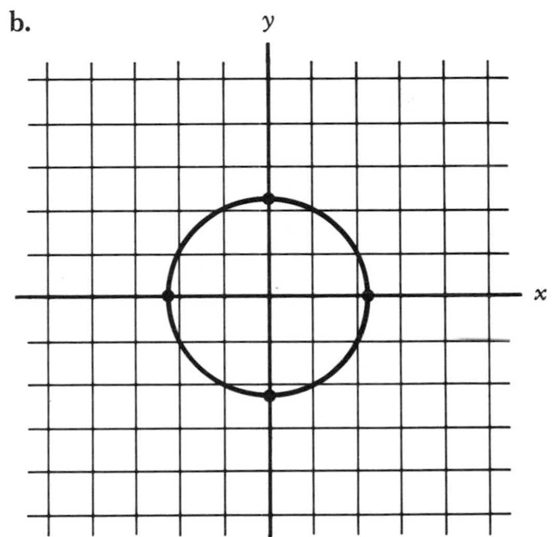
b.
y
x

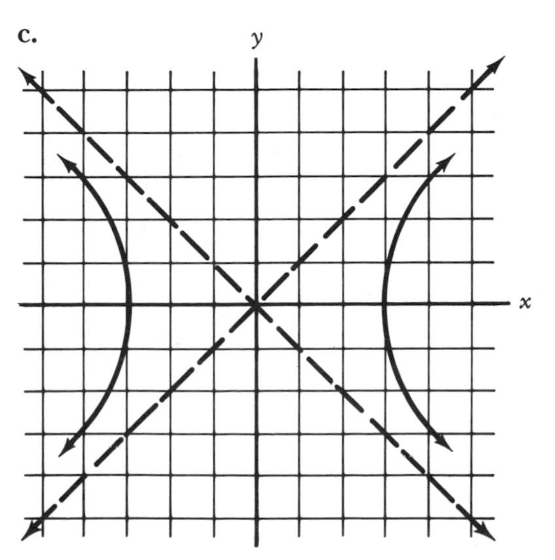
c.
y
x

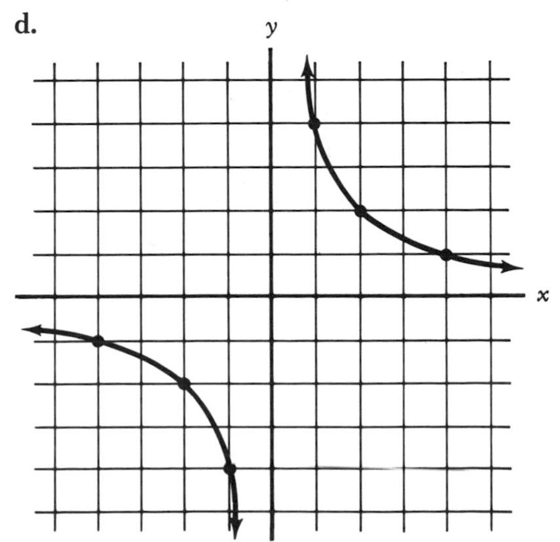
d.
y
x

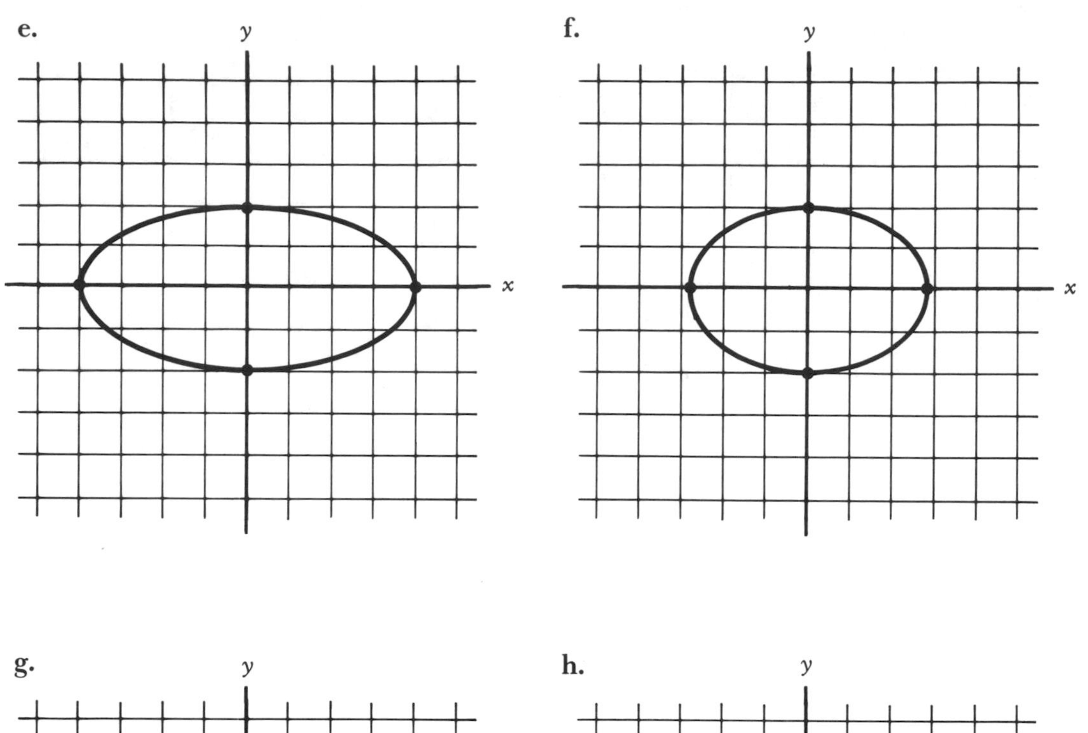
e.
y
x
f.
y
x

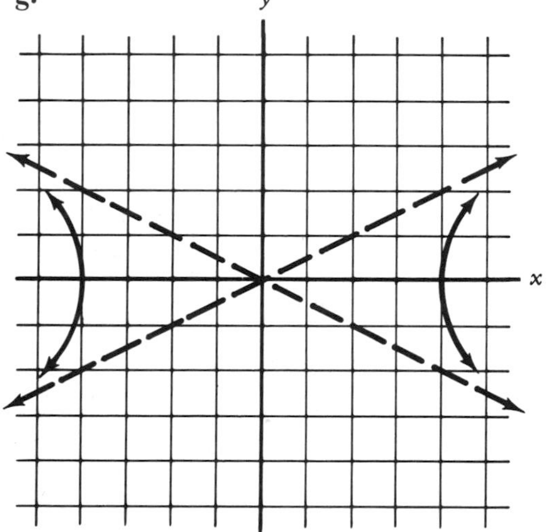
g.
y
x

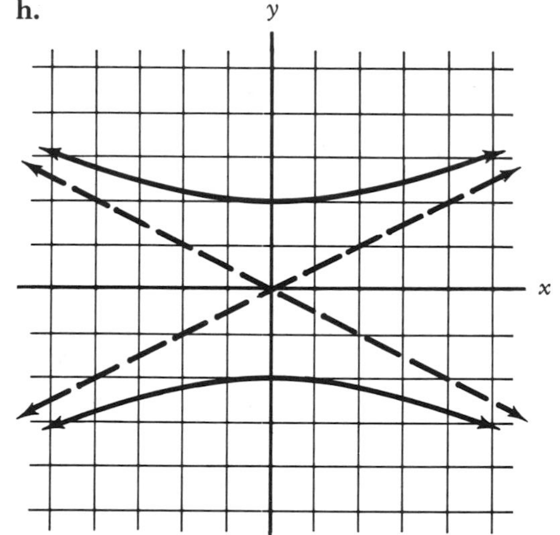
h.
y
x

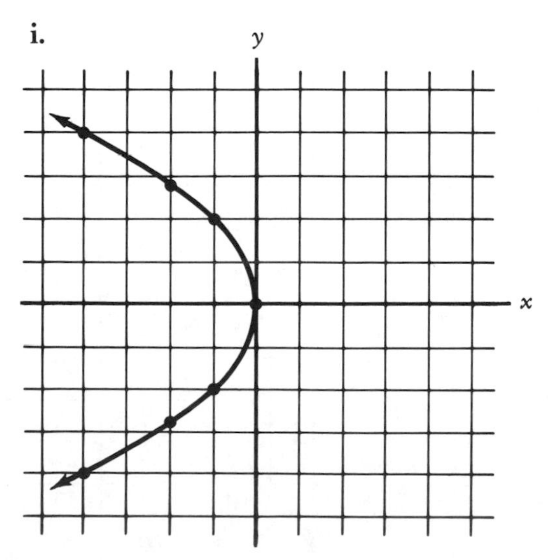
i.
y
x

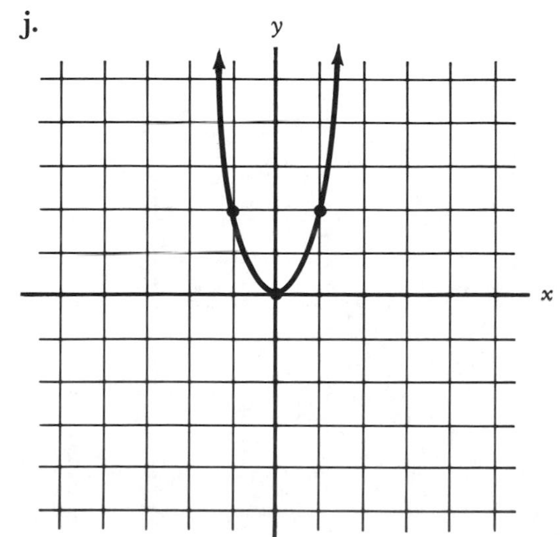
j.
y
x

4. **a.** $n = kv$ **b.** $G = kt^2$ **c.** $I = \frac{k}{t}$ **d.** $P = kv^6$

5. 8 footcandles; $\sqrt{2}$ feet
6. 400 meters
7. 160 kilograms per square meter
8. 1.105 liters

8.2 Exponential and Logarithmic Functions

1 A function of the form $f(x) = b^x$, where $b > 0$ and $b \neq 1$, is called an *exponential function*. For example, $f(x) = 2^x$, $f(x) = \left(\frac{1}{3}\right)^x$, and $f(x) = 4^x$ are exponential functions.

Example: Graph $f(x) = 2^x$.

Solution

The function can be graphed by plotting a few points and connecting them with a smooth curve. More advanced techniques are needed to show that $(\sqrt{3}, 2^{\sqrt{3}})$ and the like are actual points of the graph. However, it will be assumed that the domain is all real numbers and the graph is a smooth and unbroken curve.

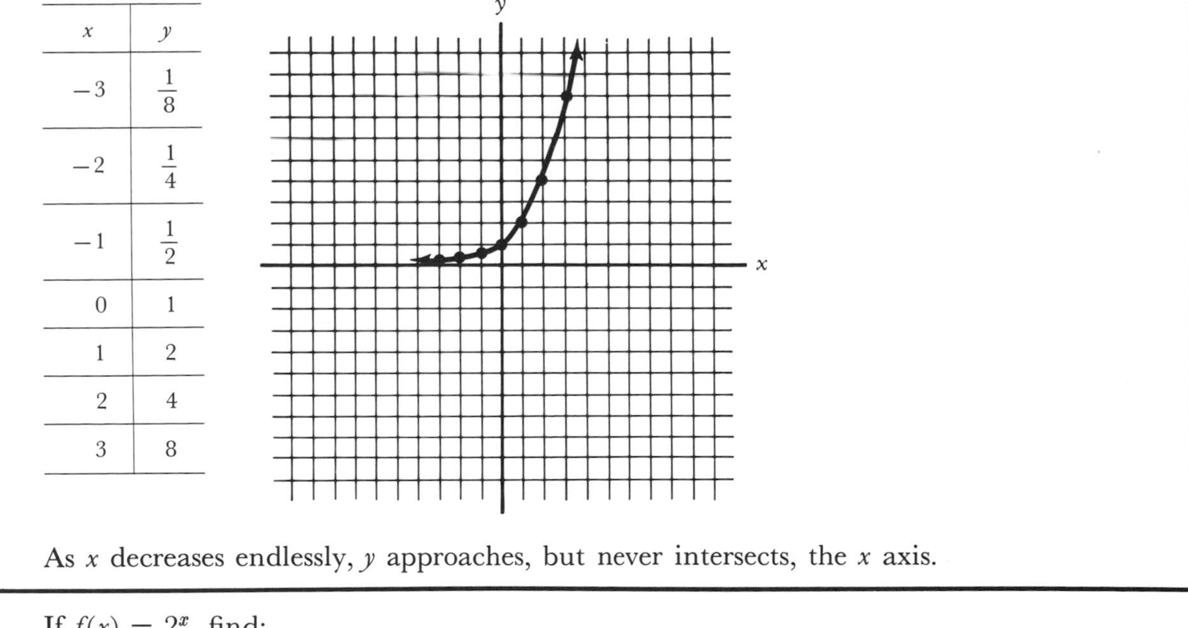

x	y
-3	$\frac{1}{8}$
-2	$\frac{1}{4}$
-1	$\frac{1}{2}$
0	1
1	2
2	4
3	8

As x decreases endlessly, y approaches, but never intersects, the x axis.

Q1 If $f(x) = 2^x$, find:

a. $f(4)$ **b.** $f(-5)$ **c.** $f\left(\frac{1}{2}\right)$

STOP • STOP • STOP • STOP • STOP • STOP • STOP • STOP • STOP

A1

a. 16: $f(4) = 2^4$

b. $\frac{1}{32}$: $f(-5) = 2^{-5} = \frac{1}{32}$

c. $\sqrt{2}$: $f\left(\frac{1}{2}\right) = 2^{1/2} = \sqrt{2}$

Q2

If $f(x) = 3^x$, find:

a. $f(0)$

b. $f(-1)$

c. $f(3)$

STOP • STOP • STOP • STOP • STOP • STOP • STOP • STOP • STOP

A2

a. 1: $f(0) = 3^0$

b. $\frac{1}{3}$: $f(-1) = 3^{-1} = \frac{1}{3}$

c. 27: $f(3) = 3^3 = 27$

Q3

If $f(x) = 3^x$, find:

a. $f(-2)$

b. $f(5)$

c. $f(8)$

STOP • STOP • STOP • STOP • STOP • STOP • STOP • STOP • STOP

A3

a. $\frac{1}{9}$: $f(-2) = 3^{-2} = \frac{1}{3^2} = \frac{1}{9}$

b. 243: $f(5) = 3^5 = 243$

c. 6,561

Q4

Complete the table and graph:

a. $f(x) = 3^x$

x	y
-1	
0	
1	
2	

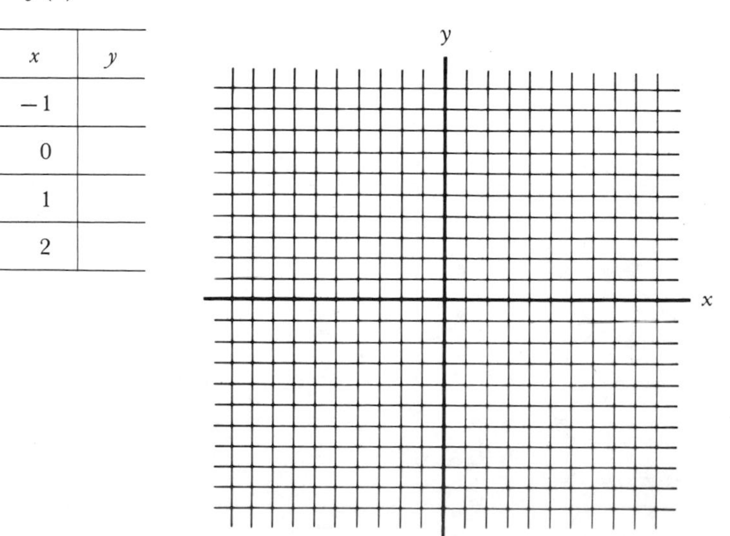

b. 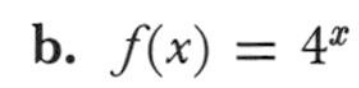
$f(x) = 4^x$

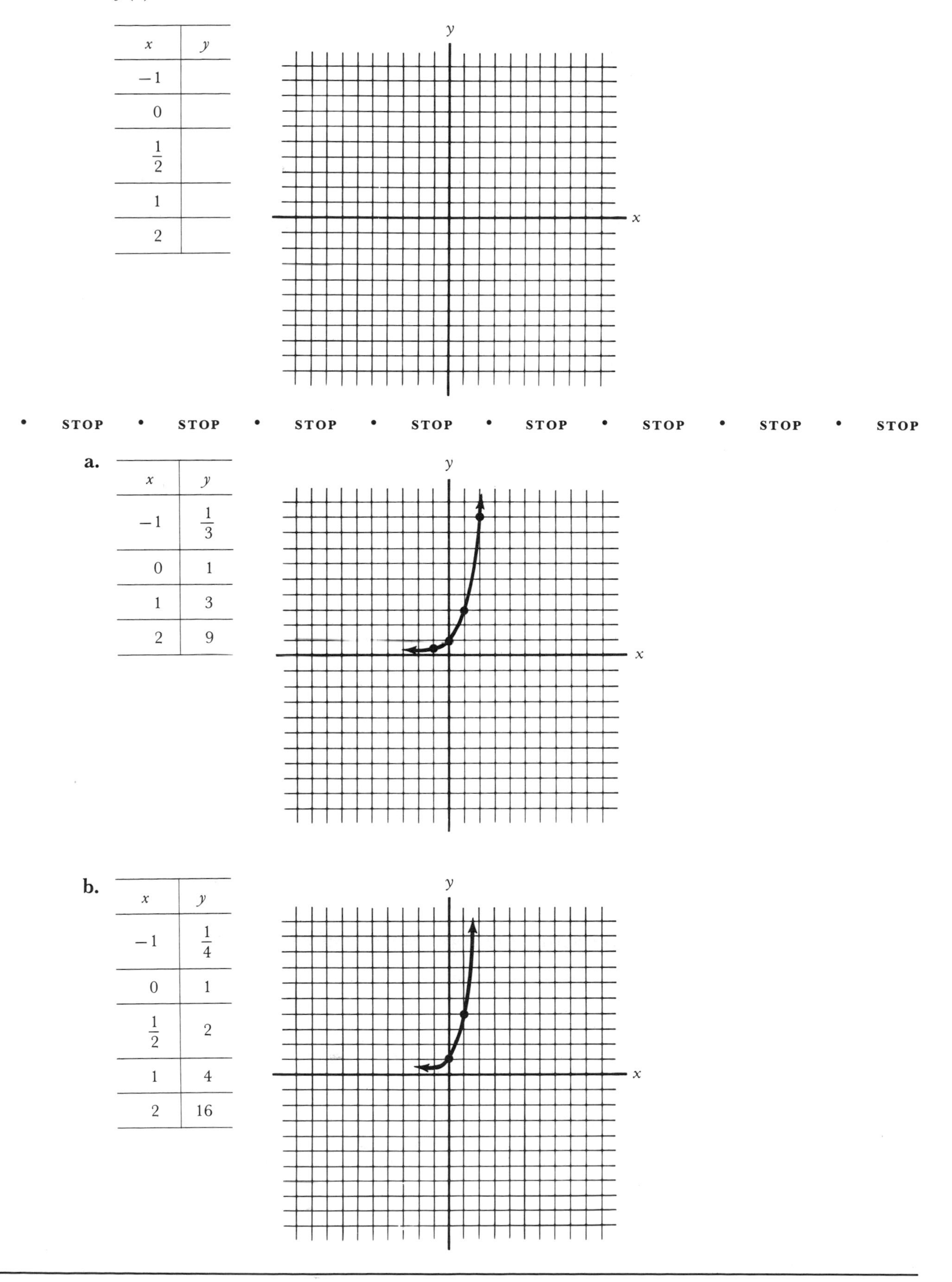

x	y
-1	
0	
$\frac{1}{2}$	
1	
2	

STOP • STOP • STOP • STOP • STOP • STOP • STOP • STOP • STOP

A4 **a.**

x	y
-1	$\frac{1}{3}$
0	1
1	3
2	9

b.

x	y
-1	$\frac{1}{4}$
0	1
$\frac{1}{2}$	2
1	4
2	16

Q5 If $f(x) = \left(\frac{1}{2}\right)^x$, find:

a. $f(-2)$ b. $f(0)$ c. $f(3)$

STOP • STOP • STOP • STOP • STOP • STOP • STOP • STOP • STOP

A5 a. 4: $f(-2) = \left(\frac{1}{2}\right)^{-2} = \frac{1^{-2}}{2^{-2}} = 2^2$

b. 1: $f(0) = \left(\frac{1}{2}\right)^0 = 1$

c. $\frac{1}{8}$: $f(3) = \left(\frac{1}{2}\right)^3 = \frac{1^3}{2^3} = \frac{1}{8}$

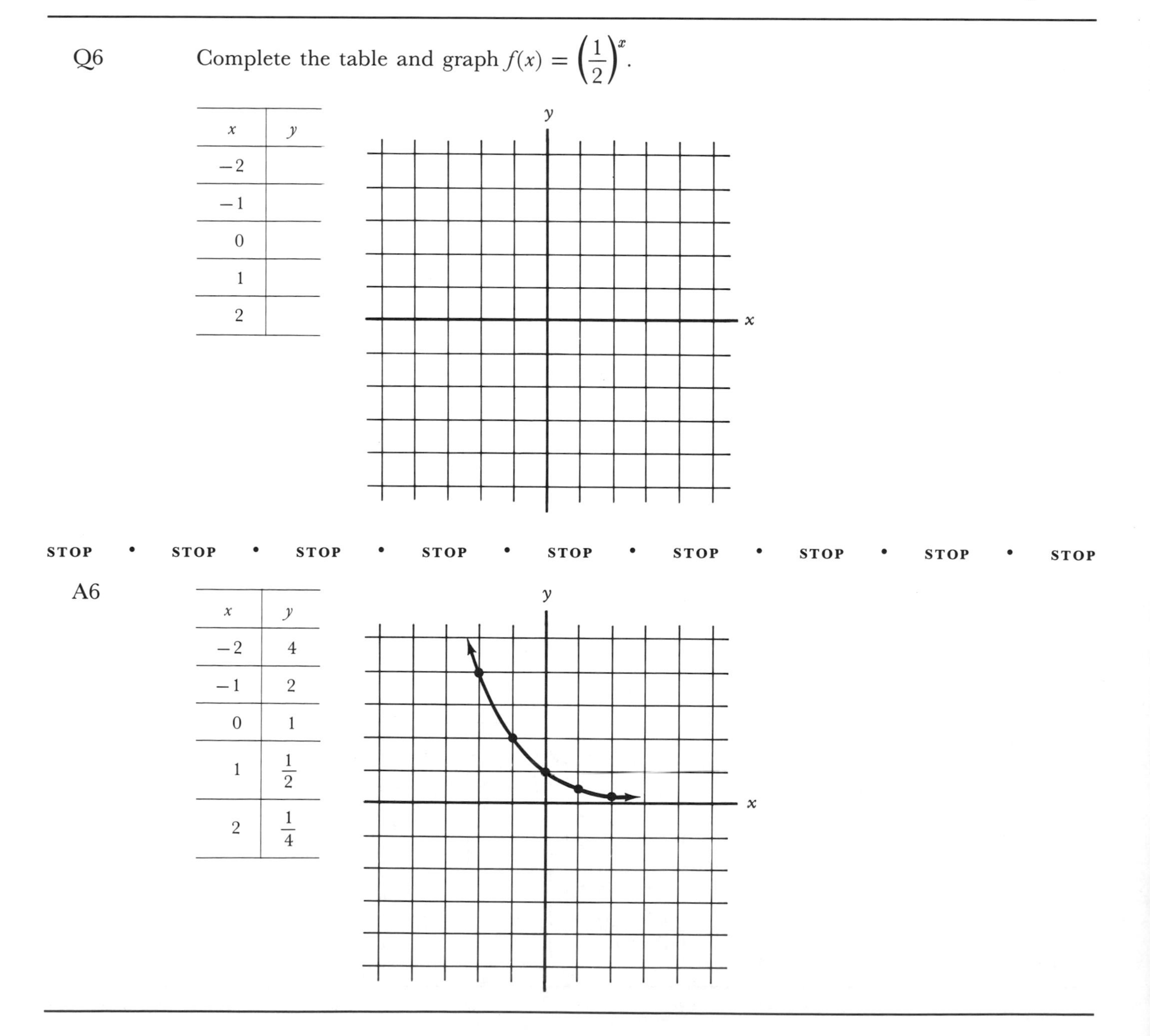

Q6 Complete the table and graph $f(x) = \left(\frac{1}{2}\right)^x$.

x	y
-2	
-1	
0	
1	
2	

STOP • STOP • STOP • STOP • STOP • STOP • STOP • STOP • STOP

A6

x	y
-2	4
-1	2
0	1
1	$\frac{1}{2}$
2	$\frac{1}{4}$

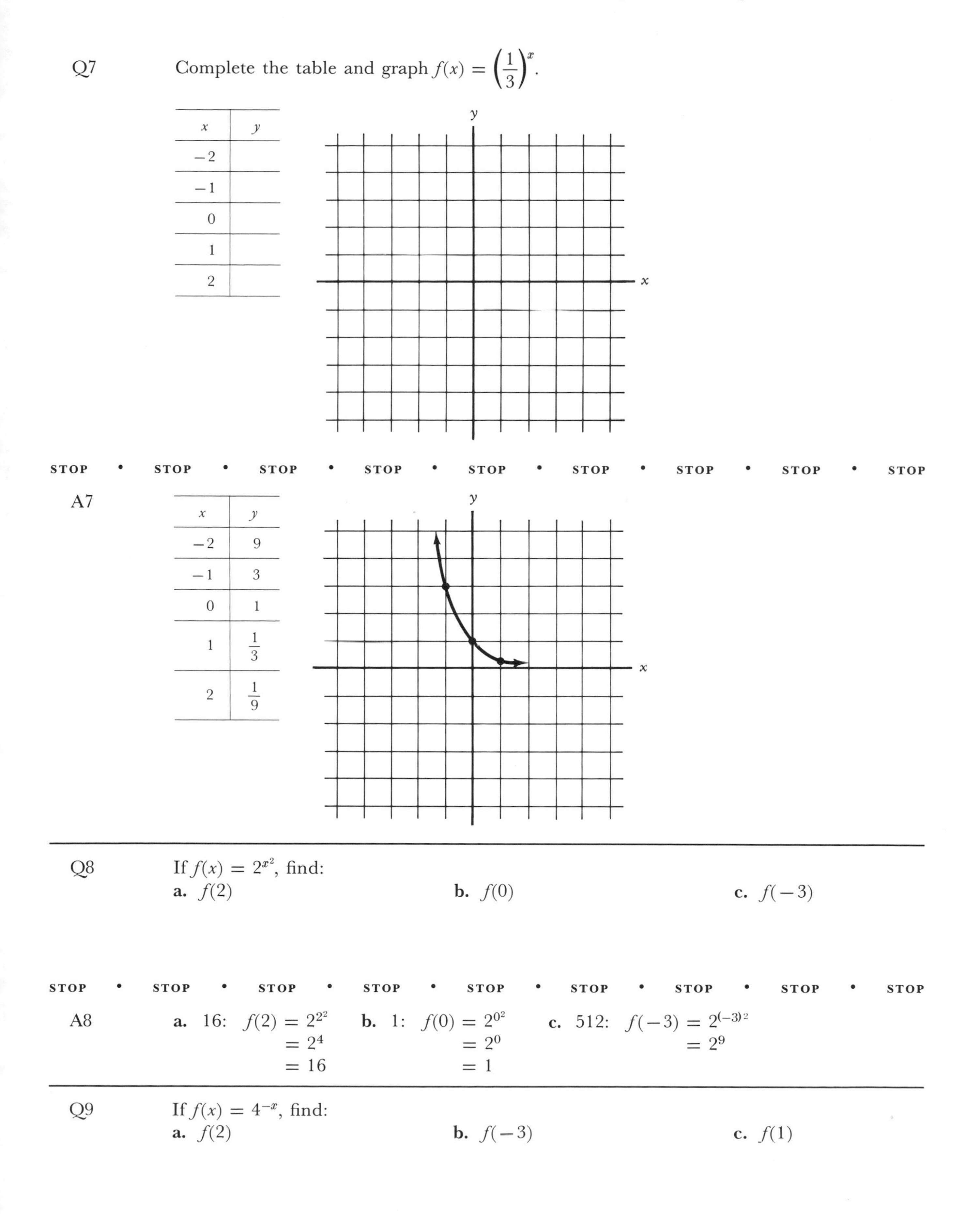

Q7 Complete the table and graph $f(x) = \left(\frac{1}{3}\right)^x$.

x	y
-2	
-1	
0	
1	
2	

STOP • STOP • STOP • STOP • STOP • STOP • STOP • STOP • STOP

A7

x	y
-2	9
-1	3
0	1
1	$\frac{1}{3}$
2	$\frac{1}{9}$

Q8 If $f(x) = 2^{x^2}$, find:

a. $f(2)$ **b.** $f(0)$ **c.** $f(-3)$

STOP • STOP • STOP • STOP • STOP • STOP • STOP • STOP • STOP

A8 **a.** 16: $f(2) = 2^{2^2} = 2^4 = 16$ **b.** 1: $f(0) = 2^{0^2} = 2^0 = 1$ **c.** 512: $f(-3) = 2^{(-3)^2} = 2^9$

Q9 If $f(x) = 4^{-x}$, find:

a. $f(2)$ **b.** $f(-3)$ **c.** $f(1)$

STOP • STOP • STOP • STOP • STOP • STOP • STOP • STOP • STOP

A9 **a.** $\frac{1}{16}$: $f(2) = 4^{-2}$
$= \frac{1}{4^2}$

b. 64: $f(-3) = 4^{-(-3)}$
$= 4^3$

c. $\frac{1}{4}$

2 You can see from the previous examples that for exponential functions, relatively small changes in x produce very large changes in y. This is what is meant by exponential growth. For example, it is said that the earth's population is growing at an exponential rate. This means that in the next few years the population will increase rapidly (unless checked by other means). Malthus's law states that while the population is growing exponentially, the food supply is increasing linearally.

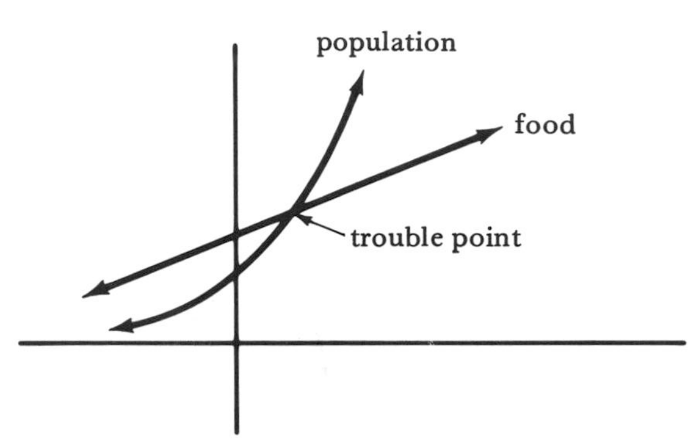

Example: Bacteria in a pure nutrient solution tend to increase according to the law

$$N = N_0 2^{kt}$$

where t is time, k a constant, and N_0 the number of bacteria present at $t = 0$. If initially 10^4 bacteria are present and $k = 1$, find the number of bacteria at the end of 2 hours.

Solution

$N_0 = 10^4 \quad k = 1, \quad t = 2$

$$\begin{aligned} N &= N_0 2^{kt} \\ &= 10^4 2^{1(2)} \\ &= 10^4(2^2) \\ &= 40{,}000 \text{ bacteria} \end{aligned}$$

Q10 In an experiment, the number of bacteria present after t days is given by $N = N_0 3^t$, where N_0 represents the number of bacteria present when $t = 0$. If $N_0 = 30{,}000$, find the bacteria present after 3 days.

STOP • STOP • STOP • STOP • STOP • STOP • STOP • STOP • STOP

A10 810,000: $N = 30{,}000(3)^3$

Q11 Galton's law of heredity states that the influence of ancestors on an individual is as follows: each parent, $\frac{1}{2}$; each grandparent, $\left(\frac{1}{2}\right)^2$; each great-grandparent, $\left(\frac{1}{2}\right)^3$; and so on. How much hereditary influence would an ancestor eight generations back have on you?

STOP • STOP • STOP • STOP • STOP • STOP • STOP • STOP • STOP

A11 $\left(\frac{1}{2}\right)^8 \doteq 0.0039$: very little influence

3 The inverse of an exponential function defined by $f = \{(x, y)|y = b^x\}$ is formed by "reversing" the coordinates of the ordered pairs of f. That is, the inverse of f is defined by

$f^{-1} = \{(x, y)|x = b^y\}$

where $b > 0$ and $b \neq 1$. The inverse of an exponential function is called a *logarithmic function.*

Example: Find the inverse of $f = \{(x, y)|y = 2^x\}$.

Solution

$\{(x, y)|x = 2^y\}$

Q12 Find f^{-1}:

a. $f = \{(x, y)|y = 3^x\}$

b. $f(x) = 5^x$

STOP • STOP • STOP • STOP • STOP • STOP • STOP • STOP • STOP

A12 a. $f^{-1} = \{(x, y)|x = 3^y\}$

b. $f^{-1} = \{(x, y)|x = 5^y\}$: $f(x) = 5^x$ implies that $y = 5^x$

Q13 Find f^{-1}:

a. $f = \left\{(x, y)|y = \left(\frac{1}{3}\right)^x\right\}$

b. $f(x) = \left(\frac{1}{2}\right)^x$

STOP • STOP • STOP • STOP • STOP • STOP • STOP • STOP • STOP

A13 a. $f^{-1} = \left\{(x, y)|x = \left(\frac{1}{3}\right)^y\right\}$ b. $f^{-1} = \left\{(x, y)|x = \left(\frac{1}{2}\right)^y\right\}$

4 The logarithmic function $\{(x, y)|x = b^y\}$ can be graphed by choosing values of y to determine corresponding values of x.

Example: Graph $x = 2^y$ (set builder notation is not necessary).

Solution

Let $y = -2, -1, 0, 1, 2, 3$, and so on, and find the corresponding x values. Then plot the ordered pairs and connect them with a smooth curve.

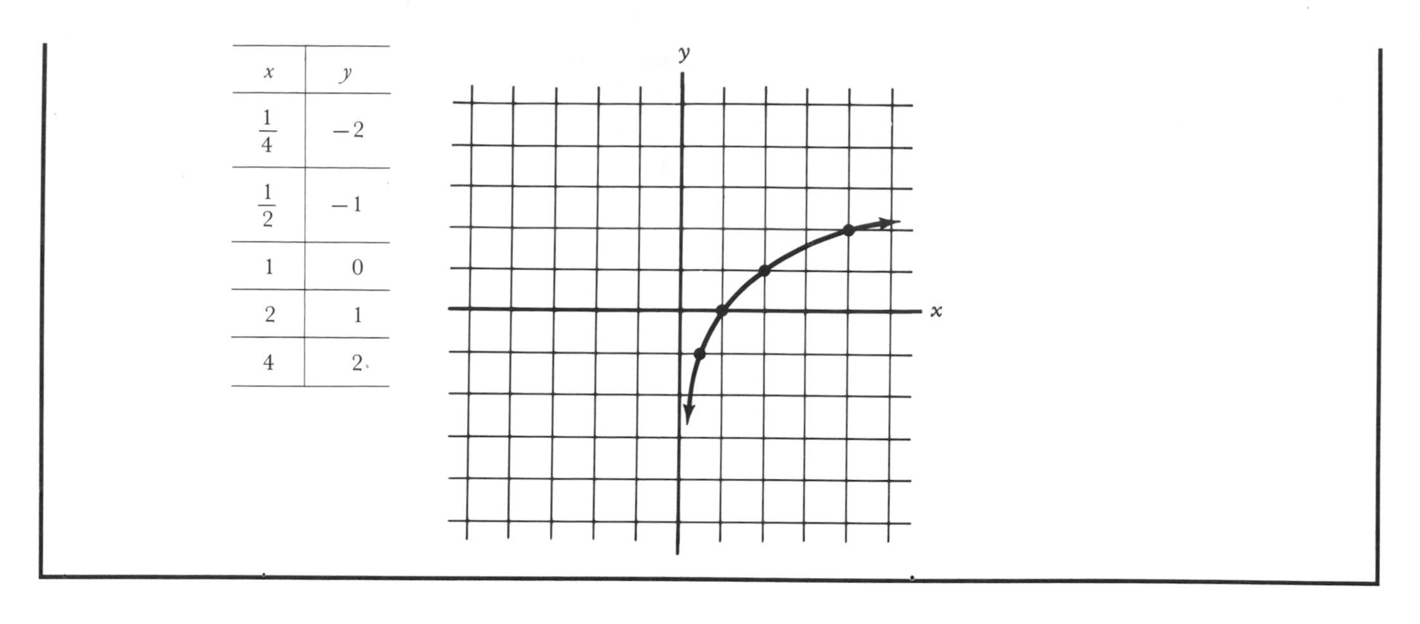

x	y
$\frac{1}{4}$	-2
$\frac{1}{2}$	-1
1	0
2	1
4	2

Q14 Complete the table and graph $x = 3^y$.

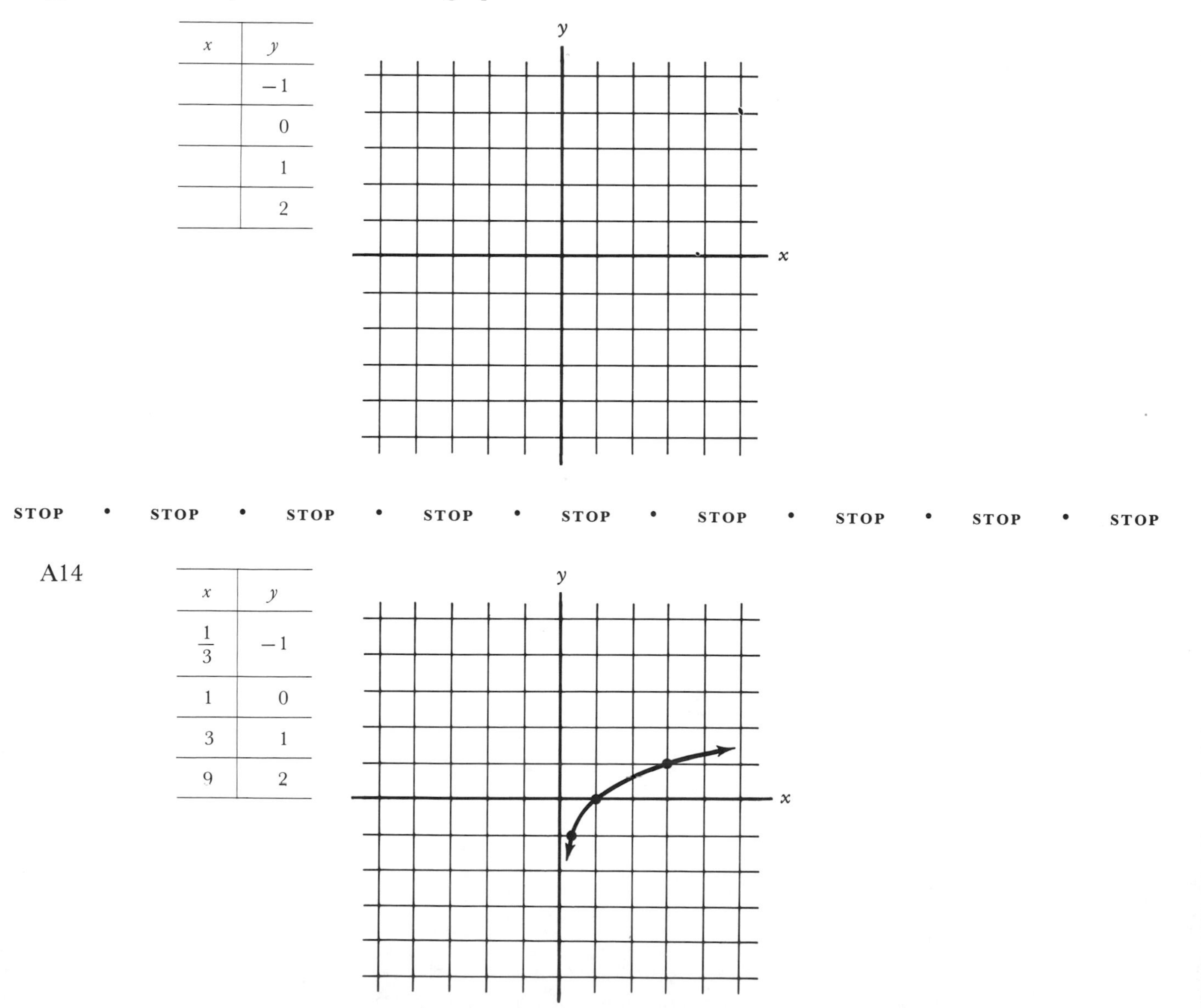

x	y
	-1
	0
	1
	2

STOP • STOP • STOP • STOP • STOP • STOP • STOP • STOP • STOP

A14

x	y
$\frac{1}{3}$	-1
1	0
3	1
9	2

Q15 Complete the table and graph $x = \left(\frac{1}{2}\right)^y$.

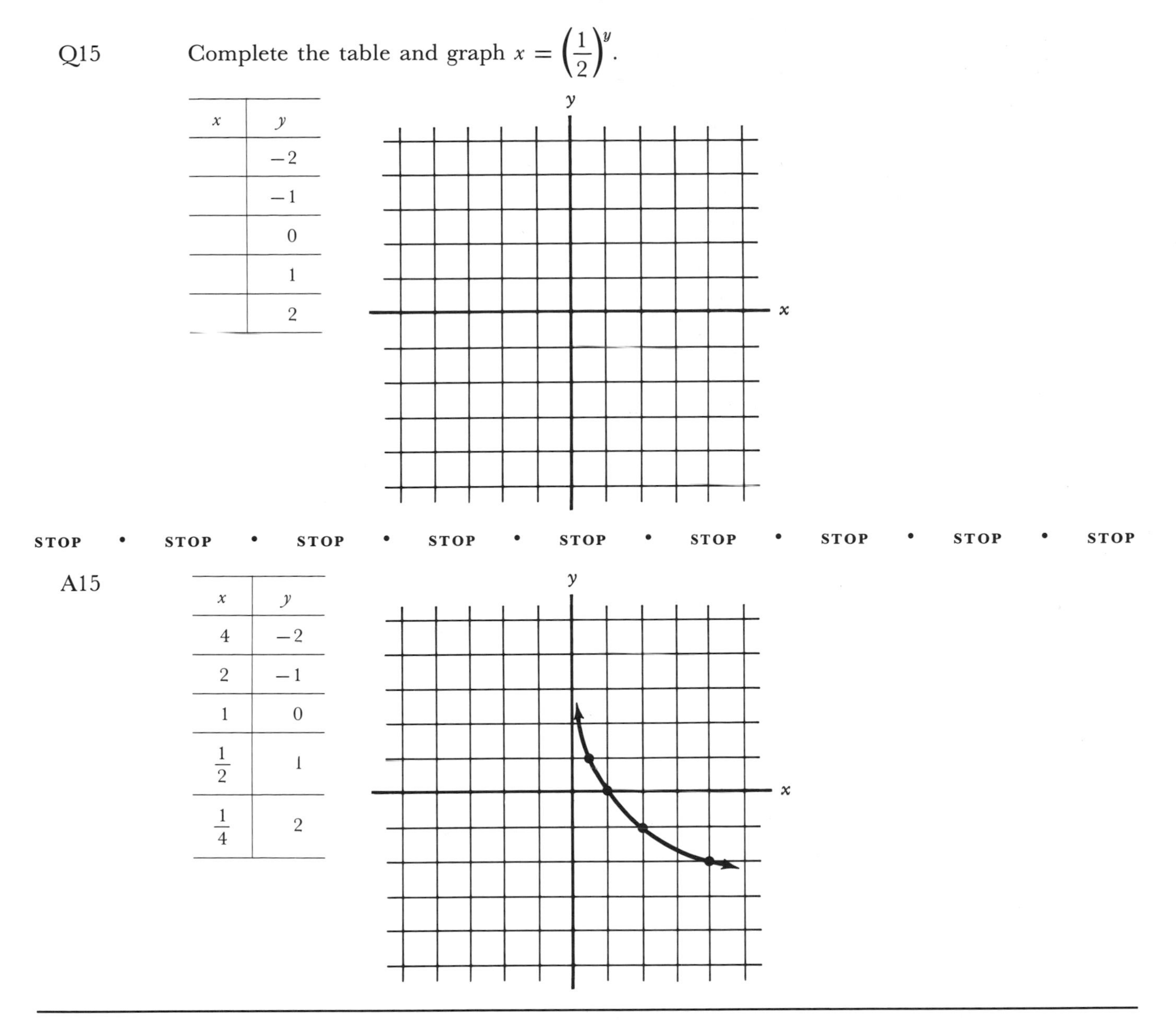

x	y
	-2
	-1
	0
	1
	2

STOP • STOP • STOP • STOP • STOP • STOP • STOP • STOP • STOP

A15

x	y
4	-2
2	-1
1	0
$\frac{1}{2}$	1
$\frac{1}{4}$	2

Q16 Graph $f(x) = 2^x$ and f^{-1} on the same coordinate system (the line $y = x$ is graphed dashed).

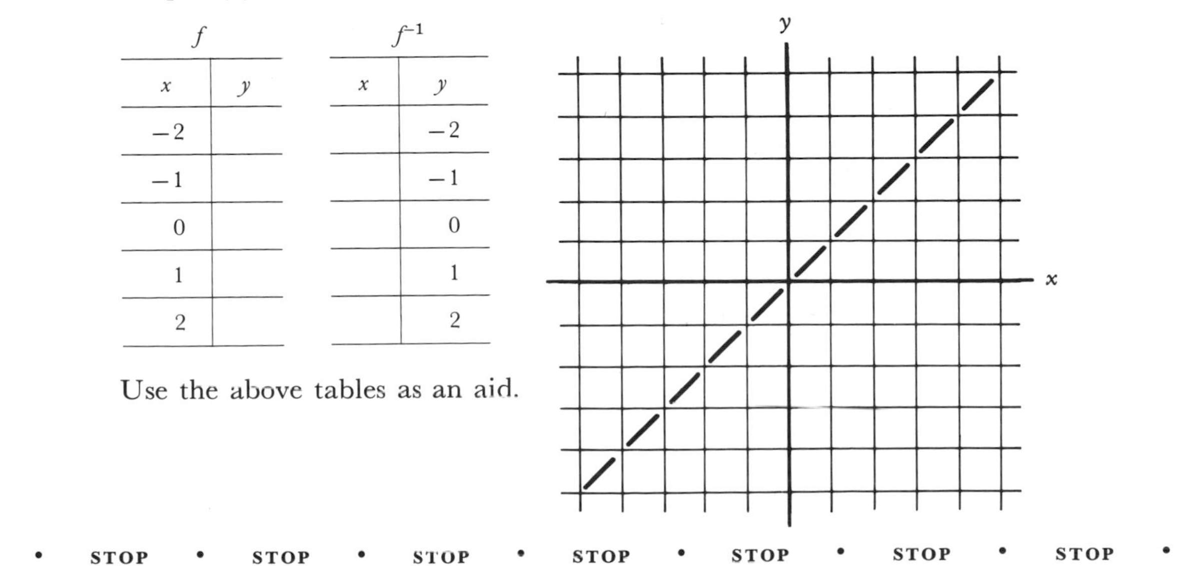

f

x	y
-2	
-1	
0	
1	
2	

f^{-1}

x	y
	-2
	-1
	0
	1
	2

Use the above tables as an aid.

STOP • STOP • STOP • STOP • STOP • STOP • STOP • STOP • STOP

A16

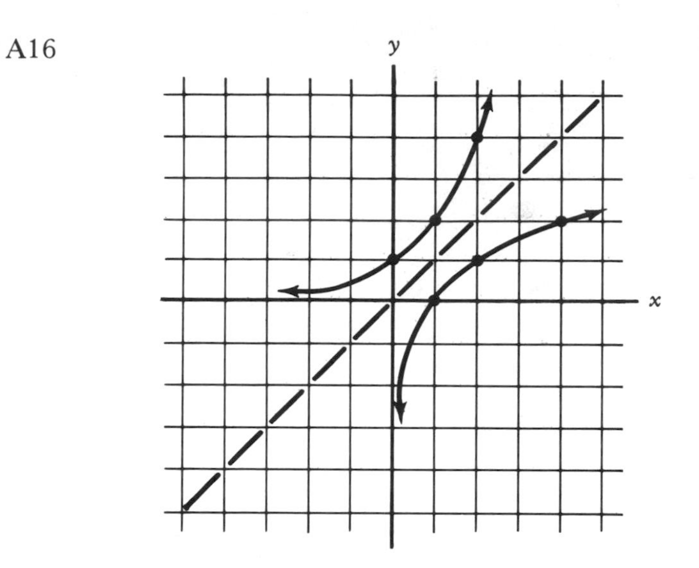

5	Q16 illustrates the fact that an exponential function and its inverse logarithmic function are symmetric to the line $y = x$. Recall from Section 6.6 on inverse functions that f and f^{-1} will always be symmetric to the line $y = x$ for any function f.

Q17 Graph $f(x) = \left(\frac{1}{2}\right)^x$ and f^{-1} on the same coordinate system.

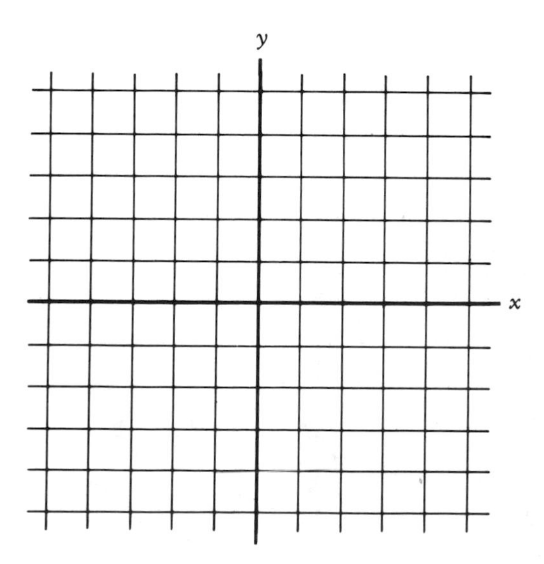

STOP • STOP • STOP • STOP • STOP • STOP • STOP • STOP • STOP

A17

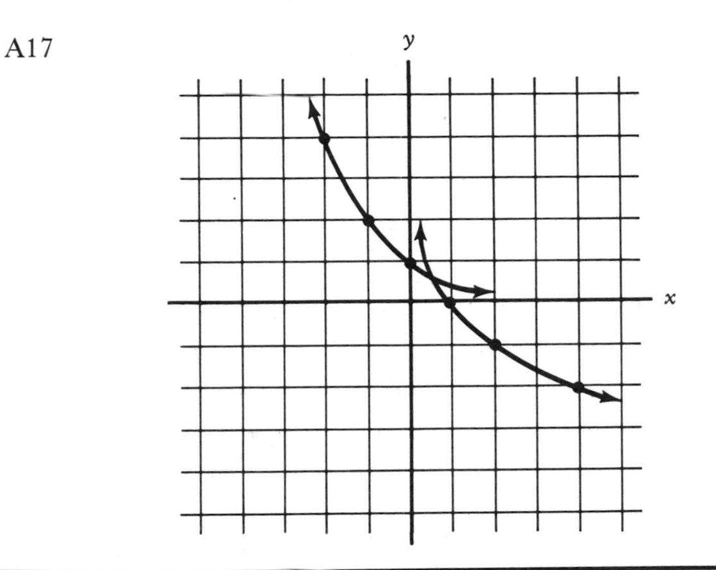

6

To express the logarithmic function $g = \{(x, y) | x = b^y\}$ by use of function notation, y must be written in terms of x. This is accomplished using logarithmic notation as follows:

$x = b^y$ is equivalent to $y = \log_b x$

read "y equals the log of x to the base b." The function g can now be expressed as $g(x) = \log_b x$.

Note that y is an exponent in the equation $x = b^y$ and y is the logarithm, to the base b, of x; hence it is said that *logarithms are exponents.*

Example: Express the inverse of $f(x) = 2^x$ in exponential form and logarithmic form.

Solution

Function	Inverse in exponential form	Inverse in logarithmic form
$f(x) = 2^x$ or $y = 2^x$	$x = 2^y$	$y = \log_2 x$

Q18

Express the inverse of f in exponential form and logarithmic form:

		Exponential form	Logarithmic form
a. $f(x) = 3^x$	f^{-1}:	$x = 3^y$	$y = \log_{(\ \ )}(\ \ \ \ )$
b. $f(x) = 4^x$	f^{-1}:	________	________
c. $f(x) = \left(\frac{1}{2}\right)^x$	f^{-1}:	________	________
d. $f(x) = 10^x$	f^{-1}:	________	________

STOP • STOP • STOP • STOP • STOP • STOP • STOP • STOP • STOP

A18

a. f^{-1}:	$x = 3^y$	$y = \log_3 x$
b. f^{-1}:	$x = 4^y$	$y = \log_4 x$
c. f^{-1}:	$x = \left(\frac{1}{2}\right)^y$	$y = \log_{1/2} x$
d. f^{-1}:	$x = 10^y$	$y = \log_{10} x$

Q19 Express the logarithmic form in equivalent exponential form:

	Logarithmic form	Exponential form
a.	$y = \log_{10}100$	$100 = 10^{(\quad)}$
b.	$y = \log_2 8$	$8 = (\quad)^y$
c.	$y = \log_{10}1$	$1 = (\quad)^{(\quad)}$
d.	$y = \log_{10}5$	________

STOP • STOP • STOP • STOP • STOP • STOP • STOP • STOP • STOP

A19 a. $100 = 10^y$ b. $8 = 2^y$ c. $1 = 10^y$ d. $5 = 10^y$

---

7 Logarithmic equations can often be evaluated by writing the logarithmic form as an equivalent exponential form. This procedure may be aided by recalling that logarithms are exponents and that the base is the base in either form. That is,

$$\log_b a = c \quad \text{implies that} \quad a = b^c$$

(logarithm: c ↔ exponent c; base: b ↔ b)

(*Note:* The base, b, will be assumed to always be positive.)

Example 1: Write $\log_{10}100 = t$ in exponential form.

Solution

$$\log_{10}100 = t \quad \text{implies that} \quad 100 = 10^t$$

(logarithm: t ↔ t; base: 10 ↔ 10)

Example 2: Write $2^3 = 8$ in logarithmic form.

Solution

$$2^3 = 8 \quad \text{implies that} \quad \log_2 8 = 3$$

(logarithm: 3 ↔ 3; base: 2 ↔ 2)

Q20 Write in exponential form:
a. $\log_{10}1{,}000 = a$ b. $\log_2 16 = 4$

STOP • STOP • STOP • STOP • STOP • STOP • STOP • STOP • STOP

A20 a. $1{,}000 = 10^a$ b. $16 = 2^4$

Q21 Write in exponential form:
a. $\log_{10}1 = 0$ b. $\log_b n = a$

STOP • STOP • STOP • STOP • STOP • STOP • STOP • STOP • STOP

A21 a. $1 = 10^0$ b. $n = b^a$

Q22 Write in logarithmic form:

a. $2^x = 64$ b. $8^0 = 1$

STOP • STOP • STOP • STOP • STOP • STOP • STOP • STOP • STOP

A22 a. $\log_2 64 = x$ b. $\log_8 1 = 0$

Q23 Write in logarithmic form:

a. $8^{4/3} = 16$ b. $2^{-3} = \frac{1}{8}$

STOP • STOP • STOP • STOP • STOP • STOP • STOP • STOP • STOP

A23 a. $\log_8 16 = \frac{4}{3}$ b. $\log_2 \frac{1}{8} = -3$

Q24 Write in exponential form:

a. $\log_8 16 = x$ b. $\log_4 x = \frac{3}{4}$

STOP • STOP • STOP • STOP • STOP • STOP • STOP • STOP • STOP

A24 a. $16 = 8^x$ b. $x = 4^{3/4}$

---

8 Logarithmic equations can often be solved by writing them in equivalent exponential form and applying properties of exponents. The unknown variable may occur in one of three places.

1. $\log_2 x = 3$

2. $\log_2 8 = x$

or

3. $\log_x 8 = 3$

The first case is solved as follows:

$$\log_2 x = 3$$
$$x = 2^3$$
$$x = 8$$

Q25 Solve $\log_{10} x = 2$.

STOP • STOP • STOP • STOP • STOP • STOP • STOP • STOP • STOP

A25 100: $x = 10^2$

Q26 Solve $\log_4 x = \frac{3}{2}$.

STOP • STOP • STOP • STOP • STOP • STOP • STOP • STOP • STOP

A26 8: $x = 4^{3/2}$
$= (4^{1/2})^3$
$= 2^3$

Q27 Solve:

a. $\log_2 x = 4$ b. $\log_8 x = \frac{1}{3}$ c. $\log_{1/8} x = \frac{-4}{3}$

STOP • STOP • STOP • STOP • STOP • STOP • STOP • STOP • STOP

A27 a. 16:
$x = 2^4$

b. 2:
$x = 8^{1/3}$

c. 16:

$$\begin{aligned} x &= \left(\frac{1}{8}\right)^{-4/3} \\ &= \frac{1^{-4/3}}{8^{-4/3}} \\ &= 8^{4/3} \\ &= (8^{1/3})^4 \\ &= 2^4 \end{aligned}$$

9 The second case, where the variable is the logarithm, is solved by applying the principle of exponents, which states that if

$x^m = x^n, x > 0, x \neq 1,$ then $m = n$

Example 1: $2^3 = 2^x$ implies that $3 = x$.

Example 2: Solve $4^2 = 8^x$ for x.

Solution

Both members of the equation have bases that are powers of 2; hence

$$\begin{aligned} 4^2 &= 8^x \\ (2^2)^2 &= (2^3)^x \\ 2^4 &= 2^{3x} \end{aligned}$$

Therefore,

$$4 = 3x$$

$$\frac{4}{3} = x$$

Example 3: Solve $\log_{10}100 = x$ for an integer value of x.

Solution

$$100 = 10^x$$
$$10^2 = 10^x$$
$$2 = x$$

Q28 Solve $\log_3 27 = x$.

STOP • STOP • STOP • STOP • STOP • STOP • STOP • STOP • STOP

A28 3: $27 = 3^x$

$$3^3 = 3^x$$
$$3 = x$$

Q29 **a.** $\log_{25}125 = x$ **b.** $\log_{10}\frac{1}{100} = x$ **c.** $\log_{3/2}\frac{4}{9} = x$

STOP • STOP • STOP • STOP • STOP • STOP • STOP • STOP • STOP

A29 **a.** $\frac{3}{2}$:

$$125 = 25^x$$
$$5^3 = (5^2)^x$$
$$5^3 = 5^{2x}$$
$$3 = 2x$$

b. -2:

$$\frac{1}{100} = 10^x$$
$$\frac{1}{10^2} = 10^x$$
$$10^{-2} = 10^x$$
$$-2 = x$$

c. -2:

$$\frac{4}{9} = \left(\frac{3}{2}\right)^x$$
$$\left(\frac{2}{3}\right)^2 = \left(\frac{3}{2}\right)^x$$
$$\left(\frac{3}{2}\right)^{-2} = \left(\frac{3}{2}\right)^x$$
$$-2 = x$$

10 The third case, where the variable is the base, is solved by applying the principle of exponents, which states that if

$a = b$ then $a^n = b^n$

If $x^2 = 4$, then $(x^2)^{1/2} = 4^{1/2}$.
If $x^{3/2} = 8$, then $(x^{3/2})^{2/3} = 8^{2/3}$.

Example 1: Solve $x^{3/2} = 27$, by raising both members to a power that yields the first power of x for the left member.

Solution

$$x^{3/2} = 27$$
$$(x^{3/2})^{2/3} = 27^{2/3}$$
$$x^{(3/2)\cdot(2/3)} = (27^{1/3})^2$$
$$x = 3^2$$
$$x = 9$$

Example 2: Solve $\log_x 128 = 7$.

Solution

$$128 = x^7$$
$$128^{1/7} = (x^7)^{1/7}$$
$$2 = x$$

Q30 Solve $\log_x 27 = 3$.

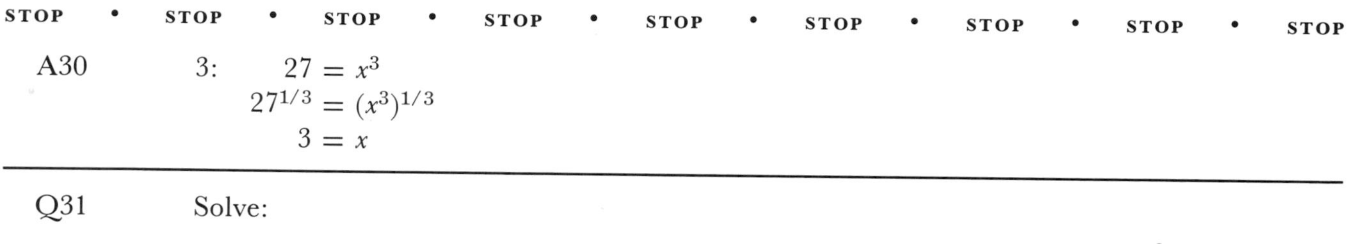

STOP • STOP • STOP • STOP • STOP • STOP • STOP • STOP • STOP

A30 3:
$$27 = x^3$$
$$27^{1/3} = (x^3)^{1/3}$$
$$3 = x$$

Q31 Solve:

a. $\log_x 81 = 4$ **b.** $\log_x 10^2 = 2$ **c.** $\log_x \frac{9}{4} = \frac{-2}{3}$

STOP • STOP • STOP • STOP • STOP • STOP • STOP • STOP • STOP

A31 **a.** 3:
$$81 = x^4$$
$$81^{1/4} = (x^4)^{1/4}$$

b. 10:
$$10^2 = x^2$$
$$(10^2)^{1/2} = (x^2)^{1/2}$$

c. $\frac{8}{27}$:

$$\frac{9}{4} = x^{-2/3}$$

$$\left(\frac{9}{4}\right)^{-3/2} = (x^{-2/3})^{-3/2}$$

$$\left(\frac{4}{9}\right)^{3/2} = x$$

$$\left[\left(\frac{4}{9}\right)^{1/2}\right]^3 = x$$

$$\left(\frac{2}{3}\right)^3 = x$$

Q32 Solve:

a. $\log_4 x = 0$ b. $\log_x 4 = \frac{1}{2}$ c. $\log_2 \frac{1}{32} = x$

STOP • STOP • STOP • STOP • STOP • STOP • STOP • STOP • STOP

A32 a. 1 b. 16 c. −5

This completes the instruction for this section.

8.2 Exercises

1. Graph:
   a. $f(x) = 3^x$ b. $f(x) = 2^{-x}$ c. $f(x) = 2^x - 1$
2. Graph $f(x) = 4^x$ and $f(x) = \left(\frac{1}{4}\right)^x$ on the same coordinate system.
3. Graph $f = \{(x, y) | y = 3^x\}$ and $f^{-1} = \{(x, y) | x = 3^y\}$ on the same coordinate system.
4. Graph:
   a. $x = 2^y$ b. $f(x) = \log_{10} x$
5. The number of bacteria present after t days is given by the formula $N = N_0 2^{(1/2)t}$, where N_0 is the number of bacteria present when $t = 0$. If there are approximately 1,000 bacteria present at the start, how many are present after
   a. 1 day? b. 2 days? c. 10 days?
6. Find the inverse of the function and express the inverse in function notation:
   a. $f(x) = 10^x$ b. $f(x) = \log_4 x$ c. $f = \left\{(x, y) | y = \left(\frac{1}{2}\right)^x\right\}$
7. Write in logarithmic form:
   a. $10^2 = 100$ b. $10^x = y$ c. $2^k = t$ d. $E^t = m$
8. Write in exponential form:
   a. $\log_2 32 = 5$ b. $\log_{10} 3 = 0.4771$ c. $\log_a 1 = 0$
   d. $\log_m n = k$

9. Solve:

a. $\log_{10}x = 5$ **b.** $\log_2 16 = x$ **c.** $\log_4 x = \frac{3}{2}$ **d.** $\log_x 4 = \frac{2}{5}$

e. $\log_3 3^x = 1$ **f.** $\log_3 x = -2$ **g.** $\log_2 64 = x$ **h.** $\log_3 x = -4$

i. $\log_{10}x = 0$ **j.** $\log_{10}0.1 = x$ **k.** $\log_{10}0.0001 = x$ **l.** $\log_{10}1{,}000 = x$

8.2 Exercise Answers

1. a. **b.**

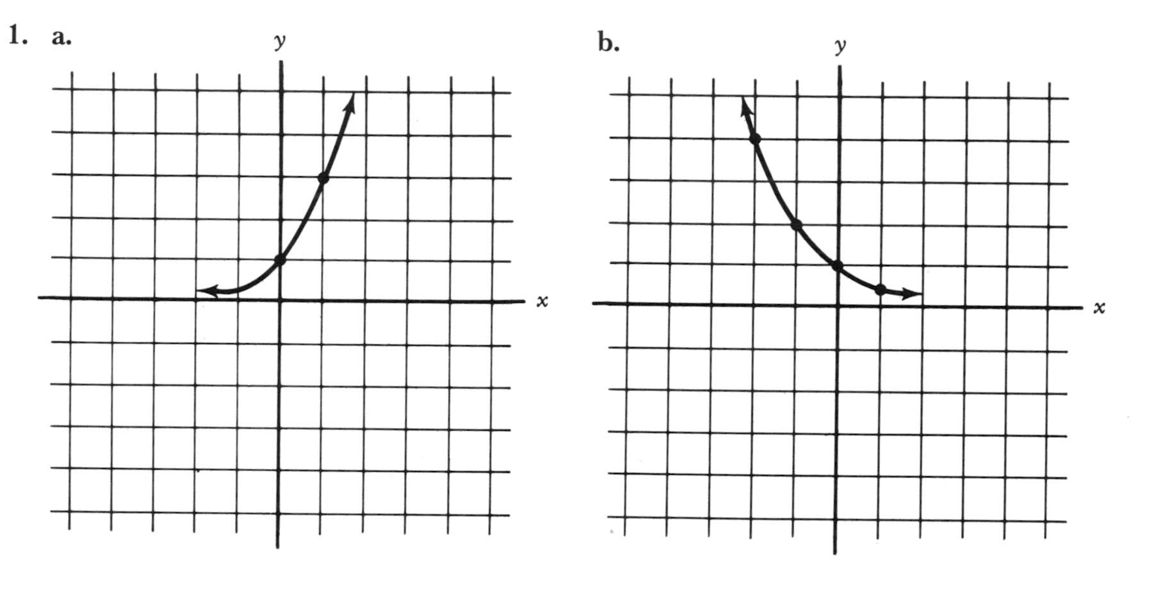

c.

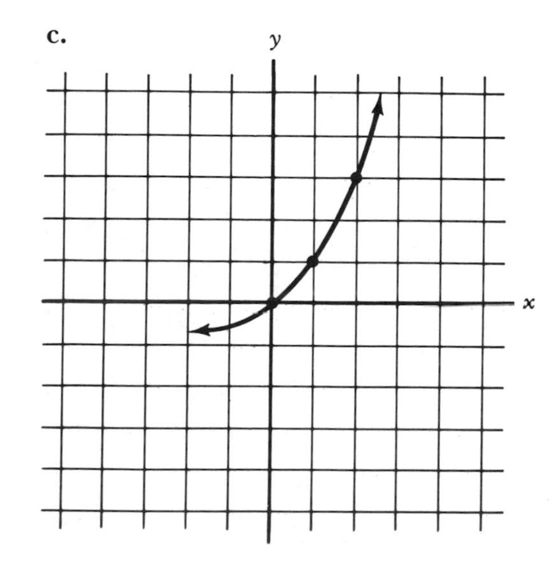

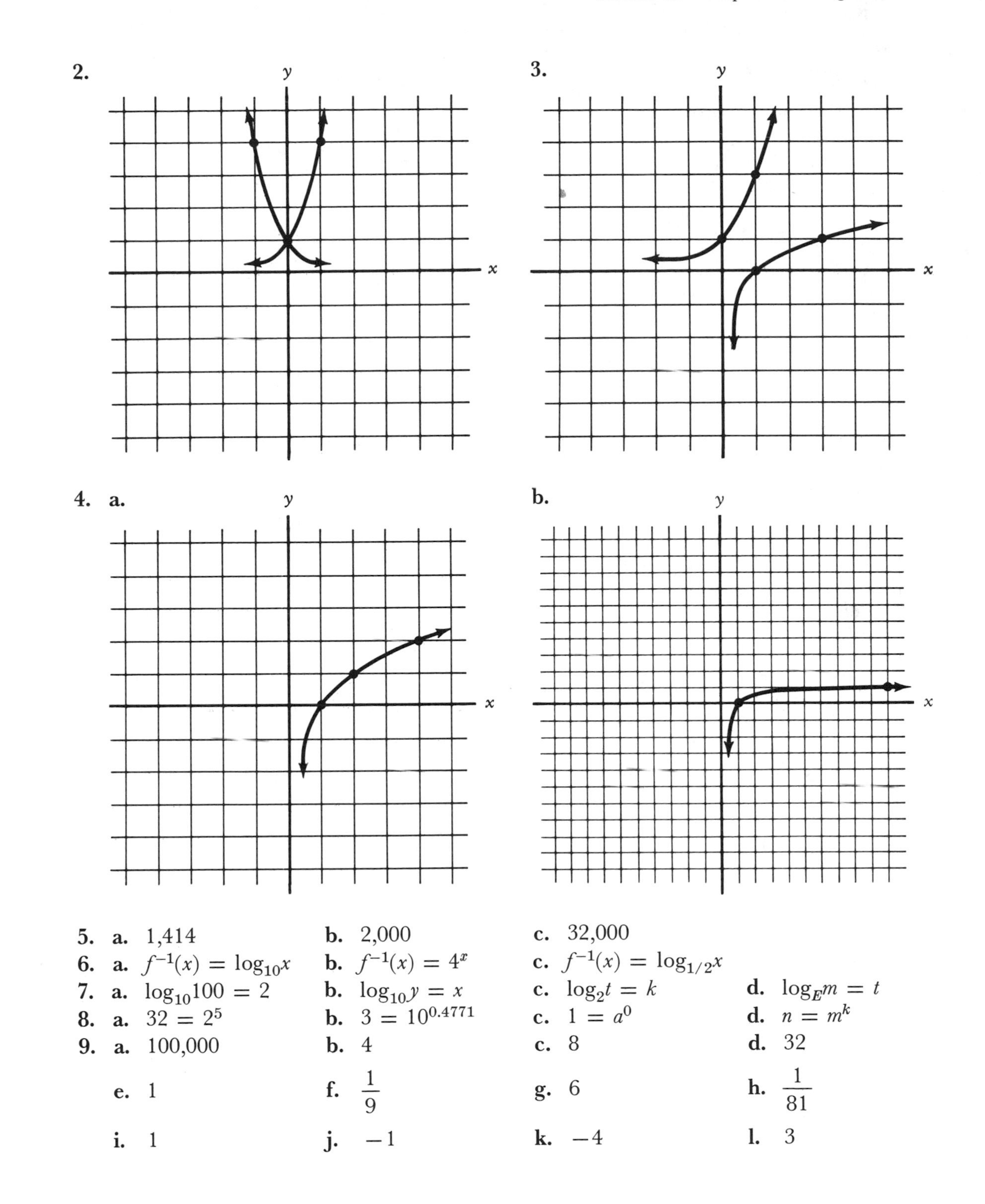

5. a. 1,414 b. 2,000 c. 32,000

6. a. $f^{-1}(x) = \log_{10}x$ b. $f^{-1}(x) = 4^x$ c. $f^{-1}(x) = \log_{1/2}x$

7. a. $\log_{10}100 = 2$ b. $\log_{10}y = x$ c. $\log_2 t = k$ d. $\log_E m = t$

8. a. $32 = 2^5$ b. $3 = 10^{0.4771}$ c. $1 = a^0$ d. $n = m^k$

9. a. 100,000 b. 4 c. 8 d. 32
e. 1 f. $\frac{1}{9}$ g. 6 h. $\frac{1}{81}$
i. 1 j. -1 k. -4 l. 3

8.3 Properties of Logarithms

1

There are three basic properties of logarithmic functions that give logarithms their many applications.

Property 1: The logarithm of a *product* of positive numbers is the *sum* of the logarithms of the factors. That is,

$\log_b xy = \log_b x + \log_b y$*

(The proof is a starred exercise.)
Some examples are:

$\log_{10} 3 \cdot 7 = \log_{10} 3 + \log_{10} 7$

$\log_{10} 12xy = \log_{10} 12 + \log_{10} x + \log_{10} y$

$\log_b k + \log_b j = \log_b kj$

*Assume all variables to be positive real numbers.

Q1 Express as a sum of logarithms:

a. $\log_{10} 2 \cdot 3$ __________

b. $\log_b mn$ __________

STOP • STOP • STOP • STOP • STOP • STOP • STOP • STOP • STOP

A1 a. $\log_{10} 2 + \log_{10} 3$ b. $\log_b m + \log_b n$

Q2 Express as the logarithm of a product:

a. $\log_{10} 10 + \log_{10} 2.36$ __________

b. $\log_b x^2 + \log_b y$ __________

STOP • STOP • STOP • STOP • STOP • STOP • STOP • STOP • STOP

A2 a. $\log_{10}(10 \cdot 2.36) = \log_{10} 23.6$ b. $\log_b x^2 y$

2 Property 2: The logarithm of a *quotient* of positive numbers is the logarithm of the numerator *minus* the logarithm of the denominator. That is,

$$\log_b \frac{x}{y} = \log_b x - \log_b y$$

Some examples are:

$$\log_{10} \frac{2.8}{10} = \log_{10} 2.8 - \log_{10} 10$$

$$\log_b \frac{x^2}{y} = \log_b x^2 - \log_b y$$

$$\log_b m - \log_b 3n = \log_b \frac{m}{3n}$$

Q3 Express as a difference of logarithms:

a. $\log_{10} \frac{2}{3}$ __________

b. $\log_b \frac{2x}{y}$ __________

STOP • STOP • STOP • STOP • STOP • STOP • STOP • STOP • STOP

A3 a. $\log_{10} 2 - \log_{10} 3$ b. $\log_b 2x - \log_b y$

Q4 Express as the logarithm of a quotient:

a. $\log_{10}7 - \log_{10}2$ ______

b. $\log_b xy - \log_b z$ ______

STOP • STOP • STOP • STOP • STOP • STOP • STOP • STOP • STOP

A4 a. $\log_{10}\frac{7}{2}$ b. $\log_b \frac{xy}{z}$

---

3 Property 3: The logarithm of a *power* of a positive number is the exponent of that power times the logarithm of the number. That is,

$$\log_b x^k = k \log_b x$$

Examples are:

$$\log_{10}2^{50} = 50 \log_{10}2$$

$$\log_b \sqrt{y} = \log_b y^{1/2} = \frac{1}{2} \log_b y$$

$$\frac{2}{3} \log_b x = \log_b x^{2/3}$$

Q5 Express without exponents or radical signs:

a. $\log_{10}7^{12}$ ______

b. $\log_b \sqrt[3]{x}$ ______

STOP • STOP • STOP • STOP • STOP • STOP • STOP • STOP • STOP

A5 a. $12 \log_{10}7$ b. $\frac{1}{3} \log_b x$

Q6 Express without exponents or radical signs:

a. $\log_{10} \sqrt{2.8}$ ______

b. $\log_b \sqrt[5]{x^3}$ ______

STOP • STOP • STOP • STOP • STOP • STOP • STOP • STOP • STOP

A6 a. $\frac{1}{2} \log_{10}2.8$ b. $\frac{3}{5}\log_b x$

Q7 Express using exponents:

a. $3 \log_{10}5$ ______

b. $4 \log_b x$ ______

STOP • STOP • STOP • STOP • STOP • STOP • STOP • STOP • STOP

A7 a. $\log_{10}5^3$ b. $\log_b x^4$

---

4 It will often be necessary to use more than one of the properties when working with logarithms. Two examples:

$$\log_b x^2 y = \log_b x^2 + \log_b y \quad \text{(Property 1)}$$

$$= 2 \log_b x + \log_b y \quad \text{(Property 3)}$$

$$\log_b \frac{xy}{\sqrt{z}} = \log_b xy - \log_b \sqrt{z} \quad \text{(Property 2)}$$

$$= \log_b x + \log_b y - \log_b z^{1/2} \quad \text{(Property 1)}$$

$$= \log_b x + \log_b y - \frac{1}{2}\log_b z \quad \text{(Property 3)}$$

Q8 Express as the logarithm of a sum or difference:

a. $\log_{10}\frac{5x}{y}$ **b.** $\log_2 \sqrt{mn}$

STOP • STOP • STOP • STOP • STOP • STOP • STOP • STOP • STOP

A8 **a.** $\log_{10}5 + \log_{10}x - \log_{10}y$ **b.** $\frac{1}{2}\log_2 m + \frac{1}{2}\log_2 n$: $\log_2 \sqrt{mn} = \log_2(mn)^{1/2}$

$$= \frac{1}{2}\log_2 mn$$

Q9 Complete the steps:

$$\log_{10}\frac{\sqrt[3]{0.007}\cdot\sqrt{231}}{5.01} = \log_{10}\frac{(0.007)^{1/3}(231)^{1/2}}{5.01}$$

= ______________________ (Properties 1, 2)

= ______________________ (Property 3)

STOP • STOP • STOP • STOP • STOP • STOP • STOP • STOP • STOP

A9 $\log_{10}(0.007)^{1/3} + \log_{10}(231)^{1/2} - \log_{10}5.01$

$\frac{1}{3}\log_{10}0.007 + \frac{1}{2}\log_{10}231 - \log_{10}5.01$

Q10 Complete the steps:

$$\log_b \sqrt[3]{3x^4y^2} = \log_b(3x^4y^2)^{1/3}$$

$$= \log_b 3^{1/3}x^{4/3}y^{2/3}$$

= ______________________ (Property 1)

= ______________________ (Property 3)

STOP • STOP • STOP • STOP • STOP • STOP • STOP • STOP • STOP

A10 $\log_b 3^{1/3} + \log_b x^{4/3} + \log_b y^{2/3}$

$\frac{1}{3}\log_b 3 + \frac{4}{3}\log_b x + \frac{2}{3}\log_b y$

Q11 Write an equivalent expression in terms of $\log_b x$, $\log_b y$, and $\log_b z$:

a. $\log_b x^2yz$ **b.** $\log_b \frac{\sqrt{x}}{yz^2}$

STOP • STOP • STOP • STOP • STOP • STOP • STOP • STOP • STOP

A11

a. $2 \log_b x + \log_b y + \log_b z$

b. $\frac{1}{2} \log_b x - \log_b y - 2 \log_b z$:

$\log_b \frac{\sqrt{x}}{yz^2}$

$\log_b x^{1/2} - (\log_b y + \log_b z^2)$

$\frac{1}{2} \log_b x - \log_b y - 2 \log_b z$

Q12 Express as the logarithm of a sum or difference:

a. $\log_b \sqrt{xy^3}$

b. $\log_{10} \frac{\sqrt[3]{1.023}}{(9.83)^{15}}$

STOP • STOP • STOP • STOP • STOP • STOP • STOP • STOP • STOP

A12

a. $\frac{1}{2} \log_b x + \frac{3}{2} \log_b y$

b. $\frac{1}{3} \log_{10} 1.023 - 15 \log_{10} 9.83$

Q13 Use the properties of logarithms to express as the logarithm of a single number:

a. $2 \log_4 \sqrt{3}$

b. $\log_b x + 2 \log_b y$

STOP • STOP • STOP • STOP • STOP • STOP • STOP • STOP • STOP

A13

a. $\log_4 3$:

$2 \log_4 \sqrt{3} = \log_4 (\sqrt{3})^2$

b. $\log_b xy^2$

Q14 Express as a single logarithm:

a. $\frac{1}{2} \log_2 5 - \log_2 7 + 2 \log_2 8$

b. $\log_b 2x + 3(\log_b x - \log_b y)$

STOP • STOP • STOP • STOP • STOP • STOP • STOP • STOP • STOP

A14

a. $\log_2 \frac{64\sqrt{5}}{7}$

b. $\log_b \frac{2x^4}{y^3}$

Q15 Express as a single logarithm:

a. $\frac{1}{2} \log_b m - \frac{3}{2} \log_b n$

b. $\log_b (x + 1) + \log_b (x - 1)$

STOP • STOP • STOP • STOP • STOP • STOP • STOP • STOP • STOP

A15 **a.** $\log_b \frac{\sqrt{m}}{\sqrt{n^3}}$ **b.** $\log_b(x^2 - 1)$

Q16 Each of the following statements is false. Correct each statement.

a. $\log_b x + \log_b y = \log_b(x + y)$ ________________

b. $\log_b \frac{x}{y} = \frac{\log_b x}{\log_b y}$ ________________

c. $\log_b 2x = 2 \log_b x$ ________________

d. $\log_b x^3 = \log_b x \cdot \log_b x \cdot \log_b x$ ________________

STOP • STOP • STOP • STOP • STOP • STOP • STOP • STOP • STOP

A16 **a.** $\log_b x + \log_b y = \log_b xy$ **b.** $\log_b \frac{x}{y} = \log_b x - \log_b y$

c. $\log_b 2x = \log_b 2 + \log_b x$ **d.** $\log_b x^3 = 3 \log_b x$

---

5 Logarithms of certain numbers can be evaluated if a few logarithms are known.

Example: If $\log_{10}2 = 0.3010$ and $\log_{10}3 = 0.4771$, find $\log_{10}6$.

Solution

$$\begin{aligned} \log_{10}6 &= \log_{10}2 \cdot 3 \\ &= \log_{10}2 + \log_{10}3 \\ &= 0.3010 + 0.4771 \\ &= 0.7781 \end{aligned}$$

Q17 If $\log_{10}2 = 0.3010$, $\log_{10}3 = 0.4771$, and $\log_{10}10 = 1$, find:

a. $\log_{10}4$ **b.** $\log_{10}5 = \log_{10}\frac{10}{2}$

STOP • STOP • STOP • STOP • STOP • STOP • STOP • STOP • STOP

A17 **a.** 0.6020:

$$\begin{aligned} \log_{10}4 &= \log_{10}2^2 \\ &= 2 \log_{10}2 \\ &= 2(0.3010) \end{aligned}$$

b. 0.6990:

$$\begin{aligned} \log_{10}\frac{10}{2} &= \log_{10}10 - \log_{10}2 \\ &= 1 - 0.3010 \end{aligned}$$

Q18 Find:

a. $\log_{10}\sqrt[5]{2}$ **b.** $\log_{10}90$

STOP • STOP • STOP • STOP • STOP • STOP • STOP • STOP • STOP

A18

a. 0.0602:

$$\log_{10}\sqrt[5]{2} = \log_{10}2^{1/5} = \frac{1}{5}\log_{10}2 = \frac{1}{5}(0.3010)$$

b. 1.9542:

$$\begin{aligned}\log_{10}90 &= \log_{10}10 \cdot 9\\ &= \log_{10}10 + \log_{10}9\\ &= 1 + \log_{10}3^2\\ &= 1 + 2\log_{10}3\\ &= 1 + 2(0.4771)\end{aligned}$$

Q19

Find:

a. $\log_{10}8$

b. $\log_{10}2{,}000$

STOP • STOP • STOP • STOP • STOP • STOP • STOP • STOP • STOP

A19

a. 0.9030:

$$\log_{10}8 = \log_{10}2^3 = 3\log_{10}2$$

b. 3.3010:

$$\begin{aligned}\log_{10}2{,}000 &= \log_{10}2 \cdot 1{,}000\\ &= \log_{10}2 + \log_{10}1{,}000\\ &= 0.3010 + \log_{10}10^3\\ &= 0.3010 + 3\log_{10}10\\ &= 0.3010 + 3(1)\end{aligned}$$

This completes the instruction for this section.

8.3 Exercises

1. Express as a sum or difference of logarithms without exponents:

   a. $\log_2\frac{3}{7}$ b. $\log_{10}\sqrt{8}$ c. $\log_b\sqrt{x^3}$ d. $\log_{10}\sqrt[5]{1.067}$

   e. $\log_b\frac{x^3}{y}$ f. $\log_b\frac{1}{\sqrt{x}}$

2. Express in terms of $\log_b x$ and $\log_b y$:

   a. $\log_b 2x\sqrt[5]{y}$ b. $\log_b\frac{x^3}{y^2}$ c. $\log_b\frac{\sqrt{x}}{y}$

3. Express as a single logarithm:

   a. $\log_{10}5 + \log_{10}7$ b. $\log_b 28 - \log_b 2$

   c. $\log_b x - 2\log_b y + \log_b x + \log_b y$

   d. $\log_b(x^2 - xy) - \log_b(2x - 2y)$

   e. $3\log_b\frac{x^2}{y} + 2\log_b xy^2$

4. If $\log_{10}2 = 0.3010$, $\log_{10}3 = 0.4771$, and $\log_{10}10 = 1$, find:

   a. $\log_{10}12$ b. $\log_{10}\frac{9}{8}$ c. $\log_{10}\sqrt[5]{12}$ d. $\log_{10}32$

   e. $\log_{10}20$ f. $\log_{10}200$ g. $\log_{10}2{,}000$ h. $\log_{10}20{,}000$

*5. Prove $\log_b xy = \log_b x + \log_b y$. (*Hint:* Assume that $\log_b x = m$ and $\log_b y = n$ and write using exponents.)

8.3 Exercise Answers

1. a. $\log_2 3 - \log_2 7$ b. $\frac{1}{2}\log_{10}8$ c. $\frac{3}{2}\log_b x$
   d. $\frac{1}{5}\log_{10}1.067$ e. $3\log_b x - \log_b y$ f. $\log_b 1 - \frac{1}{2}\log_b x$
2. a. $\log_b 2 + \log_b x + \frac{1}{5}\log_b y$ b. $3\log_b x - 2\log_b y$
   c. $\frac{1}{2}\log_b x - \log_b y$
3. a. $\log_{10}35$ b. $\log_b 14$ c. $\log_b \frac{x^2}{y}$ d. $\log_b \frac{x}{2}$
   e. $\log_b x^8 y$
4. a. 1.0791 b. 0.0512 c. 0.2158 d. 1.5050
   e. 1.3010 f. 2.3010 g. 3.3010 h. 4.3010

*5. Let $\log_b x = m$ and $\log_b y = n$
   Hence $x = b^m$ $y = b^n$
   $xy = b^{m+n}$
   Now, write by use of logarithms:
   $\log_b xy = m + n$
   Therefore,
   $\log_b xy = \log_b x + \log_b y$

8.4 Computation with Logarithms

1

Using the properties of logarithms discussed in Section 8.3, logarithms can be used to perform complicated arithmetic operations. Logarithms used for computation are to the base 10 and are called *common logarithms*. Common logarithms of powers of 10 can be found by writing the logarithmic statement using exponents.

Example: Find $\log_{10}100$.

Solution

Let

$$\begin{aligned} \log_{10}100 &= n \\ 100 &= 10^n \\ 10^2 &= 10^n \\ 2 &= n \end{aligned}$$

Hence $\log_{10}100 = 2$.

Q1 Find $\log_{10}1{,}000$ by completing the steps.
Let $\log_{10}1{,}000 = n$

$$1{,}000 = \_\_\_\_$$
$$\_\_\_\_ = \_\_\_\_$$
$$\_\_\_\_ = \_\_\_\_$$

STOP • STOP • STOP • STOP • STOP • STOP • STOP • STOP • STOP

A1 3: $1{,}000 = 10^n$
$10^3 = 10^n$
$3 = n$

Q2 Find:

a. $\log_{10}10$ **b.** $\log_{10}1{,}000$ **c.** $\log_{10}1$

STOP • STOP • STOP • STOP • STOP • STOP • STOP • STOP • STOP

A2 **a.** 1:
$\log_{10}10 = n$
$10 = 10^n$

b. 3

c. 0:
$\log_{10}1 = n$
$1 = 10^n$
$10^0 = 10^n$

Q3 Find:

a. $\log_{10}\dfrac{1}{10}$ **b.** $\log_{10}\dfrac{1}{100}$

STOP • STOP • STOP • STOP • STOP • STOP • STOP • STOP • STOP

A3 **a.** -1:

$$\log_{10}\frac{1}{10} = n$$
$$\frac{1}{10} = 10^n$$
$$10^{-1} = 10^n$$

b. -2:

$$\log_{10}\frac{1}{100} = n$$
$$\frac{1}{100} = 10^n$$
$$10^{-2} = 10^n$$

2 In general, the common logarithm of a power of 10 is the power to which 10 is raised. That is,

$\log_{10}10^c = c$

Common logarithms are usually written $\log x$, $x > 0$, omitting the subscript denoting the base.

Examples:

$$\log 1 = 0$$
$$\log 10^1 = 1$$
$$\log 10^2 = 2$$
$$\log 1{,}000 = \log 10^3 = 3$$
$$\log 10{,}000 = \log 10^4 = 4$$
$$\log \frac{1}{10} = \log 10^{-1} = -1$$
$$\log 0.01 = \log \frac{1}{100} = \log 10^{-2} = -2$$

Q4 Find the logarithms by writing the number as a power of 10:
a. log 100,000 **b.** log 0.001

STOP • STOP • STOP • STOP • STOP • STOP • STOP • STOP • STOP

A4 **a.** 5: $\log 100{,}000 = \log 10^5$ **b.** -3: $\log 0.001 = \log \frac{1}{1{,}000} = \log 10^{-3}$

3 In order to be able to compute with logarithms it will be necessary to find the logarithms of numbers that are not integer powers of 10. For example, to find log 3.24 one would need to solve the equation $3.24 = 10^n$. The logarithm n is clearly not an integer. However, Table V in the Appendix gives the approximate value of the logarithm to four decimal places. Table V gives the logarithms of numbers N from 1.00 to 9.99. That is, $1 \leqslant N < 10$. Since $\log 1 = 0$ and $\log 10 = 1$, the logarithm of N will vary from 0 to 1. That is, $0 \leqslant \log N < 1$.

To find the logarithm of a number such as 3.24 in Table V, find the first two digits (3.2) in the first column (titled N) and the third digit (4) in the top row. At the intersection of this row and column is the log 3.24.

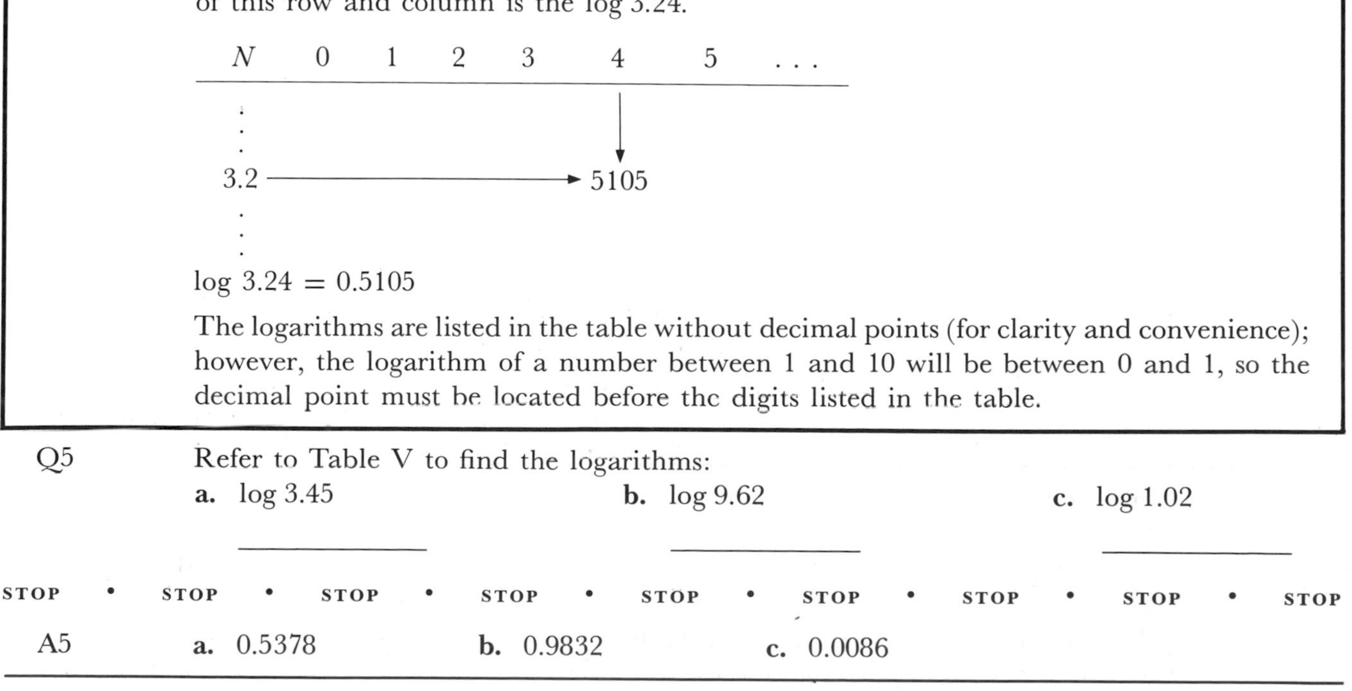

$\log 3.24 = 0.5105$

The logarithms are listed in the table without decimal points (for clarity and convenience); however, the logarithm of a number between 1 and 10 will be between 0 and 1, so the decimal point must be located before the digits listed in the table.

Q5 Refer to Table V to find the logarithms:
a. log 3.45 **b.** log 9.62 **c.** log 1.02

STOP • STOP • STOP • STOP • STOP • STOP • STOP • STOP • STOP

A5 **a.** 0.5378 **b.** 0.9832 **c.** 0.0086

Q6 Find:

a. $\log 1.11$ ______ **b.** $\log 3$ ______ **c.** $\log 6.7$ ______

STOP • STOP • STOP • STOP • STOP • STOP • STOP • STOP • STOP

A6 **a.** 0.0453 **b.** 0.4771: $\log 3 = \log 3.00$ **c.** 0.8261: $\log 6.7 = \log 6.70$

---

4 Logarithms of numbers with more than three digits can be approximated by rounding off to three digits.

Examples:

$\log 2.476 = \log 2.48 = 0.3945$
$\log 7.232 = \log 7.23 = 0.8591$

Note that even though equal signs are used, most logarithms are only approximations.

Q7 Find:

a. $\log 6.492$ ______ **b.** $\log 5.555$ ______ **c.** $\log 8.001$ ______

STOP • STOP • STOP • STOP • STOP • STOP • STOP • STOP • STOP

A7 **a.** 0.8122 **b.** 0.7451 **c.** 0.9031

---

5 If $\log N$ is known, Table V can be used to find N.

Example: If $\log N = 0.8525$, find N.

Solution

Search the logarithm values in Table V until 8525 is located. The number at the head of the row is 7.1 and at the head of the column above 8525 is 2; hence $N = 7.12$.

Q8 Refer to Table V to find N:

a. $\log N = 0.4393$ ______ **b.** $\log N = 0.7210$ ______ **c.** $\log N = 0.8710$ ______

STOP • STOP • STOP • STOP • STOP • STOP • STOP • STOP • STOP

A8 **a.** 2.75 **b.** 5.26 **c.** 7.43

Q9 Find N:

a. $\log N = 0.9031$ ______ **b.** $\log N = 0.3010$ ______ **c.** $\log N = 0$ ______

STOP • STOP • STOP • STOP • STOP • STOP • STOP • STOP • STOP

A9 **a.** 8 **b.** 2 **c.** 1

---

6 If the value of the $\log N$ cannot be located in the table, choose the logarithm nearest the given value.

Example: If $\log N = 0.7733$, find N.

Solution

$\log N$ is between 0.7731 and 0.7738 but nearer 0.7731. Hence $N = 5.93$.

Q10 Find N:

a. $\log N = 0.4037$ ________ b. $\log N = 0.8789$ ________

STOP • STOP • STOP • STOP • STOP • STOP • STOP • STOP • STOP

A10 a. 2.53 b. 7.57

7 If the given logarithm is halfway between two logarithms, choose the number that corresponds to the larger logarithm.

Q11 Find N:

a. $\log N = 0.8417$ ________ b. $\log N = 0.3865$ ________

STOP • STOP • STOP • STOP • STOP • STOP • STOP • STOP • STOP

A11 a. 6.95: 0.8417 is halfway between 0.8414 and 0.8420.
b. 2.44

Q12 Find N:

a. $\log N = 0.7194$ ________ b. $\log N = 0.8449$ ________

STOP • STOP • STOP • STOP • STOP • STOP • STOP • STOP • STOP

A12 a. 5.24 b. 7.00

8 To find the logarithms of numbers $N \geqslant 10$ or $N < 1$, it will be necessary to write the numbers in scientific notation. Recall from Section 4.3 that a positive number N is written in scientific notation as follows:

$$N = n \times 10^c$$

where $1 \leqslant n < 10$ and c is an integer.

Examples:

The following numbers are expressed in scientific notation:

$273 = 2.73 \times 10^2$ $7 = 7 \times 10^0$
$40 = 4.0 \times 10$ $0.15 = 1.5 \times 10^{-1}$

Q13 Write in scientific notation:

a. 305 ________ b. 0.0053 ________ c. 81,400 ________ d. 6 ________

STOP • STOP • STOP • STOP • STOP • STOP • STOP • STOP • STOP

A13 a. 3.05×10^2 b. 5.3×10^{-3} c. 8.14×10^4 d. 6×10^0

Q14 Write without using scientific notation:

a. 3.18×10^2 ________ b. 7.03×10^{-3} ________ c. 2×10^0 ________ d. 4.18×10^5 ________

STOP • STOP • STOP • STOP • STOP • STOP • STOP • STOP • STOP

A14 a. 318 b. 0.00703 c. 2 d. 418,000

9 The logarithm of a number N written in scientific notation is determined as follows: If $N = n \times 10^c$, where $1 \leqslant n < 10$ and $c \in I$,

$$\begin{aligned}\log N &= \log (n \times 10^c)\\ &= \log n + \log 10^c\\ &= \log n + c \log 10\\ &= \log n + c\end{aligned}$$

Since n is between 1 and 10, find the log n in Table V and then add the value of c. The log n is called the *mantissa* and c is called the *characteristic* of $\log N$. Hence $\log N =$ mantissa + characteristic, where the mantissa is found in Table V and the characteristic is the power of 10 when N is written in scientific notation.

Example: Find log 316.

Solution

$$\begin{aligned}\log 316 &= \log (3.16 \times 10^2)\\ &= \log 3.16 + \log 10^2\\ &= 0.4997 + 2\\ &= 2.4997\end{aligned}$$

Q15 Find $\log N$ by completing the steps:

$\log 827 = \log (8.27 \times 10^2)$

$=$ ________ $+$ ________

$=$ ________ $+$ ________

$=$ ________

STOP • STOP • STOP • STOP • STOP • STOP • STOP • STOP • STOP

A15 2.9175: $\log 8.27 + \log 10^2$

$0.9175 + 2$

2.9175

Q16 For $\log N = 2.9175$ the characteristic is ____ and the mantissa is ________.

STOP • STOP • STOP • STOP • STOP • STOP • STOP • STOP • STOP

A16 2; 0.9175

Q17 Find $\log N$:

a. log 826 b. log 21.7

STOP • STOP • STOP • STOP • STOP • STOP • STOP • STOP • STOP

A17

a. 2.9170:

$\log 826 = \log(8.26 \times 10^2)$
$= \log 8.26 + \log 10^2$

b. 1.3365:

$\log 21.7 = \log(2.17 \times 10)$
$= \log 2.17 + \log 10$

Q18

Find $\log N$:

a. log 197

b. log 2,070

STOP • STOP • STOP • STOP • STOP • STOP • STOP

A18

a. 2.2945:

$\log 197 = \log 1.97 + \log 10^2$

b. 3.3160:

$\log 2{,}070 = \log 2.07 + \log 10^3$

Q19

Find $\log N$:

a. log 7.15

b. log 71.5

STOP • STOP • STOP • STOP • STOP • STOP • STOP • STOP • STOP

A19

a. 0.8543 **b.** 1.8543

10

The mantissa is the portion of the logarithm found in Table V and all values in Table V are between 0 and 1. When working with logarithms it is important to keep the mantissa positive. When the characteristic of a logarithm is negative, it must not be added to the positive mantissa.

Example: Find log 0.0316.

Solution

$\log 0.0316 = \log(3.16 \times 10^{-2})$
$= \log 3.16 + \log 10^{-2}$
$= 0.4997 + {}^{-}2$

The mantissa is 0.4997 and the characteristic is -2. Since the sum would result in a negative matissa, they are not combined.

Q20

Find log 0.652.

STOP • STOP • STOP • STOP • STOP • STOP • STOP • STOP • STOP

A20

$0.8142 + {}^{-}1$: $\log 0.652 = \log(6.52 \times 10^{-1})$
$= \log 6.52 + \log 10^{-1}$
$= 0.8142 + {}^{-}1$

Q21 Find log N:

a. $\log 0.0023$

b. $\log 0.000875$

STOP • STOP • STOP • STOP • STOP • STOP • STOP • STOP • STOP

A21 **a.** $0.3617 + {}^{-}3$:

$\log 0.0023 = \log(2.3 \times 10^{-3})$

b. $0.9420 + {}^{-}4$

11 When dividing logarithms by integers it will be necessary to express the negative characteristic as a multiple of the divisor. This can be accomplished by adding and subtracting the same integer to the logarithm.

Examples:

Express the negative characteristic as a multiple of 10.

$$\begin{aligned} \log 0.0023 &= 0.3617 + {}^{-}3 \\ &= 7.3617 + {}^{-}3 - 7 \qquad \text{(add and subtract 7)} \\ &= 7.3617 - 10 \end{aligned}$$

$$\begin{aligned} \log 0.000875 &= 0.9420 + {}^{-}4 \\ &= 6.9420 + {}^{-}4 - 6 \\ &= 6.9420 - 10 \end{aligned}$$

Q22 Find log N writing negative characteristics as a positive integer before the mantissa minus a multiple of 10:

a. $\log 0.0316$

b. $\log 0.872$

STOP • STOP • STOP • STOP • STOP • STOP • STOP • STOP • STOP

A22 **a.** $8.4997 - 10$:

$$\begin{aligned} \log 0.0316 &= \log(3.16 \times 10^{-2}) \\ &= \log 3.16 + \log 10^{-2} \\ &= 0.4997 + {}^{-}2 \\ &= 8.4997 + {}^{-}2 - 8 \\ &= 8.4997 - 10 \end{aligned}$$

b. $9.9405 - 10$:

$$\begin{aligned} \log 0.872 &= \log(8.72 \times 10^{-1}) \\ &= \log 8.72 + \log 10^{-1} \\ &= 0.9405 + {}^{-}1 \\ &= 9.9405 - 10 \end{aligned}$$

Q23 Find $\frac{1}{2}\log 0.00528$.

STOP • STOP • STOP • STOP • STOP • STOP • STOP • STOP • STOP

A23 $3.8613 - 5$: $\frac{1}{2}\log 0.00528 = \frac{1}{2}\log(5.28 \times 10^{-3})$

$$= \frac{1}{2}(0.7226 + {}^{-}3)$$

$$= \frac{1}{2}(7.7226 - 10)$$

$$= (3.8613 - 5)$$

12 The value of N can be determined if $\log N$ is known, keeping in mind that only the mantissa is found in Table V.

Example: Find N if $\log N = 2.5105$.

Solution

characteristic ↘ ↙ mantissa

$$\log N = 2.5105$$
$$N = n \times 10^c$$

n is found in Table V and c is the characteristic.

$$N = 3.24 \times 10^2$$
$$= 324$$

This process is called finding the antilogarithm. That is, antilog $2.5105 = 324$.

Q24 Refer to Table V and find N:

a. $\log N = 1.8762$ ______

b. $\log N = 2.9304$ ______

STOP • STOP • STOP • STOP • STOP • STOP • STOP • STOP • STOP

A24 **a.** 75.2: $N = 7.52 \times 10^1$ **b.** 852: $N = 8.52 \times 10^2$

Q25 Find N:

a. $\log N = 3.9566$ ______

b. $\log N = 8.8274 - 10$ ______

STOP • STOP • STOP • STOP • STOP • STOP • STOP • STOP • STOP

A25 **a.** 9,050: $N = 9.05 \times 10^3$ **b.** 0.0672: $N = 6.72 \times 10^{-2}$

Q26 Find:

a. antilog 0.4742 ____________

b. antilog 7.7218 − 10 ____________

STOP • STOP • STOP • STOP • STOP • STOP • STOP • STOP • STOP

A26 **a.** 2.98 **b.** 0.00527

Q27 **a.** antilog 4.4520 ____________

b. antilog 9.9033 − 10 ____________

STOP • STOP • STOP • STOP • STOP • STOP • STOP • STOP • STOP

A27 **a.** 28,300 **b.** 0.8

Q28 Find N:

a. $\log N = 6.4771$ ____________

b. $\log N = 0.3010$ ____________

STOP • STOP • STOP • STOP • STOP • STOP • STOP • STOP • STOP

A28 **a.** 3,000,000 **b.** 2: $N = 2 \times 10^0$

---

13 With the increasing use of high-speed calculators and computers, logarithms are being used less and less to perform complicated arithmetic computations; however, logarithms can be conviently used to solve certain types of problems and equations. One additional property is necessary in order to use logarithms in computation. If $a = b$, then $\log a = \log b$; and conversely, if $\log m = \log n$, then $m = n$. Three-digit accuracy will be required for all computations.

Example: Find $\sqrt{417}$.

Solution

Let

$$N = \sqrt{417}$$
$$\log N = \log \sqrt{417}$$
$$\log N = \log 417^{1/2}$$
$$\log N = \frac{1}{2} \log 417$$
$$\log N = \frac{1}{2}(2.6201)$$
$$\log N = 1.3101$$
$$N = 20.4$$

Q29 Find $\sqrt[3]{4.17}$.

STOP • STOP • STOP • STOP • STOP • STOP • STOP • STOP • STOP

A29 1.61:

$$
\begin{aligned}
N &= \sqrt[3]{4.17} \\
\log N &= \log \sqrt[3]{4.17} \\
\log N &= \log 4.17^{1/3} \\
\log N &= \frac{1}{3} \log 4.17 \\
\log N &= \frac{1}{3} \log 4.17 \\
\log N &= \frac{1}{3}(0.6201) \\
\log N &= 0.2067 \\
N &= 1.61
\end{aligned}
$$

Q30 Find $\sqrt[3]{71.5}$.

STOP • STOP • STOP • STOP • STOP • STOP • STOP • STOP • STOP

A30 4.15:

$$
\begin{aligned}
N &= \sqrt[3]{71.5} \\
\log N &= \log \sqrt[3]{71.5} \\
\log N &= \log 71.5^{1/3} \\
\log N &= \frac{1}{3} \log 71.5 \\
\log N &= \frac{1}{3}(1.8543) \\
\log N &= 0.6181 \\
N &= 4.15
\end{aligned}
$$

Q31 Find 2^{50}.

STOP • STOP • STOP • STOP • STOP • STOP • STOP • STOP • STOP

A31 1.12×10^{15}:

$$\begin{aligned} N &= 2^{50} \\ \log N &= \log 2^{50} \\ \log N &= 50 \log 2 \\ \log N &= 50(0.3010) \\ \log N &= 15.0500 \\ N &= 1.12 \times 10^{15} \end{aligned}$$

Remember that operations with logarithms are approximations (three-digit accuracy in this text).

Q32 Use logarithms to approximate the product (53.8)(0.739).

[*Hint:* $N = (53.8)(0.739)$

$\log N = \log (53.8) + \log 0.739$]

STOP • STOP • STOP • STOP • STOP • STOP • STOP • STOP • STOP

A32 39.8:

$$\begin{aligned} \log N &= \log 53.8 + \log 0.739 \\ \log N &= 1.7308 + 9.8686 - 10 \\ \log N &= 11.5994 - 10 \\ \log N &= 1.5994 \\ N &= 3.98 \times 10^{1} \end{aligned}$$

14 Equations that involve exponents can be solved (approximately) by use of logarithms.

Example: Solve $4^x = 7$.

Solution

$$\begin{aligned} \log 4^x &= \log 7 \\ x \log 4 &= \log 7 \\ x &= \frac{\log 7}{\log 4} \\ x &= \frac{0.8451}{0.6021} \end{aligned}$$

(This step can be performed by division or by again using logarithms.)

$x = 1.40$

Note that it was not necessary to find an antilogarithm because x was the quotient of the logarithms.

Q33 Solve $2^x = 5$.

STOP • STOP • STOP • STOP • STOP • STOP • STOP • STOP • STOP

A33 2.32:

$$\log 2^x = \log 5$$
$$x \log 2 = \log 5$$
$$x = \frac{\log 5}{\log 2}$$
$$x = \frac{0.6990}{0.3010}$$
$$x = 2.32$$

Q34 Solve $5^{x+1} = 9$.

STOP • STOP • STOP • STOP • STOP • STOP • STOP • STOP • STOP

A34 0.365:

$$\log 5^{x+1} = \log 9$$
$$(x + 1) \log 5 = \log 9$$
$$x + 1 = \frac{\log 9}{\log 5}$$
$$x + 1 = \frac{0.9542}{0.6990}$$
$$x + 1 = 1.365$$
$$x = 0.365$$

Q35 Solve:

a. $N = \dfrac{\sqrt[3]{62.5}}{(0.038)^4}$

b. $10^{3x} = 92$

STOP • STOP • STOP • STOP • STOP • STOP • STOP • STOP • STOP

A35 **a.** 1.90×10^6:

$$\log N = \frac{1}{3}\log 62.5 - 4\log 0.038$$

b. 0.655:

$$\log 10^{3x} = \log 92$$
$$3x = \frac{\log 92}{\log 10}$$
$$3x = \frac{1.9638}{1}$$
$$x = 0.655$$

Q36 The compound-interest formula is $A = P(1 + r)^n$, where A is the amount of money compounded annually after n years for an initial investment of P dollars at r rate of interest. How much money would you have after 20 years if you invested \$100 at 6 percent interest?

STOP • STOP • STOP • STOP • STOP • STOP • STOP • STOP • STOP

A36 \$321:

$$A = 100(1 + .06)^{20}$$
$$A = 100(1.06)^{20}$$
$$\log A = \log 100 + \log 1.06^{20}$$
$$\log A = \log 100 + 20\log 1.06$$
$$\log A = 2 + 20(0.0253)$$
$$\log A = 2.506$$
$$A = 3.21 \times 10^2$$

Q37 Suppose that a distant ancestor of yours at the time of the birth of Christ had deposited \$1 in a bank at 3 percent interest compounded annually. How much would you collect in 1980 if you are the designated heir?

STOP • STOP • STOP • STOP • STOP • STOP • STOP • STOP • STOP

A37 $\$2.21 \times 10^{25}$:

$$\begin{aligned} A &= 1(1.03)^{1980} \\ \log A &= 1{,}980 \log 1.03 \\ \log A &= 1{,}980(0.0128) \\ \log A &= 25.344 \\ A &= 2.21 \times 10^{25} \end{aligned}$$

Q38 The time t in seconds that it takes a simple pendulum L feet long to complete one oscillation (through a small arc) is given by the formula $t = 2\pi\sqrt{\dfrac{L}{g}}$. Find t for a pendulum 4 feet long (use $\pi = 3.14$ and $g = 32.2$).

STOP • STOP • STOP • STOP • STOP • STOP • STOP • STOP • STOP

A38 2.21 seconds.

---

15 Table V has been used to find logarithms of numbers with three digits; however, using the technique of *interpolation,* logarithms of numbers with four digits can be found.

Example: Find log 2.348.

Solution

The log 2.348 is between log 2.34 and log 2.35 and can be found by assuming that logarithms are in the same ratio as the numbers N (a slightly incorrect assumption).

$$0.01\left[0.008\left[\begin{array}{l}\log 2.340 = 0.3692 \\ \log 2.348 = x\end{array}\right]d\right.\quad \begin{array}{l} \\ \\ \log 2.350 = 0.3711\end{array}\quad 0.0019$$

The value x can be determined by finding and adding d to the log 2.34 (d is added because the logarithms are increasing.)

$$\frac{d}{0.0019} = \frac{0.008}{0.01}$$

$d = 0.0015$ (rounded off to four decimal places)

Hence

$$\begin{aligned}\log 2.348 &= 0.3692 + d\\ &= 0.3692 + 0.0015\\ &= 0.3707\end{aligned}$$

The same technique can be used to find N to four digits given $\log N$. Logarithms are useful for computational purposes only if they can be used conveniently. Generally, if greater accuracy is desired, five-decimal-place common logarithms tables should be acquired.

This completes the instruction for this section.

8.4 Exercises

1. Find:
   a. $\log 10$ b. $\log 1$ c. $\log 0.001$ d. $\log 1{,}000$
2. Find:
   a. $\log 237$ b. $\log 58.3$ c. $\log 0.0615$ d. $\log 0.00486$
3. Solve for N:
   a. $\log N = 2.6937$ b. $\log N = 4.8899$
   c. $\log N = 6.7218 - 10$ d. $\log N = 0.6990$
4. Find:
   a. antilog 1.5400 b. antilog 3.6604
   c. antilog 6.8100 − 10 d. antilog 9.9802 − 10
5. Find:
   a. $15^{3.4}$ b. $\sqrt[3]{\dfrac{(15.8)(62.1)}{12}}$ c. $\sqrt[5]{0.00382}$ d. $4.63^{2.08}$
6. Solve:
   a. $5^x = 15$ b. $2^{x+6} = 32$ c. $3^x = 12$ d. $5^x = 10$
7. Find the amount A if interest is compounded at the rate of 5 percent annually for 14 years if the initial investment is \$217.
8. The area of a triangle with sides a, b, and c, can be calculated from the formula $A = \sqrt{s(s-a)(s-b)(s-c)}$, where s is one-half the perimeter. Approximate the area of a triangle with sides 12.6, 18.2, and 14.1 cm.
9. The pulse rate p of a person h meters tall is given approximately by $p = \dfrac{94}{\sqrt{h}}$, where p is the number of heartbeats per minute. If a person is 1.7 meters tall (5 feet 6 inches), what should the pulse rate be?
10. What is the approximate height of a child with a normal pulse of 98?
11. Five chain letters are sent out the first mailing, 5^2 letters the second mailing, 5^3 letters the third mailing, and so on. How many letters are sent out the tenth mailing?

8.4 Exercise Answers

	a	b	c	d
1.	a. 1	b. 0	c. −3	d. 3
2.	2.3747	b. 1.7657	c. 8.7889 − 10	d. 7.6866 − 10
3.	a. 494	b. 77,600	c. 0.000527	d. 5
4.	a. 34.7	b. 4,580	c. 0.000646	d. 0.955
5.	a. 9,970	b. 4.34	c. 0.328	d. 24.2
6.	a. 1.68	b. −1	c. 2.26	d. 1.43

7. \$430
8. 88.7 cm^2
9. 72.1
10. 0.92 meter
11. 9,770,000 letters

Chapter 8 Sample Test

At the completion of Chapter 8 it is expected that you will be able to work the following problems.

8.1 Special Functions and the Conics

1. Graph:

   a. $f(x) = \frac{2}{x}$ b. $4x^2 + 2y^2 = 8$ c. $\frac{x^2}{9} - \frac{y^2}{16} = 1$ d. $y^2 = 8x$

2. The distance a particle falls in a certain medium is directly proportional to the square of the length of the time it falls.
   a. If d = distance and t = time, represent the statement with an equation.
   b. If the particle falls 20 meters in 4 seconds, how far will it fall in 12 seconds?

8.2 Exponential and Logarithmic Functions

3. Graph:
   a. $f(x) = 3^x$ b. $f^{-1} = \{(x, y) | x = 3^y\}$
4. Solve:

   a. $\log_6 36 = x$ b. $\log_4 x = \frac{3}{2}$ c. $\log_x 9 = \frac{2}{3}$ d. $\log_{10} 0.1 = x$

8.3 Properties of Logarithms

5. Express as a single logarithm:
   a. $\log_{10} 3 + \log_{10} 6$ b. $\log_b x - 2 \log_b y$ c. $12 \log_2 4$
   d. $2 \log_b x - (\log_b y + 2 \log_b z)$
6. Express in terms of $\log_b x$ and $\log_b y$:

   a. $\log_b \frac{x}{y}$ b. $\log_b x^n$ c. $\log_b \frac{x^2 y}{17.2}$ d. $\log_b \frac{\sqrt[5]{x}}{y^3}$

8.4 Computation with Logarithms

7. Find:
   a. $\sqrt{8.6}$ b. $8.1^{0.81}$
8. How much money must be deposited in a bank today to amount to \$1,000 in 10 years, at 5 percent compounded annually?

Chapter 8 Sample Test Answers

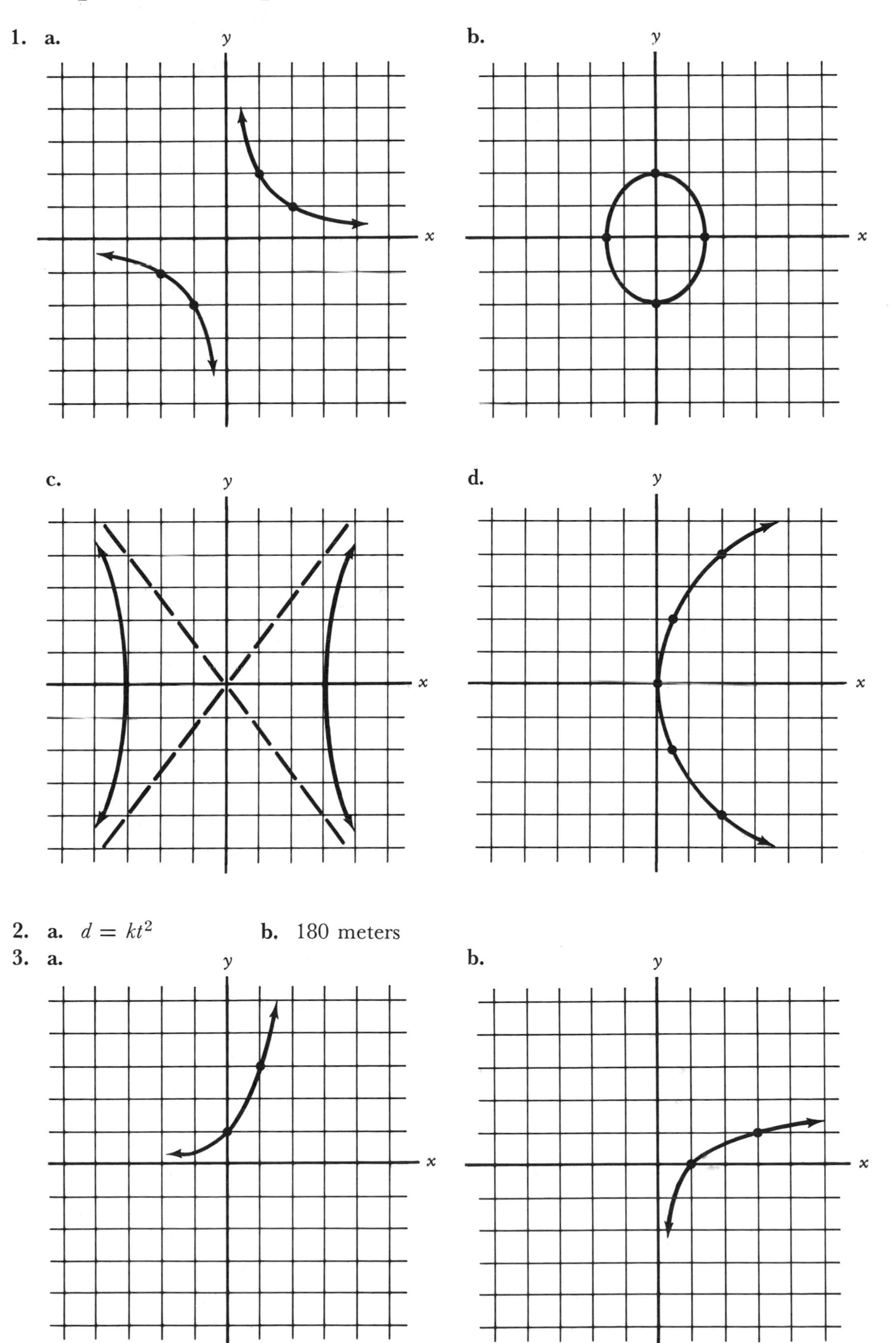

2. a. $d = kt^2$ **b.** 180 meters

4. a. 2 b. 8 c. 27 d. -1

5. a. $\log_{10}18$ b. $\log_b \frac{x}{y^2}$ c. $\log_2 4^{12}$ d. $\log_b \frac{x^2}{yz^2}$

6. a. $\log_b x - \log_b y$ b. $n \log_b x$ c. $2 \log_b x + \log_b y - \log_b 17.2$
d. $\frac{1}{5} \log_b x - 3 \log_b y$

7. a. 2.93 b. 5.44

8. $614

Chapter 9

Sequences and Series

9.1 Sequences

1

A *sequence* is a function f whose domain is a set of natural numbers, 1, 2, 3, . . . , n. The elements $f(1), f(2), f(3), \ldots, f(n)$ are called the *terms* of the sequence. The number of terms is *infinite* if the domain of the sequence is all natural numbers. In this text sequences will be assumed to be infinite unless a finite domain is given.

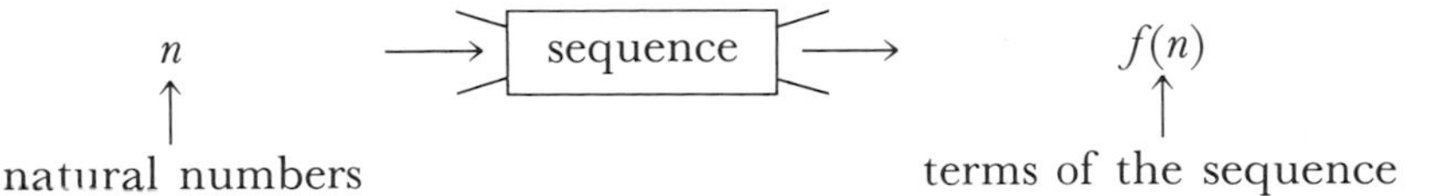

It is common practice to write the terms of the sequence as $a_1, a_2, a_3, \ldots, a_n, \ldots$ and to describe a sequence by writing a rule that shows how the nth term is formed.

Example: Find the first five terms of the sequence $a_n = 2n - 1$.

Solution

Since a_n is a sequence, the first five terms are formed by substituting 1, 2, 3, 4, 5 into the function a_n.

$a_1 = 2(1) - 1 = 1$ $\quad a_2 = 2(2) - 1 = 3$ $\quad a_3 = 2(3) - 1 = 5$ $\quad a_4 = 2(4) - 1 = 7$ $\quad a_5 = 2(5) - 1 = 9$

Hence the first five terms are 1, 3, 5, 7, and 9.

Q1 Find the first five terms of the sequence $a_n = 4n$:

$a_1 =$ _____, $a_2 =$ _____, $a_3 =$ _____, $a_4 =$ _____, $a_5 =$ _____

STOP • STOP • STOP • STOP • STOP • STOP • STOP • STOP • STOP

A1 4, 8, 12, 16, 20

Q2 Find the first five terms of the sequence:

a. $a_n = n^2$ **b.** $a_n = \frac{1}{n}$

STOP • STOP • STOP • STOP • STOP • STOP • STOP • STOP • STOP

A2 **a.** 1, 4, 9, 16, 25 **b.** $1, \frac{1}{2}, \frac{1}{3}, \frac{1}{4}, \frac{1}{5}$

Q3 Find the first five terms:

a. $a_n = \dfrac{n}{n+1}$ **b.** $a_n = 2^n$

STOP • STOP • STOP • STOP • STOP • STOP • STOP • STOP • STOP

A3 **a.** $\dfrac{1}{2}, \dfrac{2}{3}, \dfrac{3}{4}, \dfrac{4}{5}, \dfrac{5}{6}$ **b.** 2, 4, 8, 16, 32

Q4 Find the first five terms:

a. $a_n = (-1)^n$ **b.** $a_n = n^2 - n$

STOP • STOP • STOP • STOP • STOP • STOP • STOP • STOP • STOP

A4 **a.** $-1, 1, -1, 1, -1$ **b.** 0, 2, 6, 12, 20

Q5 Find a_7:

a. $a_n = \dfrac{(n-1)(n-2)(n-3)}{n}$ **b.** $a_n = (-1)^{n+1}$

STOP • STOP • STOP • STOP • STOP • STOP • STOP • STOP • STOP

A5 **a.** $\dfrac{120}{7}$: $a_7 = \dfrac{(7-1)(7-2)(7-3)}{7}$

$= \dfrac{120}{7}$

b. 1: $a_7 = (-1)^{7+1}$

2

A sequence in which each term differs from the preceding term by a constant amount is called an *arithmetic sequence* or *arithmetic progression*. The constant d is called the *common difference*.

Example: Show that 3, 6, 9, 12, . . . represents an arithmetic sequence.

Solution

$12 - 9 = 3$
$9 - 6 = 3$
$6 - 3 = 3$

Each term differs from the preceding term by the constant, 3.

Q6 Find the common difference:

a. 17, 15, 13, . . . **b.** $\dfrac{1}{2}, 1, \dfrac{3}{2}, 2, \ldots$

STOP • STOP • STOP • STOP • STOP • STOP • STOP • STOP • STOP

A6

a. -2: $13 - 15 = -2$
$15 - 17 = -2$

b. $\frac{1}{2}$: $2 - \frac{3}{2} = \frac{1}{2}$

$\frac{3}{2} - 1 = \frac{1}{2}$

$1 - \frac{1}{2} = \frac{1}{2}$

The common difference is the difference between a term and the preceding term.

Q7

Do the following represent arithmetic sequences?

a. 7, 6, 5, 4, . . . b. 2, 4, 8, 16, . . .

STOP • STOP • STOP • STOP • STOP • STOP • STOP • STOP • STOP

A7

a. yes: $d = -1$ b. no: no constant difference

3

An arithmetic sequence can be formed by adding the common difference to the first term and successive terms of the sequence.

Example: Find the first four terms of the sequence where $a_1 = 5$ and $d = 4$.

Solution

Add 4 to the first term, 5, and each successive term.

5, 9, 13, 17

Q8

Write the first four terms:

a. $a_1 = 5, d = -2$ b. $a_1 = 0, d = \frac{1}{2}$

STOP • STOP • STOP • STOP • STOP • STOP • STOP • STOP • STOP

A8

a. $5, 3, 1, -1$ b. $0, \frac{1}{2}, 1, \frac{3}{2}$

Q9

Find a_4:

a. $a_3 = 12, d = -3$ b. $a_2 = 0, d = 4$

STOP • STOP • STOP • STOP • STOP • STOP • STOP • STOP • STOP

A9

a. 9: $a_4 = a_3 + d$
$= 12 + {}^{-}3$

b. 8: $a_4 = a_3 + d$
$= a_2 + d + d$
$= 0 + 4 + 4$

4

A formula for a_n of an arithmetic sequence can be determined by recalling that any term can be obtained from the preceding term by adding a constant d. That is,

$a_2 = a_1 + d$
$a_3 = (a_1 + d) + d = a_1 + 2d$
$a_4 = (a_1 + 2d) + d = a_1 + 3d$
$a_5 = (a_1 + 3d) + d = a_1 + 4d$

Note that the coefficient of d is one less than the subscript of a; hence $a_n = a_1 + (n - 1)d$.

Example 1: Find the tenth term of the arithmetic sequence given $a_1 = 2$ and $d = 3$.

Solution

$$\begin{aligned} a_n &= a_1 + (n - 1)d \\ a_{10} &= 2 + (10 - 1)3 \\ &= 2 + 27 \\ &= 29 \end{aligned}$$

Example 2: Find a_6 given $a_1 = -2$ and $d = 5$.

Solution

$$\begin{aligned} a_n &= a_1 + (n - 1)d \\ a_6 &= -2 + (6 - 1)5 \\ &= 23 \end{aligned}$$

The formula for a_n was arrived at by establishing a pattern.

The pattern established by a few examples is assumed to always hold. This technique for finding a_n is not considered a proof.

Q10 Find a_{10}:

a. $a_1 = 1, d = 2$ **b.** $a_1 = 4, d = -2$

STOP • STOP • STOP • STOP • STOP • STOP • STOP • STOP • STOP

A10 **a.** 19: $a_{10} = 1 + (10 - 1)2$ **b.** -14: $a_{10} = 4 + (10 - 1)(-2)$

Q11 Find the indicated term:

a. $1, 5, 9, \ldots; a_{50}$ **b.** $a_1 = 1$, $d = \frac{-1}{2}$; a_{12}

STOP • STOP • STOP • STOP • STOP • STOP • STOP • STOP • STOP

A11 **a.** 197: $a_{50} = 1 + (50 - 1)4 = 1 + 196$

b. $\frac{-9}{2}$: $a_{12} = 1 + (12 - 1)\left(\frac{-1}{2}\right) = \frac{2}{2} - \frac{11}{2}$

Q12 Find the indicated term:

a. $1, 4, 7, \ldots; a_7$ **b.** $9, 3, -3, \ldots; a_{12}$

STOP • STOP • STOP • STOP • STOP • STOP • STOP • STOP • STOP

A12 **a.** 19: $a_7 = 1 + (7 - 1)3$ **b.** -57: $a_{12} = 9 + (12 - 1)(-6)$

Q13 Find the indicated term:

a. $0.5, 0.75, 1, \ldots; a_{26}$

b. $a, b, 2b - a, \ldots; a_6$

STOP • STOP • STOP • STOP • STOP • STOP • STOP • STOP • STOP

A13 **a.** 6.75

b. $5b - 4a$: $d = b - a$

$$\begin{aligned} a_6 &= a + (6 - 1)(b - a) \\ &= a + 5b - 5a \\ &= 5b - 4a \end{aligned}$$

---

5 The general term (a_n) of an arithmetic sequence can be determined if a_1 and d are known.

Example: Find the general term a_n for the arithmetic sequence where $a_1 = 7$ and $d = -1$.

Solution

$$\begin{aligned} a_n &= a_1 + (n - 1)d \\ &= 7 + (n - 1)(-1) \\ &= 7 + {}^{-}n + 1 \\ &= -n + 8 \end{aligned}$$

Hence the general term is $a_n = -n + 8$.

Q14 Find the general term, a_n:

a. $a_1 = -3, d = 2$

b. $a_1 = \frac{1}{2}, d = -3$

STOP • STOP • STOP • STOP • STOP • STOP • STOP • STOP • STOP

A14 **a.** $a_n = 2n - 5$:

$$\begin{aligned} a_n &= -3 + (n - 1)2 \\ &= -3 + 2n - 2 \\ &= 2n - 5 \end{aligned}$$

b. $a_n = -3n + \frac{7}{2}$:

$$\begin{aligned} a_n &= \frac{1}{2} + (n - 1)(-3) \\ &= \frac{1}{2} + {}^{-}3n + 3 \\ &= -3n + \frac{7}{2} \end{aligned}$$

Q15 Find a_n:

a. $3, 9, 15, \ldots$

b. $4, 14, 24, \ldots$

STOP • STOP • STOP • STOP • STOP • STOP • STOP • STOP • STOP

A15

a. $a_n = 6n - 3$:
$$a_1 = 3, \quad d = 6$$
$$a_n = 3 + (n - 1)6 = 3 + 6n - 6 = 6n - 3$$

b. $a_n = 10n - 6$:
$$a_1 = 4, \quad d = 10$$
$$a_n = 4 + (n - 1)10$$

Q16 Find a_n:

a. $5, 2, -1, \ldots$

b. $-5, -9, -13, \ldots$

STOP • STOP • STOP • STOP • STOP • STOP • STOP • STOP • STOP

A16

a. $a_n = -3n + 8$:
$$a_n = 5 + (n - 1)(-3)$$

b. $a_n = -4n - 1$:
$$a_n = -5 + (n - 1)(-4)$$

Q17 If a freely falling object falls 9.8 meters the first second, 19.6 meters the second second $(9.8 \cdot 2)$, 29.4 meters the third second $(9.8 \cdot 3)$, and so on, how many meters will it fall during the tenth second?

STOP • STOP • STOP • STOP • STOP • STOP • STOP • STOP • STOP

A17 98 meters: $a_n = 9.8, \quad d = 9.8$
$$a_{10} = 9.8 + (10 - 1)9.8 = 98$$

Q18 Mr. Mealing accepted a job that pays $8,000 the first year with yearly increases of $400. Find his salary during his tenth year on the job.

STOP • STOP • STOP • STOP • STOP • STOP • STOP • STOP • STOP

A18 $11,600: $a_{10} = 8{,}000 + (10 - 1)400 = 11{,}600$

Q19 In a theatre, the front row has 30 seats. Each row after the first row has 6 seats more than the previous row. Find the number of seats in the fifteenth row.

STOP • STOP • STOP • STOP • STOP • STOP • STOP • STOP • STOP

A19 114 seats: $a_{15} = 30 + (15 - 1)6$
$= 114$

6 A sequence in which each term divided by the preceding term is a constant is called a *geometric sequence* or *geometric progression.* The constant r is called the *common ratio* of the sequence.

Example: Show that 3, 9, 27, 81, ... represents a geometric sequence.

Solution

$\frac{81}{27} = 3 \qquad \frac{27}{9} = 3 \qquad \frac{9}{3} = 3$

Each term divided by the preceding term is the constant 3. Therefore, $r = 3$.

Q20 Find the common ratio:

a. $1, -2, 4, -8, \ldots$ **b.** $1, \frac{1}{2}, \frac{1}{4}, \frac{1}{8}, \frac{1}{16}, \ldots$

STOP • STOP • STOP • STOP • STOP • STOP • STOP • STOP • STOP

A20 **a.** -2: $\frac{-2}{1} = -2$

$\frac{4}{-2} = -2$

etc.

b. $\frac{1}{2}$: $\frac{\frac{1}{2}}{1} = \frac{1}{2}$

$\frac{\frac{1}{4}}{\frac{1}{2}} = \frac{1}{2}$

etc.

Q21 Find r:

a. $8, 12, 18, \ldots$ **b.** $\frac{1}{3}, \frac{2}{3}, \frac{4}{3}, \ldots$

STOP • STOP • STOP • STOP • STOP • STOP • STOP • STOP • STOP

A21

a. $\frac{3}{2}$: $r = \frac{12}{8} = \frac{3}{2}$

b. 2: $r = \frac{\frac{2}{3}}{\frac{1}{3}} = 2$

Q22

State whether the following sequences are arithmetic or geometric and find the common difference or ratio:

a. 1, 2, 3, 4, . . . __________, ______

b. 1, −1, 1, −1, . . . __________, ______

c. −16, 8, −4, 2, . . . __________, ______

d. 9, 6, 4, $\frac{8}{3}$, . . . __________, ______

STOP • STOP • STOP • STOP • STOP • STOP • STOP • STOP • STOP

A22

a. arithmetic, $d = 1$

b. geometric, $r = -1$

c. geometric, $r = \frac{-1}{2}$

d. geometric, $r = \frac{2}{3}$

7

A geometric sequence can be found by multiplying the first term and successive terms by the common ratio r.

Example: Find the first four terms of the geometric sequence where $a_1 = 5$ and $r = 3$.

Solution

5, 15, 45, 135

Q23

Find the first four terms of the geometric sequence where $a_1 = 12$ and $r = \frac{1}{2}$.

STOP • STOP • STOP • STOP • STOP • STOP • STOP • STOP • STOP

A23

12, 6, 3, $\frac{3}{2}$

Q24

Find the first four terms:

a. $a_1 = 2, r = -2$

b. $a_1 = 1, r = \frac{2}{3}$

STOP • STOP • STOP • STOP • STOP • STOP • STOP • STOP • STOP

A24

a. 2, −4, 8, −16

b. 1, $\frac{2}{3}, \frac{4}{9}, \frac{8}{27}$

8 The technique for finding the general term (a_n) of an arithmetic sequence can also be used for finding a_n for a geometric sequence. Recall that each term of a geometric sequence is determined by multiplying the preceding term by the common ratio, r. Thus

$$a_2 = a_1 r$$
$$a_3 = (a_1 r)r = a_1 r^2$$
$$a_4 = (a_1 r^2)r = a_1 r^3$$
$$a_5 = (a_1 r^3)r = a_1 r^4$$

Note that the power of r is one less than the subscript of a; hence

$$a_n = a_1 r^{n-1}$$

Example: Find the eighth term in the geometric sequence 3, 6, 12, ...

Solution

$a_1 = 3 \qquad r = 2$

$$\begin{aligned} a_n &= a_1 r^{n-1} \\ a_8 &= 3(2)^{8-1} \\ &= 3(2)^7 \\ &= 3(128) \\ &= 383 \end{aligned}$$

Q25 Find the indicated term:

a. $a_1 = 4, r = -1; a_6$

b. $a = -3, r = \frac{1}{3}; a_4$

STOP • STOP • STOP • STOP • STOP • STOP • STOP • STOP • STOP

A25 **a.** -4:

$$\begin{aligned} a_6 &= 4(-1)^{6-1} \\ &= 4(-1) \end{aligned}$$

b. $\frac{-1}{9}$:

$$\begin{aligned} a_4 &= (-3)\left(\frac{1}{3}\right)^{4-1} \\ &= (-3)\left(\frac{1}{3}\right)^3 \\ &= \frac{-1}{3^2} \end{aligned}$$

Q26 Find the indicated term:

a. 18, 12, 8, . . . ; a_8

b. $4, 2\sqrt{2}, 2, \ldots ; a_6$

STOP • STOP • STOP • STOP • STOP • STOP • STOP • STOP • STOP

A26 **a.** $\frac{256}{243}$:

$a = 18\left(\frac{2}{3}\right)^7$

b. $\frac{\sqrt{2}}{2}$:

$r = \frac{\sqrt{2}}{2}$

$a_6 = 4\left(\frac{\sqrt{2}}{2}\right)^5$

$= \frac{2^2(2^{1/2})^5}{2^5}$

Q27 Find the indicated term:

a. 0.01, 0.05, 0.25, . . . ; a_6

b. -8, 4, -2, . . . ; a_{10}

STOP • STOP • STOP • STOP • STOP • STOP • STOP • STOP • STOP

A27 **a.** 31.25:

$r = \frac{0.05}{0.01}$, $a_6 = 0.01(5)^5$

$= 5$ $\quad = 0.01(3{,}125)$

b. $\frac{1}{64}$:

$a_{10} = -8\left(\frac{-1}{2}\right)^9$

$= \frac{-2^3}{-2^9}$

$= \frac{1}{2^6}$

9 The general term of a geometric sequence can be expressed if a_1 and r are known.

Example: Find a_n if $a_1 = 6$ and $r = 2$.

Solution

$a_n = a_1 r^{n-1}$

$= 6(2)^{n-1}$

Hence the sequence is expressed by $a_n = 6(2)^{n-1}$.

Q28 Find a_n:

a. $a_1 = 2, r = -1$

b. $a_1 = 5, r = \frac{1}{2}$

STOP • STOP • STOP • STOP • STOP • STOP • STOP • STOP • STOP

A28 a. $a_n = 2(-1)^{n-1}$ b. $a_n = 5\left(\frac{1}{2}\right)^{n-1}$

Note: When an answer is left with an exponent of $n - 1$, the answer can be written with an exponent of n, as follows:

$$\begin{aligned} a_n &= 2(-1)^{n-1} \\ &= 2(-1)^n(-1)^{-1} \\ &= \frac{2(-1)^n}{-1} \\ &= -2(-1)^n \end{aligned}$$

Q29 Find a_n for the geometric sequences:

a. $-2, -6, \ldots$ b. $-3, \frac{3}{2}, \ldots$

STOP • STOP • STOP • STOP • STOP • STOP • STOP • STOP • STOP

A29 a. $a_n = -2(3)^{n-1}$:

$$r = \frac{-6}{-2} = 3$$

b. $a_n = -3\left(\frac{-1}{2}\right)^{n-1}$:

$$r = \frac{\frac{3}{2}}{-3} = \frac{-1}{2}$$

Q30 In a culture, there are 200 bacteria. Every hour the number of bacteria in the culture doubles. Find the number of bacteria in the culture at the end of 8 hours.

STOP • STOP • STOP • STOP • STOP • STOP • STOP • STOP • STOP

A30 25,600 bacteria: $a_1 = 200, r = 2$

$$\begin{aligned} a_8 &= 200(2)^{8-1} \\ &= 25{,}600 \end{aligned}$$

Q31 A rubber ball dropped from a height of 9 decimeters always rebounds one-third the distance of the previous fall. How far does the ball rebound the sixth time?

STOP • STOP • STOP • STOP • STOP • STOP • STOP • STOP • STOP

A31 $\frac{1}{81}$ decimeters: $a_1 = 3$ (first rebound)

$$r = \frac{1}{3}$$

This completes the instruction for this section.

9.1 Exercises

1. a. A sequence in which each term and its preceding term differ by a constant is called a(n) ______________.
   b. A sequence in which the quotient of each term and its preceding term is a constant is called a(n) ______________.
2. Write the formula that determines the nth term of a(n):
   a. arithmetic sequence b. geometric sequence
3. Find the first four terms of the sequences:
   a. $a_n = n^n$ b. $a_n = (-3)^n$
   c. $a_n = 4n - 15$ d. $a_n = n$
   e. $a_n = n(n-1)(n-2)$ f. $a_n = \frac{n-1}{n+1}$
4. Find the indicated term:
   a. $a_1 = -2, d = \frac{1}{4}; a_{10}$ b. $a_5 = 4, d = 8; a_1$
   c. $10 + \sqrt{2}, 10, 10 - \sqrt{2}, \ldots; a_{17}$ d. $2, 2.5, 3, \ldots; a_{38}$
5. Find the indicated term:
   a. $a_1 = 3, r = 4; a_6$ b. $a_1 = \frac{1}{5}, r = \frac{-1}{2}; a_8$
   c. $-16, 20, -25, \ldots; a_9$ d. $-4, 2, -1, \ldots; a_{10}$
6. Find a_n:
   a. $2, 7, 12, \ldots$ b. $3, \frac{15}{4}, \frac{18}{4}, \ldots$
   c. $-3, 0, 3, \ldots$ *d. $a_3 = 4, d = 3$
7. Find a_n:
   a. $5, 10, 20, \ldots$ b. $\frac{1}{9}, \frac{1}{3}, 1, \ldots$
   c. $10, -2, \frac{2}{5}, \ldots$ *d. $a^2 = 9, r = \frac{1}{3}$
8. If you accepted a job for \$6,000 a year and received a \$300 raise each year, how much would you be earning in 10 years?
9. A child builds with blocks, placing 35 blocks in the first row, 31 in the second row, 27 in the third row, and so on. How many blocks will be in the ninth row?
10. If you start with 1 cent and double it each day, how much would you have on the thirty-first day?
11. A certain ball when dropped from a height rebounds three-fifths of the original height. How high will the ball rebound after the fourth bounce if it was dropped from a height of 10 meters?

9.1 Exercise Answers

1. a. arithmetic sequence b. geometric sequence
2. a. $a_n = a_1 + (n-1)d$ b. $a_n = a_1 r^{n-1}$
3. a. $1, 4, 27, 256$ b. $-3, 9, -27, 81$
   c. $-11, -7, -3, 1$ d. $1, 2, 3, 4$
   e. $0, 0, 6, 24$ f. $0, \frac{1}{3}, \frac{1}{2}, \frac{3}{5}$

4. **a.** $\frac{1}{4}$ **b.** -28 **c.** $10 - 15\sqrt{2}$ **d.** 20.5

5. **a.** 3,072 **b.** $\frac{-1}{640}$ **c.** $\frac{-390{,}625}{4{,}096}$ **d.** $\frac{1}{128}$

6. **a.** $a_n = 5n - 3$ **b.** $a_n = \frac{3n + 9}{4}$

c. $a_n = 3n - 6$ ***d.** $a_n = 3n - 5$

7. **a.** $a_n = 5(2)^{n-1}$ ***b.** $a_n = \frac{1}{9}(3)^{n-1}$

c. $a_n = 10\left(\frac{-1}{5}\right)^{n-1}$ ***d.** $a_n = 27\left(\frac{1}{3}\right)^{n-1}$

8. $8,700
9. 3
10. 2^{30} cents (over $10 million)
11. $\frac{162}{125}$ meters

9.2 Series

1 The indicated sum of the terms of a sequence is called a *series*. For example, associated with the finite sequence 1, 3, 5, 7 is the series $1 + 3 + 5 + 7$. The symbol S_n denotes the sum of the first n terms of the associated series.

Example: Given $a_n = n^2$, find S_5 (do not sum).

Solution

$S_5 = 1 + 4 + 9 + 16 + 25$

Q1 From the given sequence, indicate the first four terms of the associated series (do not sum).

a. $a_n = 3n$ **b.** $a_n = \frac{1}{n}$

$S_4 =$ ________________ $S_4 =$ ________________

STOP • STOP • STOP • STOP • STOP • STOP • STOP • STOP • STOP

A1 **a.** $S_4 = 3 + 6 + 9 + 12$ **b.** $S_4 = 1 + \frac{1}{2} + \frac{1}{3} + \frac{1}{4}$

Q2 Indicate the first three terms of the associated series:

a. $a_n = \frac{n}{n + 1}$ **b.** $a_n = 2^n$

STOP • STOP • STOP • STOP • STOP • STOP • STOP • STOP • STOP

A2 a. $S_3 = \frac{1}{2} + \frac{2}{3} + \frac{3}{4}$ b. $S_3 = 2 + 4 + 8$

2 The Greek letter $\sum$ (sigma) is often used to indicate a series. For example, the series $3 + 6 + 9 + 12$ can be written

$$\sum_{n=1}^{4} 3n$$

where it is understood that n is to be replaced by each of the numbers 1, 2, 3, and 4 in the expression $3n$, and then the sum of resulting terms is to be indicated. That is,

$$\sum_{n=1}^{4} 3n = 3(1) + 3(2) + 3(3) + 3(4)$$
$$= 3 + 6 + 9 + 12$$

In general,

$$\sum_{n=1}^{k} a_n = a_1 + a_2 + a_3 + \cdots + a_k$$

The symbol $\sum$ is called the *summation symbol* and the letter n is called the *index of summation.* It is a "dummy" variable in the sense that any other letter in the same position would give the same result. That is,

$$\sum_{n=1}^{4} 3n = \sum_{i=1}^{4} 3i = \sum_{k=1}^{4} 3k$$

The symbol $\sum_{n=1}^{4} 3n$ is read "the summation of $3n$ from $n = 1$ to 4."

Examples:

$$\sum_{n=1}^{4} (2n - 1) = 2(1) - 1 + 2(2) - 1 + 2(3) - 1 + 2(4) - 1$$
$$= 1 + 3 + 5 + 7$$

$$\sum_{i=1}^{3} i^2 = 1^2 + 2^2 + 3^2 + 4^2$$
$$= 1 + 4 + 9 + 16$$

Q3 Express in expanded notation (do not sum):

a. $\sum_{n=1}^{4} 6n$ b. $\sum_{i=1}^{3} (-1)^i$

STOP • STOP • STOP • STOP • STOP • STOP • STOP • STOP • STOP

A3

a. $6 + 12 + 18 + 24$:

$$\sum_{n=1}^{4} 6n = 6(1) + 6(2) + 6(3) + 6(4)$$

b. $-1 + 1 - 1$:

$$\sum_{i=1}^{3} (-1)^i = (-1)^1 + (-1)^2 + (-1)^3$$

Q4

Express in expanded notation:

a. $\sum_{n=1}^{6} (7n - 2)$

b. $\sum_{k=1}^{7} (k^2 + 1)$

STOP • STOP • STOP • STOP • STOP • STOP • STOP • STOP • STOP

A4

a. $5 + 12 + 19 + 26 + 33 + 40$

b. $2 + 5 + 10 + 17 + 26 + 37 + 50$

Q5

Express in expanded notation:

a. $\sum_{n=3}^{6} 2n^2$

b. $\sum_{i=0}^{2} (-1)^i(i + 1)$

(*Note:* The index does not begin with 1.)

STOP • STOP • STOP • STOP • STOP • STOP • STOP • STOP • STOP

A5

a. $18 + 32 + 50 + 72$:

$$\sum_{n=3}^{6} 2n^2 = 2(3)^2 + 2(4)^2 + 2(5)^2 + 2(6)^2$$

b. $1 - 2 + 3$:

$$\sum_{i=0}^{2} (-1)^i(i + 1) = (-1)^0(0 + 1) + (-1)^1(1 + 1) + (-1)^2(2 + 1)$$

Q6

Expand:

a. $\sum_{n=1}^{4} (-1)^{n+1}n$

b. $\sum_{n=1}^{4} \frac{1}{2n - 1}$

STOP • STOP • STOP • STOP • STOP • STOP • STOP • STOP • STOP

A6

a. $1 - 2 + 3 - 4$

b. $1 + \frac{1}{3} + \frac{1}{5} + \frac{1}{7}$

Q7 Expand and evaluate:

a. $\sum_{n=1}^{4} (n^2 + 2)$ b. $\sum_{i=1}^{3} \frac{1}{i}$

STOP • STOP • STOP • STOP • STOP • STOP • STOP • STOP • STOP

A7 a. 38:
$3 + 6 + 11 + 18$

b. $\frac{11}{6}$:
$1 + \frac{1}{2} + \frac{1}{3}$

Q8 Evaluate:

a. $\sum_{n=0}^{4} \frac{n}{n+1}$ b. $\sum_{i=1}^{5} 1^i$

STOP • STOP • STOP • STOP • STOP • STOP • STOP • STOP • STOP

A8 a. $\frac{163}{60}$ b. 5

3 The indicated sum of the terms of an arithmetic sequence is an *arithmetic series*. The formula for the sum of the first n terms of an arithmetic sequence is $S_n = \frac{n}{2}(a_1 + a_n)$ and can be obtained as follows: The sum of the first n terms of an arithmetic sequence with common difference d and nth term a_n is

$$S_n = a_1 + (a_1 + d) + (a_1 + 2d) + \cdots + (a_n - d) + a_n$$

The reverse of this sum is

$$S_n = a_n + (a_n - d) + (a_n - 2d) + \cdots + (a_1 + d) + a_1$$

Adding these equations,

$$2S_n = (a_1 + a_n) + (a_1 + a_n) + (a_1 + a_n) + \cdots + (a_1 + a_n) + (a_1 + a_n)$$

Since there are n terms of $(a_1 + a_n)$,

$$2S_n = n(a_1 + a_n)$$

Therefore,

$$S_n = \frac{n}{2}(a_1 + a_n)$$

This formula for S_n can be used to find the sum of the first n terms of an arithmetic series.

Example 1: Find $\sum_{n=1}^{4} (2n - 1)$.

Solution

$a_1 = 1 \qquad a_n = a_4 = 7 \qquad n = 4$

$$S_4 = \frac{4}{2}(1 + 7)$$
$$= 2(8)$$
$$= 16$$

Example 2: Find the sum of the series $2 + 4 + 6 + \cdots + 100$.

Solution:

$a_1 = 2 \qquad a_n = a_{50} = 100 \qquad n = 50$

$$S_{50} = \frac{50}{2}(2 + 100)$$
$$= 2{,}550$$

Q9 Evaluate using the formula (Frame 3) for finding S_n:

a. $\sum_{n=1}^{10} (4n - 3)$ **b.** $\sum_{i=1}^{8} (-2i + 7)$

STOP • STOP • STOP • STOP • STOP • STOP • STOP • STOP • STOP

A9 **a.** 190:

$$S_{10} = \frac{10}{2}(1 + 37)$$
$$= 5(38)$$

b. -16:

$$S_8 = \frac{8}{2}(5 + {}^-9)$$
$$= 4(-4)$$

Q10 Evaluate:

a. $\sum_{n=1}^{10} (5n - 3)$ **b.** $1 + 2 + 3 + \cdots + 100$

STOP • STOP • STOP • STOP • STOP • STOP • STOP • STOP • STOP

A10 **a.** 245:
$a_1 = 2,\ a_n = a_{10} = 47$

b. 5,050:
$a_1 = 1,\ a_n = a_{100} = 100$

Q11 Find the sum of the first 10 terms of $12 + 8 + 4 + \cdots$ by first finding a_{10}.

STOP • STOP • STOP • STOP • STOP • STOP • STOP • STOP • STOP

A11 -60: $d = -4$, $a_n = a_1 + (n - 1)d$

$a_{10} = 12 + (10 - 1)(-4)$

$= -24$

$S_{10} = \frac{10}{2}(12 + {}^{-}24)$

Q12 Evaluate:

a. $3 + 6 + 9 + \cdots; S_{10}$ **b.** $1 + 2 + 3 + \cdots + 200$

STOP • STOP • STOP • STOP • STOP • STOP • STOP • STOP • STOP

A12 **a.** 165:

$a_{10} = 3 + (10 - 1)3$

$= 30$

$S_{10} = \frac{10}{2}(3 + 30)$

b. 20,100

Q13 If Ms. Bakich invests \$100 on her 21st birthday, \$200 on her 22nd birthday, \$300 on her 23rd birthday, and so on until she is 30 years old, how much money does she invest altogether?

STOP • STOP • STOP • STOP • STOP • STOP • STOP • STOP • STOP

A13 \$5,500: $a_1 = 100$, $a_{10} = 100 + (10 - 1)100$

$a_{10} = 1{,}000$

$S_{10} = \frac{10}{2}(100 + 1{,}000)$

Q14 Find the sum of the odd integers from 1 through 99.

STOP • STOP • STOP • STOP • STOP • STOP • STOP • STOP • STOP

A14 2,500: $a_1 = 1, a_{50} = 99$

$$a_{50} = \frac{50}{2}(1 + 99)$$

4 A *geometric series* is the indicated sum of the terms of a geometric sequence. If the nth term of a geometric sequence is

$$a_n = a_1 r^{n-1}$$

then $S_n = a_1 + a_1 r + a_1 r^2 + a_1 r^3 + \cdots + a_1 r^{n-1}$, where r is the common ratio. Using a technique similar to the one used for finding S_n for an arithmetic series, a formula can be derived for finding S_n for a geometric series. Let

$$S_n = a_1 + a_1 r + a_1 r^2 + \cdots + a_1 r^{n-2} + a_1 r^{n-1}$$

Multiply both members by r:

$$rS_n = a_1 r + a_1 r^2 + a_1 r^3 + \cdots + a_1 r^{n-1} + a_r^{\,n}$$

Subtract the second equation from the first:

$$S_n - rS_n = a_1 - a_1 r^n$$

$$S_n(1 - r) = a_1(1 - r^n)$$

$$\boxed{S_n = \frac{a_1(1 - r^n)}{1 - r}}$$

The above formula for S_n can be used to sum the first n terms of a geometric series.

Example: 1: Evaluate $\sum_{n=1}^{5} 3^n$.

Solution

$a_1 = 3 \qquad a_n = a_5 = 3^5 \qquad r = 3$

$$S_5 = \frac{3(1 - 3^5)}{-2}$$

$$= 363$$

Example 2: Sum the first four terms of $27 + 9 + 3 + \cdots$.

Solution

$a_1 = 27 \qquad r = \frac{1}{3}$

$$S_4 = \frac{27\left[1 - \left(\frac{1}{3}\right)^4\right]}{\frac{2}{3}}$$

$$= \frac{27\left(\frac{80}{81}\right)}{\frac{2}{3}}$$

$$= 40$$

Q15 Evaluate:

a. $\sum_{n=1}^{6} 5\left(\frac{1}{3}\right)^n$

b. $\sum_{n=1}^{4} 2(3)^n$

STOP • STOP • STOP • STOP • STOP • STOP • STOP • STOP • STOP

A15 **a.** $\frac{1{,}820}{729}$:

$$S_6 = \frac{\frac{5}{3}\left[1 - \left(\frac{1}{3}\right)^6\right]}{1 - \frac{1}{3}} = \frac{\frac{5}{3}\left[1 - \frac{1}{729}\right]}{\frac{2}{3}} = \frac{\frac{5}{3}\left(\frac{728}{729}\right)}{\frac{2}{3}}$$

b. 240:

$$S_4 = \frac{6(1 - 3^4)}{1 - 3} = \frac{6(1 - 81)}{-2} = -3(-80)$$

Q16 Evaluate:

a. $\sum_{n=1}^{6} \frac{2}{3}(2)^n$

b. $1 + 2 + 4 + 8 + \cdots + 32$

STOP • STOP • STOP • STOP • STOP • STOP • STOP • STOP • STOP

A16 **a.** 84:

$$S_6 = \frac{\frac{4}{3}(1 - 2^6)}{1 - 2} = \frac{\frac{4}{3}(-63)}{-1}$$

b. 63:

$a_1 = 1, n = 6, r = 2$

Q17 Evaluate:

a. $\sum_{k=1}^{8} \frac{(-1)^{k-1}}{2^k}$ **b.** $\sum_{i=1}^{6} \frac{2^i}{3}$

STOP • STOP • STOP • STOP • STOP • STOP • STOP • STOP • STOP

A17 **a.** $\frac{85}{256}$: **b.** 42:

$a_1 = \frac{1}{2}, r = \frac{-1}{2}$ $a_1 = \frac{2}{3}, r = 2$

---

5 An infinite series is denoted by $\sum_{n=1}^{+\infty} a_n$ and is expanded

$$\sum_{n=1}^{+\infty} a_n = a_1 + a_2 + a_3 + \cdots$$

An example is

$$\sum_{n=1}^{+\infty} n^2 = 1 + 4 + 9 + 16 + \cdots$$

Q18 Expand:

a. $\sum_{n=1}^{+\infty} 2^n$ **b.** $\sum_{n=1}^{+\infty} \frac{1}{n}$

STOP • STOP • STOP • STOP • STOP • STOP • STOP • STOP • STOP

A18 **a.** $2 + 4 + 8 + \cdots$ **b.** $1 + \frac{1}{2} + \frac{1}{3} + \frac{1}{4} + \cdots$

Q19 Expand:

a. $\sum_{n=1}^{+\infty} (-1)^n$ **b.** $\sum_{n=1}^{+\infty} (-1)^{n+1} \frac{n}{n+1}$

STOP • STOP • STOP • STOP • STOP • STOP • STOP • STOP • STOP

A19 **a.** $-1 + 1 - 1 + 1 - \cdots$ **b.** $\frac{1}{2} - \frac{2}{3} + \frac{3}{4} - \cdots$

6 The sum of an infinite geometric series can be determined if the absolute value of the common ratio is less than 1; that is, $|r| < 1$. The formula is developed as follows: For n terms,

$$S_n = \frac{a_1(1 - r^n)}{1 - r} \qquad |r| < 1$$

$$= \frac{a_1 - a_1 r^n}{1 - r}$$

$$= \frac{a_1}{1 - r} - \frac{a_1 r^n}{1 - r}$$

Since $|r| < 1$, the absolute value of r^n becomes very small as n becomes very large. In fact, as n approaches $+\infty$, r^n approaches zero, and in turn the expression $\frac{a_1 r^n}{1 - r}$ approaches zero. Hence the sum of an infinite geometric series can be expressed as

$$S_{+\infty} = \frac{a_1}{1 - r} \qquad |r| < 1$$

Example: Evaluate $\sum_{n=1}^{+\infty} \left(\frac{1}{3}\right)^{n-1}$.

Solution

$$a_1 = 1 \qquad r = \frac{1}{3}$$

$$S_{+\infty} = \frac{1}{1 - \frac{1}{3}}$$

$$= \frac{1}{\frac{2}{3}}$$

$$= \frac{3}{2}$$

Q20 Evaluate:

a. $\sum_{n=1}^{+\infty} \left(\frac{1}{2}\right)^{n-1}$ **b.** $\sum_{k=1}^{+\infty} \left(\frac{1}{4}\right)^{k}$

STOP • STOP • STOP • STOP • STOP • STOP • STOP • STOP • STOP

A20

a. 2:

$a_1 = 1, r = \frac{1}{2}$

$S_{+\infty} = \frac{1}{1 - \frac{1}{2}}$

b. $\frac{1}{3}$:

$a_1 = \frac{1}{4}, r = \frac{1}{4}$

$S_{+\infty} = \frac{\frac{1}{4}}{1 - \frac{1}{4}}$

Q21 Evaluate:

a. $\sum_{n=0}^{+\infty} \frac{1}{2^n}$

b. $\sum_{i=1}^{+\infty} -2\left(\frac{3}{5}\right)^i$

STOP • STOP • STOP • STOP • STOP • STOP • STOP • STOP • STOP

A21

a. 2:

$S_{+\infty} = \frac{1}{1 - \frac{1}{2}}$

b. -3:

$S_{+\infty} = \frac{\frac{-6}{5}}{1 - \frac{3}{5}}$

Q22 Evaluate:

a. $\sum_{n=1}^{+\infty} \frac{1}{2}(4)^n$

b. $\sum_{n=1}^{+\infty} -2\left(\frac{-1}{2}\right)^n$

STOP • STOP • STOP • STOP • STOP • STOP • STOP • STOP • STOP

A22

a. no sum:
$r = 4$, r must be
less than 1

b. $\frac{2}{3}$:

$S_{+\infty} = \frac{1}{1 + \frac{1}{2}}$

Q23 What distance will a ball rebound before coming to rest if it is dropped from a height of 6 feet, and after each fall it rebounds two-thirds of the distance it fell? (The distance of rebound counts only the upward bounce, not the downward motion.)

STOP • STOP • STOP • STOP • STOP • STOP • STOP • STOP • STOP

A23 12 feet: $S_{+\infty} = \sum_{n=1}^{+\infty} 6\left(\frac{2}{3}\right)^n$

Q24 The first swing of a bob on a pendulum is 10 centimeters. If on each subsequent swing it travels 0.9 as far as on the preceding swing, how far will the bob travel before coming to rest?

STOP • STOP • STOP • STOP • STOP • STOP • STOP • STOP • STOP

A24 100 centimeters: $S_{+\infty} = \frac{10}{1 - \frac{9}{10}}$

This completes the instruction for this section.

9.2 Exercises

1. Write the formula for the sum of a:
   a. geometric series b. arithmetic series c. infinite geometric series
2. Expand and evaluate:
   a. $\sum_{n=1}^{4} n^3$ b. $\sum_{n=1}^{8} 2n$
3. Express in expanded notation:
   a. $\sum_{n=5}^{8} (n^2 + n)$ b. $\sum_{k=2}^{+\infty} 2{,}400(10)^{-2k}$
4. Find the sum of the arithmetic series:
   a. $\sum_{n=1}^{10} [2 + (n - 1)7]$ b. $\sum_{k=1}^{5} (2 + 5k)$ c. $\sum_{n=1}^{50} \frac{n + 1}{2}$
   d. $\sum_{i=1}^{10} (8i - 5)$ e. $\sum_{n=1}^{12} (3n + 2)$ f. $\sum_{n=1}^{20} (2n - 5)$

5. Find the sum of the first 75 positive integers.
6. Find the sum of the geometric series:

a. $\sum_{n=1}^{4} 5(4)^{n-1}$ b. $\sum_{k=1}^{7} 4\left(\frac{-3}{2}\right)^k$ c. $\sum_{i=1}^{3} \left(\frac{3}{4}\right)^i$

d. $2 + \frac{4}{3} + \frac{8}{9} + \cdots$ e. $\frac{3}{2} + 1 + \frac{2}{3} + \cdots$ f. $\sum_{n=1}^{+\infty} 100\left(\frac{3}{4}\right)^n$

g. $\sum_{i=1}^{5} \left(\frac{1}{3}\right)^i$ h. $\sum_{i=2}^{5} \frac{1}{2}(-1)^i$ i. $\sum_{i=1}^{+\infty} 5\left(\frac{3}{2}\right)^i$

j. $\sum_{n=1}^{+\infty} \frac{1}{2^n}$ k. $\sum_{n=0}^{+\infty} \frac{1}{2^n}$ l. $\sum_{n=2}^{+\infty} \frac{1}{2^n}$

7. A particular substance decays in such a way that it loses half its weight each day. If there is 256 grams to begin with, how much is left after 10 days?
8. If you place 1 cent on the first square of a chessboard, 2 cents on the second square, 4 cents on the third, and so on, continuing to double until all 64 squares are covered, how much money will be on the 64th square? How much money will be on the whole board?
9. A pile of logs has 24 logs in the first layer, 23 in the second, 22 in the third, and so on, the last layer containing 10 logs. Find the total number of logs in the pile.
10. A vacuum pump removes one-half of the air in a container at each stroke. After 10 strokes, what percentage of the original amount of air remains in the container?
11. A culture of bacteria increases 20 percent every hour. If the original culture contains 1,000 bacteria, find a formula for the number of bacteria present after t hours.
12. A rubber ball is dropped from a height of 10 meters. If it rebounds one-half the distance after each fall, find the total distance that the ball rebounds before coming to rest.

9.2 Exercise Answers

1. a. $S_n = \frac{a_1(1 - r^n)}{1 - r}$ b. $S_n = \frac{n}{2}(a_1 + a_n)$

   c. $S_{+\infty} = \frac{a_1}{1 - r}, |r| < 1$
2. a. 100 b. 72
3. a. $30 + 42 + 56 + 72$ b. $0.24 + 0.0024 + 0.000024 + \cdots$
4. a. 335 b. 85 c. $\frac{1,325}{2}$ d. 390 e. 258 f. 320
5. 2,850
6. a. 425 b. $\frac{-1,389}{32}$ c. $\frac{111}{64}$ d. 6

   e. $\frac{9}{2}$ f. 300 g. $\frac{121}{243}$ h. 0

   i. no sum j. 1 k. 2 l. $\frac{1}{2}$

7. $256\left(\frac{1}{2}\right)^{10}$ grams
8. 9.22×10^{18} dollars, 1.84×10^{19} dollars
9. 255
10. $\frac{25}{256}$ percent
11. $1{,}000\left(\frac{6}{5}\right)^{t}$
12. 10 meters

9.3 The Binomial Theorem

1 A power of a binomial can be expanded by repeated multiplication.

Examples:

$$\begin{aligned}(x + y)^2 &= (x + y)(x + y)\\ &= x^2 + 2xy + y^2\end{aligned}$$

$$\begin{aligned}(2a - b)^3 &= (2a - b)^2(2a - b)\\ &= (4a^2 - 4ab + b^2)(2a - b)\\ &= 8a^3 - 12a^2b + 6ab^2 - b^3\end{aligned}$$

Q1 Expand:

a. $(a + b)^2$ **b.** $(a - b)^2$

STOP • STOP • STOP • STOP • STOP • STOP • STOP • STOP • STOP

A1 **a.** $a^2 + 2ab + b^2$ **b.** $a^2 - 2ab + b^2$

Q2 Expand:

a. $(3x + 2y)^2$ **b.** $(a + b)^3$

STOP • STOP • STOP • STOP • STOP • STOP • STOP • STOP • STOP

A2 **a.** $9x^2 + 12xy + 4y^2$ **b.** $a^3 + 3a^2b + 3ab^2 + b^3$

2 When expanding a power of a binomial, a series is formed. The series is not arithmetic or geometric; however, certain patterns are evident. The expansions of the following powers of the binomial $a + b$ can be obtained by repeated multiplication by $a + b$.

$$\begin{aligned}(a + b)^1 &= a + b\\ (a + b)^2 &= a^2 + 2ab + b^2\\ (a + b)^3 &= a^3 + 3a^2b + 3ab^2 + b^3\\ (a + b)^4 &= a^4 + 4a^3b + 6a^2b^2 + 4ab^3 + b^4\\ (a + b)^5 &= a^5 + 5a^4b + 10a^3b^2 + 10a^2b^3 + 5ab^4 + b^5\end{aligned}$$

In each case, it is possible to observe that when expanding $(a + b)^n$:

1. Each expansion has $n + 1$ terms.
2. The first term is a^n and the last term is b^n.
3. The exponents of a decrease by 1 in each term.
4. The exponents of b increase by 1 in each term.
5. The sum of the exponents of a and b in each term is n.
6. The coefficient of any term after the first can be obtained by multiplying the coefficient of the preceding term by its exponent of a and dividing by the number of this preceding term.

Example: Use the above technique to find the first four terms of $(x + y)^{10}$.

Solution

$$(x + y)^{10} = x^{10} + 10x^9y + 45x^8y^2 + 120x^7y^3 + \cdots$$

$$45 = \frac{10(9)}{2} \qquad 120 = \frac{45(8)}{3}$$

Q3 When expanding $(x + y)^8$:

a. The first term is ______.

b. The last term is ______.

c. The sum of the exponents in any term will be ______.

STOP • STOP • STOP • STOP • STOP • STOP • STOP • STOP • STOP

A3 a. x^8 b. y^8 c. 8

Q4 Write the powers of the variables when expanding $(x + y)^8$:
a. The second term is $8x$—y—.
b. The third term is $28x$—y—.
c. The fourth term is $56x$—y—.

STOP • STOP • STOP • STOP • STOP • STOP • STOP • STOP • STOP

A4 a. $8x^7y$ b. $28x^6y^2$ c. $56x^5y^3$
(*Note:* The sum of the exponents of x and y in each term is 8.)

3 Recall that the coefficient of the fifth term of $(x + y)^8$ is found by multiplying the coefficient of the fourth term by its exponent of x and dividing by the number of the term (4).

Q5 Complete the fifth and sixth terms of
$(x + y)^8 = x^8 + 8x^7y + 28x^6y^2 + 56x^5y^3 +$

STOP • STOP • STOP • STOP • STOP • STOP • STOP • STOP • STOP

A5 $70x^4y^4 + 56x^3y^5$: $70 = \frac{(56)(5)}{4}$, $56 = \frac{(70)(4)}{5}$

Q6 Expand $(x + y)^5$:

STOP • STOP • STOP • STOP • STOP • STOP • STOP • STOP • STOP

A6 $x^5 + 5x^4y + 10x^3y^2 + 10x^2y^3 + 5xy^4 + y^5$

Q7 Expand $(x + y)^7$:

STOP • STOP • STOP • STOP • STOP • STOP • STOP • STOP • STOP

A7 $x^7 + 7x^6y + 21x^5y^2 + 35x^4y^3 + 35x^3y^4 + 21x^2y^5 + 7xy^6 + y^7$

Q8 Find the first four terms of the expansion:

a. $(a + b)^{10}$ **b.** $(m + n)^{20}$

STOP • STOP • STOP • STOP • STOP • STOP • STOP • STOP • STOP

A8 **a.** $a^{10} + 10a^9b + 45a^8b^2 + 120a^7b^3$

b. $m^{20} + 20m^{19}n + 190m^{18}n^2 + 1{,}140m^{17}n^3$

---

4 Expansion of binomials in which the coefficients of the terms are not 1 can be completed by making appropriate substitutions for the terms.

Example: Expand $(2x - 3y)^4$.

Solution

Let $a = 2x$ and $b = -3y$. Hence,

$$\begin{aligned}(2x - 3y)^4 &= (a + b)^4 \\ &= a^4 + 4a^3b + 6a^2b^2 + 4ab^3 + b^4 \\ (a = 2x,\ b = -3y) &= (2x)^4 + 4(2x)^3(-3y) + 6(2x)^2(-3y)^2 + 4(2x)(-3y)^3 + (-3y)^4 \\ &= 16x^4 - 96x^3y + 216x^2y^2 - 216xy^3 + 81y^4\end{aligned}$$

Q9 Expand $(x - 2y)^4$. Let $a = x$, $b = -2y$.

STOP • STOP • STOP • STOP • STOP • STOP • STOP • STOP • STOP

A9 $x^4 - 8x^3y + 24x^2y^2 - 32xy^3 + 8y^4$:

$$(x - 2y)^4 = (a + b)^4$$
$$= a^4 + 4a^3b + 6a^2b^2 + 4ab^3 + b^4$$
$$= x^4 + 4x^3(-2y) + 6x^2(-2y)^2 + 4x(-2y)^3 + (-2y)^4$$

Q10 Expand $(2x + 3)^4$.

STOP • STOP • STOP • STOP • STOP • STOP • STOP • STOP • STOP

A10 $16x^4 + 96x^3 + 216x^2 + 216x + 81$: Let $a = 2x$, $b = 3$:

$$(2x + 3)^4 = (a + b)^4$$
$$= a^4 + 4a^3b + 6a^2b^2 + 4ab^3 + b^4$$
$$= (2x)^4 + 4(2x)^3(3) + 6(2x)^2(3)^2 + 4(2x)(3)^3 + (3)^4$$

Q11 Expand:

a. $(x - 3)^3$ **b.** $(x - y)^5$

STOP • STOP • STOP • STOP • STOP • STOP • STOP • STOP • STOP

A11 **a.** $x^3 - 9x^2 + 27x - 27$

b. $x^5 - 5x^4y + 10x^3y^2 - 10x^2y^3 + 5xy^4 - y^5$

Q12 Expand $(a + b)^6$.

STOP • STOP • STOP • STOP • STOP • STOP • STOP • STOP • STOP

A12 $a^6 + 6a^5b + 15a^4b^2 + 20a^3b^3 + 15a^2b^4 + 6ab^5 + b^6$

---

5 The expansion of a binomial can be expressed in terms of a formula; however, it will first be necessary to develop a convenient shorthand notation. The coefficients in a binomial expansion are often expressed in a shorter, but equivalent fashion, using notation known as *factorial notation.* If n represents a natural number,

$$n! = n(n - 1)(n - 2)(n - 3) \cdots 4 \cdot 3 \cdot 2 \cdot 1$$

The symbol $n!$ is read "n factorial."

Examples:

$1! = 1$
$2! = 2 \cdot 1$
$3! = 3 \cdot 2 \cdot 1$
$4! = 4 \cdot 3 \cdot 2 \cdot 1$
$5! = 5 \cdot 4! = 5 \cdot 4 \cdot 3! = 5 \cdot 4 \cdot 3 \cdot 2 \cdot 1$

$$\frac{6!}{4!3!} = \frac{6 \cdot 5 \cdot \overset{1}{\cancel{4!}}}{\underset{1}{\cancel{4!}}3!} = \frac{6 \cdot 5}{3!} = \frac{6 \cdot 5}{3 \cdot 2 \cdot 1} = 5$$

For purposes of completeness, $0!$ is defined to be 1.

Q13 Expand and evaluate:

a. $5!$ b. $4!0!$

STOP • STOP • STOP • STOP • STOP • STOP • STOP • STOP • STOP

A13 a. 120: $5! = 5 \cdot 4 \cdot 3 \cdot 2 \cdot 1$ b. 24: $0! = 1$

Q14 Simplify by expanding the factorial notation:

a. $\frac{9!}{11!}$ b. $\frac{9!}{4!5!}$

STOP • STOP • STOP • STOP • STOP • STOP • STOP • STOP • STOP

A14 a. $\frac{1}{110}$:

$$\frac{9!}{11!} = \frac{9!}{11 \cdot 10 \cdot 9!}$$

$$= \frac{1}{11 \cdot 10} \cdot \frac{9!}{9!}$$

b. 126:

$$\frac{9!}{4!5!} = \frac{9 \cdot 8 \cdot 7 \cdot 6 \cdot 5!}{4 \cdot 3 \cdot 2 \cdot 1 \cdot 5!}$$

Q15 Simplify:

a. $\frac{4!}{2!2!}$ b. $3!0!$

STOP • STOP • STOP • STOP • STOP • STOP • STOP • STOP • STOP

A15 a. 6 b. 6

---

6 The notation $\binom{n}{k}$ is defined as follows:

$$\binom{n}{k} = \frac{n!}{k!(n-k)!}$$

Some examples are:

$$\binom{5}{2} = \frac{5!}{2!(5-2)!} = \frac{5 \cdot 4 \cdot 3!}{2 \cdot 1 \cdot 3!} = 10$$

$$\binom{6}{3} = \frac{6!}{3!(6-3)!} = \frac{6 \cdot 5 \cdot 4 \cdot 3!}{3 \cdot 2 \cdot 1 \cdot 3!} = 20$$

$$\binom{3}{3} = \frac{3!}{3!(3-3)!} = \frac{3!}{3!0!} = 1$$

The notation $\binom{5}{3}$ is read "5 choose 3" because it represents the number of ways three objects may be chosen from five objects.

Q16 Evaluate:

a. $\binom{4}{1}$ b. $\binom{4}{2}$ c. $\binom{4}{0}$

STOP • STOP • STOP • STOP • STOP • STOP • STOP • STOP • STOP

A16 a. 4:
$$\binom{4}{1} = \frac{4!}{1!(4-1)!}$$
b. 6:
$$\binom{4}{2} = \frac{4!}{2!(4-2)!}$$
c. 1:
$$\binom{4}{0} = \frac{4!}{0!(4-0)!}$$

Q17 Evaluate:

a. $\binom{5}{2}$ b. $\binom{9}{4}$ c. $\binom{5}{0}$

STOP • STOP • STOP • STOP • STOP • STOP • STOP • STOP • STOP

A17 a. 10 b. 126 c. 1

7 The expansion of $(a+b)^n$ can now be stated by use of factorial notation.

The binomial theorem: If $a,b \in R$ and $n,k \in N$, then

$$(a+b)^n = \binom{n}{0}a^n + \binom{n}{1}a^{n-1}b + \binom{n}{2}a^{n-2}b^2 + \cdots + \binom{n}{k}a^{n-k}b^k + \cdots + \binom{n}{n}b^n$$

Example: Use the binomial theorem to expand $(a+b)^4$.

Solution

$$(a+b)^4 = \binom{4}{0}a^4 + \binom{4}{1}a^3b + \binom{4}{2}a^2b^2 + \binom{4}{3}ab^3 + \binom{4}{4}b^4$$
$$= a^4 + 4a^3b + 6a^2b^2 + 4ab^3 + b^4$$

$$\binom{4}{0} = \frac{4!}{0!(4-0)!} = \frac{4!}{1 \cdot 4!} = 1 \qquad \binom{4}{1} = \frac{4!}{1!(4-1)!} = \frac{4 \cdot 3!}{1 \cdot 3!} = 4$$

$$\binom{4}{2} = \frac{4!}{2!(4-2)!} = \frac{4!}{2!2!} = 6 \qquad \binom{4}{3} = \frac{4!}{3!(4-3)!} = 4$$

$$\binom{4}{4} = \frac{4!}{4!(4-4)!} = 1$$

Q18 Expand $(x + y)^3$ using the binomial theorem.

STOP • STOP • STOP • STOP • STOP • STOP • STOP • STOP • STOP

A18 $x^3 + 3x^2y + 3xy^2 + y^3$: $\binom{3}{0}x^3 + \binom{3}{1}x^2y + \binom{3}{2}xy^2 + \binom{3}{3}y^3$

Q19 Expand $(a - b)^8$ by use of the binomial theorem:

$\binom{8}{0}a^8 + \binom{8}{1}a^7b +$

Complete:

STOP • STOP • STOP • STOP • STOP • STOP • STOP • STOP • STOP

A19 $a^8 - 8a^7b + 28a^6b^2 - 56a^5b^3 + 70a^4b^4 - 56a^3b^5 + 28a^2b^6 - 8ab^7 + b^8$:

$$(a - b)^8 = \binom{8}{0}a^8 + \binom{8}{1}a^7(-b) + \binom{8}{2}a^6(-b)^2 + \binom{8}{3}a^5(-b)^3 + \binom{8}{4}a^4(-b)^4 + \binom{8}{5}a^3(-b)^5 + \binom{8}{6}a^2(-b)^6 + \binom{8}{7}a(-b)^7 + \binom{8}{8}(-b)^8$$

Q20 Complete the expansion of $(2x + 3y)^5$. (*Think:* $a = 2x$, $b = 3y$, but do not substitute a and b.)

$(2x + 3y)^5 = \binom{5}{0}(2x)^5 + \binom{5}{1}(2x)^4(3y) +$

STOP • STOP • STOP • STOP • STOP • STOP • STOP • STOP • STOP

A20 $32x^5 + 240x^4y + 720x^3y^2 + 1{,}080x^2y^3 + 810xy^4 + 243y^5$:

$$(2x + 3y)^5 = \binom{5}{0}(2x)^5 + \binom{5}{1}(2x)^4(3y) + \binom{5}{2}(2x)^3(3y)^2 + \binom{5}{3}(2x)^2(3y)^3 + \binom{5}{4}(2x)(3y)^4 + \binom{5}{5}(3y)^5$$

$$= 32x^5 + 5 \cdot 16x^4(3y) + 10 \cdot 8x^3(9y^2) + 10 \cdot 4x^2(27y^3) + 5 \cdot 2x(81y^4) + 243y^5$$

Q21 Expand $(2x + 5)^5$.

STOP • STOP • STOP • STOP • STOP • STOP • STOP • STOP • STOP

A21 $32x^5 + 400x^4 + 2{,}000x^3 + 5{,}000x^2 + 6{,}250x + 3{,}125$:

$$(2x + 5)^5 = \binom{5}{0}(2x)^5 + \binom{5}{1}(2x)^4(5) + \binom{5}{2}(2x)^3(5)^2 + \binom{5}{3}(2x)^2(5)^3 + \binom{5}{4}(2x)(5)^4 + \binom{5}{5}(5)^5$$

$$= 32x^5 + 5(16x^4)(5) + 10(8x^3)(25) + 10(4x^2)(125) + 5(2x)(625) + 3{,}125$$

8 The following technique can be used to find a particular term of a binomial expansion. To find the rth term of $(a + b)^n$:

1. The exponent of b is $r - 1$ (1 less than the number of the term).
2. Since the sum of the exponents is n, the exponent of a is $n - (r - 1)$.
3. The coefficient is $\binom{n}{r-1}$.

That is, the rth term of $(a + b)^n$ is

$$\binom{n}{r-1}a^{n-(r-1)}b^{r-1}$$

Example: Find the fifth term of $(x + y)^9$.

Solution

$r = 5$ and $n = 9$

$$\binom{9}{5-1}x^{9-(5-1)}y^{5-1} = \binom{9}{4}x^5y^4 = 126x^5y^4$$

Remember that the exponents on the terms must total n.

Q22 Find the fourth term of $(x + y)^7$.

$r = 4$ and $n = 7$

$$\binom{7}{4-1}x^{7-(4-1)}y^{4-1} = \_\_\_\_\_\_$$

STOP • STOP • STOP • STOP • STOP • STOP • STOP • STOP • STOP

A22 $35x^4y^3$

Q23 Find the fifth term of $(x - y)^6$.

STOP • STOP • STOP • STOP • STOP • STOP • STOP • STOP • STOP

A23 $15x^2y^4$: $\binom{6}{4}x^2y^4$

Q24 Find the sixth term of $(x + 2y)^{10}$.

STOP • STOP • STOP • STOP • STOP • STOP • STOP • STOP • STOP

A24 $8{,}064x^5y^5$: $\binom{10}{5}x^5y^5$

Q25 Find the eighth term of $(2x^2 + y)^7$.

STOP • STOP • STOP • STOP • STOP • STOP • STOP • STOP • STOP

A25 y^7

This completes the instruction for this section.

9.3 Exercises

1. Expand:

   a. $(x - y)^4$ **b.** $(2x + y)^5$ **c.** $(x - y)^7$ **d.** $\left(\frac{x}{2} + \frac{y}{2}\right)^6$

   e. $(a - b)^5$ **f.** $(2x + 3)^3$ **g.** $(mx - n^2)^3$ **h.** $\left(\frac{x}{3} + 2y\right)^5$

2. Write the first four terms of the expansion:
   a. $(x + 2y)^{12}$ **b.** $(3a - b)^{14}$
3. Find the required term:
   a. third term of $(2x^2 - y)^9$
   b. fourth term of $(x - 2y)^5$
   c. sixth term of $(x - 2y)^7$
   d. fourth term of $(2a + b)^{10}$
   e. third term of $(x - 1)^9$

9.3 Exercise Answers

1. **a.** $x^4 - 4x^3y + 6x^2y^2 - 4xy^3 + y^4$
   b. $32x^5 + 80x^4y + 80x^3y^2 + 40x^2y^3 + 10xy^4 + y^5$
   c. $x^7 - 7x^6y + 21x^5y^2 - 35x^4y^3 + 35x^3y^4 - 21x^2y^5 + 7xy^6 - y^7$

d. $\frac{x^6}{64} + \frac{3x^5y}{32} + \frac{15x^4y^2}{64} + \frac{5x^3y^3}{16} + \frac{15x^2y^4}{64} + \frac{3xy^5}{32} + \frac{y^6}{64}$
e. $a^5 - 5a^4b + 10a^3b^2 - 10a^2b^3 + 5ab^4 - b^5$
f. $8x^3 + 36x^2 + 54x + 27$
g. $m^3x^3 - 3m^2n^2x^2 + 3mn^4x - n^6$
h. $\frac{x^5}{243} + \frac{10x^4y}{81} + \frac{40x^3y^2}{27} + \frac{80x^2y^3}{9} + \frac{80xy^4}{3} + 32y^5$

2. a. $x^{12} + 24x^{11}y + 264x^{10}y^2 + 1760x^9y^3$
   b. $3^{14}a^{14} - 14(3^{13})a^{13}b + 91(3^{12})a^{12}b^2 - 364(3^{11})a^{11}b^3$
3. a. $4{,}608x^{14}y^2$ b. $-80x^2y^3$ c. $-672x^2y^5$ d. $15{,}360a^7b^3$
   e. $36x^7$

Chapter 9 Sample Test

At the completion of Chapter 9 it is expected that the student will be able to work the following problems.

9.1 Sequences

1. Find:
   a. the first 5 terms of $a_n = 4 - 3n$
   b. the eighth term of $a_n = 4 - 3n$
2. Find the fifth and nth terms of the sequence:
   a. 2, 2.3, 2.6, 2.9, . .   b. $\frac{1}{4}, \frac{1}{2}, 1, 2, \ldots$
3. William wishes to construct a ladder with nine rungs which diminish uniformly from 24 inches at the base to 18 inches at the top. Determine the length of the seven intermediate rungs.

9.2 Series

4. Find the sum:
   a. $\sum_{n=1}^{18} (3n - 2)$ b. $\sum_{i=1}^{12} (3i + 2)$ c. $\sum_{n=1}^{6} (3^{-n})$
   d. $\sum_{i=1}^{+\infty} \left(\frac{-1}{4}\right)^i$ e. $\sum_{n=1}^{5} (n^2 - n)$
5. Each length of the path of the bob of a swinging pendulum is 95 percent of the preceding arc length. If its initial arc length is 18 centimeters, find the total distance that the bob travels before the pendulum comes to rest.

9.3 The Binomial Theorem

6. Expand:
   a. $(x + y)^5$ b. $(3a - 2b)^4$
7. Find:
   a. the first four terms of $(x + y)^{50}$
   b. the seventh term of $(x - 2y)^{12}$

Chapter 9 Sample Test Answers

1. a. $1, -2, -5, -8, -11$ b. -20
2. a. $a_5 = 3.2$, $a_n = 0.3n + 1.7$ b. $a_5 = 4$, $a_n = 2^{n-3}$
3. $23\frac{1}{4}, 22\frac{1}{2}, 21\frac{3}{4}, 21, 20\frac{1}{4}, 19\frac{1}{2}, 18\frac{3}{4}$
4. a. 477 b. 258 c. $\frac{364}{729}$ d. $\frac{-1}{5}$ e. 40
5. 360 centimeters
6. a. $x^5 + 5x^4y + 10x^3y^2 + 10x^2y^3 + 5xy^4 + y^5$
   b. $81a^4 - 216a^3b + 216a^2b^2 - 96ab^3 + 16b^4$
7. a. $x^{50} + 50x^{49}y + 1{,}225x^{48}y^2 + 19{,}600x^{47}y^3$
   b. $59{,}136x^6y^6$

Appendix

Table I Metric Weights and Measures

Units of Length

1 millimeter (mm)	=	1000 micrometers (μm)
1 centimeter (cm)	=	10 millimeters
1 decimeter (dm)	=	10 centimeters
1 meter (m)	=	10 decimeters
1 dekameter (dam)	=	10 meters
1 hectometer (hm)	=	10 dekameters
1 kilometer (km)	=	10 hectometers

Units of Weight

1 milligram (mg)	=	1000 micrograms (μg)
1 centigram (cg)	=	10 milligrams
1 decigram (dg)	=	10 centigrams
1 gram (g)	=	10 decigrams
1 dekagram (dag)	=	10 grams
1 hectogram (hg)	=	10 dekagrams
1 kilogram (kg)	=	10 hectograms
1 megagram (Mg)	=	1000 kilograms (metric ton)

Units of Area

1 square centimeter	=	100 square millimeters
1 square decimeter	=	100 square centimeters
1 square meter	=	100 square decimeters
1 square kilometer	=	1 000 000 square meters
1 hectare (ha)	=	10 000 square meters
1 are (a)	=	100 square meters

Units of Capacity (Volume)

1 milliliter (ml)	=	1 cubic centimeter (cc or cm^3)
1 centiliter (cl)	=	10 milliliters
1 deciliter (dl)	=	10 centiliters
1 liter (l)	=	10 deciliters
	=	1000 cubic centimeters
1 deckaliter (dal)	=	10 liters
1 hectoliter (hl)	=	10 dekaliters
1 kiloliter (kl)	=	10 hectoliters
1 megaliter (Ml)	=	1000 kiloliters

Table II Symbols

The section number indicates the first use of the symbol.

Section	Symbol	Reading	Meaning
1.1	$\{a\}$	"set containing the element a"	Braces used to denote a set
1.1	$=$	"is equal to"	
1.1	$\{\ \}$ or $\varnothing$	"empty set"	Set with no elements
1.1	$A \subseteq B$	"A is a subset of B"	
1.1	$A \not\subseteq B$	"A is not a subset of B"	
1.1	$A \subset B$	"A is a proper subset of B"	
1.1	$\neq$	"is not equal to"	
1.1	$A \not\subset B$	"A is not a proper subset of B"	
1.1	$\in$	"is an element of"	
1.1	$\notin$	"is not an element of"	
1.1	$, \ldots$	"and so on"	Indicates that list continues
1.1	$A \cap B$	"A intersect B" or "the intersection of A and B"	
1.1	$A \cup B$	"A union B" or "the union of A and B"	
1.1	a^n	"a to the nth"	Exponential notation
1.2	N	the set of natural numbers	
1.2	W	the set of whole numbers	
1.2	$+$	"plus"	Indicates addition (sum) of two numbers
1.2	$-$	"minus"	Indicates subtraction (difference) of two numbers
1.2	$a \cdot b$	"a times b"	Raised dot to indicate the product of two numbers
1.2	$a(b) = (a)(b) = (a)b$	"a times b"	Product of two numbers
1.2	$(\ )$	"quantity"	Parentheses used as grouping symbols
1.2	$\div$	"divided by"	Indicates division (quotient) of two numbers
1.2	$\frac{a}{b}$	"a over b"	Fraction indicating a out of b equal parts
1.3	I	the set of integers	
1.3	$^{+}$	"positive"	Raised plus sign to denote a positive number
1.3	$^{-}$	"negative"	Raised minus sign to denote a negative number
1.3	^{-}x	"the opposite of x"	Raised minus sign to indicate the opposite of a number
1.4	Q	the set of rational numbers	

1.4	$0.\overline{ab}\ \ a,b \in W$	numbers under bar repeat endlessly	
1.4	LCD	"least common denominator"	
1.5	L	the set of irrational numbers	
1.5	$\sqrt{n}$	"the positive (principal) square root of n"	
1.5	$-\sqrt{n}$	"the negative square root of n"	
1.5	R	the set of real numbers ($Q \cup L = R$)	
1.5	π	pi	Constant whose value is generally given as approximately equal to 3.14 or $\frac{22}{7}$
3.2	$a > b$	"a is greater than b"	
3.2	$a < b$	"a is less than b"	
3.2	$a \leqslant b$	"a is less than or equal to b"	
3.2	$a \geqslant b$	"a is greater than or equal to b"	
3.2	$\{x \mid x = a\}$	"the set of all x such that x is equal to a"	Set builder notation
3.2	$x > 0$	"x is greater than zero"	Indicates that the number is positive
3.2	$x < 0$	"x is less than zero"	Indicates that the number is negative
3.2	$a < x < b$	"x is greater than a and less than b	
3.2	a x	open graph; indicates that $x > a$	
3.2	a x	closed graph; indicates that $x \geqslant a$	
3.2	a b x	indicates that $a < x \leqslant b$	
3.2	a b x	indicates that $x \leqslant a$ or $x \geqslant b$	
3.3	$\lvert x \rvert$	"the absolute value of x"	$\lvert x \rvert = x$ if $x \geqslant 0$ $\lvert x \rvert = -x$ if $x < 0$
4.1	$\sqrt[n]{a}$	"the principal nth root of a"	
4.3	$a \times b$	"a times b"	
4.5	i	indicates the imaginary number $\sqrt{-1}$	
4.5	C	the set of complex numbers	
4.5	J	the set of imaginary numbers	
4.5	P	the set of pure imaginary numbers	
5.1	$\pm$	"positive or negative"	
6.1	$\longrightarrow$	"corresponds to"	

6.1	(a, b)	"the ordered pair a, b"	
6.2	x_1 and x_2	"x sub-one and x sub-two"	Subscript used to distinguish between two values of x
6.2	$f(x)$	"f of x"	Function notation
6.4	m	denotes the slope of a line	
6.6	f^{-1}	"f inverse" or "the inverse of f"	
6.6	$f^{-1}(x)$	"inverse function of $f(x)$"	
7.2	%	"percent"	
7.3	$\begin{bmatrix} a & b \\ c & d \end{bmatrix}$	"matrix"	Rectangular array of numbers
7.3	$\begin{vmatrix} a & b \\ c & d \end{vmatrix}$	"determinant"	Numerical value associated with its corresponding matrix
8.1	$\doteq$	"is approximately equal to"	
8.2	log	abbreviation of "logarithm"	
8.2	$y = \log_b x$	"y equals the log of x to the base b"	Equivalent to $x = b^y$
8.4	$\log x$	common logarithm equivalent to $\log_{10} x$	
8.4	antilog	abbreviation of antilogarithm (N in the equation $\log N = a$)	
9.1	d	common difference in an arithmetic sequence	
9.1	a_n	"a sub-n"	General term of an arithmetic or geometric sequence
9.1	r	common ratio in a geometric sequence	
9.2	S_n	"S sub-n"	Indicates the sum of the first n terms of the associated series
9.2	$\sum$	Greek letter "sigma," summation symbol for a series	
9.2	$\sum_{n=1}^{k} a_n$	equivalent to $a_1 + a_2 + a_3 + \cdots + a_k$	
9.2	$+\infty$	"positive infinity"	
9.2	$S_{+\infty}$	sum of an infinite series	
9.3	$n!$	"n factorial"	
9.3	$\binom{n}{k}$	"n choose k"	

The following formulas appear in the text. The section number is indicated.

2.3 and 2.8	Properties of exponents: stated in text
4.1	Properties of radicals: stated in text

Section	Formula
4.3 and 4.4	Properties of exponents: stated in text
5.2	Quadratic formula: For the equation $ax^2 + bx + c = 0$, where $a,b,c \in R$, $a \neq 0$, $x = \frac{-b \pm \sqrt{b^2 - 4ac}}{2a}$
6.3	Slope, m, of a line containing the points (x_1, y_1) and (x_2, y_2): $m = \frac{y_2 - y_1}{x_2 - x_1} = \frac{y_1 - y_2}{x_1 - x_2}$
6.4	Point-slope form of a linear equation with slope m and containing the point (x_1, y_1): $y - y_1 = m(x - x_1)$
6.4	Slope-intercept form of a linear equation of slope m: $y = mx + b$, where $(0, b)$ is the y intercept
7.3	Determinant of the 2×2 matrix $\begin{bmatrix} A & B \\ C & D \end{bmatrix}$: $\begin{vmatrix} A & B \\ C & D \end{vmatrix} = AD - BC$
7.3	Determinant of the 3×3 matrix $\begin{bmatrix} A & B & C \\ D & E & F \\ G & H & I \end{bmatrix}$: $\begin{vmatrix} A & B & C \\ D & E & F \\ G & H & I \end{vmatrix} = AEI + BFG + CDH - GEC - HFA - IDB$
7.3	Cramer's rule (system of two equations in two variables): Given $\begin{matrix} Ax + By = C \\ Dx + Ey = F \end{matrix}$ $D = \begin{vmatrix} A & B \\ D & E \end{vmatrix}$ $D_x = \begin{vmatrix} C & B \\ F & E \end{vmatrix}$ $D_y = \begin{vmatrix} A & C \\ D & F \end{vmatrix}$ $x = \frac{D_x}{D}$ $y = \frac{D_y}{D}$
7.3	Cramer's rule (system of three equations in three variables): *Given* $\begin{matrix} Ax + By + Cz = D \\ Ex + Fy + Gz = H \\ Ix + Jy + Kz = L \end{matrix}$ $D = \begin{vmatrix} A & B & C \\ E & F & G \\ I & J & K \end{vmatrix}$ $D_x = \begin{vmatrix} D & B & C \\ H & F & G \\ L & J & K \end{vmatrix}$ $D_y = \begin{vmatrix} A & D & C \\ E & H & G \\ I & L & K \end{vmatrix}$ $D_z = \begin{vmatrix} A & B & D \\ E & F & H \\ I & J & L \end{vmatrix}$ $x = \frac{D_x}{D}$ $y = \frac{D_y}{D}$ $z = \frac{D_z}{D}$
8.1	General form of a circle with center at the origin with radius r: $x^2 + y^2 = r^2$

8.1	General form of an ellipse with center at the origin: $\frac{x^2}{a^2} + \frac{y^2}{b^2} = 1$ where $a > 0$ and $b > 0$
8.1	Standard forms of the hyperbola: 1. $\frac{x^2}{a^2} - \frac{y^2}{b^2} = 1$ 2. $\frac{y^2}{b^2} - \frac{x^2}{a^2} = 1$, where $a > 0$ and $b > 0$
8.3	Properties of logarithms: stated in text
9.1	The nth term of an arithmetic sequence: $a_n = a_1 + (n - 1)d$, where a_1 = first term, $n \in N$; d = common difference between two consecutive terms of the sequence
9.1	The nth term of a geometric sequence: $a_n = a_1 r^{n-1}$, where a_1 = first term, $n \in N$; r = common ratio between two consecutive terms of the sequence
9.2	Sum of the first n terms of an arithmetic series: $S_n = \frac{n}{2}(a_1 + a_n)$
9.2	Sum of the first n terms of a geometric series: $S_n = \frac{a_1(1 - r^n)}{1 - r}$ where r is the common ratio
9.2	Sum of an infinite geometric series: $S_{+\infty} = \frac{a_1}{1 - r}$ $\lvert r \rvert < 1$ (where r is the common ratio)
9.3	$n! = n(n-1)(n-2)(n-3) \cdots 4 \cdot 3 \cdot 2 \cdot 1$
9.3	$\binom{n}{k} = \frac{n!}{k!(n-k)!}$
9.3	The binomial theorem: If $a,b \in R$ and $n,k \in N$, then $(a + b)^n = \binom{n}{0} a^n + \binom{n}{1} a^{n-1}b + \binom{n}{2} a^{n-2}b^2 + \cdots + \binom{n}{k} a^{n-k}b^k + \cdots + \binom{n}{n} b^n$

Table III Metric and English Conversion

Length

English	Metric
1 inch	= 2.54 cm
1 foot	= 30.5 cm
1 yard	= 91.4 cm
1 mile	= 1610 m
1 mile	= 1.61 km
0.0394 in	= 1 mm
0.394 in	= 1 cm
39.4 in	= 1 m
3.28 ft	= 1 m
1.09 yd	= 1 m
0.621 mi	= 1 km

Capacity

English	Metric
1 gallon	= 3.79 l
1 quart	= 0.946 l
0.264 gal	= 1 l
1.05 qt	= 1 l

Weight

English	Metric
1 ounce	= 28.3 g
1 pound	= 454 g
1 pound	= 0.454 kg
0.0353 oz	= 1 g
0.00220 lb	= 1 g
2.20 lb	= 1 kg

Area

English	Metric
1 square inch	= 6.45 square centimeters
1 square foot	= 929 square centimeters
1 square yard	= 8361 square centimeters
1 square rod	= 25.3 square meters
1 acre	= 4047 square meters
1 square mile	= 2.59 square kilometers
10.76 square feet	= 1 square meter
1,550 square inches	= 1 square meter
0.0395 square rod	= 1 square meter
1.196 square yards	= 1 square meter
0.155 square inch	= 1 square centimeter
247.1 acres	= 1 square kilometer
0.386 square mile	= 1 square kilometer

Volume

English	Metric
1 cubic inch	= 16.39 cubic centimeters
1 cubic foot	= 28.317 cubic centimeters
1 cubic yard	= 0.7646 cubic meter
0.06102 cubic inch	= 1 cubic centimeter
35.3 cubic feet	= 1 cubic meter

Table IV Powers and Roots

No.	Square	Cube	Square Root	Cube Root	No.	Square	Cube	Square Root	Cube Root
1	1	1	1.000	1.000	51	2,601	132,651	7.141	3.708
2	4	8	1.414	1.260	52	2,704	140,608	7.211	3.732
3	9	27	1.732	1.442	53	2,809	148,877	7.280	3.756
4	16	64	2.000	1.587	54	2,916	157,464	7.348	3.780
5	25	125	2.236	1.710	55	3,025	166,375	7.416	3.803
6	36	216	2.449	1.817	56	3,136	175,616	7.483	3.826
7	49	343	2.646	1.913	57	3,249	185,193	7.550	3.848
8	64	512	2.828	2.000	58	3,364	195,112	7.616	3.871
9	81	729	3.000	.2080	59	3,481	205,379	7.681	3.893
10	100	1,000	3.162	2.154	60	3,600	216,000	7.746	3.915
11	121	1,331	3.317	2.224	61	3,721	226,981	7.810	3.936
12	144	1,728	3.464	2.289	62	3,844	238,328	7.874	3.958
13	169	2,197	3.606	2.351	63	3,969	250,047	7.937	3.979
14	196	2,744	3.742	2.410	64	4,096	262,144	8.000	4.000
15	225	3,375	3.873	2.466	65	4,225	274,625	8.062	4.021
16	256	4,096	4.000	2.520	66	4,356	287,496	8.124	4.041
17	289	4,913	4.123	2.571	67	4,489	300,763	8.185	4.062
18	324	5,832	4.243	2.621	68	4,624	314,432	8.246	4.082
19	361	6,859	4.359	2.668	69	4,761	328,509	8.307	4.102
20	400	8,000	4.472	2.714	70	4,900	343,000	8.367	4.121
21	441	9,261	4.583	2.759	71	5,041	357,911	8.426	4.141
22	484	10,648	4.690	2.802	72	5,184	373,248	8.485	4.160
23	529	12,167	4.796	2.844	73	5,329	389,017	8.544	4.179
24	576	13,824	4.899	2.884	74	5,476	405,224	8.602	4.198
25	625	15,625	5.000	2.924	75	5,625	421,875	8.660	4.217
26	676	17,576	5.099	2.962	76	5,776	438,976	8.718	4.236
27	729	19,683	5.196	3.000	77	5,929	456,533	8.775	4.254
28	784	21,952	5.292	3.037	78	6,084	474,552	8.832	4.273
29	841	24,389	5.385	3.072	79	6,241	493,039	8.888	4.291
30	900	27,000	5.477	3.107	80	6,400	512,000	8.944	4.309
31	961	29,791	5.568	3.141	81	6,561	531,441	9.000	4.327
32	1,024	32,768	5.657	3.175	82	6,724	551,368	9.055	4.344
33	1,089	35,937	5.745	3.208	83	6,889	571,787	9.110	4.362
34	1,156	39,304	5.831	3.240	84	7,056	592,704	9.165	4.380
35	1,225	42,875	5.916	3.271	85	5,225	614,125	9.220	4.397
36	1,296	46,656	6.000	3.302	86	7,396	636,056	9.274	4.414
37	1,369	50,653	6.083	3.332	87	7,569	658,503	9.327	4.431
38	1,444	54,872	6.164	3.362	88	7,744	681,472	9.381	4.448
39	1,521	59,319	6.245	3.391	89	7,921	704,969	9.434	4.465
40	1,600	64,000	6.325	3.420	90	8,100	729,000	9.487	4.481
41	1,681	68,921	6.403	3.448	91	8,281	753,571	9.539	4.498
42	1,764	74,088	6.481	3.476	92	8,464	778,688	9.592	4.514
43	1,849	79,507	6.557	3.503	93	8,649	804,357	9.644	4.531
44	1,936	85,184	6.633	3.530	94	8,836	830,584	9.695	4.547
45	2,025	91,125	6.708	3.577	95	9,025	857,375	9.747	4.563
46	2,116	97,336	6.782	3.583	96	9,216	884,736	9.798	4.579
47	2,209	103,823	6.856	3.609	97	9,409	912,673	9.849	4.595
48	2,304	110,592	6.928	3.634	98	9,604	941,192	9.899	4.610
49	2,401	117,649	7.000	3.659	99	9,801	970,299	9.950	4.626
50	2,500	125,000	7.071	3.684	100	10,000	1,000,000	10.000	4.642

Table V Common Logarithms

N	0	1	2	3	4	5	6	7	8	9
1.0	0000	0043	0086	0128	0170	0212	0253	0294	0334	0374
1.1	0414	0453	0492	0531	0569	0607	0645	0682	0719	0755
1.2	0792	0828	0864	0899	0934	0969	1004	1038	1072	1106
1.3	1139	1173	1206	1239	1271	1303	1335	1367	1399	1430
1.4	1461	1492	1523	1553	1584	1614	1644	1673	1703	1732
1.5	1761	1790	1818	1847	1875	1903	1931	1959	1987	2014
1.6	2041	2068	2095	2122	2148	2175	2201	2227	2253	2279
1.7	2304	2330	2355	2380	2405	2430	2455	2480	2504	2529
1.8	2553	2577	2601	2625	2648	2672	2695	2718	2742	2765
1.9	2788	2810	2833	2856	2878	2900	2923	2945	2967	2989
2.0	3010	3032	3054	3075	3096	3118	3139	3160	3181	3201
2.1	3222	3243	3263	3284	3304	3324	3345	3365	3385	3404
2.2	3424	3444	3464	3483	3502	3522	3541	3560	3579	3598
2.3	3617	3636	3655	3674	3692	3711	3729	3747	3766	3784
2.4	3802	3820	3838	3856	3874	3892	3909	3927	3945	3962
2.5	3979	3997	4014	4031	4048	4065	4082	4099	4116	4133
2.6	4150	4166	4183	4200	4216	4232	4249	4265	4281	4298
2.7	4314	4330	4346	4362	4378	4393	4409	4425	4440	4456
2.8	4472	4487	4502	4518	4533	4548	4564	4579	4594	4609
2.9	4624	4639	4654	4669	4683	4698	4713	4728	4742	4757
3.0	4771	4786	4800	4814	4829	4843	4857	4871	4886	4900
3.1	4914	4928	4942	4955	4969	4983	4997	5011	5024	5038
3.2	5051	5065	5079	5092	5105	5119	5132	5145	5159	5172
3.3	5185	5198	5211	5224	5237	5250	5263	5276	5289	5302
3.4	5315	5328	5340	5353	5366	5378	5391	5403	5416	5428
3.5	5441	5453	5465	5478	5490	5502	5514	5527	5539	5551
3.6	5563	5575	5587	5599	5611	5623	5635	5647	5658	5670
3.7	5682	5694	5705	5717	5729	5740	5752	5763	5775	5786
3.8	5798	5809	5821	5832	5843	5855	5866	5877	5888	5899
3.9	5911	5922	5933	5944	5955	5966	5977	5988	5999	6010
4.0	6021	6031	6042	6053	6064	6075	6085	6096	6107	6117
4.1	6128	6138	6149	6160	6170	6180	6191	6201	6212	6222
4.2	6232	6243	6253	6263	6274	6284	6294	6304	6314	6325
4.3	6335	6345	6355	6365	6375	6385	6395	6405	6415	6425
4.4	6435	6444	6454	6464	6474	6484	6493	6503	6513	6522
4.5	6532	6542	6551	6561	6571	6580	6590	6599	6609	6618
4.6	6628	6637	6646	6656	6665	6675	6684	6693	6702	6712
4.7	6721	6730	6739	6749	6758	6767	6776	6785	6794	6803
4.8	6812	6821	6830	6839	6848	6857	6866	6875	6884	6893
4.9	6902	6911	6920	6928	6937	6946	6955	6964	6972	6981
5.0	6990	6998	7007	7016	7024	7033	7042	7050	7059	7067
5.1	7076	7084	7093	7101	7110	7118	7126	7135	7143	7152
5.2	7160	7168	7177	7185	7193	7202	7210	7218	7226	7235
5.3	7243	7251	7259	7267	7275	7284	7292	7300	7308	7316
5.4	7324	7332	7340	7348	7356	7364	7372	7380	7388	7396
5.5	7404	7412	7419	7427	7435	7443	7451	7459	7466	7474
5.6	7482	7490	7497	7505	7513	7520	7528	7536	7543	7551
5.7	7559	7566	7574	7582	7589	7597	7604	7612	7619	7627
5.8	7634	7642	7649	7657	7664	7672	7679	7686	7694	7701
5.9	7709	7716	7723	7731	7738	7745	7752	7760	7767	7774

N	0	1	2	3	4	5	6	7	8	9
6.0	7782	7789	7796	7803	7810	7818	7825	7832	7839	7846
6.1	7853	7860	7868	7875	7882	7889	7896	7903	7910	7917
6.2	7924	7931	7938	7945	7952	7959	7966	7973	7980	7987
6.3	7993	8000	8007	8014	8021	8028	8035	8041	8048	8055
6.4	8062	8069	8075	8082	8089	8096	8102	8109	8116	8122
6.5	8129	8136	8142	8149	8156	8162	8169	8176	8182	8189
6.6	8195	8202	8209	8215	8222	8228	8235	8241	8248	8254
6.7	8261	8267	8274	8280	8287	8293	8299	8306	8312	8319
6.8	8325	8331	8338	8344	8351	8357	8363	8370	8376	8382
6.9	8388	8395	8401	8407	8414	8420	8426	8432	8439	8445
7.0	8451	8457	8463	8470	8476	8482	8488	8494	8500	8506
7.1	8513	8519	8525	8531	8537	8543	8549	8555	8561	8567
7.2	8673	8579	8585	8591	8597	8603	8609	8615	8621	8627
7.3	8633	8639	8645	8651	8657	8663	8669	8675	8681	8686
7.4	8692	8698	8704	8710	8716	8722	8727	8733	8739	8745
7.5	8751	8756	8762	8768	8774	8779	8785	8791	8797	8802
7.6	8808	8814	8820	8825	8831	8837	8842	8848	8854	8859
7.7	8865	8871	8876	8882	8887	8893	8899	8904	8910	8915
7.8	8921	8927	8932	8938	8943	8949	8954	8960	8965	8971
7.9	8976	8982	8987	8993	8998	9004	9009	9015	9020	9025
8.0	9031	9036	9042	9047	9053	9058	9063	9069	9074	9079
8.1	9085	9090	9096	9101	9106	9112	9117	9122	9128	9133
8.2	9138	9143	9149	9154	9159	9165	9170	9175	9180	9186
8.3	9191	9196	9201	9206	9212	9217	9222	9227	9232	9238
8.4	9243	9248	9253	9258	9263	9269	9274	9279	9284	9289
8.5	9294	9299	9304	9309	9315	9320	9325	9330	9335	9240
8.6	9345	9350	9355	9360	9365	9370	9375	9380	9385	9390
8.7	9395	9400	9405	9410	9415	9420	9425	9430	9435	9440
8.8	9445	9450	9455	9460	9465	9469	9474	9479	9484	9489
8.9	9494	9499	9504	9509	9513	9518	9523	9528	9533	9538
9.0	9542	9547	9552	9557	9562	9566	9571	9576	9581	9586
9.1	9590	9595	9600	9605	9609	9614	9619	9624	9628	9633
9.2	9638	9643	9647	9652	9657	9661	9666	9671	9675	9680
9.3	9685	9689	9694	9699	9703	9708	9713	9717	9722	9727
9.4	9731	9736	9741	9745	9750	9754	9759	9763	9768	9773
9.5	9777	9782	9786	9791	9795	9800	9805	9809	9814	9818
9.6	9823	9827	9832	9836	9841	9845	9850	9854	9859	9863
9.7	9868	9872	9877	9881	9886	9890	9894	9899	9903	9908
9.8	9912	9917	9921	9926	9930	9934	9939	9943	9948	9952
9.9	9956	9961	9965	9969	9974	9978	9983	9987	9991	9996

Glossary

1. When a bar is placed over a vowel, the vowel says its own name.
2. A curved mark over a vowel indicates the following sounds:
   - **a.** ă as in at
   - **b.** ĕ as in bed
   - **c.** ĭ as in it
   - **d.** ŏ as in ox
   - **e.** ŭ as in rug

Abscissa (ăb sĭs′ sŭ) First member of an ordered pair, such as x in (x,y).

Absolute value (ăb′ sŭ loot văl′ ū) **(of a number)** Distance between the number and zero.

Acute (ŭ kūt′)
(angle) Angle whose measure is less than 90°.
(triangle) Triangle with three angles less than 90°.

Add (ăd) To combine into one sum or quantity.

Addend (ăd′ ĕnd) Number or quantity to be added to another.

Addition (ă dĭ′ shŭn) Act of adding.

Additive (ăd′ ĭ tĭv) **(inverses)** (*See* Opposites.)

Algebraic (ăl′ jŭ brā′ ĭk) **(expression)** Open expressions containing variables, such as $a + 3$, $5y - 2x$, $9(b + 7)$, $(x - 2)(x + 3)$, $\frac{x}{y} - x$.

Altitude (ăl′ tĭ to͞od)
(of a parallelogram) Perpendicular distance between two parallel sides.
(of a trapezoid) Perpendicular distance between the two parallel sides.
(of a triangle) Line segment from the vertex of a triangle perpendicular to the base.

Analytic (ăn′ ŭ lĭt′ ĭk) **(method)** Developed by algebraic means.

Angle (ăng′ g′l) Two rays with a common endpoint.

Antilogarithm (ăn′ tĭ log′ ŭ rĭth′m) (*See* Section 8.4, Frame 12.)

Area (air′ ē ŭ) Measure (in square units) of the region within a closed curve (including polygons) in a plane.

Arithmetic (ŭ rĭth′ mŭ tĭk) **(sequence or progression)** Sequence in which each term differs from the preceding term by a constant amount.

Ascending (ă sĕnd′ ĭng) **(order of the terms of a polynomial with respect to a particular variable)** Terms of a polynomial arranged so that the degree of each term increases from left to right with respect to a particular variable.

Associative (ŭ sō′ shŭ tĭv) Indicates that the grouping of three numbers in an addition or multiplication can be changed without affecting the sum or product.

Asymptotes (ăs′ ĭm tōts) (*See* Section 8.1, Frame 10.)

Base (bās)
(of a cylinder) Two circular regions in a right-circular cylinder.
(of a logarithm) Value of b in $\log_b x$.
(of a power) Number being raised to a power, such as 3 in 3^5.
(of a triangle) Side opposite a vertex.

Bases (bās′ ĭz) **(of a trapezoid)** Parallel sides.

Binomial (bī nō′ mē ŭl) Polynomial that has exactly two terms.

Braces (brās′ ĭz) Symbols, { }, used for grouping expressions or to indicate a set, such as 1, 2, 3, . . .}.

Characteristic (kăr′ ăk tur is′ tĭk) **(of a logarithm)** (*See* Section 8.4, Frame 9.)

Circle (sur′ kŭl) Set of all points in a plane whose distance from a given point (center) is equal to a positive number, r (radius).

Circumference (sur kŭm′ fur ĕns) **(of a circle)** Distance around.

Closed (set) A set is said to be closed under a particular operation when you can perform the operation on any two numbers of the set and get another number in the set.

Closure (klō′ zhur) **(property)** Property which indicates that a set is closed under a particular operation. (*See* Closed.)

Coefficient (kō′ ĕ fĭsh′ ĕnt) (*See* Numerical coefficient.)

Commutative (kŭ mū′ tŭ tĭv) Indicates that the order of two numbers in an addition or multiplication can be changed without affecting the sum or product.

Complementary (kŏm plŭ mĕn′ tŭ rē) **(angles)** Two or more angles whose measures added together result in a sum of 90°.

Complex (kŏm′ plĕks)

(fraction) Rational expression that has a fraction in its numerator or its denominator, or both.

(number) Number that can be put in the form $a + bi$, where a and b are real numbers and $i = \sqrt{-1}$.

Composite (kŏm pŏz′ ĭt) **(number)** Natural number that has more than two natural-number factors.

Cone (kōn) **(right-circular)** Circular region (base) and the surface made up of line segments that connect the circle with a point (vertex) located on a line through the center of the circle and perpendicular to the plane of the circle.

Conic (sections) (kŏn′ ĭk) Circle, ellipse, parabola, and hyperbola.

Conjugates (kŏn′ jo͞o gŭts) Two binomials that differ only in the sign of the second term, such as $a + b$ and $a - b$.

Consecutive (kŏn sĕk′ ū tĭv) **(integers)** Integers that differ by 1.

Consistent (kŏn sĭs′ tĕnt) **(equations)** Two equations in two unknowns (variables) which have only one solution.

Constant (kŏn′ stănt) Numbers without a literal coefficient, such as 5, -7, $\frac{7}{8}$, 0.

Coordinate axes (kō or′ dĭ nĭt ăk′ sēz) x axis and y axis in a (rectangular) coordinate system.

Coordinates (kō or′ dĭ nĭts) **(of a point)** Two numbers that locate a point on a (rectangular) coordinate system, such as (3, 2), which indicates that the x coordinate is 3 and the y coordinate is 2.

Correspondence (kŏr′ rĕs pŏn′ dĕns) Manner in which elements of two sets are matched.

Cube (kūb) Solid figure of six equal faces.

(of a number) Indication that a number is being used three times as a factor, such as 2^3 $(2 \cdot 2 \cdot 2)$.

(root) The value b such that $\sqrt[3]{a} = b$ if $b^3 = a$.

Cylinder (sĭl′ ĭn dur) **(right-circular)** Two circular regions with the same radius in parallel planes (bases) connected by line segments perpendicular to the planes of the two circles.

Cylindrical (sĭ lĭn′ drĭ kŭl) Relating to, or having the properties of, a cylinder.

Decimal (dĕs′ ĭ mŭl) **(fraction)** Fraction whose denominator is a power (multiple) of 10, such as $0.5 = \frac{5}{10}$, $2.67 = \frac{267}{100}$.

Degree (dŭ grē′)
(measurement of an angle) 1/360 of a complete revolution of a ray around its endpoint.
(of a polynomial) Greatest degree of any of its terms.
(of a term) The degree of a constant is zero. The number zero has no degree. The degree of a term of one variable agrees with the exponent on the variable (the degree of a term with more than one variable is equal to the sum of all exponents on individual variables).

Denominator (dŭ nŏm′ ĭ nā tur) Bottom number in a fraction, such as 3 in $\frac{2}{3}$.

Dependent (dē pĕn′ dĕnt) **(equations)** Two equations in two unknowns (variables) which have an infinite number of solutions.

Descending (dē sĕnd′ ĭng) **(order of the terms of a polynomial with respect to a particular variable)** Terms of a polynomial arranged so that the degree of each term decreases from left to right with respect to a particular variable.

Determinant (dŭ tur′ mĭ nŭnt) **(of a matrix)** Numerical value associated with a matrix.

Diameter (dī ăm′ ŭ tur) **(of a circle)** Twice the radius (of the circle).

Difference (dĭf′ ur ĕns) Result of a subtraction of two numbers, such as 5 in $7 - 2 = 5$.

Digit (dĭj′ ĭt) Any one of the symbols 0, 1, 2, 3, 4, 5, 6, 7, 8, 9.

Directrix (dĭ rĕk′ trĭks) (*See* Section 8.1, Frame 13.)

Discriminant (dĭs krĭm′ ĭ nŭnt) **(of a quadratic equation)** Value of the number $b^2 - 4ac$ if $ax^2 + bx + c = 0$, where $a,b,c \in R$, $a \neq 0$.

Disjoint (dĭs joint′) Two or more sets that have no common elements (the intersection of disjoint sets is empty).

Distributive (dĭs trĭb′ ū tĭv) **(property of multiplication over addition or subtraction)** Property that permits changing sums or differences to products (or vice versa); that is,
$a(b + c) = ab + ac$
$a(b - c) = ab - ac$
$(a + b)c = ac + bc$
$(a - b)c = ac - bc$

Divide (dĭ vīd′) Process of determining the number of equal parts in a number or quantity.

Dividend (dĭv′ ĭ dĕnd) In a division problem, the number that is being divided, such as 15 in $15 \div 3$.

Division (dĭ vĭ′ zhŭn) Act of dividing.

Divisor (dĭ vī′ zur) Number by which another number (dividend) is divided, such as 3 in $15 \div 3$.

Domain (dŭ mān) **(of a relation)** Set consisting of the first members of the ordered pairs of the relation.

Elements (ĕl′ ŭ ments) Things that make up a set.

Ellipse (ĕ lĭps′) Set of all points in a plane the sum of whose distances from two fixed points (foci) in the plane is a constant.

Empty (ĕmp′ tē) **(set)** Set with no elements.

Equation (ē kwā′ zhŭn) Statement that the expressions on opposite sides of an equal sign represent the same number, such as $2 + 3 = \frac{12 - 2}{2}$ (mathematical statement of equality).

Equilateral (ē′ kwĭ lăt′ ur ŭl) **(triangle)** Triangle with three equal sides.

Equivalent (ŭ kwĭv′ ŭ lent)
(equations) Two or more equations that have the same solution set.
(expressions) Two or more expressions that have the same evaluation for all replacements of the variable(s).

Exponent (ĕks′ pō nŭnt) n in the expression x^n.

Exponential (ĕks′ pŭ nĕn′ shŭl) **(function)** Function of the form $f(x) = b^x$, where $b > 0$ and $b \neq 1$.

Extremes (ĕks trēmz′) **(of a proportion)** First and fourth terms of a proportion, such as 2 and 35 in $2:5 = 14:35$ or $\frac{2}{5} = \frac{14}{35}$.

Factor (făk′ tur) One of the values in a multiplication expression, such as 3 in (3)(5).

Factorial (făk tō′ rĭ ăl) **(notation)** (*See* Section 9.3, Frame 5.)

Finite (fī′ nīt) **(set)** Set in which the elements can be counted and the count has a last number.

Foci (fō′ sī) Plural of focus. (*See* Ellipse or Hyperbola.)

Focus (fō′kŭs) Singular of foci. (*See* Ellipse, Hyperbola, or Parabola.)

Fraction (frăk′ shŭn) Number which indicates that some whole has been divided into a number of equal parts and that a portion of the equal parts is represented.

Function (fŭngk′ shŭn) Relation in which no two ordered pairs have the same first coordinate (each domain element of the relation corresponds to exactly one range element).

Geometric (jē′ ŭ mĕt′ rĭk) **(sequence or progression)** Sequence in which each term divided by the preceding term is a constant.

Hexagon (hĕk′sŭ gŏn) Polygon with six sides.

Horizontal (hŏr′ ĭ zŏn′ tŭl) **(line)** Line parallel to the plane of the horizon.

Hyperbola (hī pur′ bŭ lŭ) Set of all points in a plane such that the difference of their distances from two fixed points (foci) in the plane is a constant.

Hypotenuse (hī pŏt′ ŭ nōōs) **(of a right triangle)** Side opposite the right angle.

Imaginary (ĭ măj′ ĭ nĕr′ ē) **(part of a complex number)** Value of b in a complex number $a + bi$.

Improper (ĭm prŏp′ ur) **(fraction)** Fraction that has a value greater than or equal to 1.

Inconsistent (ĭn kŏn sĭs′ tĕnt) **(equations)** Two equations in two unknowns (variables) that have no common solution.

Index (ĭn′ dĕks) **(of a radical)** Number that indicates the desired root, such as 3 in $\sqrt[3]{8}$ (the index 2 is understood when the radical sign is being used to indicate square root).

Inequality (ĭn′ ē kwăl′ ĭ tē) Expression consisting of two unequal quantities, with the sign of inequality ($>$, $<$, $\neq$) between them.

Infinite (ĭn′ fĭ nĭt) **(set)** Set whose count in unending.

Integer (ĭn′ tŭ jur) Any number that is in the following list: $\ldots, -2, -1, 0, 1, 2, \ldots$.

Interpolation (ĭn tur′ pŭ lā′ shŭn) (*See* Section 8.4, Frame 15.)

Intersection (ĭn′ tur sĕk′ shŭn) **(of two sets)** Third set that contains those elements, and only those elements, that belong to both (one and the other) of the original sets.

Inverse (ĭn vurs′)
(of a function) Relation formed by interchanging the coordinates of each ordered pair of the function (the inverse of a function is not necessarily a function).

(operations) Addition and subtraction, multiplication and division are said to be inverse operations because one undoes the effect of the other; that is, $5 + 2 - 2 = 5$ or $5 - 2 + 2 = 5$ and $7(2) \div 2 = 7$ or $7 \div 2(2) = 7$.

Irrational (ĭr răsh′ ŭn ŭl) **(numbers)** Infinite nonrepeating decimals. (square roots of positive nonperfect squares are but one class of irrational numbers).

Isosceles (ī sŏs′ ĕ lēz) **(triangle)** Triangle with at least two equal sides.

Kilogram (kĭl′ ŭ grăm) Base (standard) unit of weight in the metric system (1 kilogram = 1000 grams.

Lateral surface (lăt′ ur ŭl sur′ fĭs) **(of a cylinder)** Curved surface that connects the bases.

Least common denominator (LCD) Smallest number that is exactly divisible by each of the original denominators of two or more fractions.

Like terms Terms of an algebraic expression which have exactly the same literal coefficients (including exponents).

Line (līn) Set of points represented by a picture, such as $\longleftrightarrow$ (A line is always straight, has no thickness, and extends forever in both directions.)

Linear (lĭn′ ē ur)

(equation) Equation that can be written in the form $Ax + By + C = 0$, where A, B, and C are real numbers, A and B both not zero.

(function) Function of the form $f(x) = mx + b$, $m \neq 0$, where m and b are constants (graph is a straight line).

(measurement) Measurement along a (straight) line.

Liter (lē′ tur) Base (standard) unit of volume or capacity in the metric system.

Literal coefficient (lĭt′ ur ŭl kō′ ĕ fĭsh′ ŭnt) Letter factor of an indicated product of a number and one or more variables.

Logarithmic (function) Inverse of an exponential function.

Mantissa (*See* Section 8.4, Frame 9.)

Matrices Plural of matrix.

Matrix Rectangular array of numbers.

Means (mēnz) **(of a proportion)** Second and third terms of a proportion, such as 5 and 14 in $2:5 = 14:35$ or $\frac{2}{5} = \frac{14}{35}$.

Meter (mē′ tur) Base (standard) unit of length in the metric system.

Minuend (mĭn′ ū ĕnd) Number or quantity from which another (subtrahend) is to be subtracted, such as 17 in $17 - 3$.

Mixed number Understood sum of a whole number and a proper fraction.

Monomial (mō nō′ mē ŭl) Polynomial that contains only one term.

Multiple (mŭl′ tĭ p'l) Product of a quantity by an integer.

Multiplication (mŭl′ tĭ plĭ kā′ shŭn) Act of multiplying.

Multiplicative (mŭl′ tĭ plĭk′ ŭ tĭv) **(inverses)** (*See* Reciprocals.)

Multiply (mŭl′ tĭ plī) To take by addition a certain number of times.

Natural (năt′ jŭ rŭl) **(or counting number)** Any number that is in the following list: 1, 2, 3, 4, ...

Negative (nĕg′ ŭ tĭv) **(number)** Any number that is located to the left of zero on a horizontal number line.

Numeral (nōō′ mur ŭl) A symbol that represents a number, such as V, 𝍸, and 5, which are all numerals naming the number five.

Numerator (nōō′ mur ā′ tur) Top number in a fraction, such as 2 in $\frac{2}{3}$.

Numerical (nōō mĕr′ ĭ kŭl) **(coefficient)** Number factor of an indicated product of a number and a variable.

Obtuse (ŏb tōōs′)
(angle) Angle whose measure is between 90° and 180°.
(triangle) Triangle that has one angle greater than 90°.

Octagon (ŏk′ tŭ gŏn) Polygon with eight sides.

Open expression Expression in which the position of an unknown number is held by a letter (variable). (*See* Algebraic expression.)

Open sentence Equation that does not contain enough information to be judged as either true or false, such as $x + 2 = 7$.

Opposite (ŏp′ ŭ zĭt) **(sides of a quadrilateral)** Pairs of sides that do not intersect.

Opposites (ŏp′ ŭ zĭtz) **(additive inverses)** Two numbers on a number line that are the same distance from zero, such as -5 and 5, $6\frac{2}{3}$ and $-6\frac{2}{3}$. (The sum of opposites is zero; that is, $a + {}^{-}a = 0$.)

Ordered pair (of numbers) Expression (x, y) that locates a point on a rectangular coordinate system (x indicates the distance and direction in a horizontal direction from the origin; y indicates the distance and direction in a vertical direction).

Ordinate (or′ dĭ nŭt) Second member of an ordered pair, such as y in (x, y).

Origin (or′ ĭ jĭn) Point of intersection of the coordinate axes in a rectangular coordinate system, identified by the ordered pair $(0, 0)$.

Parabola (pŭ răb′ ŭ lŭ) Graph of the quadratic function $f(x) = ax^2 + bx + c$, where a, b, and c are constants and $a \neq 0$ (set of all points in a plane that are equidistant from a fixed point (focus) and a fixed line (directrix) in the plane).

Parallel (păr′ ŭ lĕl) **(lines)** Two or more lines in the same plane that have no points in common.

Parallelogram (păr′ ŭ lĕl′ ŭ gram) Quadrilateral with opposite sides parallel.

Parentheses (pŭ rĕn′ thŭ sēz) Symbols, ( ), used for grouping expressions, such as $5 + (a + 3)$ or $2(x - 1)$; or to indicate an ordered pair of numbers, such as (2, 3).

Pentagon (pĕn′ tŭ gŏn) Polygon with five sides.

Percent (pur sĕnt′) Fraction with a denominator of 100, such as $\frac{7}{100} = 0.07 = 7$ percent.

Perfect square
(integer) Integer obtained by squaring an integer.
(monomial) Square of a monomial.
(rational number) Square of a rational number.
(trinomial) Square of a binomial (trinomial of the form $a^2 \pm 2ab + b^2$).

Perimeter (pŭ rĭm′ ŭ tur) **(of a polygon)** Total length of all the sides of the polygon.

Perpendicular (pur′ pĕn dĭk′ ū lur) **(lines)** Two intersecting lines that form a right angle.

Plane (plān) Flat surface (such as a tabletop) that extends infinitely in every direction.

Point Location in space that has no thickness (represented on paper by a dot).

Polygon (pŏl′ ĭ gŏn) Plane closed figure of three or more angles.

Polynomial (pŏl′ ĭ nō′ mē ŭl) Algebraic expression made up of sums, differences, and products of variables and numbers.

Polynomial (function) Defined by $P(x) = a_nx^n + a_{n-1}x^{n-1} + \cdots + a_0$, where $n \in W$ and the a's are constants.

Positive (pŏz′ ĭ tĭv) **(number)** Any number that is located to the right of zero on a horizontal number line.

Power (pow′ ur) Expression used when referring to numbers with exponents, such as 3^5, "the fifth power of three."

Prime (prīm)

(number) Natural number other than 1 divisible by exactly two natural number factors, itself and 1.

(polynomial over the set of integers) Polynomial that does not contain an integer factor other than 1 or -1 and cannot be factored into two other polynomials with integer coefficients.

Principal (prĭn′sĭ p′l) Amount of money invested or borrowed.

Product (prŏd′ ŭkt) Result of a multiplication problem.

Proper (prŏp′ ur)

(fraction) Fraction that has a value less than 1.

(subset) Subset of another set in which at least one element appears in the other set which does not appear in the subset.

Proportion (prŭ pōr′ shŭn) Statement of equality between two ratios.

Quadrant (kwŏd′ rănt) Any of the four parts into which a plane is divided by rectangular coordinate axes lying in that plane.

Quadratic (kwŏd răt′ ĭk) **(equation)** Equation that can be written in the form $ax^2 + bx + c = 0$, where a, b, and c are real numbers and $a \neq 0$.

Quadrilateral (kwŏd′ rĭ lăt′ ur ûl) Polygon with four sides.

Quantity (kwŏn′ tĭ tē) Indicates an expression that has been placed within parentheses (can represent an entire expression, such as the quantity $x + 7$).

Quotient (kwō′ shŭnt) Number or quantity that results from the division of one number or quantity (dividend) by another (divisor).

Radical sign (răd′ ĭ kŭl sīn) Symbol $\sqrt{\ }$ ($\sqrt[n]{a}$ is the nth root of a).

Radicand (răd′ ĭ kănd) Number under the radical sign, such as 2 in $\sqrt[3]{2}$.

Radii (rā′ dē ī) Plural of radius.

Radius (rā′ dē ŭs) **(of a circle)** Line segment connecting the center of a circle to any point on the circle (measure of this distance).

Range (rānj) **(of a relation)** Set consisting of the second members of the ordered pairs of the relation.

Ratio (rā′shō) Quotient of two quantities or numbers.

Rational (răsh′ ŭn ŭl)

(expression) Expression in which both numerator and denominator are polynomials, excluding all possible values of the variables that would produce a zero denominator.

(function) Defined by $R(x) = \dfrac{P(x)}{Q(x)}$, $Q(x) \neq 0$, where $P(x)$ and $Q(x)$ are polynomials.

(number) Any number that can be written in the form $\dfrac{p}{q}$, where p and q are integers and $q \neq 0$ (terminating or infinite-repeating decimals).

Ray (rā) Part of a line on one side of a point, which includes the point (endpoint of the ray).

Real (rē ŭl)

(number) Either a rational or irrational number (the union of the set of rational numbers and the set of irrational numbers is the set of real numbers).

(part of a complex number) Value of a in a complex number $a + bi$.

Reciprocal (rŭ sĭp′ rŭ kăl) **(function)** $f(x) = \frac{1}{x}, x \neq 0$.

Reciprocals (rŭ sĭp′ rŭ kălz) **(multiplicative inverses)** Two numbers whose product is 1, such as 5 and $\frac{1}{5}$, $\frac{-5}{8}$ and $\frac{-8}{5}$, x and $\frac{1}{x}(x \neq 0)$.

Rectangle (rĕk′ tăng g'l) Parallelogram whose sides meet at right angles.

Rectangular (rĕk tăng′ gū lur) **(coordinate system)** Formed by crossing (intersecting) vertical and horizontal number lines.

Reduce (rŭ doos′) Process of dividing both numerator and denominator of a fraction by a common factor.

Relation (rŭ lā′ shŭn) Set of ordered pairs of numbers.

Replacement set Set of permissible values of a variable in an open (algebraic) expression or equation.

Right

(angle) Any of the angles formed by two intersecting lines where all four angles are equal (the measure of a right angle is 90°).

(triangle) Triangle with one angle of 90°.

Scalene (skā lēn′) **(triangle)** Triangle that has no pair of sides equal.

Scientific (sī′ ĕn tĭf′ ĭk) **(notation)** Number written as a number greater than or equal to 1 and less than 10 multiplied by 10 raised to some power.

Segment (sĕg′ mĕnt) **(line)** Portion of a line between two points, including its endpoints.

Sequence (sē′ kwĕns) (*See* Section 9.1, Frame 1.)

Series (sēr′ ēz) Indicated sum of the terms of a sequence.

Set (sĕt) Well-defined (membership in the set is clear) collection of things.

Set notation The use of braces to indicate a set, such as {2, 3, 7} indicates the set containing the numbers 2, 3, and 7.

Signed numbers Number expressing both directional and quantitative values, such as $+14\frac{1}{2}$, $-\pi$.

Slope (slōp) **(of a line)** Computed between any two points (x_1,y_1) and (x_2,y_2) of the line and is defined as the "rise" $(y_1 - y_2)$ divided by the "run" $(x_1 - x_2)$ between the two points.

Solution (sŭ lū′ shŭn) **(truth set)** Set of all values from the replacement set which converts the open sentence into a true statement.

Solutions (sŭ lū′ shŭnz) **(to an equation)** All values of the variable from the replacement set which convert the open sentence (equation) into a true statement.

Square (skwair) Rectangle whose sides are of equal length.

(of a number) Indication that a number is being used twice as a factor, such as $3^2(3 \cdot 3)$.

(roots) The square roots of a number are the values which when squared will produce the number, such as the square roots of 16 are 4 and −4.

Subset (sŭb′ sĕt) Set in which each element is also an element of another set.

Subtract (sŭb trăkt′) To withdraw or take away, as one number from another.

Subtraction (sŭb trăk′ shŭn) Act of subtracting.

Subtrahend (sŭb′ trŭ hĕnd′) Number or quantity to be subtracted from another (minuend), such as 3 in 17 − 3.

Sum (sŭm) Result of an addition of two or more numbers.

Supplementary (sup′ lu mĕn′ tŭ rē) **(angles)** Two or more angles whose measures added together result in a sum of 180°.

Terms (turmz) Parts of an expression separated by an addition symbol

Trapezoid (trăp′ ŭ zoid) Quadrilateral in which only one pair of opposite sides is parallel.

Triangle (trī′ ăng g'l) Polygon with three sides.

Trinomial (trī nō′ mē ŭl) Polynomial that has exactly three terms.

Union (ūn′ yŭn) **(of two sets)** Third set, which contains all those elements that belong to either (one or the other) of the original sets.

Universal (ū′ nĭ vur′ sŭl) **(set)** (*See* Replacement set.)

Variable (vair′ ĭ ŭ b'l) Letter in an algebraic expression which can be replaced by any one of a set of many numbers.

Vertex (vur′ tĕks)
(of an angle) Common endpoint of two rays which form an angle.
(of a parabola) Highest or lowest point on the graph of a quadratic function.

Vertical (vur′ tĭ kŭl) **(line)** Line perpendicular to the plane of the horizon.

Vertices (vur′ tĭ sēz) **(of a polygon)** Plural of vertex.

Volume (vŏl′ yŭm) Measure (in cubic units) of the space within a closed solid figure.

Whole (hōl) **(number)** Any number that is in the following list: 0, 1, 2, 3, . . . (the next whole number is formed by adding 1 to the previous whole number).

***x* axis** (x ăk′ sĭs) Horizontal number line in a (rectangular) coordinate system.

***x* intercept(s)** (x ĭn′ tur sĕpt′) **(of a graph)** Point(s) where graph intersect(s) the x axis.

***y* axis** (y ăk′ sĭs) Vertical number line in a (rectangular) coordinate system.

***y* intercept(s)** (y ĭn′ tur sĕpt′) **(of a graph)** Point(s) where graph intersect(s) the y axis.

Index